A. Ambikapathy

Control System

Control System

Prof (Dr.) A. AMBIKAPATHY
B.E., M.Tech., PhD
HOD Electricals & Electronics Engineering
Galgotias College of Engineering and Technology
Greater Noida

KHANNA BOOK PUBLISHING CO. (P) LTD.

Publisher of Engineering and Computer Books

4C/4344, Ansari Road, Darya Ganj, New Delhi-110002
Phone : 011-23244447-48 **Mobile:** +91-9910909320
E-mail : contact@khannabooks.com
Website : www.khannabooks.com

Dear Readers,

To prevent the piracy, this book is secured with HIGH SECURITY HOLOGRAM on the front title cover. In case you don't find the hologram on the front cover title, please write us to at contact@khannabooks.com or whatsapp us at +91-99109 09320 and avail special gift voucher for yourself.

Specimen of Hologram on front Cover title:

Moreover, there is a SPECIAL DISCOUNT COUPON for you with EVERY HOLOGRAM.

How to avail this SPECIAL DISCOUNT:

Step 1: Scratch the hologram
Step 2: Under the scratch area, your "coupon code" is available
Step 3: Logon to www.khannabooks.com
Step 4: Use your "coupon code" in the shopping cart and get your copy at a special discount
Step 5: Enjoy your reading!

ISBN: 978-93-80016-94-8

Control System
by A. Ambikapathy

Edition: 2026

To view complete list of books, Please scan the QR Code:

Published by:
Khanna Book Publishing Co. (P) Ltd.
CIN: U22110DL1998PTC095547

Visit us at: **www.khannabooks.com**
Write us at: **contact@khannabooks.com**

Printed in Bharat.

Dedicated to
Parents
Husband
Mother-in-law
Brother and Sisters
Daughter
Friend

Preface

The textbook on "Automatic Control System" has been written for the undergraduate students of ECE, EIE, EEE, EE and IC departments. The book is also useful to the teachers those who teach this course. The effort has been taken in preparation of this book such that the students using this book acquire the knowledge. The material of this book is very useful to the students for understanding the concepts as well as from examination point of view. A number of examples are solved in a detailed manner in every unit so that the students can practice more on numerical part. Every unit contains exercise problem with answers this helps the students from examination point of view. The more effort has been taken to explain the procedure for drawing different plots such as Bode plot, Polar plot and root locus. The content of this book is explained in a simple language so that understanding will be much easy. I will be thankful to the students and teachers who pinpoint the errors and suggestions for the improvement of the book.

Please send your feedback to **ami_ty21@yahoo.com**

A.AMBIKAPATHY

Acknowledgement

I thank my parents Mr. Aladiyan and Mrs.Vasantha for their continuous support. I thank my husband for his guidance and motivation. K.E.T. Arasu B.E, MBA. I am greatful to my mother-in-law Mrs. Ellammal for her support.

I thank my seniors who supported me by their guidance.

Dr. Pradeep Kumar, Chairman, Delhi College of Technology and Management.

I would like to thank firstly, the God for his blessings, my parants for their contribution in my life, my husband for his support, my mother-in-law and all my family member for their encouragement.

I also would like to thank Dr. V.K. Dwivedi, Director, Galgotias College of Engineering and Technology.

Dr. Pardeep kumar, Chairman, Delhi College of Technology and Management.

Dr. Gajendra Singh, Dean Academics, ABES.

Dr. Prabhakar Tiwari, Associate Profession, MMMUT.

A.AMBIKAPATHY

15th February, 2019

Prof (Dr).A. Ambikapathy

I wish to congratulate Prof (Dr).A. Ambikapathy for her efforts in writing book on "Control System" for UG (Electrical and Electronics Engineering) students. The entire book is written with simple language with large number example numerical are solved and large number of unsolved numerical with answer key are given for students to solve themselves. The special feature of this book is the presentation. This book covers all the topics connected to Control Systems including Lag, Lead , Lag lead compensators and state space analysis.

I forward this book to the future engineering graduates.

Prof(Dr).Pardeep Kumar

Pro VC, Galgotias University

Greater Noida

UP, India.

Plot No. 2, Sector-17A,
Yamuna Expressway,
Distt. Gautam Budh Nagar
0120 - 4370000

Contents

1

Basics of Control System

1.0 INTRODUCTION

Control system theory evolved as an engineering discipline and due to universality of the principles involved, it is extended to various fields like economy, sociology, biology, medicine, etc. The automatic control has become an integral part of modern manufacturing industries, controlling pressure, temperature, humidity, viscosity and flow in process industry.

1.1 CONTROL SYSTEM CLASSIFICATION

What is system?

When a number of elements (or) components are connected in a sequence to perform a specified function, the group thus formed is called a system.

What is control system?

In a system when the output quantity is controlled by varying the input quantity, the system is called control system. The output quantity is called controlled variable (or) response and input quantity is called command signal (or) excitation.

With respect to feedback signal, control system can be sub divided into two types (i) Open loop system and (ii) Closed loop system.

Control system can be classified in a number of ways. They are classified either as an open loop system (or) as a closed loop system. According to the differential equation describing the system, control systems are classified as linear or non linear, time varying or time invariant. control systems are classified as continuous data and discrete data systems according to the types of the signal found in the system. Control systems are also classified according to the components connected in the system. Electrical systems, mechanical systems, Hydraulic systems, Pneumatic systems, electromechanical systems etc. Depending upon the equations describing the behaviour, systems are classified as lumped or distributed system. System which have lumped parameters are described by differential equations whereas distributed system is described by partial differential equations. If the systems have fixed parameters and inputs they are called deterministic systems. On other hand systems which have randomness in parameters or inputs are called stochastic systems. Fig. (1.1) shows classification of control systems.

Let we see about open loop and closed loop system in brief.

1.2 OPEN LOOP SYSTEM

Any physical system which does not automatically correct the variation in its output, is called an open loop system, or system in which the output quantity has no effect upon the input quantity is called open loop control system. This means that the output is not feedback to the input for correction.

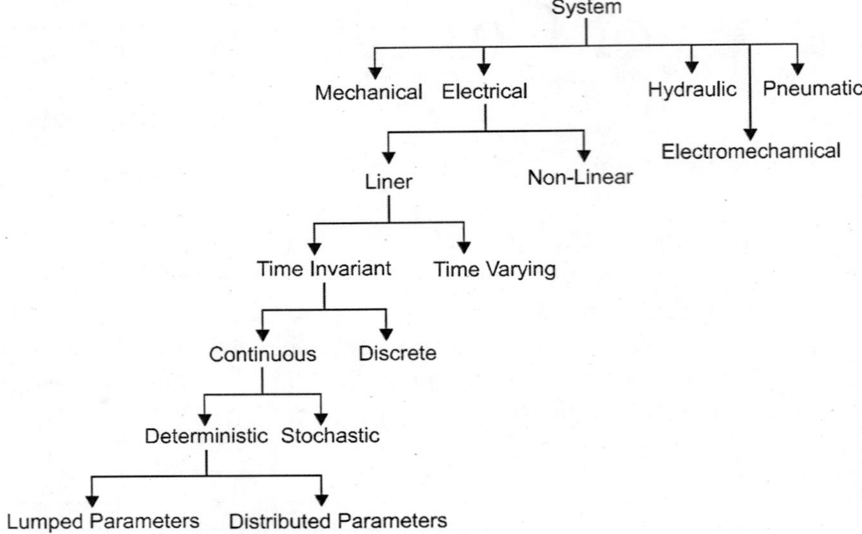

Fig. 1.1: Classification of control system.

In open loop control system the output can be varied by varying the input. But due to external disturbances the system output may change. When the output changes due to disturbances, it is not followed by changes in input to correct the output. The schematic block diagram of open loop control system shown in Fig. (1.2a).

I/P — Open Loop System (Plant) — O/P

Fig. 1.2(a): Block diagram of open loop system.

To understand the principle of operation of open loop system, consider the action of a blind man who attempts to take a book from a table. The blind man gives command to his brain to take the book. The brain senses his command and gives out order to the arms to take the book from the table. The blind man stretches out his hands and tries to take the book from the table. In his single attempt to get the book, he may succeed (or) fail.

In case he fails he has to change the reference input. The blind man may or may not succeed in his attempt. The blind man's action is an example of open loop system. There is no automatic correction of the error. The block diagram of the above action is shown in Fig. (1.2b)

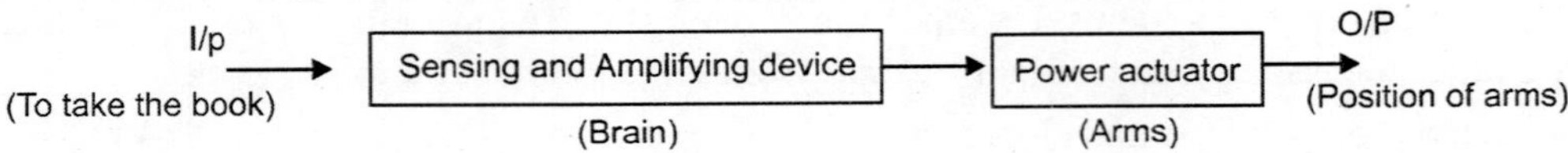

Fig. 1.2(b): Block diagram of open loop system with example.

1.3 CLOSED LOOP SYSTEM

Control system in which the output has an effect upon the input quantity so as to maintain the desired output are called closed loop system.

The open loop system can be modified as closed loop system by providing a feedback. The provision of feedback is that, it can automatically corrects the changes in output due to disturbances. Hence the closed loop system is also called automatic control system. The general block diagram of an automatic control system is shown in Fig. (1.3a). It consists of an error detector, controller, plant and feedback path elements.

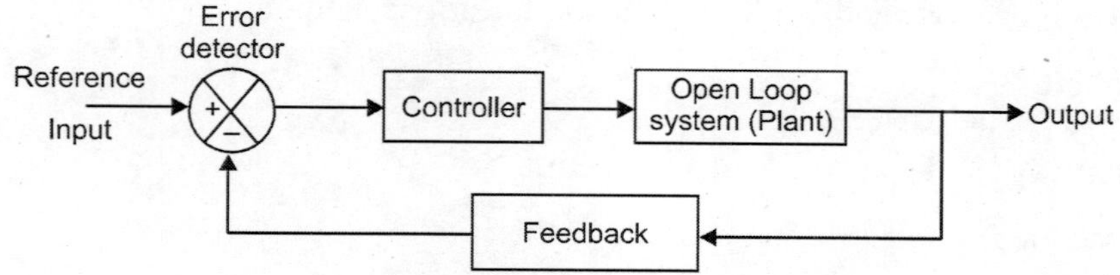

Fig. 1.3(a): Block diagram of closed loop system.

The feedback path elements samples the output and converts in to a signal of same types as that of reference signal. The feedback signal is proportional to output signal and it is fed to the error detector. The error signal which is generated by the error detector is the difference between reference signal and feedback signal. The controller modifies and amplifies the error signal to produce better control action. The modified error signal is fed to the plant to correct its output.

To understand the principle of operation of a closed-loop system, consider, the action of a normal man in picking up a book from the table. He gives command to the brain and the brain in turn commands and actuates the arms. The arms try to take the book If there is any error; suddenly the eyes detect that error and sends the signal to the brain. The brain gets aware of this error and moves the arms such that he pick up the book without miss. The eyes compare the input and output and tell the brain the amount of error between actual output and the desired output. Correction is automatically made by the brain until the desired result is achieved. This is an example of closed loop system and its block diagram is shown in Fig. (1.3b).

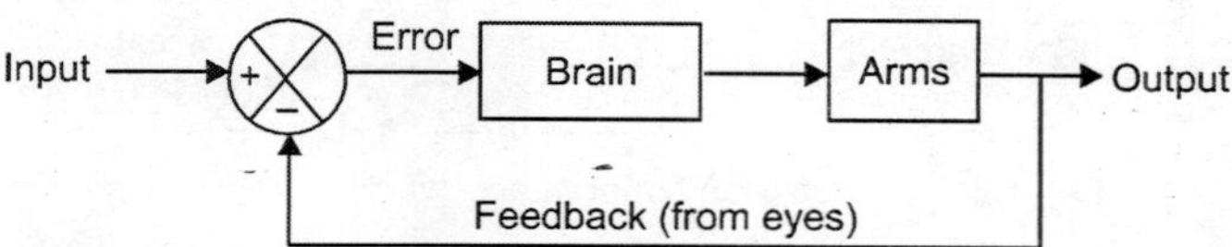

Fig. 1.3(b): Block diagram of closed loop system with example.

1.4 ADVANTAGES AND DISADVANTAGES OF OPEN AND CLOSED LOOP SYSTEM

1.4.1 Advantages of Open Loop System

1. The open loop systems is simple and economical.
2. The open loop system is easier to construct.
3. Generally the open loop systems are stable.

1.4.2 Disadvantages of Open Loop Systems

1. The open loop systems are inaccurate and unreliable.
2. The changes in the output due to external disturbances are not corrected automatically.

1.4.3 Advantages of Closed Loop Systems

1. The closed loop system outputs are more accurate to desired output.
2. The output is accurate both for linear and non-linearity conditions.
3. Less noise.

1.4.4 Disadvantages of Closed Loop Systems

1. The closed loop systems are complex and costlier
2. The feedback in closed loop system may lead to oscillatory response.
3. The feedback reduces the overall gain of the system.
4. Stability is a major problem in closed loop system and more care is needed to design a stable closed loop system.

1.5 EXAMPLES OF CONTROL SYSTEMS

1.5.1 Example 1: Temperature Control

Open loop system

The electric furnace shown in Fig. (1.5a) is an example for open loop system. The output in the system is desired temperature. The temperature of the system is raised by heat generated by the heating element. The output temperature depends on the time during which the supply to heater remains ON.

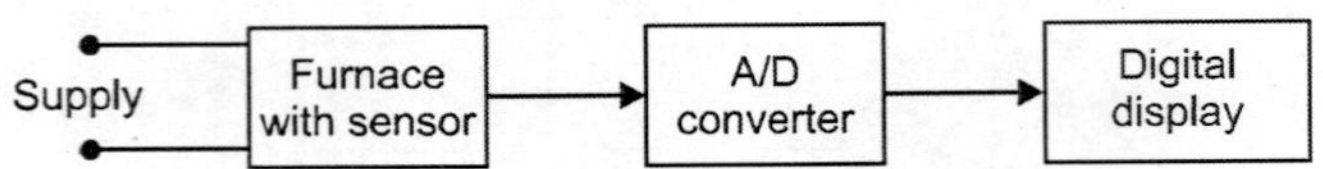

Fig. 1.5(a): Open loop temperature control system.

The ON and OFF of the supply is governed by the time setting. The temperature is measured by a sensor, which gives an analog voltage corresponding to the temperature of the furnace. The analog signal is converted to digital signal by an analog to digital converter (A/D converter). The digital signal is given to the digital display device to display the temperature.

Closed loop system

The electric furnace shown in Fig. (1.5b) is example for closed loop system. The output of the system is desired temperature and it depends on the time during which the supply remains ON.

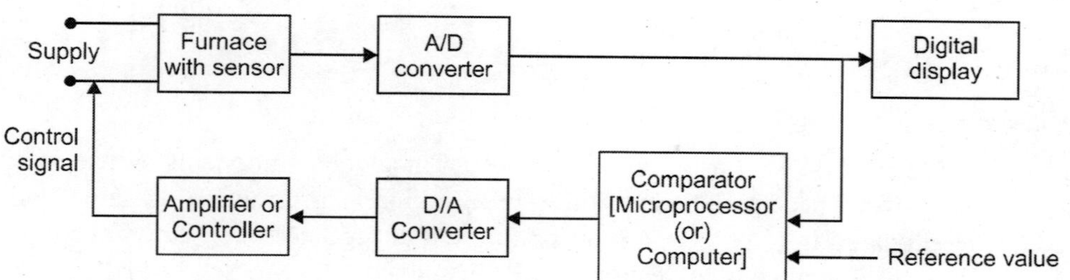

Fig. 1.5.1(b): Closed loop temperature control system.

The desired temperature is set to the system through keyboard or as a signal corresponding to desired temperature via ports. The actual temperature is sensed by sensor and converted to digital signal by the A/D converter. The comparator compares with desired temperature. If it finds any difference then it sends signals through D/A converter and Amplifier. Thus the system automatically corrects any change in output. Hence it is called closed loop system.

1.5.2 Example 2: Traffic Control

Open loop system

In traffic control, traffic signals which are operated on a time basis constitutes an open loop control system. The sequence of control signals are based on a time slot given for each signal. The time slots are decided based on a traffic study. The system will not measure the density of the traffic before giving the signal. Since the time slot does not change according to traffic density, the system is open loop system.

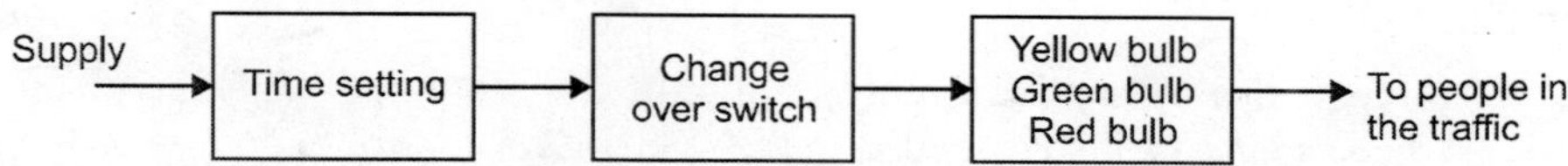

Fig. 1.5.2(a): Open loop traffic control system.

Closed loop system

Traffic control system can be made as a closed loop system if the time slots of the signals are decided based on the density of vehicles. In closed loop traffic control system, the density of the traffic is measured on all the sides and the information is fed to a computer. The timings of the control signals are decided by the computer based on the density of traffic. Since the closed loop system dynamically changes the timings according to the flow of vehicles, it will be better then open loop system.

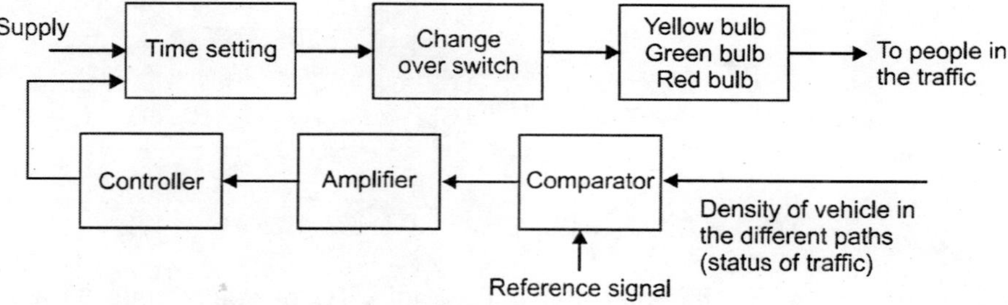

Fig. 1.5.2(b): Closed loop traffic control system.

1.5.3 Example 3: Numerical Control System

Open loop system

Numerical control is a method of controlling the motion of machine components by use of numbers. Here, the position of work head tool is controlled by the binary information contained on a disc. A magnetic disc is prepared in binary form representing the desired part P (P is the metal part to be machined). The tool will operate on the desired Part P. To start the system, the disc is fed through the reader to the digital to analog converter (D/A converter) The D/A converter converts the FM [frequency modulated] output of the reader to a analog signal. It is amplified and fed to servomotor which positions the cutter on the desired part P. The position of the cutter head is controlled by the angular motion of the servomotor. This is an open loop system since no feedback path exists between the output and input. The system positions the tool for a given input command. Any deviation in the desired position is not checked and corrected automatically.

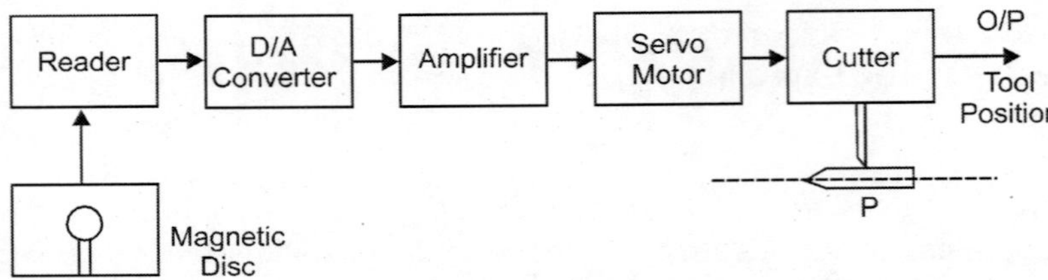

Fig. 1.5(c): Open loop numerical control system.

Closed loop system

A magnetic disc is prepared in binary form representing the desired part *P* (*P* is the metal part to be machined). To start the system, the disc is loaded in the reader. The controller compares the frequency modulated input pulse signal with the feedback pulse signal. The controller is a computer (or) micro-processor system. The controller carries out mathematical operations on the difference in the pulse signals and generates an error signal.

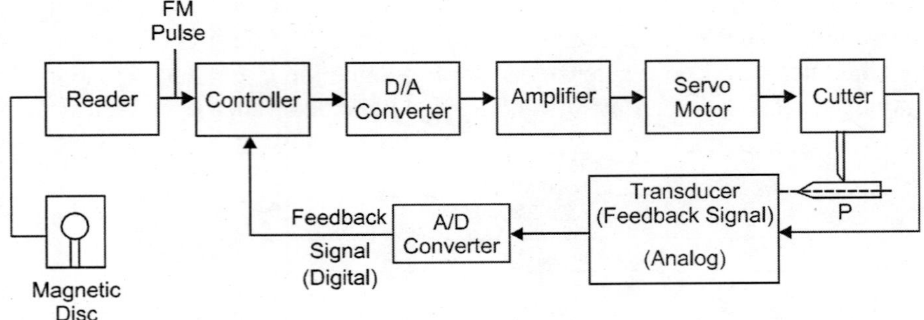

Fig. 1.5(d): Closed loop numerical control system.

The D/A converter converts the controller output pulse (error signal) into an analog signal. The amplified analog signal rotates the servomotor to position the tool on the job. The position of the cutter head is controlled according to the input of the servomotor. The transducer attached to the cutter head converts the motion into an electrical signal. The analog electrical signal is converted to the digital pulse signal by the A/D converter. Then this signal is compared with the input pulse signal. If there is any difference between these two, the controller sends a signal to the servomotor. Thus the system automatically corrects any deviation in the desired output position. An advantage of numerical control is that complex parts can be produced with uniform tolerances at the maximum milling speed.

1.5.4 Example 4: Position Control System using Servomotor

The position control system shown in Fig. (1.5e) is a closed loop system. The system consists of a servomotor powered by a generator. The load whose position has to be controlled is connected to motor shaft through gear wheels. Potentiometers are used to convert the mechanical motion to electrical signals. The desired load position (θ_R) is set on the input potentiometer and the actual load position (θ_C) is fed to feedback potentiometer. The difference between the two angular positions generates an error signal, which is amplified and fed to generator field circuit. The induced emf of the generator drives the motor. The rotation of the motor stops when the error signal is zero. i.e. when the desired load position is reached.

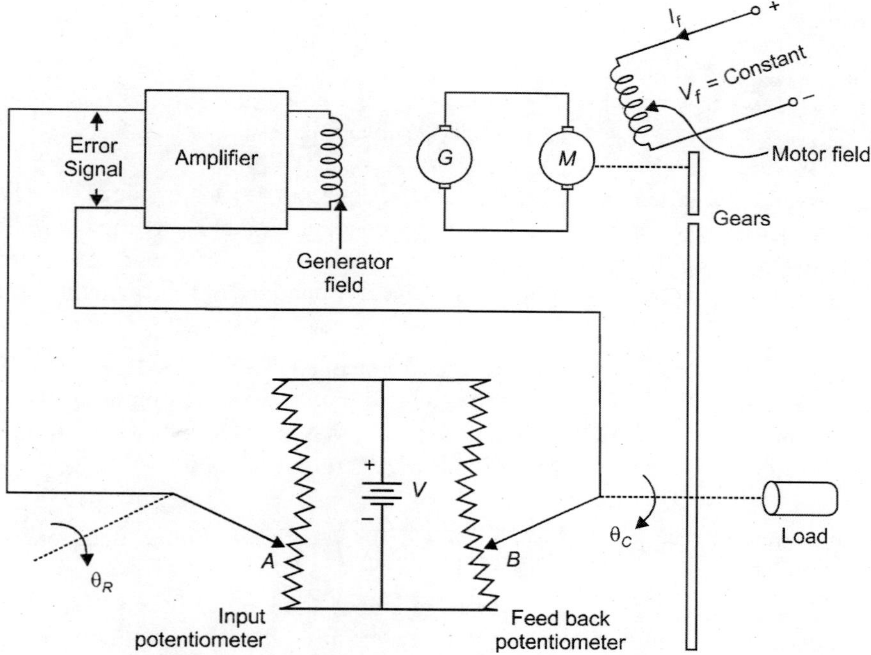

Fig. 1.5(e): A position control system (Servomechanism).

1.6 MATHEMATICAL MODEL OF CONTROL SYSTEM

A control system is a collection of physical objects (components) connected together to serve an objective. The input and output relations of various physical components of a system are governed by differential equations. The mathematical model of a control system constitutes a set of differential equations.

The response or output of the system can be studied by solving the differential equations for various input conditions.

The mathematical model of a system is linear if it obeys the principle of superposition and homogenity. The principle implies that if a system model has responses $y_1(t)$ and $y_2(t)$ to any inputs $x_1(t)$ and $x_2(t)$ respectively, then the system response to the linear combination of these inputs $a_1 x_1(t) + a_2 x_2(t)$ is given by linear combination of the individual outputs $a_1 y_1(t) + a_2 y_2(t)$ where a_1 and a_2 are constants. The principle of superposition can be explained diagrammatically let H be a linear system as shown in Fig. (1.6).

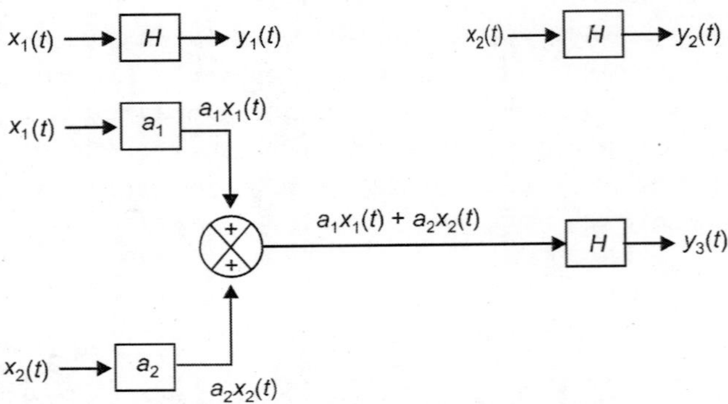

Fig. 1.6: Principle of linear and superposition.

A mathematical model will be linear if the differential equation describing the system has constant coefficients (or the coefficients may be function of independent variables). If the coefficients of the differential equation describing the system are constants then the model is linear time invariant. If the coefficients of differential equations governing the system are functions of time then the model is linear time varying.

The differential equations of a linear time invariant system can be reshaped into different form for the convenience of analysis. One such model for single input and single output system analysis is transfer function of the system. "The transfer function of a system is defined as the ratio of Laplace transform of output to the Laplace transform of input with zero initial conditions.

$$\text{Transfer function} = \frac{\text{Laplace transform of output}}{\text{Laplace transform of input}} \Big/ \text{with zero initial condition.}$$

The transfer function can be obtained by taking Laplace transform of the differential equations governing the system with zero initial conditions and rearranging the resulting algebraic equations to get the ratio of output to input.

1.7 THERMAL SYSTEM

List of symbols used in thermal system

$q \Rightarrow$ Heat flow rate, Kcal/sec

$\theta_1 \Rightarrow$ Absolute temperature of emitter, °K

$\theta_2 \Rightarrow$ Absolute temperature of receiver, °K

$\Delta\theta \Rightarrow$ Temperature difference, °C

$A \Rightarrow$ Area normal to heat flow, m^2

$K \Rightarrow$ Conduction coefficient, kcal/sec–°C

$K_f \Rightarrow$ Radiation coefficient, kcal/sec–°C

$K = HA \Rightarrow$ Convection coefficient, kcal/m^2–sec–°C

$K \Rightarrow$ Thermal conductivity, kcal/m–sec–°C

$\Delta x \Rightarrow$ Thickness of conductor, m

$R \Rightarrow$ Thermal resistance, °C–sec/kcal

$C \Rightarrow$ Thermal capacitance, kcal/°C

Heat flow rate

Thermal systems are those that involve the transfer of heat from one substance to another. There are different ways of heat flow from one substance to another. They are conduction, convection and radiation.

For conduction,

$$\text{Heat flow rate, } q = K\Delta\theta = KA/\Delta x$$

For Convection,

$$\text{Heat flow rate, } q = K\Delta\theta = HA\,\Delta\theta$$

For Radiation,

$$\text{Heat flow rate, } q = K_r\,(\theta^4{}_1 - \theta^4{}_2)$$

If $\theta_1 >> \theta_2$ then, $q = K_r\,\overline{\theta}^4$

where $$\overline{\theta}_4 = (\theta_1^4 - \theta_2^4)^{1/4}$$

Note: $\overline{\theta}_4$ is called effective temperature difference of the emitter and receiver.

The resistance and capacitance are basic elements of thermal system.

The thermal resistance, R is defined as the ratio of change in temperature and change in heat flow rate (when heat is transfered between two substances).

$$\text{Thermal resistance, } R = \frac{\text{Change in temperature, °C}}{\text{Change in heat flow rate, kcal/sec}} = \frac{d(\Delta\theta)}{dq}$$

For conduction or convection,

Heat flow rate, $q = K\Delta\theta$

On differentiating we get,

$$dq = Kd(\Delta\theta)$$

$$\therefore \quad \frac{d(\Delta\theta)}{dq} = \frac{1}{K}$$

But thermal resistance, $$R = \frac{d(\Delta\theta)}{dq}$$

$\therefore$ Thermal resistance, $R = \dfrac{1}{K}$ for conduction

$= \dfrac{1}{K} = \dfrac{1}{HA}$ for convection

For radiation,

Heat flow rate, $q = k_r\, \overline{\theta}_4$

on differentiating we get $dq = K_r,\, 4\overline{\theta}^3\, d\overline{\theta}$

$$\therefore \quad \frac{d\overline{\theta}}{dq} = \frac{1}{K_r 4\overline{\theta}^3}$$

But thermal resistance, $R = \dfrac{d\overline{\theta}}{dq}$

$\therefore$ Thermal resistance (for radiation)

$$R = \frac{1}{4K_r\overline{\theta}^3}$$

Thermal capacitance, C is defined as the ratio of change in heat stored and change in temperature.

Thermal capacitance, $C = \dfrac{\text{Change in heat stored, kcal}}{\text{Change in temperature, °C}}$

Let,

$M \rightarrow$ Mass of substance considered, kg

$C_p \rightarrow$ Specific heat of substance, kcal/ kg - °C

Now,

Thermal capacitance, $C = MC_p$

Consider a simple thermal system shown in Fig. (1.7). Let us assume that the tank is insulated to eliminate heat loss to the surrounding air, there is no heat storage in the insulation and liquid in the tank is kept at uniform temperature by perfect mixing. Let derive the transfer function for the Fig. (1.7).

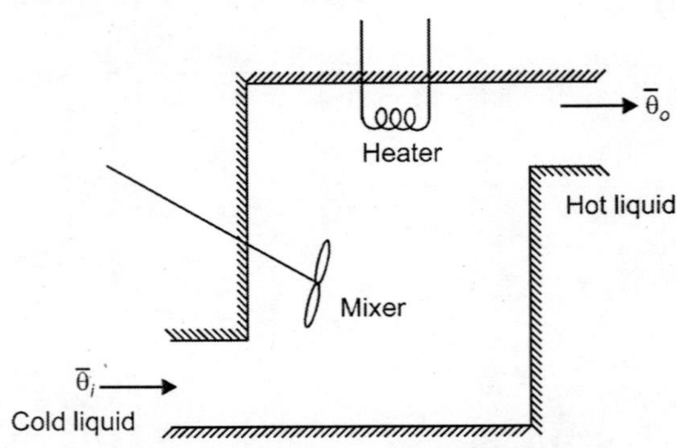

Fig. 1.7: Thermal system cold liquid.

Let,

$\overline{\theta}_i$ ⇒ Steady state temperature of inflowing liquid, °C.

$\overline{\theta}_0$ ⇒ Steady state temperature of outflowing liquid, °C.

G ⇒ Steady state liquid flow rate, kg/sec.

M ⇒ Mass of liquid in tank, kg.

c ⇒ Specific heat of liquid kcal/kg °C.

R ⇒ Thermal resistance, °C - sec/ kcal

C ⇒ Thermal capacitance, kcal/ °C

Q ⇒ Steady state heat input rate. kcal/sec.

Let us assume that the temperature of inflowing liquid is kept constant. Let the heat input rate to the system supplied by the heater is suddenly changed from $\overline{Q}$ to $\overline{Q} + q_i$.

Due to this, the heat output flow rate will gradually change from $\overline{Q}$ to $\overline{Q} + q_0$. The temperature of the outflowing liquid will also be changed from $\overline{Q}_0$ to $\overline{Q}_0 + Q$.

For this system the equation for q_0, C and R are obtained as follows,

Change in output heat flow rate q_0 = {Liquid flow rate, G} × {Specific heat of liquid, c} × {Change in temperature θ}

$$q_0 = Gc\theta \tag{1.1}$$

Thermal capacitance C = Mass, (M) × {Specific heat of liquid (c)}

$$C = Mc \tag{1.2}$$

$$\text{Thermal resistance, } R = \frac{\text{Change in temperature, } \theta}{\text{Change in heat flow rate, } q_0}$$

$$R = \frac{\theta}{q_0} \tag{1.3}$$

On substituting for q_0 from equation (1.1) in equation (1.3) we get,

$$R = \frac{\theta}{Gc\theta} = \frac{1}{Gc}$$

$$R = \frac{1}{Gc} \tag{1.4}$$

In this system, the rate of change of temperature is directly proportional to change in heat input rate.

$$\therefore \quad \frac{d\theta}{dt} \propto q_i - q_0 \tag{1.5}$$

Let the constant of proportionality is the capacitance C of the system.

$$\therefore \quad \frac{C\,d\theta}{dq} = q_i - q_o \tag{1.6}$$

Equation (1.6) is the differential equation governing the system. Since equation (1.6) is of first order equation, the system is first order system.

From equation (1.3), $R = \dfrac{\theta}{q_0}$,

$$\therefore \quad q_o = \frac{\theta}{R} \qquad (1.7)$$

On substituting for q_0 from equation (1.7) in equation (1.6) we get

$$C\frac{d\theta}{dt} = q_i - \frac{\theta}{R}$$

$$C\frac{d\theta}{dt} = \frac{Rq_i - \theta}{R}$$

$$RC\frac{d\theta}{dt} = Rq_i - \theta$$

$$RC\frac{d\theta}{dt} + \theta = Rq_i \qquad (1.8)$$

Let, $L(\theta) = \theta(s)$; $L\left[\dfrac{d\theta}{dt}\right] = s\theta(s)$ and $L[q_i] = Q_i(s)$

On taking Laplace transform of equation (1.8) we get,

$$RC\, s\theta\,(s) + \theta(s) = RQ_i\,(s)$$

$$\theta(s)\,[sRC + 1] = RQ_i\,(s)$$

$\dfrac{\theta(s)}{Q_i(s)}$ is the required transfer function of the system.

$$\frac{\theta(s)}{Q_i(s)} = \frac{R}{sRC+1} = \frac{R}{RC\left[s+\dfrac{1}{RC}\right]} = \frac{\dfrac{1}{C}}{s+\dfrac{1}{RC}}$$

$$\frac{\theta(s)}{Q_i(s)} = \frac{\dfrac{1}{C}}{s+\dfrac{1}{RC}} \qquad (1.9)$$

The above equation (1.9) is the transfer function of thermal system of example which is taken for analysis.

The transfer function of thermal system of Fig. (1.8) is

$$\frac{\theta(s)}{Q_i(s)} = \frac{1/C}{s+\dfrac{1}{RC}}$$

In equation (1.9), the term RC is time constant of the system.

1.8 HYDRAULIC SYSTEM

The hydraulic system may be classified into

1. Liquid level system
2. Hydraulic devices

The liquid level system consists of storage tanks and connecting pipes. The variables to be controlled are liquid height in tanks and flow rate in pipes. The driving force is the relative difference of the liquid heights in the tanks.

The hydraulic devices are using incompressible oil as their working medium. These devices are used for controlling the forces and motions. The driving force is the high pressure oil supplied by the hydraulic pumps.

The variables of hydraulic system are volumetric flow rate, q and pressure, P. The volumetric flow rate, q is through variable and it is analogous to current. The pressure, P is across variable and it is analogous to voltage.

Three basic elements of hydraulic systems are resistance, capacitance and Inertance. The liquid flowing out of a tank can meet the resistance in several ways. Liquid while flowing through a pipe meet with resistance due to the friction between pipe walls and liquid. Presence of valves, bends, coupling of pipes of different diameter also offer resistance to liquid flow.

The capacitance is an energy storage element and it represents storage in gravity field. The inertance represents fluid inertia and is derived from the inertia forces required to accelerate the fluid in a pipe. It is also an energy storage element. But the energy storage due to inertance element is negligible compared to that of capacitance element.

Consider the flow through a short pipe connecting two tanks. The resistance for liquid flow in such a pipe or restriction, is defined as the change in the level difference, necessary to cause a unit change in the flow rate.

$$R = \frac{\text{Change in level difference, m}}{\text{Change in flow rate, m}^3\text{/sec}}$$

Capacitance C of a tank is defined as the change in quantity of stored liquid, necessary to cause a unit change in the potential (head).

$$C = \frac{\text{Change in liquid stored, m}^3}{\text{Change in head, m}}$$

A simple liquid level system shown in Fig. (1.8a) with the steady state flow rate, $\bar{Q}$ and steady state head, $\bar{H}$

Let,

$\bar{Q} \rightarrow$ Steady state flow rate (before any change occurred)

$q_i \rightarrow$ Small deviation of inflow rate from its steady state valve, m^3/sec

$q_o \rightarrow$ Small deviation of outflow rate from its steady state valve, m^3/sec

$\bar{H} \rightarrow$ Steady state head (before any change has occurred), m

$h \rightarrow$ Small deviation of head from its steady-state valve, m.

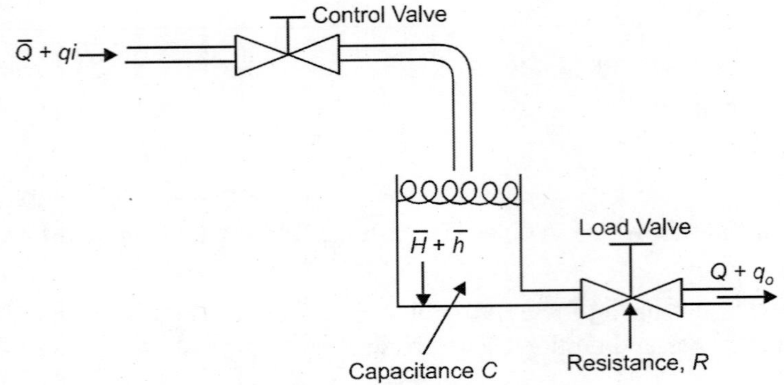

Fig. 1.8(a): Liquid level system.

Let the system be considered linear. The differential equation governing the system is obtained by equating, the change in flow rate to the amount stored in the tank. In a small time interval dt, Let the change in flow rate be $(q_i - q_o)$, and the change in height be dh.

Now, change in storage = change in flow rate

$$\therefore \quad C\,dh = (q_i - q_o)\,dt \tag{1.10}$$

The resistance,

$$R = \frac{\text{Change in head}}{\text{Change in outflow rate}}$$

$$= \frac{h}{q_o} \tag{1.11}$$

$$\therefore \quad q_o = \frac{h}{R} \tag{1.12}$$

On substituting for q_0 from equation (1.12) in equation (1.10) we get,

$$C\,dh = \left(q_i - \frac{h}{R}\right) dt$$

$$C\,dh = \left(\frac{q_i R - h}{R}\right) dt$$

$$RC\frac{dh}{dt} = q_i R - h$$

$$RC\frac{dh}{dt} + h = q_i R \tag{1.13}$$

The equation (1.13) is the differential equation governing the system. The term RC is the time constant of the system. On taking Laplace transform of equation (1.13) we get,

$$RC\,sH(s) + H(s) = Q_i(s)\,R$$

$$(sRC + 1)\,H(s) = Q_i(s)\,R$$

$$\frac{H(s)}{Q_i(s)} = \frac{R}{(sRC+1)} = \frac{R}{RC\left(s+\frac{1}{RC}\right)} = \frac{\frac{1}{C}}{s+\frac{1}{RC}}$$

$$\boxed{\frac{H(s)}{Q_i(s)} = \frac{\frac{1}{C}}{s+\frac{1}{RC}}} \qquad (1.14)$$

The equation (1.14) is the required transfer function of the system.

The hydraulic devices are used in hydraulic feedback systems and in combined electro-mechanical-hydraulic systems. In hydraulic devices, power is transmitted through the action of fluid flow under pressure and the fluid is in compressible. The fluid used are petroleum based oils or non-inflammable synthetic oils.

The most frequently used hydraulic device in control system is hydraulic motor-pump set. It consists of a variable stroke, hydraulic pump and a fixed stroke hydraulic motor as shown in Fig. (1.8b). The device accepts a linear displacement (stroke length) as input and delivers a large output torque.

The hydraulic motor is controlled by amount of oil delivered by the pump. By mechanically changing the pump stroke, the oil delivered by the pump is controlled. In a pump the input is mechanical power and output is hydraulic power.

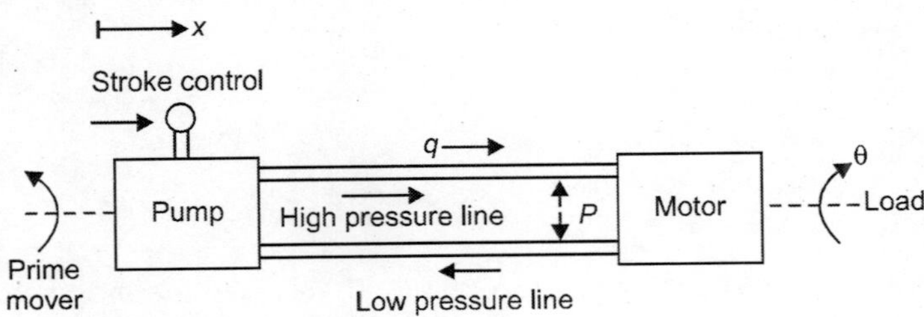

Fig. 1.8(b): Hydraulic motor-pump set.

Let,

$q_p \Rightarrow$ Rate at which the oil flows from the pump

$q_m \Rightarrow$ Oil flow rate through the motor

$q_i \Rightarrow$ Leakage flow rate

$q_c \Rightarrow$ Compressibility flow rate

$x \Rightarrow$ Input stroke length

$\theta \Rightarrow$ Output angular displacement of motor

$P \Rightarrow$ Pressure drop across motor.

The rate at which the oil flow from the pump is proportional to stroke angle, i.e., $q_p \alpha\, x$.

$\therefore$ oil flow rate from the pump, $q_p = K_p x$ (1.15)

where K_p is a constant and is equal to the ratio of rate of oil flow to unit stroke angle.

The rate of oil flow through the motor is proportional to motor speed i.e., $q_m \, \alpha \, \frac{d\theta}{dt}$

$\therefore$ oil flow rate through motor} $q_m = K_m \frac{d\theta}{dt}$ (1.16)

where $K_m \rightarrow$ motor displacement constant.

All the oil from the pump does not flow through the motor in the proper channels. Due to back pressure in the motor, a portion of the ideal flow from the pump leaks back in the pistons of motor and pump. The back pressure is the pressure that is built up by the hydraulic flow to overcome the resistance to free movement offered by load on motor shaft.

It is usually assumed that the leakage flow is proportional to motor pressure, i.e. $q_i \, \alpha \, p$

$\therefore$ Leakage flow rate, $q_i = K_i \, P$

where, $K_i \rightarrow$ constant

The back pressure built up by the motor not only causes leakage flow in the motor and pump but also causes the oil in the lines to compress. Volume compressibility flow is essentially proportional to pressure and therefore the rate of flow is proportional to the rate of change of pressure, i.e., $q_c \, \alpha \, \frac{dp}{dt}$

$\therefore$ Compressibility flow rate, $q_c = K_c \frac{dp}{dt}$ (1.17)

when $K_c \rightarrow$ coefficient of compressibility.

The rate at which oil flows from the pump is given by sum of oil flow rate through the motor, leakage flow rate and compressibility flow rate.

$$q_p = q_m + q_i + q_c \quad (1.18)$$

on substituting from equation q_m, q_i and q_c values in equation (1.18) we get,

$$K_p \, x = K_m \frac{d\theta}{dt} + K_i \, P + K_c \frac{dp}{dt} \quad (1.19)$$

The torque T_m developed by the motor's proportional to the pressure drop and balances the load torque.

$\therefore$ Hydraulic motor torque, $T_m \propto P \Rightarrow T_m = K_t P$ (1.20)

where K_t is motor torque constant.

If the load is consist of moment of inertia J and viscous friction with coefficient B,

Then, Load torque, $T_l = J \frac{d^2\theta}{dt^2} + B \frac{d\theta}{dt}$ (1.21)

Hydraulic power input $= q_m P$ (1.22)

Substitute for q_m, from equation (1.16) in equation (1.22)

$$\text{Hydraulic power input} = K_m \frac{d\theta}{dt} P \quad (1.23)$$

$$\text{Mechanical power output} = T_m \frac{d\theta}{dt} \qquad (1.24)$$

Substitute for T_m from equation (1.20) in equation (1.24)

$$\therefore \quad \text{Mechanical power output} = K_t P \frac{d\theta}{dt} \qquad (1.25)$$

It hydraulic motor losses are neglected then the hydraulic motor input is equal to mechanical power output of hydraulic motor.

$$\therefore \quad K_m \frac{d\theta}{dt} P = K_t \, P \frac{d\theta}{dt} \qquad (1.26)$$

From equation (1.26), it is clear that $K_m = K_t$ Hence, equation (1.20) can be written as

$$T_m = K_t p = K_m P$$

Since the motor toque equals to load torque

$$T_m = T_l$$

$$\therefore \quad K_m \, P = J \frac{d^2\theta}{dt^2} + B \frac{d\theta}{dt}$$

$$P = \frac{J}{K_m} \frac{d^2\theta}{dt^2} + \frac{B}{K_m} \frac{d\theta}{dt} \qquad (1.27)$$

On differentiating with respect to t, we get

$$\frac{dP}{dt} = \frac{J}{K_m} \frac{d^3\theta}{dt^3} + \frac{B}{K_m} \frac{d^2\theta}{dt} \qquad (1.28)$$

On substituting for P and $\frac{dP}{dt}$ from equation (1.27) and equation (1.28) in equation (1.19) we get,

$$K_p x = K_m \frac{d\theta}{dt} + K_i \left[\frac{J}{K_m} \frac{d^2\theta}{dt^2} + \frac{B}{K_m} \frac{d\theta}{dt} \right] + K_c \left[\frac{J}{K_m} \frac{d^3\theta}{dt^3} + \frac{B}{K_m} \frac{d^2\theta}{dt^2} \right]$$

$$K_p x = \frac{K_c J}{K_m} \frac{d^3\theta}{dt^2} + \left[\frac{K_i J}{K_m} + \frac{K_c B}{K_m} \right] \frac{d^2\theta}{dt^2} + \left[K_m + \frac{K_i B}{K_m} \right] \frac{d\theta}{dt}$$

On taking Laplace transform with zero initial conditions, we get,

$$K_p X(s) = \frac{K_c J}{K_m} s^3 \theta(s) + \left[\frac{K_i J}{K_m} + \frac{K_c B}{K_m} \right] s^2 \theta(s) + \left[K_m + \frac{K_i B}{K_m} \right] s\theta(s)$$

$$\frac{\theta(s)}{X(s)} = \frac{K_p}{s \left[\frac{K_c J}{K_m} s^2 + \left[\frac{K_i J + K_c B}{K_m} \right] s + \frac{K_{m^2} + K_i B}{K_m} \right]} \qquad (1.29)$$

In hydraulic systems, normally $K_c << K_m$,
therefore, Put $K_c = 0$ in equation (1.29)

$$\frac{\theta(s)}{X(s)} = \frac{K_p}{s\left[\frac{K_i J}{K_m} s + \frac{K_{m^2} + K_i B}{K_m}\right]} \tag{1.30}$$

The equation (1.30) is the required transfer function of the system

1.9 PNEUMATIC SYSTEM

Pneumatic system uses compressible fluid as working medium usually air. In pneumatic systems, compressibility effects of gas cannot be neglected and hence dynamic equations are obtained using conservation of mass. In pneumatic systems, change in fluid inertia energy and the fluid's internal thermal energy are assumed negligible.

The variables of pneumatic system are mass flow rate, q_m and pressure, P. The mass flow rate is through variable and it is analogous to current. The pressure is across variable and it is analogous to voltage. The two basic elements of pneumatic system are the resistance and capacitance. The restrictions in the pipes and valves offers resistance to gas flow.

The gas flow resistance, R is defined as the rate of change in gas pressure difference for a change in gas flow rate.

$$R = \frac{\text{Change in gas pressure difference, N/m}^2}{\text{Change in gas flow rate, kg/sec}}$$

The pneumatic capacitance is defined for a pressure vessel and depends on the type of expansion process involved. The capacitance of a pressure vessel may be defined as the ratio of change in gas stored for a change in gas pressure.

$$C = \frac{\text{Change in gas stored, kg}}{\text{Change in gas pressure, N/m}^2}$$

Pneumatic devices are employed in guided missiles, air craft systems, automation of production machinery and in many other fields as automatic controllers.

A simple pneumatic system shown in Fig. (1.9) and it consists of a pneumatic bellows in line with the restriction. The pneumatic bellows consists of a hollow chamber with thin pneumatic walls. The side walls of bellows are corrugated and the input and output surface are flat. An increase in pressure within the bellows results in an increase in separation between the input and output surfaces.

Let

$\bar{P}_i \Rightarrow$ Steady state value of input air pressure.

$p_i \Rightarrow$ Increase in the pressure of air-source.

$\bar{P} \Rightarrow$ Steady state value of pressure inside the bellows.

$p \Rightarrow$ Increase in pressure inside the bellow.

$\bar{Q}_m \Rightarrow$ Steady-state valve of air flow rate.

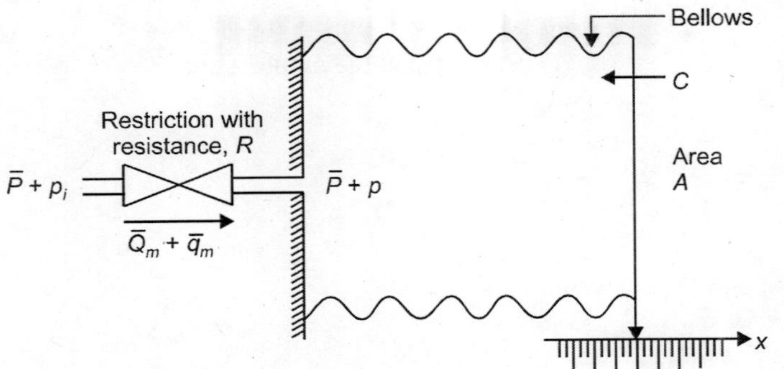

Fig. 1.9: Pneumatic system.

$q_m \Rightarrow$ Increase in air flow rate.

$A \Rightarrow$ Area of each flat surface of the bellows.

$R \Rightarrow$ Resistance of the restriction.

$C \Rightarrow$ Capacitance of the bellows.

$x \Rightarrow$ Displacement of the movable surface of the bellows.

Let the pressure of air source be increased from its steady state valve by an amount p_i. Thus results in an increase in air flow by q_m and increase in the pressure inside the bellows by p. Due to increase in pressure, there will be a displacement of the movable surface of the bellows, by an amount x. Here, the terms p_i, q_m, p and x are all functions of time t and therefore can be expressed as $p_i(t)$, $q_m(t)$ $p(t)$ and $x(t)$.

The force exerted on the movable surface of the bellows is proportional to increase in pressure inside the bellows, i.e., $f_b \propto p(t)$

∴ Force exerted on the movable surface of the bellow,

$$f_b = Ap(t) \tag{1.31}$$

The force opposing the movement of the flat surface of bellow walls is proportional to the displacement i.e., $f_0 \propto x(t)$.

∴ Force opposing the motion,

$$f_0 = Kx(t) \tag{1.32}$$

where $K \rightarrow a$ constant representing the stiffness of the bellows.

At steady state the above two forces are balanced,

∴ $$f_b = f_0 \quad \therefore\ Ap(t) = Kx(t) \tag{1.33}$$

$$\text{The resistance, } R = \frac{\text{Difference between change in pressure}}{\text{Change in air flow rate}}$$

$$= \frac{p_i(t) - p(t)}{q_m(t)}$$

∴ $$q_m(t) = \frac{p_i(t) - p(t)}{R} \tag{1.34}$$

The capacitance, $$C = \frac{\text{Change in air flow rate}}{\text{Rate of change of pressure}}$$

$$= \frac{q_m(t)}{dp(t)/dt}$$

$$\therefore \quad q_m(t) = C\frac{dp(t)}{dt} \tag{1.35}$$

On equating (1.34) and (1.35) we get

$$C\frac{dp(t)}{dt} = \frac{p_i(t) - p(t)}{R} \tag{1.36}$$

$$RC\frac{dp(t)}{dt} + p(t) = p_i(t) \tag{1.37}$$

From the equation (1.33), we get

$$p(t) = \frac{K}{A}x(t) \tag{1.38}$$

On differentiating equation (1.38) with respect to t, we get

$$\frac{dp(t)}{dt} = \frac{K}{A}\frac{dx(t)}{dt} \tag{1.39}$$

On substituting for $p(t)$ and $\frac{dp(t)}{dt}$ from equation (1.38) and (1.39) in equation (1.37) we get,

$$RC\left[\frac{K}{A}\frac{dx(t)}{dt}\right] + \frac{K}{A}x(t) = p_i(t) \tag{1.40}$$

On taking Laplace transform with zero initial conditions we get

$$RC\left[\frac{K}{A}s\,X(s)\right] + \frac{K}{A}X(s) = P_i(s)$$

$$\frac{X(s)}{P_i(s)} = \frac{A/K}{RCs+1} = \frac{A/K}{\tau s+1}$$

where $\tau = RC \rightarrow$ time constant of the system

$$\frac{X(s)}{P_i(s)} = \frac{A/K}{\tau s+1} \tag{1.41}$$

The equation (1.41) is the required transfer function of the system.

1.10 ELECTRICAL SYSTEM

The models of electrical systems can be obtained by using resistor, capacitor and inductor. The current-voltage relation of resistor, inductor and capacitor are given in table 1. For modelling electrical systems the electrical network or equivalent circuit is formed by using R, L and C and voltage or current source.

The differential equations governing the electrical systems can be formed by writing Kirchoff's current law and Kirchoff's voltage Law equations. The transfer function can be obtained by taking Laplace transform of the differential equations and rearranging them as a ratio of output to input.

Table 1

Element	Voltage across the element	Current through the element
i(t) R + V(t) −	$V(t) = Ri(t)$	$i(t) = V(t)/R$
i(t) L + V(t) −	$V(t) = L\frac{d}{dt}i(t)$	$i(t) = \frac{1}{L}\int V(t)\,dt$
i(t) C + − V(t)	$V(t) = \frac{1}{C}\int i(t)\,dt$	$i(t) = C\frac{dV(t)}{dt}$

Let consider one electrical network and derive transfer function for the Fig. (1.10a).

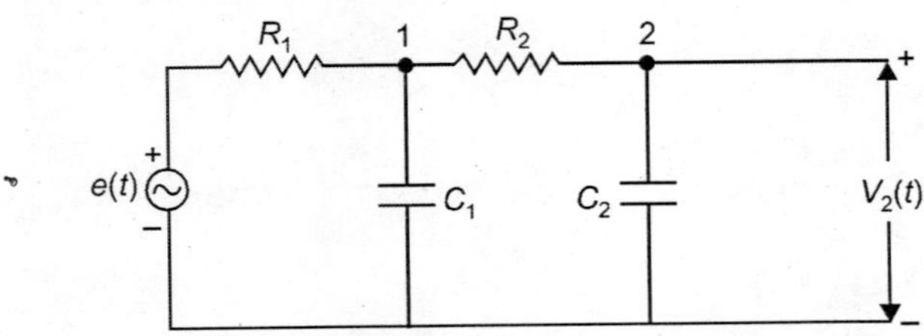

Fig. 1.10(a): Electrical network.

$e(t) \rightarrow$ Input voltage

$V_2(t) \rightarrow$ Output voltage

Laplace transform of $e(t) = L(e(t)) = E(s)$

Laplace transform of $V_2(t) = V_2(s)$

The transfer function of the network is $\frac{V_2(s)}{E(s)}$

Convert the voltage source which is in series with resistance R_1 into equivalent current source as shown in Fig. (1.10b). The network has two nodes. Let the node voltages be V_1 and V_2.

The laplace transform of node voltages V_1 and V_2 are $V_1(s)$ and $V_2(s)$ respectively. The differential equations governing the network are given by the Kirchoff's Current Law equations at these nodes.

At node 1, by Kirchoff's current law (Fig. 1.10b)

$$\frac{V_1}{R_1} + C_1 \frac{dV_1}{dt} + \frac{V_1 - V_2}{R_2} = \frac{e}{R_1} \tag{1.42}$$

on taking laplace transform with zero initial conditions.

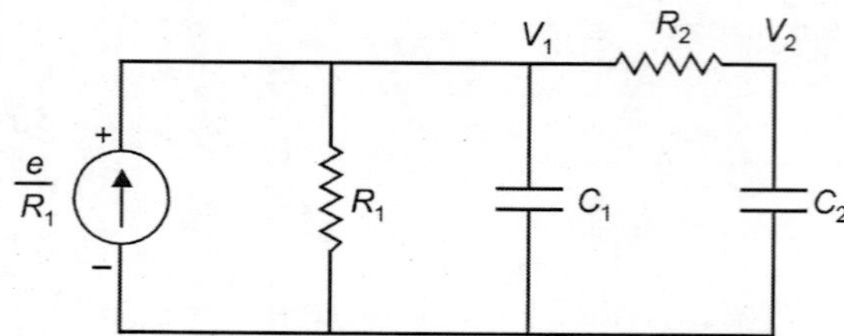

Fig. 1.10(b): Electrical network (After converting voltage source into current source).

$$\frac{V_1(s)}{R_1} + C_1\, sV_1(s) + \frac{V_1(s)}{R_2} - \frac{V_2(s)}{R_2} = \frac{E(s)}{R_1} \tag{1.43}$$

$$V_1(s)\left[\frac{1}{R_1} + sC_1 + \frac{1}{R_2}\right] - \frac{V_2(s)}{R_2} = \frac{E(s)}{R_1}$$

At node 2, by Kirchoff's current law (Fig. (1.10b)

$$\frac{V_2 - V_1}{R_2} + C_2 \frac{dV_2}{dt} = 0 \tag{1.44}$$

on taking laplace transform of equation (1.43) with zero initial conditions.

$$\frac{V_2(s)}{R_2} - \frac{V_1(s)}{R_2} + C_2\, s\, V_2(s) = 0$$

$$\frac{V_1(s)}{R_2} = \frac{V_2(s)}{R_2} + C_2\, s\, V_2(s) = \left[\frac{1}{R_2} + s\, C_2\right] V_2(s)$$

$$V_1(s) = [1 + sC_2R_2]\; V_2(s) \tag{1.45}$$

substituting for $V_1(s)$ from equation (1.45) in equation (1.43) we get,

$$(1 + s\, R_2\, C_2)\; V_2(s)\left[\frac{1}{R_1} + s\, C_1 + \frac{1}{R_2}\right] - \frac{V_2(s)}{R_2} = \frac{E(s)}{R_1}$$

$$\left[\frac{(1 + s\, R_2\, C_2)(R_2 + R_1 + s\, C_1\, R_1\, R_2) - R_1}{R_1 R_2}\right] V_2(s) = \frac{E(s)}{R_1}$$

$$\frac{V_2(s)}{E(s)} = \frac{R_2}{[(1 + s\, R_2\, C_2)(R_1 + R_2 + s\, C_1\, R_1\, R_2) - R_1]} \tag{1.46}$$

The equation (1.46) is the required transfer function of the system.

Difference between Hydraulic and Pneumatic system

Pneumatic	Hydraulic
1. Working fluid is compressible	1. Working fluid is incompressible
2. Working fluid lack lubricating property	2. Working fluid acts as a lubricant
3. Operating pressure is lower	3. Higher operating pressure
4. Output power is less	4. More output power
5. Accuracy of actuator is poor	5. More accuracy can be made
6. External leakage is permissible but internal leakage must to avoided	6. Internal leakage is permissible and external leakage must be avoided
7. Non-return pipes are required	7. Return pipes are required
8. In sensitive to temperature changes	8. Sensitive to changes in temperature.
9. Fire and explosion proof	9. Not a fire and explosion proff.

1.11 ELECTROMECHANICAL SYSTEM

Electro mechanical system consists of both electrical system and mechanical system. For example generator and motor both are called as electromechanical system. In case of generator input is mechanical and output is electrical then in case of motor input is electrical and output is mechanical.

The differential equation governing the electrical system can be formed by Kirchoff's laws and for mechanical system, the equation is formed by newton's second law.

Example 1: Transfer function of Armature Controlled DC motor

The speed of DC motor is directly proportional to armature voltage and inversely proportional to flux. In armature controlled DC motor, the desired speed is obtained by varying the armature voltage. This is an electromechanical control system. The electrical system consists of armature and field circuits but for analysis purpose, only the armature circuit is considered because the field is excited by constant voltage. The mechanical system consists of the rotating part and load connected to the shaft of the motor. The armature controlled DC motor system is shown in Fig. (1.11a).

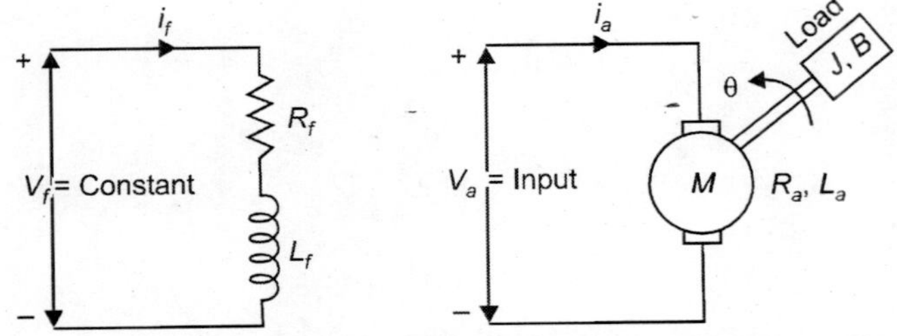

Fig. 1.11(a): Armature controlled DC motor.

Let,

$R_a \rightarrow$ Armature Resistance, Ω

$L_a \rightarrow$ Armature Inductance, H

$i_a \rightarrow$ Armature Current, A

$V_a \rightarrow$ Armature voltage, V

$e_b \rightarrow$ Back emf, V

$K_t \rightarrow$ Torque constant, $N-m/A$

$T \rightarrow$ Torque developed by motor, $N-m$

$\theta \rightarrow$ Angular displacement of shaft, Radian

$J \rightarrow$ Moment of inertia of motor and load, kgm^2/Rad

$B \rightarrow$ Frictional co-efficient of motor and load, $N-m/$ (rad - sec)

$K_b \rightarrow$ Back emf constant, $V/$(rad - sec)

The equivalent circuit of armature is shown in Fig. (1.11b)

By applying KVL to Fig. (1.11b), we get,

$$i_a R_a + L_a \frac{di_a}{dt} + e_b = V_a \tag{1.47}$$

Torque of DC motor is proportional to the product of flux and current. Since flux is constant in this system the torque is proportional current alone.

$$T \alpha\, i_a$$

∴ Torque

$$T = K_t\, i_a \tag{1.48}$$

Fig. 1.11(b): Equivalent circuit of Armature.

The mechanical system of the motor is shown in Fig. (1.11c). the differential equation governing the mechanical system of motor is given by

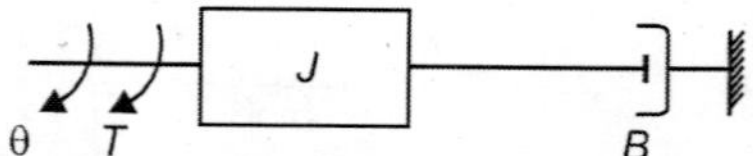

Fig. 1.11(c): Mechanical system of the motor.

$$J\frac{d^2\theta}{dt^2} + B\frac{d\theta}{dt} = T \tag{1.49}$$

The back emf of DC machine is proportional to speed of shaft.

∴
$$e_b \,\alpha \frac{d\theta}{dt};\; e_b = K_b \frac{d\theta}{dt} \tag{1.50}$$

The laplace transform of various time domain signals involved in this system are shown below

$$L(V_a) = V_a(s); \quad L[e_b] = E_b(s)$$

$$L[i_a] = I_a(s); \quad L[\theta] = \theta(s)$$
$$L(T) = T(s)$$

The differential equations governing the armature controlled DC system are

$$i_a R_a + L_a \frac{di_a}{dt} + e_b = V_a; \qquad J\frac{d^2\theta}{dt^2} + B\frac{d\theta}{dt} = T$$

$$T = K_t i_a; \qquad e_b = K_b \frac{d\theta}{dt}$$

on taking laplace transform of the differential equations with zero initial conditions for above equations, we get,

$$I_a(s)\, R_a + L_a\, sI_a(s) + E_b(s) = V_a(s) \tag{1.51}$$

$$T(s) = K_t\, I_a(s) \tag{1.52}$$

$$Js^2\, \theta(s) + Bs\, \theta(s) = T(s) \tag{1.53}$$

$$E_b(s) = K_b s \cdot \theta(s) \tag{1.54}$$

on equating (1.52) and (1.53)

$$K_t\, I_a(s) = J^2 s\theta(s) + Bs\theta(s)$$

$$K_t\, I_a(s) = \theta(s)\,[Js^2 + Bs] \tag{1.55}$$

$$I_a(s) = \theta(s)\left[\frac{Js^2 + Bs}{K_t}\right] \tag{1.56}$$

Equation (1.51) can be written as

$$(R_a + sL_a)\, I_a(s) + E_b(s) = V_a(s) \tag{1.57}$$

Substituting $E_b(s)$ and $I_a(s)$ from equation (1.54) and (1.56) in equation (1.57)

$$(R_a + sL_a)\frac{(Js^2 + Bs)}{K_t}\,\theta(s) + K_b\, s\theta(s) = V_a(s)$$

$$\left[\frac{(R_a + sL_a)(Js^2 + Bs) + K_b K_t s}{K_t}\right]\theta(s) = V_a(s) \tag{1.58}$$

The required transfer function is $\theta(s)/V_a(s)$

From equation (1.58)

$$\frac{\theta(s)}{V_a(s)} = \frac{K_t}{[(R_a + sL_a)(Js^2 + Bs) + K_b\, K_t s]} \tag{1.59}$$

$$= \frac{K_t}{R_a Js^2 + R_a Bs + L_a Js^3 + L_a Bs^2 + K_b K_t s}$$

$$= \frac{K_t}{s[JL_a s^2 + (JR_a + BL_a)s + (BR_a + K_b K_t)]}$$

$$\frac{\theta(s)}{V_a(s)} = \frac{K_t/JL_a}{s\left[s^2 + \left(\frac{JR_a + BL_a}{JL_a}\right)s + \left(\frac{BR_a + K_bK_t}{JL_a}\right)\right]} \tag{1.60}$$

This also can be written in another form i.e.

From equation (1.59)

$$\frac{\theta(s)}{V_a(s)} = \frac{K_t}{\left[(R_a + sL_a)(Js^2 + Bs) + (K_bK_ts)\right]}$$

$$\frac{\theta(s)}{V_a(s)} = \frac{K_t}{R_a\left[\frac{sL_a}{R_a} + 1\right]Bs\left[1 + \frac{Js^2}{Bs}\right] + K_b\ K_ts}$$

$$\frac{\theta(s)}{V_a(s)} = \frac{K_t/R_aB}{s\left[(1 + s\,T_a)(1 + s\,T_m) + \left(\frac{K_bK_t}{R_aB}\right)\right]} \tag{1.61}$$

where, $\frac{L_a}{R_a} = T_a \rightarrow$ Electrical time constant

$\frac{J}{B} = T_m \rightarrow$ Mechanical time constant

The equation (1.61) is the required transfer function.

Example 2: Transfer function of field controlled DC Motor

The speed of a DC motor is directly proportional to armature voltage and inversely proportional to flux. In field controlled DC motor the armature voltage is kept constant and the speed is varied by varying the flux of the field. Since flux is directly proportional to field current, the flux is varied by varying field current. The speed control is an electromechanical control system. The electrical system consists of armature and field excited by a constant voltage. The mechanical system consists of the rotating part and the load connected to the shaft of the motor. The field controlled DC motor shown in Fig. (1.11d).

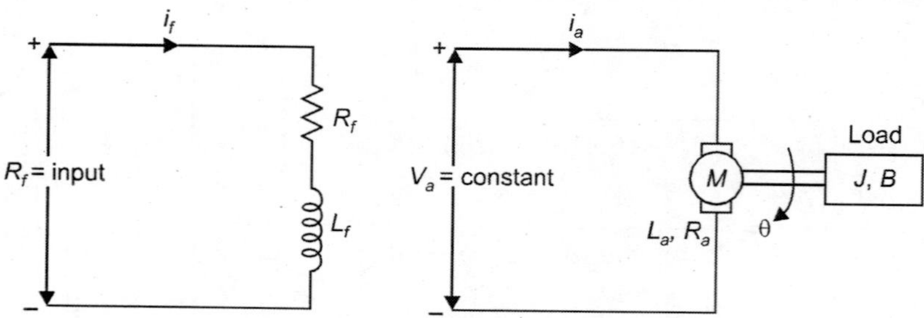

Fig. 1.11(d): Field controlled DC motor.

Let,

$R_f \rightarrow$ Field resistance, Ω

$L_f \rightarrow$ Field inductance, H

$i_f \rightarrow$ Field current, A

$V_f \rightarrow$ Field voltage, V

$T \rightarrow$ Torque developed by motor, N–m

$K_{tf} \rightarrow$ Torque constant, N–m/A

$J \rightarrow$ Moment of inertia of motor and load, kg-m^2/rad

$B \rightarrow$ Frictional coefficient of motor and load, N–m/(rad/sec)

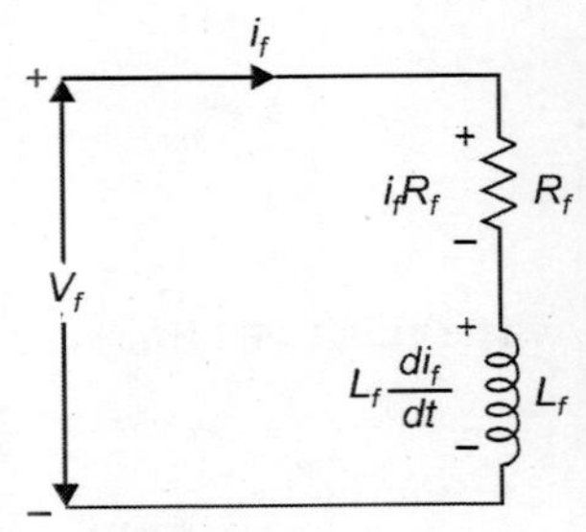

Fig. 1.11(e): Equivalent circuit of field.

The equivalent circuit of field in shown in Fig. (1.11e).

By Kirchoff's voltage law, we can write,

$$R_f\, i_f + L_f \frac{di_f}{dt} = V_f \tag{1.62}$$

The Torque of DC motor is proportional to product of flux and armature current. Since armature current is constant in this system, the torque is proportional to flux alone, but flux is proportional to field current.

$$T \,\alpha\, i_f \;\therefore\; \text{Torque, } T = Kt_f\, i_f \tag{1.63}$$

The mechanical system of the motor is shown in Fig. (1.11f).

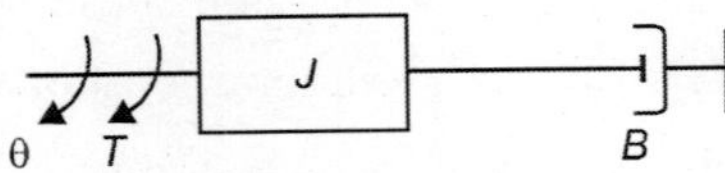

Fig. 1.11(f)

The differential equation governing the mechanical system of the motor is given by

$$J\frac{d^2\theta}{dt^2} + B\frac{d\theta}{dt} = T \tag{1.64}$$

The laplace transform of various time domain signals involved in this system are shown below.

$$L\,[i_f] = I_f\,(s);\; L(T) = T(s);\; L(V_f) = V_f(s)$$

$$L[\theta] = \theta(s)$$

The differential equations governing the field controlled DC motor are

$$R_f i_f + L_f\frac{di_f}{dt} = V_f;\quad T_f = Kt_f\, i_f;\quad J\frac{d^2\theta}{dt^2} + B\frac{d\theta}{dt} = T$$

on taking laplace transform of the system differential equation, we get

$$R_f\, I_f\,(s) + L_f\; sI_f(s) = V_f(s) \tag{1.65}$$

$$T(s) = K_{tf}\, I_f(s) \tag{1.66}$$

$$Js^2\,\theta(s) + Bs\,\theta(s) = T(s) \tag{1.67}$$

on equating (1.66) and (1.67)

$$K_{tf}\, I_f(s) = Js^2\theta\,(s) + Bs\,\theta(s)$$

$$K_{tf}\, I_f(s) = \theta(s)\,[Js^2 + Bs]$$

$$I_f(s) = s\,\frac{(Js + B)}{K_{tf}}\,\theta(s) \tag{1.68}$$

The equation (1.65) can be written as

$$(R_f + sL_f)\, I_f(s) = V_f(s) \tag{1.69}$$

on substituting for $I_f(s)$ from equation (1.68) in equation (1.69) we get,

$$(R_f + sL_f)\; s\,\frac{(Js + B)}{K_{tf}}\,\theta(s) = V_f(s)$$

$$\frac{\theta(s)}{V_f(s)} = \frac{K_{tf}}{s(R_f + sL_f)\,(B + s\,J)}$$

$$= \frac{K_{tf}}{s\,R_f\left(1 + \frac{s\,L_f}{R_f}\right) B\left(1 + \frac{s\,J}{B}\right)}$$

$$\frac{\theta(s)}{V_f(s)} = \frac{K_m}{s\,(1 + s\,T_f)\,(1 + s\,T_m)} \tag{1.70}$$

where
$K_m = K_{tf}/R_f B \rightarrow$ Motor gain constant
$T_f = L_f/R_f \rightarrow$ Field time constant
$T_m = J/B \rightarrow$ Mechanical time constant

1.12 FEED BACK CHARACTERISTICS OF CONTROL SYSTEM

1.12.1 Feedback and Non-Feedback Systems

Feedback system plays an important role in modern engineering practice because they have the possibility for being adopted to perform their assigned tasks automatically. A non-feedback (open loop) system represented by the block diagram and signal flow graph in Fig. (1.12a) and (1.12b) is achieved by a single signal at the input (for single input systems). There is no provision within this system for supervision of the output and no mechanism is provided to correct the system behaviour for any lack of proper performance of system components, changing environment, loading or ignorance of the exact value of process parameters. On the other hand, a feedback (closed loop) system represented by the block diagram and signal flow graph is driven by two signals (more signals could be employed), one the input signal and the other, a signal called the feedback signal derived from the output of the system. The feedback signal act as self-correcting mechanism for feedback system which is explained below.

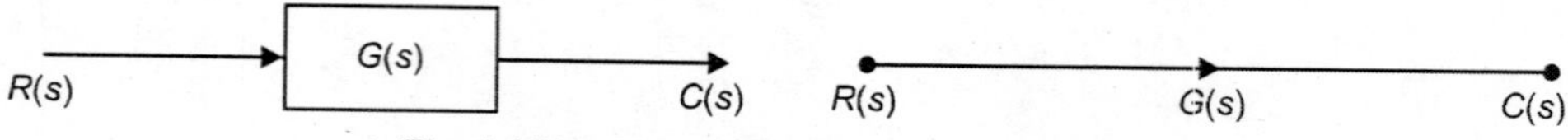

Fig. 1.12(a): Non-feedback system (open loop).

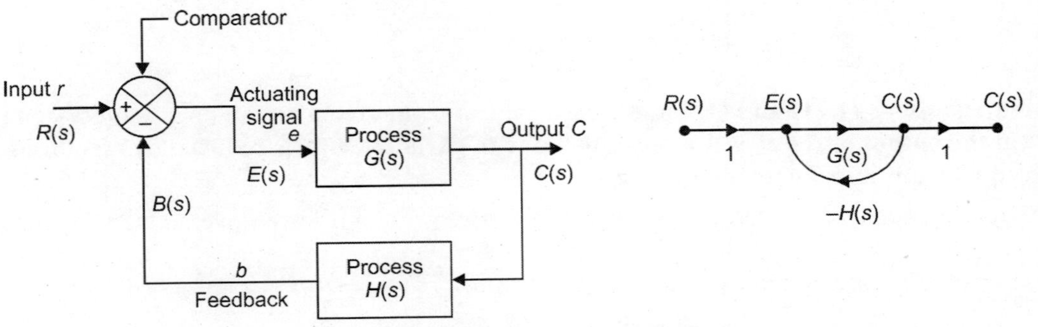

Fig. 1.12(b): Feedback system (closed loop).

The output signal c is measured by a sensor $H(s)$, which produces a feedback signal b with the input signal r generating the actuating signal e, which is measure of discrepancy between r and b. The actuating signal is applied to the process $G(s)$ so as to influence the output c and to reduce the error e.

The beneficial effects of feedback in control system.

1. The controlled variable accurately follows the desired value.
2. External disturbances are greatly reduced.
3. Accurate control
4. It improves the speed of system response.

1.12.2 Reduction of Parameter Variation by use of Feedback

The main motive of using feedback in the control system is to reduce the sensitivity of the system of parameter variation.

Let us define sensitivity on a quantitative basis,

In the open loop case

$C(s) = G(s)\ R(s)$ from Fig. (1.12a) suppose due to parameter variations $G(s)$ changes to $[G(s) + \Delta G(s)]$ where $|G(s) >>| \Delta G(s)$. The output of the open-loop system then changes to

$$C(s) + \Delta C(s) = [G(s) + \Delta G(s)]\ R(s)$$

or
$$\Delta C(s) = \Delta G(s)\ R(s) \qquad (1.71)$$

Similarly, In the closed loop case, the output

$$C(s) = \frac{G(s)}{1 + G(s)\ H(s)} R(s) \text{ from Fig. (1.12b)}$$

changes to
$$C(s) + \Delta C(s) = \frac{G(s) + \Delta G(s)}{1 + G(s)\ H(s) + \Delta G(s)\ H(s)} R(s)$$

Due to variation $\Delta G(s)$ in $G(s)$,

where $|G(s)| >> |\Delta G(s)|$, the ouput of the closed-loop system then changes to

$$\Delta C(s) = \frac{\Delta G(s)}{1 + G(s)\,H(s)} R(s) \tag{1.72}$$

From equations (1.71) and (1.72) it is seen that in comparison to the open loop system, the change in the output of the closed-loop system due to variation *G*(*s*) is reduced bya factor of [1 + *G*(*s*) *H*(*s*)] which is much greater than unity.

The term sensitivity is used to describe the relative variation in the overall transfer function $T(s) = \frac{C(s)}{R(s)}$ due to variation in *G*(*s*) and is defined below,

$$\text{Sensitivity} = \frac{\text{Percentage change in } T(s)}{\text{Percentage change in } G(s)}$$

For small incremental variation in *G*(*s*), the sensitivity is written in the quantitative form as,

$$S_G^T = \frac{\partial T / T}{\partial G / G} = \frac{\partial Ln\, T}{\partial Ln\, G} \tag{1.73}$$

where $S_G^T \rightarrow$ Sensitivity of *T* with respect to *G*

The use of feedback in reducing sensitivity to parameter variation is an important advantage of control system. To have a highly accurate open-loop system, the components of *G*(*s*) must be selected to meet the specifications rigidly in order to fulfil the overall goals of the system. On the other hand, in a closed loop system *G*(*s*) may be less rigidly specified, Since the effects of parameter variations are mitigated by the use of feedback. However a closed loop system requires careful selection of the components of the feedback sensor *H*(*s*) since *G*(*s*) is made up of power elements and *H*(*s*) is made up of measuring elements which operate at low power levels, the selection of accurate *H*(*s*) in far less costly than that of *G*(*s*) to meet the exact specification.

1.12.3 Control Over System Dynamics by use of Feedback

Let us consider the elementary single-loop feedback system shown in Fig. (1.12c)

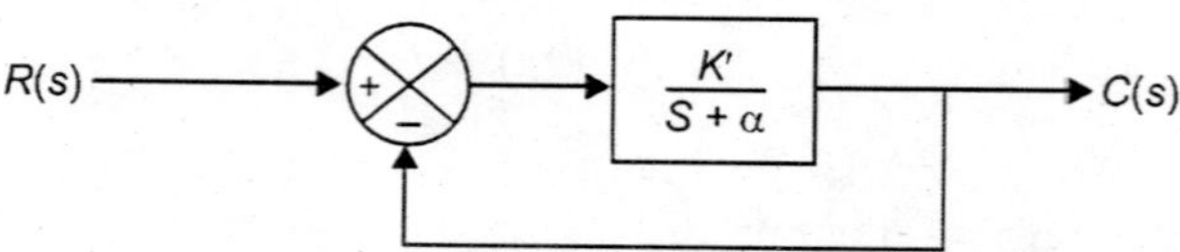

Fig. 1.12(c): A simple feedback system.

The open loop transfer function of the system is

$$\frac{C(s)}{R(s)} = \frac{K'}{s + \alpha} \tag{1.76}$$

We can write equation (1.76) as $\frac{C(s)}{R(s)} = \frac{K}{\tau s + 1}$ if $K = \frac{K'}{\alpha}$ and $\tau = \frac{1}{\alpha}$ (1.77)

There are two alternative forms of expressing transfer function

ie., The closed loop transfer function is,

$$\frac{C(s)}{R(s)} = \frac{K}{\tau s + (1+K)} = \frac{K'}{s + (\alpha + K')} \quad (1.78)$$

$$= \frac{K/(1+K)}{\tau_c s + 1}; \quad \tau_c = \frac{t}{(1+K)} \quad (1.79)$$

From equations (1.78) and (1.79) it is understood that the introduction of negative feedback is to shift the systems's pole from $-\alpha$ to $-(\alpha + K')$ (or) $-\alpha(1+K)$, alternatively to reduce the system time constant from τ to $\tau/(1+K)$.

Now we shall examine the effect of these changes in system transfer function on its dynamics response.

For this purpose we shall assume that the system is excited by an impulse input $r(t) = \delta(t)$. The laplace transform of an impulse excitation is $R(s) = 1$.

Taking inverse laplace transform of equation (1.76) and (1.78) with $R(s) = 1$, we have

$$C(t) = L^{-1}\left[\frac{K'}{s+\alpha}\right] = K'e^{-\alpha t} = K'e^{-t/\tau} \quad \text{(for open loop)} \quad (1.80)$$

and

$$C(t) = L^{-1}\left[\frac{K'}{s+\alpha(1+K)}\right] = K'e^{-\alpha(1+K)t}$$

$$C(t) = K'e^{-t/\tau_c} \quad \text{(for closed loop)} \quad (1.81)$$

The location of pole and the dynamic response of non feedback and feedback systems are shown in Fig. (1.12d). These responses delays in accordance with respective time constants. As the closed loop time constant $\tau_c = \tau/(1+k)$. Its response delays much faster which means that the speed of system's response with loop closed is faster by a factor of $(1+k)$ compared to the open-loop system.

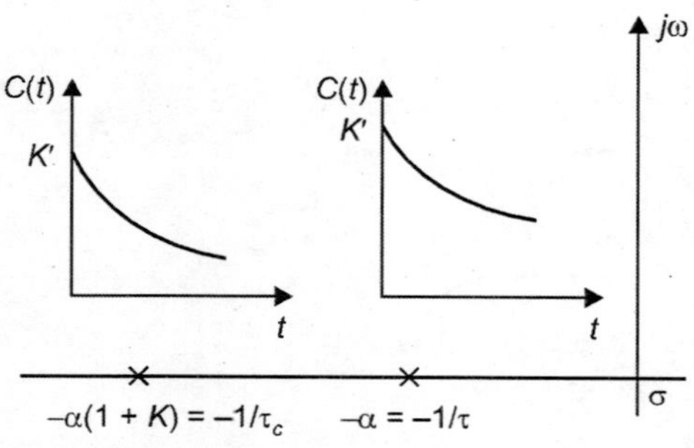

Fig. 1.12(d): Impulse response of open and closed loop system.

From this, it is concluded that feedback controls the dynamics of the system by adjusting the location of its poles. It is however, important to note here that feedback introduces the possibility of instability. A closed loop system may be unstable. but the feedback is a powerful techniques for control of system dynamics.

1.12.4 Effects of Feedback on Bandwidth

A control system is said to be a low-pass filter-if it responds to frequency ω_b at which the gain drops to $1/\sqrt{2}$ of its dc value. This frequency ω_b is the bandwidth of the system. A large bandwidth implies that the system responds accurately to higher frequencies i.e. fast changing signals, which is another way of looking at the speed of response of a control system.

Consider once again the Fig. (1.12.c) the open and closed loop transfer functionscan be expressed in frequency domain as

$$G(s) = \frac{K}{j\omega\tau + 1}; \quad \text{open loop [from equation (1.76)]} \tag{1.82}$$

$$G(s) = \frac{K/(1+K)}{j\omega\tau_c + 1}; \quad \text{closed loop [from equation (1.79)]} \tag{1.83}$$

Their bandwidth is determined as below,
from equation (1.82)

$$(\omega_b\tau)^2 + 1 = (\sqrt{2})^2$$

$$(\omega_b\tau)^2 + 1 = 2 - 1 = 1$$

$$\omega_b^2 = 1/\tau^2$$

$$\omega_b = 1/\tau \quad \text{(for open loop)} \tag{1.84}$$

Similarly, from equation (1.83)

$$(\omega_b\tau_c)^2 + 1 = \left(\sqrt{2}\right)^2$$

$$(\omega_b\tau_c)^2 + 1 = 2 - 1 = 1$$

$$(\omega_b\tau_c)^2 = 1$$

$$\omega_b^2 = \frac{1}{\tau_c^{\ 2}}$$

$$\omega_b = \frac{1}{\tau_c} \quad \text{for closed loop} \tag{1.85}$$

$$\frac{\omega_b \text{ (closed loop)}}{\omega_b \text{ (open loop)}} = \frac{\tau}{\tau_c} = (1 + K)$$

Thus the closed loop system has a band width (1 + *k*) times the bandwidth of the open loop system. This implies increased speed response.

1.12.5 Linearizing Effect of Feedback

The another property of feedback is linearizing effect which is illustrated by means of single and simple closed loop system as shown in Fig. (1.12e).

In a static system the gains are independent of time. We shall assume that the forward block function is non-linear and it is expressed as

$$e = f(e) = e^2 \quad \text{(By square law function)}$$

when the feedback loop is open

$$e = r \Rightarrow c = r^2$$

which is plotted in graph (1) of Fig. (1.12e) on the other hand when the loop is closes, we have

$$e = f(e) = (r - c)^2$$

which is plotted in graph (2) of Fig. (1.12e)

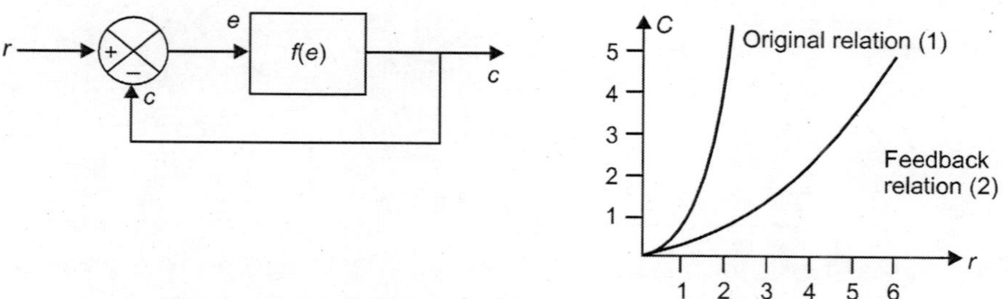

Fig. 1.12(e): Simple closed loop system with its response curve.

It is easily seen by comparison of the graphs (1) and (2) that the input-output relation (*c*/*r*) is approximately linear over much wider range for the closed loop system.

1.12.6 Regenerative Feedback

In regenerative feedback, the output is feedback to input with positive sign as shown in Fig. (1.12f).

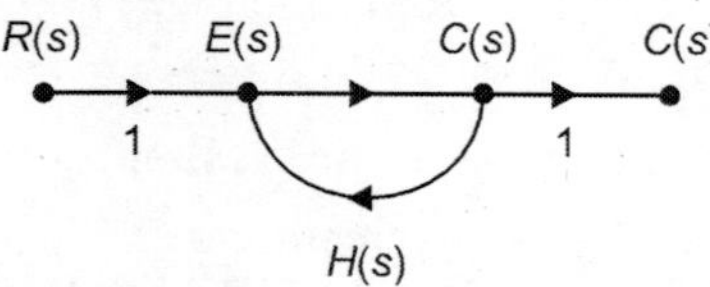

Fig. 1.12(f): A regenerative feedback control system.

In this case, the transfer function is given by,

$$\frac{C(s)}{R(s)} = \frac{G(s)}{1 - G(s)\,H(s)} \tag{1.86}$$

There is a negative sign in the denominator of equation (1.86) which indicates the possibility of denominator becoming equal to zero thereby giving an infinite output for a infinite input which is the condition of instability.

The regenerative feedback is sometimes used for increasing the loop gain of feedback system.

Conclusion: The feedback characteristics of control system can be written in short way that, when compared to non-feedback system, the feedback control systems are having many advantages.

1. The controlled variable accurately follows the desired value
2. External disturbances are greatly reduced.

3. Accurate control.
4. High speed of response.
5. The use of feedback in reducing sensitivity to parameter variation is an important advantage.
6. Feedback is used to control the dynamics of the system by adjusting the location of its poles.
7. The closed loop system a bandwidth $(1 + K)$ times the bandwidth of open loop system. This implies increased speed of response.
8. The linear characteristics of closed loop system.
9. Since, the inner regenerative feedback provides high loop gain, the closed loop transfer function becomes insensitive to $G(s)$.

1.13 MECHANICAL TRANSLATION SYSTEMS

The model of mechanical translational system can be obtained by using three basic elements like mass, spring and dash-pot. These three elements are essential for mechanical system.

The weight of the mechanical system is represented by the element mass and it is assumed to be concentrated at the centre of the body. The elastic deformation of the body can be represented by a spring. The friction existing in mechanical system can be represented by the dash-pot. The dash pot is a piston moving inside a cylinder filled with viscous fluid.

When a force is applied to a translational mechanical system, it is opposed by opposing forces due to mass, friction and elasticity of the system. The force acting on a mechanical body are governed by newton's second law of motion. For translational system it states that the sum of forces acting on a body is zero. (or Newton's second law states that the sum of applied forces is equal to the sum of opposing forces on a body.

List of symbols used in mechanical translational system.

$x \Rightarrow$ Displacement, m. (or) $y \Rightarrow$ Displacement, m.

$v \Rightarrow \frac{dx}{dt} \Rightarrow$ velocity, m/sec.

$a \Rightarrow \frac{d^2x}{dt^2} \Rightarrow$ Acceleration, m/sec^2.

$f \Rightarrow$ Applied force, N (Newtons)

$f_m \Rightarrow$ Opposing force offered by mass of the body, N

$f_k \Rightarrow$ Opposing force offered by the elasticity of the body (spring), N.

$f_b \Rightarrow$ Opposing force offered by the friction of the body (dash-pot), N.

$M \Rightarrow$ Mass, kg

$K \Rightarrow$ Stiffness of spring, N/m.

$B \Rightarrow$ Viscous friction co-efficient, N-sec/m

Note: Lower case letters are functions of time.

Force balance equations of Idealized elements

1. Consider an ideal mass elements shown in Fig. (1.13a) which has negligible friction and elasticity. Let a force be applied on it. The mass will offer an opposing force which is proportional to acceleration of the body.

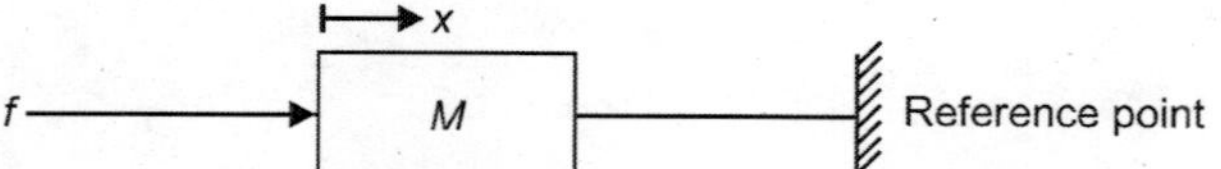

Fig. 1.13(a): Ideal mass element.

Let,

$f \rightarrow$ Applied force

$f_m \rightarrow$ Opposing force due to mass

Here $$f_m \,\alpha\, \frac{d^2x}{dt^2} \text{ (or) } f_m = \frac{Md^2x}{dt^2}$$

By Newton's second law,

$$f = f_m = \frac{Md^2x}{dt^2} \quad (1.87)$$

2. Consider an ideal frictional element dashpot shown in Fig. (1.13b) which has negligible mass and elasticity. Let a force be applied on it. The dashpot will offer an opposing force which is proportional to velocity of the body.

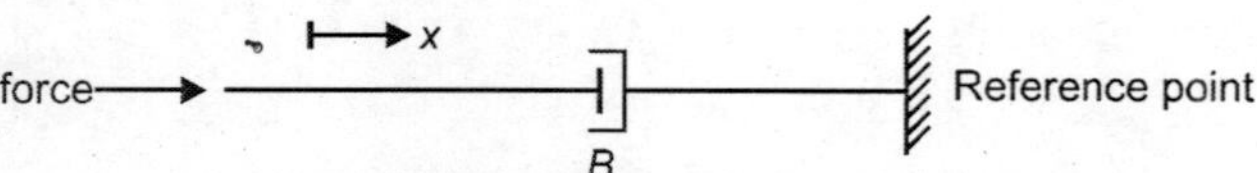

Fig. 1.13(b): Ideal dashpot element.

Let,

f = Applied force

f_b = Opposing force due to friction

Here, $$f_b \propto \frac{dx}{dt} \quad \text{or} \quad f_b = -B\frac{dx}{dt}$$

By Newton's second law,

$$f = f_b = B\frac{dx}{dt} \quad (1.88)$$

3. When the dashpot has displacement at both ends as shown in Fig. (1.13c), the opposing force is proportional to differential velocity.

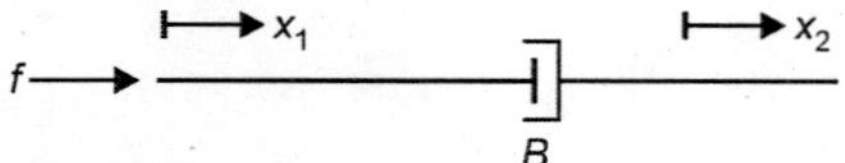

Fig. 1.13(c): Ideal dashpot with displacement at both ends.

$$f_b \propto \frac{d}{dt}(x_1 - x_2);$$

$$f_b = B\frac{d}{dt}(x_1 - x_2)$$

$$f = f_b = B\frac{d}{dt}(x_1 - x_2) \tag{1.89}$$

4. Consider an ideal elastic element spring shown in Fig. (1.13d), which has negligible mass and friction. Let a force applied on it. The spring will offer an opposing force which is proportional to displacement of the body

Fig. 1.13(d): Ideal spring element.

Let,

f = Applied force

f_k = Opposing force due to elasticity

Here, $f_k \alpha x$ (or) $f_k = Kx$

$$f = f_k = Kx \tag{1.90}$$

5. When the spring has displacement at both ends as shown in Fig. (1.13e), the opposing force is proportional to differential displacement.

Fig. 1.13(e): Ideal spring with displacement at both ends.

$$f_k \propto (x_1 - x_2)$$

$$f_k = K(x_1 - x_2)$$

By Newton's second law.

$$f = f_k = K(x_1 - x_2) \tag{1.91}$$

6. Consider a liner arrangement is shown in Fig. [1.13f (a) and (b)].

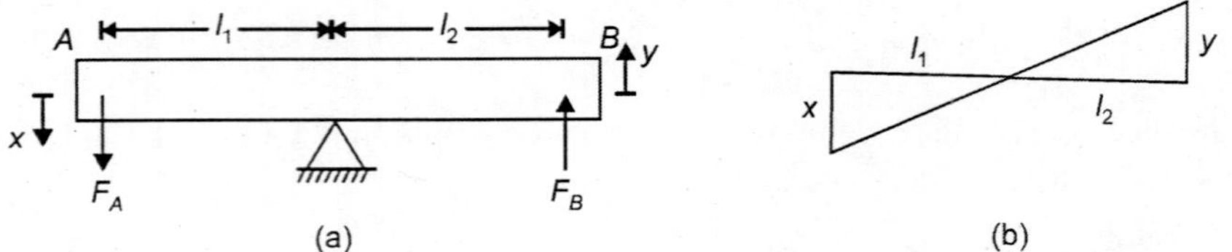

Fig. 1.13(f): A lever arrangement.

For Fig. [1.13f(a)], the following equation can be written

$$F_A l_1 + F_B l_2 = 0 \tag{1.92}$$

From Fig. [1.13f(b)], the following equation is written

$$\frac{x}{l_1} = \frac{y}{l_2} \tag{1.93}$$

7. Consider the gear arrangement shown in Fig. (1.13g) the primary gear has N_1 teeth and the secondary gear has N_2 teeth. the torque applied to the shaft of N_1 gear is $T_1(t)$ and the torque available at the shaft of N_2 gear will be $T_2(t)$.

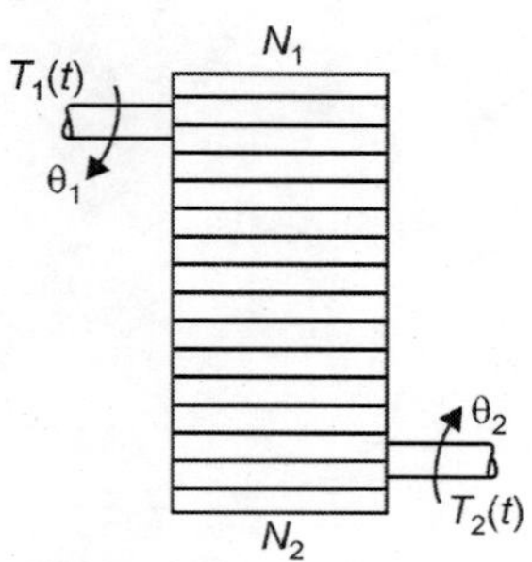

Fig. 1.13(g): Gear arrangement.

The following equations are written for Fig. (1.13g)

$$T_1\theta_1 = T_2\theta_2$$

$$N_1\theta_1 = N_2\theta_2$$

$$\therefore \quad \frac{T_1}{T_2} = \frac{\theta_2}{\theta_1} = \frac{N_1}{N_2} \tag{1.94}$$

The gear arrangement is similar to lever arrangement the dynamic equations for rotational systems are written on similar line for translational system. The linear displacement x meter is replaced by angular displacement θ radians, The force $f(t)$ is replaced by the torque $T(t)$ kg-m.

Guidelines to determine the transfer function of mechanical translational system

1. In mechanical translational system, the differential equations governing the system are obtained by writing force balance equations at every nodes in the system. The nodes are meeting point of elements. Generally the nodes are mass elements in the system. In some cases the nodes may be without mass element.
2. The linear displacement of the masses (nodes) are assumed x_1, x_2, x_3, etc. and assign a displacement to each mass (node). The first derivative of the displacement is velocity and the second derivative of the displacement is acceleration.
3. Draw the free body diagrams of the system. The free body diagram is obtained by drawing each mass separately and then marking all the forces acting on that mass (nodes). Always the opposing force acts in a direction opposite to applied force. The mass has to move in the direction of the applied force. Hence the displacement, velocity and acceleration of the mass will be in the direction of the applied force. If there is no applied force then the displacement, velocity and acceleration of the mass is in a direction opposite to that of opposing force.

4. For each free body diagram write one differential equation by equating the sum of applied forces to the sum of opposing forces.
5. Take laplace transform of differential equations to convert them to algebraic equations. Then rearrange the s-domain equations to eliminate the unwanted variables and obtain the ratio between output variable to input variable. This ratio is the transfer function of the system.

Note:

Laplace transform of $x(t) = L[x(t)] = X(s)$

Laplace transform of $\frac{dx(t)}{dt} = L\left[\frac{d}{dt}[x(t)]\right] = sX(s)$

Laplace transform of $\frac{d^2x(t)}{dt^2} = L\left[\frac{d^2}{dt^2}[x(t)]\right] = s^2\,X(s)$

EXAMPLE 1

Write the differential equations governing the mechanical system shown in Fig. (Ex. 1.1) and determine the transfer function.

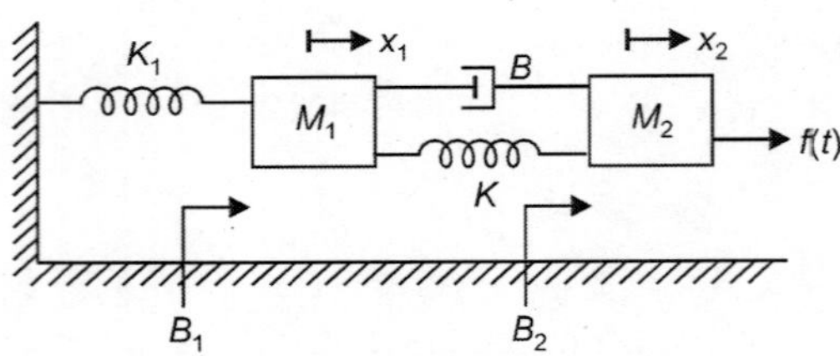

Fig. Ex.1

Solution: In the given system, applied force $f(t)$ is the input and displacement x is the output.

Let laplace transform of $f(t) = L[f(t)] = F(s)$

and laplace transform of $x = L[x] = X(s)$

The required transfer function is $\frac{X_2(s)}{F(s)}$

The system has two nodes and they are mass M_1 and M_2. The differential equations governing the system are given by force balance equations at these nodes.

Let the displacement of mass M_1 be x_1. The free body diagram of mass M_1 is shown in Fig. (Ex. 1.1.1). The opposing forces acting on mass M_1 are marked as $f_{m1}, f_{b1}, f_b, f_{k1}$ and f_k

$$f_{m1} = M_1\frac{d^2x_1}{dt^2};\quad f_{b1} = B_1\frac{dx_1}{dt};\quad f_{k1} = K_1x_1$$

$$f_b = B\frac{d}{dt}(x_1 - x_2);\quad f_k = K(x_1 - x_2)$$

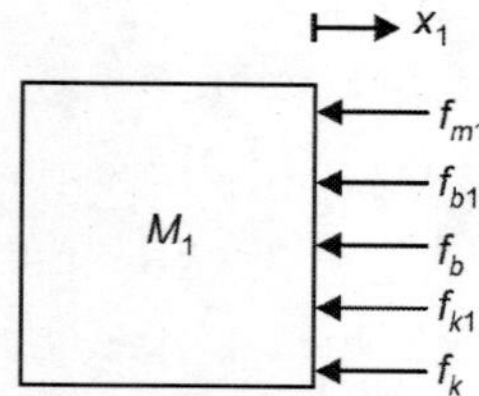

Fig. Ex. 1.1.1: Freebody diagram of mass M_1.

By newton's second law,

$$f_{m1} + f_{b1} + f_b + f_{k1} + f_k = 0$$

$$M_1 \frac{d^2x_1}{dt^2} + B_1 \frac{dx_1}{dt} + B\frac{d(x_1 - x_2)}{dt} + K_1x_1 + K(x_1 - x_2) = 0 \quad (1.1.1)$$

On taking laplace transform with zero initial conditions

$$M_1s^2X_1(s) + B_1sX_1(s) + Bs[X_1(s) - X_2(s)] + K_1X_1(s) + K[X_1(s) - X(s)] = 0$$

$$X_1(s)\,[M_1s^2 + (B_1 + B)\,s + (K_1 + K)] - X_2(s)\,[Bs + K] = 0$$

$$X_1(s)\,[M_1s^2 + (B_1 + B)\,s + (K_1 + K)] = X_2(s)\,[Bs + K]$$

$$X_1(s) = X_2(s)\frac{Bs + K}{M_1s^2 + (B_1 + B)s + (K_1 + K)} \quad (1.1.2)$$

Note: Laplace transform of $x_1 = L(x_1) = X_1(s)$

The freebody diagram of mass M_2 is shown in Fig. (Ex. 1.1.2). The opposing forces acting on M_2 are marked as f_{m2}, f_{b2}, f_b and f_k.

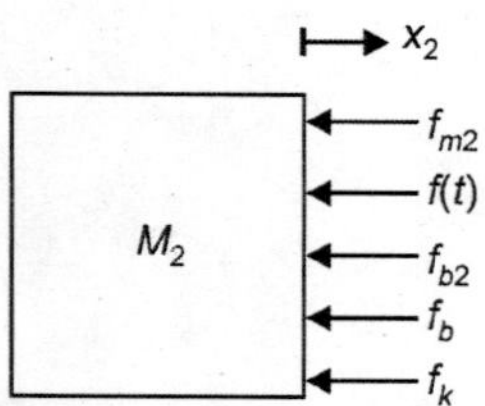

Fig. Ex. 1.1.2

$$f_{m2} = M_2\frac{d^2x_2}{dt^2}; \quad f_{b2} = B_2\frac{dx_2}{dt}$$

$$f_b = B\frac{d}{dt}(x_2 - x_1); \quad f_k = K(x_2 - x_1)$$

By newton's second law,

$$f_{m2} + f_{b2} + f_b + f_k = f(t)$$

$$M_2\frac{d^2x_2}{dt^2} + B_2\frac{dx_2}{dt} + B\frac{d}{dt}(x_2 - x_1) + K(x_2 - x_1) = f(t) \quad (1.1.3)$$

On taking laplace transform with zero initial conditions.

$$M_2 s^2 X_2(s) + B_2\, s\, X_2(s) + Bs[X_2(s) - X_1(s)] + K[X_2(s) - X_1(s)] = F(s)$$

$$X_2(s)\,[M_2 s^2 + (B_2 + B)\, s + K] - X_1(s)\,[Bs + K] = F(s) \qquad (1.1.4)$$

Substituting for $x_1(s)$ from equation (1.1.2) in (1.1.4) we get,

$$X_2(s)\,[M_2 s^2 + (B_2 + B)s + K] - X_2(s)\,\frac{(Bs + K)^2}{M_1 s^2 + (B_1 + B)s + (K_1 + K)} = F(s)$$

$$X_2(s)\left[\frac{[M_1 s^2 + (B_1 + B)\, s + (K_1 + K)\,(M_2 s^2 + (B_2 + B)s + K) - (Bs + K)^2]}{M_1 s^2 + (B_1 + B)s + (K_1 + K)}\right] = F(s)$$

$$\frac{X_2(s)}{F(s)} = \frac{M_1 s^2 + (B_1 + B)s + (K_1 + K)}{[M_1 s^2 + (B_1 + B)s + (K_1 + K)\,[M_2 s^2 + (B_2 + B)s + K] - (Bs + K)^2]} \qquad (1.1.5)$$

The equation (1.1.5) is the required transfer function.

Result: The differential equations governing the system are

1. $M_1 \dfrac{d^2x_1}{dt^2} + B_1 \dfrac{dx_1}{dt} + B\dfrac{d}{dt}\,(x_1 - x_2) + K_1 x_1 + K(x_1 - x_2) = 0$

2. $M_2 \dfrac{d^2x_2}{dt^2} + B_2 \dfrac{dx_2}{dt} + B\dfrac{d}{dt}\,(x_2 - x_1) + K(x_2 - x_1) = f(t)$

The transfer function of the system is

$$\frac{X(s)}{F(s)} = \frac{M_1 s^2 + (B_1 + B)s + (K_1 + K)}{[M_1 s^2 + (B_1 + B)s + (K_1 + K)]\,[M_2 s^2 + (B_2 + B)s + K]\cdot(Bs + K)^2}$$

EXAMPLE 2

Determine the transfer function $\dfrac{Y_2(s)}{F(s)}$ of the system shown in Fig. (Ex. 1.2).

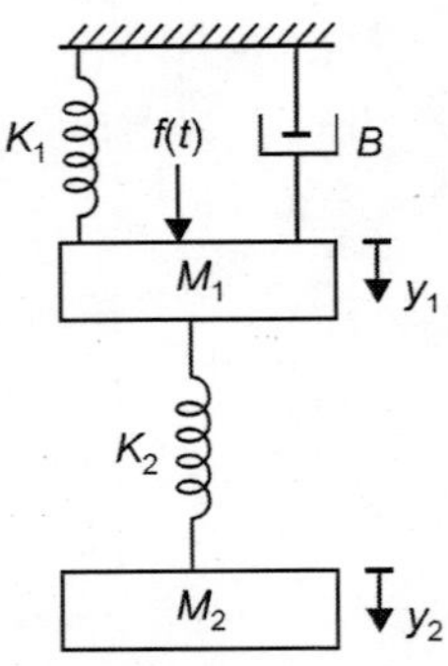

Fig. Ex. 1.2

Solution: Let, laplace transform of $f(t) = L[f(t)] = F(s)$

Laplace transform of $y_1 = L(y_1) = Y_1(s)$

Laplace transform of $y_2 = L(y_2) = Y_2(s)$

The system has two nodes and they are mass M_1 and M_2, the differential equations governing the system and the force balance equations at these nodes are,

The freebody diagram of mass M_1 is shown in Fig. (Ex. 1.2.1). The opposing forces are marked as f_{m1}, f_b, f_{k1} and f_{k2}.

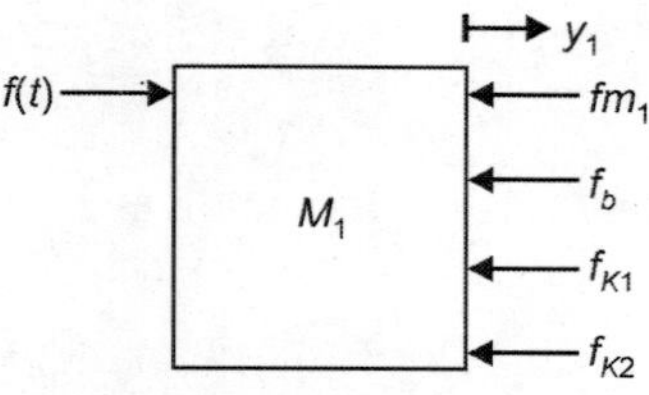

Fig. Ex. 1.2.1

$$f_{m1} = M_1 \frac{d^2 y_1}{dt^2}; \quad f_b = B\frac{dy_1}{dt}; \quad f_{k1} = K_1 y_1$$

$$f_{k2} = K_2 (y_1 - y_2)$$

By newton's second law,

$$f_{m1} + f_b + f_{k1} + f_{k2} = f(t)$$

$$M_1 \frac{d^2 y_1}{dt^2} + B\frac{dy_1}{dt} + K_1 y_1 + K_2(y_1 - y_2) = f(t) \quad (1.2.1)$$

On taking laplace transform with zero initial condition

$$M_1 s^2 Y_1(s) + sBY_1(s) + K_1 Y_1(s) + K_2 [Y_1(s) - Y_2(s)] = F(s)$$

$$Y_1(s) [M_1 s^2 + Bs + K_1 + K_2] - Y_2(s) K_2 = F(s) \quad (1.2.2)$$

The free body diagram of mass M_2 is shown in Fig. (Ex. 1.2.2). The opposing forces acting on M_2 are f_{m2} and f_{k2}

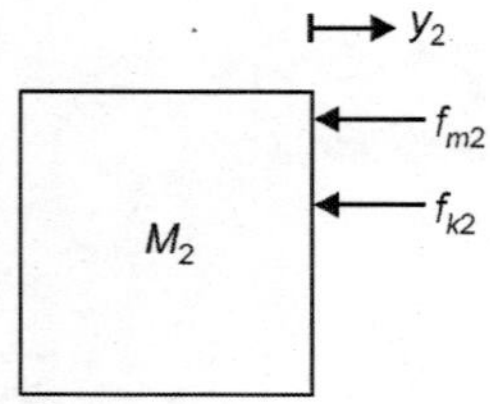

Fig. Ex. 1.2.2

$$f_{m2} = M_2 \frac{d^2 y_2}{dt^2}; \quad f_{k2} = K_2(y_2 - y_1)$$

By newton's second law, $f_{m2} + f_{k2} = 0$

$$M_2 \frac{d^2 y_2}{dt^2} + K_2 (y_2 - y_1) = 0 \tag{1.2.3}$$

On taking laplace transform,

$$M_2 s^2 Y_2(s) + K_2[Y_2(s) - Y_1(s)] = 0$$

$$Y_2(s)\,[M_2 s^2 + K_2] - Y_1(s)\,K_2 = 0$$

$$Y_1(s) = Y_2(s)\left[\frac{M_2 s^2 + K_2}{K_2}\right] \tag{1.2.4}$$

Substituting for $Y_1(s)$ from equation (1.2.4) in equation (1.2.2)

$$Y_2(s)\left\{\left[\frac{M_2 s^2 + K_2}{K_2}\right][M_1 s^2 + Bs + (K_1 + K_2)]\right\} - Y_2(s) K_2 = F(s)$$

$$Y_2(s)\left[\frac{(M_2 s^2 + K_2)(M_1 s^2 + Bs + (K_1 + K_2) - K_2^2}{K_2}\right] = F(s)$$

$$\therefore \quad \frac{Y_2(s)}{F(s)} = \frac{K_2}{[M_1 s^2 + Bs + (K_1 + K_2)][M_2 s^2 + K_2] - K_2^2}$$

Result: The differential equations governing the systems are

1. $M_1 \frac{d^2 y_1}{dt^2} + B \frac{dy_1}{dt} + K_1 y_1 + K_2(y_1 - y_2) = f(t)$

2. $M_2 \frac{d^2 y_2}{dt^2} + K_2 (y_2 - y_1) = 0$

The transfer function of the system is

$$\frac{Y_2(s)}{F(s)} = \frac{K_2}{[M_1 s^2 + Bs + (K_1 + K_2)][M_2 s^2 + K_2] - K_2^2}$$

EXAMPLE 3

Write the equations in s-domain for the system shown in Fig. (Ex. 1.3). Determine the transfer function of the system.

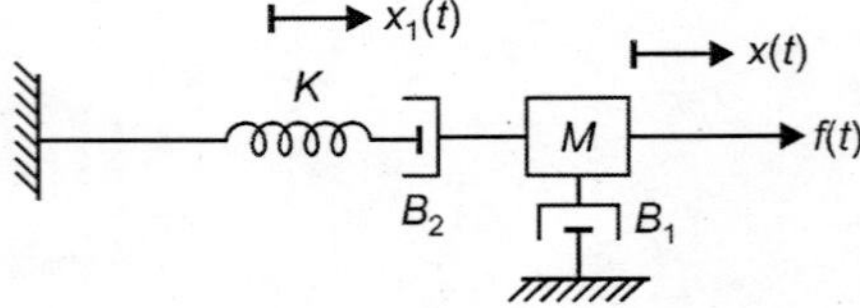

Fig. Ex. 1.3

Solution: Let,

Laplace transform of $x(t) = X(s)$

Lpalace transform of $f(t) = F(s)$

Let x_1 be the displacement at the meeting point of spring and dashpot. Laplace transform of x_1 is $X_1(s)$.

The system has two nodes and they are mass M and the meeting point of spring and dashpot. The differential equations governing the system are the force balance equations at these nodes. The equations of motion in the s-domain are obtained by taking laplace transform of the differential equations

The freebody diagram of mass M is shown in Fig. (Ex. 1.3.1). The opposing forces are marked as f_m, f_{b1} and f_{b2}

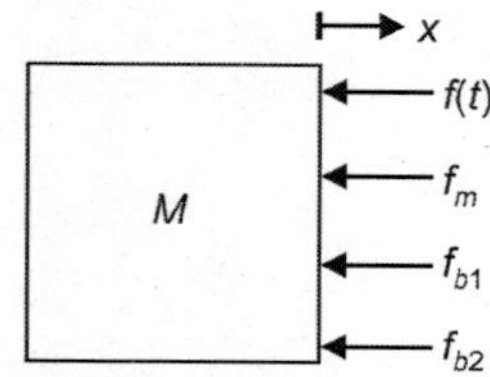

Fig. Ex. 1.3.1

$$f_m = M\frac{d^2x}{dt^2}; \quad f_{b1} = B_1\frac{dx}{dt}$$

$$f_{b2} = B_2\frac{d}{dt}(x - x_1)$$

By newton's second law the force balance equation is

$$f_m + f_{b1} + f_{b2} = f(t)$$

$$\therefore \quad M\frac{d^2x}{dt^2} + B_1\frac{dx}{dt} + B_2\frac{d}{dt}(x - x_1) = f(t) \quad (1.3.1)$$

on taking laplace transform

$$Ms^2 X(s) + B_1 s\, X(s) + B_2 s\,[X(s) - X_1(s)] = F(s)$$

$$[Ms^2 + (B_1 + B_2)\, s]\, X(s) - B_2 s\, X_1(s) = F(s) \quad (1.3.2)$$

The free body diagram at the meeting point of spring and dashpot is shown in Fig. (Ex. 1.3.2) the opposing forces are marked as f_k and f_{b2}.

$$f_{b2} = B_2\frac{d}{dt}(x_1 - x); \quad f_k = Kx_1$$

By newton's second law, $f_{b2} + f_k = 0$

$$\therefore \quad B_2\frac{d}{dt}(x_1 - x) + Kx_1 = 0 \quad (1.3.3)$$

on taking laplace transform

$$B_2 s[X_1(s) - X(s)] + KX_1(s) = 0$$

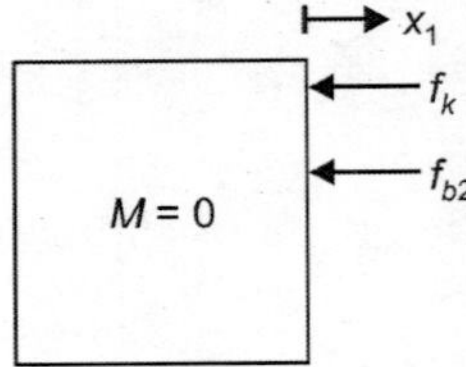

Fig. Ex. 1.3.2

$$(B_2\,s + K)\,X_1(s) - B_2\,sX(s) = 0 \tag{1.3.4}$$

$$X_1(s) = \frac{B_2 s}{(B_2 s + K)}\,X(s) \tag{1.3.5}$$

Substituting for $x_1(s)$ from equation (1.3.5) in equation (1.3.2)

$$[Ms^2 + (B_1 + B)s]\,X(s) - B_2 s\left[\frac{B_2 s}{B_2 s + K}\right]X(s) = F(s)$$

$$X(s)\left[\frac{[Ms^2 + (B_1 + B_2)s][B_2 s + K] - (B_2 s)^2}{(B_2 s + K)}\right] = F(s)$$

$$\therefore \quad \frac{X(s)}{F(s)} = \frac{(B_2 s + K)}{[Ms^2 + (B_1 + B_2)s][B_2 s + K] - (B_2 s)^2} \tag{1.3.6}$$

Result: The differential equations governing the system are,

1. $\dfrac{Md^2x}{dt^2} + B_1\dfrac{dx}{dt} + B_2\dfrac{d}{dt}(x - x_1) = f(t)$

2. $B_2\dfrac{d}{dt}(x_1 - x) + Kx_1 = 0$

The transfer function of the system is

$$\frac{X(s)}{F(s)} = \frac{B_2 s + K}{[Ms^2 + (B_1 + B_2)\,s][B_2 s + K] - (B_2 s)^2}$$

EXAMPLE 4

Determine the transfer function $x_1(s)/F(s)$ and $\dfrac{X_2(s)}{F(s)}$, for the system shown in Fig. (Ex. 1.4)

Solution: Let,

Laplace transform of $f(t) = F(s)$

Laplace transform of $x_1 = X_1(s)$

Laplace transform of $x_2 = X_2(s)$

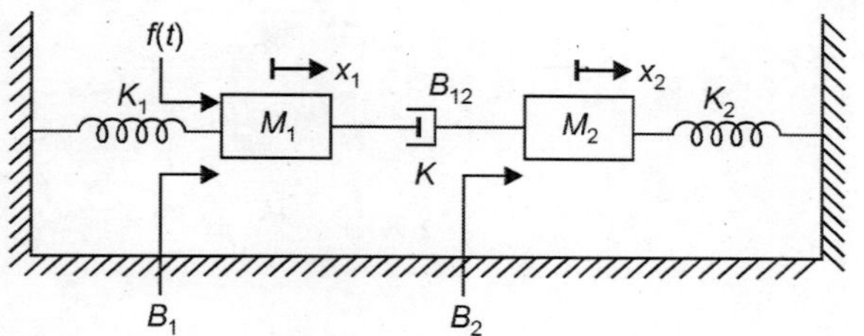

Fig. Ex. 1.4

The system has two nodes and they are mass M_1 and M_2. The differential equations governing the system are the force balance equations at these nodes. The free body diagram of mass M_1 is shown in Fig. (Ex. 1.4.1). The opposing forces are marked as f_{m1}, f_{b1}, f_{b12} and f_{k1}.

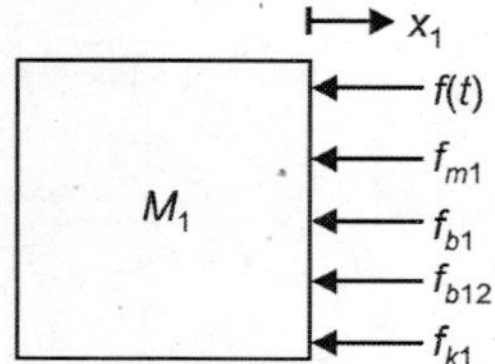

Fig. Ex. 1.4.1

$$f_{m1} = M_1 \frac{d^2x_1}{dt^2}; \quad f_{b1} = B_1 \frac{dx_1}{dt}; \quad f_{b12} = B_{12} \frac{d}{dt}(x_1 - x_2) \quad f_{k1} = K_1 x_1$$

By newton's second law, $f_{m1} + f_{b1} + f_{b12} + f_{k1} = f(t)$

$$M_1 \frac{d^2x_1}{dt^2} + B_1 \frac{dx_1}{dt} + B_{12} \frac{d}{dt}(x_1 - x_2) + K_1 x_1 = f(t) \tag{1.4.1}$$

on taking laplace transform with zero initial conditions

$$M_1 s^2 X_1(s) + B_1 s X_1(s) + B_{12} [X_1(s) - X_2(s)] + K_1 X_1(s) = F(s)$$

$$X_1(s) [M_1 s^2 + (B_1 + B_{12})s + K_1] - B_{12}\, sX_2(s) = F(s) \tag{1.4.2}$$

The free body diagram of mass M_2 is shown in Fig. (Ex. 1.4.2) the opposing forces are marked as f_{m2}, f_{b2}, f_{b12} and f_{k2}

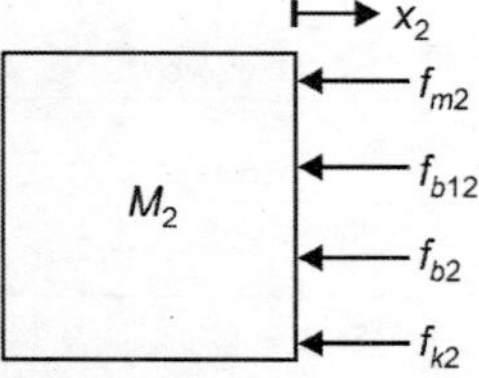

Fig. Ex. 1.4.2

$$f_{m2} = M_2 \frac{d^2x_2}{dt^2}; \quad f_{b2} = B_2 \frac{dx_2}{dt}; \quad f_{b12} = B_{12} \frac{d}{dt}(x_2 - x_1)$$

$$f_{k2} = K_2 x_2$$

By newton's second law,

$$f_{m2} + f_{b2} + f_{b12} + f_{k2} = 0$$

$$M_2 \frac{d^2x_2}{dt^2} + B_2 \frac{dx_2}{dt} + B_{12} \frac{d}{dt}(x_2 - x_1) + K_2 x_2 = 0 \qquad (1.4.3)$$

on taking laplace transform

$$M_2 s^2 X_2(s) + B_2 s X_2(s) + B_{12} s[X_2(s) - X_1(s)] + K_2 X_2(s) = 0$$

$$X_2(s)[M_2 s^2 + B_2 s + B_{12} s + K_2] - X_1(s) B_{12} s = 0$$

$$X_2(s)[M_2 s^2 + s[B_2 + B_{12}] + K_2] - B_{12} s X_1(s) = 0$$

$$X_2(s) = \frac{B_{12} s X_1(s)}{[M_2 s^2 + (B_2 + B_{12}) s + K_2]} \qquad (1.4.4)$$

Substituting $X_2(s)$ from equation (1.4.4) in equation (1.4.2)

$$X_1(s)[M_1 s^2 + (B_1 + B_{12}) s + K_1] - \frac{(B_{12} s)^2 X_1(s)}{(M_2 s^2 + (B_2 + B_{12}) s + K_2} = F(s)$$

$$\frac{X_1(s)\left[[M_1 s^2 + (B_1 + B_{12})s + K_1][M_2 s^2 + (B_2 + B_{12})s + K_2] - (B_1 s)^2\right]}{M_2 s^2 + (B_2 + B_{12} s) + K_2} = F(s)$$

$$\therefore \quad \frac{X_1(s)}{F(s)} = \frac{M_2 s^2 + (B_2 + B_{12})s + K_2}{[M_1 s^2 + (B_1 + B_{12})s + K_1][M_2 s^2 + (B_2 + B_{12})s + K_2] - [B_1 s]^2} \qquad (1.4.5)$$

From equation (1.4.4),

$$X_1(s) = \frac{\left[[M_2 s^2 + (B_2 + B_{12})s + K_2]\right] X_2(s)}{B_{12} s} \qquad (1.4.6)$$

Substituting $X_1(s)$ from equation (1.4.6) in equation (1.4.2)

$$\frac{X_2(s)[M_2 s^2 + (B_2 + B_{12})s + K_2}{B_{12} s}[M_1 s^2 + (B_1 + B_{12})s + K_1] - B_{12} s X_2(s) = F(s)$$

$$X_2(s)\left[\frac{[M_2 s^2 + (B_2 + B_{12}) s + K_2][M_1 s^2 + (B_1 + B_{12} s + K_1)] - (B_{12} s)^2]}{B_{12} s}\right] = F(s)$$

$$\therefore \quad \frac{X_2(s)}{F(s)} = \frac{B_{12} s}{[M_2 s^2 + (B_2 + B_{12}) s + K_2][M_1 s^2 + (B_1 + B_{12})s + K_1] - [B_{12} s]^2}$$

Result: The differential equation governing the system are,

1. $M_1 \frac{d^2x_1}{dt^2} + B_1 \frac{dx_1}{dt} + B_{12} \frac{d}{dt}(x_1 - x_2) + K_1 x_1 = f(t)$

2. $M_2 \frac{d^2x_2}{dt^2} + B_2 \frac{dx_2}{dt} + B_{12} \frac{d}{dt}(x_2 - x_1) + K_2 x_2 = 0$

The transfer function of the system are,

1. $\frac{X_1(s)}{F(s)} = \frac{M_2 s^2 + (B_2 + B_{12})s + K_2}{[M_1 s^2 + (B_1 + B_{12})s + K_1][M_2 s^2 + (B_2 + B_{12})s + K_2] - (B_{12}s)^2}$

2. $\frac{X_2(s)}{F(s)} = \frac{B_{12}s}{[M_2 s^2 + (B_2 + B_{12})s + K_2][M_1 s^2 + (B_1 + B_{12})s + K_1] - (B_{12}s)^2}$

1.14 MECHANICAL ROTATIONAL SYSTEM

The model of rotational mechanical systems can be obtained by using three elements. Moment of inertial [J] of mass, dash-pot with rotational frictional coefficient [B] and torsional spring with stiffness [K].

The weight of the rotational mechanical system is represented by the moment of inertia of the mass. The moment of inertia of the system or body is considered to be concentrated at the centre of gravity of the body. The elastic deformation of the body can be represented by a spring. The friction existing in rotational mechanical system can be represented by the dash-pot. The dash-pot is a piston rotating inside a cylinder filled with viscous fluid.

When a torque is applied to a rotational mechanical system, it is opposed by opposing torques due to moment of inertia, friction and elasticity of the system. The torque acting on a rotational mechanical body are governed by newtons's second law of motion of rotational systems. It states that the sum of torques acting on a body is zero.

List of Symbols Used in Mechanical Rotational System

θ = Angular displacement, rad

$\frac{d\theta}{dt}$ = Angular velocity, rad/sec

$\frac{d^2\theta}{dt^2}$ = Angular acceleration, rad/sec^2

T = Applied Torque, N–m

J = Moment of inertia, kg-m^2/rad

B = Rotational frictional coefficient, N-m/(rad/sec)

K = Stiffness of the spring, N-m/rad

Torque Balance Equations of Idealised Elements

Consider an ideal mass element shown in Fig. (1.14a) which has negligible friction and elasticity. The opposing torque due to moment of inertia is proportional to the angular acceleration.

Let,

T = Applied torque

T_j = Opposing torque due to moment of inertia of the body.

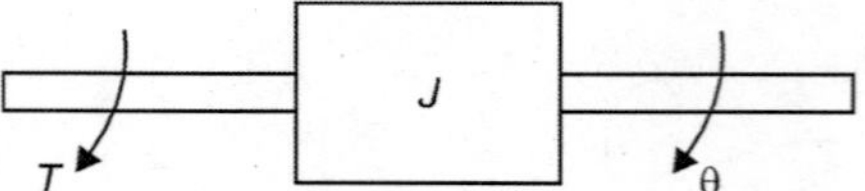

Fig. 1.14(a): Ideal rotational mass element.

Hence, $T_j \; \alpha \; \frac{d^2\theta}{dt^2}$ or $T_j = J\frac{d^2\theta}{dt^2}$

By newton's second law

$$T = T_j = J\frac{d^2\theta}{dt^2}$$

Consider an ideal frictional element dash-pot shown in Fig. (1.14b) which has negligible moment of inertia and elasticity. Let a torque be applied on it. The dash-pot will offer an opposing torque which is proportional to the angular velocity of the body.

Fig. 1.14(b): Ideal rotational dash-pot with one and fixed to reference.

Let, T = Applied torque.

T_b = Opposing torque due to friction.

$$T_b \; \alpha \; \frac{d\theta}{dt} \quad \text{(or)} \quad T_b = B\frac{d\theta}{dt}$$

By newton's second law, $T = T_b = B\frac{d\theta}{dt}$

When the dash-pot has angular displacement at both ends as shown in Fig. (1.14c), the opposing torque is proportional to the differential angular velocity.

Fig. 1.14(c): Ideal dash-pot with angular displacement at both sides.

$$T_b \; \alpha \; \frac{d}{dt}(\theta_1 - \theta_2)$$

$$T_b = B\frac{d}{dt}(\theta_1 - \theta_2)$$

By newton's second law

$$T = T_b = B\frac{d}{dt}(\theta_1 - \theta_2)$$

Consider an ideal elastic element, torsional spring as shown in Fig. (1.14d), which has negligible moment of inertia and friction. Let a torque be applied on it. The torsional spring will offer an using torque which is proportional to angular displacement of the body.

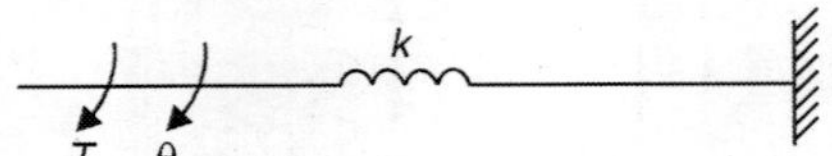

Fig. 1.14(d): Ideal spring with one end fixed to reference.

Let,

T = Applied torque

T_k = Opposing torque due to elasticity

$T_k \propto \theta$

$T_k = K\theta$

By newton's second law

$$T = T_k = K\theta$$

when the spring has angular displacement at both ends as shown in Fig. (1.14e) the opposing torque is proportional to differential angular displacement.

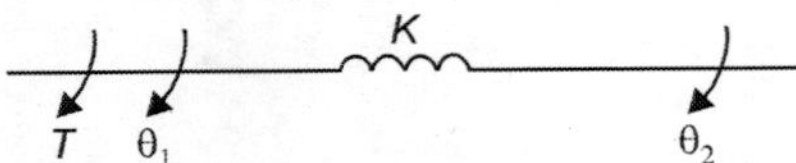

Fig. 1.14(e): Ideal spring with angular displacement at both ends.

$$T_k \propto (\theta_1 - \theta_2)$$
$$T_k = K(\theta_1 - \theta_2)$$
$$T = T_k = K(\theta_1 - \theta_2)$$

Guideline to determine the transfer function of Mechanical Rotational System

1. In mechanical rotational system, the differential equations governing the system are obtained by writing torque equations at nodes in the system. The nodes are meeting point of elements. Generally the nodes are mass elements with moment of inertia in the system. In some cases the nodes may be without mass element.
2. The angular displacement of the moment of inertia of the masses (nodes) are assumed as θ_1, θ_2, θ_3 etc. and assign a displacement to each mass (nodes). The first derivative of angular displacement is angular velocity and the second derivative of the angular displacement is angular acceleration.
3. Draw the free body diagrams of the system. The free body diagram is obtained by drawing each moment of inertia of mass separately and then sum all the torques acting on that body. Always the opposing torques act in a direction opposite to applied torque.
4. The mass has to rotate in the direction of the applied torque. Hence the angular displacement, velocity and acceleration of the mass will be in the direction of the applied torque. If there is no applied torque then the angular displacement, velocity and acceleration of the mass is in a direction opposite to that of opposing torque.
5. For each free body diagram write one differential equation by equating the sum of applied torque to the sum of opposing torques.

6. Take laplace transform of differential equation to convert them to algebraic equations. Then rearrange the s-domain equations to eliminate the unwanted variables and obtain the relation between output variables and input variable. This ratio is the transfer function of the system.

Note: Laplace transform of $\theta = L(\theta) = \theta(s)$

Laplace transform of $\frac{d\theta}{dt}$ with zero initial conditions

$$L\left[\frac{d\theta}{dt}\right] = s\theta(s)$$

EXAMPLE 5

Write the differential equations governing the mechanical rotational system shown in Fig. (Ex. 1.5). Obtain the transfer function of the system

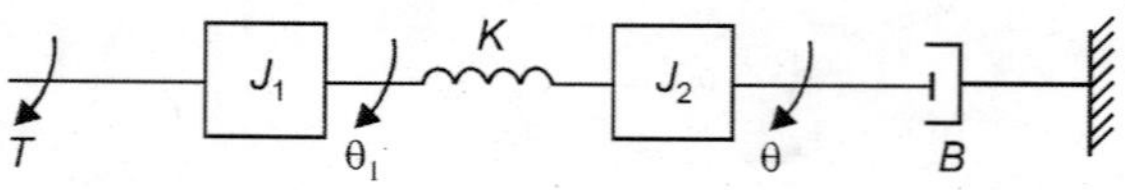

Fig. Ex. 1.5

Solution: In the given system, applied torque T is the input and angular displacement θ is the output.

Let

Laplace transform of $T = L(T) = T(s)$

Laplace transform of $\theta = L(\theta) = \theta(s)$

Hence the required transfer function is $\frac{\theta(s)}{T(s)}$

The system has two nodes and they are masses with moment of inertial J_1 and J_2. The differential equations governing the system are given by torque balance equations at these nodes.

Let the angular displacement of mass with moment of inertial J_1 be θ_1. The Laplace transform of $\theta_1 = L[\theta_1] = \theta_1(s)$. The free body diagram of J_1 is shown in Fig. (Ex. 1.5.1). The opposing torques acting on J_1 are marked as T_{j1} and T_k.

Fig. Ex. 1.5.1: Freebody diagram of mass with moment of inertial J_1.

$$T_{j1} \alpha \frac{d^2\theta_1}{dt^2} \quad \text{(or)} \quad T_{j1} = J_1 \frac{d^2\theta_1}{dt^2}$$

$$T_k \alpha (\theta_1 - \theta_2) \quad \text{(or)} \quad T_k = K(\theta_1 - \theta)$$

By newton's second law,

$$T_{j1} + T_k = T$$

$$J_1 \frac{d^2\theta}{dt^2} + K(\theta_1 - \theta) = T \quad (1.5.1)$$

On taking laplace transform with zero initial conditions,

$$J_1 s^2 \theta_1(s) + K_1 \theta_1(s) - K_1 \theta(s) = T(s)$$

$$\theta_1(s) [J_1 s^2 + K_1] - \theta(s) K_1 = T(s) \quad (1.5.2)$$

The free body diagram of mass with moment of inertia J_2 is shown in Fig. (Ex.1.5.2) the opposing torque acting on J_2 are marked as T_{j2}, T_b and T_k.

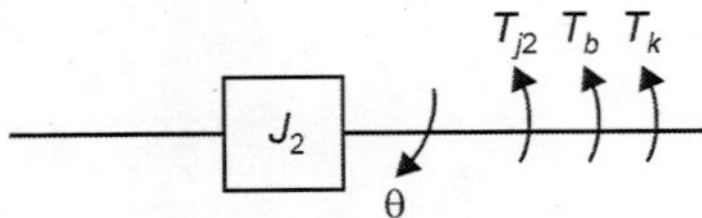

Fig. Ex. 1.5.2: Free body diagram of mass with moment of inertia J_2.

$$T_{j2} \frac{d^2\theta}{dt^2}, \quad T_b \propto \frac{d\theta}{dt}$$

$$T_k \propto (\theta - \theta_1)$$

$$\therefore \quad T_{j2} = J_2 \frac{d^2\theta}{dt^2}; \quad T_b = B \frac{d\theta}{dt}$$

and $$T_k = K(\theta - \theta_1)$$

By newton's second law,

$$T_{j2} + T_b + T_k = 0$$

$$J_2 \frac{d^2\theta}{dt^2} + B \frac{d\theta}{dt} + K(\theta - \theta_1) = 0$$

$$J_2 \frac{d^2\theta}{dt^2} + B \frac{d\theta}{dt} + K\theta - K\theta_1 = 0 \quad (1.5.3)$$

on taking laplace transform with zero initial conditions,

$$J_2 s^2 \theta(s) + Bs\,\theta(s) + K\,\theta(s) - K\theta_1(s) = 0$$

$$\theta(s) [J_2 s^2 + Bs + K] - K\theta_1(s) = 0$$

$$\theta_1(s) = \frac{(J_2 s^2 + Bs + K)}{K} \theta(s) \quad (1.5.4)$$

Substituting for $\theta_1(s)$ from equation (1.5.4) in equation (1.6.2) we get,

$$(J_1 s^2 + K) \frac{(J_2 s^2 + Bs + K)}{K} \theta(s) - K\theta(s) = T(s)$$

$$\left[\frac{(J_1 s^2 + K)(J_2 s^2 + Bs + K) - K^2}{K} \right] \theta(s) = T(s)$$

$$\therefore \qquad \frac{\theta(s)}{T(s)} = \frac{K}{(J_1 s^2 + K)(J_2 s^2 + Bs + K) - K^2} \qquad (1.5.5)$$

Results: The differential equations governing the system are

1. $J_1 \dfrac{d^2\theta_1}{dt^2} + K\theta_1 - K\theta = T$

2. $J_2 \dfrac{d^2\theta_2}{dt^2} + B\dfrac{d\theta}{dt} + K\theta - K\theta_1 = 0$

The transfer function of the system is

$$\frac{\theta(s)}{T(s)} = \frac{K}{(J_1 s^2 + K)(J_2 s^2 + Bs + K) - K^2}$$

EXAMPLE 6

Write the differential equations governing the mechanical rotational system shown in Fig. (Ex. 1.6) and determine the transfer function $\theta(s)/T(s)$.

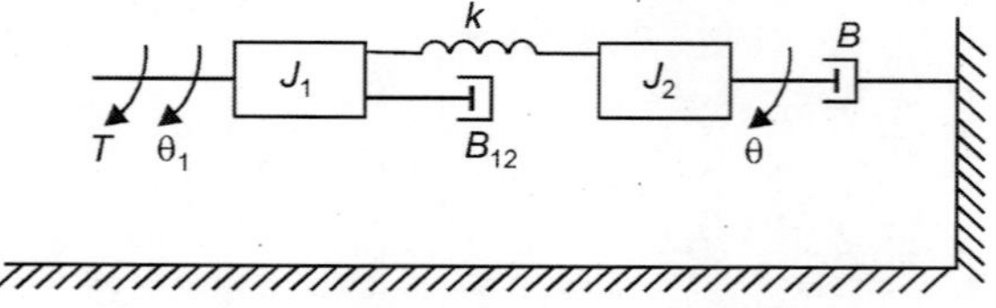

Fig. Ex. 1.6

Solution: In the given system the torque T is the input and the angular displacement θ is the output.

Let,

Laplace transform of $T = L(T) = T(s)$

Laplace transform of $\theta = L(\theta) = \theta(s)$

Hence the required transfer function is $\theta(s)/T(s)$ the system has two nodes and they are masses with moment of inertia J_1 and J_2 the differential equations governing the system are given by torque balance equations at these nodes.

Let the angular displacement of mass with moment of inertia J_1 be θ_1. The laplace transform of $\theta_1 = L(\theta_1) = \theta_1(s)$ the free body diagram of J_1 is shown in Fig. (Ex. 1.6.1). The opposing torques acting on J_1 are marked as T_{j1}, T_{b12} and T_k.

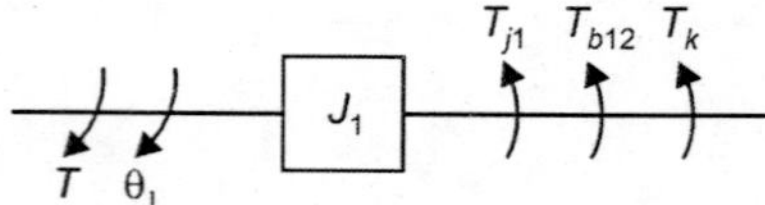

Fig. Ex. 1.6.1: Free body diagram of mass with moment of inertia J_1.

$$T_{j1} = J_1 \frac{d^2\theta}{dt^2}$$

$$T_{b12} = B_{12} \frac{d}{dt}(\theta_1 - \theta)$$

$$T_k = K(\theta_1 - \theta)$$

By newton's second law

$$T_{j1} + T_{b12} + T_k = T$$

$$J_1 \frac{d^2\theta_1}{dt^2} + B_{12}\frac{d}{dt}(\theta_1 - \theta) + K(\theta_1 - \theta) = T \qquad (1.6.1)$$

on taking laplace transform with zero initial conditions

$$J_1 s^2\theta_1(s) + B_{12}\, s[\theta_1(s) - \theta(s)] + K[\theta_1(s) - \theta(s)] = T(s)$$

$$\theta_1(s)\,[J_1 s^2 + B_{12}s + K] - \theta(s)\,[B_{12}s + K] = T(s) \qquad (1.6.2)$$

The free body diagram of mass with moment of inertia J_2 as shown in Fig. (Ex. 1.6.2). The opposing torques are marked as T_{j2}, T_{b12}, T_b and T_k.

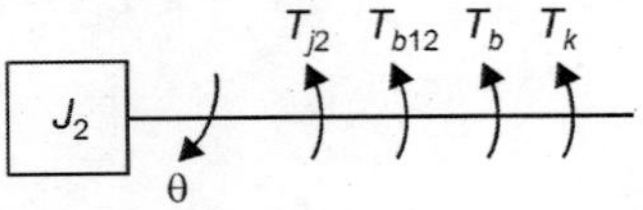

T_{j2} T_{b12} T_b T_k

J_2

θ

Fig. Ex. 1.6.2: Free body diagram of mass with moment of inertia J_2

$$T_{j2} = J_2 \frac{d^2\theta}{dt^2}; \quad T_{12} = B_{12}\frac{d}{dt}(\theta - \theta_1)$$

$$T_b = B\frac{d\theta}{dt} \quad \text{and} \quad T_k = K(\theta - \theta_1)$$

By newton's second law

$$J_{j2} + T_{b12} + T_b + T_k = 0$$

$$J_2 \frac{d^2\theta}{dt^2} + B_{12}\frac{d}{dt}(\theta - \theta_1) + B\frac{d\theta}{dt} + K(\theta - \theta_1) = 0 \qquad (1.6.3)$$

on taking laplace transform with zero initial conditions,

$$J_2 s^2\,\theta(s) - B_{12}s\,\theta_1(s) + s\theta(s)\,[B_{12} + B] + K\theta(s) - K\theta_1(s) = 0$$

$$\theta(s)\,[s^2 J_2 + s(B_{12} + B) + K] - \theta_1(s)\,[sB_{12} + K] = 0$$

$$\theta_1(s) = \frac{s^2 J_2 + s(B_{12} + B) + K}{(sB_{12} + K)}\,\theta(s) \qquad (1.6.4)$$

substituting for $\theta_1(s)$ from equation (1.7.4) in equation (1.7.2) we get

$$[J_1 s^2 + sB_{12} + K]\,\frac{[J_2 s^2 + s(B_{12} + B) + K]\,\theta(s)}{[sB_{12} + K]} - (sB_{12} + K)\,\theta(s) = T(s)$$

$$\left[\frac{(J_1s^2 + sB_{12} + K)(J_2s^2 + s(B_{12} + B) + K) - (sB_{12} + K)^2}{(sB_{12} + K)}\right]\theta(s) = T(s)$$

$$\therefore \quad \frac{\theta(s)}{T(s)} = \frac{(sB_{12} + K)}{(J_1s^2 + sB_{12} + K)(J_2s^2 + s(B_{12} + B) + K) - (sB_{12} + K)^2} \quad (1.6.5)$$

Results: The differential equations governing the system are

1. $J_1 \frac{d^2\theta_1}{dt^2} + B_{12}\frac{d}{dt}(\theta_1 - \theta) + K(\theta_1 - \theta) = T$

2. $J_2 \frac{d^2\theta}{dt^2} + B_{12}\frac{d\theta}{dt} + \frac{d\theta}{dt}(B_{12} + B) + K\theta - K\theta_1 = 0$

The transfer function of the system is

$$\frac{\theta(s)}{T(s)} = \frac{(sB_{12} + K)}{(J_1s^2 + sB_{12} + K)(J_2s^2 + s(B_{12} + B) + K) - (sB_{12} + K)^2}$$

EXAMPLE 7

For the electro-mechanical system shown in Fig. (Ex. 1.7) determine the transfer function $\frac{Y(s)}{V(s)}$.

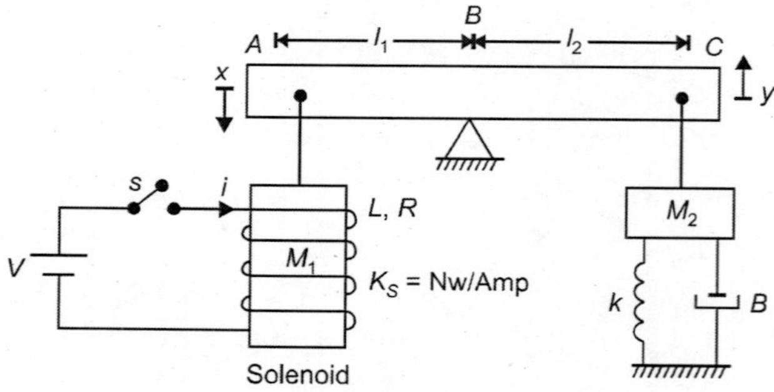

Fig. Ex. 1.7

Solution: In the given system, applied input is V and the output y is the linear displacement.

Let

Laplace transfer of $V = L(V) = V(s)$

Laplace transform of $y = L(y) = Y(s)$

Hence the required transfer function is $\frac{Y(s)}{V(s)}$

The system has two nodes and they are masses M_1 and M_2. The differential equations governing the system are given by force balance equations at these nodes.

Let we first write the loop equation for electrical circuit. The equivalent circuit of that electrical circuit as shown in Fig. (Ex. 1.7.1).

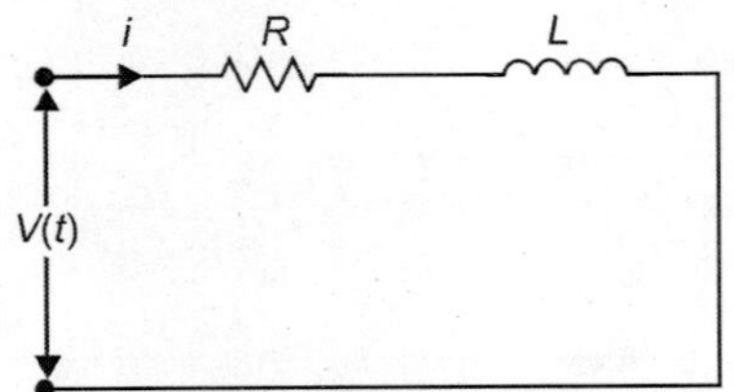

Fig. Ex. 1.7.1: Equivalent circuit of electrical circuit for given system.

By applying KVL, for Fig. (Ex. 1.7.1).

$$V(t) = Ri + L\frac{di}{dt} \qquad (1.7.1)$$

on taking Laplace transform of equation (1.8.1)

$$V(s) = R\,I(s) + Ls\,I(s)$$

$$V(s) = I(s)\,[R + Ls]$$

$$I(s) = \frac{V(s)}{[R + Ls]} \qquad (1.7.2)$$

The mechanical elements present in mass is K_s (i.e. spring)

∴ The force on the solenoid is $F_s = K_s\,I(s)$ (1.7.3)

Let F_a be the reaction force at point A. Then substitute $I(s)$ from equation (1.7.2) in equation (1.7.3).

$$F_s(s) = \frac{K_s\,V(s)}{(R + Ls)} = M_1 s^2\,X(s) + F_a \qquad (1.7.4)$$

The force acting at point c is

$$F_c = M_2\frac{d^2y}{dt^2} + B\frac{dy}{dt} + Ky$$

on taking laplace transform

$$F_c(s) = M_2 s^2\,Y(s) +\text{-}Bs\,Y(s) + K\,Y(s)$$

$$Fc\,(s) = Y(s)\,[M_2 s^2 + Bs + K] \qquad (1.7.5)$$

Taking moment at the point, the following equation is obtained

$$-F_a l_1 + F_c l_2 = 0$$

$$F_a = F_c\left(\frac{l_2}{l_1}\right) = \left(\frac{l_2}{l_1}\right)(M_2 s^2 + Bs + K)\,Y(s) \qquad (1.7.6)$$

When the force is pulling the solenoid down, the following equation may be written,

$$\frac{x}{l_1} = \frac{y}{l_2}$$

$$\therefore \qquad x = \left(\frac{l_1}{l_2}\right) y$$

Substituting (1.7.6) in equation (1.7.4)

$$\frac{K_s V(s)}{R + Ls} = \left[M_1 s^2 \frac{l_1}{l_2} + \frac{l_2}{l_1}(M_2 s^2 + Bs + K)\right] Y(s) \qquad \left[\because X(s) = \frac{l_1}{l_2} Y(s)\right]$$

$$\frac{Y(s)}{V(s)} = \frac{Ks}{(R + Ls)\left(M_1 s^2 \frac{l_1}{l_2} + \frac{l_2}{l_1}(M_2 s^2 + Bs + K)\right)}$$

Results: The differential equations governing the system are,

1. $V(t) = R_i + L\dfrac{di}{dt}$

2. $F_s = \dfrac{Ks\,V(t)}{R + L\dfrac{di}{dt}} = M_1 \dfrac{d^2x}{dt^2} + F_a$

3. $F_c = M_2 \dfrac{d^2y}{dt^2} + B\dfrac{dy}{dt} + Ky$

4. $\dfrac{x}{l_1} = \dfrac{y}{l_2}$

The transfer function of the system is

$$\frac{Y(s)}{V(s)} = \frac{Ks}{[R + Ls]\left[M_1 s^2 \left(\frac{l_1}{l_2}\right) + \left(\frac{l_2}{l_1}\right)(M_1 s^2 + Bs + K)\right]}$$

EXAMPLE 8

Consider the electro-mechanical system shown in Fig. (Ex. 1.8). Assuming that the electrical force acting on the core of mass M is proportional to the current i in the inductor obtain the transfer function of the system.

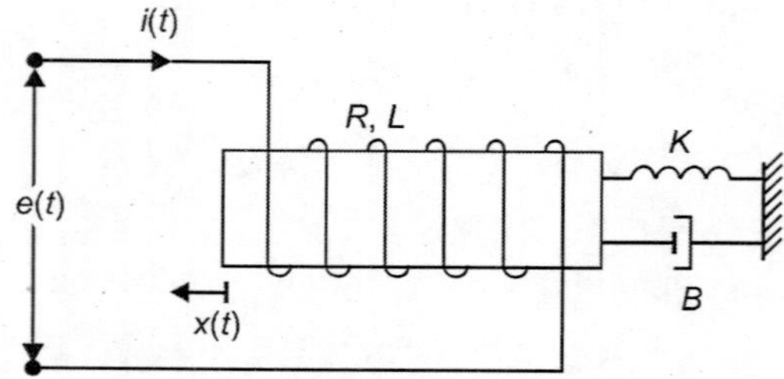

Fig. Ex. 1.8

Solution: In the given system the emf $e(t)$ is the input and the linear displacement $x(t)$ is the output.

Let,

Laplace transform of $x(t) = L[x(t)] = X(s)$

Laplace transform of $e(t) = L[e(t)] = E(s)$

Hence required transfer function in $\dfrac{X(s)}{E(s)}$.

The system has only one node, that is M. The differential equation are formed or mechanical system and also for electrical system.

First, Let write differential equation for mechanical system. The free body diagram of M is shown in Fig. (Ex. 1.8.1). The opposing forces acting on M are marked as f_m, f_b and f_k.

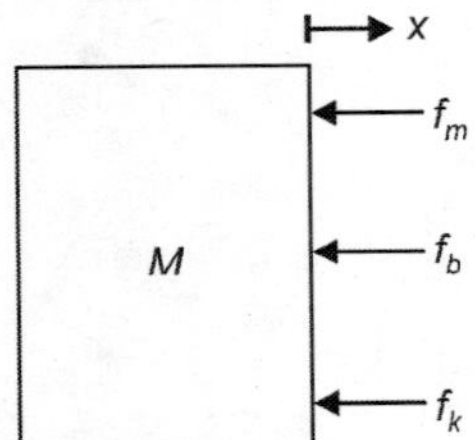

Fig. Ex. 1.8.1: Free body diagram of mass.

$$f_m = M\frac{d^2x}{dt^2}; \quad f_b = B\frac{dx}{dt}; \quad f_k = Kx$$

By newton's second law,

$$f(t) = f_m + f_b + f_k$$

$$f(t) = M\frac{d^2x}{dt^2} + B\frac{dx}{dt} + Kx \tag{1.8.1}$$

$f(t)$ is the electrical force acting on the mass

on taking Laplace transform of equation (1.8.1)

$$F(s) = Ms^2 X(s) + Bs\,X(s) + K\,X(s)$$

$$F(s) = X(s)\,[Ms^2 + Bs + K] \tag{1.8.2}$$

The equivalent circuit for the electrical circuit as shown in Fig. (Ex. 1.8.2)

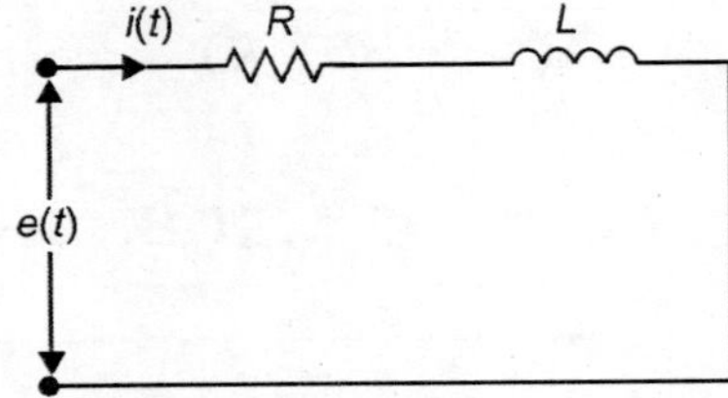

Fig. Ex. 1.8.2

By applying KVL,

$$e(t) = Ri\,(t) + L\frac{di(t)}{dt} \tag{1.8.3}$$

on taking laplace transform of equation (1.8.3)

$$E(s) = R\,I(s) + Ls\,I(s)$$

$$E(s) = I(s)\,[R + Ls] \Rightarrow I(s) = \frac{E(s)}{(R + Ls)} \tag{1.8.4}$$

and also, $f(t) \propto i(t); f(t) = K_s\, i(t)$

where, $K_s \rightarrow$ Proportionality constant

$$\therefore \quad F(s) = K_s\, I(s) \tag{1.8.5}$$

Substitute $I(s)$ from equation (1.8.4) in equation (1.8.5)

$$F(s) = K_s\left[\frac{E(s)}{R + Ls}\right] \tag{1.8.6}$$

By equating equations (1.8.6) and (1.8.2)

$$x(s)\,[Ms^2 + Bs + K] = K_s\left[\frac{E(s)}{R + Ls}\right]$$

$$\frac{X(s)}{E(s)} = \frac{Ks}{(R + Ls)\,(Ms^2 + Bs + K)} \tag{1.8.7}$$

Result: The differential equations governing the system case,

1. $e(t) = iR + L\dfrac{di}{dt}$
2. $f(t) = M\dfrac{d^2x}{dt^2} + B\dfrac{dx}{dt} + Kx$
3. $f(t) = K_s\, i(t)$

The transfer function of the system is

$$\frac{X(s)}{E(s)} = \frac{K_s}{(R + Ls)\,(Ms^2 + Bs + K)}$$

1.15 ELECTRICAL ANALOGOUS OF MECHANICAL TRANSLATIONAL SYSTEMS

Systems remain analogous as long as the differential equations governing the system or transfer functions are of identical form. The electric analogue of any other kind of system is of great importance since it is easier to construct electrical models and analyse them.

The three basic elements mass, dash-pot and spring that are used in modelling mechanical translational systems are analogous to resistance, inductance and capacitance of electrical system. The input force in mechanical system is analogous to either voltage source or current source in electrical systems.

The output velocity in mechanical system is analogous to either current (or) voltage in an element in electrical system. Since the electrical systems has two types of inputs either voltage or current source, there are two types of analogies: force-voltage analogy and force-current analogy.

Force-voltage Analogy

The force balance equations of mechanical elements and their analogous electrical elements in force-voltage analogy are shown in Table 2.

Table 2

Mechanical System	Electrical system
Input: Force Output: Velocity	Input: Voltage source Output: Current through the element.
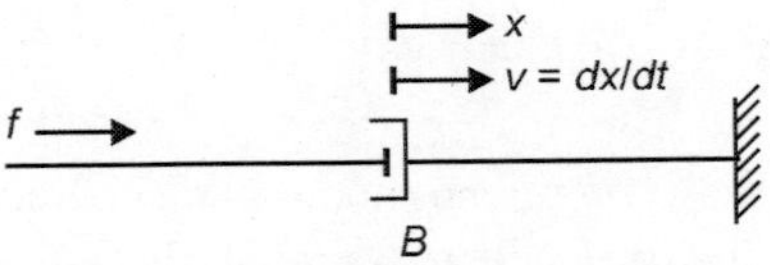 x, $v = dx/dt$, f, B $f = \frac{Bdx}{dt} = Bv$	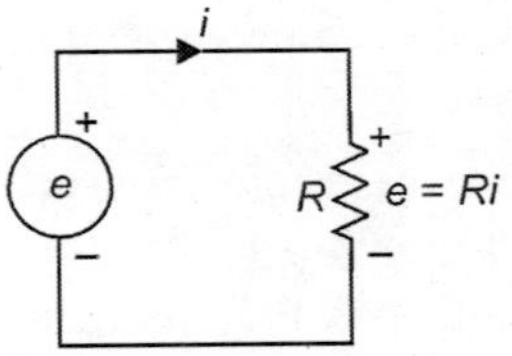 i, e, R $e = Ri$
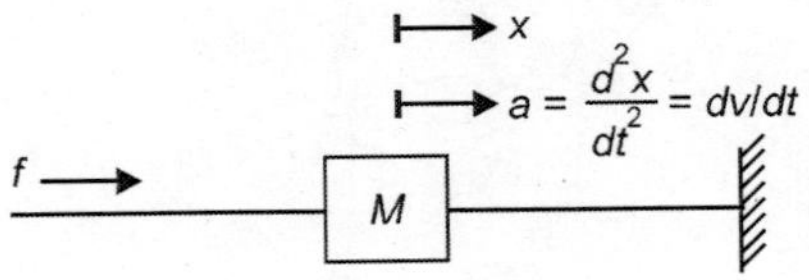 x, $a = \frac{d^2x}{dt^2} = dv/dt$, f, M $f = \frac{Md^2x}{dt^2} = \frac{Mdv}{dt}$	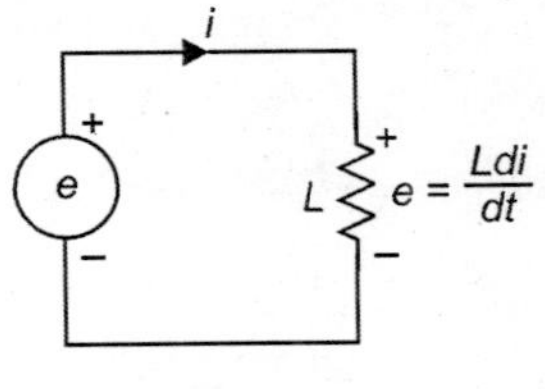 i, e, L $e = \frac{Ldi}{dt}$
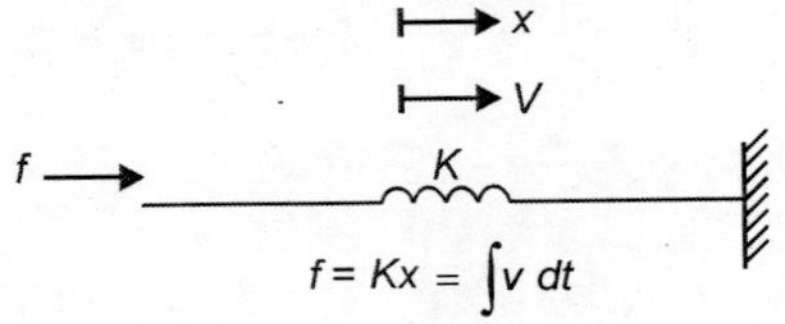 x, V, f, K $f = Kx = \int v\,dt$	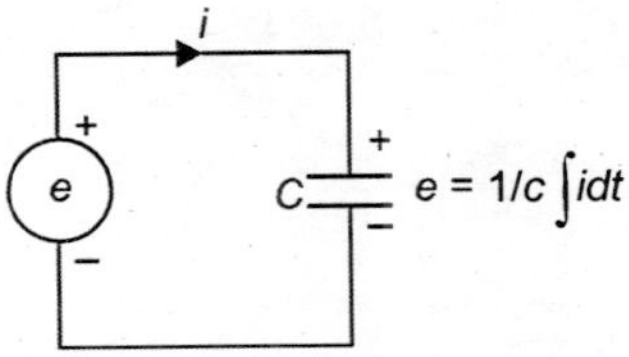 i, e, C $e = 1/c \int idt$

Table 3		
Item	**Mechanical system**	**Electrical system**
Independent variable (Input)	Force, f	Voltage, e
Dependent variable (Output)	Velocity, v	Current, i
Dissipative element	Frictional coefficient of dash-pot, B	Resistance, R
Storage element	Mass, M	Inductance, L
	Stiffness of spring, K	Inverse of capacitance, $\frac{1}{c}$
Physical law	Newton's second law $\Sigma f = 0$	KVL $\Sigma V = 0$
Changing the level of independent variable	Lever $\frac{f_1}{f_2} = \frac{l_1}{l_2}$	Transformer $\frac{e_1}{e_2} = \frac{N_1}{N_2}$

The following points serve as guide lines to obtain electrical analogous of mechanical systems based on force-voltage analogy.

1. In electrical system the elements which are in series will have same current, likewise in mechanical system, the elements having same velocity are said to be in series.
2. The elements having same velocity in mechanical system should have the same current in electrical analogous system.
3. Each node in the mechanical system corresponds to a closed loop in electrical system. A mass is considered as a node.
4. The number of meshes in electrical analogous is same as that of the number of nodes in mechanical system. Hence the number of mesh currents and system equations will be same as that of the number of velocities of nodes in mechanical system (number of masses = number of loops).
5. The mechanical driving sources and passive elements connected to the node in mechanical system should be represented as closed loop in electrical analogous system.
6. The elements connected between two masses in mechanical system is represented as a common elements between two meshes in electrical analogous system.

Note: Force voltage analogy is based on KVL.

Force Current Analogy

Table 4	
Mechanical System	**Electrical system**
Input: Force Output: Velocity 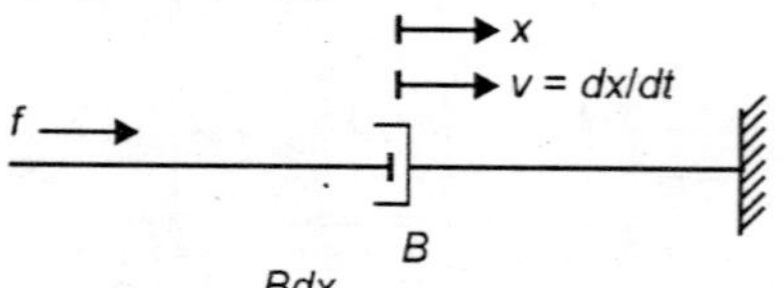 x $v = dx/dt$ f B $f = \frac{Bdx}{dt^2} = Bv$	Input: Current source Output: Voltage across the element 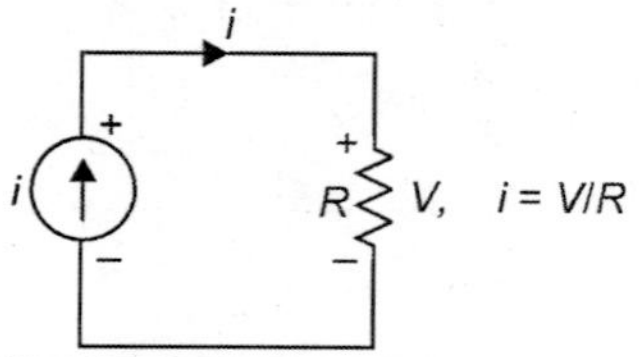 i i + − R + − V, $i = V/R$

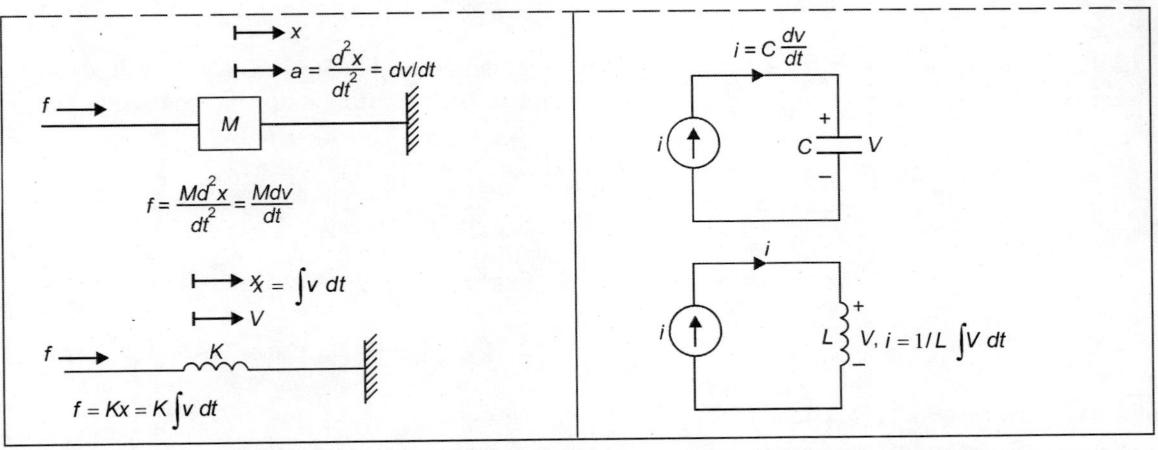

Table 5		
Item	**Mechanical system**	**Electrical system**
Independent variable (Input)	Force, f	Current, i
Dependent variable (Output)	Velocity, v	Voltage, V
Dissipative element	Frictional coefficient of dash-pot, B	Conductance, $G = 1/R$
Storage element	Mass, M Stiffness of spring, K	Capacitance, C Inverse of inductance, $1/L$
Physical law	Newton's second law $\Sigma f = 0$	KCL $\Sigma i = 0$
Changing the level of independent variable	Lever $\frac{f_1}{f_2} = \frac{l_1}{l_2}$	Transformer $\frac{i_1}{i_2} = \frac{N_1}{N_2}$

The following points serve as guidelines to obtain electrical analogous of mechanical system based on force-current analogy.

1. In electrical system elements which are in parallel will have same voltage, likewise in mechanical system, the elements having same force are said to be in parallel.
2. The elements having same velocity in mechanical system can have analogous to voltage in electrical analogous system.
3. Each mass in the mechanical system corresponds to a node in electrical system.
4. The number of node voltages and system equations will be same as that of the number of velocities of masses in mechanical system.
5. The mechanical driving sources and passive elements connected to the node in mechanical system should be represented by analogous elements connected to a node in electrical system.
6. The element connected between two nodes in mechanical system is represented as a common element between two nodes in electrical analogous system.

Note: Force current analogy based on KCL.

EXAMPLE 9

Write the differential equations governing the mechanical system shown in Fig. (Ex. 1.9) Draw the force voltage and force-current electrical analogous circuits and verify by writing mesh and node equations.

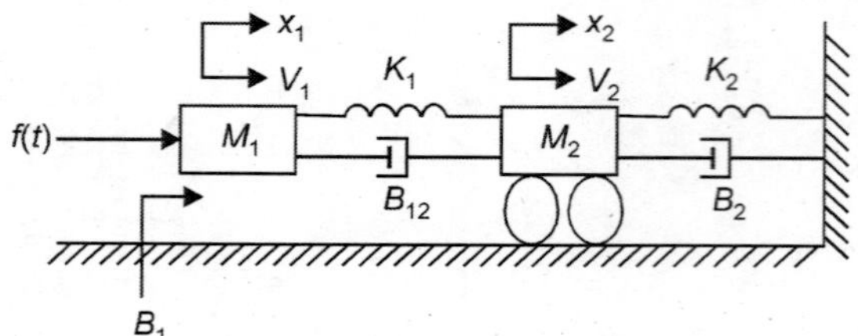

Fig. Ex. 1.9

Solution: Write force balance equation for given mechanical system.

Free body diagram of mass M_1 as shown in Fig. (Ex. 1.9.1). Forces are marked as f_{m1}, f_{b1}, f_{b12}, f_{k1} and $f(t)$

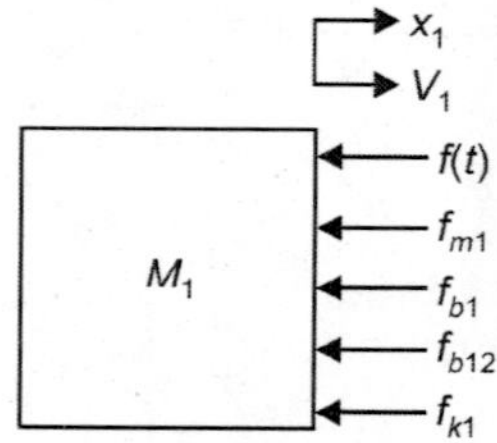

Fig. Ex. 1.9.1: Free body diagram of mass M_1.

By newton's second law,

$$f_{m1} + f_{b1} + f_{b12} + f_{k1} = f(t)$$

$$M_1 \frac{d^2x_1}{dt^2} + B_1 \frac{dx_1}{dt} + B_{12} \frac{d}{dt}(x_1 - x_2) + K_1 (x_1 - x_2) = f(t) \quad (1.9.1)$$

The freebody diagram of mass M_2 as shown in Fig. (Ex. 1.9.2). Forces are marked as, f_{m2}, f_{b2}, f_{b12}, f_{k1} and f_{k2}.

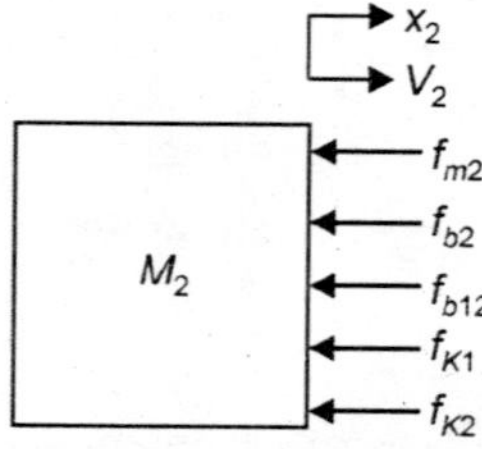

Fig. Ex. 1.9.2: Frebody diagram of mass M_2.

By newton's second law,

$$f_{m2} + f_{b2} + f_{b12} + f_{k1} + f_{k2} = 0$$

$$M_2 \frac{d^2x_2}{dt^2} + B_2 \frac{dx_2}{dt} + K_2x_2 + B_{12} \frac{d}{dt}(x_2 - x_1) + K_1 (x_2 - x_1) = 0 \quad (1.9.2)$$

on replacing the displacement by velocity in the differential equations (1.9.1) and (1.9.2) of the mechanical system we get,

$$\left[\text{i.e.,} \frac{d^2x}{dt^2} = \frac{dv}{dt}; \quad \frac{dx}{dt} = v \quad \text{and} \quad x = \int v dt\right]$$

equation (1.9.1) become,

$$M_1 \frac{dv_1}{dt} + B_1v_1 + B_{12} (v_1 - v_2) + K_1 \int (v_1 - v_2)\, dt = f(t) \quad (1.9.3)$$

equation (1.9.2) become,

$$M_2 \frac{dv_2}{dt} + B_2v_2 + K_2 \int v_2 dt + B_{12}(v_2 - v_1) + K_1 \int (v_2 - v_2)\, dt = 0 \quad (1.9.4)$$

Force-voltage Analogous Circuit

The given mechanical system has two nodes. Hence the force-voltage analogous electrical circuit will have two meshes.

The force applied to mass M_1 is represented by a voltage source in first mesh. The elements M_1, B_1, K_1 and B_{12} are connected to first node. Hence they are represented by analogous element in mesh 1 forming a closed path. The elements K_1, B_{12}, M_2, K_2 and B_2 are connected to second node. Hence they are represented by analogous element in mesh 2 forming a closed path.

The elements K_1 and B_{12} are common between node 1 and 2 and so they are represented by analogous element as common elements between two meshes. The force-voltage electrical analogous circuit is shown in Fig. (Ex. 1.9.3). The electrical analogous elements for the elements of mechanical system are given below.

$f(t) \rightarrow e(t)$	$M_1 \rightarrow L_1$	$B_1 \rightarrow R_1$	$K_1 \rightarrow 1/C_1$
$x_1 \rightarrow i_1$	$M_2 \rightarrow L_2$	$B_2 \rightarrow R_2$	$K_2 \rightarrow 1/C_2$
$x_2 \rightarrow i_2$		$B_{12} \rightarrow R_{12}$	

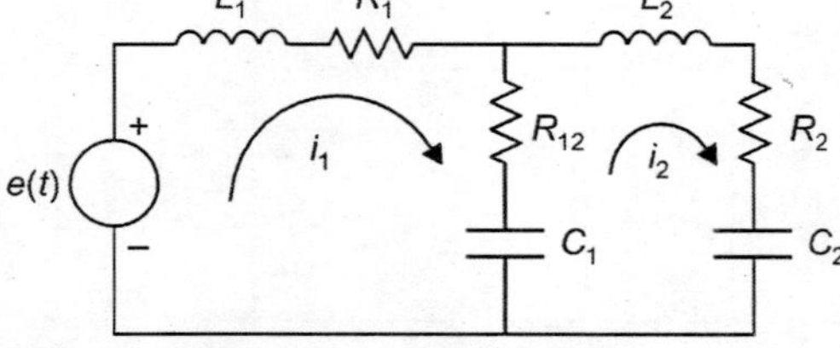

Fig. Ex. 1.9.3: Force voltage electrical analogous circuit.

The mesh equations using KVL for the circuit shown in Fig. (Ex. 1.9.3) are given below.

For first loop, apply Kirchoff's voltage law,

$$L_1 \frac{di_1}{dt} + R_1 i_1 + R_{12}(i_1 - i_2) + \frac{1}{C_1}\int (i_1 - i_2)\, dt = e(t) \quad (1.9.5)$$

For second loop, apply Kirchoff's voltage law,

$$L_2 \frac{di_2}{dt} + R_2 i_2 + \frac{1}{c_2}\int i_2\, dt + R_{12}(i_2 - i_1) + \frac{1}{c_1}\int (i_2 - i_1)\, dt = 0 \quad (1.9.6)$$

It is observed that the mesh basis equations (1.10.5) and (1.10.6) are similar to the differential equations (1.9.3) and (1.9.4) governing the mechanical system.

Force Current Analogous Circuit

The given mechanical system has two nodes. Hence the force current analogous electrical circuit will have two nodes

The force applied to mass M_1 is represented as a current source connected to node 1 in analogous electrical circuit. The elements M_1, B_1, K_1 and B_{12} are connected to first node. Hence they are represented by analogous elements connected to node 1 in analogous electrical circuit. The elements K_1, B_{12} M_2, K_2 are B_2 are connected to second node. Hence they are represented by analogous electrical circuit.

The elements K_1 and B_{12} are common between 1 and 2 so they are represented as common element between two nodes in electrical analogous circuit. The force current electrical analogous circuit is shown in Fig. (1.10.4)

The electrical analogous elements for the elements of mechanical system are given below

$f(t) \rightarrow i(t)$	$M_1 \rightarrow C_1$	$B_1 \rightarrow 1/R_1$	$K_1 \rightarrow 1/L_1$
$x_1 \rightarrow V_1$	$M_2 \rightarrow C_2$	$B_2 \rightarrow 1/R_2$	$K_2 \rightarrow 1/L_2$
$x_2 \rightarrow V_2$			

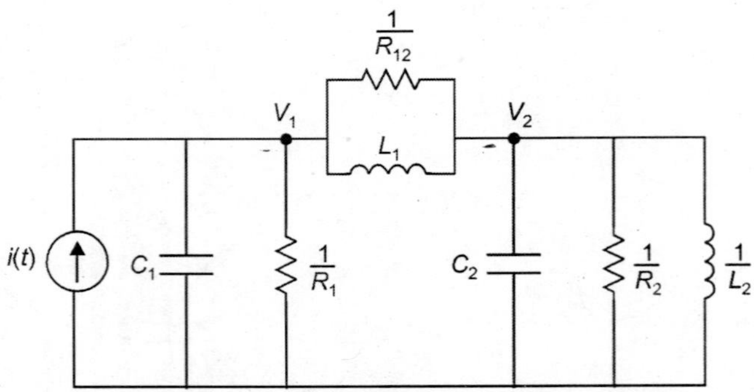

Fig. 1.9.4: Force-current electrical analogous circuit.

The node basis equations using Kirchoff's current Law for the circuit shown in Fig. (Ex. 1.9.4) are given below.

For node 1,
$$C_1 \frac{dV_1}{dt} + \frac{1}{R_1} V_1 + \frac{1}{R_{12}} (V_1 - V_2) + \frac{1}{L_1} \int (V_1 - V_2)\, dt = i(t) \quad (1.9.7)$$

For node 2,
$$C_2 \frac{dV_2}{dt} + \frac{1}{R_2} V_2 + \frac{1}{L_2} \int V_2 dt + \frac{1}{R_{12}} (V_2 - V_1) + \frac{1}{L_1} \int (V_2 - V_1)\, dt = 0 \quad (1.9.8)$$

It is observed that the node basis equations (1.9.7) and (1.9.8) are similar to the differential equations (1.9.3) and (1.9.4) governing the mechanical system.

EXAMPLE 10

Write the differential equations governing the mechanical system shown in Fig. (Ex. 1.10). Draw the force-voltage and force current electrical analogous circuits and verify by writing mesh and node equations.

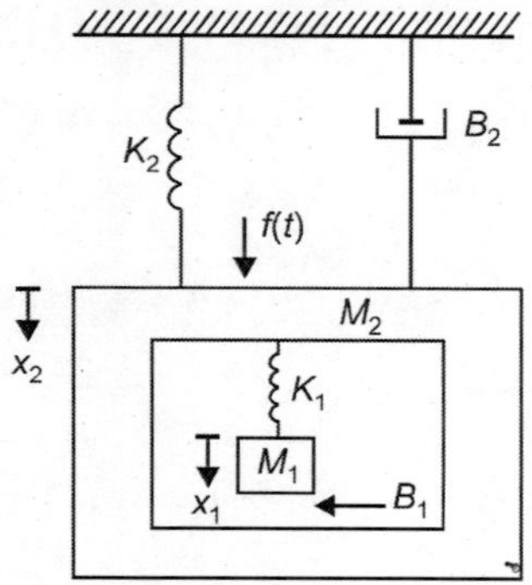

Fig. Ex. 1.10

Solution: First let write force balance equation for given mechanical system.

Free body diagram of mass M_1 as shown in Fig. (Ex. 1.10.1). Forces are marked as f_{m1}, f_{b1}, and f_{k1}.

By newton's second law

$$f_{m1} + f_{b1} + f_{k1} = 0$$

$$M_1 \frac{d^2 x_1}{dt^2} + B_1 \frac{d}{dt}(x_1 - x_2) + K_1 (x_1 - x_2) = 0 \quad (1.10.1)$$

Fig. Ex. 1.10.1: Free body diagram of mass M_1.

The free body diagram of mass M_2 shown in Fig. (Ex. 1.10.2). The opposing forces are marked as $f_{m2}, f_{b2}, f_{b1}, f_{k2}$ and f_{k1}

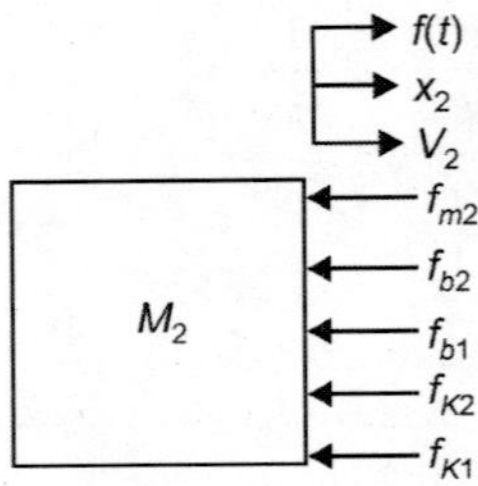

Fig. Ex. 1.10.2: Free body diagram of mass M_2.

By newton's second law

$$f_{m2} + f_{b2} + f_{b1} + f_{k2} + f_{k1} = f(t)$$

$$M_2 \frac{d^2x_2}{dt^2} + B_2 \frac{dx_2}{dt} + K_2x_2 + B_1 \frac{d}{dt}(x_2 - x_1) + K_1 (x_2 - x_1) = f(t) \tag{1.10.2}$$

on replacing the displacements by velocity in the differential equation (1.10.1) and (1.10.2) governing the mechanical system we get,

$$\left[\text{i.e., } \frac{d^2x}{dt^2} = \frac{dv}{dt}; \frac{dx}{dt} = v \text{ and } x = \int v dt\right]$$

The equation (1.10.1) and (1.10.2) become

$$M_1 \frac{dv_1}{dt} + B_1 (v_1 - v_2) + K_1 \int (v_1 - v_2)\, dt = 0 \tag{1.10.3}$$

$$M_2 \frac{dv_2}{dt} + B_2v_2 + K_2 \int v_2\, dt + B_1 (v_2 - v_1) + K_1 \int (v_2 - v_1)\, dt = f(t) \tag{1.10.4}$$

Force-voltage Analogous Circuit

The given mechanical system has two nodes. Hence the force voltage analogous electrical circuit will have two meshes. The force applied to mass, M_2 is represented by a voltage source is second mesh.

The elements M_1, K_1 and B_1 are connected to first node. Hence they are represented by analogous element in mesh 1 forming closed path. The elements M_2, K_2, B_2, B_1 and K_1 are connected to second node. Hence they are represented by analogous element in mesh 2 forming a closed path.

The element B_1 and K_1 are common between node 1 and 2 and so they are represented as common elements between mesh 1 and mesh 2. The force-voltage electrical analogous circuit shown in Fig. (Ex. 1.10.3)

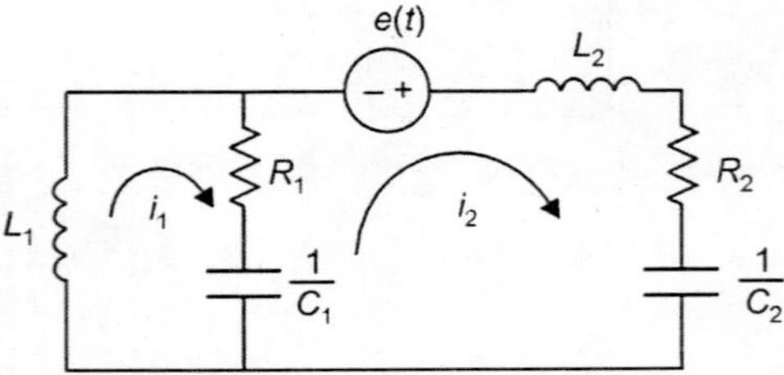

Fig. Ex. 1.10.3: Force-voltage electrical analogous circuit.

The electrical analogous elements for the elements of mechanical system are given below,

$f(t) \rightarrow e(t)$ $\quad x_1 \rightarrow i_1$ $\quad M_1 \rightarrow L_1$ $\quad K_1 \rightarrow 1/C_1$ $\quad B_1 \rightarrow R_1$

$x_2 \rightarrow i_2$ $\quad M_2 \rightarrow L_2$ $\quad K_2 \rightarrow 1/C_2$ $\quad B_2 \rightarrow R_2$

The mesh basis equations using KVL for the circuit shown in Fig. (1.10.3) are given below.

$$L_1 \frac{di_1}{dt} + R_1 (i_1 - i_2) + \frac{1}{C_1} \int (i_1 - i_2)\, dt = 0 \tag{1.10.5}$$

$$L_2 \frac{di_2}{dt} + R_2 i_2 + \frac{1}{C_2} \int i_2 dt + \frac{1}{C_1} \int (i_2 - i_1)\, dt + R_1 (i_2 - i_1) = i(t) \tag{1.10.6}$$

It is observed that the mesh basis equations (1.10.5) and (1.10.6) are similar to the differential equations (1.10.3) and (1.10.4) governing the mechanical system.

Force-Current Analogous Circuit

The given mechanical system has two nodes. Hence the force-current analogous electrical circuit will have two nodes. The force applied to mass M_2 is represented as a current source connected to node 2 in analogous electrical circuit.

The elements M_1 K_1 and B_1 are connected to first node. Hence they are represented by analogous elements as connected to node 1 in analogous electrical circuit. The elements M_2, K_2, B_2, B_1 and K_1 are connected to second node. Hence they are represented by analogous elements as elements connected to node 2 in analogous electrical circuits.

The elements K_1 and B_1 is common to node 1 and 2 and so they are represented by analogous elements as common elements between two nodes in analogous circuit. The force current electrical analogous circuit shown in Fig. (Ex. 1.10.4).

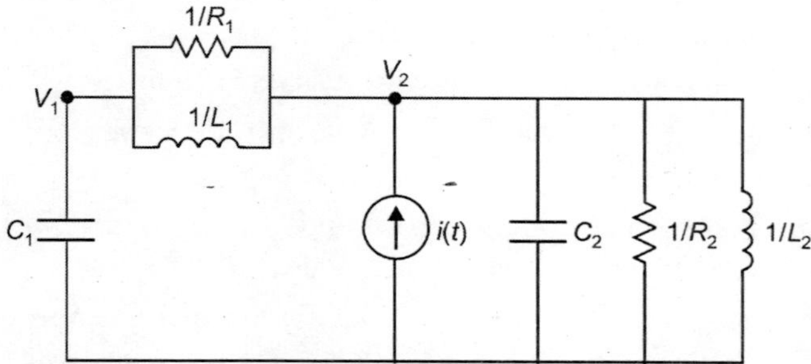

Fig. Ex. 1.10.4: Force current electrical analogous circuit.

The electrical analogous elements for the elements of mechanical system are given below

$f(t) \rightarrow i(t)$ $\quad x_1 \rightarrow V_1$ $\quad M_1 \rightarrow C_1$ $\quad B_1 \rightarrow 1/R_1$ $\quad K_1 \rightarrow 1/L_1$

$x_2 \rightarrow V_2$ $\quad M_2 \rightarrow C_2$ $\quad B_2 \rightarrow 1/R_2$ $\quad K_2 \rightarrow 1/L_2$

The node basis equations using Kirchoff's current law for the circuit shown in Fig. (Ex. 1.10.4)

$$C_1 \frac{dv_1}{dt} + \frac{1}{R_1}(v_1 - v_2) + \frac{1}{L_1}\int (v_1 - v_2)\, dt = 0 \tag{1.10.7}$$

$$C_2 \frac{dv_2}{dt} + \frac{1}{R_2} v_2 + \frac{1}{L_1}\int v_2\, dt + \frac{1}{R_1}(v_2 - v_1) + \frac{1}{L_1}\int (v_2 - v_1)\, dt = i(t) \tag{1.10.8}$$

It is observed that the node basis equations (1.10.7) and (1.10.8) are similar to the differential equations (1.10.3) and (1.10.4) governing the mechanical system.

1.16 ELECTRICAL ANALOGOUS OF MECHANICAL ROTATIONAL SYSTEM

The three basic elements such as moment of inertia, rotational dashpot and torsional spring that are used in mechanical rotational systems are analogous to resistance, inductance and capacitance of electrical system. The input torque in mechanical system is analogous to either voltage source or current source in electrical systems. the output angular velocity in mechanical rotational system is analogous to either current (or) voltage in electrical system. Since the electrical systems has two types of inputs either voltage source or current source so, there are two types of analogies such as Torque-voltage analogy and torque-current analogy.

Torque-Voltage Analogy

The torque balance equations of mechanical rotational elements and their analogous electrical elements in torque-voltage analogy are shown in Table 6. The table 7 shows the List of analogous quantities in torque-voltage analogy.

The following points serve as guidelines to obtain electrical analogous of mechanical rotational systems based on torque voltage analogy.

1. In electrical system the elements which are in series will have same current, likewise in mechanical system, the elements having same angular velocity said to be in series.
2. The elements having same angular velocity in mechanical system should have same current in electrical analogous system.
3. Each node in the mechanical system corresponds to a closed loop in electrical system. he moment of inertia of mass is considered as a node.
4. The number of meshes in electrical analogous is same as that of the number of nodes in mechanical system. Hence the number of mesh currents and system equations will be same as that of the number of angular velocities of nodes in mechanical system.
5. The mechanical driving sources and passive elements connected to the node in mechanical system should be represented by a closed loop in electrical analogous system.
6. The elements connected between two nodes in mechanical system is represented as a common element between two meshes in electrical analogous system.

Table 1.6

Mechanical System	Electrical system
Input: Torque Output: Angular velocity	Input: Voltage source Output: Current through the element
$T = \frac{Bd\theta}{dt} = B\omega$, $\omega = \frac{d\theta}{dt}$	$e = Ri$
$T = J\frac{d^2\theta}{dt^2} = J\frac{d\omega}{dt}$, $a = \frac{d^2\theta}{dt^2} = \frac{d\omega}{dt}$	$e = \frac{Ldi}{dt}$
$T = K\theta = K\int \omega dt$, $\theta = \int \omega dt$	$e = 1/C \int i\, dt$

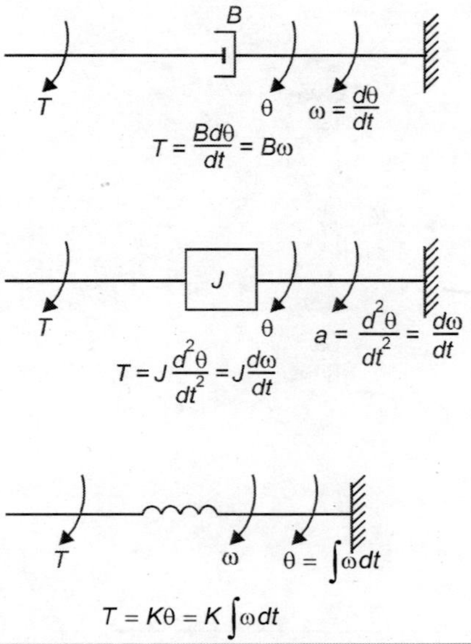

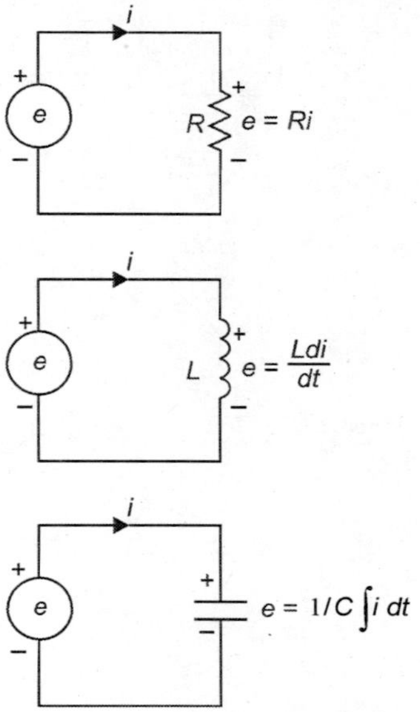

Table 1.7

Item	Mechanical system	Electrical system
Independent variable (Input)	Torque, T	Voltage, e
Dependent variable (Output)	Angular velocity, ω	Current, i
Dissipative element	Rotational coefficient of dash-pot, B	Resistance, R
Storage element	Moment of inertia, J Stiffness of spring, K	Inductance, L Inverse of capacitance, $1/C$
Physical law	Newton's second law $\Sigma T = 0$	Kirchoff's Voltage Law $\Sigma V = 0$
Changing the level of independent varia	Lever $\frac{T_1}{T_2} = \frac{N_1}{N_2}$	Transformer $\frac{e_1}{e_2} = \frac{N_1}{N_2}$

Torque-Current Analogy

The torque balance equations of mechanical elements and their analogous electrical elements in torque-current analogy are shown in Table 1.8. The Table 1.9 shows the list of analogous quantities in torque current analogy.

Table 8	
Mechanical system	**Electrical system**
Input: Torque Output: Angular velocity	Input: Current source Output: Voltage across the element
$T = \frac{Bd\theta}{dt} = B\omega$, $\omega = \frac{d\theta}{dt}$	V, $i = V/R$
$T = K\theta = K\int \omega dt$, $\theta = \int \omega dt$	V, $i = \frac{1}{L}\int V\,dt$
$T = J\frac{d^2\theta}{dt^2} = J\frac{d\omega}{dt}$, $a = \frac{d^2\theta}{dt^2} = \frac{d\omega}{dt}$	V, $i = C\frac{dv}{dt}$

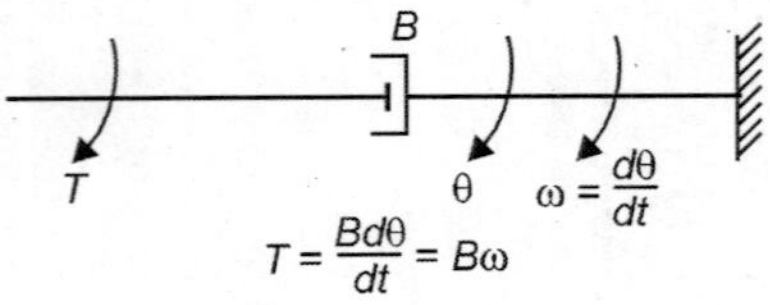

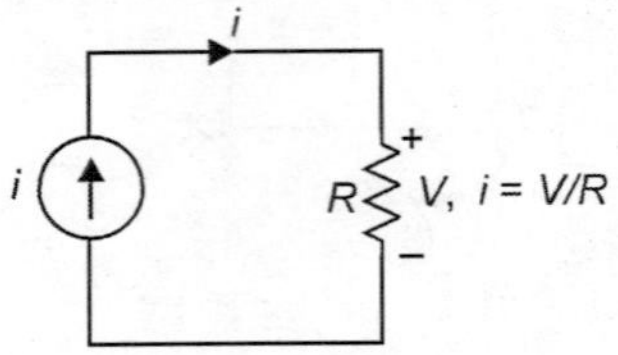

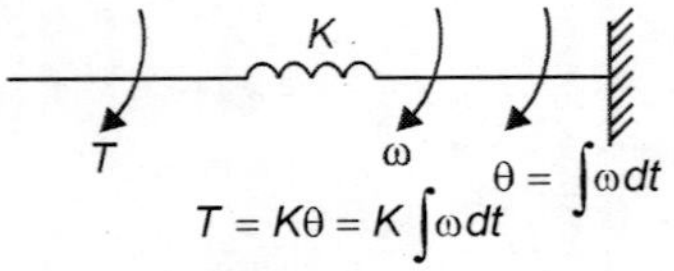

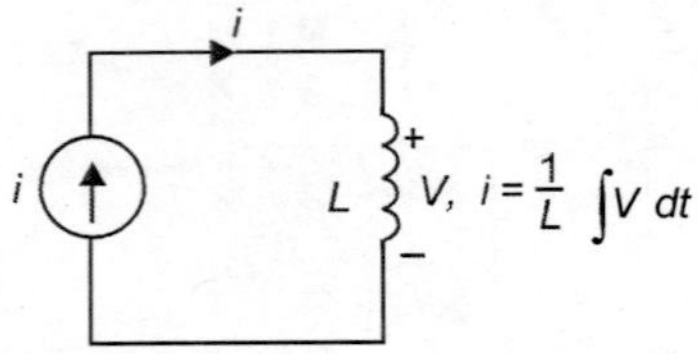

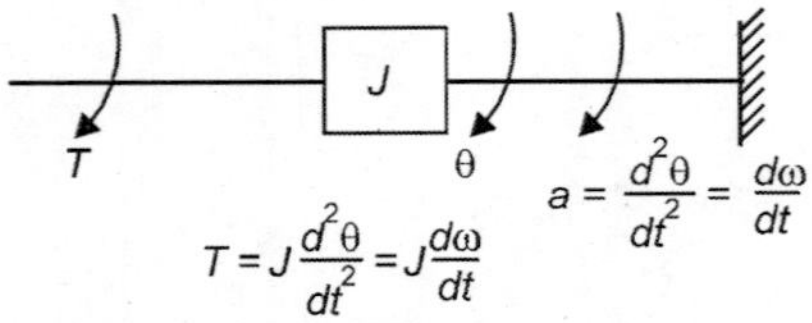

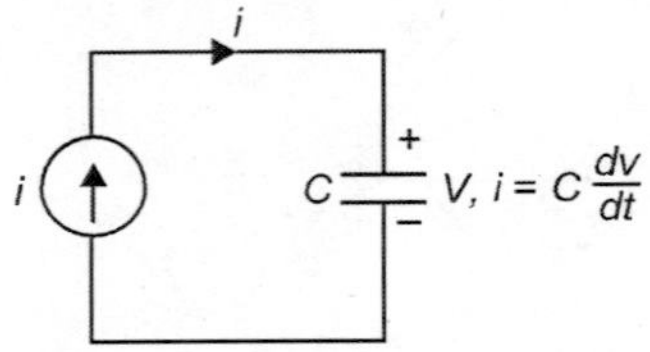

Table 9		
Item	**Mechanical system**	**Electrical system**
Independent variable (Input)	Torque, T	Current, i
Dependent variable (Output)	Angular velocity, ω Angular displacement, θ	Voltage, v Charge, q
Dissipative element	Rotational coefficient of dash-pot, B	Conductance, $G = 1/R$
Storage element	Moment of inertia, J Stiffness of spring, K	Capacitance, C Inverse of inductance, 1/L
Physical law	Newton's second law $\Sigma T = 0$	Kirchoff's current law $\Sigma i = 0$
Changing the level of independent variable	Gear $\frac{T_1}{T_2} = \frac{N_1}{N_2}$	Transformer $\frac{i_1}{i_2} = \frac{N_2}{N_1}$

The following points serve as guidelines to obtain electrical analogous of mechanical rotational systems based on torque-current analogy.

1. In electrical system the elements which are in parallel will have same voltage, similarly in mechanical systems, the elements having same torque are said to be in parallel.
2. The elements having same angular velocity in mechanical system should have same voltage in electrical analogous system.

3. Each node in the mechanical system corresponds to a node in electrical system. The moment of inertia of mass is considered as a node.
4. The number of nodes in electrical analogous is same as that of the number of nodes in mechanical system. Hence the number of node voltages and system equations will be same as that of the number of angular velocities of nodes in mechanical system.
5. The mechanical driving sources and passive elements connected to the node in mechanical system should be represented by analogous element connected to a node in analogous electrical system.
6. The elements connected between two nodes in mechanical system is represented as a common element between two nodes in electrical analogous system.

EXAMPLE 11

Write the differential equations governing the mechanical rotational system shown in Fig. (Ex. 1.11) the torque-voltage and torque-current electrical analogous circuits and verify by writing mesh and node equations.

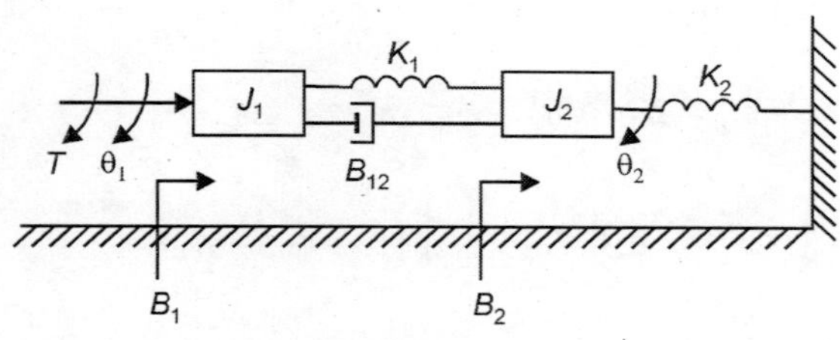

Fig. Ex. 1.11

Solution: The given mechanical rotational systems has two nodes. The differential equations governing the mechanical rotational system are given by torque balance equations at these nodes.

Let the angular displacements of J_1 and J_2 be θ_1 and θ_2.

The free body diagram of J_1 is shown in Fig. (Ex. 1.11.1). the opposing torques are marked as T_{j1}, T_{b1} and T_{k1}

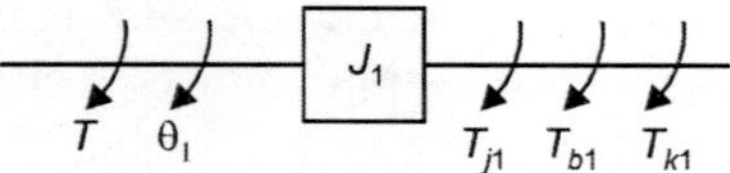

Fig. Ex. 1.11.1: Free body diagram of mass with moment of inertia J_1.

By newton's second law,

$$T_{j1} + T_{b1} + T_{k1} = T$$

$$J_1 \frac{d^2\theta_1}{dt^2} + B_1 \frac{d\theta_1}{dt} + K_1(\theta_1 - \theta_2) = T \qquad (1.11.1)$$

The freebody diagram of J_2 is shown in Fig. (Ex. 1.11.2). The opposing torques are marked as Tj_2, T_{b2}, T_{k2} and T_{k1}.

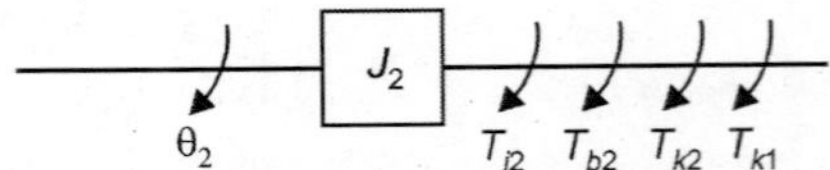

Fig. Ex. 1.11.2: Free body diagram of mass with moment of inertia J_2.

By newton's second law

$$T_{j2} + T_{b2} + T_{k2} + T_{k1} = 0$$

$$J_2 \frac{d^2\theta_2}{dt^2} + B_2 \frac{d\theta_2}{dt} + K_2\theta_2 + K_1(\theta_2 - \theta_1) = 0 \tag{1.11.2}$$

on replacing the angular displacements by angular velocity in the differential equations (1.11.1) and (1.11.2) governing the mechanical rotational system we get,

$$\left[\text{i.e., } \frac{d^2\theta}{dt^2} = \frac{d\omega}{dt}; \frac{d\theta}{dt} = \omega; \; q = \int \omega \, dt\right]$$

The equations (1.11.1) and (1.11.2) become

$$J_1 \frac{d\omega_1}{dt} + B_1\omega_1 + K_1 \int (\omega_1 - \omega_2)\, dt = T \tag{1.11.3}$$

$$J_2 \frac{d\omega_2}{dt} + B_2\omega_2 + K_2 \int \omega_2 dt + K_1 \int (\omega_2 - \omega_1)\, dt = 0 \tag{1.11.4}$$

Torque-Voltage Analogous Circuit

The given mechanical system has two nodes (J_1 and J_2) hence the torque-voltage analogous electrical circuit also will have two meshes. The torque applied to J_1 is represented by a voltage source in first mesh.

The elements J_1, B_1 and K_1 are connected to first node. Hence they are represented by analogous element in mesh, forming a closed path. The elements J_2, B_2 and K_1 are connected to second node. Hence they are represented by analogous element in mesh 2 forming a closed path.

The element K_1 is common between node 1 and 2 and so it is represented by analogous element as common element between two meshes. The torque-voltage electrical analogous circuit is shown in Fig. (Ex. 1.11.3).

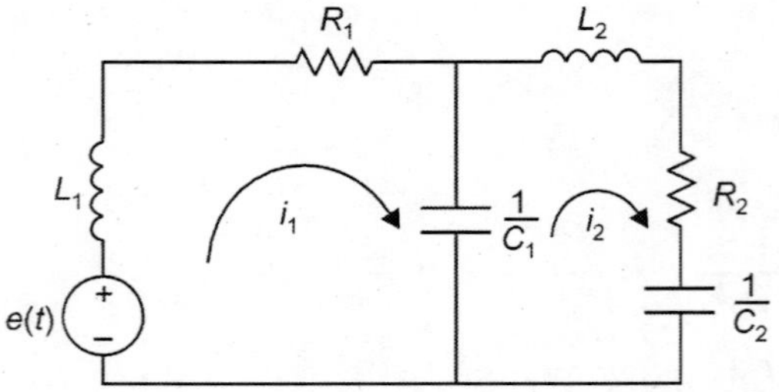

Fig. Ex. 1.11.3: Torque-voltage electrical analogous circuit.

The electrical analogous elements for the elements of mechanical rotational system are given below

$T \rightarrow e(t)$ $\quad J_1 \rightarrow L_1$ $\quad B_1 \rightarrow R_1$ $\quad K_1 \rightarrow 1/C_1$

$\theta_1 \rightarrow i_1$ $\quad J_2 \rightarrow L_2$ $\quad B_2 \rightarrow R_2$ $\quad K_2 \rightarrow 1/C_2$

$\theta_2 \sim i_2$

The mesh basis equations using KVL for the circuit shown in Fig. (Ex. 1.11.3)

$$L_1 \frac{di_1}{dt} + R_1 i_1 + \frac{1}{c_1} \int i_2 dt + \frac{1}{c_1} \int (i_2 - i_1)\, dt = 0 \qquad (1.11.6)$$

It is observed that the mesh basis equations (1.11.5) and (1.11.6) are similar to the differential equations (1.11.3) and (1.11.4) governing the mechanical system.

Torque-Current Analogous Circuit

The given mechanical system has two nodes. Hence the torque-current analogous electrical circuit will have two nodes. The torque applied to J_1 is represented as a current source connected to node 1 in analogous electrical circuit.

The element J_1, B_1 and K_1 are connected to first node. Hence they are represented by analogous elements as elements connected to node 1 in analogous electrical circuit. The elements J_2, B_2, K_2 and K_1 are connected to second node. Hence they are represented by analogous elements connected to node 2 in analogous electrical circuit.

The element K_1 is common between node 1 and 2. So it is represented by analogous element as common element between node 1 and 2. The torque current electrical analogous circuit is shown in Fig. (Ex. 1.11.4).

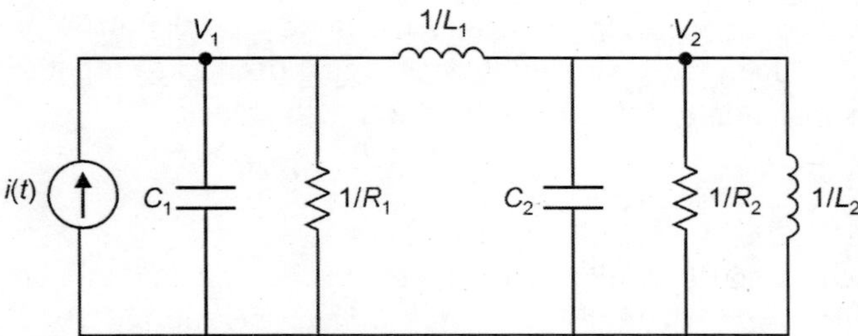

Fig. Ex. 1.11.4: Torque-current electrical analogous circuit.

The electrical analogous elements for the elements of mechanical rotational system are given below.

$T \rightarrow i(t)$ $\quad B_1 \rightarrow 1/R_1$ $\quad \theta_1 \rightarrow v_1$ $\quad J_1 \rightarrow C_1$ $\quad K_1 \rightarrow 1/L_1$

$B_2 \rightarrow 1/R_2$ $\quad \theta_2 \rightarrow v_2$ $\quad J_2 \rightarrow C_2$ $\quad K_2 \rightarrow 1/L_2$

The node basis equations using KCL for the circuit shown in Fig. (Ex. 1.11.4) are given below.

$$C_1 \frac{dv_1}{dt} + \frac{1}{R_1} v_1 + \frac{1}{L_1} \int (v_1 - v_2)\, dt = i(t) \qquad (1.11.7)$$

$$C_2 \frac{dv_2}{dt} + \frac{1}{R_2} v_2 + \frac{1}{L_2} \int v_2 \, dt + \frac{1}{L_1} \int (v_2 - v_1) \, dt = 0 \tag{1.11.8}$$

It is observed that the node basis equations (1.12.7) and (1.12.8) are similar to the differential equations (1.12.3) and (1.12.4) governing the mechanical system.

1.17 BLOCK DIAGRAM

A control system may consist of a number of components. In control engineering to show the functions performed by each component, we commonly use a diagram called the block diagram. **A block diagram is a pictorial representation of the functions performed by each component and of the flow of signals. which shows interrelationships that exist among the various components. The elements of a block diagram are block, branch point and summing point.**

Block

In a block diagram all system variables are linked to each other through functional blocks. The functional block is a symbol for the mathematical operation on the input signal to the block that produces the output. The transfer functions of the components are usually entered in the corresponding blocks, which are connected by arrows to indicate the direction of the flow of signals. Fig. (1.17.1) shows the functional block of the block diagram.

Fig. 1.17.1: Functional block.

The arrow pointing towards the block indicates the input, and the arrow leading away from the block represents the output. Such arrows are referred as signal flow. The output signal from the block is given by the product of input signal and transfer function in the block.

Summing Point

Summing points are used to add two (or) more signals in the system. Referring to Figure (1.17.2). A circle with a cross is the symbol that indicates the summing operation.

The plus or minus sign at each arrow indicates whether the signal is to be added or subtracted.

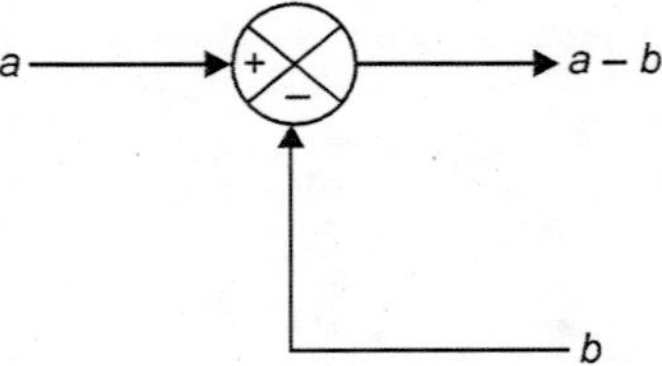

Fig. 1.17.2: Summing Point.

Branch Point

A branch point is a point from which the signal goes from one block to other blocks (or) to summing points. A branch point is shown in Fig. (1.17.3).

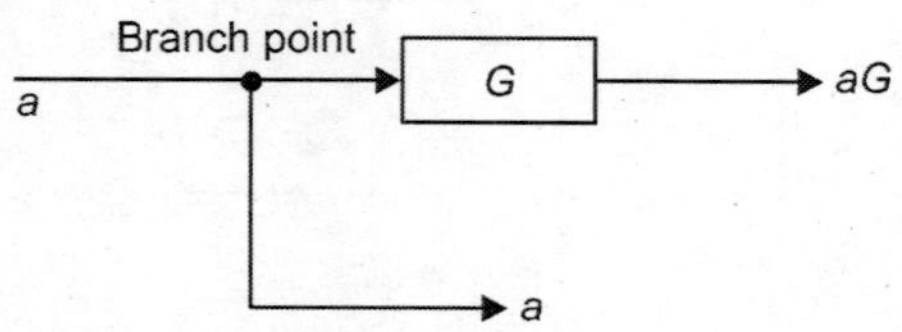

Fig. 1.17.3: Branch point.

Construction of blocK diagram for control systems

A control system can be represented diagrammatically by block diagram. The differential equations governing the system are used to construct the block diagram. By taking laplace transform the differential equations are converted to algebraic equations. The equations will have variables and constants. From the working knowledge of the system the input and output variables are identified and the block diagram for each equation can be drawn. Each equation gives one section of block diagram. The output of one section will be input for another section. The various sections are interconnected to obtain the overall block diagram of the system.

Basic connections for blocks

There are three basic connections namely (1) Cascade (2) Parallel (3) Feedback.

1. **Cascade connection:** Two blocks are said to be connected in cascade if the ouput of the first-block becomes input to the second block. The cascade connected blocks are shown in Fig. (1.17.4).

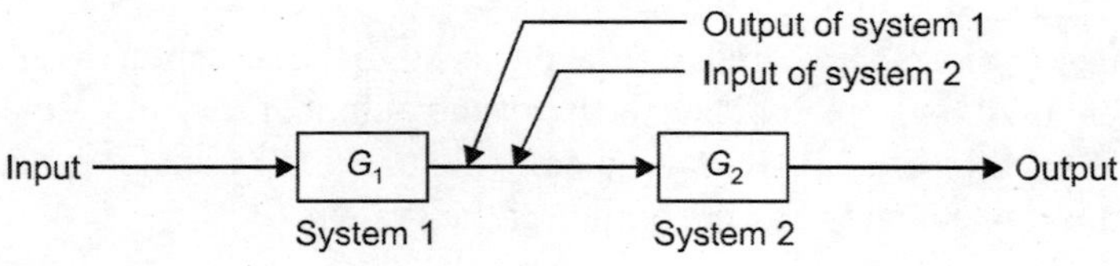

Fig. 1.17.4

The same definition may be extended to *n* number of blocks.

Note: It also can be named as series connection.

2. **Parallel connection:** Two blocks are said to be connected in parallel if the same input is applied to both the blocks and the output from the blocks are added using summer. The same definition may be extended to n blocks in parallel.

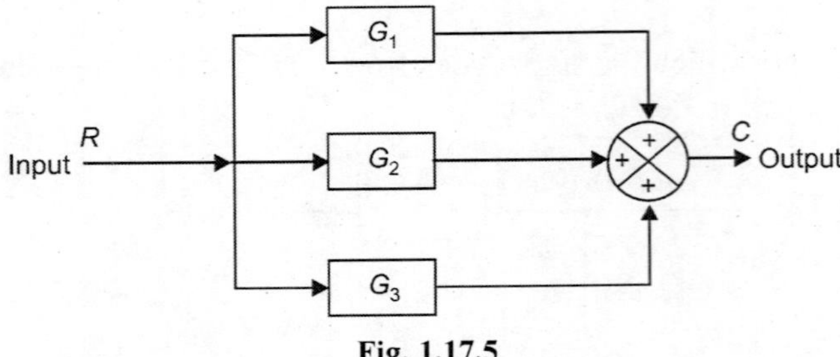

Fig. 1.17.5

3. **Feedback connection:** Two blocks are said to be connected in feedback if the output of the first block is feedback and becomes input to the second block. The output of the second block is compared with the reference input and becomes input to the first block. If the output of the second block aids the reference input to the feedback then it is known as positive feedback. Usually the first block is called the forward path transfer function or open loop transfer function and is represented by G and feedback transfer function is usually represented by H.

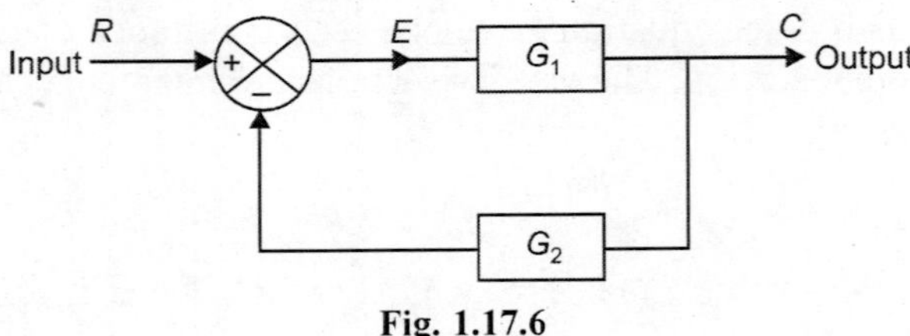

Fig. 1.17.6

EXAMPLE 12

Consider the electromechanical system shown in Fig. (Ex. 1.12). Assuming the magnetic pull produced by the solenoid is equal to K_0 times the current i, flowing through the solenoid, (a) draw a block diagram showing each component of the system. Indicate the input and output of each block clearly. (b) Determine the transfer function of each block. (c) Determine equivalent overall transfer function of the system. Take E as the input to the system and x_1 as the output of the system.

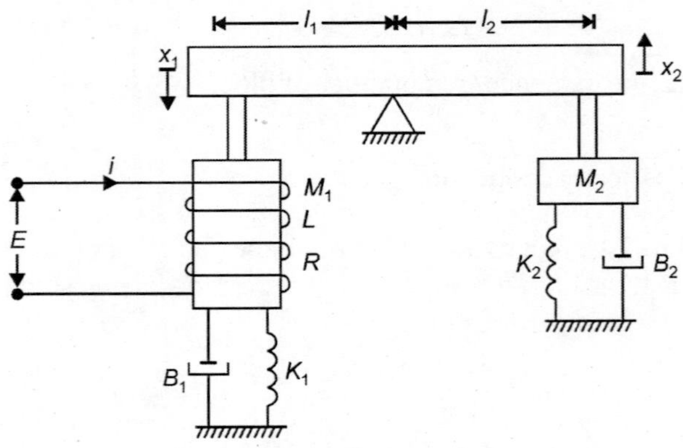

Fig. Ex. 1.12

Solution:

Step 1: Let F_1 is the force acting at x_1 and F_2 is the force acting at x_2, the following equation can be written

$$F_1 l_1 = F_2 l_2$$

$$\frac{F_2}{F_1} = \frac{l_1}{l_2} \tag{1.12.1}$$

Step 2: Let write equation for electrical circuit,

$$e(t) = Ri + L\frac{di}{dt}$$

$$E(s) = (R + sL)\, I(s)$$

$$I(s) = E(s)/(R + sL) \tag{1.12.2}$$

Step 3: $F(s) = K_0\, I(s)$

$$\frac{F(s)}{I(s)} = K_0 \tag{1.12.3}$$

Step 4: At point x_2, the following equation is written

$$F_2(t) = M_2\frac{d^2x_2}{dt^2} + B_2\frac{dx_2}{dt} + K_2\, x_2$$

on taking laplace transform,

$$F_2(s) = (M_2 s^2 + B_2 s + K_2)\, X_2(s) \tag{1.12.4}$$

Step 5:

$$\frac{x_1}{l_1} = \frac{x_2}{l_2}$$

$$\frac{x_1}{x_2} = \frac{l_1}{l_2} \tag{1.12.5}$$

Step 6: The following equation is written for mass M_1.

$$F_m = M_1\frac{d^2x_1}{dt^2} + B_1\frac{dx_1}{dt} + K_1\, x_1$$

on taking laplace transform,

$$F_m(s) = M_1 s^2\, X_1(s) + B_1 s\, X_1(s) + K_1 X_1(s)$$

$$F_m(s) = X_1(s)[M_1 s^2 + B_1 s + K_1]$$

$$\frac{F_m(s)}{X_1(s)} = (M_1 s^2 + B_1 s + K_1) \tag{1.12.6}$$

Step 7: The following equation is written at x_1

$$F(s) - F_m(s) = F_1(s) \tag{1.12.7}$$

By combining all the equations, the block diagram can be drawn as shown in Fig. (Ex. 1.12.1) and (Ex. 1.12.2)

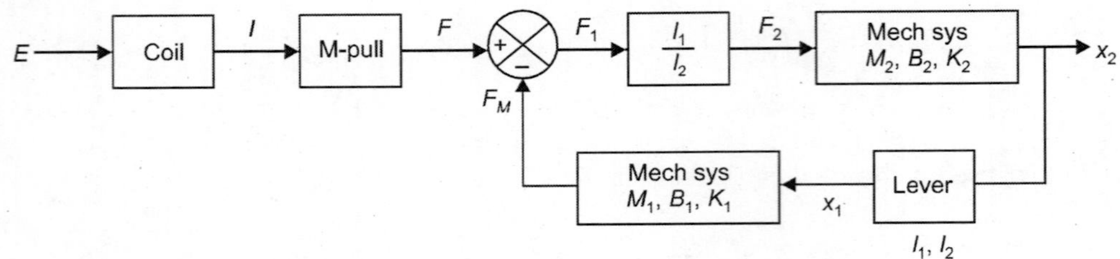

Fig. Ex. 1.12.1

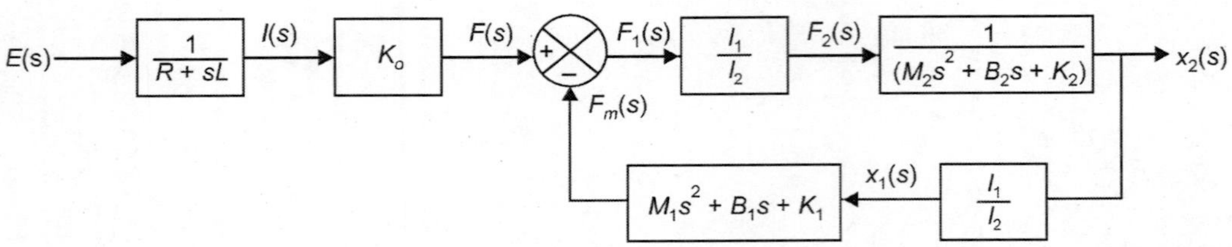

Fig. Ex. 1.12.2

using the block diagram reduction technique, we may obtain

$$\frac{X_1(s)}{E(s)} = \frac{K_0}{(R + sL)\left[\frac{l_1}{l_2}(M_1s^2 + B_1s + K_1) + (M_2s^2 + B_2s + K_2)\left(\frac{l_2}{l_1}\right)\right]}$$

EXAMPLE 13

Construct the block diagram of armature controlled dc motor.

Solution: The differential equations governing the armature controlled dc motor are (refer section 1.11)

$$V_a = i_a R_a + L_a \frac{di_a}{dt} + e_b \tag{1.13.1}$$

$$T = K_t \, i_a \tag{1.13.2}$$

$$T = J\frac{d\omega}{dt} + B\omega \tag{1.13.3}$$

$$e_b = K_b\omega \tag{1.13.4}$$

where,

$$\omega = \frac{d\theta}{dt} \tag{1.13.5}$$

on taking laplace transform of equation (1.13.1) we get,

$$V_a(s) = I_a(s)\, R_a + L_a\, sI_a(s) + E_b(s)$$

$$V_a(s) - E_b(s) = I_a(s)\,[R_a + sL_a]$$

$$I_a(s) = \frac{[V_a(s) - E_b(s)]}{[R_a + sL_a]}$$

$$I_a(s) = \frac{1}{(R_a + sL_a)}\,[V_a(s) - E_b(s)] \qquad (1.13.6)$$

from equation (1.13.6)

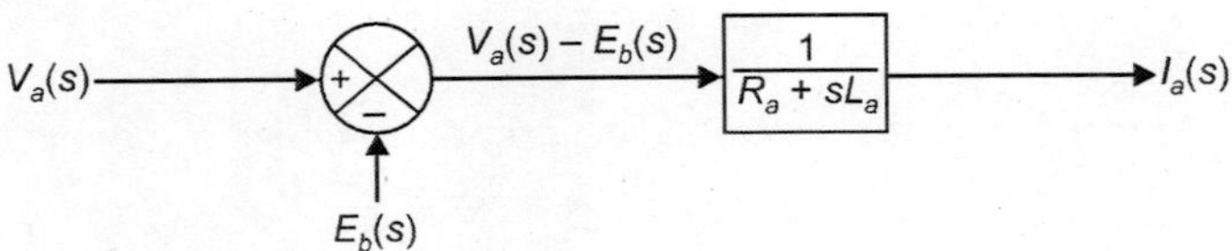

Fig. Ex. 1.13.1

on taking laplace transform of equation (1.13.2) we get

$$T(s) = K_t\, I_a(s)$$

from equation (1.13.7)

$I_a(s)$ → K_t → $T(s)$

Fig. Ex. 1.13.2

on taking laplace transform of equation (1.13.3) we get,

$$T(s) = \omega(s)\,[Js + B]$$

$$\omega(s) = \left(\frac{1}{Js + B}\right) T(s) \qquad (1.13.8)$$

from equation (1.13.8)

$T(s)$ → $\frac{1}{Js + B}$ → $\omega(s)$

Fig. Ex. 1.13.3

on taking laplace transform of equation (1.13.4) we get,

$$E_b(s) = K_b\, \omega(s) \qquad (1.13.9)$$

$\omega(s)$ → K_b → $E_b(s)$

Fig. Ex. 1.13.4

on taking laplace transform of equation (1.13.5)

$$\omega(s) = s\theta(s)$$

$$\theta(s) = \frac{1}{s}\,\omega(s)$$

ω(s)

1/s

θ(s)

Fig. Ex. 1.13.5

The overall block diagram of armature controlled dc motor is obtained by connecting various sections of Fig. (Ex. 1.13.1 to 1.13.5).

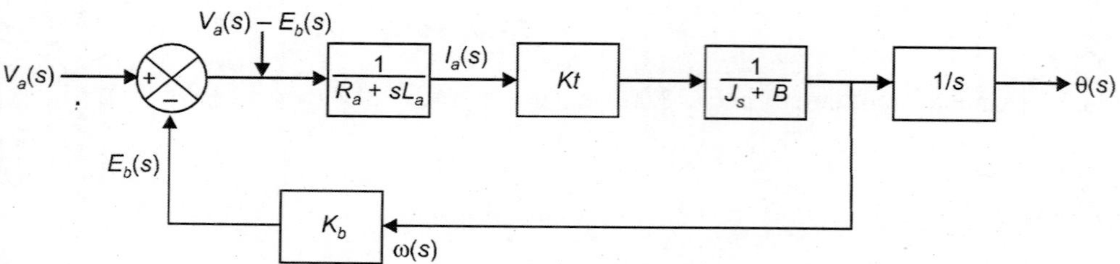

Fig. Ex. 1.13.6: Complete block diagram of armature controlled dc motor.

EXAMPLE 14

Construct the block diagram of field controlled dc motor.

Solution: The differential equations governing the field controlled dc motor are (refer section 1.11 example 2)

$$V_f = R_f i_f + L_f \frac{di_f}{dt} \quad (1.14.1)$$

$$T = Kt_f\, i_f \quad (1.14.2)$$

$$T = J\frac{d^2\theta}{dt} + B\frac{d\theta}{dt} \quad (1.14.3)$$

on taking laplace transform of equation (1.14.1)

$$V_f(s) = R_f\, I_f(s) + L_f s\, I_f(s)$$

$$V_f(s) = I_f(s)\,[R_f + s\,L_f]$$

$$I_f(s) = \frac{V_f(s)}{[R_f + s\,L_f]}$$

$$I_f(s) = \frac{1}{(R_f + s\,L_f)} V_f(s) \tag{1.14.4}$$

From equation (1.14.4)

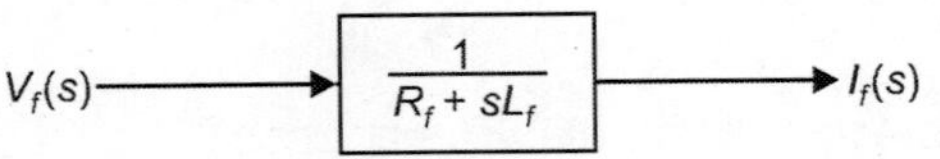

Fig. Ex. 1.14.1

on taking laplace transform of equation (1.14.2),

$$T(s) = K_{tf}\, I_f(s) \tag{1.14.5}$$

From equation (1.14.5)

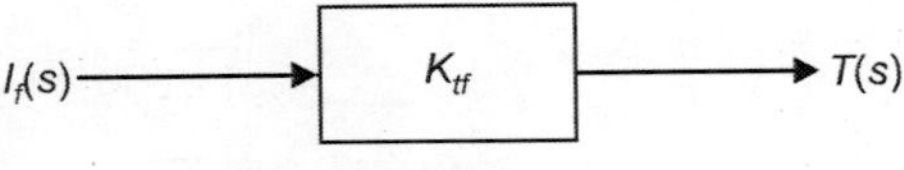

Fig. Ex. 1.14.2

on taking laplace transform of equation (1.14.3) we get,

$$T(s) = Js^2\,\theta(s) + Bs\,\theta(s)$$

$$T(s) = \theta(s)\,[Js^2 + Bs]$$

$$\therefore \quad \theta(s) = \frac{1}{(Js^2 + Bs)} T(s) \tag{1.14.6}$$

From equation (1.14.6)

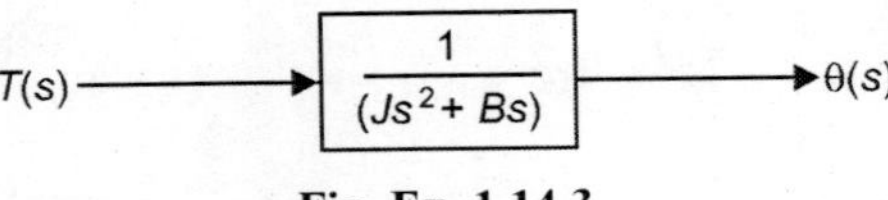

Fig. Ex. 1.14.3

By combining Fig. (1.14.1), (1.14.2) and (1.14.3) the overall block diagram of field control DC motor can be obtained.

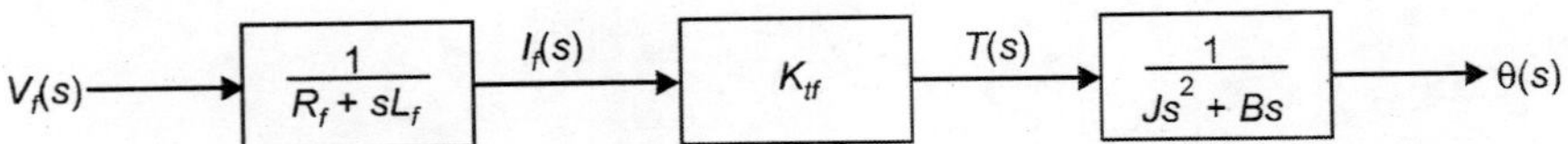

Fig. Ex. 1.14.4: Block diagram of field controlled DC motor.

Block diagram reduction

The block diagram can be reduced to find the overall transfer function of the system. The following rules can be used for block diagram reduction. The rules are framed such that any modification made on the diagram does not alter the input and output relation.

Rules of Block Diagram Algebra

Rule 1: Combining the blocks in Cascade (Multiplication)

$A \rightarrow$ [G_1] $\xrightarrow{A\,G_1}$ [G_2] $\rightarrow A\,G_1G_2 \Rightarrow A \rightarrow$ [G_1G_2] $\rightarrow A\,G_1G_2$

Rule 2: Combining parallel blocks (Addition or Subtraction)

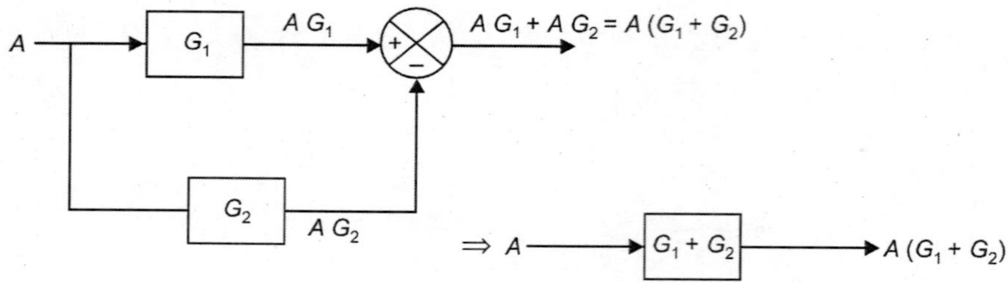

Rule 3: Moving the branch point ahead of the block [Branch point to right of gain (*BR* 1/*G*)]

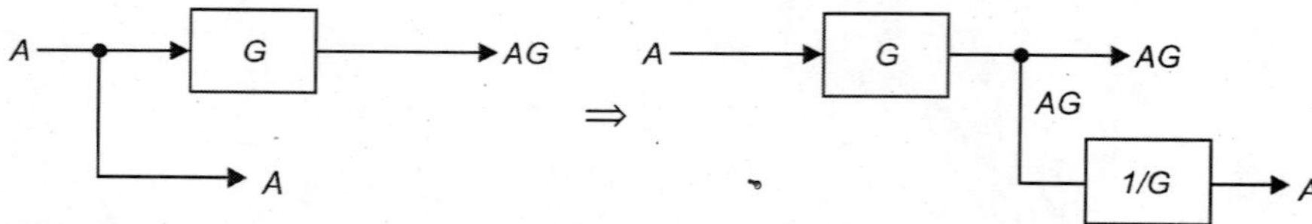

Rule 4: Moving the branch point before the block [Branch point to left of gain (BLG)]

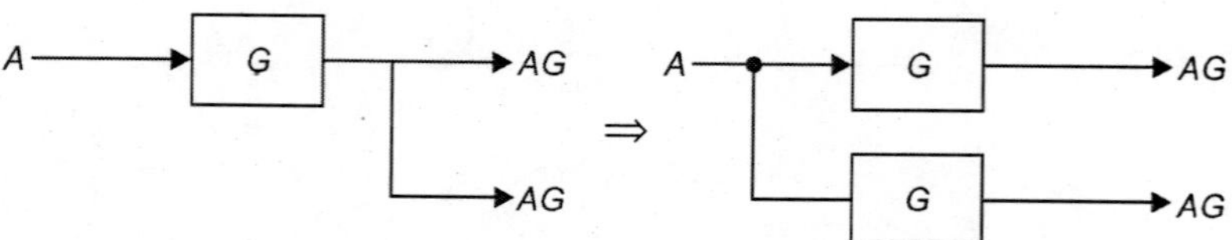

Rule 5: Moving the summing point ahead of the block [summing point to right of gain (*SRG*)]

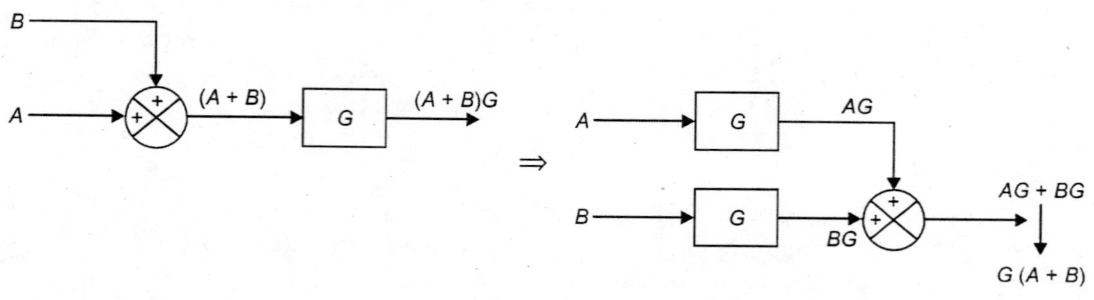

Rule 6: Moving the summing point before the block [Summing point to left of gain (*SL* 1/*G*)]

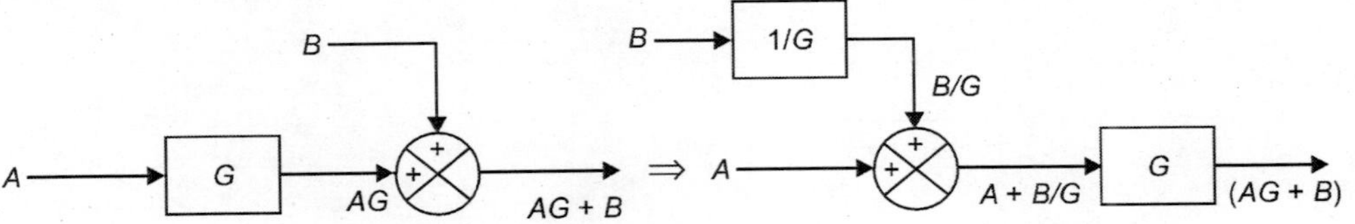

Rule 7: Interchanging summing point [No need to add any block]

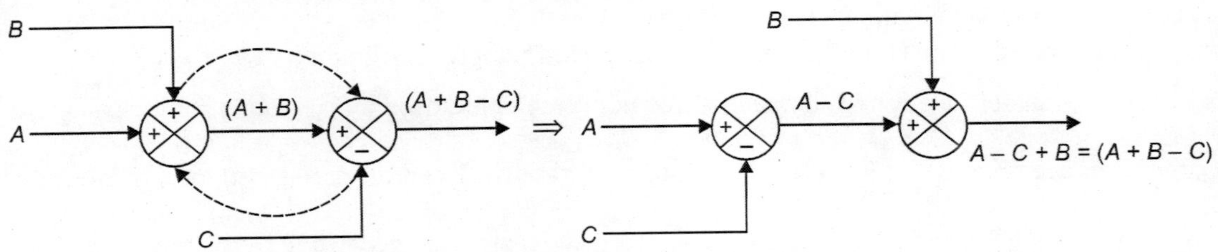

Rule 8: Splitting summing points

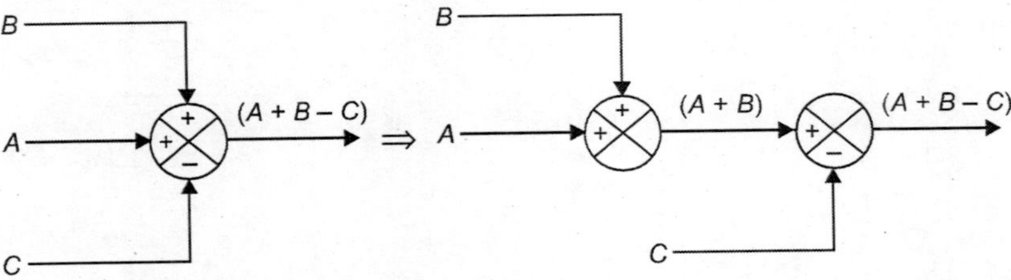

Rule 9: Combining summing points

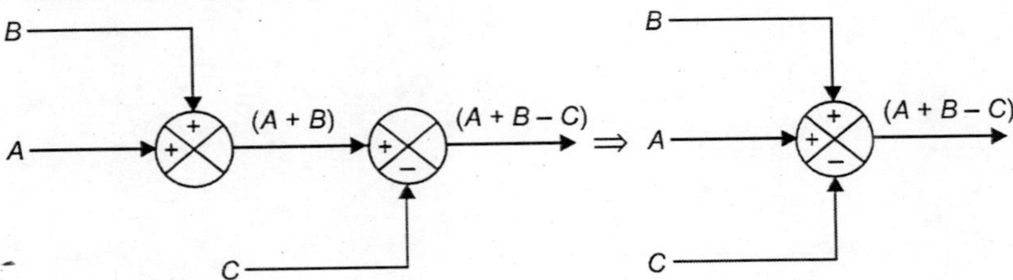

Rule 10: Elimination of feedback loop [Finding transfer function for simple loop]

(i) Negative feed back

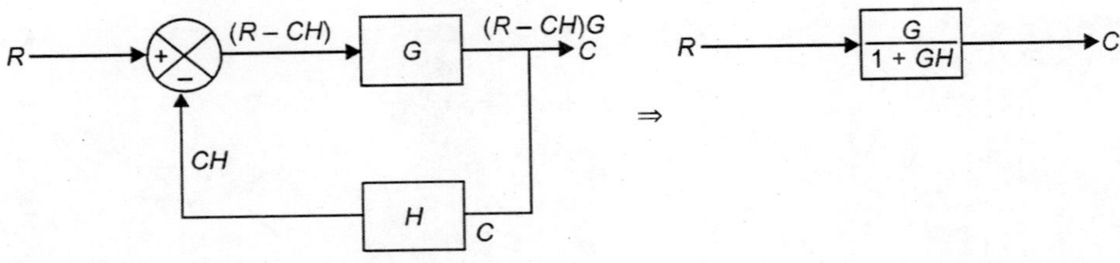

Proof:

$$C = (R - CH)\,G \Rightarrow C = RG - CGH$$

$$C + CGH = RG$$

$$C\,[1 + GH] = RG$$

$$\boxed{\frac{C}{R} = \frac{G}{1 + GH}}$$

Note: In rule 3, 4, 5 and 6, G and $1/G$ implies the compensation block which is to be a added to the block diagram so as to maintain the original signal while interchanging the brnach point and summing point. G means $\Rightarrow$ add gain (G) block and $1/G$ means add reciprocal of gain block ($1/G$).

* The rule 10 is transfer function of simple loop. Whenever simple closed loop is found in block diagram, that can be replaced by $\frac{G}{1+GH}$ or $\frac{G}{1-GH}$ form. It means the loop can be reduced in standard form which is proved in the proof of 10th rule. The + and − signs are used according to the type of feedback is used.

(ii) Positive feedback

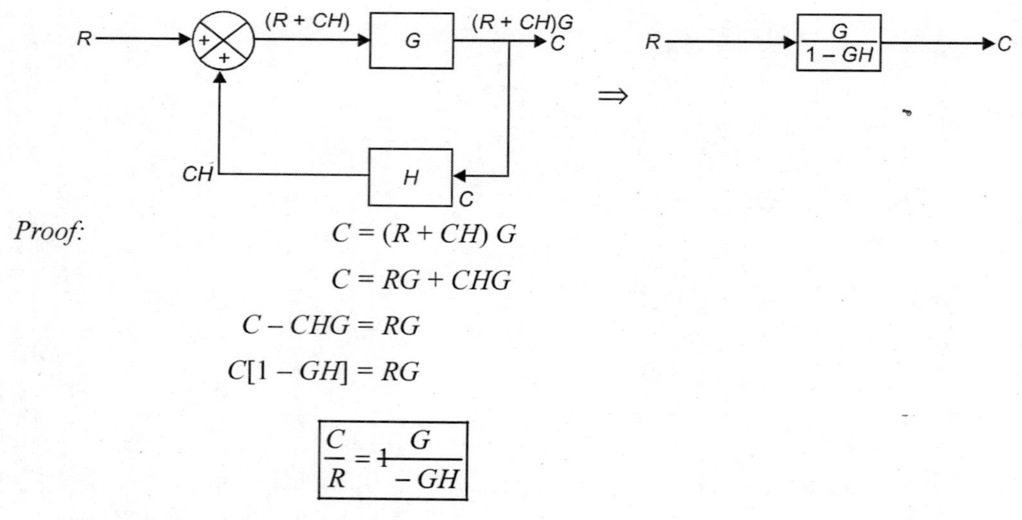

Proof:

$$C = (R + CH)\,G$$

$$C = RG + CHG$$

$$C - CHG = RG$$

$$C[1 - GH] = RG$$

$$\boxed{\frac{C}{R} = \frac{G}{1 - GH}}$$

EXAMPLE 15

Reduce the block diagram shown in Fig. (Ex. 1.15) and find C/R.

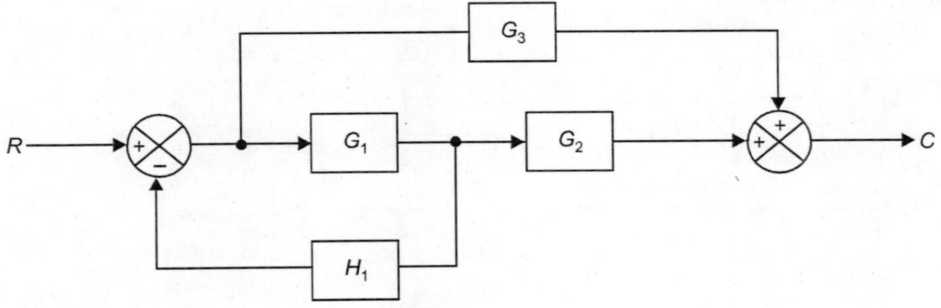

Fig. Ex. 1.15

Solution:

Step 1: Move the branch point after the block G_1 (Rule no. 3) [BR $1/G$]

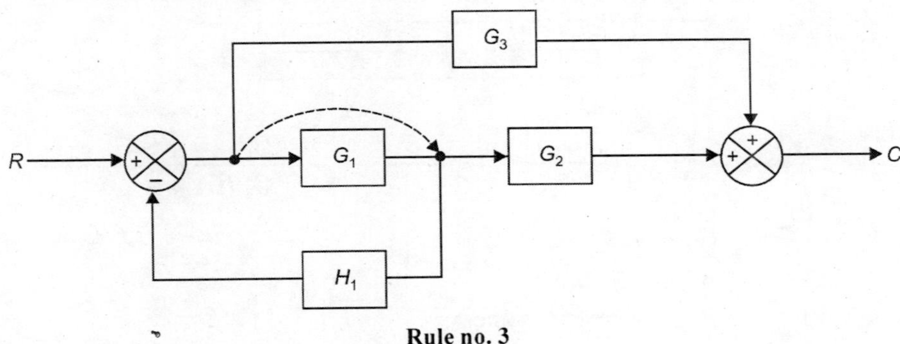

Rule no. 3

Step 2: Eliminate the feedback path and combine blocks in cascade. (Rule 1 and Rule 10)

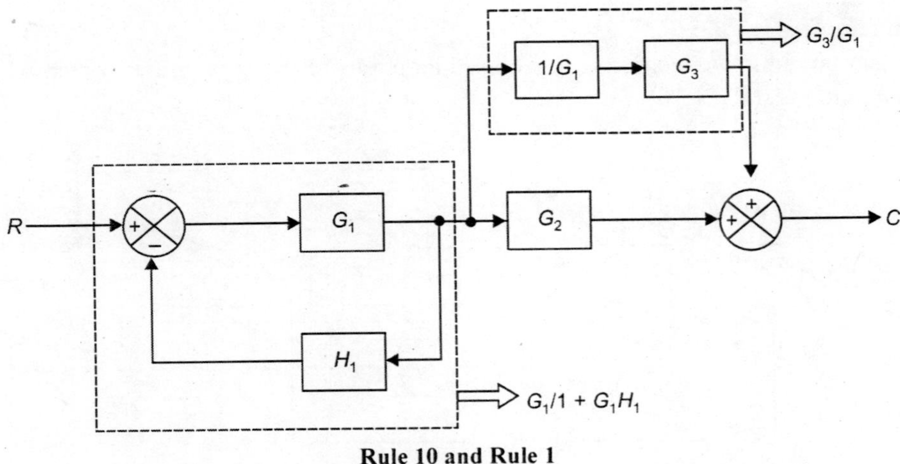

Rule 10 and Rule 1

Step 3: Combining parallel blocks [Rule 2] [Addition]

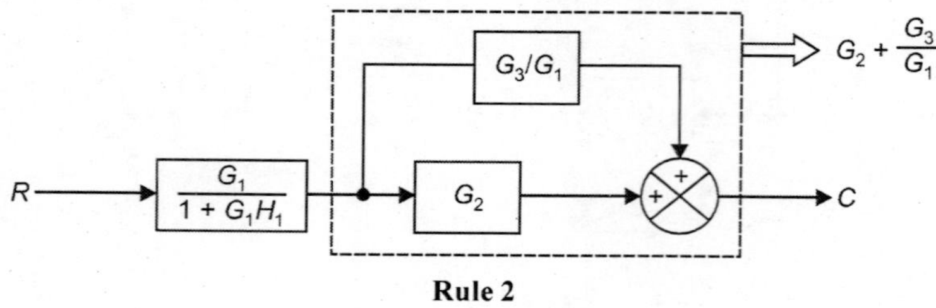

Rule 2

Step 4: Combining blocks in Cascade (Rule 1)

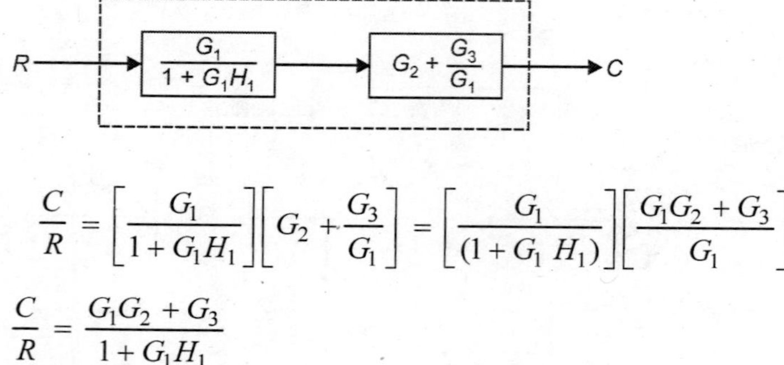

$$\frac{C}{R} = \left[\frac{G_1}{1+G_1H_1}\right]\left[G_2 + \frac{G_3}{G_1}\right] = \left[\frac{G_1}{(1+G_1\ H_1)}\right]\left[\frac{G_1G_2+G_3}{G_1}\right]$$

$$\frac{C}{R} = \frac{G_1G_2+G_3}{1+G_1H_1}$$

Results: The overall transfer function of the system,

$$\frac{C}{R} = \frac{G_1G_2+G_3}{1+G_1H_1}$$

EXAMPLE 16

Using block diagram reduction technique find closed loop transfer function of the system whose block diagram is shown in Fig. (Ex. 1.16).

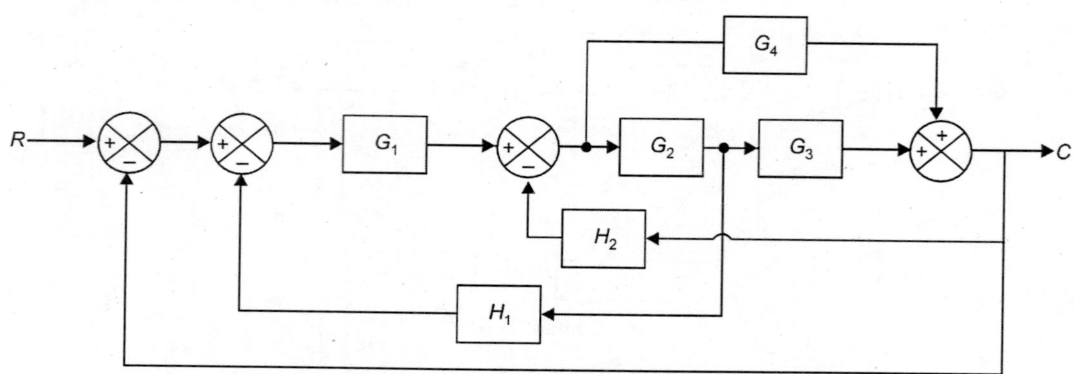

Fig. Ex. 1.16

Solution:

Step 1: Moving the branch point before the block G_2 (Rule 4) [*BLG*]

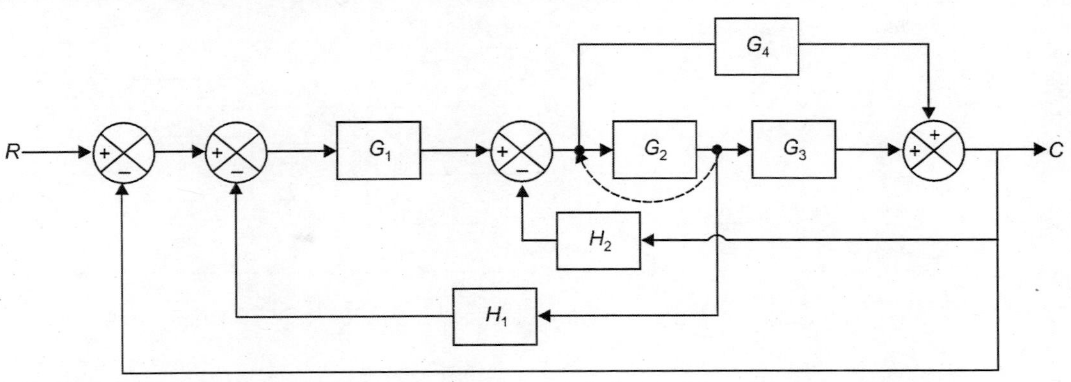

Note: According to rule 4, while changing the branch point before block G_2 [*BLG*], compensation block should added [G_2] with the corresponding branch poiint.

Step 2: Combining the blocks in cascade and eliminating parallel blocks [Rule 1 and 2]

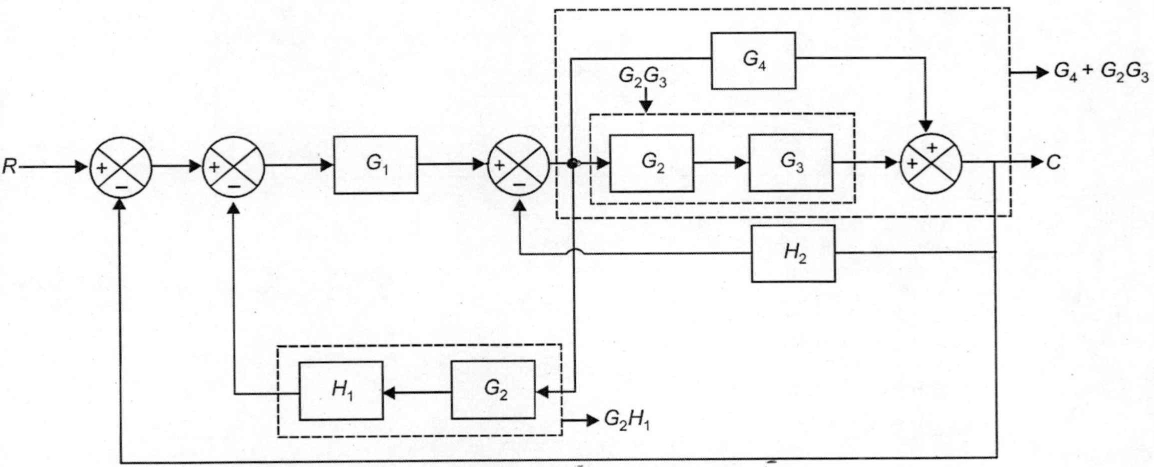

Step 3: Moving summing point before the block (Rule 6) [*SL* $1/G$]

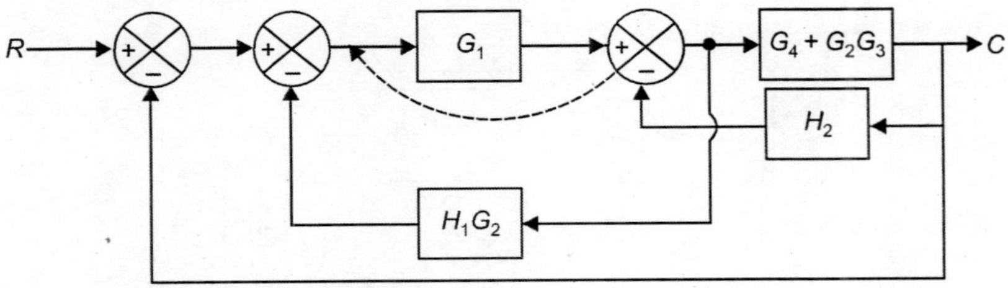

Note: According to rule 6, while moving the summing point before block G_1, the compensation block should be added ($1/G_1$) with the summing point.

Step 4: Interchanging summing points (Rule 7) [According to rule 7, interchange summing point 2 and 3 without compensation)

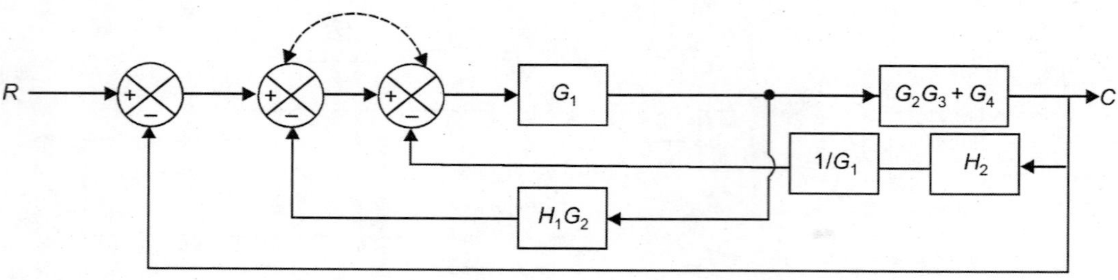

Step 5: Eliminating feedback path and combining blocks in cascade (Rule 10 and 1)

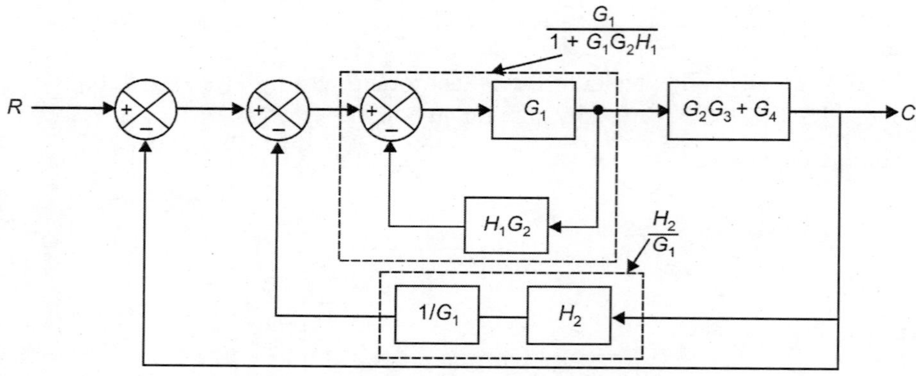

Step 6: Combining blocks in cascade (Rule 1)

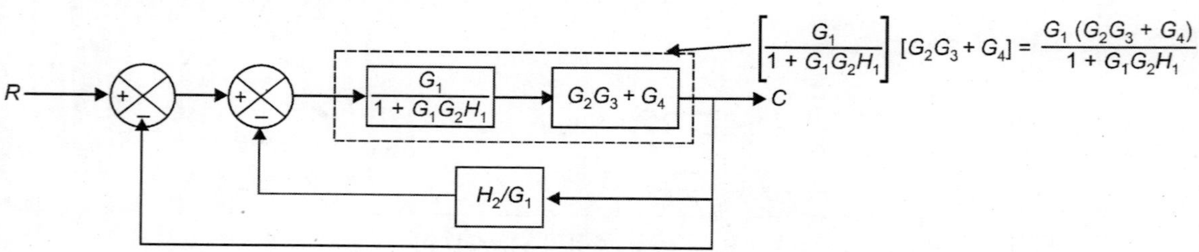

Step 7: Eliminating the feedback path (Rule 10)

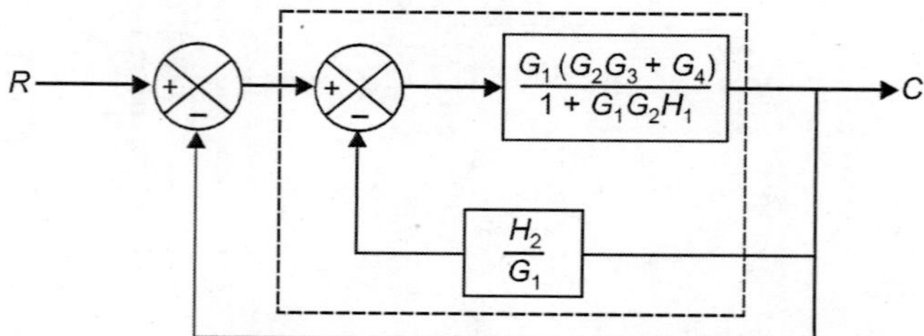

Step 8: Eliminating the feedback path

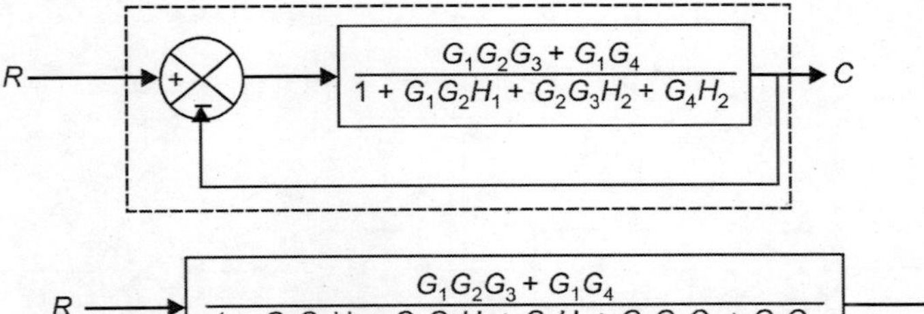

Step 9:

$$R \longrightarrow \boxed{\frac{G_1G_2G_3 + G_1G_4}{1 + G_1G_2H_1 + G_2G_3H_2 + G_4H_2 + G_1G_2G_3 + G_1G_4}} \longrightarrow C$$

Results: The overall transfer function is given by

$$\frac{C}{R} = \frac{G_1\,G_2\,G_3 + G_1\,G_4}{1 + G_1\,G_2\,H_1 + G_2\,G_3\,H_2 + G_4\,H_2 + G_1\,G_2\,G_3 + G_1\,G_4}$$

Proof for step 7

Rule 10 is

$$\frac{G}{1+GH} = \frac{\dfrac{G_1(G_2G_3+G_4)}{1+G_1G_2H_1}}{1+\left(\dfrac{G_1(G_2G_3+G_4)}{1+G_1G_2H_1}\right)\left(\dfrac{H_2}{G_1}\right)}$$

$$\Rightarrow \frac{G_1(G_2G_3+G_4)}{1+G_1G_2H_1+G_2G_3H_2+G_4H_2}$$

$$\Rightarrow \frac{G_1G_2G_3+G_1G_4}{1+G_1G_2H_1+G_2G_3H_2+G_4H_2}$$

Proof for step 8: Here $H = 1$

$$\frac{G}{1+GH} \Rightarrow \frac{\dfrac{G_1\,G_2\,G_3+G_1\,G_4}{1+G_1\,G_2\,H_1+G_2\,G_3\,H_2+G_4\,H_2}}{1+\dfrac{G_1\,G_2\,G_3+G_1\,G_4}{1+G_1\,G_2\,H_1+G_2\,G_3\,H_2+G_4\,H_2}}$$

$$\Rightarrow \frac{\dfrac{G_1\,G_2\,G_3+G_1\,G_4}{(1+G_1\,G_2\,H_1+G_2\,G_3\,H_2+G_4\,H_2)}}{\dfrac{(1+G_1\,G_2\,H_1+G_2\,G_3\,H_2+G_4\,H_2)+(G_1\,G_2\,G_3+G_1\,G_4)}{(1+G_1\,G_2\,H_1+G_2\,G_3\,H_2+G_4\,H_2)}}$$

$$\Rightarrow \frac{G_1\,G_2\,G_3+G_1\,G_4}{(1+G_1\,G_2\,H_1+G_2\,G_3\,H_2+G_4\,H_2+G_1\,G_2\,G_3+G_1\,G_4)}$$

EXAMPLE 17

Consider the block diagram shown in Fig. (Ex. 1.17) using the block diagram reduction method find C/R.

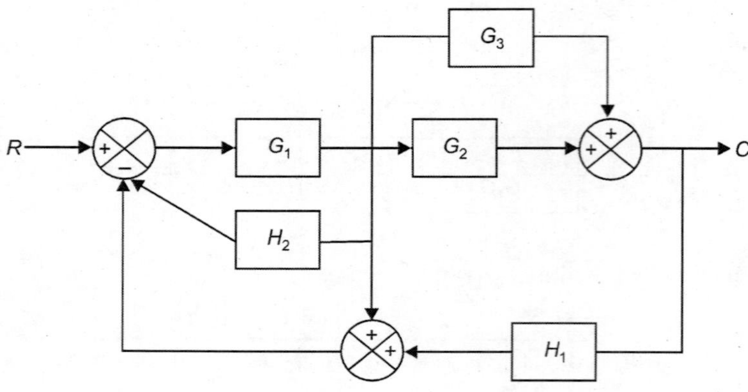

Fig. Ex. 1.17

Solution:

Step 1: Combining the blocks G_2 and G_3 [since two blocks are in parallel (Rule 2)]

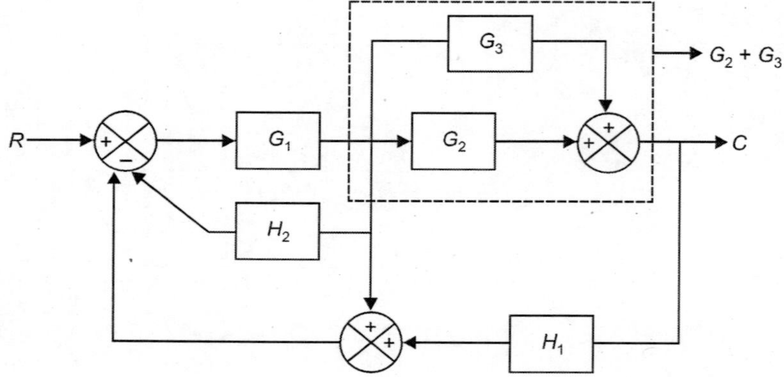

Step 2: Moving the branch point after the block $(G_2 + G_3)$ (Rule 3) $\left[BR\dfrac{1}{G}\right]$

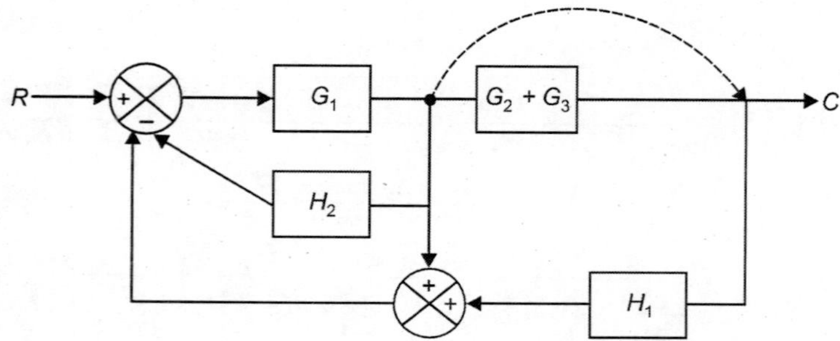

Step 3: Reducing parallel blocks and eliminating feedback path (Rule 2 and 10)

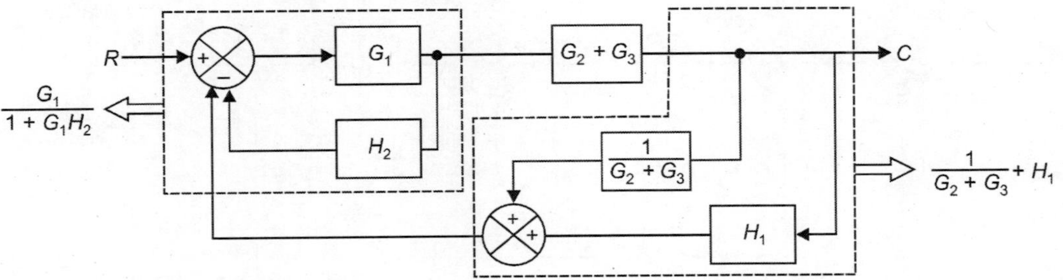

Step 4: Combining cascade blocks (Rule 1)

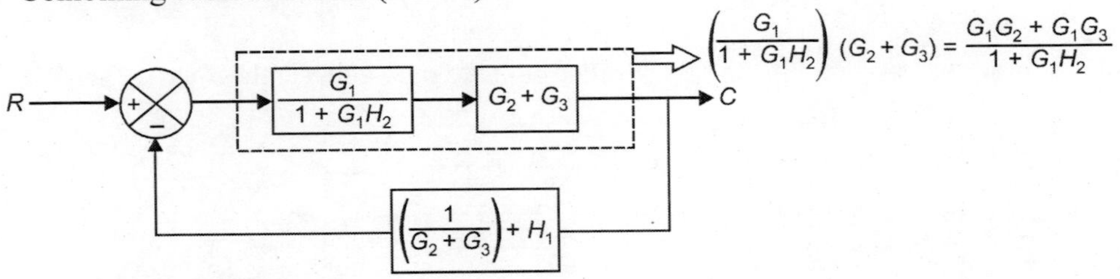

Step 5: Eliminating feedback path (Rule 10)

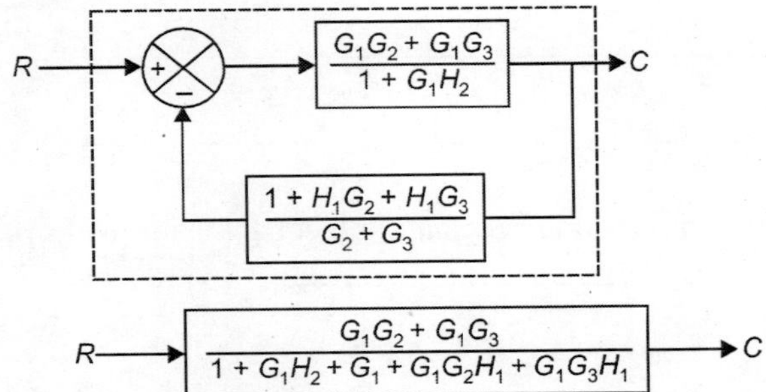

Step 6:

$$R \rightarrow \boxed{\frac{G_1G_2 + G_1G_3}{1 + G_1H_2 + G_1 + G_1G_2H_1 + G_1G_3H_1}} \rightarrow C$$

Result: The overall transfer function (*C*/*R*) is

$$C/R = \frac{G_1 G_2 + G_1 G_3}{1 + G_1 H_2 + G_1 + G_1 G_2 H_1 + G_1 G_3 H_1}$$

EXAMPLE 18

Find *C*/*R* for the block shown in Fig. (Ex. 1.18).

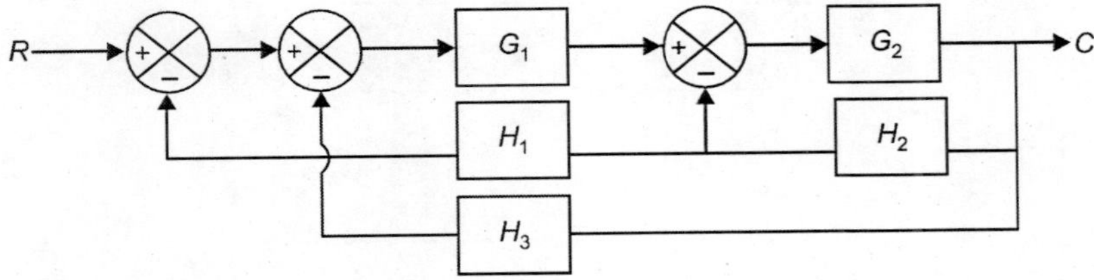

Fig. Ex. 1.18

Step 1: Rearrange the feedback paths. Then eliminate feedback path (Rule 10)

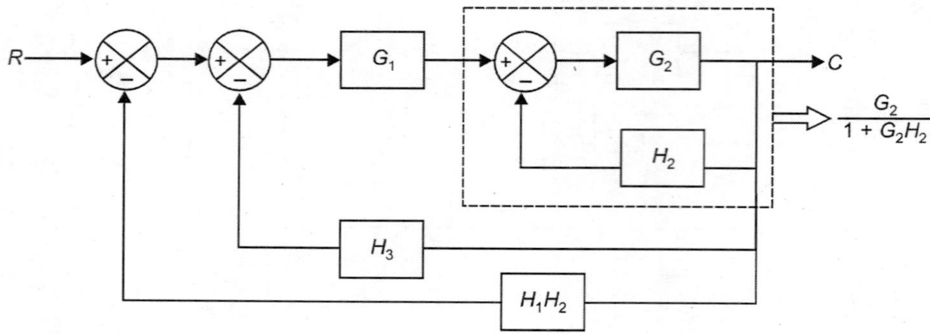

Step 2: Combining the cascade blocks and eliminating feedback path (Rule 1 and 10)

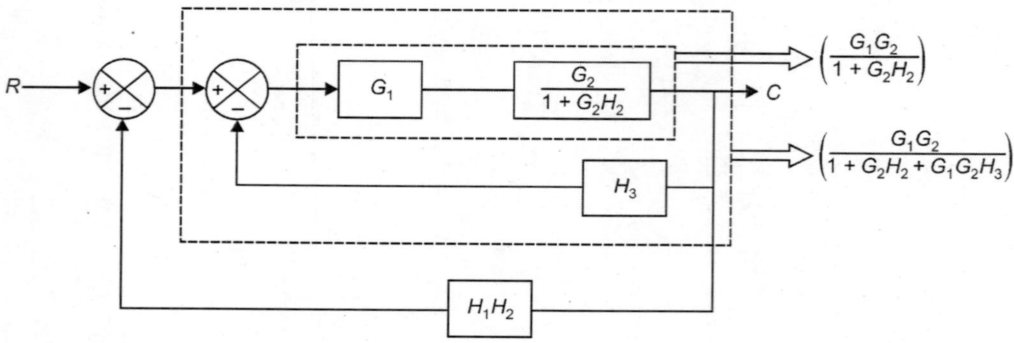

Step 3: Reducing parallel blocks and eliminating feedback path (Rule 10)

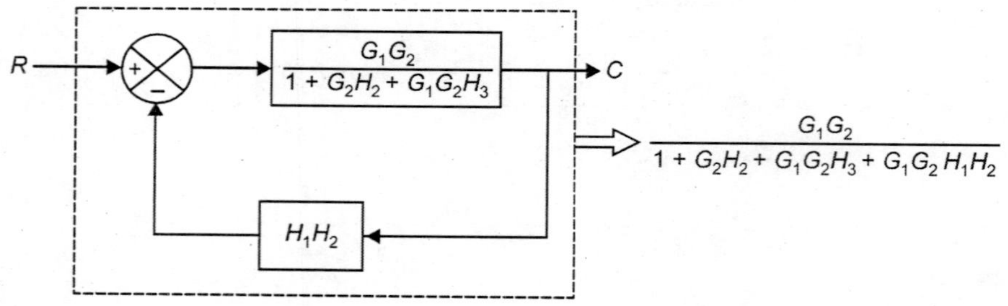

Step 4:

R → $\frac{G_1G_2}{1 + G_2H_2 + G_1G_2H_3 + G_1G_2H_1H_2}$ → C

Result: The overall transfer function (C/R) in

$$C/R = \frac{G_1 G_2}{1 + G_2 H_2 + G_1 G_2 H_3 + G_1 G_2 H_1 H_2}$$

EXAMPLE 19

For the block diagram shown in Fig. (Ex. 1.19) find *C/R* by block diagram reduction method.

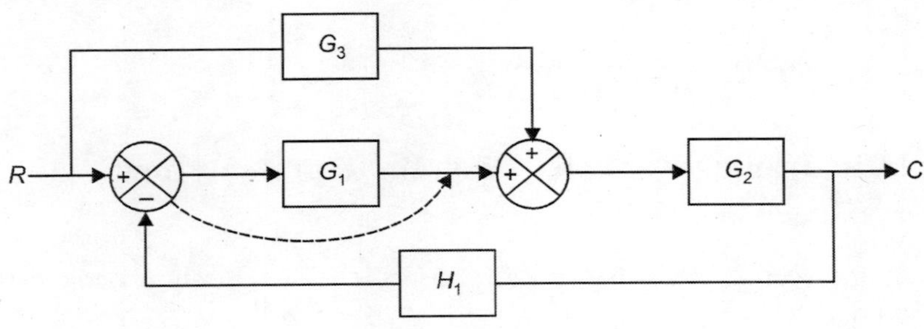

Fig. Ex. 1.19

Solution:

Step 1: Move the summing point after block G_1 (Rule 5) (*SRG*) and interchanging the summing point (Rule 7)

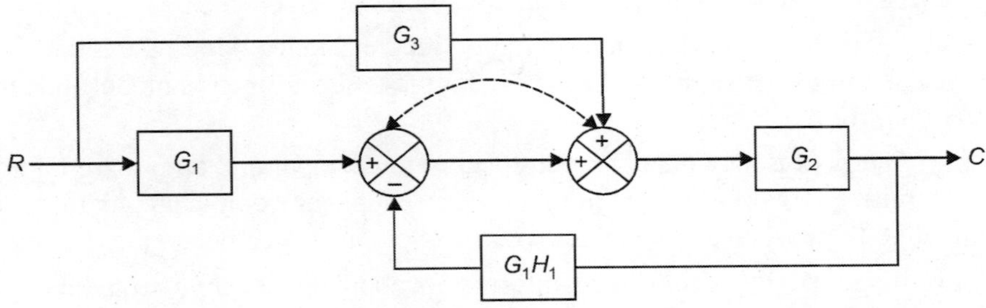

Step 2: Eliminate feedback path and parallel path (Rule 10 and 12)

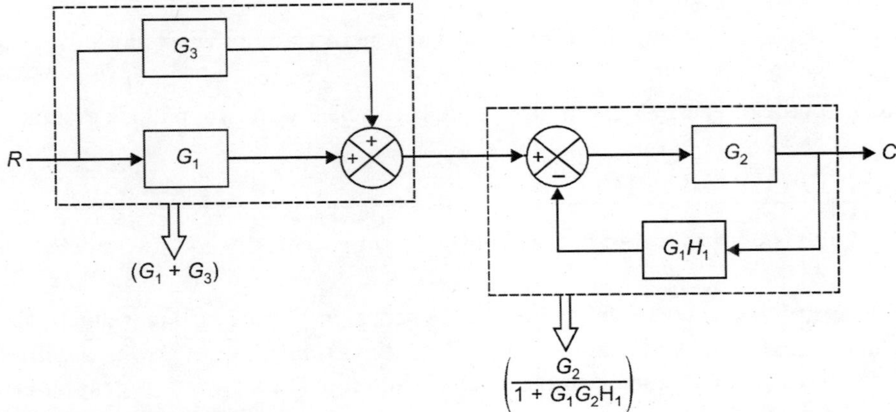

Step 3: Combine the blocks (Rule 1)

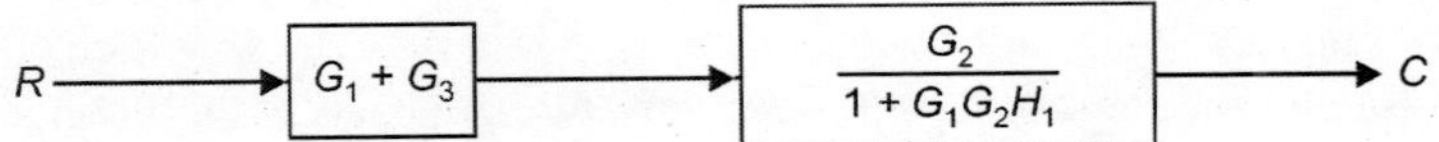

$$R \rightarrow \boxed{\frac{G_1G_2 + G_2G_3}{1 + G_1G_2H_1}} \rightarrow C$$

Result: The overall transfer function (C/R) is

$$C/R = \frac{G_1\, G_2 + G_2\, G_3}{1 + G_1\, G_2\, H_1}$$

Advantages and Disadvantages of Block Diagram Representation

Advantages

1. Easy to form the over all block diagram for the whole system by simply connecting the blocks of each component.
2. The functional operation of the system can be readily visualized by examining the block diagram rather than the physical system itself.

Disadvantages

1. The basic assumption is that, there is no interaction between blocks.
2. The possibility of important functions being omitted or hidden within the block. This may be due to manipulations of the physical equation for the transfer function model and putting in the block diagram form.
3. Passive symmetrical π networks are reciprocal because input and output are interchangeable and any amplifier network is non-reciprocal because input and output are not inter-changeable, but both will be represented by block diagram without any reference to the above property.
4. From block diagram it is possible to get information about the dynamic behaviour of the system and no information about the construction of the system is known. Thus a mechanical system, electrical system and hydraulic system having the same dynamic equations will be represented by the same block diagram.
5. The block diagram is not unique. From analysis point of view, different block diagrams may be drawn for the same system.
6. The main source of energy flow in the system is not clearly shown in block diagram.

1.18 SIGNAL FLOW GRAPHS

The signal flow graph is used to represent the control system graphically and it was developed by S.J. Mason.

A signal flow graph is a diagram that represents a set of simultaneous linear algebraic equation by taking laplace transform, the time domain differential equations governing a control system can be transferred to a set of algebraic equations in s-domain. The signal flow graph of the system can be constructed using these equations.

The signal flow graph depicts the flow of signals from one point of a system to another and gives the relationships among the signals. A signal flow graph consists of a network in which nodes are connected by directed branches. Each node represents a system variable and each branch connected between two nodes acts as a signal multiplier. Each branch has a gain or transmittance when the signal pass through a branch, it gets multiplied by the gain of the branch. Note that the signal flow is only one direction. The direction of signal flow is directed by an arrow placed on the branch and the multiplication is indicated along the branch.

Definition of Terms Used in Signal Flow Graph

Node: A node is a point representing a variable or signal.

Example: *a* •

Branch: A branch is directed line segment joining two nodes. The arrow on the branch indicates the direction of signal flow and the gain of a branch is known as transmittance.

Example: *a* • ——→—— • *b*

Transmittance: The gain acquired by the signal when it travels from one node to another is called transmittance. The transmittance can be real (or) complex.

Example: *a* • ——→—— • *b* (G_1)

Here the transmittance is G_1

Input node (or) Source: It is a node that has only outgoing branches.

Example:

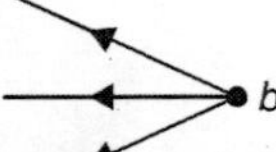

Output node (or) Sink: It is a node that has only incoming branches.

Example: *b*

Mixed node: It is a node that has both incoming and outgoing branches.

Path: A path is a traversal of connected branches in the directions of the branch arrows. The path should not cross a node more than once.

Open path: A open path starts at a node and ends at another node.

Example:

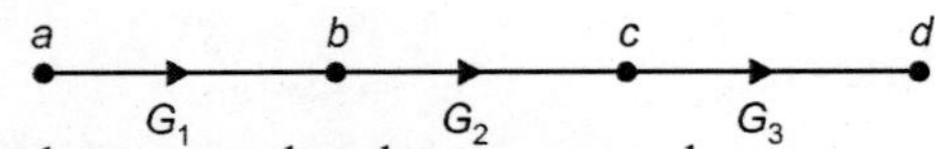

Closed path: Closed path starts and ends at same node.

Example:

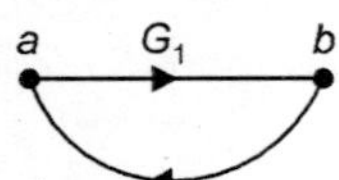

Forward path: It is a path from an input node to an output node that does not cross any node more than once.

Forward path gain: It is the product of the branch transmittances (gains) of a forward path.

Example:

$a \xrightarrow{G_1} b \xrightarrow{G_2} c \xrightarrow{G_3} d$

its forward path gain $= G_1G_2G_3$

Individual loop: It is a closed path starting from a node and after passing through a certain part of a graph arrives at the same node without crossing any node more than once.

Example: a G_1 b, H_1

Individual Loop gain: It is the product of the branch transmittances (gains) of a loop.

Example: a G_1 b, H_1

its loop gain $= G_1H_1$

Non-touching loops: Number of loops does not have a common node, then they are said to be nontouching loops.

Example:

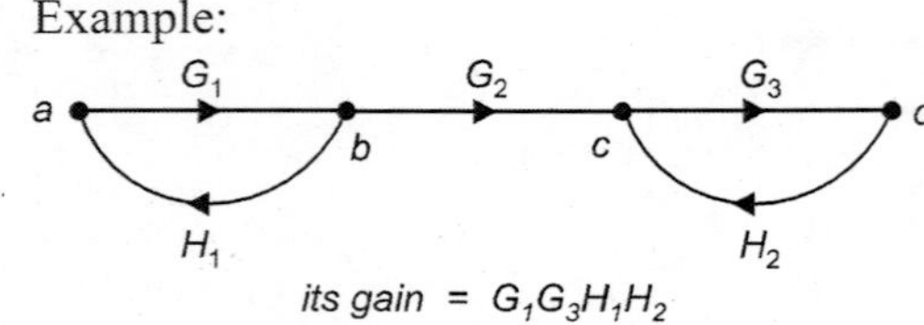

its gain $= G_1G_3H_1H_2$

Properties of Signal Flow Graph

The basic properties of signal flow graph are the following:

1. The algebraic equations which are used to construct signal flow graph must be in the form of cause and effect relationship.
2. It is applicable to linear system only.
3. A node represents the variable (or) signal.
4. A node adds the signals of all incoming branches and transmits the sum to all outgoing branches.
5. A mixed node which has both incoming and outgoing signals can be treated as output node by taking on outgoing branch of unity transmittance.
6. A branch indicates functional dependence of one signal on the other.
7. The signal travel along branches only in the marked direction and when it travels it gets multiplied by the gain (or) transmittance of the branch.
8. The signal flow graph of system is not unique. By rearranging the system equations different types of signal flow graphs can be drawn for a given system.

Signal Flow Graph Algebra

Signal flow graph for a system can be reduced to obtain the transfer function of the system using the following rules. Let see the rules to obtain transfer function from signal flow graph.

Rule 1: In coming signal to a node through a branch is given by the product of a signal at previous node and the gain of the branch.

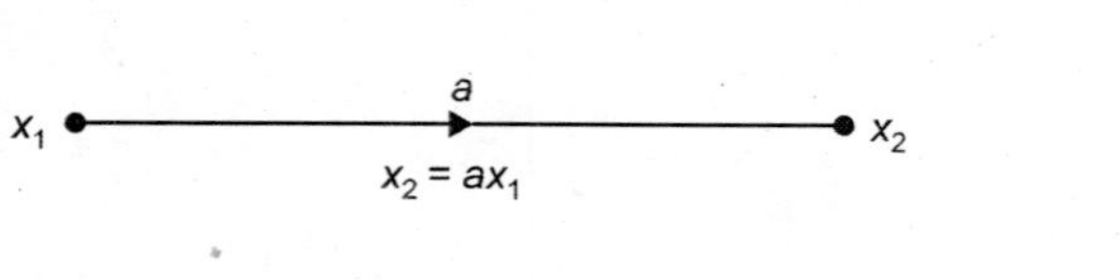

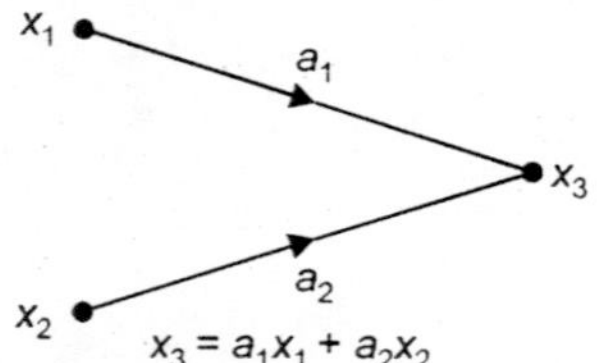

Rule 2: Cascaded branches can be combined to give a singal branch whose transmittance is equal to the product of individual branch transmittance.

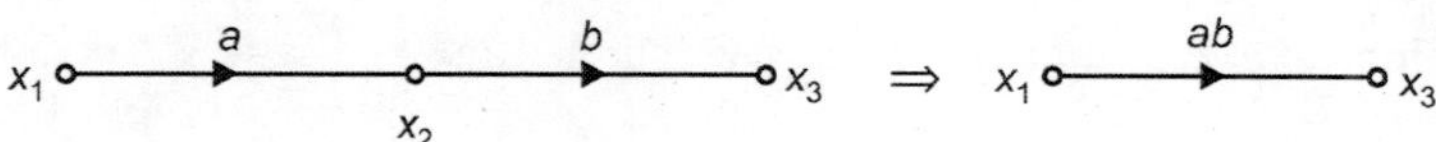

Rule 3: Parallel branches may be represented by single branch whose transmittance is the sum of individual branch transmittances.

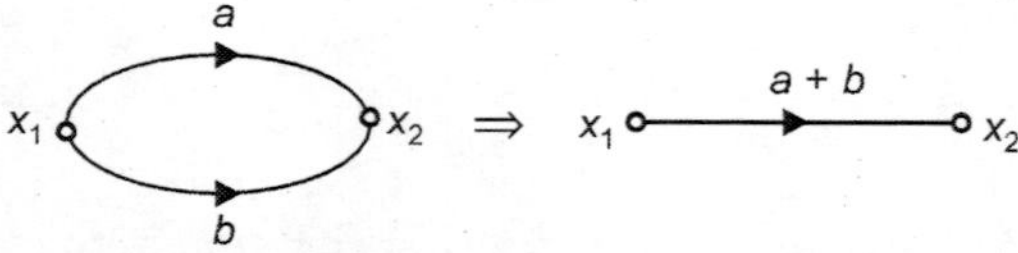

Rule 4: A mixed node can be eliminated by multiplying transmittance of outgoing branch (from the mixed node) to the transmittance of all incoming branches to the mixed node.

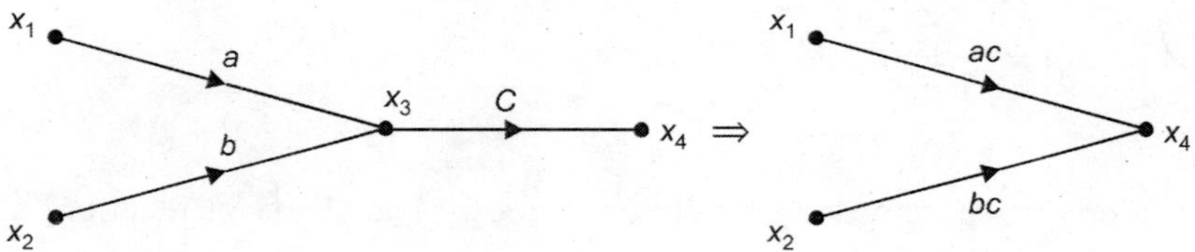

Rule 5: A loop may be eliminated by writing equations at the input and output node and rearranging the equations to find the ratio of output to input. This ratio gives the gain of resultant branch.

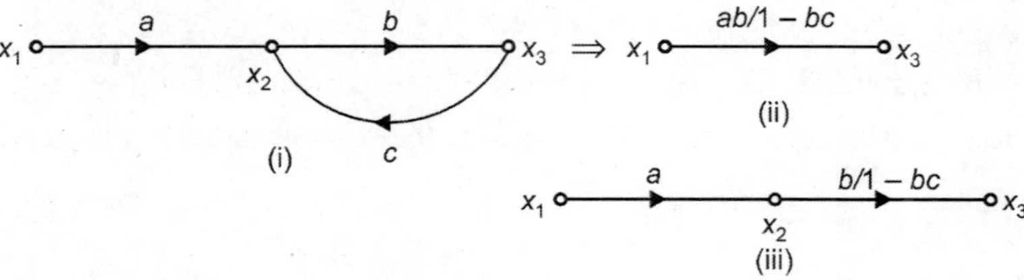

Proof:

$$x_2 = ax_1 + cx_3$$

$$x_3 = bx_2$$

Substitute x_2 in equation for x_3

$\therefore$

$$x_3 = b\,[ax_1 + cx_3]$$

$$x_3 = abx_1 + bcx_3$$

$$x_3 - bcx_3 = abx_1$$

$$x_3[1 - bc] = abx_1$$

$$\frac{x_3}{x_1} = \frac{ab}{1-bc} \text{ or } x_3 = x_1\left[\frac{ab}{1-bc}\right]$$

Signal Flow Graph Reduction

The signal flow graph of a system can be reduced either by using the rules of a signal flow graph algebra (i.e.) by writing equations at every node and then rearranging these equations to get the ratio of output and input (Transfer function).

The signal flow graph reduction by above method will be time consuming and tedious S.J. Mason have developed a simple procedure to determine the transfer function of the system represented as a signal flow graph. He has developed a formula called by his name Mason's gain formula which can be directly used to find the transfer function of the system.

Mason's Gain Formula

The Mason's gain formula is used to determine the transfer function of the system from the signal flow graph of the system.

Let, $R(s)$ = Input of the system

and $C(s)$ = Output of the system

Transfer function of the system

$$T(s) = \frac{C(s)}{R(s)}$$

The Mason's gain formula states the overall gain of the system [Transfer function] as follows

Overall gain, $$T = 1/\Delta \sum_k P_k \Delta_k$$

where, $T = T(s)$ = Transfer function of the system

P_K = Forward path gain of Kth forward path

Δ = 1 – (sum of individual loop gains) + [sum of gain products of all possible combinations of two non-touching loops] – [Sum of gain products of all possible combinations of three non-touching loops] + ...

Construction of Signal Flow Graph for Control Systems

A control system can be represented diagrammatically by signal flow graph. The differential equations governing the system are used to construct the signal flow graph. By taking laplace transform the differential equations can be converted to algebraic equations. The constants and variables of the equations are identified. From the working knowledge of the system, the variables are identified as input, output and intermediate variables. For each variable a node is assigned in signal flow graph and constants are the gain or transmittance of the branches connecting the nodes. For each equation a signal flow graph is drawn and then they are interconnected to give overall signal flow graph of the system

EXAMPLE 20

Construct a signal flow graph for armature controlled dc motor.

$\Delta_K = \Delta$ of the part of the graph which is not touching K^{th} forward path.

Solution: The differential equations governing the armature controlled dc motor (Refer section 1.17 example 14)

$$V_a = i_a R_a + L_a \frac{di_a}{dt} + e_b \tag{1.20.1}$$

$$T = K_t i_a \tag{1.20.2}$$

$$T = J\frac{d\omega}{dt} + B\omega \tag{1.20.3}$$

$$e_b = K_b\omega \tag{1.20.4}$$

where, $$\omega = \frac{d\theta}{dt} \tag{1.20.5}$$

on taking laplace transform of equation (1.20.1 to 1.20. 5)

$$V_a(s) = I_a(s)\,[R_a + s\,La] + E_b\,(s) \tag{1.20.6}$$

$$T(s) = K_t\, I_a(s) \tag{1.20.7}$$

$$T(s) = J_s\,\omega(s) + B\omega(s) \tag{1.20.8}$$

$$E_b(s) = K_b\,\omega(s) \tag{1.20.9}$$

$$\omega(s) = s\,\theta(s) \tag{1.20.10}$$

The input and output variables of armature controlled dc motor are armature voltage $V_a(s)$ and angular displacement $\theta(s)$ respectively. The variables $I_a(s)$, $T(s)$, $E_b(s)$ and $\omega(s)$ are intermediate variables.

The equations (1.20.6) to (1.20.10) are rearranged and individual signal flow graph are shown in Fig. (1.20.4 to 1.20.5)

From eqn. (1.20.6)

$$V_a(s) = E_b(s) = I_a(s)\,[R_a + s\,L_a]$$

$$I_a(s) = \frac{1}{[R_a + s\,L_a]}[V_a(s) - E_b(s)]$$

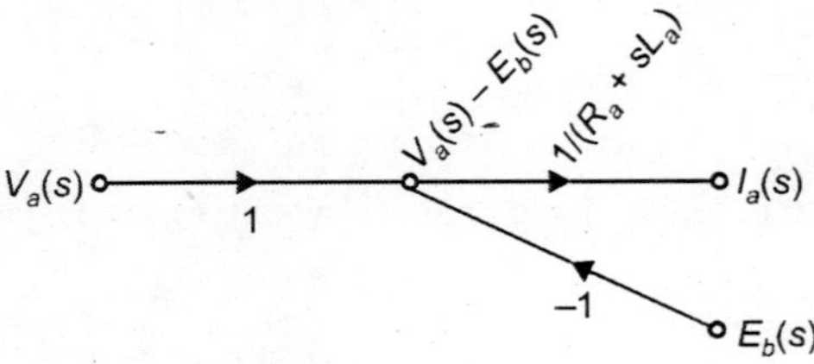

Fig. Ex. 1.20.1

from eqn (1.20.7)

$$T(s) = K_t\, I_a(s)$$

$I_a(s)$ —— Kt ——> $T(s)$

Fig. Ex. 1.20.2

from eqn. (1.20.8)

$$T(s) = \omega(s)\,[Js + B]$$

$$\omega(s) = \frac{1}{Js + B}\, T(s)$$

T(s) → 1/Js + B → ω(s)

Fig. Ex. 1.20.3

from equation (1.20.9)

$$E_b(s) = K_b\, \omega(s)$$

ω(s) → K_b → $E_b(s)$

Fig. Ex. 1.20.4

from equation (1.20.10)

$$\theta(s) = \frac{1}{s}\, \omega(s)$$

ω(s) → 1/s → θ(s)

Fig. Ex. 1.20.5

The overall signal flow graph of armature controlled dc motor is obtained by interconnecting the individual signal flow graphs shown in Fig. 1.20.1 to 1.20.5. The overall signal flow graph is shown in Fig. 1.20.6

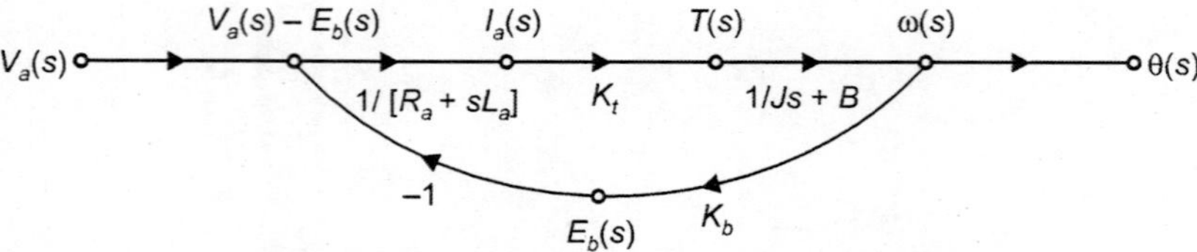

Fig. Ex. 1.20.6: Signal flow graph of armature controlled dc motor.

EXAMPLE 21

Construct a signal flow graph for field controlled dc motor.

Solution: The differential equations governing the field controlled dc motor (Refer Section 1.17 example 15)

$$V_f = R_f\, i_f + L_f \frac{di_f}{dt} \tag{1.21.1}$$

$$T = K_{tf}\, i_f \tag{1.21.2}$$

$$T = J\frac{d^2\theta}{dt^2} + B\frac{d\theta}{dt} \quad (1.21.3)$$

on taking laplace transform of equation (1.21.1 to 1.21.3) we get,

from equation (1.21.1)

$$V_f(s) = R_f I_f(s) + L_f s\, I_f(s)$$

$$V_f(s) = V_f(s)/\ [R_f + s\, L_f]$$

$$I_f(s) = V_f(s)/\ (R_f + s\, L_f) \quad (1.21.4)$$

from equation (1.21.2)

$$T(s) = K_{tf} I_f(s) \quad (1.21.5)$$

from equation (1.21.3)

$$T(s) = Js^2\, \theta(s) + Bs\, \theta(s)$$

$$\theta(s) = \frac{1}{Js^2 + B} T(s) \quad (1.21.6)$$

From equation (1.21.4, 1.21.5 and 1.21.6)

$V_f(s)$ —— $1/[R_f + sL_f]$ —→ $I_f(s)$ —— K_{tf} —→ $T(s)$ —— $1/Js^2 + B$ —→ $\theta(s)$

Fig. Ex. 1.21.1

The Fig. (1.21.1) shows the overall signal flow graph of field controlled dc motor.

EXAMPLE 22

Find the overall transfer function of the system whose signal flow graph is shown in Fig. (Ex. 1.22).

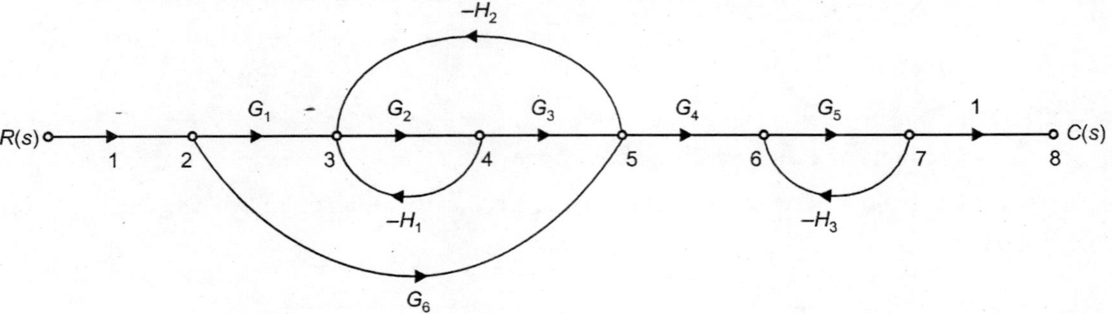

Fig. Ex. 1.22

Solution:

Step 1: Forward path gains

There are two forward paths. $\therefore\ K = 2$

Let forward path gains be P_1 and P_2

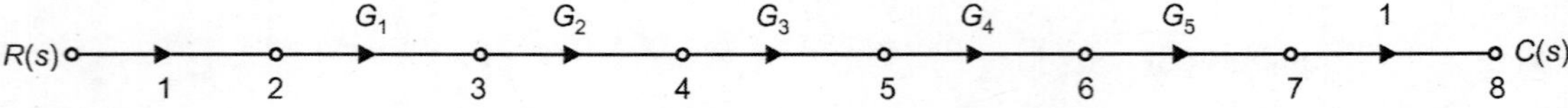

Fig. Ex. 1.22.1: Forward path-1

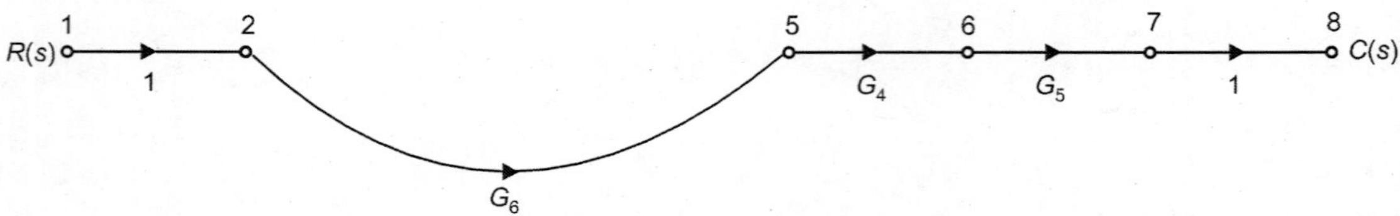

Fig. Ex. 1.22.2: Forward path – 2.

Gain of forward path – 1, $P_1 = G_1\ G_2\ G_3\ G_4\ G_5$

Gain of forward path – 2, $P_2 = G_4\ G_5\ G_6$

Step 2: Individual Loop Gain

There are three individual loops. Let individual loop gains be P_{11}, P_{21} and P_{31}.

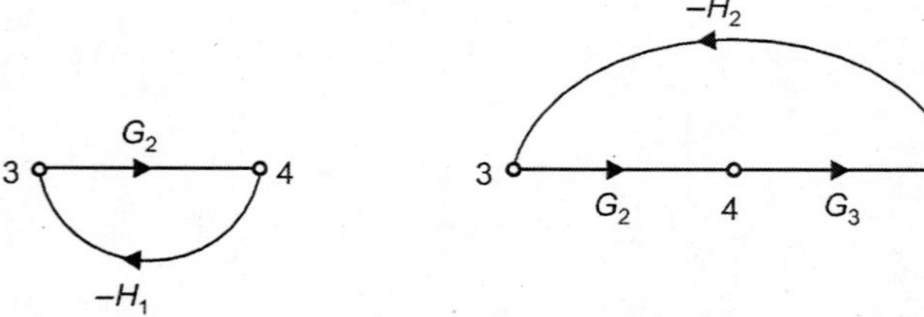

Fig. Ex. 1.22.3: Loop 1.

Fig. Ex. 1.22.4: Loop 2.

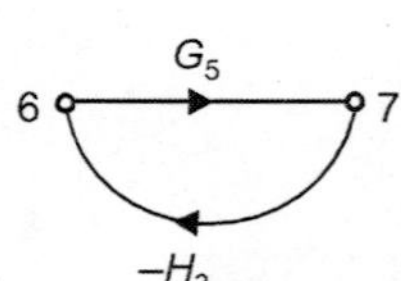

Fig. Ex. 1.23.5: Loop 3.

Loop gain of individual loop – 1, $L_1 = -\ G_2\ H_1$

Loop gain of individual loop – 2, $L_2 = -\ G_2\ G_3\ H_2$

Loop gain of individual loop – 3, $L_3 = -\ G_5\ H_3$

Step 3: Gain Products of Two Non-Touching Loops

There are two combinations of two non-touching loops. Let the gain products of two non-touching loops be L_{13} and L_{23}

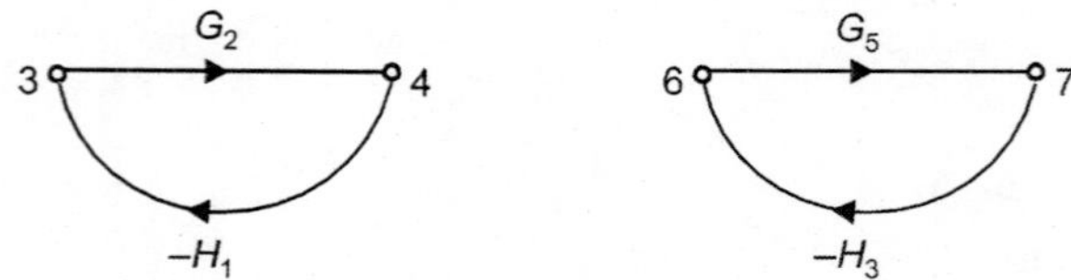

Fig. Ex.1.22.6: First combination of two non-touching loops

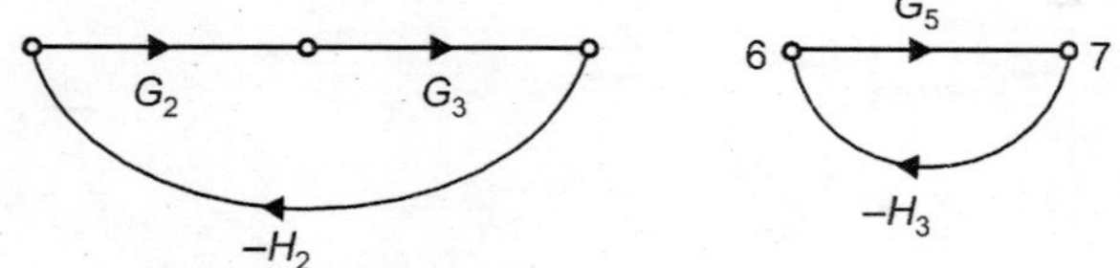

Fig. Ex. 1.22.7: Second combination of two non-touching loops

Gain product of first combination of two non touching loops

$$L_{13} = L_1L_3 = (-G_2H_1)(-G_5H_3) = G_2G_5H_1H_3$$

Gain product of second combination of two non touching loops

$$L_{23} = L_2L_3 = (-G_2G_3H_2)(-G_5H_3) = G_2G_3G_5H_2H_3$$

Calculation of Δ and ΔK

$$\Delta = 1 - (L_1 + L_2 + L_3) + (L_{13} + L_{23})$$

$$\Delta = 1 + G_2H_1 + G_2G_3H_2 + G_5H_3 + G_2G_5H_1H_3 + G_2G_3G_5H_2H_3$$

$\Delta_1 = 1$ [Since there is no part of graph which is not touching with first forward path]

i.e. $\Delta_1 = 1\text{-}0$

The part of the graph which is non touching with second forward path is shown in Fig. (Ex. 1.22.8)

$$\Delta_2 = 1 - L_1 = 1 - (-G_2H_1)$$

$$\Delta_2 = 1 + G_2H_1$$

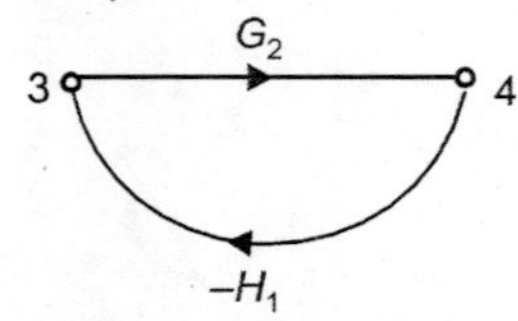

Fig. Ex. 1.22.8

Transfer Function, *T*

By Mason's gain formula the transfer function

$$T = \frac{1}{\Delta}\sum_k P_k\,\Delta_k$$

$$= \frac{1}{\Delta}[P_1\Delta_1 + P_2\Delta_2] \qquad \text{[Here } K = 2\text{, since we have only two forward paths]}$$

$$\therefore \quad T = \frac{G_1G_2G_3G_4G_5 + G_4G_5G_6(1 + G_2H_1)}{1 + G_2H_1 + G_2G_3H_2 + G_5H_3 + G_2G_5H_1H_3 + G_2G_3G_5H_2H_3}$$

EXAMPLE 23

Find overall gain of the system whose signal flow graph in shown in Fig. (Ex. 1.23)

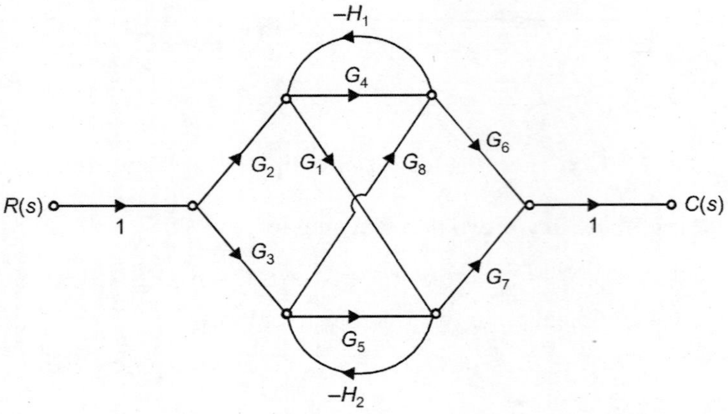

Fig. Ex. 1.23

Solution: Let us number the node as shown in Fig. (Ex. 1.23.1)

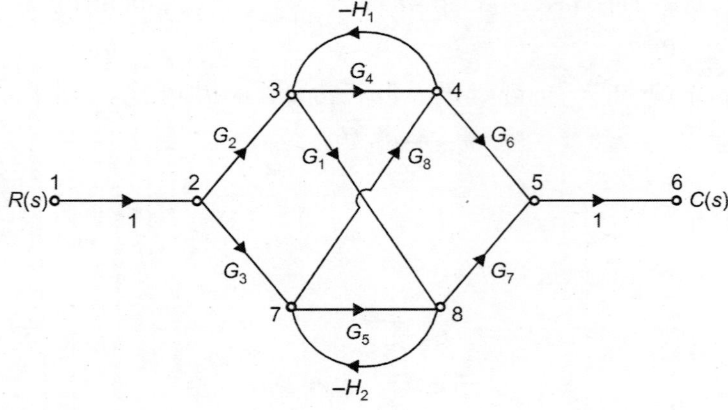

Fig. Ex. 1.23.1

Forward Path Gains

There are six forward path. $\therefore$ $K = 6$.

Let the forward path gains be P_1, P_2, P_3, P_4, P_5 and P_6.

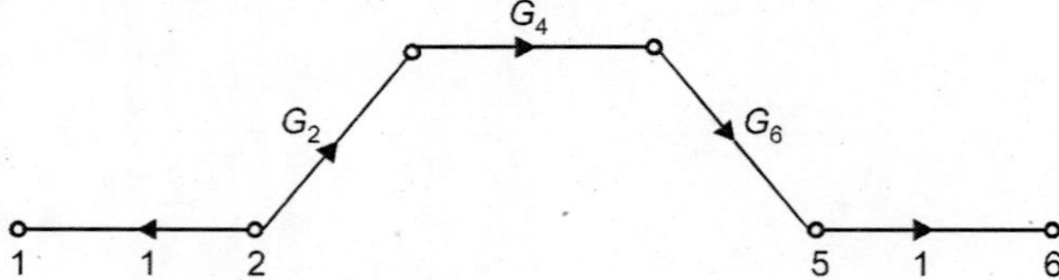

Fig. Ex. 1.23.2

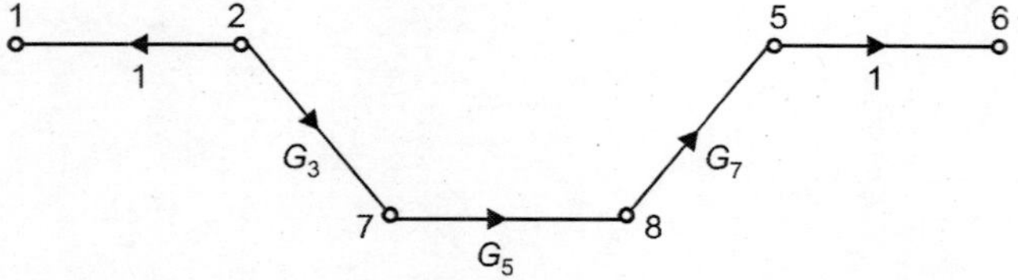

Fig. Ex. 1.23.3

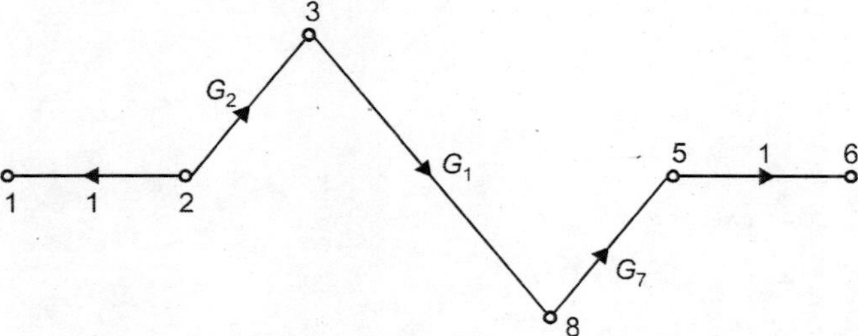

Fig. Ex. 1.23.4

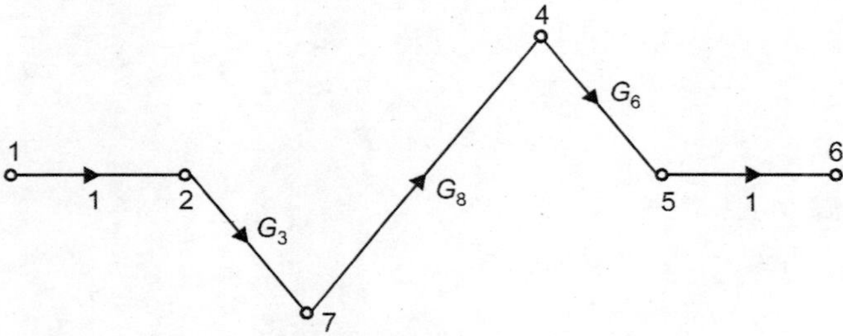

Fig. Ex. 1.23.5

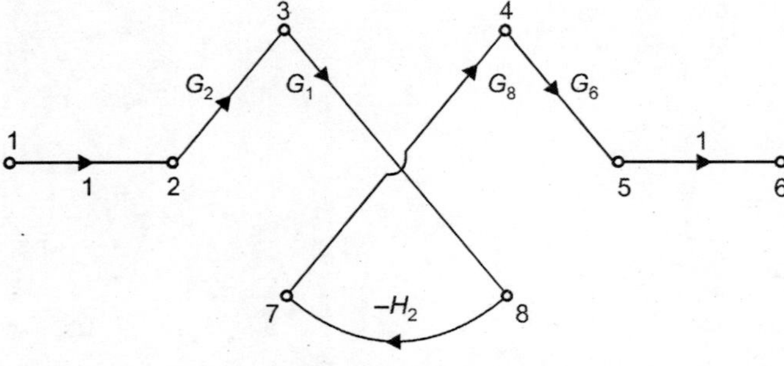

Fig. Ex. 1.23.6

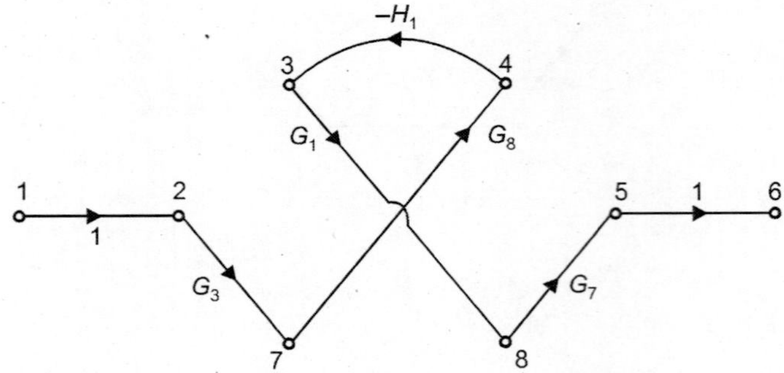

Fig. Ex. 1.23.7

Gain of forward path – 1, $P_1 = G_2\, G_4\, G_6$

Gain of forward path – 2, $P_2 = G_3\, G_5\, G_7$

Gain of forward path – 3, $P_3 = G_1\, G_2\, G_7$

Gain of forward path – 4, $P_4 = G_3\, G_8\, G_6$

Gain of forward path – 5, $P_5 = -G_1\, G_2\, G_6\, G_8\, H_2$

Gain of forward path – 6, $P_6 = -G_1\, G_3\, G_7\, G_8\, H_1$

Individual Loop Gain

There are three individual loops.

Let individual loop gains be, P_{11}, P_{21} and P_{31}

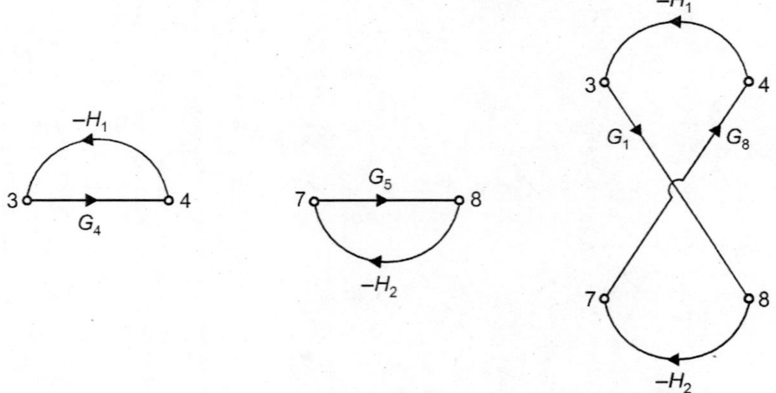

Fig. Ex. 1.23.8: Loop 1. **Fig. Ex. 1.23.9:** Loop 2. **Fig. Ex. 1.23.10:**Loop 3.

Loop gain of individual loop-1, $L_1 = -\,G_4\, H_1$

Loop gain of individual loop-2, $L_2 = -\,G_5\, H_2$

Loop gain of individual loop-3, $L_3 = G_1\, G_8\, H_1\, H_2$

Gain Products of Two non-Touching Loops

There is only one combination of two non-touching loops. Let the gain product of two non-touching loops be L_{12}.

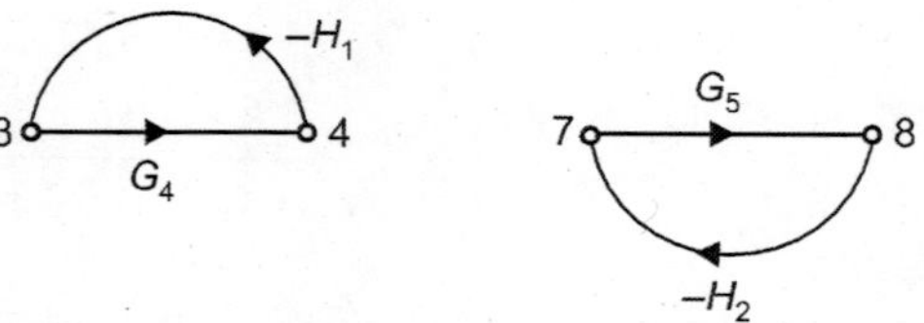

Fig. Ex. 1.23.11: First combination of two non-touching loops.

Gain product of first combination of two non-touching loops

$$L_{12} = L_1 L_2 = (-G_4 H_1)(-G_5 H_2) \Rightarrow G_4 G_5 H_1 H_2$$

Calculation of Δ and ΔK

$$\Delta = 1 - (L_1 + L_2 + L_3) + L_{12}$$

$$= 1 - (-G_4 H_1 - G_5 H_2 + G_1 G_8 H_1 H_2) + (G_4 G_5 H_1 H_2)$$

$$= 1 + G_4 H_1 + G_5 H_2 - G_1 G_8 H_1 H_2 + G_4 G_5 H_1 H_2$$

The part of the graph non-touching forward path –1 is shown in Fig. (Ex. 1.23.12)

$\therefore$ $$\Delta_1 = 1 - (-G_5 H_2) = 1 + G_5 H_2$$

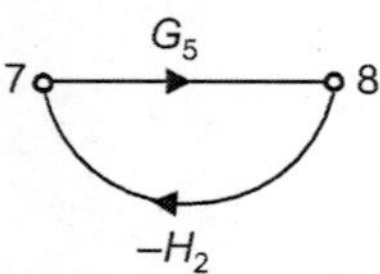

Fig. Ex. 1.23.12

The part of the graph non-touching forward path–2 is shown in Fig. (Ex. 1.23.13)

$\therefore$ $$\Delta_2 = 1 - (-G_4 H_1) = 1 + G_4 H_1$$

Fig. Ex. 1.23.13

There is no part of the graph which is non-touching with forward paths 3, 4, 5, and 6

$\therefore$ $$\Delta_3 = \Delta_4 = \Delta_5 = \Delta_6 = 1$$

Transfer Function

By Mason's gain formula, $T = \dfrac{1}{\Delta}\left[\sum_k P_k \, \Delta_k\right]$

$$= \frac{1}{\Delta}\ [P_1\Delta_1 + P_2\Delta_2 + P_3\Delta_3 + P_4\Delta_4 + P_5\Delta_5 + P_6\Delta_6]$$

(Here $K = 6$, since we have six forward paths)

$$T = \frac{(G_2\,G_4\,G_6)(1 + G_5\,H_2) + (G_3\,G_5\,G_7)(1 + G_4\,H_1) + G_1\,G_2G_7 + G_3\,G_8\,G_6 + (-G_1\,G_2\,G_6\,G_8\,H_2) + (-G_1\,G_3\,G_7\,G_8\,H_1)}{1 + G_4\,H_1 + G_5\,H_2 - G_1\,G_8\,H_1\,H_2 + G_4\,G_5\,H_1\,H_2}$$

EXAMPLE 24

Find the overall gain $C(s)/R(s)$ for the signal flow graph shown in Fig. (Ex. 1.24).
Mason gain formula:

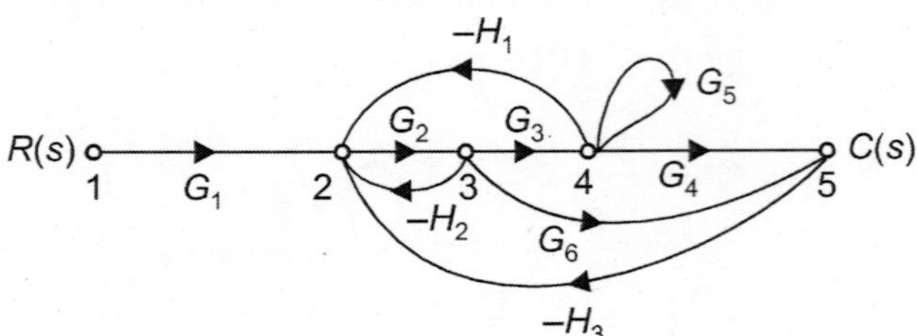

Fig. Ex. 1.24

Solution:

Forward Path Gains

There are two forward paths. $\therefore\ K = 2$

Let the forward path gains be P_1 and P_2

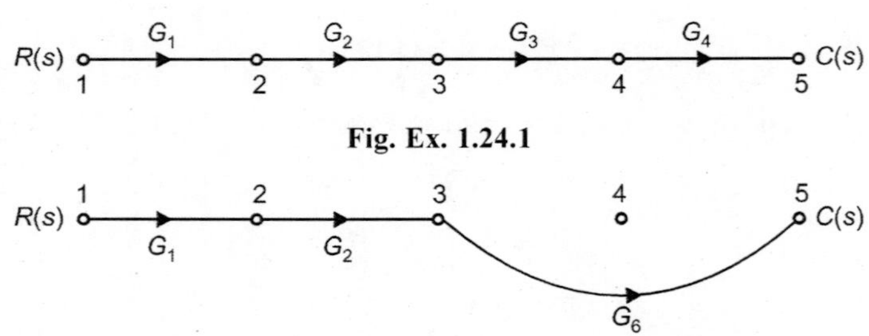

Fig. Ex. 1.24.1

Fig. Ex. 1.24.2

Gain of forward path-1 $P_1 = G_1\,G_2\,G_3\,G_4$

Gain of forward path-2 $P_2 = G_1\,G_2\,G_6$

Individual Loop Gain

There are five individual loops. Let the individual loop gains be P_{11}, P_{21}, P_{31}, P_{41} and P_{51}

Loop gain of individual loop-1, $L_1 = -\,G_2\,G_3\,H_1$

Loop gain of individual loop-2, $L_2 = -\,H_2\,G_2$

Loop gain of individual loop-3, $L_3 = -G_2 G_6 H_3$

Loop gain of individual loop-4, $L_4 = -G_2 G_3 G_4 H_3$

Loop gain of individual loop-5, $L_5 = G_5$

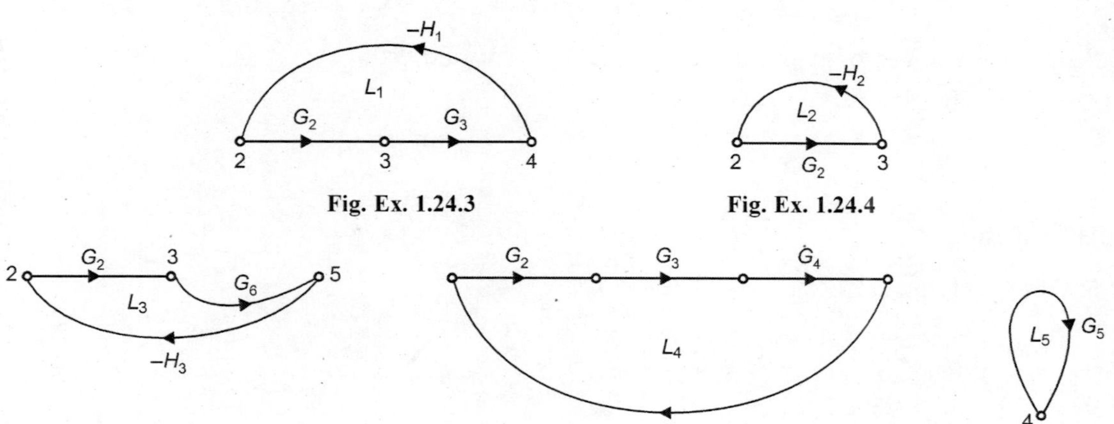

Fig. Ex. 1.24.3 Fig. Ex. 1.24.4

Fig. Ex. 1.24.4 Fig. Ex. 1.24.5 Fig. Ex. 1.24.6

Gain Products of Two Non-touching Loops (Combinations of pair of loops)

There are two combinations of two non-touching loops.

Let the gain products of two non-touching loops be L_{25} and L_{35}.

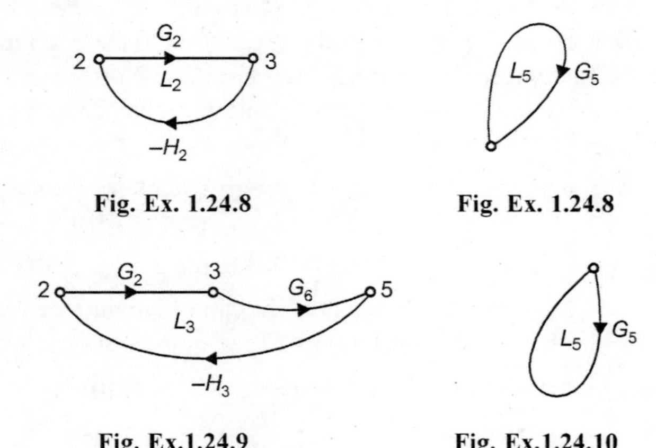

Fig. Ex. 1.24.8 Fig. Ex. 1.24.8

Fig. Ex.1.24.9 Fig. Ex.1.24.10

Gain product of first combination of two non-touching loops

$$L_{25} = L_2 L_5 = (-G_2 H_2)(G_5) = -G_2 G_5 H_2$$

Gain product of second combination of two non-touching loops

$$L_{35} = L_3 L_5 = (-G_2 G_6 H_3)$$
$$= -G_2 G_5 G_6 H_3$$

Calculation of Δ and Δ*K*

$$\Delta = 1 - (L_1 + L_2 + L_3 + L_4 + L_5) + (L_{25} + L_{35})$$

$$= 1 - (-G_2 G_3 H_1 - H_2 G_2 - G_2 G_3 G_4 H_3 + G_5 - G_2 G_6 H_3) + (- G_2 H_2 G_5 - G_2 G_5 G_6 H_3)$$

Since there is no part of graph which is not touching forward Path-1.

$$\therefore \quad \Delta = 1$$

$$\Delta_2 = 1 - G_5$$

Transfer Function

By Mason's gain formula

The transfer functions $T = \frac{1}{\Delta} \sum_k P_k \Delta_k = \frac{1}{\Delta} [P_1 \Delta_1 + P_2 \Delta_2]$

$$T = \frac{G_1 G_2 G_3 G_4 + G_1 G_2 G_6 - G_1 G_2 G_5 G_6}{1 + G_2 G_3 H_1 + H_2 G_2 + G_2 G_3 G_4 H_3 - G_5 - G_2 H_2 G_5 + G_2 G_6 H_3 - G_2 G_5 G_6 H_3}$$

Procedure for Converting Block Diagram to Signal Flow Graph

The signal flow graph and block diagram of a system provides the same information but there is no standard procedure for reducing the block diagram to find the transfer function of the system. Also the block diagram reduction technique will be tedious and it is difficult to choose the rule to be applied for simplification. Hence it will be easier if the block diagram is converted to signal flow graph and Mason's gain formula is applied to find the transfer function. The following procedure can be used to convert block diagram to signal flow graph.

1. Assume nodes at input, output, at every summing point, at every branch point and in between cascade blocks.
2. Draw the nodes separately as small circles and number the circles in the order, 1, 2, 3, 4 ... etc.
3. From block diagram find the gain between each node in the main forward path and connect all the corresponding circles by straight line and mark the gain between the nodes.
4. Draw the feed forward path between various nodes and mark the gain of feed forward path along with sign.
5. Draw the feedback paths between various nodes and mark the gain of feedback path along with sign.

EXAMPLE 25

Convert the block diagram to signal flow graph and determine the transfer function using Mason's gain formula.

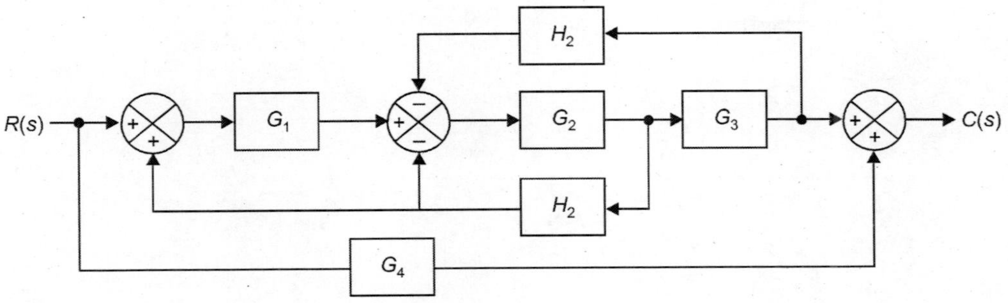

Fig. Ex. 1.25

Solution: The nodes are assigned at input/output, at every summing point and branch point as shown in Fig. Ex. 1.25.1.

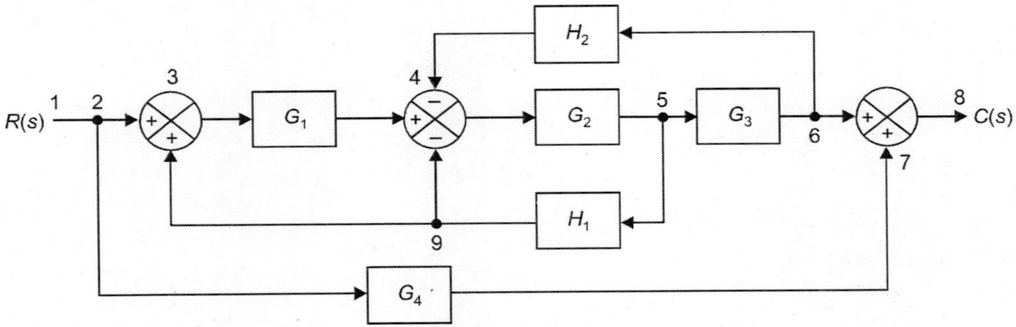

Fig. Ex. 1.25.1

The signal flow graph of above block diagram shown in Fig. (Ex. 1.25.2)

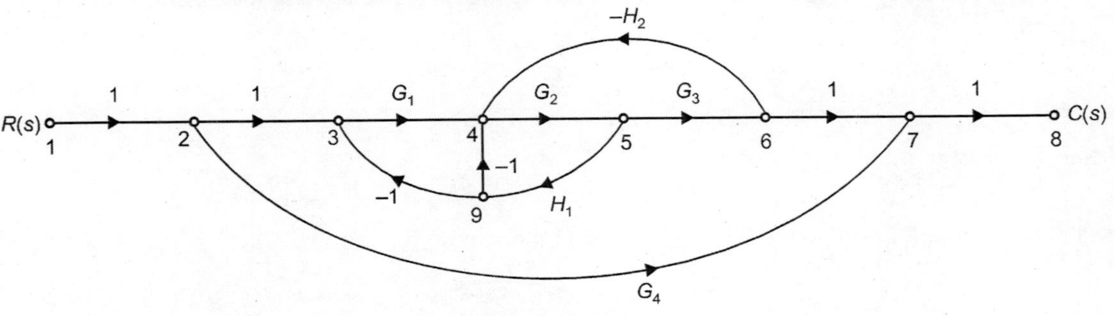

Fig. Ex. 1.25.2

Forward Path Gain

There are two forward path, $\therefore$ $K = 2$

Let the forward path gains be P_1 and P_2.

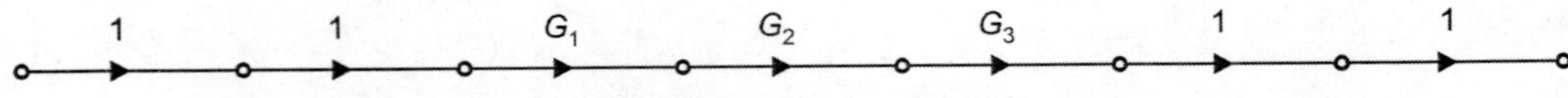

Fig. Ex. 1.25.3: Forward path-1

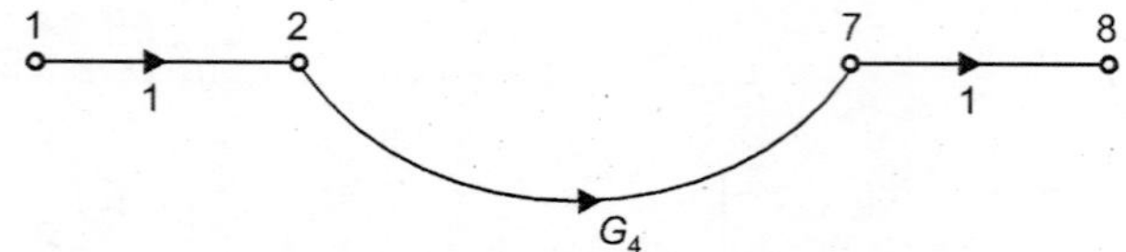

Fig. Ex. 1.25.4: Forward path-2

Gain of forward path-1, $P_1 = G_1\, G_2\, G_3$

Gain of forward path-2, $P_2 = G_4$

Individual Loop Gain

There are three individual loops with gains P_{11}, P_{21} and P_{31}

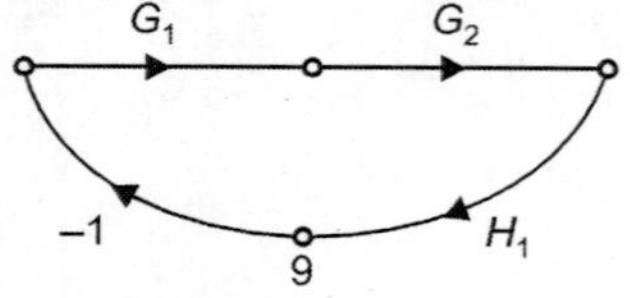

Fig. Ex. 1.25.5: Loop 1.

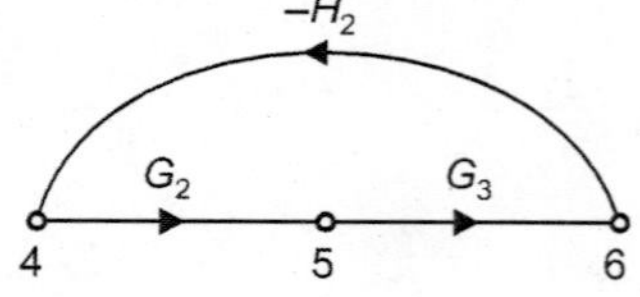

Fig. Ex. 1.25.6: Loop 2.

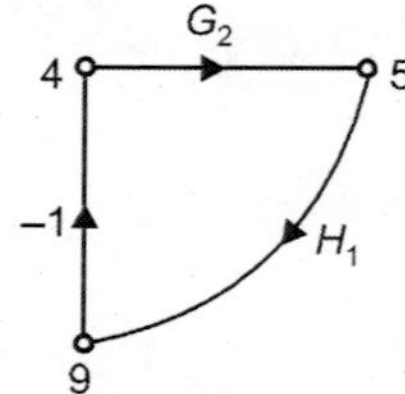

Fig. Ex. 1.257: Loop 3.

Gain of individual loop-1, $L_1 = -\, G_1\, G_2\, H_1$

Gain of individual loop-2, $L_2 = -\, G_2\, G_3\, H_2$

Gain of individual loop-3, $L_3 = -\, G_2\, H_1$

Gain Products of Two Non-touching Loops

There are no possible combinations of two non-touching. Three non-touching loops, etc.

Calculation of Δ and ΔK

$$\Delta = 1 - [L_1 + L_2 + L_3]$$

$$\Delta = 1 + G_1\, G_2\, H_1 + G_2\, G_3\, H_2 + G_2\, H_1$$

Since no part of graph touches forward path-1

$$\Delta_1 = 1$$

The part of graph non-touching forward Path-2 is shown in Fig. (Ex. 1.25.8)

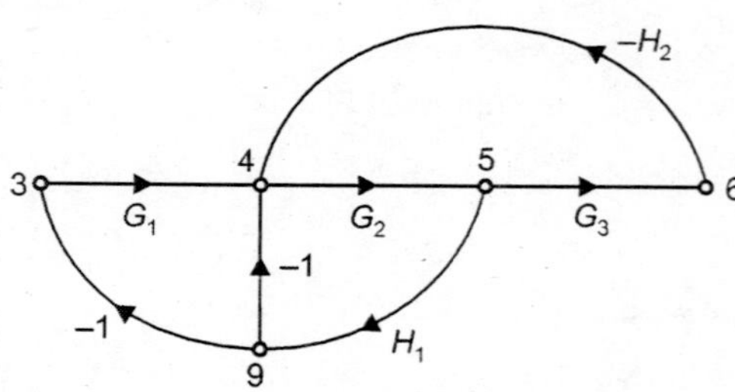

Fig. Ex. 1.25.8

$$\Delta_2 = 1 - [G_1 G_2 H_1 - G_2 G_3 H_2 - G_2 H_1]$$
$$\Delta_2 = 1 + G_1 G_2 H_1 + G_2 G_3 H_2 + G_2 H_1$$

Transfer Function, *T*

$$T = \frac{1}{\Delta} \sum_k P_k \Delta_k$$

$$= \frac{1}{\Delta} [P_1 \Delta_1 + P_2 \Delta_2]$$

$$T = \frac{G_1 G_2 G_3 + G_4 + G_1 G_2 G_4 H_1 + G_2 G_3 G_4 H_2 + G_2 G_4 H_1}{1 + G_1 G_2 H_1 + G_2 G_3 H_2 + G_2 H_1}$$

EXAMPLE 26

Convert the block diagram to signal flow graph and determine the transfer function using Mason's gain formula.

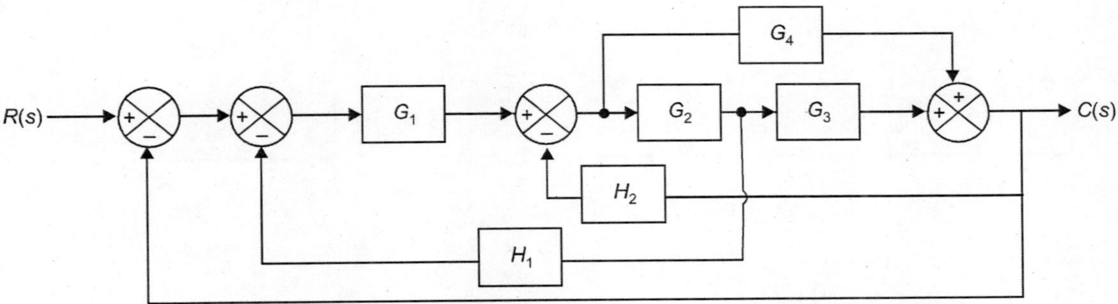

Fig. Ex. 1.26

Solution: The nodes are assigned at input, output, at every summing point and branch point as shown in Fig. (Ex. 1.26.1)

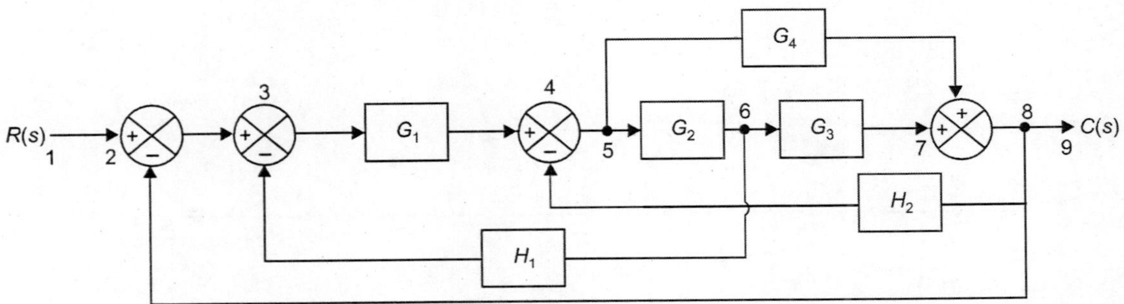

Fig. Ex. 1.26.1

The signal flow graph for the above block diagram shown in Fig. (Ex. 1.26.2)

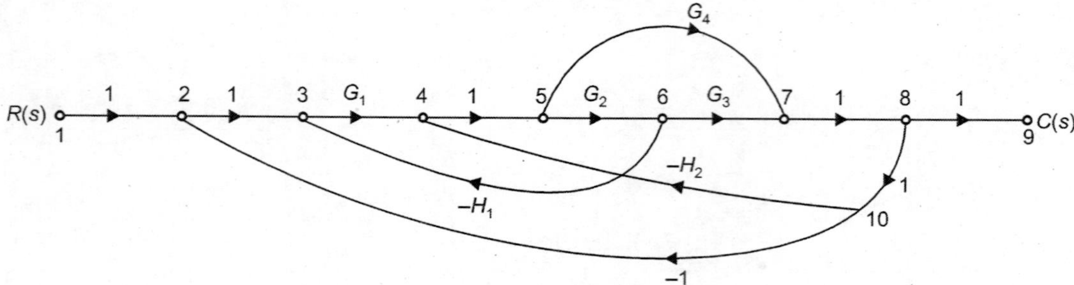

Fig. Ex. 1.26.2

I. Forward Path Gain

There are two forward paths $\therefore K = 2$

Let the gain of the forward paths be P_1 and P_2

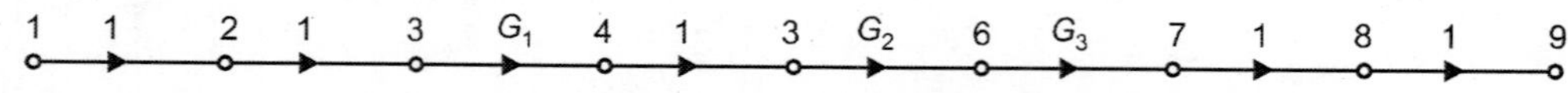

Fig. Ex. 1.26.3: Forward path-1.

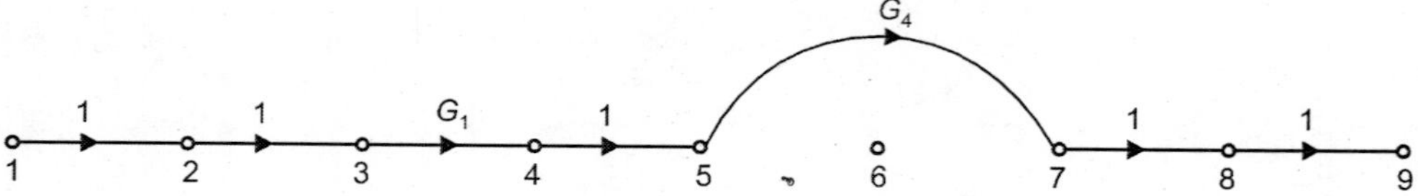

Fig. Ex. 1.26.4: Forward path-2.

Gain of forward path-1, $P_1 = G_1\, G_2\, G_3$

Gain of forward path-2 $P_2 = G_1\, G_4$

II. Individual Loop Gain

There are five individual loops let the individual loop gain be P_{11}, P_{21}, P_{31}, P_{41} and P_{51}.

Loop gain of individual loop-1, $L_1 = -\,G_1\, G_2\, G_3$

Loop gain of individual loop-2, $L_2 = -\,G_2\, G_1\, H_1 = -G_1\, G_2\, H_1$

Loop gain of individual loop-3, $L_3 = -\,G_2\, G_3\, H_2$

Loop gain of individual loop-4, $L_4 = -\,G_1\, G_4$

Loop gain of individual loop-5, $L_5 = -\,G_4\, H_2$

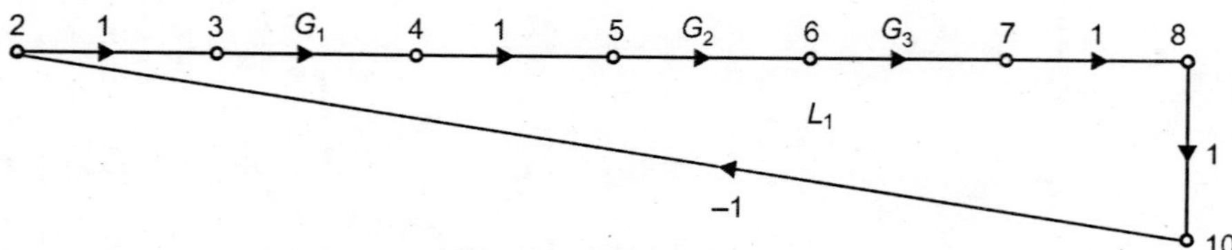

Fig. Ex. 1.26.5: Loop-1.

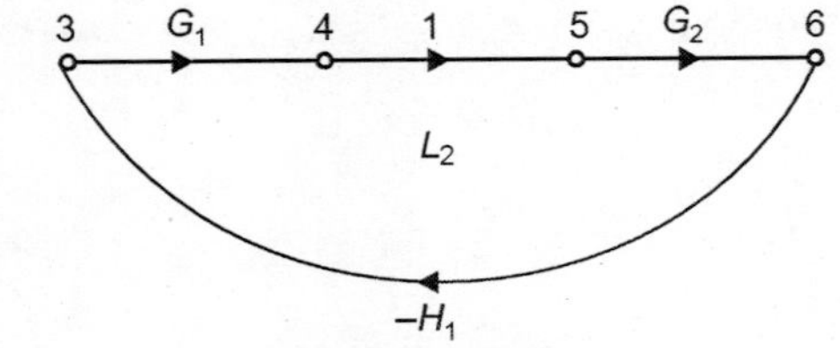

Fig. Ex. 1.26.6: Loop-2.

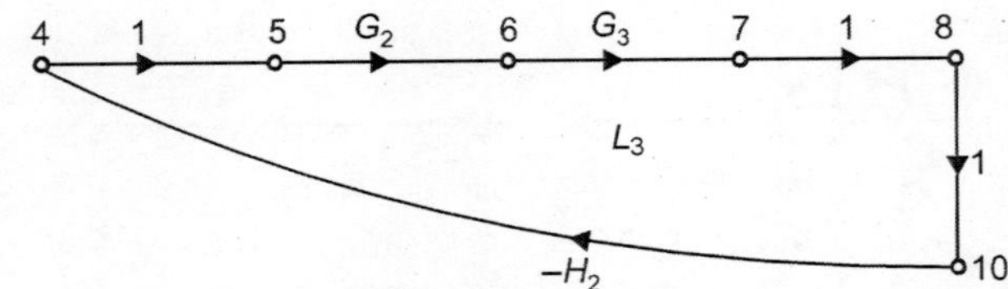

Fig. Ex. 1.26.7: Loop-3.

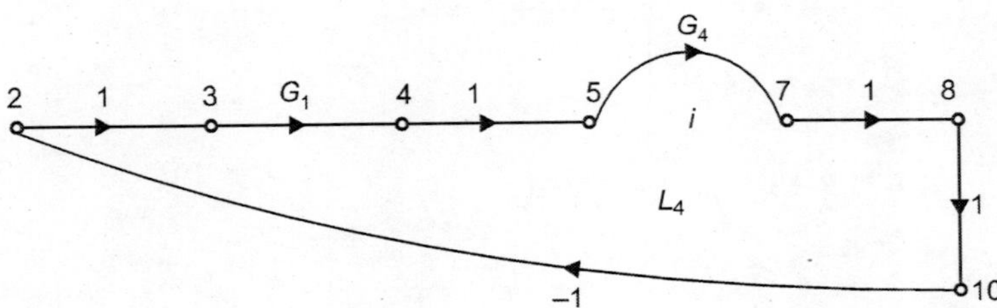

Fig. Ex. 1.26.8: Loop-4.

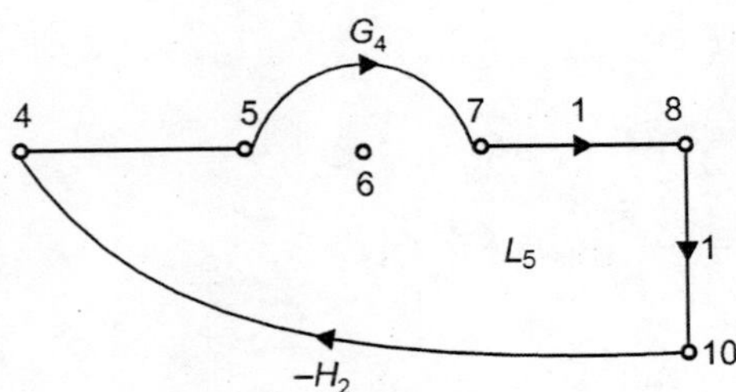

Fig. Ex. 1.26.9: Loop-5.

III. Gain Products of Two Non-touching Loops

There are no possible combinations of two non-touching loops, threen-touching loops, etc.

IV. Calculation of Δ and Δ*K*

$$\Delta = 1 - [L_1 + L_2 + L_3 + L_4 + L_5]$$
$$= 1 + G_1\, G_2\, G_3 + G_1\, G_2\, H_1 + G_2\, G_3\, H_2 + G_4\, G_1 + G_4\, H_2$$

Since no part of graph is non-touching with forward paths 1 and 2, $\Delta_1 = \Delta_2 = 1$

V. Transfer Function, *T*

By Mason's gain formula

$$T = \frac{1}{\Delta}\sum_K P_K \Delta_K = \frac{1}{\Delta}\,[P_1\Delta_1 + P_2\Delta_2]$$

$$T = \frac{G_1 G_2 G_3 + G_1 G_4}{1 + G_1 G_2 G_3 + G_1 G_2 H_1 + G_2 G_3 H_2 + G_1 G_4 + G_4 H_2}$$

SOME WORKED EXAMPLES

EXAMPLE 27

Draw the block diagram of the circuit shown in Fig. (Ex. 1.27).

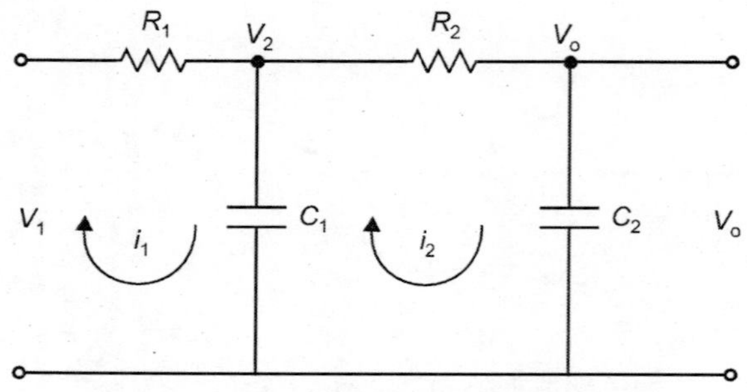

Fig. Ex. 1.27

Solution:

$$i_1(t) = V_1(t) - V_2(t)/R_1 \quad (1.27.1)$$

$$V_2(t) = \frac{1}{C_1}\int [i_1(t) - i_2(t)]\,dt \quad (1.27.2)$$

$$i_2(t) = \frac{1}{R_2}[V_2(t) - V_0(t)] \quad (1.27.3)$$

$$V_0(t) = \frac{1}{C_2}\int i_2\,dt \quad (1.27.4)$$

By taking laplace transform of equations (1.27.1 to 1.27.4)

$$I_1(s) = \frac{1}{R_1}[V_1(s) - V_2(s)] \quad (1.27.5)$$

$$V_2(s) = \frac{1}{sC_1}[I_1(s) - I_2(s)] \quad (1.27.6)$$

$$I_2(s) = \frac{1}{R_2}[V_2(s) - V_0(s)] \tag{1.27.7}$$

$$V_0(s) = \frac{1}{sC_2} I_2(s) \tag{1.27.8}$$

from equation (1.27.5)

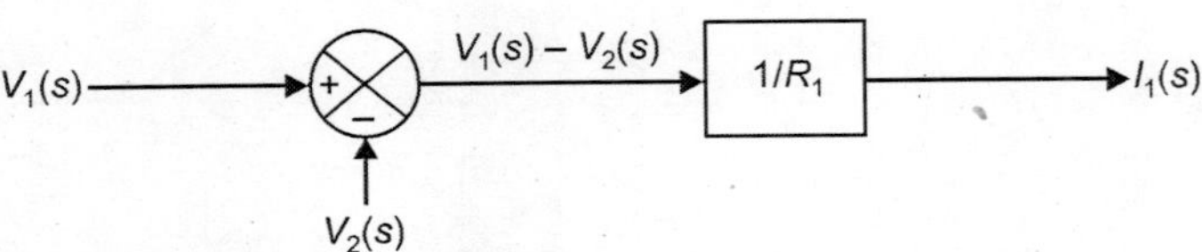

Fig. Ex. 1.27.1

from equation (1.27.6)

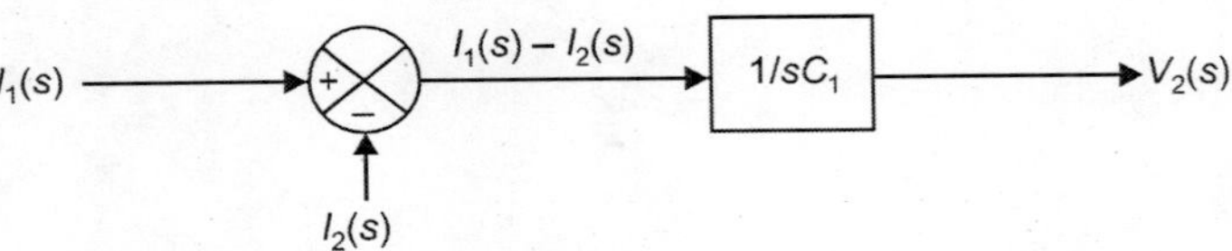

Fig. Ex. 1.27.2

from equation (1.27.7)

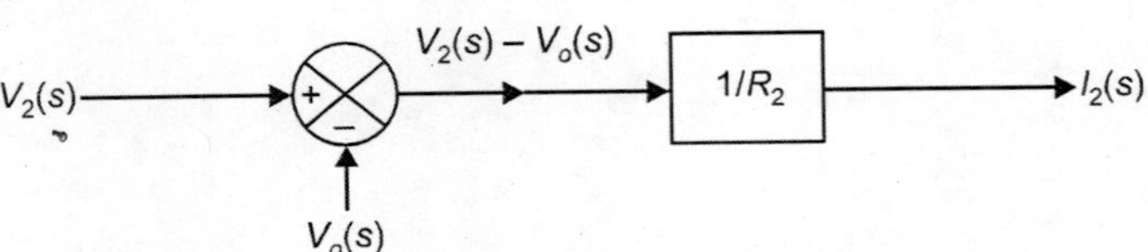

Fig. Ex. 1.27.3

from equation (1.28.8)

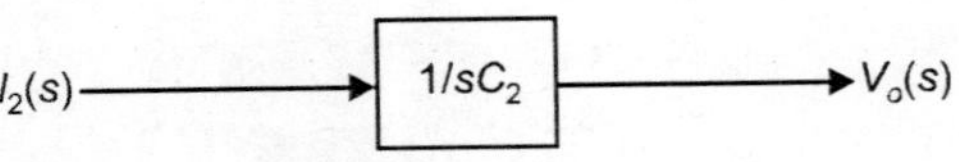

Fig. Ex. 1.27.4

Combining all the Fig. (1.27.1 to 1.27.4)

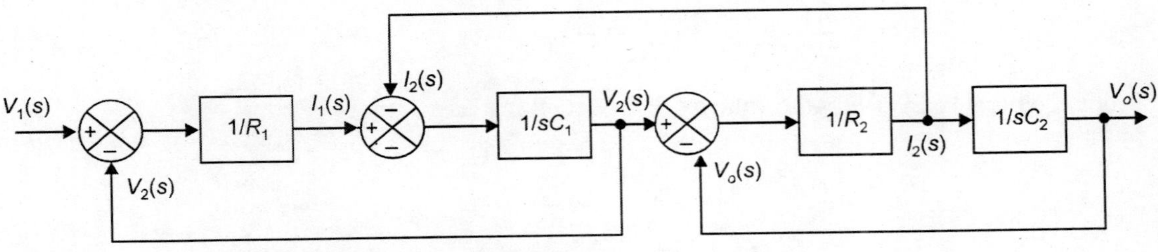

Fig. Ex. 1.27.5

EXAMPLE 28

Determine the ratio C/R

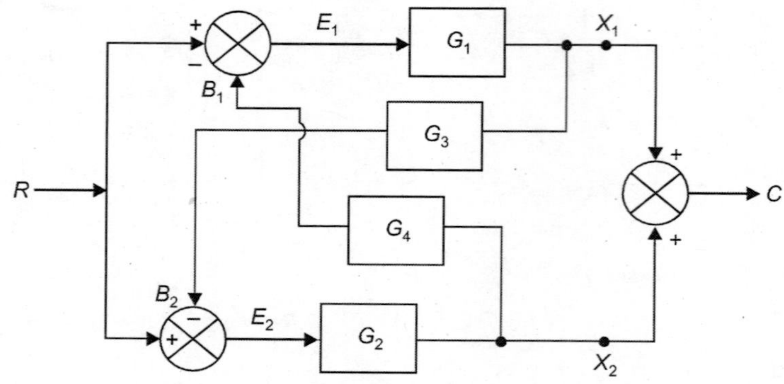

Fig. Ex. 1.28

Solution: From above block diagram

$$E_2 = R - B_2$$

$$B_2 = X_1 G_3 \qquad (1.28.1)$$

$$X_2 = E_2 G_2$$

$$X_2 = G_2(R - B_2)$$

$$X_2 = G_2 (R - X_1 G_3) \qquad (1.28.2)$$

$$B_1 = X_2 G_4$$

$$B_1 = G_4 G_2 (R - X_1 G_3)$$

$$E_1 = R - B_1 = R - G_2 G_4 (R - X_1 G_3) \qquad (1.28.3)$$

$$X_1 = E_1 G_1$$

$$X_1 = G_1 [R - G_2 G_4 (R - X_1 G_3)]$$

$$X_1 = G_1 R - G_1 G_2 G_4 R + G_1 G_2 G_3 G_4 X_1$$

$$X_1 = R (G_1 - G_1 G_2 G_4) + G_1 G_2 G_3 G_4 X_1 \qquad (1.28.4)$$

$$X_1 [1 - G_1 G_2 G_3 G_4] = R (G_1 - G_1 G_2 G_4)$$

$$\frac{X_1}{R} = \frac{G_1 - G_1 G_2 G_4}{1 - G_1 G_2 G_3 G_4} \qquad (1.28.5)$$

From equation (1.29.5) put the value of X_1 in equation (1.29.2) in X_2

$$X_2 = G_2 \left[R - \frac{G_1 (1 - G_2 G_4) G_3 R}{1 - G_1 G_2 G_3 G_4} \right]$$

$$= \frac{G_2 R [1 - G_1 G_3]}{1 - G_1 G_2 G_3 G_4}$$

$$\frac{X_2}{R} = \frac{G_2\,(1 - G_1\,G_3)}{1 - G_1\,G_2\,G_3\,G_4}$$

$$\frac{C}{R} = \frac{X_1}{R} + \frac{X_2}{R}$$

$$\frac{C}{R} = \frac{G_1 + G_2 - G_1\,G_2\,G_4 - G_1\,G_2\,G_3}{1 - G_1\,G_2\,G_3\,G_4}$$

Result: The ratio of C/R is

$$C/R = \frac{G_1 + G_2 - G_1\,G_2\,G_4 - G_1\,G_2\,G_3}{1 - G_1\,G_2\,G_3\,G_4}$$

EXAMPLE 29

Determine the transfer function C_1/R_1, C_2/R_2, C_1/R_2 and C_2/R_1 from the block diagram shown in Fig. (1.29).

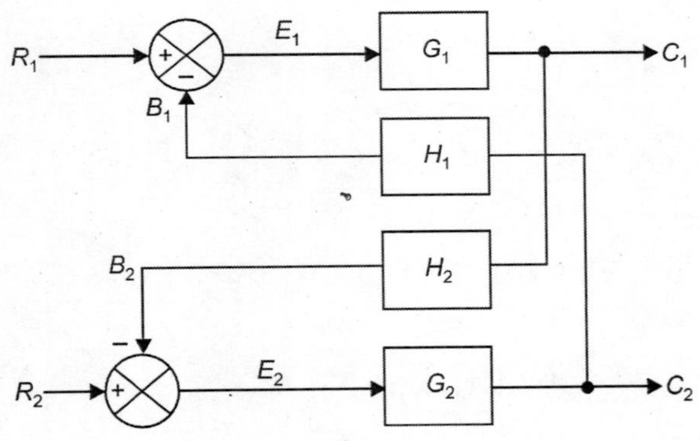

Fig. 1.29

Solution: From above block diagram

$$E_2 = R_2 - B_2 \tag{1.29.1}$$

$$B_2 = C_1\,H_2 \tag{1.29.2}$$

$$C_2 = E_2\,G_2 \tag{1.29.3}$$

Put the value of E_2 from equation (1.29.1) in (1.29.3)

$$C_2 = G_2\,[R_2 - B_2] \tag{1.29.4}$$

from equation (1.29.2) put the value of B_2 in equation (1.29.4)

$$C_2 = G_2\,(R_2 - C_1\,H_2) \tag{1.29.5}$$

$$B_1 = C_2\,H_1 \tag{1.29.6}$$

Put the value of C_2 from (1.29.5) in (1.29.6)

$$B_1 = G_2 H_1 (R_2 - C_1 H_2) \quad (1.29.7)$$

$$E_1 = R_1 - B_1 \quad (1.29.8)$$

$$E_1 = R_1 - G_2 H_1 (R_2 - C_1 H_2) \quad (1.29.9)$$

$$C_1 = E_1 G_1 \quad (1.29.10)$$

Also, from (1.29.8) put the value of E_1 in (1.29.9)

$$C_1 = G_1[R_1 - G_2 H_1 [R_2 - C_1 H_2]] \quad (1.29.11)$$

$$C_1 [1 - G_1 G_2 H_1 H_2] = G_1 R_1 - G_1 G_2 H_1 R_2 \quad (1.29.12)$$

when $R_1 = 0$ from equation (1.29.12)

$$\frac{C_1}{R_2} = \frac{-G_1 G_2 H_1}{1 - G_1 G_2 H_1 H_2} \quad (1.29.13)$$

from equation (1.29.13) put the value of C_1 in equation (1.29.5)

$$C_2 = G_2 \left[R_2 + \frac{G_1 G_2 H_1 H_2}{1 - G_1 G_2 H_1 H_2} \right]$$

$$\frac{C_2}{R_2} = \frac{G_2}{1 - G_1 G_2 H_1 H_2} \quad (1.29.14)$$

when $R_2 = 0$, (1.29.12) becomes

$$C_1 = G_1 [R_1 + C_1 G_2 H_1 H_2]$$

$$C_1 = G_1 R_1 + C_1 G_1 G_2 H_1 H_2$$

$$\frac{C_1}{R_1} = \frac{G_1}{1 - G_1 G_2 H_1 H_2} \quad (1.29.15)$$

when $R_2 = 0$, (1.29.5) becomes

$$C_2 = - G_2 C_1 H_2 \quad (1.29.16)$$

from equation (1.29.15) put the value of C_1 in (1.29.16)

$$C_2 = \frac{-G_1 G_2 R_1 H_2}{1 - G_1 G_2 H_1 H_2}$$

$$\frac{C_2}{R_1} = \frac{-G_1 G_2 H_2}{1 - G_1 G_2 H_1 H_2} \quad (1.29.17)$$

Result:

1. $\frac{C_1}{R_1} = \frac{G_1}{1 - G_1 G_2 H_1 H_2}$

2. $\dfrac{C_2}{R_2} = \dfrac{G_2}{1 - G_1 G_2 H_1 H_2}$

3. $\dfrac{C_1}{R_2} = \dfrac{-G_1 G_2 H_1}{1 - G_1 G_2 H_1 H_2}$

4. $\dfrac{C_2}{R_1} = \dfrac{-G_1 G_2 H_2}{1 - G_1 G_2 H_1 H_2}$

EXAMPLE 30

Draw the signal flow graph for the network shown in Fig. (Ex. 1.30) take V_3 as output node.

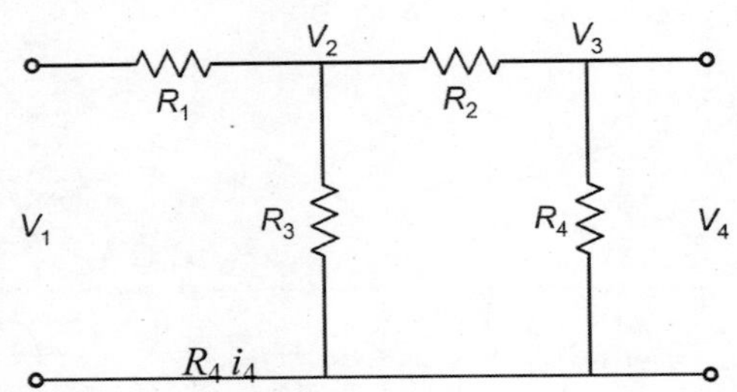

Fig. Ex. 1.30

Solution: The circuit can be redrawn with current directions.
From Fig. (Ex. 1.30.1)

$$i_1 = \frac{V_1 - V_2}{R_1} = \frac{V_1}{R_1} - \frac{V_2}{R_1}$$

$$V_2 = I_3 R_3 = R_3 (i_1 - i_2) = R_3 i_1 - R_3 i_2$$

$$i_2 = \frac{V_2 - V_3}{R_2} = \frac{V_2}{R_2} - \frac{V_3}{R_2}$$

$$V_3 = R_4 i_4 = R_4 i_2$$

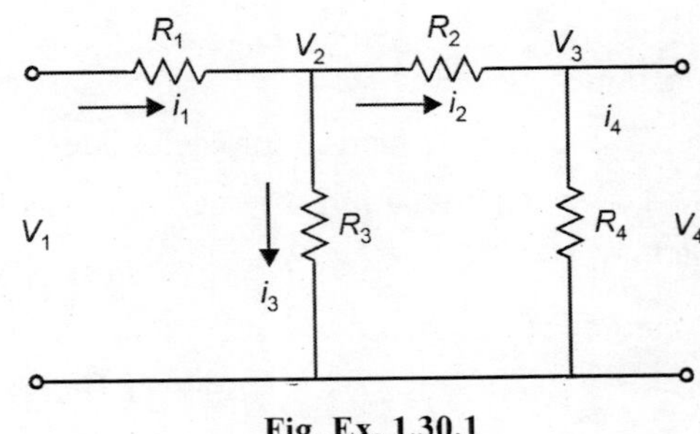

Fig. Ex. 1.30.1

Note: $i_4 = i_2$

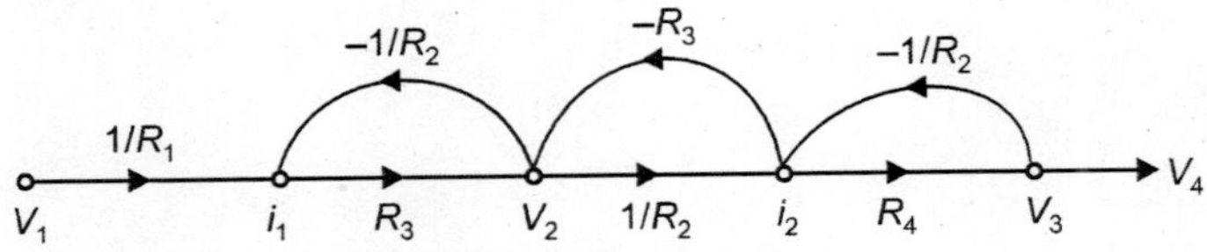

Fig. Ex. 1.30.2: Signal flow graph

1.19PROCEDURE FOR CONVERTING SIGNAL FLOW GRAPH IN TO BLOCK DIAGRAM

Step 1: For given signal flow graph, write the system equations.

Step 2: At each node consider the incoming branches only.

Step 3: Add all incoming signals algebraically at a node.

Step 4: For +ve (or) – sign in system equations use a summing point

Step 5: For the gain of each branch of signal flow graph draw the block having the same transfer function as the gain of the branch.

EXAMPLE 31

Draw the block diagram from the given signal flow graph.

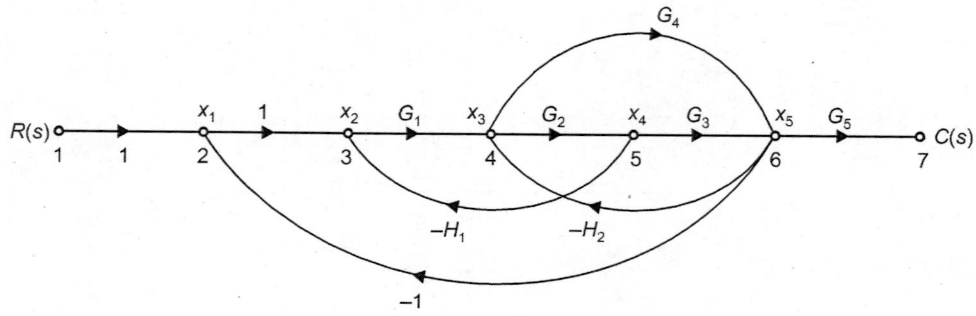

Fig. Ex. 1.31

Solution:

Step 1: System equation for each node.

The system equation at node 2, the incoming branches are from $R(s)$ and x_5

$$x_1 = 1R(s) - 1x_5 = R(s) - x_5 \qquad (1.31.1)$$

at node 3, there are two incoming branches

$$x_2 = 1x_1 - H_1\, x_4 = x_1 - H_1 x_4 \qquad (1.31.2)$$

at node 4, there are two incoming branches

$$x_3 = G_1 x_2 - H_2 x_5 \qquad (1.31.3)$$

Similarly at node 5 and 6 the system equation are

$$x_4 = G_2 x_3 \qquad (1.31.4)$$

$$x_5 = G_3 x_4 + G_4 x_3 \qquad (1.31.5)$$

Step 2: Draw the block diagram for each system equation

for equation (1.31.1)

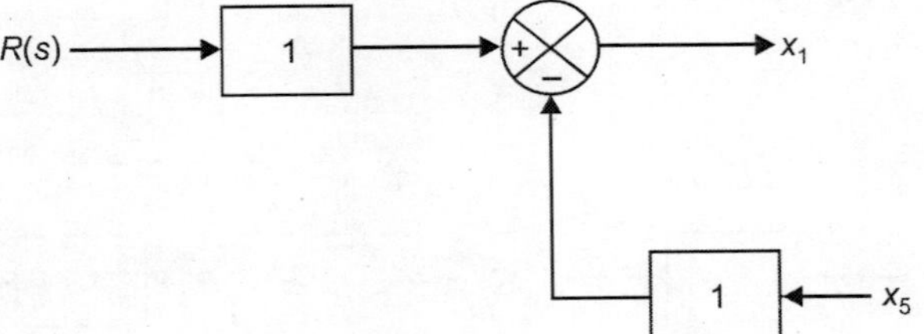

Fig. Ex. 1.31.1

for equation (1.31.2)

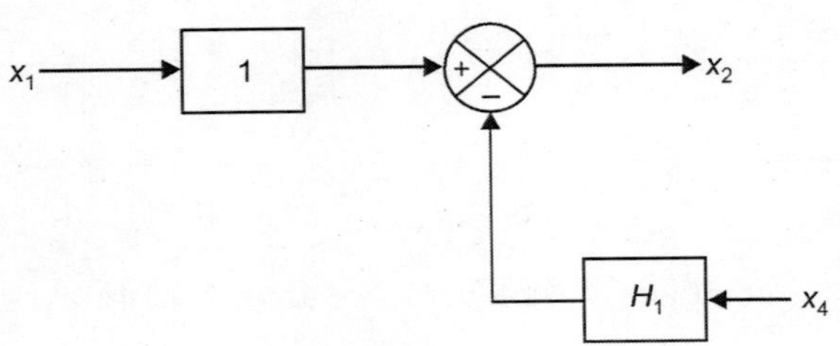

Fig. Ex. 1.31.2

for equation (1.31.3)

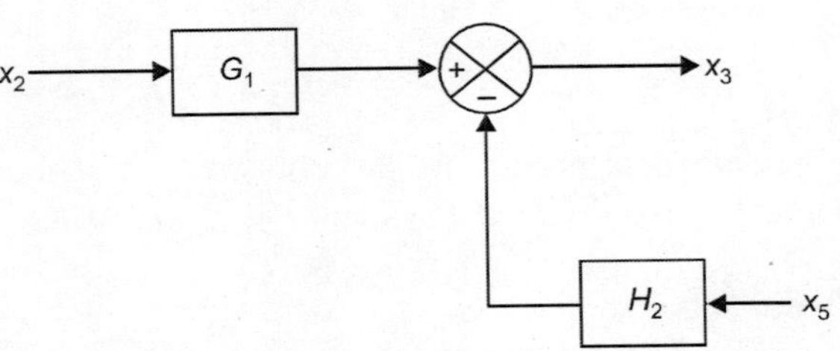

Fig. Ex. 1.31.3

for equation (1.31.4)

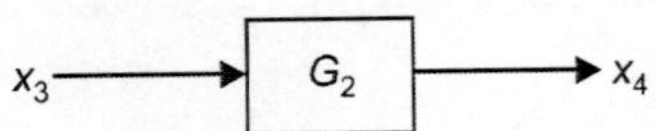

Fig. Ex. 1.31.4

for equation (1.31.5)

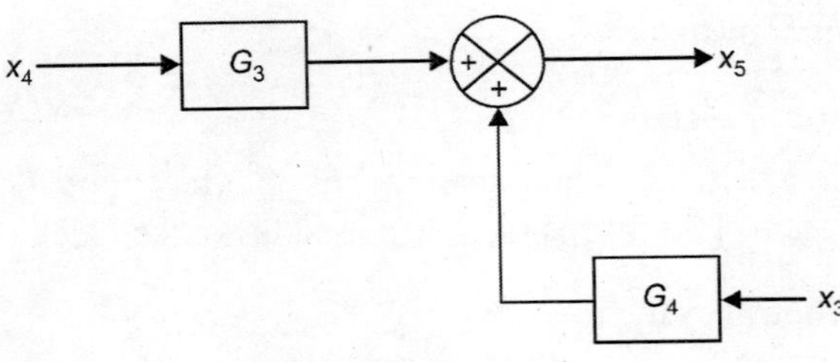

Fig. Ex. 1.31.5

Step 3: Combine all the block diagrams i.e. from (Ex. 1.31.1) to Fig. (Ex. 1.31.7)

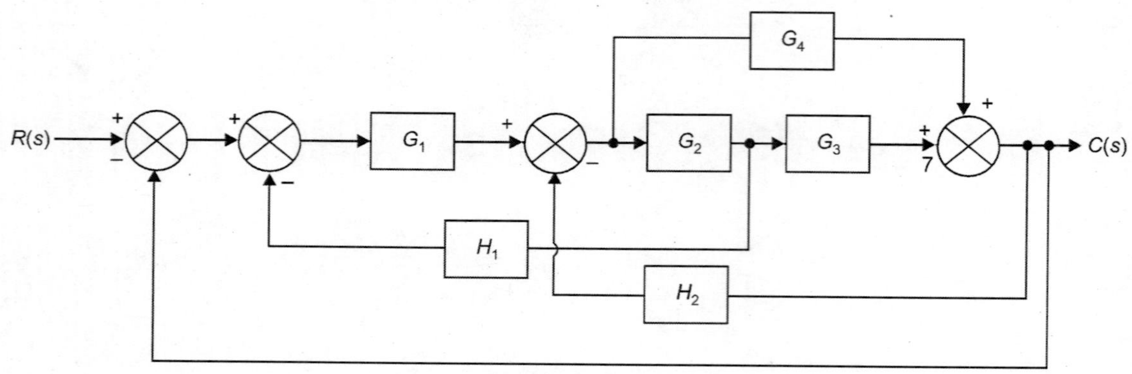

Fig. Ex. 1.31.7

Result: The Fig. (1.31.6) is the complete block diagram of given signal flow graph.

QUESTIONS AND ANSWERS

Q1.1 What is system?

Ans. When a number of elements (or) components are connected in a sequence to perform a specified function, the group thus formed is called a system.

Q1.2 What is control system?

Ans. A system consists of a number of components connected together to perform a specified function. In a system when the output quantity is controlled by varying the input quantity, then the system is called control system.

The output quantity is called controlled variable or response and input quantity is called command signal (or) excitation.

Q1.3 What are the two major type of control system?

Ans.
1. Open loop control system
2. Closed loop control system.

Q1.4 Define open loop control system.

Ans. The control system in which the output quantity has no effect upon the input quantity are called open loop control system. This means that the output is not feedback to the input for correction.

Q1.5 Define closed loop control system.

Ans. The control system in which the output has an effect upon the input quantity so as to maintain the desired output value are called closed loop control system.

Q1.6 What is feedback? What type of feedback is employed in control system?

Ans. The feedback is a control action in which the output is sampled and a proportional signal is given to input for automatic correction of any changes in desired output.

Negative feedback is employed in control system.

Q1.7 What are the components of feedback control system?

Ans. The components of feedback control system are plants, feedback path elements, error detector and controller.

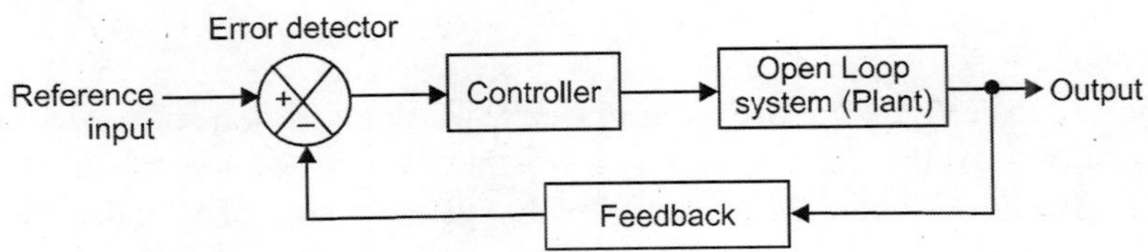

Q1.8 Why negative feedback is preferred in a closed loop system?

Ans. The negative feedback results in better stability in steady state and rejects any disturbance signals. It also has low sensitivity to parameter variations. Hence negative feedback is preferred in closed loop systems.

Q1.9 What are the characteristics of negative feedback?

Ans. The characteristics of negative feedback are,

(a) Accuracy in tracking steady state value

(b) Rejection of disturbance signals.

(c) Low sensitivity to parameter variations.

(d) Reduction in gain at the expense of better stability.

Q1.10 What is the effect of positive feedback on stability?

Ans. The positive feedback increases the error signal and drives the output to instability. But sometimes the positive feedback is used in minor loops in control system to amplify certain internal signals or parameters.

Q1.11 Distinguish between open loop and closed loop system?

Ans.

Open loop	Closed loop
(a) Inaccurate and unreliable	(a) Accurate and reliable
(b) Simple and economical	(b) Complex and costlier
(c) The changes in output due to external disturbances are not corrected automatically	(c) The changes in output due to external disturbances are corrected automatically.
(d) They are generally stable	(d) Great efforts are neede to design a stable system

Q1.12 What is servomechanism?

Ans. The servomechanism is a feedback control system component which corrects the position of mechanical load to desired output.

Q1.13 State the principle of superposition and homogeneity (or) state the principle of superposition?

Ans. The principle of superposition and homogeneity states that if the system has responses $y_1(t)$ and $y_2(t)$ for the inputs $X_1(t)$ and $X_2(t)$ respectively then the system responses to the linear combination of these input $a_1X_1(t) + a_2X_2(t)$ is given by linear combination of the individual outputs $a_1\,y_1(t) + a_2\,y_2\,(t)$, where a_1 ande a_2 are constants.

Q1.14 Define linear system?

Ans. A system is said to be linear if it obeys the principle of superposition and homogeneity principle of superposition states that the response of a system to a weighed sum of signals is equal to the corresponding weighed sum of the responses of the system to each of the individual input signals.

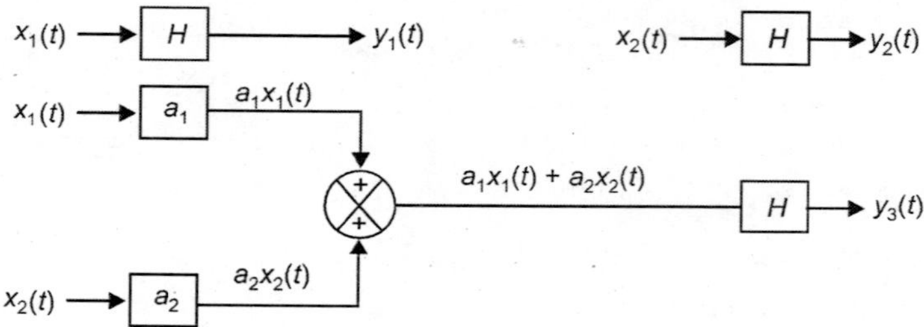

Q1.15 What is time invariant system?

Ans. A system is said to be time invariant if its input-output characteristics do not change with time. A linear time invariant system can be represented by constant coefficient differential equations. (In linear time varying system the coefficients of the differential equations governing the system are function of time).

Q1.16 Define transfer function?

Ans. The transfer function of a system is defined as the ratio of Laplace transform of output to Laplace transform of input with zero initial conditions. It is also defined of the Laplace transform of the impulse response of system with zero initial conditions.

Q1.17 State whether transfer function technique is applicable to non-linear system and whether the transfer function is independent of the input of a system?

Ans.

(i) The transfer function technique is not applicable to non-linear system.

(ii) The transfer function of a system is independent of input and depends only on system parameters but the output of a system depends on input.

Q1.18 What are the basic elements used for modelling mechanical translational system?

Ans. The model of mechanical translational system can be obtained by using three basic elements mass, spring and dashpot.

Q1.19 Write the force balance equation of ideal mass element?

Ans. Let a force be applied to an ideal mass M. The mass will offer an opposing force, f_m which is proportional to acceleration.

$$\text{Therefore, } f = f_m = M\frac{d^2x}{dt^2}$$

Q1.20 Write the force balance equation of ideal dashpot.

Ans. Let a force f be applied to an ideal dashpot, with viscous frictional coefficient B. The dashpot will offer an opposing force, f_b which is proportional to velocity.

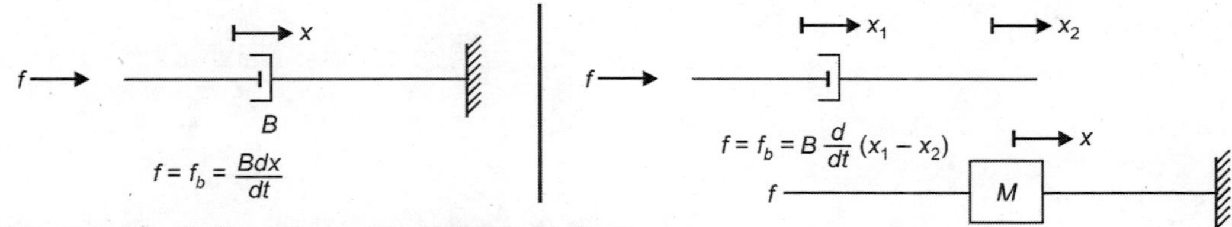

Q1.21 Write the force balance equation of ideal spring?

Ans. Let a force f be applied to an ideal spring with spring constant K. The spring will offer as opposing force f_K which is proportional to displacement.

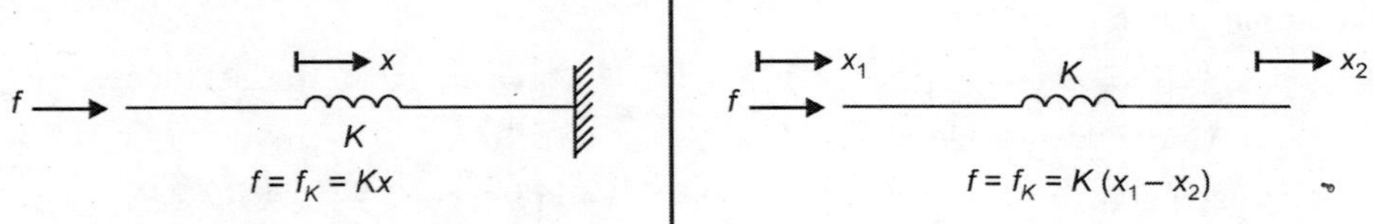

Q1.22 What are the basic elements used for modelling mechanical rotational system?

Ans. The model of mechanical rotational system can be obtained by using three basic elements mass with moment of inertia, J dashpot with rotational frictional coefficient, B and spring with stiffness, K.

Q1.23 Write the torque balance equation of an ideal rotational mass element.

Ans.

J
T θ B

Let a torque T be applied to an ideal mass with moment of inertial, J. The mass will offer an opposing torque T_j which is proportional to angular acceleration.

$$T = T_j = J\frac{d^2\theta}{dt^2}$$

Q1.24 Write the torque balance equation of an ideal rotational dashpot.

Ans. Let a torque T be applied to a rotational dash-pot with frictional coefficient B. The dashpot will offer an opposing torque which is proportional to angular velocity.

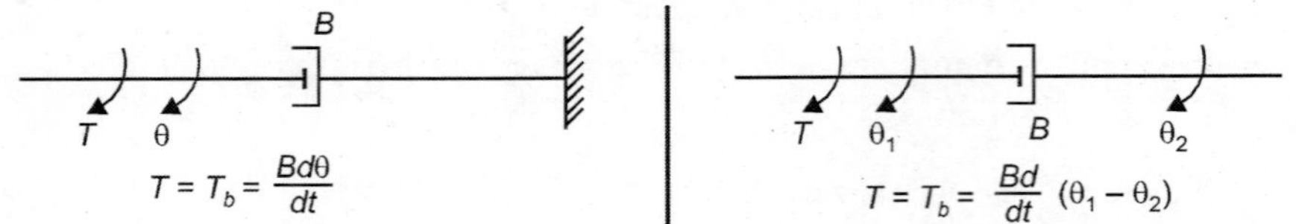

Q1.25 Write the torque balance equation of ideal rotational spring.

Ans. Let a torque T be applied to an ideal to an ideal rotational spring with constant K. The spring will offer an opposing torque T_k which is proportional to angular displacement.

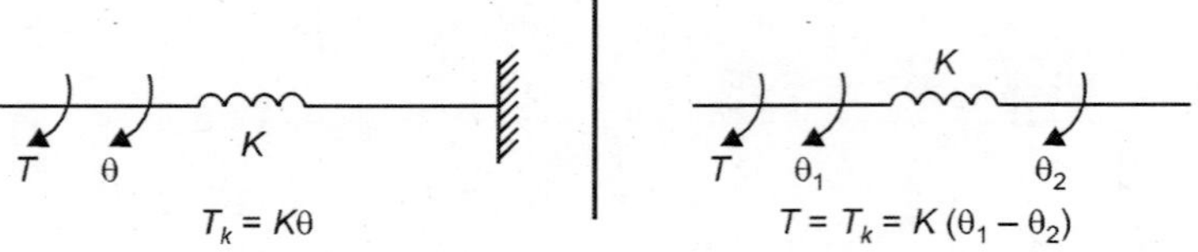

Q1.26 Name the two types of electrical analogous for mechanical system.

Ans. The two types of analogies for the mechanical system are force-voltage and force current analogy.

Q1.27 Write the analogous electrical elements in force-voltage analogy for the elements of mechanical translational system

Ans.

Force, $f \rightarrow$ voltage, e

Velocity, $v \rightarrow$ Current, i

Displacement, $x \rightarrow$ Charge, q

Frictional coefficient $B \rightarrow$ Resistance, R

Mass, $M \rightarrow$ Inductance, L

Stiffness, $K \rightarrow$ Inverse of capacitance, $1/C$

Newton's second $\rightarrow$ KVL

$\sum f = 0 \quad \sum V = 0$

Q1.28 Write the analogous electrical elements in force current analogy for the elements of mechanical translational system.

Ans.

Force, $f \rightarrow$ current, i

Velocity, $v \rightarrow$ Voltage, v

Displacement, $x \rightarrow$ Flux, ϕ

frictional coefficient, $B \rightarrow$ Conductance

$G = 1/R$

Mass, $M \rightarrow$ Capacitance, C

Stiffness, $K \rightarrow$ Inverse of inductance, $1/C$

Newton's second $\rightarrow$ KCL

$\sum f = 0 \quad \sum i = 0$

Q1.29 Write the analogous electrical elements in torque-voltage analogy for the elements of mechanical rotational system.

Ans.

Torque, $T \rightarrow$ Voltage, e

Angular velocity, $\omega \rightarrow$ Current, i

Angular displacement, $\theta \rightarrow$ Charge, θ

Frictional coefficient, $B \rightarrow$ Resistance, R

Moments of inertia, $J \rightarrow$ Inductance, L

Stiffness of spring, $K \rightarrow$ Inverse of capacitance, $1/C$

Newton's second law $\sum T = 0 \qquad \sum V = 0$

Q1.30 Write the analogous electrical elements in torque-current analogy for the elements of mechanical rotational system..

Ans.

Torque, $T \rightarrow$ Current, c

Angular velocity, $\omega \rightarrow$ Voltage, v

Angular displacement, $\theta \rightarrow$ Flux, ϕ

Frictional coefficient, $B \rightarrow$ Conductance $G = 1/R$

Moment of inertia, $J \rightarrow$ Capacitance, C

Stiffness of spring, $K \rightarrow$ Inverse of inductance, $1/L$

Newton's second law, $\sum T = 0, \qquad \sum i = 0$

Q1.31 What is block diagram? What are the basic components of block diagram?

Ans. A block diagram of system is pictorial representation of the functions performed by each component of the system and shows the flow of signals. The basic elements of block diagram are block, branch point and summing point.

Q1.32 What is the basic for framing the rules of block diagram reduction technique?

Ans. The ruless for block diagram reduction techniques are framed such that any modification made on the diagram does not alter the input output relation.

Q1.33 Write the rule for moving the summing point ahead of a block.

Ans.

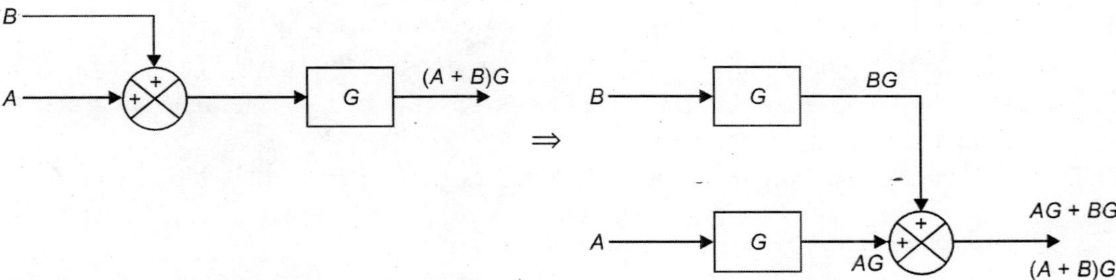

Q1.34 Write the rule for eliminating negative feedback loop?

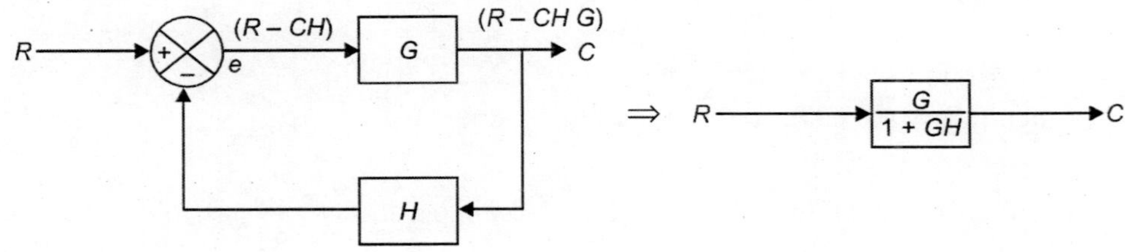

Ans. ***Proof:***

$$e = (R - CH)G \Rightarrow C + CHG = RG$$
$$C = RG - CHG \Rightarrow C\,(1 + GH) = RC$$
$$\frac{C}{R} = \frac{G}{1 + GH}$$

Q1.35 What is a signal flow graph?

Ans. A signal flow graph is a diagram that represents a set of simultaneous linear algebraic equations. By taking Laplace transform, the time domain differential equations governing a control system can be transfereed to a set of algebraic equations in S-domain. The signal flow graph of the system can be constructed using these equations.

Q1.36 What is transmittance?

Ans. The transmittance is the gain acquired by the signal when it travels from one node to another node in signal flow graph.

Q1.37 What is Sink and Source?

Ans. Source is the input node in the signal flow graph and it has only outgoing branches. Sink is a output node in the signal flow graph and it has only incoming branches.

Q1.38 Define non-touching loop.

Ans. The loops are said to be non-touching if they do not have common nodes.

Q1.39 What are the basic proporties of signal flow graph?

Ans.
1. Signal flow graph is applicable to linear system.
2. It consists of nodes and branches. A node is a point representing a variable on signal. A branch indicates functional dependence of one signal on the other.
3. A node adds the signals of all incoming branches and transmits this sum to all outgoing branches.
4. Signals travel along branches only in the marked direction and gets multiplied by the gain of the branch.
5. The algebraic equations must be in the form of cause and effect relationship.

Q1.40 Write the Mason's gain formula.

Ans. Mason's gain formula states that the overall gain of the system [Transfer function] as follows,

$$\text{overall gain, } T = \frac{1}{\Delta}\sum_{K} P_k\,\Delta_k$$

$T = T(s)$ = Transfer function of the system

K = Number of forward paths in the signal flow graph.

P_R = Forward path gain of Kth forward path.

$\Delta = 1 -$ [Sum of individual loop gains] + [Sum of gain products of all possible combinations of two non-touching loops] – [Sum of gain products of all possible combinations of three non-touching loops] + ...

ΔK = [Δ for that part of the graph which is not touching K^{th} forward path].

Q1.41 For the given signal flow graph, identify the number of forward path and number of individual loop.

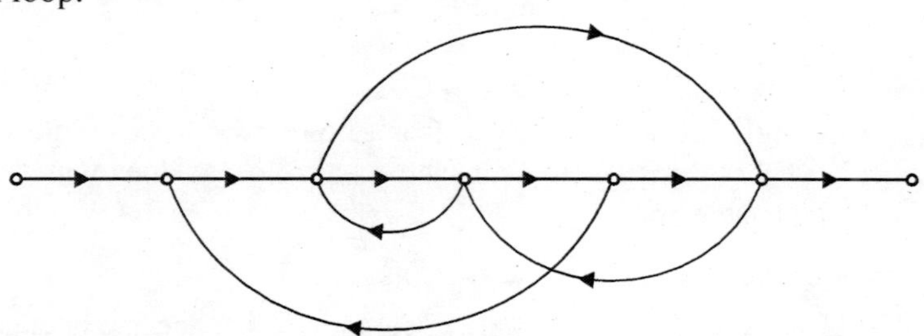

Ans. No. of forward paths = 2

No. of individual loop = 3.

Q1.42 What are the basic elements of thermal system?

Ans. The basic elements of thermal system are thermal resistance and thermal capacitance.

Q1.43 Define thermal resistance?

Ans. The thermal resistance for heat transfer between two substances is defined as the ratio of change in temperature and change in heat flow rate.

$$\text{Thermal resistance, } R = \frac{\text{Change in temperature, °C}}{\text{Change in heat flow rate, Kcal/sec.}}$$

Q1.44 Define thermal capacitance?

Ans. Thermal capacitance is defined as the ratio of change in heat stored and change in temperature.

$$\text{Thermal capacitance, } C = \frac{\text{Change in heat stored, Kcal}}{\text{Change in temperature, °C}}$$

Q1.45 Mention the electrical analogous of simple thermal system?

Ans. The electrical analogous of simple Ist order thermal system is R–C parallel circuit.

Q1.46 What are the basic elements of hydraulic system?

Ans. The basic elements of hydraulic system are resistance, inductance and capacitance.

Q1.47 Define hydraulic resistance?

Ans. The resistance for liquid flow is defined as the change in the level difference necessary to cause the unit change in flow rate.

$$R = \frac{\text{Change in level difference, } m}{\text{Change in flow node, m}^2\text{/sec}}$$

Q1.48 Define hydraulic capacitance?

Ans. The capacitance C of tank is defined as the change in quantity of stored liquid necessary to cause a unit change in head (height).

$$C = \frac{\text{Change in liquid stored, m}^3}{\text{Change in head, m.}}$$

Q1.49 What is inertance?

Ans. The inertance represents fluid inertia drived from the inertial forces required to accelerate a fluid in a pipe. It is a energy storing element. The energy storage due to intertance is negligible compared to capacitance element.

Q1.50 What are the basic elements of pneumatic systems?

Ans. The basic elements of pneumatic systems are pneumatic resistance and pneumatic capacitance.

Q1.51 Define pneumatic resistance?

Ans. The gas flow resistance, R may be defined as follows:

$$R = \frac{\text{Change in gas pressure difference N/m}^2}{\text{Change in gas flow rate, m}^2\text{/sec}}$$

EXERCISES

Q.1 Write the circuit differential equations governing the mechanical system. Also draw the force-voltage and force-current analogous circuit for Fig. 1.

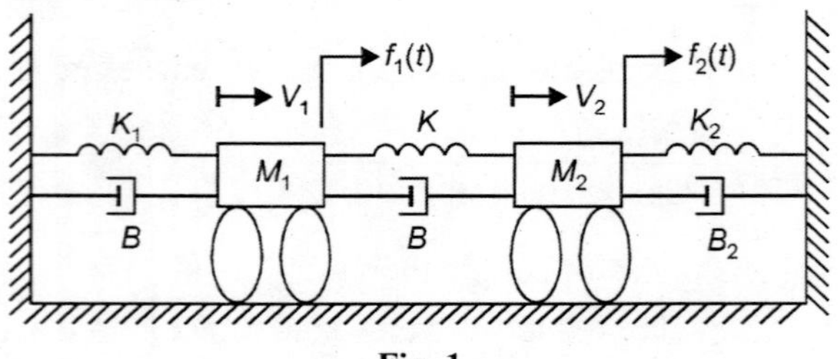

Fig. 1

Ans. $$f_1(t) = M_1 \frac{dV_1}{dt} + B_1V_1 + B(V_1 - V_2) + K_1 \int V_1\, dt + K \int (V_1 - V_2)\, dt$$

$$f_2(t) = M_2 \frac{dV_2}{dt} + B_2 V_2 + B\,(V_2 - V_1) + K_2 \int V_2\, dt + K \int (V_2 - V_1)\, dt$$

Note: $V = \frac{dx}{dt}$.

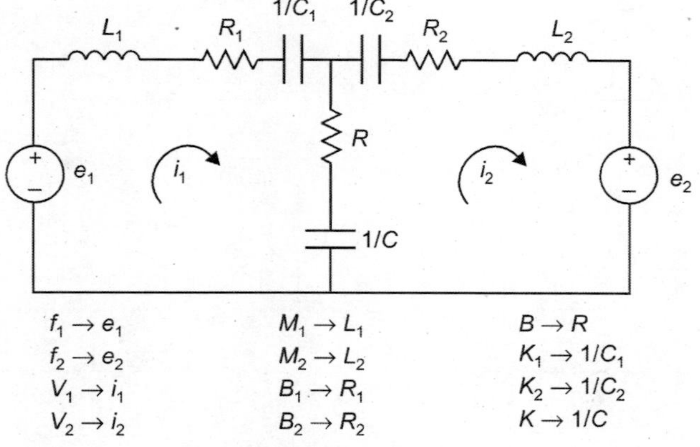

Fig. 2: Voltage analogous circuit.

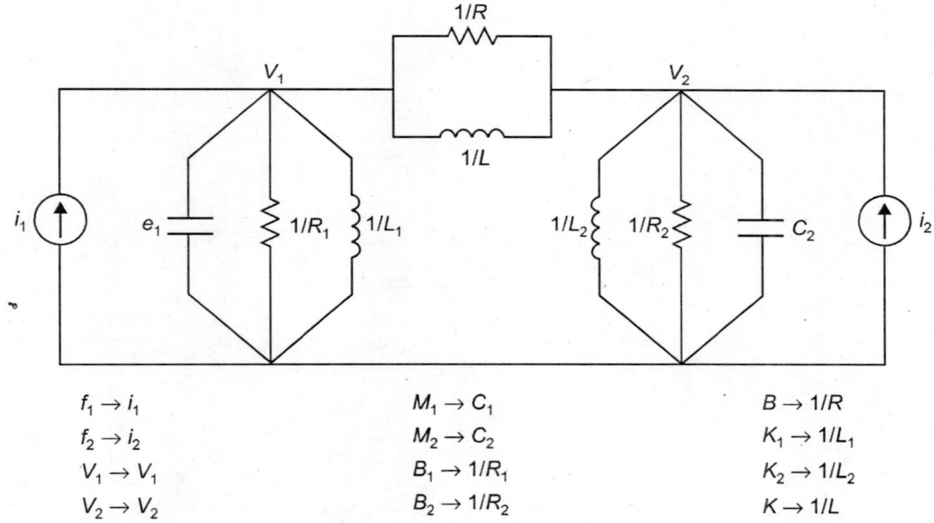

Fig. 3: Force-current analogous circuit.

Q.2 Write the differential equations governing the mechanical system. Draw the force-voltage and force-current analogous circuits for Fig. (4).

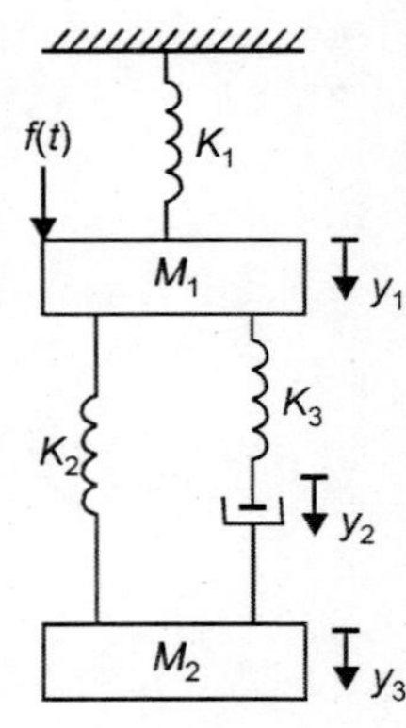

Fig. 4

Ans.

$$M_1 \frac{dV_1}{dt} + K_1 \int V_1\, dt + K_2 \int (V_1 - V_3)\, dt + K_3 \int (V_1 - V_2)\, dt = f(t)$$

$$K_3 \int (V_2 - V_1)\, dt + B\,(V_2 - V_3) = 0$$

$$M_2 \frac{dV_3}{dt} + B\,(V_3 - V_2) + K_2\,(V_3 - V_1)\, dt = 0$$

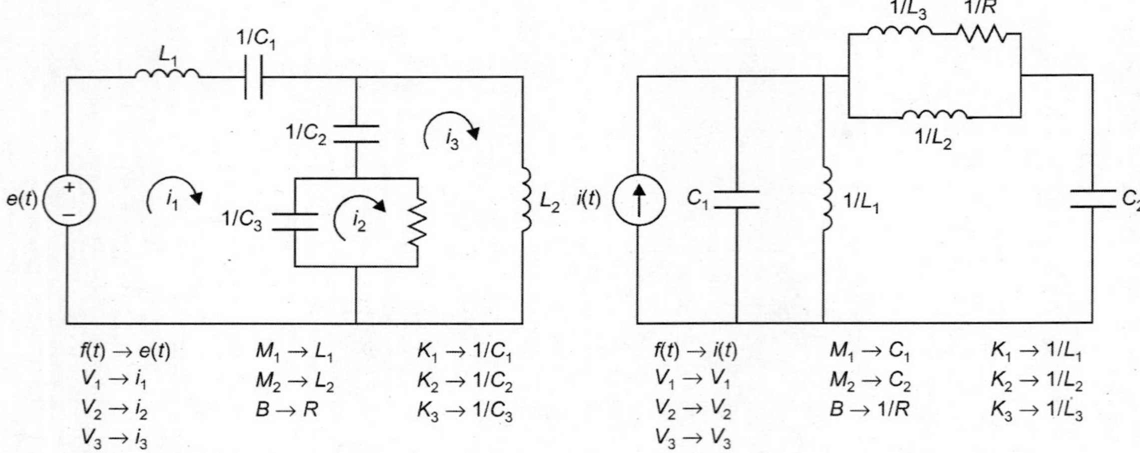

Fig. 5: Force-voltage analogous circuit. **Fig. 6**: Force-current analogous circuit.

Q.3 Consider the mechanical translational system. Draw force-voltage and force current analogous circuits for Fig. (7).

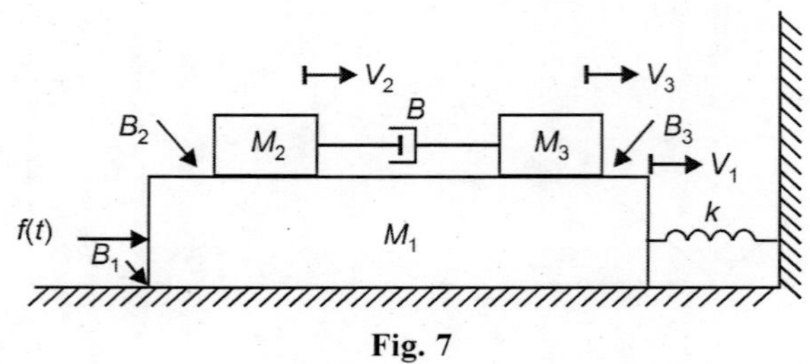

Fig. 7

Ans.

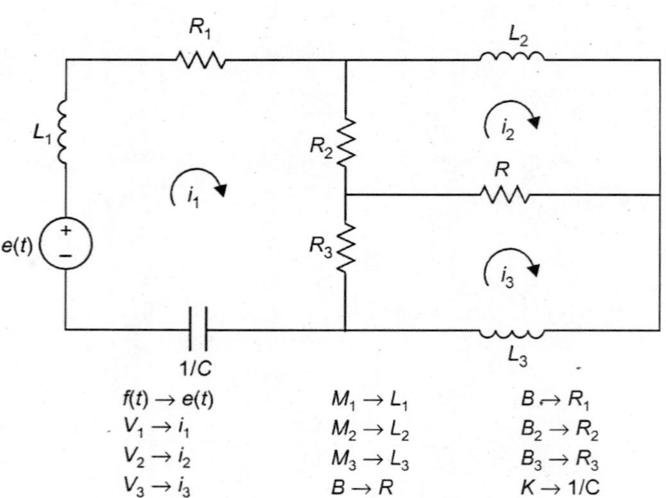

Fig. 8: Force-voltage analogous circuit.

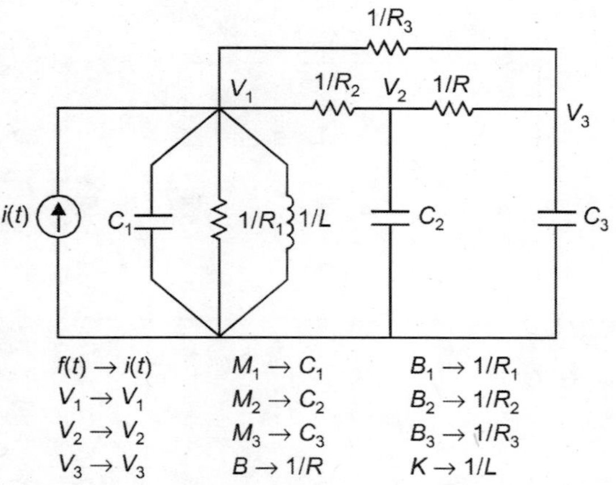

Fig. 9: Force-current analogous circuit.

Q.4 Write the differential equations governing the rotational mechanical system. Also draw the torque-voltage and torque-current analogous circuits for Fig. (10).

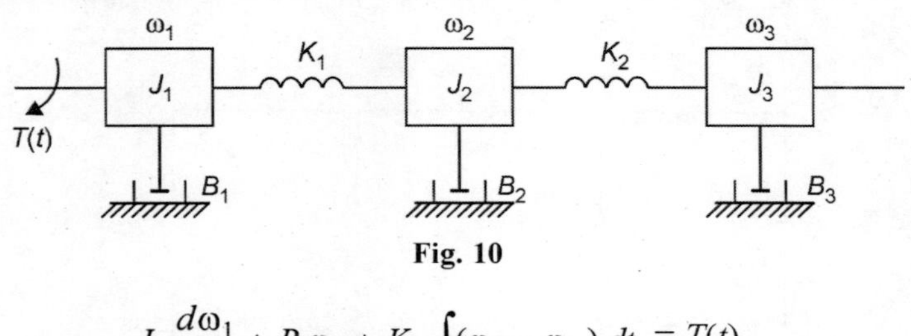

Fig. 10

Ans.

$$J_1 \frac{d\omega_1}{dt} + B_1\omega_1 + K_1 \int(\omega_1 - \omega_2)\, dt = T(t)$$

$$J_2 \frac{d\omega_2}{dt} + B_2\,\omega_2 + K_1 \int(\omega_2 - \omega_1)\, dt + K_2 \int(\omega_2 - \omega_3)\, dt = 0$$

$$J_3 \frac{d\omega_3}{dt} + B_3\,\omega_3 + K_2 \int(\omega_3 - \omega_2)\, dt = 0$$

$T(t) \rightarrow e(t)$
$\omega_1 \rightarrow i_1$
$\omega_2 \rightarrow i_2$
$\omega_3 \rightarrow i_3$

$J_1 \rightarrow L_1$
$J_2 \rightarrow L_2$
$J_3 \rightarrow L_3$

$B_1 \rightarrow R_1$
$B_2 \rightarrow R_2$
$B_2 \rightarrow R_3$
$K_1 \rightarrow 1/C_1$
$K_2 \rightarrow 1/C_2$

Fig. 11: Torque-voltage analogous

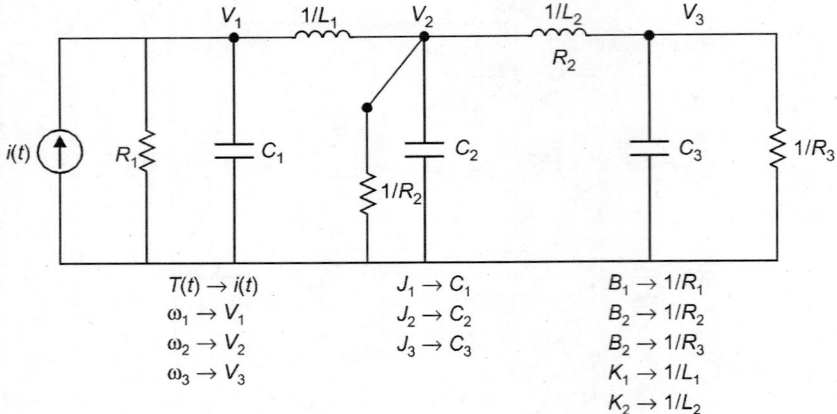

Fig. 12: Torque-current analogous circuit

Q.5 In an electrical circuit the elements resistance, capacitance, and inductance and connected in parallel across the voltage source E. Draw translation mechanical analogous system and rotational mechanical analogous system for Fig. (13).

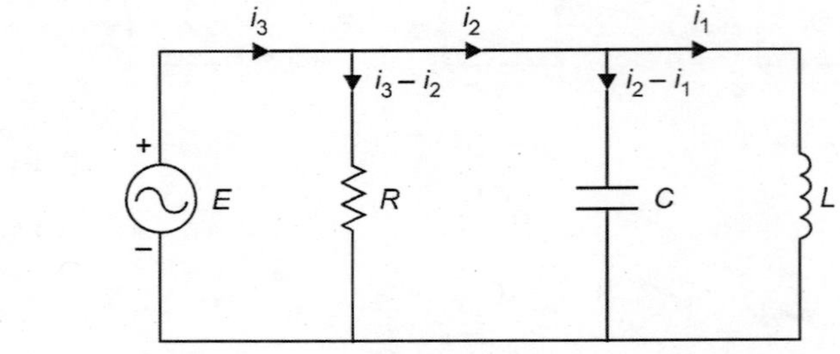

Fig. 13

Ans.

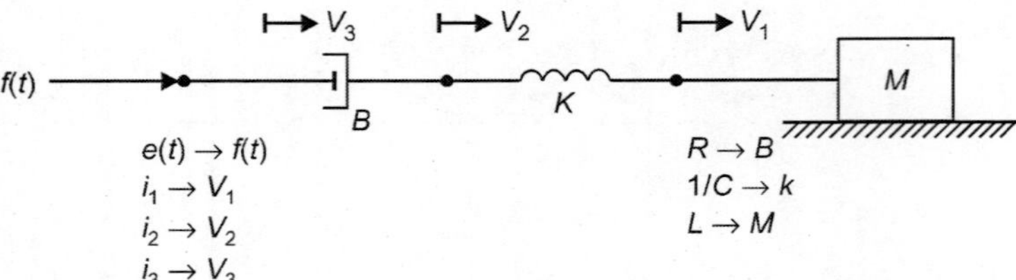

Fig. 14: Analogous mechanical traslational system.

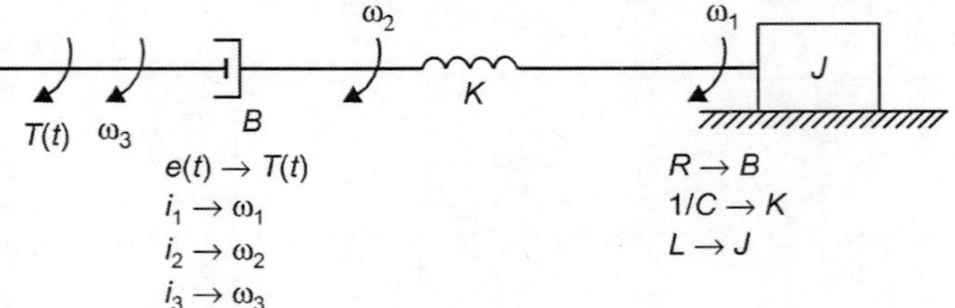

Fig. 15: Analogous mechanical rotational system

Q.6 Use reduction method to find out *C*/*R* by for the following diagram for Fig. (16).

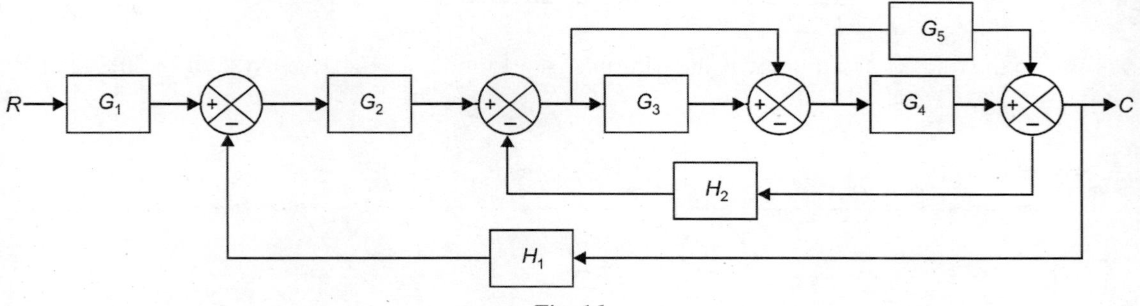

Fig. 16

Ans. $$\frac{C}{R} = \frac{G_1 G_2 (1 + G_3)(G_4 + G_5)}{[1 + (1 + G_3)(G_4 + G_5) H_2 + (1 + G_3)(G_4 + G_5) G_2 H_1]}$$

Q.7 By using the following block diagram obtain *C*/*R* for Fig. (17).

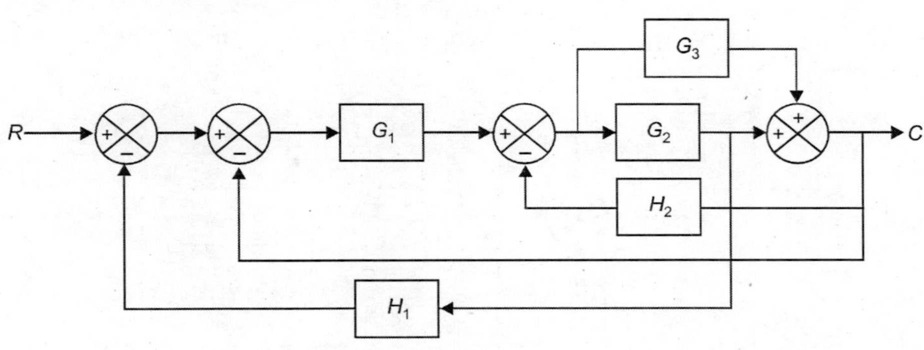

Fig. 17

Ans. $$\frac{C}{R} = \frac{G_1 G_2 + G_1 G_3}{[1 + G_1 G_2 H_1 + G_1 G_2 + G_1 G_2 + G_1 G_3 + G_2 H_2 + G_3 H_2]}$$

Q.8 Obtain *C*/*R* for Fig. (18).

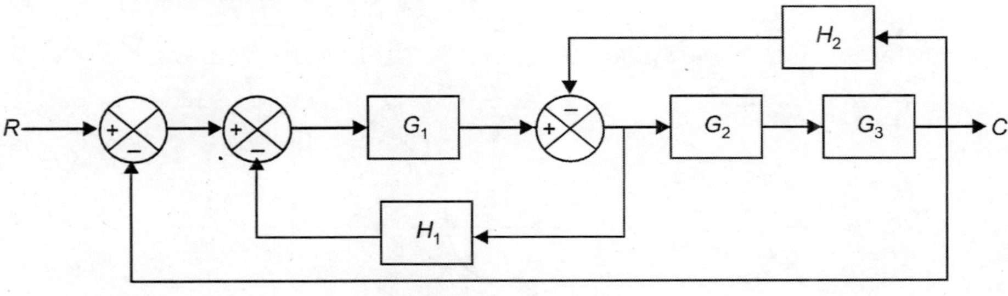

Fig. 18

Ans.
$$\frac{C}{R} = \frac{G_1 G_2 G_3}{1 + G_2 G_3 H_2 + G_1 G_2 H_1 + G_1 G_2 G_3}$$

Q.9 Obtain the transfer function by using Mason's gain formula for the given system for Fig. (19).

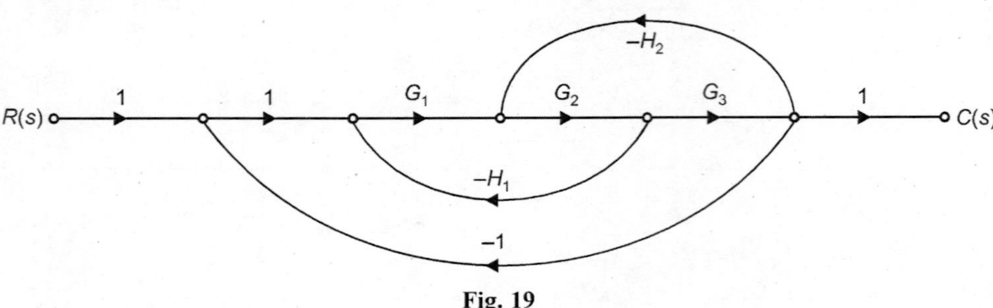

Fig. 19

Ans.
$$\frac{C(s)}{R(s)} = \frac{G_1 G_2 G_3}{1 + G_1 G_2 H_1 + G_2 G_3 H_2 + G_1 G_2 G_3}$$

Q.10 Convert the block diagram into signal flow graph and find the transfer function of the system shown in the Fig. (20).

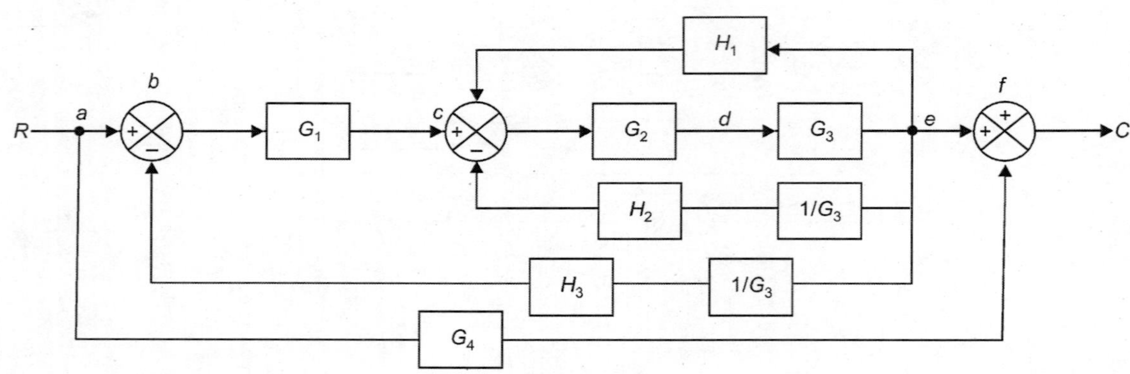

Fig. 20

Ans.
$$\frac{G_1 G_2 G_3 + G_4 + G_2 G_3 G_4 H_1 + G_2 G_4 H_2 + G_1 G_2 G_4 H_3}{1 + G_2 G_3 H_1 + G_2 H_2 + G_1 G_2 H_3}$$

Q.11. Find the transfer function of the system whose signal flow graph as shown in Fig. (22).

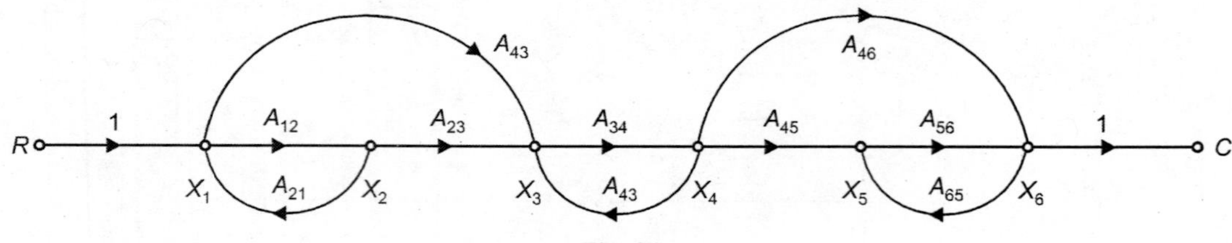

Fig. 21

Ans. $$\frac{C}{R} = \frac{A_{12}A_{23}A_{34}A_{45}A_{56} + A_{13}A_{34}A_{45}A_{56} + A_{12}A_{23}A_{34}A_{46} + A_{13}A_{34}A_{46}}{1-(A_{12}A_{21} + A_{34}A_{43} + A_{56}A_{65}) + (A_{12}A_{21}A_{34}A_{43} + A_{12}A_{21}A_{56}A_{65} + A_{34}A_{43}A_{56}A_{56}) - (A_{12}A_{21}A_{34}A_{43}A_{56}A_{65})}$$

Q.12 Obtain the transfer function of the system shown in Fig. (22).

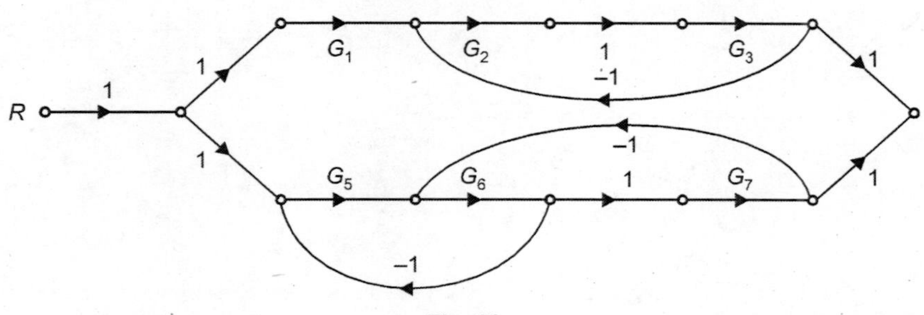

Fig. 22

Ans.

$$\frac{C}{R} = \frac{G_1G_2G_3G_4(1+G_5G_6+G_6G_7) + G_4G_5G_6G_7(1+G_1G_2+G_2G_3)}{1+G_1G_2+G_2G_3+G_5G_6+G_6G_7+G_1G_2G_3G_6+G_5G_6G_2G_3+G_1G_2G_6G_7+G_2G_3G_6G_7}$$

Q.13 Find X_8/X_1 for Fig. (23).

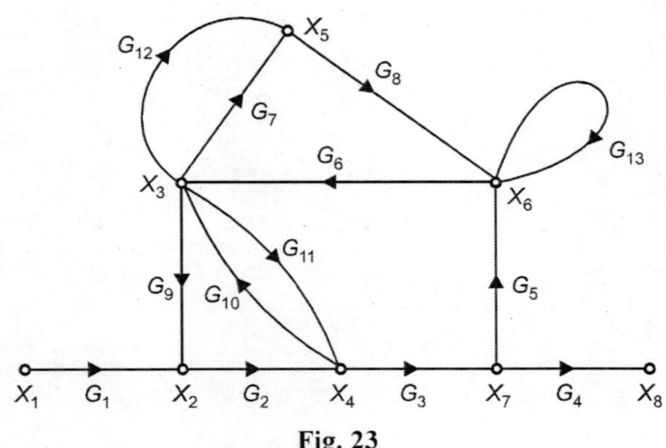

Fig. 23

Ans. $$\frac{X_8}{X_1} = \frac{[G_1G_2G_3G_4][1-(G_7G_{12}+G_6G_7G_8G_{13})+G_7G_{12}G_{13}]}{1-\{G_2G_9G_{10}+G_{10}G_{11}+G_2G_3G_5G_6G_9+G_3G_5G_6G_{11}+G_7G_{12}+G_6G_7G_8+G_{13}\}+G_2G_9G_{10}G_{13}+G_{10}G_{11}G_{13}+G_7G_{12}G_{13}}$$

Q.14 Obtain $\frac{X_8}{X_1}$ and $\frac{X_8}{X_2}$ from the signal flow graph shown in Fig. 24.

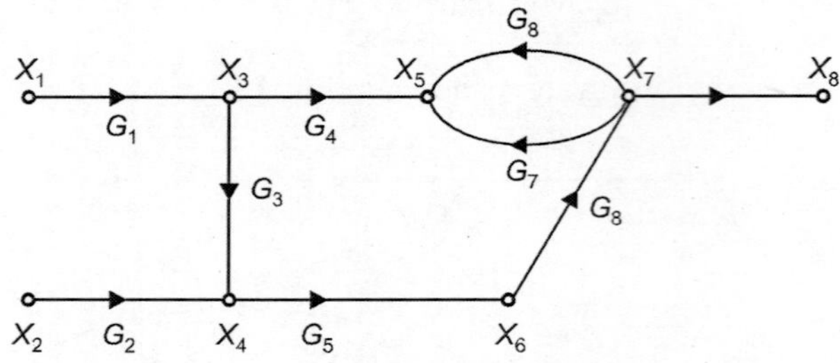

Fig. 24

Ans.

$$\frac{X_8}{X_1} = G_1\, G_4\, G_6\, G_9 + G_1\, G_4\, G_7\, G_9 + G_9 + G_1\, G_3\, G_5\, G_8\, G_9$$

$$\frac{X_8}{X_2} = G_2\, G_5\, G_8\, G_9$$

Q.15 Find the transfer functions of the given networks for the Fig. (25, 26, 27 and 28).

(a)

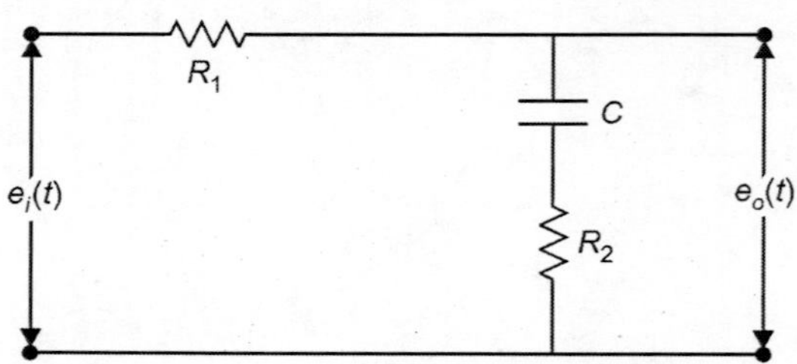

Fig. 25

Ans.

$$\frac{E_0(s)}{E_i(s)} = \frac{1 + s\,R_2\,C}{1 + s\,(R_1 + R_2)\,C}$$

(b)

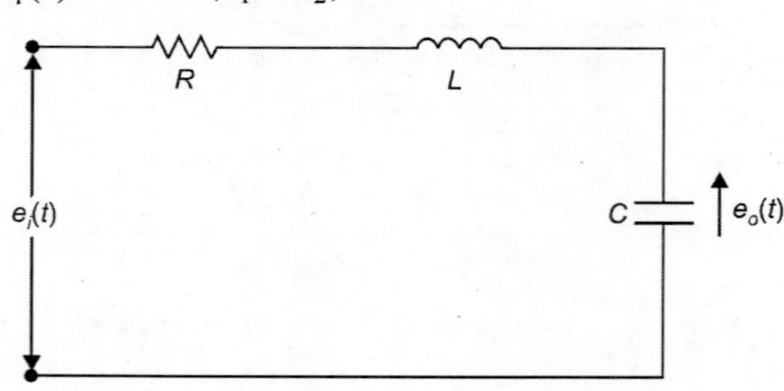

Fig. 26

Ans.
$$\frac{E_0(s)}{E_i(s)} = \frac{1}{s^2\ LC + sRC + 1}$$

(c)

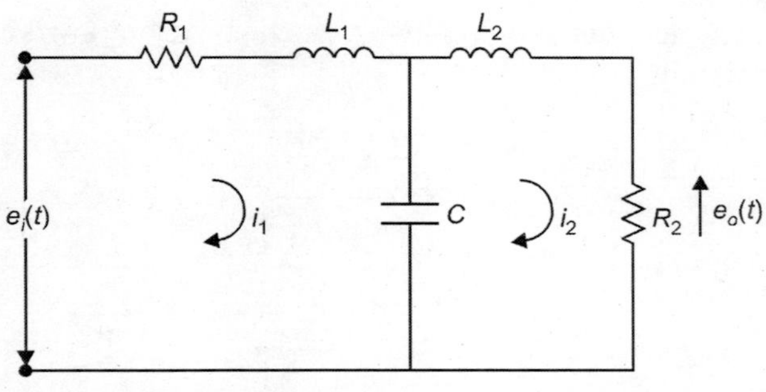

Fig. 27

Ans.
$$\frac{E_0(s)}{E_i(s)} = \frac{sR_2C}{(s^2L_1\,C + sR_1\,C + 1)\,(s^2\,L_2\,C + s\,R_2\,C + 1) - 1}$$

(d)

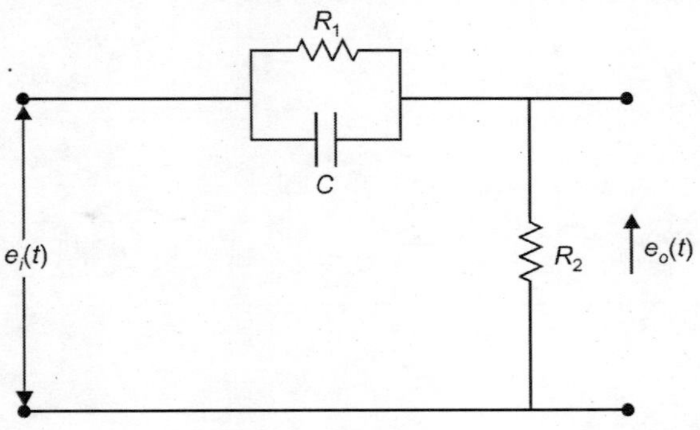

Fig. 28

Ans.
$$\frac{E_0(s)}{E_i(s)} = \frac{s\,R_1R_2\,C + R_2}{s\,R_1\,R_2\,C + (R_1 + R_2)}$$

Q.16 Find the transfer function of the circuit given below?

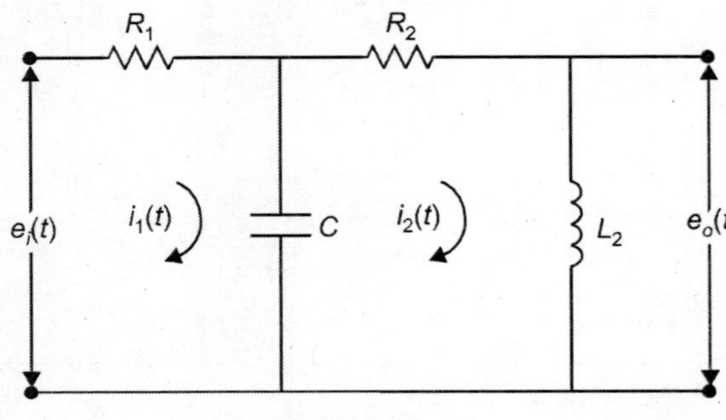

Fig. 29

Ans.

$$\frac{E_o(s)}{E_i(s)} = \frac{s^2 L_2 C}{[sR_1C+1][s^2(L_1+L_2)C\,s\,R_2\,C+1]-1}$$

Q.17 For the mechanical system derive the transfer function. Also draw the force-voltage and force current analogous circuits.

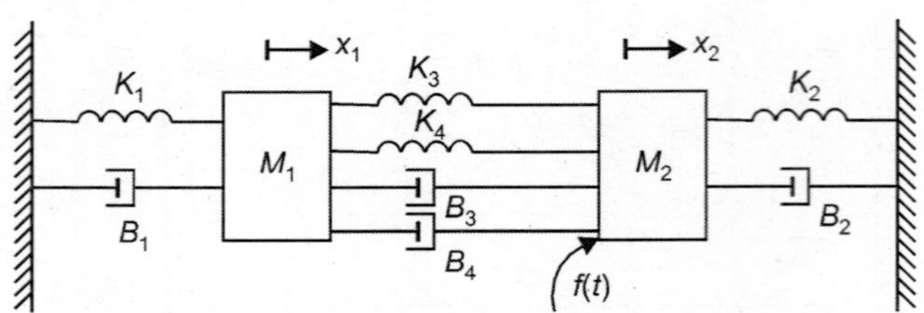

Fig. 30

Ans.

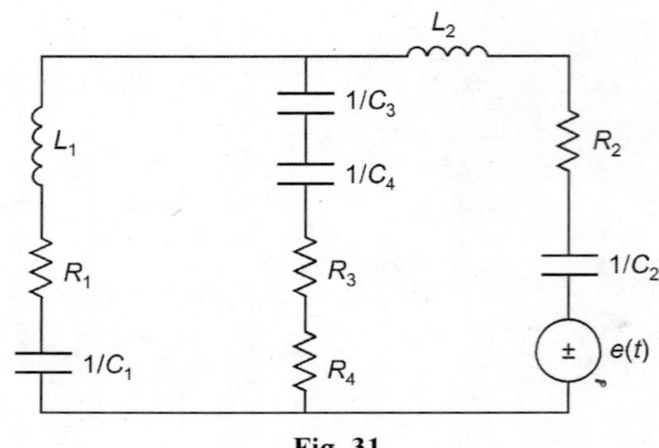

Fig. 31

$$\frac{X_2(s)}{F(s)} = \frac{M_1s^2 + s(B_1+B_3+B_4)+(K_1+K_3)}{(M_1s^2+s(B_1+B_3+B_4)+(K_1+K_3)(M_2s^2+B_2s+K_2+K_3+K_4)}$$

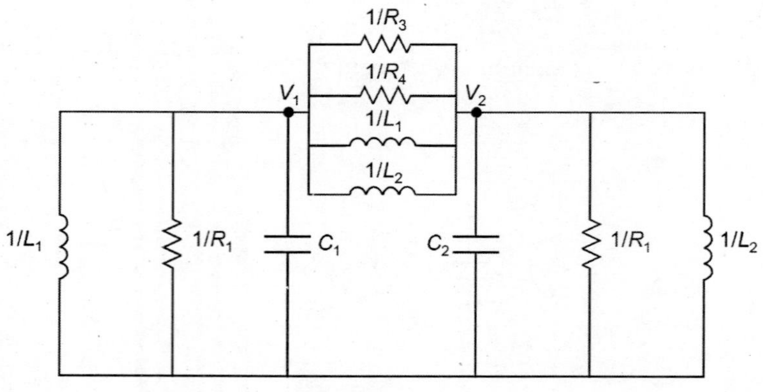

Fig. 32

Q.18 Write the differential equations for the rotational mechanical system. Also draw the torque-voltage and torque-current analogous circuits.

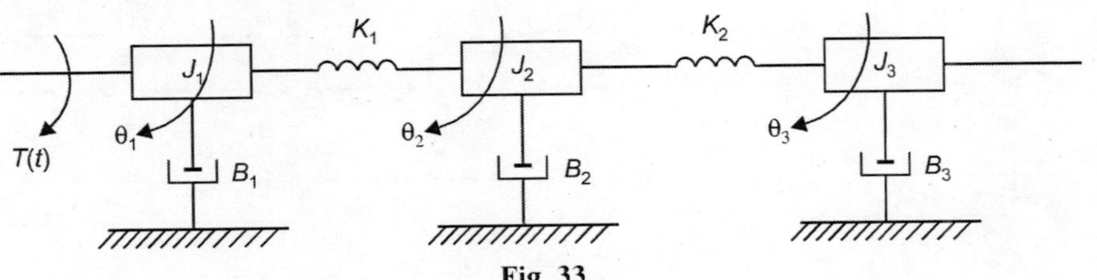

Fig. 33

Ans. (i) The differential equations are,

$$\theta_1(s)\,[J_1s^2 + B_1s + K_1] - K_1\,\theta_2(s) = T(s) \quad (1)$$

$$\theta_2(s)\,[J_2s^2 + B_2s + (K_1 + K_2] - K_1\,\theta_1(s) - K_2\theta_3(s) = 0 \quad (2)$$

$$\theta_3(s)\,[J_3s^2 + B_3s + K_2] - K_2\,\theta_2(s) = 0 \quad (3)$$

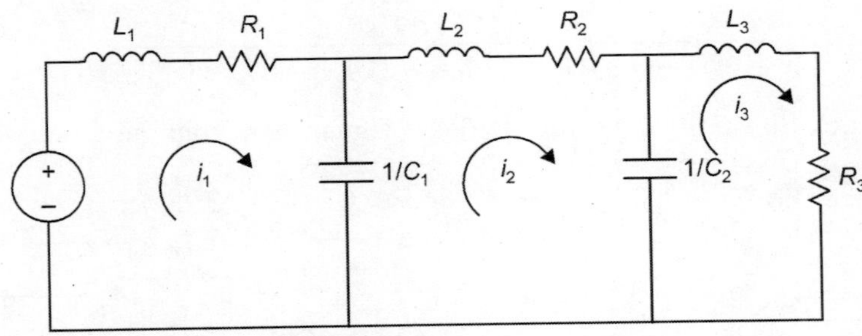

Fig. 34: Force analogous circuit voltage.

(ii) Force-current analogous circuit

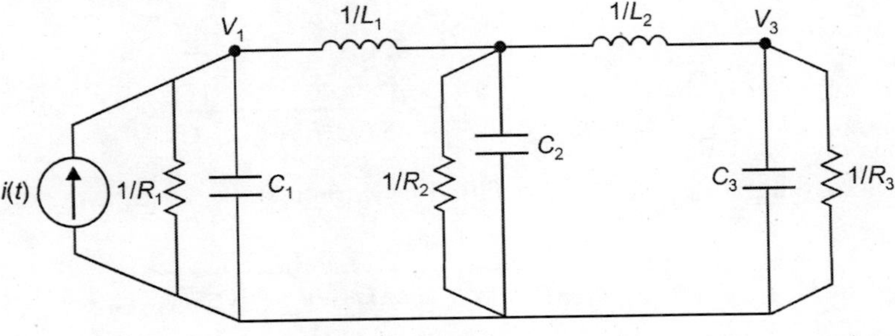

Fig. 35

Q.19 Use block diagram reduction technique to find out *C*/*R*.

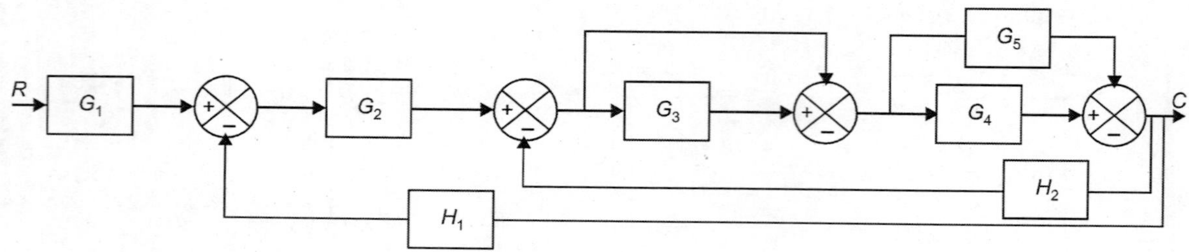

Fig. 36

Ans.
$$\frac{G_1G_2(1+G_3)(G_4+G_5)}{[1+(1+G_3)(G_4+G_5)H_2+(1+G_3)(G_4+G_5)G_2H_2]}$$

Q.20 Convert the block diagram shown in Fig. (36) to signal flow graph and determine the transfer function.

Ans.
$$\frac{G_1G_2(1+G_3)(G_4+G_5)}{[1+(1+G_3)(G_4+G_5)H_2+(1+G_3)(G_4+G_5)G_2H_1]}$$

Q.21 Convert the block diagram shown in Fig. 37 to signal flow graph and determine the transfer function.

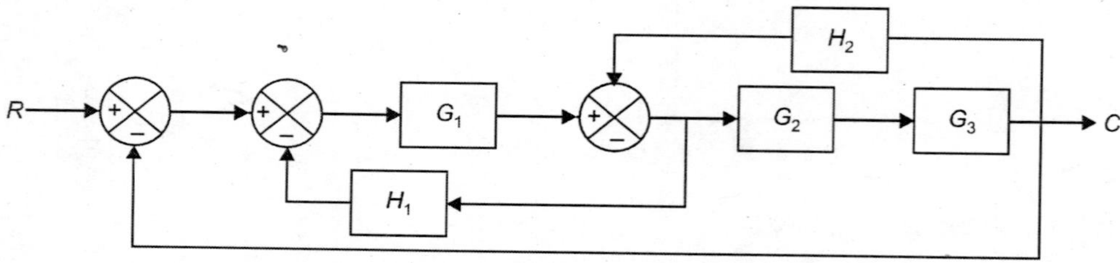

Fig. 37

Ans.
$$C/R=\frac{G_1G_2G_3}{1+G_2G_3H_2+G_1G_2H_1+G_1G_2G_3}$$

Q.22 Convert the block diagram in Fig. (37) to signal flowgraph and determine the transfer function.

Ans.
$$C/R=\frac{G_1G_2G_3}{1+G_2G_3H_2+G_1G_2H_1+G_1G_2G_3}$$

Q.23 Convert the signal flowgraph shown in Fig. (38) into block diagram and determine the transfer function.

Ans.
$$\frac{C(s)}{R(s)}=\frac{[G_1G_2G_3G_4][1-(G_7G_{12}+G_6G_7G_8G_{13})+G_7G_{12}G_{13}]}{1-\{G_2G_7G_{10}+G_{10}G_{11}+G_2G_3G_5G_6G_9+G_3G_5G_6G_{11}+G_7G_{12}+G_6G_7G_8+G_{13}\}+ G_2G_9G_{10}G_{13}+G_{10}G_{13}G_{11}G_{13}+G_7G_{12}G_{13}}$$

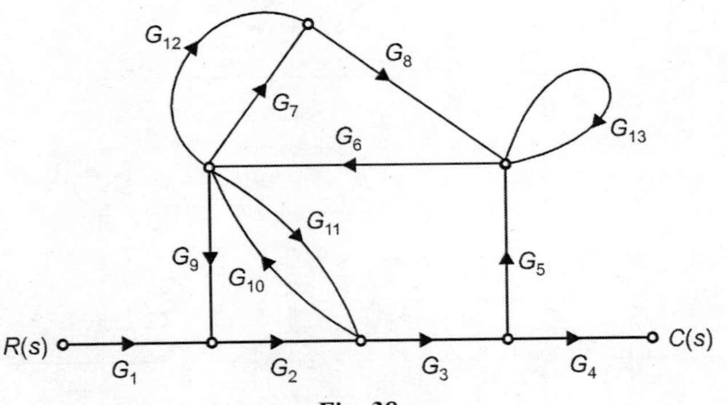

Fig. 38

Q.24 Find transfer function of network shown in Fig. 9.39.

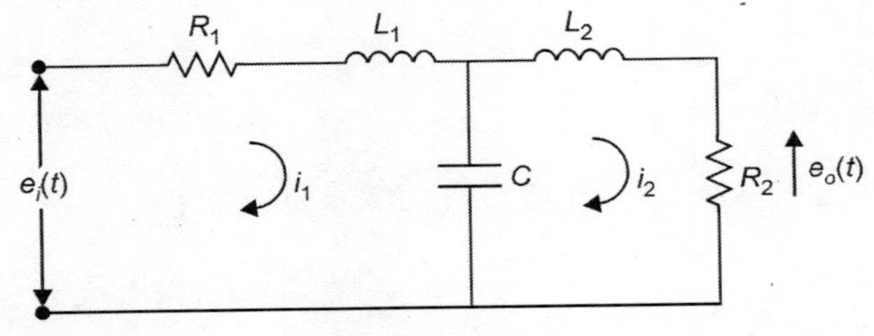

Fig. 39

Ans.
$$\frac{E_o(s)}{E_i(s)} = \frac{sR_2C}{(s^2L_1C + sR_1C + 1)\,Cs^2L_2C + sR_2C + 1) - 1}$$

Q.25 Obtain $E_o(s)/E_i(s)$

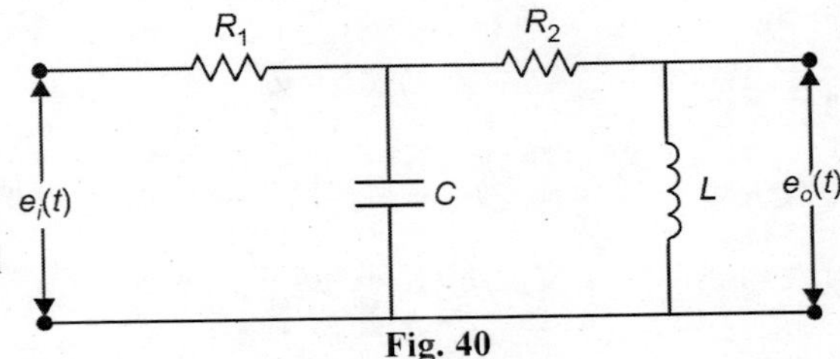

Fig. 40

Ans.
$$\frac{E_o(s)}{E_i(s)} = \frac{sL_2C}{(sR_1C + 1)\,[s^2(L_1 + L_2)]}$$

Q.26 Draw the block diagram for the given signal flowgraph.

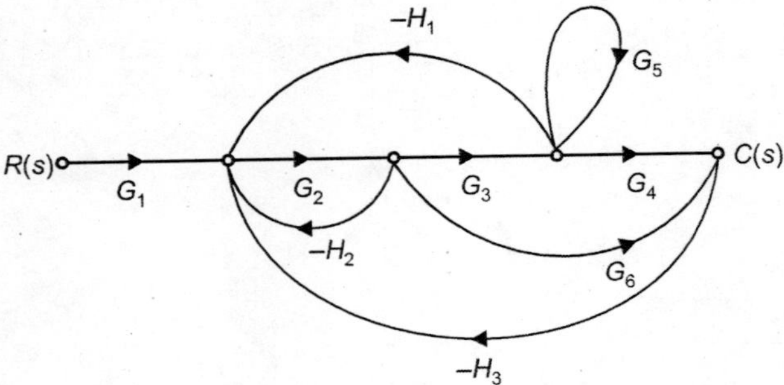

Fig. 41

Ans.

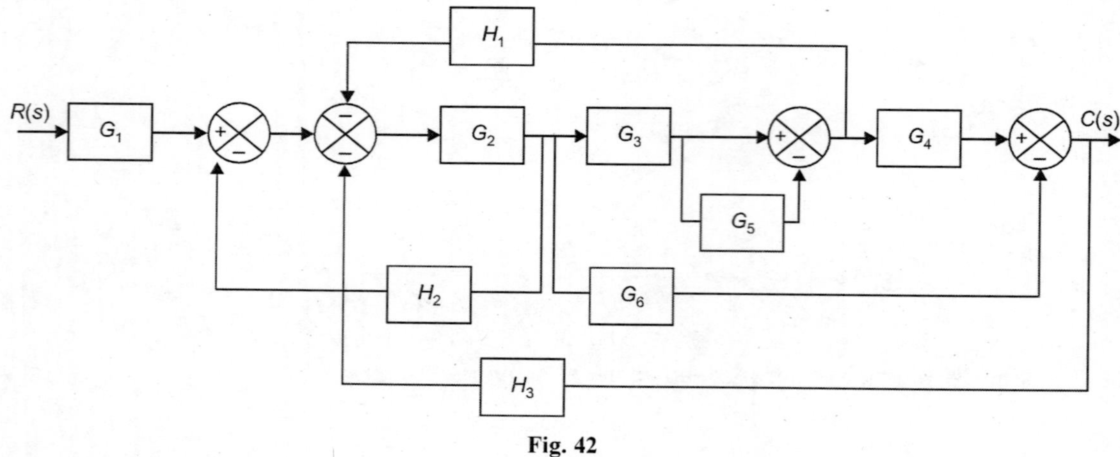

Fig. 42

SOME WORKED EXAMPLES

EXAMPLE 1

For the mechanical system shown in Fig. (1) below find $\frac{X(s)}{F(s)}$.

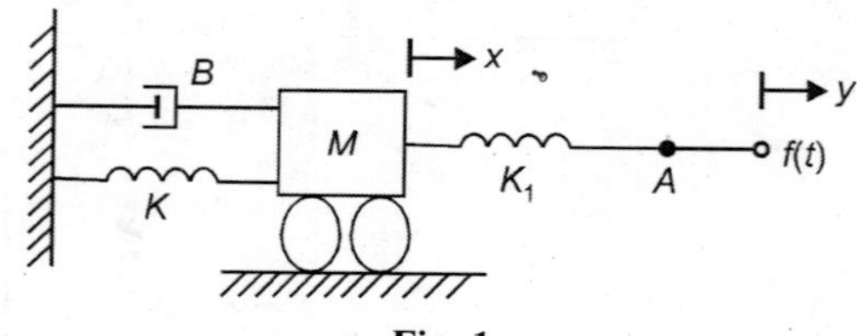

Fig. 1

Solution: The forces which are acting on mass M is f_m, f_b, f_k and f_{k1}.

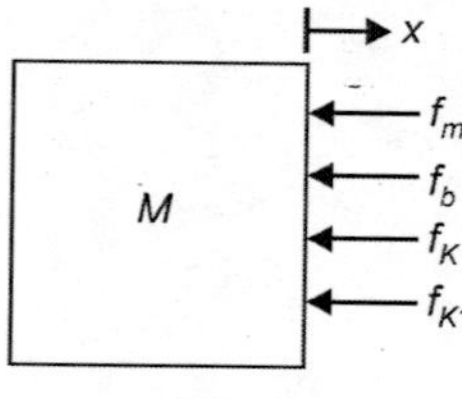

Fig. 2

$$f_m \propto d^2x/dt^2 \Rightarrow f_m = M\frac{d^2x}{dt^2}$$

$$f_b \propto dx/dt \Rightarrow f_b = B\frac{dx}{dt}$$

$$f_k \propto x \Rightarrow f_k = K_x$$

$$f_{k1} \propto (x - y) \Rightarrow f_{k1} = K_1(x - y)$$

Applied force on M is zero, so the force balance equation becomes.

$$0 = f_m + f_b + f_k + f_{k1}$$

$$0 = M\frac{d^2x}{dt^2} + B\frac{dx}{dt} + Kx + K_1(x - y)$$

on taking laplace transform,

$$0 = Ms^2X(s) + Bs\,X(s) + KX(s) + K_1\,X(s) - K_1\,Y(s) \qquad (1)$$

At point A, the force $f(t)$ is applied so we can draw the diagram at point A,

$$f_{k1} \longleftarrow \overset{A}{\bullet} \longrightarrow f(t) \qquad \vdash\!\to x \qquad \vdash\!\to y$$

Fig. 3

For the above Fig. (3), the force balance equation is,

$$f_{k1} \propto (y - x)$$

$$f_{k1} = K_1(y - x)$$

We know applied force at point A is $f(t)$

$$\therefore \quad f(t) = f_{k1}$$

$$f(t) = K_1(y - x)$$

on taking laplace transform,

$$F(s) = K_1\,Y(s) - K_1\,X(s) \qquad (2)$$

Substituting equation (2) in (1) we obtain,

$$\boxed{\frac{X(s)}{F(s)} = \frac{1}{(Ms^2 + Bs + K)}} \qquad (3)$$

Result: The equation (3) is the required transfer function.

EXAMPLE 2

Determine the transfer function of the circuit given below.

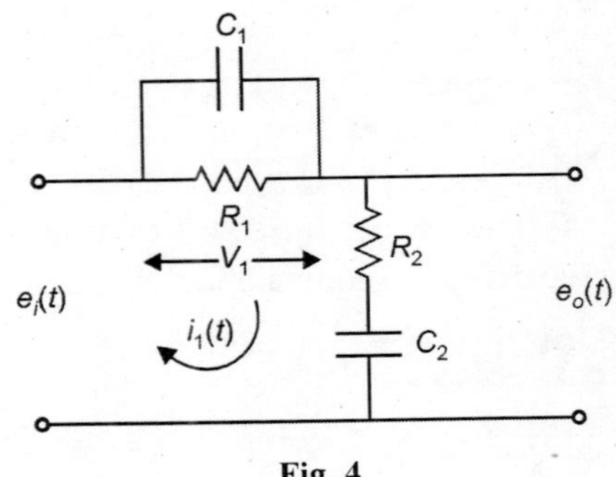

Fig. 4

Solution: The required transfer function is $\dfrac{E_o(s)}{E_i(s)}$

From above Fig. (4), we can write,

$$e_o(t) = R_2 i_1(t) + \frac{1}{C_2}\int i_2(t)$$

on taking laplace transform

$$E_o(s) = R_2\, I_1(s) + \frac{I_1(s)}{s\,C_2}$$

$$E_o(s) = I_1(s)\left[R_2 + \frac{1}{sC_2}\right]$$

$$E_o(s) = I_1(s)\left[\frac{sR_2\,C_2 + 1}{s\,C_2}\right] \quad (1)$$

Similarly for $e_i(t)$, $$E_i(s) = I_1(s)\left[\left(\frac{R_1}{R_1\,C_1\,s+1}\right) + \left(\frac{R_2\,C_2 s + 1}{s\,C_2}\right)\right] \quad (2)$$

Divide equation (1) by (2)

$$\boxed{\frac{E_o(s)}{E_i(s)} = \frac{(1+R_1C_1s)\,(1+R_2C_2s)}{(1+R_1C_1s)\,(1+R_2C_2s)+R_1C_2s}} \quad (3)$$

Result: The equation (3) is the required transfer function.

EXAMPLE 3

In an electrical circuit, the elements resistance, capacitance and inductance are connected in parallel across the voltage source E. Draw the translational mechanical analogous circuit.

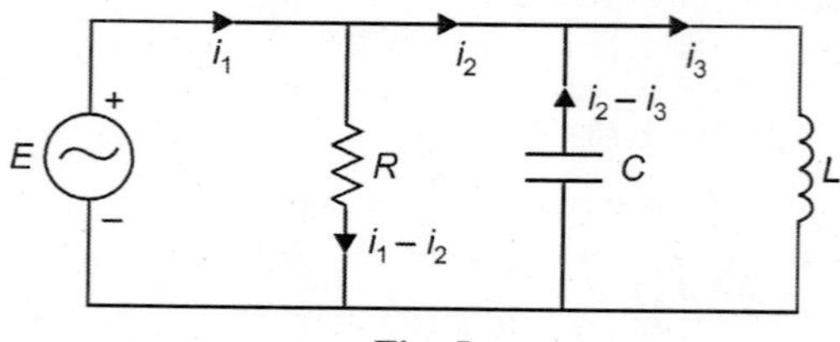

Fig. 5

Solution: This is the reverse procedure of force-voltage analogous (i.e.,) voltage-force analogous.

- E is analogous to $f \rightarrow$ (i.e) force
- R is analogous to $B \rightarrow$ (i.e) dash pot
- C is analogous to $1/K \rightarrow$ (i.e) spring
- L is analogous to $M \rightarrow$ (i.e) mass
- i_1 is analogous to $x_1 \rightarrow$ (i.e) displacement
- i_2 is analogous to $x_2 \rightarrow$ (i.e) displacement
- i_3 is analogous to $x_3 \rightarrow$ (i.e) displacement

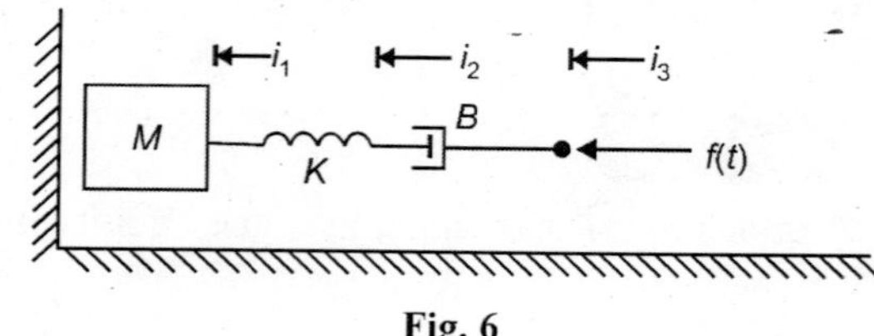

Fig. 6

by replacing the equivalent electrical to mechanical analogous, the following circuit to obtained.

EXAMPLE 4

Convert the below signal flow graph into block diagrams.

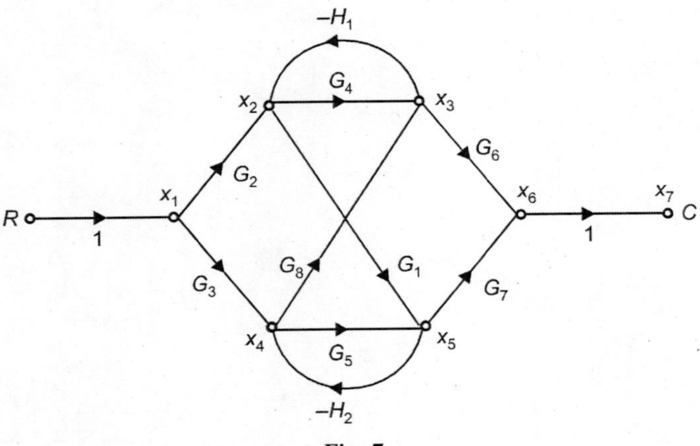

Fig. 7

Solution:

Step 1: Write the system equation for every node.

at node x_1	$x_1 = R(1)$	(1)
at node x_2	$x_2 = -x_3H_1 + x_1G_2$	(2)
at node x_3	$x_3 = x_2G_4 + x_4G_8$	(3)
at node x_4	$x_4 = x_1G_3 - x_5H_2$	(4)
at node x_5	$x_5 = x_2G_1 + x_4G_5$	(5)
at node x_6	$x_6 = x_3G_6 + x_5G_7$	(6)
at node x_7	$x_7 = x_6(1)$	(7)

Step 2: Combine all the equations for drawing block diagram by combining the equations (1) to (7), let draw the block diagram.

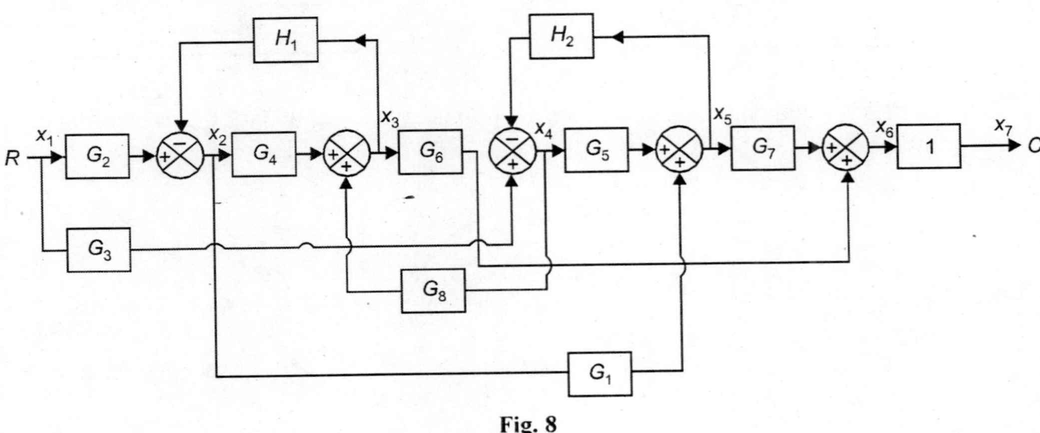

Fig. 8

The above Fig. (8) is the complete block diagram for the given signal flow graph.

EXAMPLE 5

Convert the given block diagram into signal flow graph:

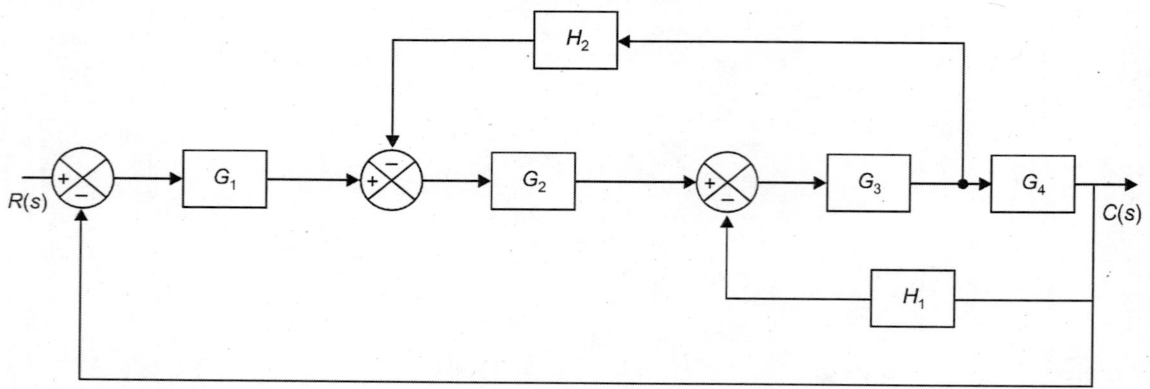

Fig. 9

Solution:

Step 1: Assign node number of input, output, summing points and branch points.

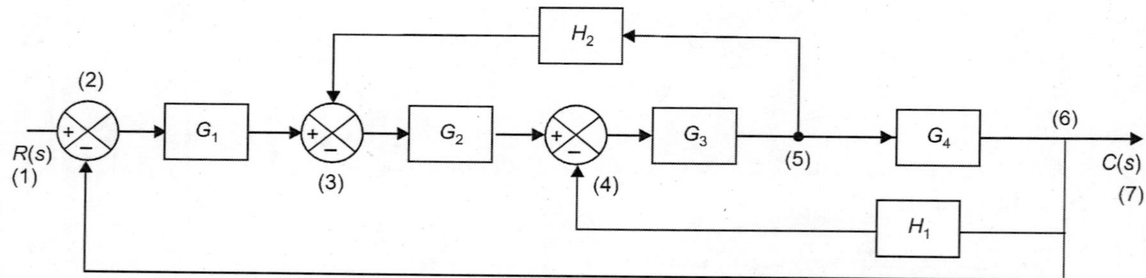

Fig. 10

Step 2: Replace i/p, o/p, branch points and summing points by node and mark the transmittance from one node to another node.

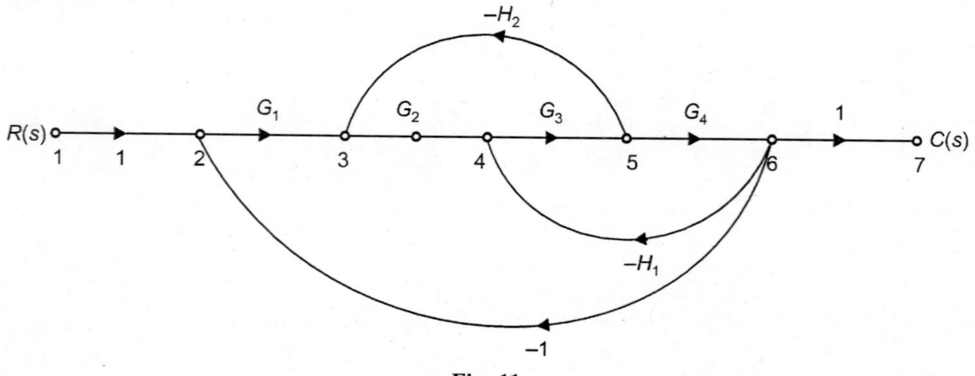

Fig. 11

EXAMPLE 6

Derive the transfer function of the given signal flow graph.

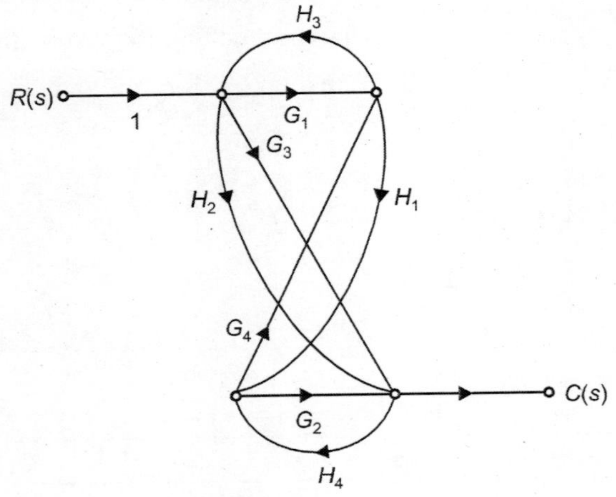

Fig. 12

Solution:

Step 1: The number of forward paths, $K = 2$

$$P_1 = G_3 \qquad\qquad P_2 = G_1\, G_2\, H_1$$

Fig. 13

Fig. 14

Step 2: The total number of individual loops are 4.

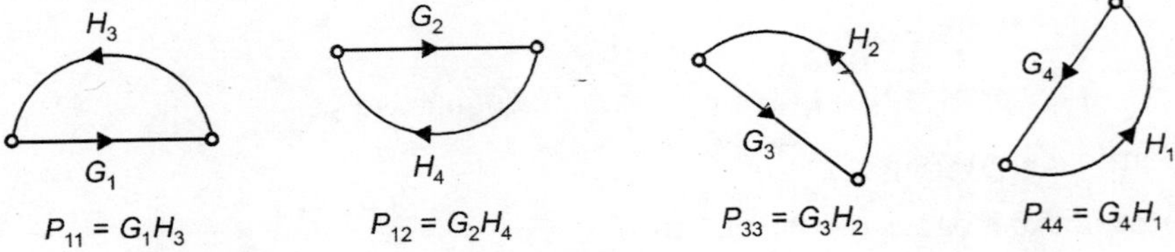

Fig. 15

Step 3: The total number of non touching loops are

$$P_{12} = G_1 G_2\, H_3\, H_4 \text{[combine 1st and 2nd loops]}$$

$$P_{34} = G_3\, G_4\, H_1\, H_2 \text{ [combine 3}^{\text{rd}}\text{ and 4}^{\text{th}}\text{ loop]}$$

Step 4: Find Δ

$$\Delta = 1 - \text{[sum of individual loop gain]} + \text{[sum of two non touching loop pairs]} \ldots$$

$$\Delta = 1 - [P_{11} + P_{22} + P_{33} + P_{44}] + [P_{12} + P_{34}]$$

$$\therefore \quad \Delta = 1 - [G_1H_3 + G_2H_2 + G_3H_2 + G_4H_1] + [G_1G_2H_3H_4 + G_3G_4H_1H_2]$$

Step 5: Find Δ_K

$\Delta_2 = 1 - 0 = 1$ [No non touching loop for path 2]

$\Delta_1 = 1 - G_4H_1$ [The loop G_1H_1 is not touching with path 1]

The mason's gain formula,

$$T(s) = \frac{1}{\Delta}\Sigma P_k \Delta_k$$

where $K = 2$

$$T(s) = \frac{1}{\Delta}[P_1\Delta_1 + P_2\Delta_2]$$

$$\boxed{T(s) = \frac{1[G_3(1 - G_4H_1) + G_1G_2H_1(1)}{[1 - (G_1H_3 + G_2H_4 + G_3H_2 + G_4H_1) + (G_1G_2H_3H_4 + G_3G_4H_1H_2)]}}$$

EXAMPLE 7

For the mechanical system shown in below Fig. (16), draw the force-voltage analogous circuit.

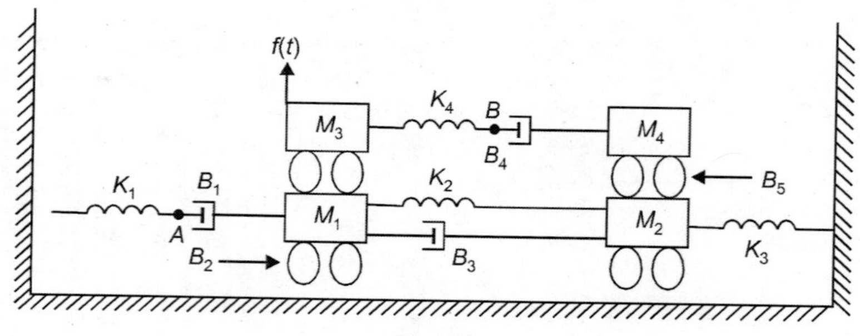

Fig. 16

Solution: Force-voltage analogous: In force voltage analogous we know

→ force is analogous to voltage

→ Displacement is analogous to loop currents

→ Mass is analogous to Inductance

→ Dash potis analogous to Resistance

→ Spring is analogous to reciprocal of capacitance

by use of above points,

$M_1 \rightarrow L_1$	$B_1 \rightarrow R_1$	$K_1 \rightarrow 1/C_1$	$x_1 \rightarrow i_1$	
$M_2 \rightarrow L_2$	$B_2 \rightarrow R_2$	$K_3 \rightarrow 1/C_2$	$x_2 \rightarrow i_2$	
$M_3 \rightarrow L_3$	$B_3 \rightarrow R_3$	$K_3 \rightarrow 1/C_3$	$x_3 \rightarrow i_3$	$f(t) = 0(t)$
$M_4 \rightarrow L_4$	$B_4 \rightarrow R_4$	$K_4 \rightarrow 1/C_4$	$x_4 \rightarrow i_4$	
	$B_5 \rightarrow R_5$		$x_5 \rightarrow i_5$	
			$x_6 \rightarrow i_6$	

Let draw the circuit,

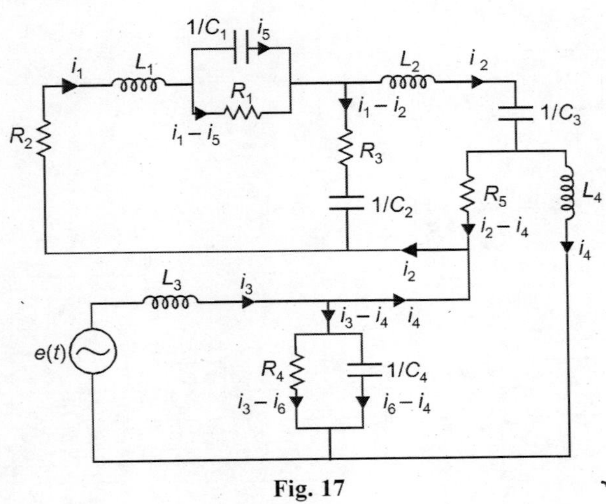

Fig. 17

EXAMPLE 8

Find the transfer function of the system using block diagram reduction for the Fig. shown below.

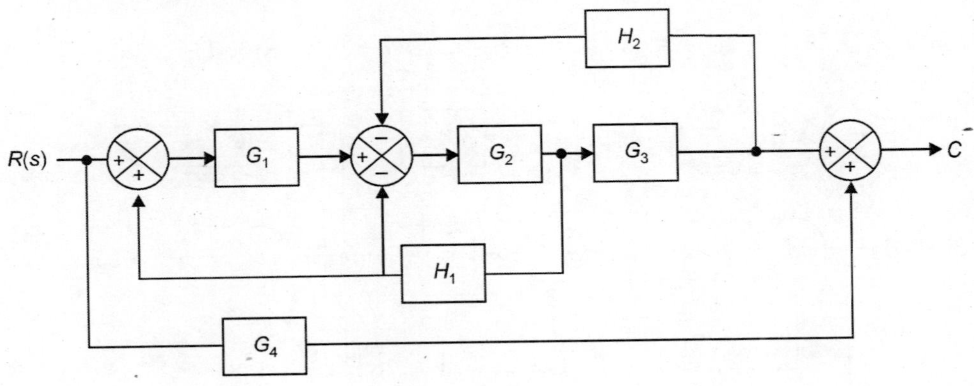

Fig. 18

Solution:

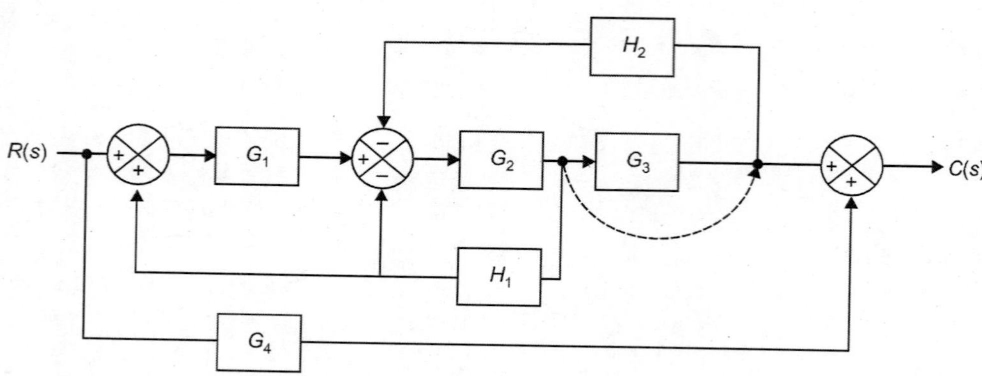

Fig. 19

Step 1: Shift the branch point to right of gain G_3. (Rule 3) (*BR* $1/G$)

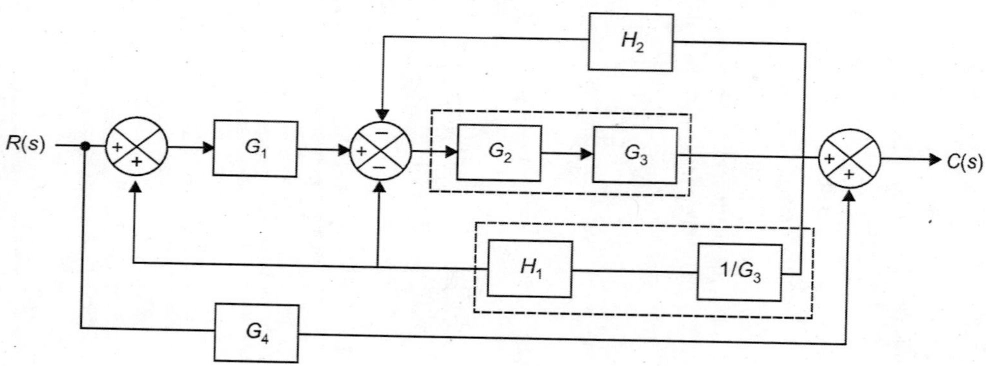

Fig. 20

Step 2: Multiply G_2 and G_3 block then multiply H_1 and $1/G_3$ blocks (Rule 1).

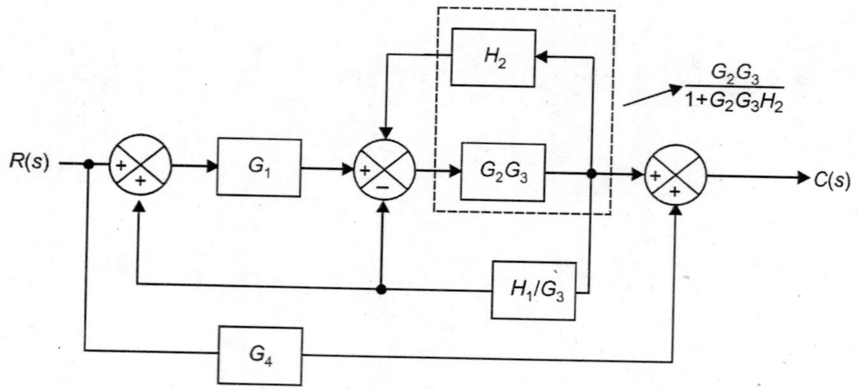

Fig. 21

Step 3: Combine G_2G_3 and H_2 blocks by substituting in equation $\frac{G}{1+GH}$ (Rule 10) ($\because$ closed loop).

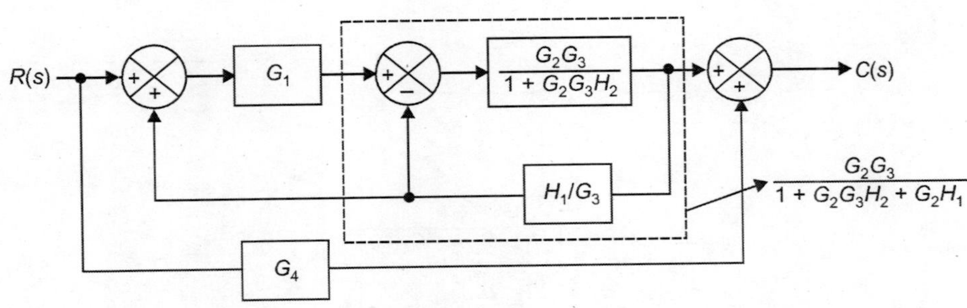

Fig. 22

Step 4: Combine $\frac{G_2G_3}{1+G_2G_3H_2+G_2H_1}$ and G_1 block (Rule 1)

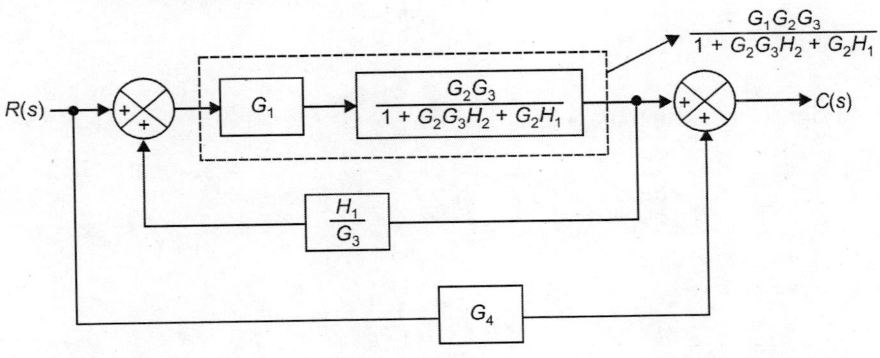

Fig. 23

Step 5: Combine the blocks H_1/G_3 and $\frac{G_2G_3}{1+G_2G_3H_2+G_2H_1}$ by rule 10.

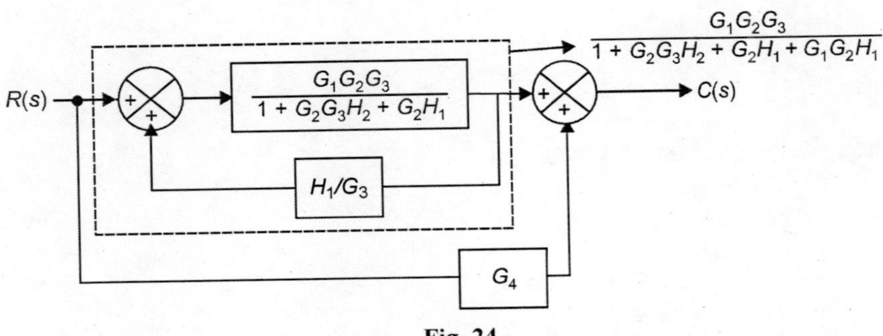

Fig. 24

Step 6: Combine the blocks using rule 2.

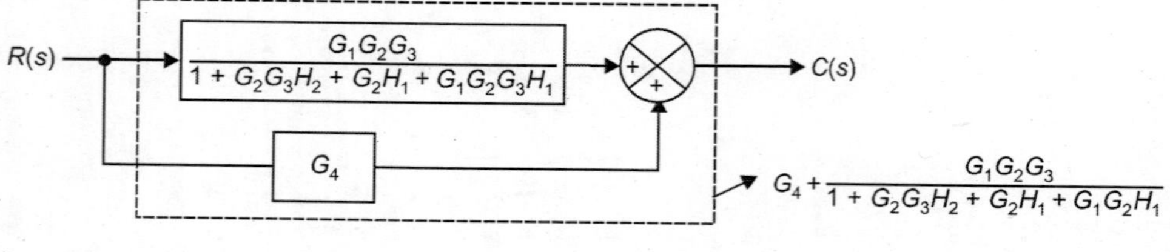

Fig. 25

R(s) → $G_4 + \frac{G_1G_2G_3}{1 + G_2G_3H_2 + G_2H_1 + G_1G_2H_1}$ → C(s)

Fig. 26

the required transfer function is,

$$\frac{C(s)}{R(s)} = G_4 + \left[\frac{G_1G_2G_3}{1 + G_2G_3H_2 + G_2H_1 + G_1G_2H_1}\right]$$

EXAMPLE 9

Convert the given signal flow graph into block diagram.

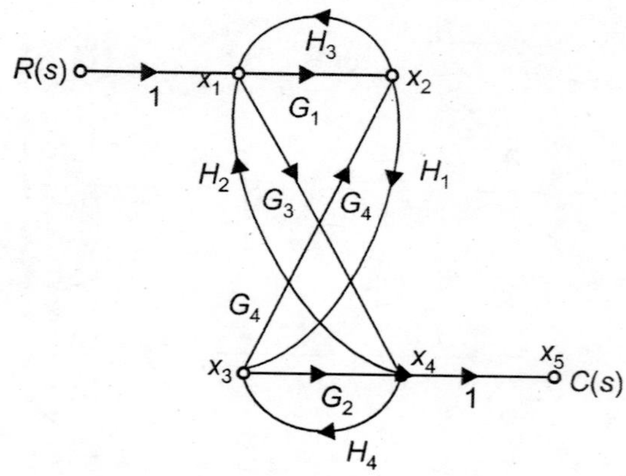

Fig. 27

Solution:

Step 1: Write the system equation for every node.

at node x_1 $\quad x_1 = R(s) + x_2H_3 + x_4H_2 \quad$ (1)

at node x_2 $\quad x_2 = x_1G_1 + x_3G_4 \quad$ (2)

at node x_3 $\quad x_3 = x_2H_1 + x_4H_4 \quad$ (3)

at node x_4 $\quad x_4 = x_1G_3 + x_3G_2 \quad$ (4)

at node x_5 $\quad x_5 = x_4(1) \quad$ (5)

Step 2: By combining the equations (1) to (5) draw the block diagram.

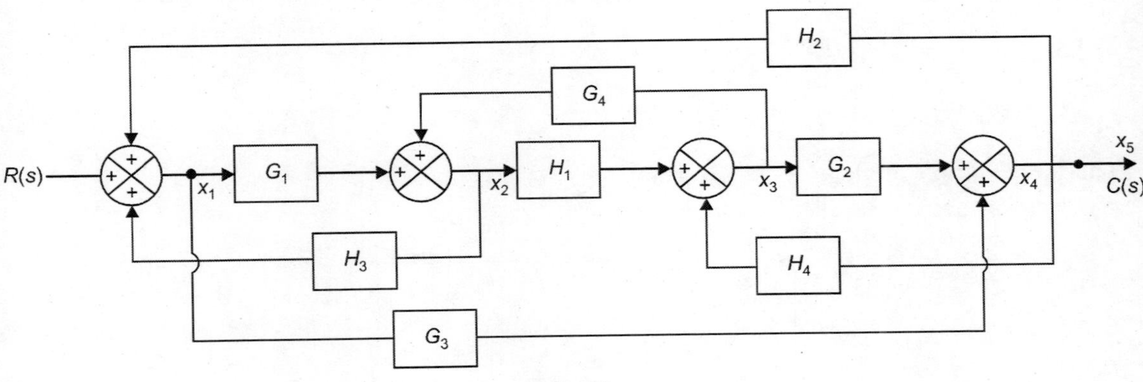

Fig. 28

The above block diagram is the required block diagram which is derived from the given signal flow graph.

EXAMPLE 10

For the mechanical system shown in Fig. (29), derive the transfer function. Also draw the force-voltage and force-current analogous circuits.

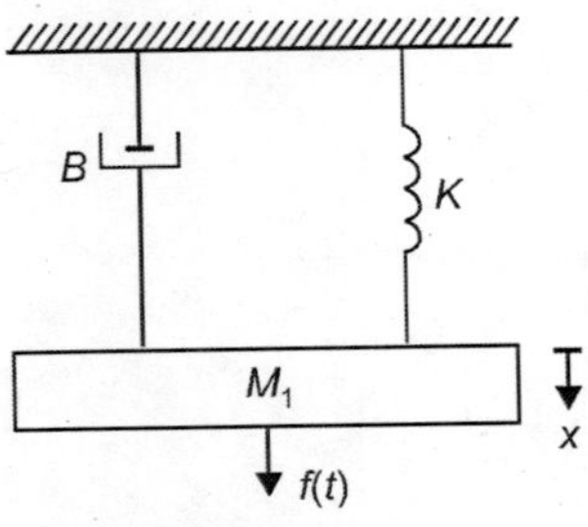

Fig. 29

Solution:

$$f_{m1} \propto d^2x/dt^2 = M_1 \frac{d^2x}{dt^2}$$

$$f_b \propto dx/dt = B\frac{dx}{dt}$$

$$f_k \propto x = Kx$$

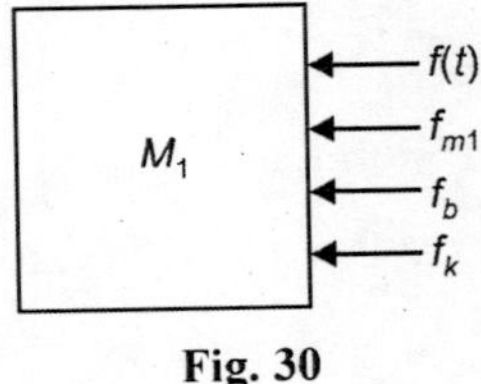

Fig. 30

By Newtons' second law,

the force balance equation is,

$$f(t) = f_{m1} + f_b + f_k$$

$$f(t) = M_1 \frac{d^2x}{dt^2} + B\frac{dx}{dt} + K_1 s$$

on taking laplace transform,

$$F(s) = M_1 s^2 X(s) + Bs\, X(s) + K\, X(s)$$

$$F(s) = X(s)\, [M_1 s^2 + Bs + K]$$

$$\therefore \quad \frac{X(s)}{F(s)} = \frac{1}{[M_1 s^2 + Bs + K]} \quad (1)$$

The force voltage analogous circuit (refer Fig. 31)

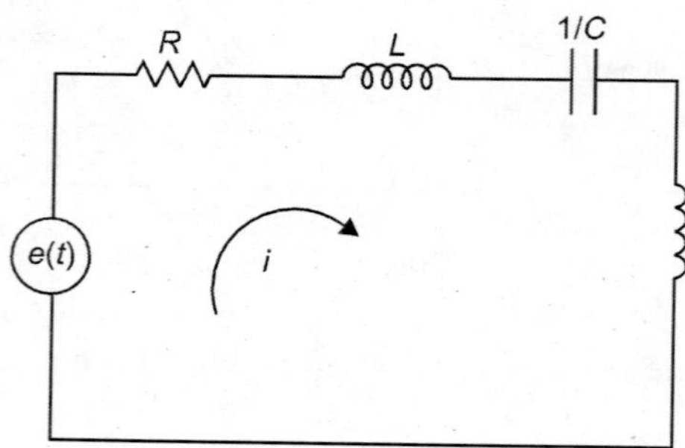

Fig. 31

In force-voltage analogy;

$$M \rightarrow L$$
$$B \rightarrow R$$
$$K \rightarrow 1/C$$
$$f(t) \rightarrow V$$
$$x \text{ (or) } y \rightarrow i$$

The force current analogous circuit (refer Fig. 32)

In force current analogy.

$$M \rightarrow C$$
$$B \rightarrow 1/R$$
$$K \rightarrow 1/L$$
$$f(t) \rightarrow i$$
$$x \text{ (or) } y \rightarrow V$$

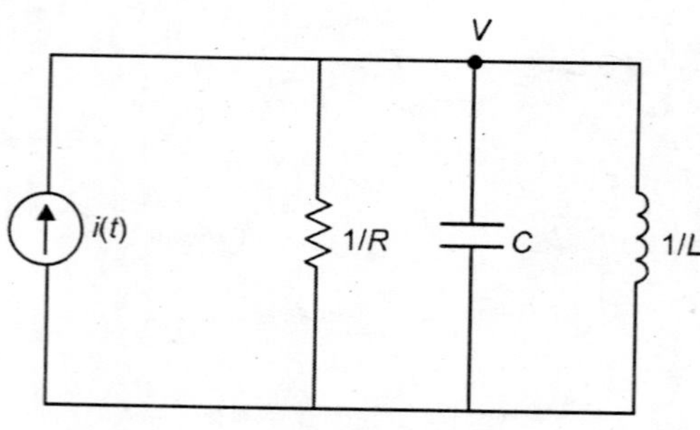

Fig. 32

EXAMPLE 11

Write the circuit differential equations governing the mechanical system. Also draw the force voltage and force current analogous circuit.

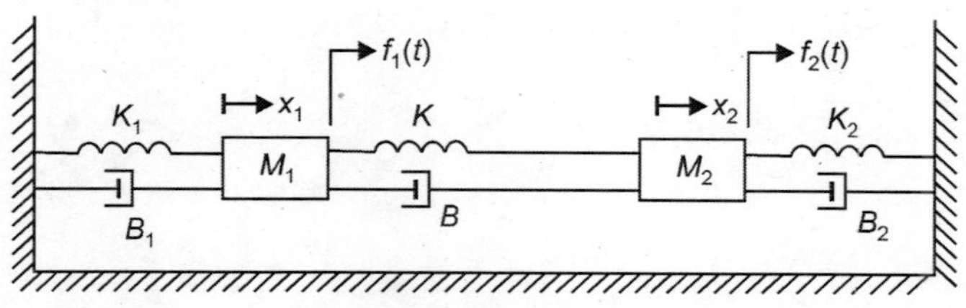

Fig. 33

Solution: (i) Differential equation

For mass M,

Fig. 34

where,

$$f_{m1} \propto \frac{d^2x_1}{dt^2} \Rightarrow M_1\frac{d^2x_1}{dt^2}$$

$$f_{k1} \propto x_1 \Rightarrow K_1x_1$$

$$f_{b1} \propto \frac{dx_1}{dt} \Rightarrow B_1\frac{dx_1}{dt}$$

$$f_k \propto (x_1 - x_2) \Rightarrow K(x_1 - x_2)$$

$$f_b \propto \frac{d}{dt}(x_1 - x_2) \Rightarrow B\frac{d}{dt}(x_1 - x_2)$$

The force balance equation is,

$$f_1(t) = f_{m1} + f_{b1} + f_{k1} + f_b + f_k$$

$$\therefore \quad f_1(t) = M_1\frac{d^2x_1}{dt^2} + B_1\frac{dx_1}{dt^2} + K_1x_1 + B\frac{d}{dt}(x_1 - x_2) + K(x_1 - x_2) \quad (1)$$

For mass M_2

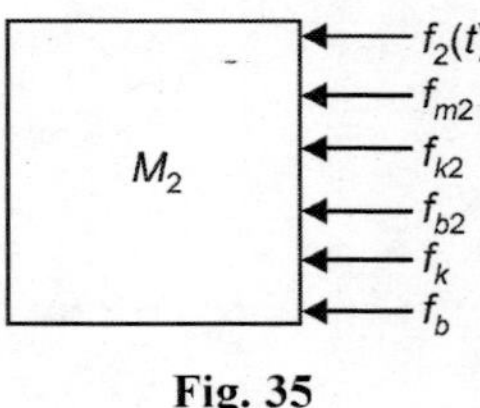

Fig. 35

where,

$$f_{m2} \propto \frac{d^2x_2}{dt^2} \Rightarrow M_2\frac{d^2x_2}{dt^2}$$

$$f_{b2} \propto dx_2/dt \Rightarrow B_2\frac{dx_2}{dt}$$

$$f_{k2} \propto x_2 \Rightarrow K_2x_2$$

$$f_b \propto \frac{d}{dt}(x_2 - x_1) \Rightarrow B\frac{d}{dt}(x_2 - x_1)$$

$$f_K \propto (x_2 - x_1) \Rightarrow K(x_2 - x_l)$$

The force balance equation is,

$$f_2(t) = f_{m2} + f_{b2} + f_{K2} + f_b + f_K$$

$$\therefore \qquad f_2(t) = M_2 \frac{d^2x_2}{dt^2} + B_2 \frac{dx_2}{dt^2} + K_2x_2 + B\frac{d}{dt}(x_2 - x_1) + K(x_2 - x_1) \qquad (2)$$

(ii) Force-voltage analogous circuit:

$$M_1 \rightarrow L_1$$
$$M_2 \rightarrow L_2$$
$$K_1 \rightarrow 1/C_1$$
$$B_1 \rightarrow R_1$$
$$K_2 \rightarrow 1/C_2$$
$$B_2 \rightarrow R_2$$
$$K \rightarrow 1/C$$
$$B \rightarrow R$$
$$f_1(t) \rightarrow V_1(t)$$
$$f_2(t) \rightarrow V_2(t)$$
$$x_1 \rightarrow i_1(t)$$
$$x_2 \rightarrow i_2(t)$$

so, let draw the force-voltage analogous circuit for the Fig. (36).

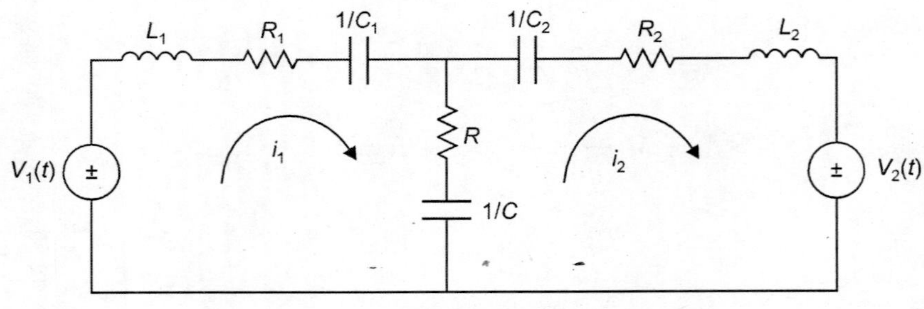

Fig. 36

(iii) Force-current analogous circuit

$M_1 \rightarrow C_1$	$K_2 \rightarrow 1/L_2$	$f_1(t) \rightarrow i_1(t)$
$M_2 \rightarrow C_2$	$B_2 \rightarrow 1/R_2$	$f_2(t) \rightarrow i_2(t)$
$K_1 \rightarrow 1/L_1$	$K \rightarrow 1/L$	$x_1 \rightarrow V_1(t)$
$B_1 \rightarrow 1/R_1$	$B \rightarrow 1/R$	$x_2 \rightarrow V_2(t)$

so, let draw the force-current analogous

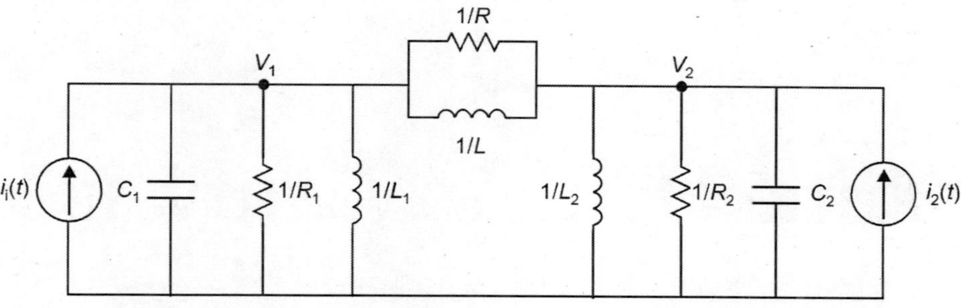

2

Time Domain Analysis

2.1 INTRODUCTION

Any system containing energy storing element like inductor, capacitor, mass and inertia etc. Posses certain energy. These energy storing elements are the part of the control system and cannot be avoided. If the energy state of the system is distributed then it takes a certain time to change from one state to another state. This disturbance sometimes occurs at input, sometimes occurs at output and sometimes at both ends. The time required to change from one state to another state is known as transient time and the values of current and voltages during this period is called transient response. These transient may have oscillations which may be either sustained or decaying in nature. This will depend upon the parameters of the system. For any system we obtain a linear differential equation. The solution of linear differential equation gives the response of the system.

The time response of the system is the output of the closed loop system as a function of time. It is denoted by *C*(*t*). The time response can be obtained by solving the differential equation governing the system. Alternatively, the response *C*(*t*) can be obtained from the transfer function of the system and the input to the system.

The output in *s*-domain, *C*(*s*) is given by the product of the transfer function and the input, *R*(*s*). On taking inverse Laplace transform of this product the time domain response, *C*(*t*) can be obtained.

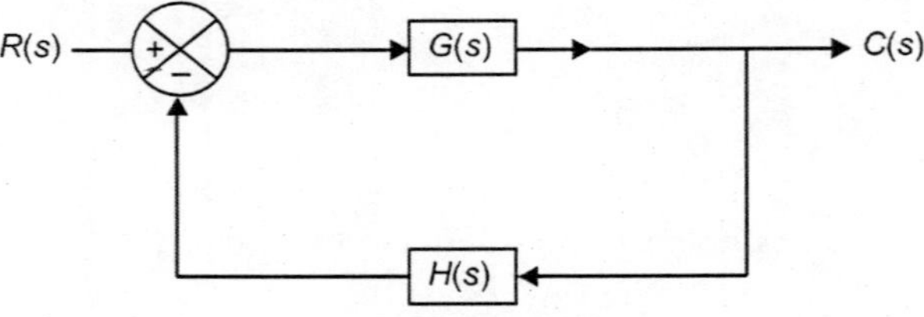

Fig. 2.1: Closed loop.

The closed loop transferfunction,

$$\frac{C(s)}{R(s)} = \frac{G(s)}{1 + G(s)\,H(s)} \tag{2.1}$$

Response in s-domain, $C(s) = R(s)\dfrac{G(s)}{1+G(s)\,H(s)}$

$$C(t) = L^{-1}\left[R(s)\frac{C(s)}{1+G(s)\,H(s)}\right] \tag{2.2}$$

The time response of a control system consists of two parts: the transient and the steady state response. The transient response of the system shows the response of the system when the input changes from one state to another state.

2.2 TEST SIGNALS

The knowledge of input signal is required to predict the response of a system. In most of the system the input signals are not known ahead of time and also it is difficult to express the input signal mathematically by simple equation. The characteristics of actual input signals are a sudden shock, a sudden changes a constant velocity and a constant acceleration. Hence test signals which resembles these characteristics are used as input signals to predict the performance of the system. The commonly used test input signals are impulse, step, ramp, and parabolic signals.

The standard test signals are

1. (a) Step Signal (b) Unit Step Signal
2. (a) Ramp Signal (b) Unit Ramp Signal
3. (a) Parabolic Signal (b) Unit Parabolic Signal
4. (a) Impulse Signal (b) Unit Impulse Signal

Since the test signals are simple functions of time, they can be easily generated in laboratories. The mathematical and experimental analysis of control system using these signals can be carried out easily. The use of the test signals can be justified because of a correlation existing between the response characteristics of a system to a test input signal and capability of the system to cope with actual input signals.

Step Signal: The step signal is a signal whose value changes from zero to A at $t = 0$ and remains constant at A for $t > 0$. The step signal resembles an actual steady input to a system. A special case of step signal is unit step in which A is unity. The mathematical representation of the step signal is

$$r(t) = Au(t)$$

where

$$u(t) = 1; \quad t \geq 0$$

$$u(t) = 0; \quad t < 0$$

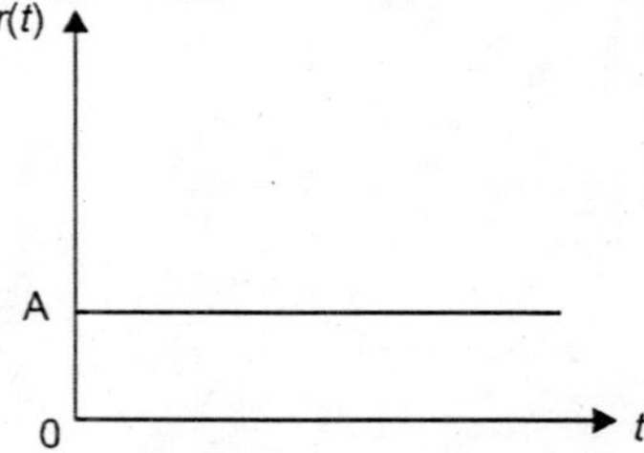

Fig. 2.2: Ramp signal.

Let $u(t)$ be defined in the interval $0 \leq t \leq \infty$, Laplace transform is obtained by multiplying $u(t)$ by e^{-st} and integrate between the limit 0 to ∞.

$$L[u(t)] = \int_0^{\infty} u(t)\, e^{-st}\, dt$$

$$= \int_0^\infty 1 \cdot e^{-st}\, dt = \left[\frac{e^{-st}}{-s}\right]_0^\infty = \frac{1}{s}$$

Step function is also called displacement function. Step function can be described as sudden application of input signal to a system.

If input is $R(s)$; then

$$R(s) = \frac{1}{s}$$

Ramp signal: The ramp signal is a signal whose value increase linearly with time from an initial value of zero at $t = 0$. The ramp signal reasembles a constant velocity input to the signal in which the value of A is unity. The mathematical representation of the ramp signal is,

$$r(t) = At; \quad \text{when } t \geq 0$$

$$r(t) = 0, \quad \text{when } t < 0$$

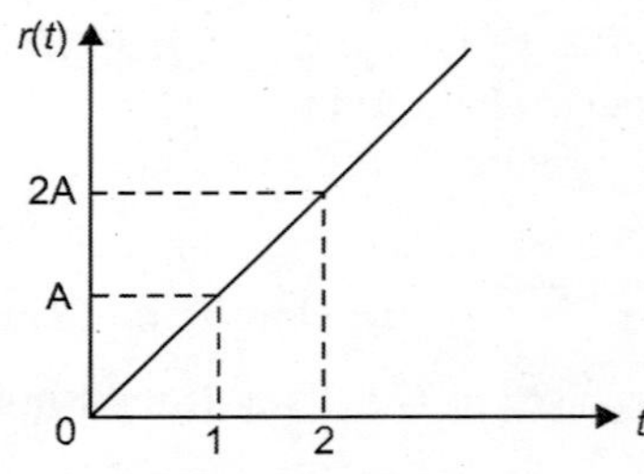

Fig. 2.3: Ramp signal.

Let $r(t)$ be defined in the interval $0 \leq t \leq \infty$ Laplace transform is obtained by multiplying $r(t)$ by e^{-st} and integrate between the limit 0 to ∞.

$$Lr(t) = \int_0^\infty r(t)\, e^{-st}\, dt \qquad \text{where } r(t) = At$$

$$Lr(t) = \int_0^\infty At\, e^{-st}\, dt = \frac{A}{s^2}$$

$$R(s) = \frac{A}{s^2}$$

If A is unity then [i.e. $A = 1$]

$$R(s) = \frac{1}{s^2}$$

Parabolic signal

In parabolic signal, the instantaneous value varies as square of the time from an initial value of zero at $t = 0$. The sketch of the signal with respect to time resembles a parabolas. The parabolic signal resembles a constant acceleration input to the system. A special case of parabolic is unit parabolic signal in which A is unity. the mathematical representation of the parabolic signal is

$$r(t) = \frac{At^2}{2}; \quad t \geq 0$$

$$r(t) = 0; \quad t < 0$$

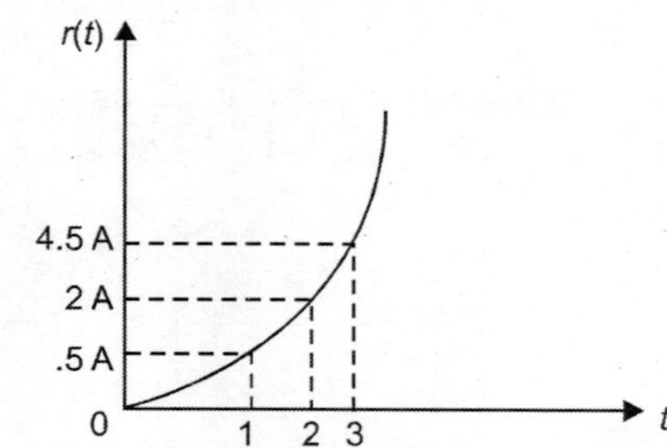

Fig. 2.4: Parabolic signal.

$$Lr(t) = \int_0^{\infty} r(t)\, e^{-st}$$

where $r(t) = \dfrac{At^2}{2}$

$$Lr(t) = \int_0^{\infty} \frac{At^2}{2} e^{-st}\, dt$$

$$Lr(t) = \frac{A}{s^3}$$

$$R(s) = \frac{A}{s^3}$$

For unit parabolic function $A = 1$

So $$R(s) = \frac{1}{s^3}$$

Impulse signal

A signal which is variable for very short duration is called impulse signal Ideal impulse signal is a unit impulse signal which is defined as a signal having zero values at all times expect at $t = 0$. At $t = 0$, the magnitude is infinite. It is denoted by $\delta(t)$ and mathematically it is expressed as

$$\delta(t) = 0 \quad \text{for } t \neq 0$$

and $$\underset{t_1 \to 0}{Lt} \int_{-t_1}^{+t_1} \delta(t)\, dt = 1$$

or $$\delta(t) = \begin{bmatrix} 0; \ t \neq 0 \\ \infty; t = 0 \end{bmatrix} \int_{-\infty}^{\infty} \delta(t)\, dt = 1$$

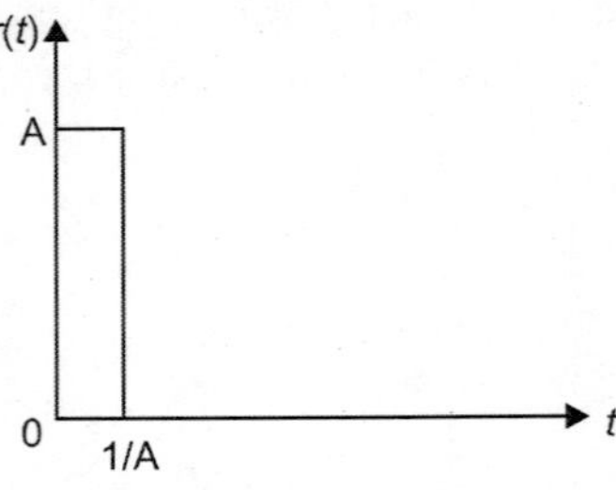

Fig. 2.5: Impulse signal.

Since a perfect impulse cannot be achieved in practice it is usually approximated by a pulse of small width but mathematically an impulse function is the derivative of a step function. Laplace transform of the impulse function is unity.

The response of the system, with input impulse signal is called weighting function of the system. It is given by the inverse Laplace transform of the system transfer function.

$$C(t) = L^{-1}\left[R(s)\frac{G(s)}{1+G(s)\,H(s)}\right] = L^{-1}\left[\frac{G(s)}{1+G(s)\,H(s)}\right]$$

$[\because R(s) = 1 \text{ for impulse}]$

Table 2.1 Standard test signals

Name of the signal	**Time domain equation of signal, $r(t)$**	**Laplace transform of the signal, $R(s)$**
Step *i/p*	A	A/s
Unit step *i/p*	1	$1/s$
Ramp *i/p*	At	A/s^2
Unit Ramp *i/p*	t	$1/s^2$
Parabolic	$At^2/2$	A/s^3
Unit parabolic	t^2/t	$1/s^3$
Impulse	$\delta(t)$	1

2.3 ORDER OF A SYSTEM

The input and output relationship of a control system can be expressed by a differential equation. The order of the system is given by the order of the differential equation governing the system. If the system is governed by *n*th order differential equation, then the system is called *n*th order system. Alternatively, the order can be determined from the transfer function of the system. The transfer function of the system can be obtained by taking Laplace transform of the differential equation governing the system and rearranging them as a ratio of two polynomials in s.

Transfer function, $$T(s) = K \cdot \frac{P(s)}{Q(s)}$$

where,

K = Constant

$P(s)$ = Numerator Polynomial

$Q(s)$ = Denominator Polynomial

The order of the system is given by the maximum power of s in the denominator polynomial, $Q(s)$. If

$$Q(s) = a_0 \cdot s^n + a_1 s^{n-1} + a_2 s^{n-2} + \ldots + a_{n-1} s + a_n$$

then, n is the order of the system.

if $n = 0$, then the system is zero order system.

if $n = 1$, then the system is first order system.

if $n = 2$, then the system is second order system and so on.

The order can be specified for both open loop system and closed loop system.

Also the value of n gives the number of poles in the transfer function. Hence the order is given by the number of poles of the transfer function.

2.4 INTRODUCTION TO PARTIAL FRACTION EXPANSION

The time response of the system is obtained by taking the inverse Laplace transform of the product of input signal and transfer function of the system. Taking inverse Laplace transform requires the knowledge of partial fraction expansion. In control System three different types of transfer function are available and they are,

Case 1: Functions with separate poles

Case 2: Functions with multiples poles

Case 3: Functions with Complex Conjugate poles.

The partial fraction of all the three cases are explained with an example,

Case 1: When the transfer functions has distinct poles (Functions with separate poles)

$$T(s) = \frac{K}{s\,(s + P_1)\,(s + P_2)}$$

By partial fraction expansion, $T(s)$ can be expressed as

$$T(s) = \frac{K}{s\,(s + P_1)\,(s + P_2)} = \frac{A}{s} + \frac{B}{s + P_1} + \frac{C}{s + P_2}$$

The residues A, B and C are given by,

$$A = T(s) \cdot s\big|_{s=0}$$

$$B = T(s)\,(s + P_1)\big|_{s=-P_1}$$

$$C = T(s)\,(s + P_2)\big|_{s=-P_2}$$

By using partial fraction find out the values of A, B and C then substitute those values in the equation of $T(s)$.

Case 2: When the transfer function has multiples poles (functions with multiple poles)

$$T(s) = \frac{K}{s\,(s + P_1)\,(s + P_2)^2}$$

By partial fraction expansion, $T(s)$ can be expressed as

$$T(s) = \frac{K}{s\,(s + P_1)\,(s + P_2)^2} = \frac{A}{s} + \frac{B}{s + P_1} + \frac{C}{(s + P_2)^2} + \frac{D}{s + P_2}$$

The residues A, B, C and D are given by

$$A = T(s) \cdot s\big|_{s=0}$$

$$B = T(s)\,(s + P_1)\big|_{s=-P_1}$$

$$C = T(s)\,(s + P_2)^2\big|_{s=-P_2}$$

$$D = \frac{d}{ds}[T(s)\,(s + P_2)^2]\Big|_{s=-P_2}$$

Find out A, B, C and D by using partial fraction then substitute in the $T(s)$ equation

Case 3: When the transfer function has complex conjugate poles (functions with complex conjugate poles)

$$T(s) = \frac{K}{(s + P_1)\,(s^2 + bs + c)}$$

By partial fraction expansion, $T(s)$ can be expressed as

$$T(s) = \frac{K}{(s + P_1)\,(s^2 + bs + s)} = \frac{A}{s + P_1} = \frac{Bs + C}{s^2 + bs + c}$$

The residue A given by, $A = T(s) \cdot (s + P_1)$

The residues B and C are solved by cross multiplying the equation A. and then equating the coefficient of s and power of s. Finally express $T(s)$ as shown below,

$$T(s) = \frac{A}{s + P_1} + \frac{Bs}{s^2 + bs + c} + \frac{C}{s^2 + bs + c}$$

$$= \frac{A}{s + P_1} + \frac{BS}{\left(s + \frac{b}{2}\right)^2 + \left(c - \frac{b^2}{4}\right)} + \frac{C}{\left(s + \frac{b}{2}\right)^2 + \left(c - \frac{b}{4}\right)^2}$$

Find out A, B and C by using partial fractioin then substitute in the $T(s)$ equation.

Note: After substituting A, B, C ... (residues) in $T(s)$ equation, the inverse lapalace transform will give the time response of a system for all the cases [case (1), case (2) and case (3)].

2.5 TIME RESPONSE OF FIRST ORDER SYSTEM FOR UNIT STEP INPUT

The closed loop transfer function of first order system,

$$\frac{C(s)}{R(s)} = \frac{1}{1+T_s} \tag{2.3}$$

If the input is unit step then, $r(t) = 1$ and $R(s) = 1/s$. [Refer table (1.2.1)]

The response in S-domain,

$$C(s) = R(s)\frac{1}{(1+Ts)}$$

$$= \frac{1}{s}\cdot\frac{1}{(1+Ts)}$$

$$= \frac{1}{sT\left(\frac{1}{T}+s\right)} = \frac{1/T}{s\,(s+1/T)} \tag{2.4}$$

By partial fraction expansion,

$$C(s) = \frac{1/T}{s\left(s+\frac{1}{T}\right)} = \frac{A}{s} + \frac{B}{\left(s+\frac{1}{T}\right)} \tag{2.5}$$

A is obtained by multiplying $C(s)$ by s and letting $s = 0$.

$$A = C(s)\cdot s = \frac{1/T}{s\left(s+\frac{1}{T}\right)}\cdot s = \left.\frac{1/T}{s+\frac{1}{T}}\right|_{s=0}$$

$$A = \frac{1/T}{1/T} = 1 \text{ ie. } A = 1$$

B is obtained by multiplying $C(s)$ by $(s + 1/T)$ and letting $s = -1/T$

$$B = C(s)\cdot\left(S+\frac{1}{T}\right) = \frac{1/T}{s\,(s+1/T)}\left(s+\frac{1}{T}\right) = \left.\frac{1/T}{s}\right|_{s=-1/T}$$

$$= \frac{1/T}{-1/T} = -1 \quad \text{ie., } B = -1$$

Substitute A and B values in equation (2.5)

$$C(s) = \frac{1}{s} - \frac{1}{s+1/T} \tag{2.6}$$

The response in time domain is obtained by taking inverse laplace transform of equation (2.6),

$$C(t) = L^{-1}[C(s)] = L^{-1}\left[\frac{1}{s} - \frac{1}{s+\frac{1}{T}}\right] = 1 - e^{-t/T} \tag{2.7}$$

Substitute various values for time (t) in equation (2.5.5)

when	$t = 0,$	$C(t) = 1 - e^{0}$	$= 0$
when	$t = T,$	$C(t) = 1 - e^{-1}$	$= 0.632$
when	$t = 2T,$	$C(t) = 1 - e^{-2}$	$= 0.865$
When	$t = 3T,$	$C(t) = 1 - e^{-3}$	$= 0.95$
When	$t = 4T,$	$C(t) = 1 - e^{-4}$	$= 0.9817$
when	$t = 5T,$	$C(t) = 1 - e^{-5}$	$= 0.993$
When	$t = \infty,$	$C(t) = 1 - e^{-\infty}$	$= 1$

Here T is called time constant of the system. After a time of $5T$, the system is assumed to have attained steady state. the input and output of the system is given below.

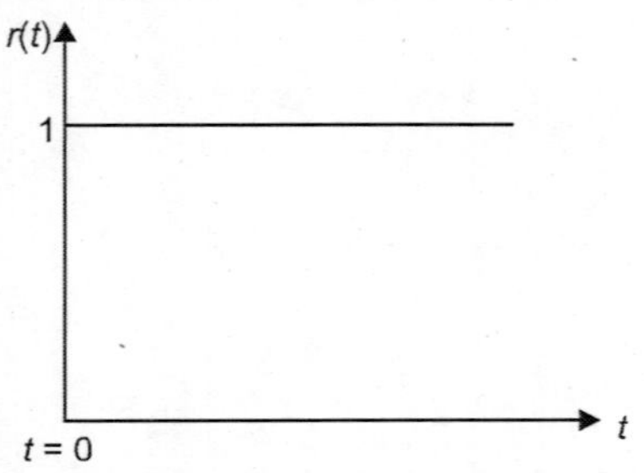

Fig. 2.6: Unit step input.

⇒

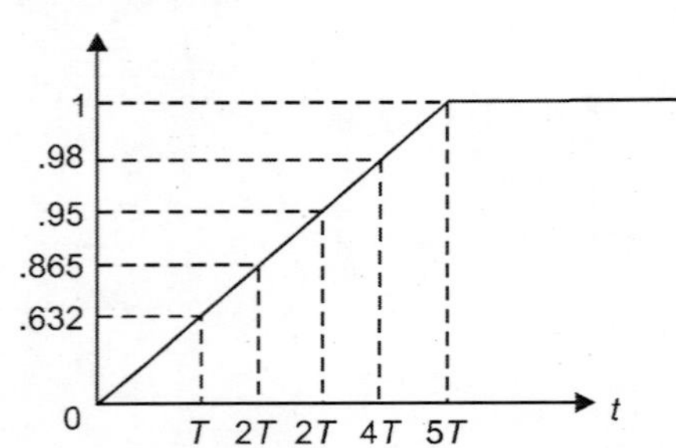

Fig. 2.7: Response of 1st order system to unit step input.

Conclusion: From fig. 2.7 the response of first order system with unit step input in stable.

2.6 TIME RESPONSE OF THE FIRST ORDER SYSTEM WITH UNIT RAMP FUNCTION

The closed loop transfer function of first order system,

$$\frac{C(s)}{R(s)} = \frac{1}{1+Ts} \qquad (2.8)$$

In the input is unit ramp then,

$$r(t) = t, \quad R(s) = 1/s^2 \text{ [refer Table (1.2.1)]}$$

∴ The response in s-domain,

$$C(s) = R(s)\frac{1}{(1+Ts)}$$

$$C(s) = \frac{1}{s^2}\cdot\frac{1}{(1+Ts)}$$

$$C(s) = \frac{1}{s^2\,(1+Ts)} = \frac{1}{s^2T\,[s+1/T]}$$

$$C(s) = \frac{1/T}{s^2\,[s+1/T]} \qquad (2.9)$$

By Partial fraction

$$C(s) = \frac{1/T}{s^2\,[s+1/T]} = \frac{A}{s} + \frac{B}{s^2} + \frac{C}{(s+1/T)}$$

$\therefore$ $$A(s)\,(s + 1/T) + B(s + 1/T) + Cs^2 = 1 \quad (2.10)$$

$$As^2 + \frac{As}{T} + Bs + \frac{B}{T} + s^2C = 1 \quad (2.11)$$

$$s^2[A + C] + s\,[B + A/T] + B/T = 1 \quad (2.12)$$

Equate s^2, s and constant terms from both L.H.S and R.H.S

$$(s^2 \text{ term})\; A + C = 0 \quad (2.13)$$

$$(s \text{ term})\; B + A/T = 0 \quad (2.14)$$

$$\text{constant term)}\; B/T = 1 \quad \therefore \quad B = T \quad (2.15)$$

Substitute (2.15) in equation (2.14)

$$T + A/T = 0$$

$$A/T = -T \quad A = -T^2 \quad (2.16)$$

Substitute (2.16) in equation (2.13)

$$-T^2 + C = 0$$

and $$C = T^2 \quad (2.17)$$

Substitute the values of A, B and C in equation $C(s)$

$$C(s) = \frac{1}{T}\left[\frac{-T^2}{s} + \frac{T}{s^2} + \frac{T^2}{s+1/T}\right]$$

$\therefore$ $$C(s) = \left[\frac{-T}{s} + \frac{1}{s^2} + \frac{T}{s + \frac{1}{T}}\right]$$

The response in time domain is obtained by taking inverse laplace transform of $C(s)$,

$$C(t) = L^{-1}\,[C(s)] = L^{-1}\left[\frac{-T}{s} + \frac{1}{s^2} + \frac{T}{s+1/T}\right]$$

$$C(t) = -T + t + T \times e^{-t/T} \quad (2.18)$$

The equation (2.18) in the time response of a first order system when unit ramp function is given as input.

$$C(t) = -T + t + T \cdot e^{-t/T}$$

when

t = 0;	$C(t) = -T + O + T{\cdot}e^{O/T} = -T$
t = T;	$C(t) = -T + T + T{\cdot}e^{-T/T} = -.632T$
t = 2T;	$C(t) = -T + 2T + T{\cdot}e^{-2T/T} = T + .865T = 1.865T$
t = 3T;	$C(t) = -T + 3T + T{\cdot}e^{-3T/T} = 2T + .95T = 2.95T$
t = 4T;	$C(t) = -T + 4T + T{\cdot}e^{-4T/T} = 3T + .993T = 3.993T$
t = ∞;	$C(t) = -T + \infty + T{\cdot}e^{-\infty/T} = \infty$

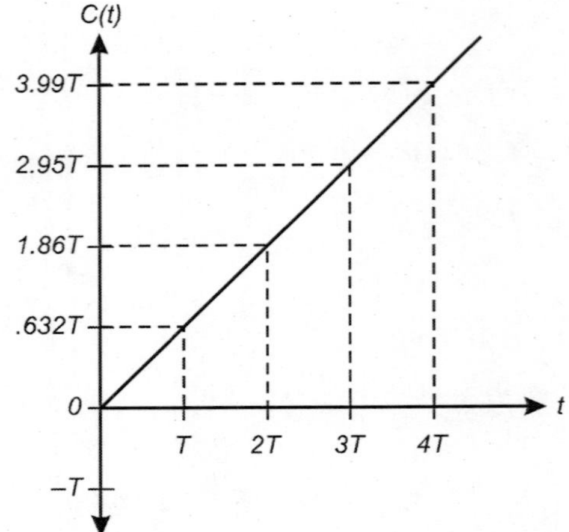

Conclusion: Marginally stable.

2.7 TIME RESPONSE OF FIRST ORDER SYSTEM WITH UNIT PARABOLIC FUNCTION

The closed loop transfer function of first order system,

$$\frac{C(s)}{R(s)} = \frac{1}{1+Ts} \tag{2.19}$$

If the input is unit parabolic then,

$$r(t) = \frac{t^2}{2}, \quad R(s) = 1/s^3 \text{ [refer Table 1.2.1)]}$$

∴ The response in s-domain,

$$C(s) = R(s)\cdot\frac{1}{1+Ts}$$

$$C(s) = \frac{1}{s^3}\cdot\frac{1}{1+Ts}$$

$$C(s) = \frac{1}{s^3\,(1+Ts)}$$

$$C(s) = \frac{1}{s^3\,T(s+1/T)}$$

$$C(s) = \frac{1/T}{s^3\,(s+1/T)} \tag{2.20}$$

By taking partial fraction,

$$\frac{A}{s}+\frac{B}{s^2}+\frac{C}{s^3}+\frac{D}{s+1/T} = 1 \tag{2.21}$$

$$As^2\,(s + 1/T) + Bs(s + 1/T) + C(s + 1/T) + Ds^3 = 1$$

$$As^3 + As^2/T + Bs^2 + Bs/T + Cs + C/T + Ds^3 = 1 \tag{2.22}$$

Equate the coefficients from both L.H.S and R.H.S of s^3, s^2, s and constant terms

$$A + D = 0 \quad (2.23)\ [\text{equate } s^3]$$

$$A/T + B = 0 \quad (2.24)\ [\text{equate } s^2]$$

$$B/T + C = 0 \quad (2.25)\ [\text{equate } s]$$

$$C/T = 1 \Rightarrow C = T \quad (2.26)\ [\text{equate constant}]$$

Substitue C in equation (2.25)

$$B/T + T = 0, \quad B = -T^2$$

Substitute B values in equation (2.24)

$$A/T + (-T)^2 = 0$$

$$A/T = T^2 \Rightarrow A = T^3$$

Substiturte A in equation (2.23)

$$T^3 + D = 0 \Rightarrow D = -T^3$$

Substitute A, B, C and D values in equation (2.21),

$$\left[\frac{T^3}{s} + \frac{-T^2}{s^2} + \frac{T}{s^3} + -\frac{T^3}{s+1/T}\right] \times \frac{1}{T} = C(s)$$

$$\frac{T^2}{s} + \frac{-T}{s^2} + \frac{1}{s^3} + \frac{-T^2}{s+1/T} = C(s)$$

By taking laplace transform,

$$C(t) = T^2 - Tt + t^2 - T^2 e^{-t/T} \quad (2.27)$$

The equation (2.27) is the time response of first order system when unit parabolic signal is given as input.

2.8 TIME RESPONSE OF FIRST ORDER SYSTEM WITH UNIT IMPULSE FUNCTION

The closed loop transfer function of first order system,

$$\frac{C(s)}{R(s)} = \frac{1}{1+Ts} \quad (2.28)$$

If the input is unit impulse function then,

$$R(s) = 1 \quad \text{(Refer Table 1.2.1)}$$

$\therefore$ The response in s-domain $C(s) = \dfrac{R(s)\times 1}{1+T_s} = 1 \times \dfrac{1}{1+Ts} = \dfrac{1}{1+Ts}$ $\quad [\;R(s) = 1]$

$$C(s) = \frac{1}{T[s+1/T]} = \frac{1/T}{(s+1/T)} \quad (2.29)$$

By applying partial fraction in equation (2.29)

$$C(s) = \frac{A}{(s+1/T)} \quad (2.30)$$

Substitute $s = -1/T$ in equation (2.30)

$$1 = \frac{A}{s+1/T} \Rightarrow 1 = \frac{A}{-1/T+1/T} \Rightarrow A = 1$$

$$\therefore \quad C(s) = \frac{1}{s+1/T} \times \frac{1}{T}$$

By taking laplace transform,

$$L^{-1}[C(s)] = L^{-1}\left[\frac{1}{T} \times \frac{1}{s+1/T}\right]$$

$$\therefore \quad C(t) = \frac{1}{T} e^{-t/T} \tag{2.31}$$

The equation (2.31) is the time response of first order system when unit impulse is given as input.

2.9 COMPARISON OF FIRST ORDER RESPONSE WITH UNIT STEP, RAMP, PARABOLIC AND IMPULSE FUNCTIONS

The time response of first order system with unit step input,

$$C(t) = 1 - e^{-t/T} \tag{2.32}$$

The time response of first order system with unit ramp input,

$$C(t) = t - T + te^{-t/T} \tag{2.33}$$

The time response of first order system with unit parabolic input,

$$C(t) = T^2 - Tt + t^2 - Te^{-t/T} \tag{2.34}$$

The time response of first order system with unit impulse input,

$$C(t) = \frac{1}{T} e^{-t/T} \tag{2.35}$$

From equations (2.32) to (2.35) it is clear that the unit step input is the derivative of unit ramp input and unit impulse input is the derivative of unit step input and vice versa. Thus, the response to the derivative of an input signal can be obtained by differentiating the response of the system to the original signal. This is the property of the linear time-invariant systems.

2.10 SECOND ORDER SYSTEM

The standard form of closed loop transfer function of second order system is given by

$$\frac{C(s)}{R(s)} = \frac{\omega_n^2}{s^2 + 2\zeta\omega_n s + \omega_n^2} \tag{2.36}$$

where,

$\omega_n \rightarrow$ Undamped natural frequency, rad/sec

$\zeta \rightarrow$ Damping ratio

The damping ratio is defined as the ratio of actual damping to the critical damping. The response *C*(*t*) of second order system depends on the value of damping ratio. Depending on the value of ζ, the system can be classified in to the following four cases,

Case 1: Undamped system, $\zeta = 0$

Case 2: Under damped system, $0 < \zeta < 1$

Case 3: Critically damped system, $\zeta = 1$

Case 4: Over damped system, $\zeta > 1$

The characteristics equation of the 2nd order system is

$$s^2 + 2\zeta\omega_n s + \omega_n^2 = 0 \tag{2.37}$$

It is a quadratic equation and the roots of this equation is given by

$$s_1, s_2 = \frac{-2\omega\zeta_n \pm \sqrt{4\zeta^2\omega_n^2 - 4\omega_n^2}}{2} = \frac{-2\zeta\omega_n \pm \sqrt{4\omega_n^2(\zeta^2-1)}}{2}$$

$$s_1, s_2 = -\zeta\omega_n \pm \omega_n\sqrt{\zeta^2 - 1} \tag{2.38}$$

when $\zeta = 0$, $s_1\, s_2 = \pm j\omega_n$; {Roots are purely imaginary and the system is undamped}

when $\zeta = 1$, $s_1, s_2 = -\omega_n$; {Roots are real and equal and the system is critically damped}

when $\zeta > 1$, $s_1, s_2 = -\zeta\omega_n \pm \omega_n\sqrt{\zeta^2 - 1}$; {Roots are real and unequal and the system is over damped}

when $0 < \zeta < 1$, $s_1, s_2 = -\zeta\omega_n \pm \omega_n\sqrt{\zeta^2 - 1}$

$$= -\zeta\omega_n \pm \omega_n\sqrt{(-1)(1-\zeta^2)} = -\zeta\omega_n \pm j\omega_n\sqrt{1-\zeta^2}$$

$$= -\zeta\omega_n \pm j\omega_d$$ {Roots are complex conjugate the system is under damped}

where, $\omega_d = \omega_n\sqrt{1-\zeta^2}$

Here ω_d called damped frequency of oscillation of the system and its unit is radian/second.

2.10.1 Response of Undamped Second Order System For Unit Step Input

The standard form of closed loop transfer function of second order system is

$$\frac{C(s)}{R(s)} = \frac{\omega_n^2}{s^2 + 2\zeta\omega_n s + \omega_n^2} \tag{2.39}$$

For undamped system, $\zeta = 0$

$$\therefore \quad \frac{C(s)}{R(s)} = \frac{\omega_n^2}{s^2 + \omega_n^2} \tag{2.40}$$

when the input is unit step, $r(t) = 1$

and $R(s) = 1/s$

$\therefore$ The response in s-domain,

$$C(s) = R(s)\cdot\frac{\omega_n^2}{s^2 + \omega_n^2}$$

$$C(s) = \frac{1}{s}\cdot\frac{\omega_n^2}{s^2 + \omega_n^2} \tag{2.41}$$

By partial fraction expansion,

$$C(s) = \frac{\omega_n^2}{s(s^2+\omega_n^2)} = \frac{A}{s} + \frac{B}{s^2+\omega_n^2} \tag{2.42}$$

A is obtained by multiplying $C(s)$ by s and letting $s = 0$

$$A = C(s) \times s \Rightarrow \frac{\omega_n^2}{\not{s}(s^2+\omega_n^2)} \times \not{s} \quad \left.\frac{\omega_n^2}{s^2+\omega_n^2}\right|_{s=0}$$

$$= \frac{\omega_n^2}{\omega_n^2} = 1$$

$$\boxed{A = 1}$$

B is obtained by multiplying $C(s)$ by $(s^2 + \omega_n^2)$ and letting $s^2 = -\ \omega_n^2$ (or) $s = j\omega_n$

$$B = C(s)\,(s^2 + \omega_n^2)$$

$$= \frac{\omega_n^2}{s(s^2+\omega_n^2)} \times (s^2+\omega_n^2) \Rightarrow \left.\frac{\omega_n^2}{s}\right|_{s=-\omega_n^2}$$

$\therefore$

$$B = \frac{\omega_n^2}{j\omega_n} = -\ j\omega_n = -\ s$$

$$\boxed{B = -s}$$

Substitute A and B values in equation (2.42)

$$C(s) = \frac{A}{s} + \frac{s}{(s^2+\omega_n^2)} = \frac{1}{s} - \frac{s}{(s^2+\omega_n^2)} \tag{2.43}$$

The time domain response of equation (2.43) is obtained by taking inverse laplace transform,

$$L^{-1}\,[C(s)] = L^{-1}\left[\frac{1}{s} - \frac{s}{s^2+\omega_n^2}\right] = 1 - \cos\omega_n t \tag{2.10.1.6}$$

The equation (2.44) is the time response of second order system when unit step input is given under $\zeta = 0$.

Substitute various time for time in equation (2.44)

when, $t = 0$	$\Rightarrow$	$C(t) = 1 - \cos\omega_n t$	$\Rightarrow$	$1 - \cos 0$	$\Rightarrow$	$1 - 1 = 0$
$t = \pi/2$	$\Rightarrow$	$C(t) = 1 - \cos\omega_n t$	$\Rightarrow$	$1 - \cos \pi/2$	$\Rightarrow$	$1 - 0 = 1$
$t = \pi$	$\Rightarrow$	$C(t) = 1 - \cos\omega_n t$	$\Rightarrow$	$1 - \cos \pi$	$\Rightarrow$	$1 - (-1) = 2$
$t = 3\pi/2$	$\Rightarrow$	$C(t) = 1 - \cos\omega_n t$	$\Rightarrow$	$1 - \cos (3\pi/2)$	$\Rightarrow$	$1 - 0 = 1$
$t = 2\pi$	$\Rightarrow$	$C(t) = 1 - \cos\omega_n t$	$\Rightarrow$	$1 - \cos 2\pi$	$\Rightarrow$	$1 - 1 = 0$

The input and output of the system shown in Fig. (2.8) and (2.9)

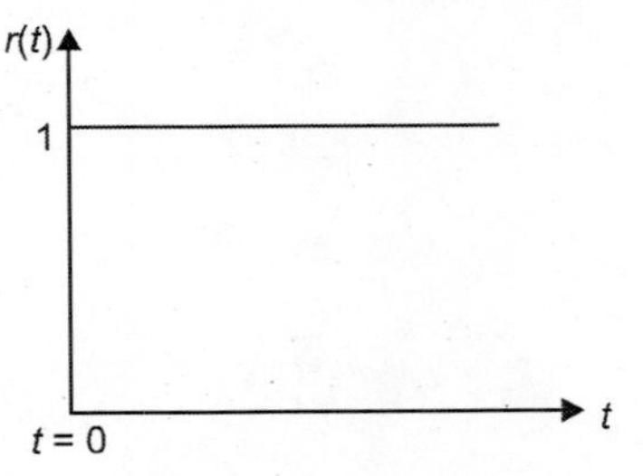

Fig. 2.8: Unit step i/p

$\Rightarrow$

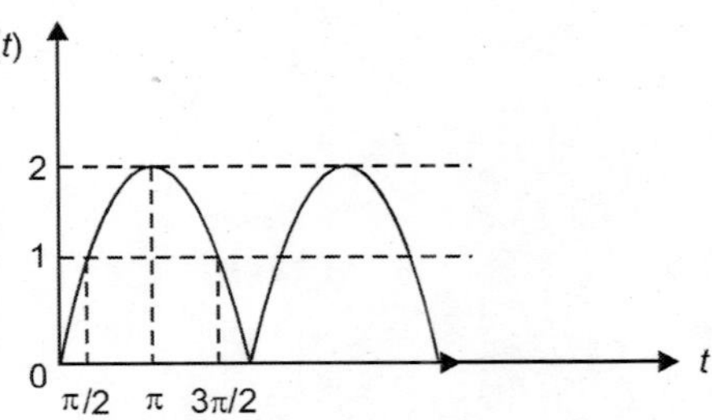

Fig. 2.9: Response of undamped second order system for unit step input.

Conclusion: The response of undamped second order system for unit step input is completely oscillatory.

Note: Every practical system has some amount of damping. Hence undamped system does not exist in practice.

2.10.2 Time Response of Second Order System For Under damped Case and When the input is Unit Step

The standard form of closed loop transfer function of second order system is

$$\frac{C(s)}{R(s)} = \frac{\omega_n^2}{s^2 + 2\zeta\omega_n s + \omega_n^2} \tag{2.45}$$

For underdamped system, $0 < \zeta < 1$. and roots of the denominator [characteristic equation] are complex conjugate.

The roots of the denominator are,

$$s = -\zeta\omega_n \pm \omega_n\sqrt{\zeta^2 - 1}$$

since $\quad \zeta < 1,\ \zeta^2 < 1$

$$B = -\zeta\omega_n \pm \omega_n\sqrt{(-1)(1-\zeta^2)} = -\zeta\omega_n \pm j\omega_n\sqrt{1-\zeta^2}$$

The damped frequency of oscillation,

$$\omega_d = \omega_n\sqrt{1-\zeta^2}$$

$$\therefore \quad s = -\zeta\omega_n \pm j\omega_d$$

The response in s-domain,

$$C(s) = R(s)\frac{\omega_n^2}{s^2 + 2\zeta\omega_n s + \omega_n^2}$$

For unit step input, $r(t) = 1$ and $R(s) = 1/s$

$$\therefore \quad C(s) = \frac{\omega_n^2}{s\,(s^2 + 2\zeta\omega_n\, s + \omega_n^2)} \tag{2.46}$$

By partial fraction expansion,

$$C(s) = \frac{A}{s} + \frac{Bs + C}{s^2 + 2\,\zeta\omega_n s + \omega_n^2}$$

$$A = s\,C(s)\big|_{s=0}$$

$$= \frac{\omega_n^2}{\omega_n^2} = 1$$

To solve B and C

$$\frac{\omega_n^2}{s(s^2 + 2\zeta\omega_n\, s + \omega_n^2)} = \frac{1}{s} + \frac{Bs + C}{s^2 + 2\zeta\omega_n s + \omega_n^2}$$

on cross multiplication and equating the coefficients of like power of s we get,

$$\omega_n^2 = s^2 + 2\,\zeta\omega_n s + \omega_n + (Bs + C)s$$

$$\omega_n^2 = s^2 + 2\zeta\omega_n s + \omega_n^2 + Bs^2 + Cs$$

Equating coefficients of s^2 we get,

$$0 = 1 + B$$

$$\therefore \quad B = -1$$

Equating coefficients of s we get,

$$0 = 2\,\zeta\omega_n + C$$

$$\therefore \quad C = 2\,\zeta\omega_n$$

Substitute A, B and C values in equation (2.10.2.3)

$$C(s) = \frac{1}{s} - \frac{(s + 2\,\zeta\omega_n)}{s^2 + 2\,\zeta\omega_n s + \omega_n^2} \tag{2.47}$$

Adding and subtracting $s^2\ \omega_n^2$ to the denominator of second term in equation (2.47)

$$= \frac{1}{s} - \frac{(s + 2\,\zeta\omega_n)}{s^2 + 2\,\zeta\omega_n s + \omega_n^2 + \zeta^2\omega_n^2 - \zeta^2\omega_n^2}$$

$$= \frac{1}{s} - \frac{(s + 2\,\zeta\omega_n)}{(s^2 + 2\zeta\omega_n\, s + \zeta^2\omega_n^2) + (\omega_n^2 - \zeta^2\omega_n^2)}$$

$$= \frac{1}{s} - \frac{(s + 2\zeta\omega_n)}{(s + \zeta\omega_n)^2 + \omega_n^2\,(1 - \zeta^2)}$$

as we know $$\omega_d = \omega_n \sqrt{1-\zeta^2}$$

$$C(s) = \frac{1}{s} - \frac{(s+2\zeta\omega_n)}{(s+\zeta\omega_n)^2 + \omega_d^2}$$

$$= \frac{1}{s} - \frac{(s+\zeta\omega_n)}{(s+\zeta\omega_n)^2 + \omega_d^2} - \frac{(\zeta\omega_n)}{(s+\zeta\omega_n)^2 + \omega_d^2} \quad (2.48)$$

Multiplying and dividing by ω_d in the third term of the equation (2.48)

$$C(s) = \frac{1}{s} - \frac{(s+\zeta\omega_n)}{(s+\zeta\omega_n)^2 + \omega_d^2} - \left(\frac{\zeta\omega_n}{\omega_d}\right)\left(\frac{\omega_d}{(s+\zeta\omega_n)^2 + \omega_d^2}\right) \quad (2.49)$$

on taking inverse Laplace transform

$$C(t) = 1 - e^{-\zeta\omega_n t}\cos\omega_d t - \frac{\zeta\omega_n}{\omega_d} e^{-\zeta\omega_n t}\sin\omega_d t$$

$$C(t) = 1 - e^{-\zeta\omega_n t}\left[+\cos\omega_d t + \frac{\zeta\omega_n}{\omega_n\sqrt{1-\zeta^2}}\sin\omega_d t\right]$$

$$= 1 - e^{-\zeta\omega_n t}\left[\cos\omega_d t + \frac{\zeta}{\sqrt{1-\zeta^2}}\sin\omega_d t\right]$$

$$= 1 - \frac{e^{-\zeta\omega_n t}}{\sqrt{1-\zeta^2}}\left[\sqrt{1-\zeta^2}\cos\omega_d t + \zeta\sin\omega_d t\right]$$

$$C(t) = \frac{e^{-\zeta\omega_n t}}{\sqrt{1-\zeta^2}}\left[\sin\omega_d t\cdot\zeta + \cos\omega_d t\cdot\sqrt{1-\zeta^2}\right]$$

$$C(t) = 1 - \frac{e^{-\zeta\omega_n t}}{\sqrt{1-\zeta^2}}\left[\sin\omega_d t\cos\theta + \cos\omega_d t\sin\theta\right]$$

(Refer note)

$$C(t) = 1 - \frac{e^{-\zeta\omega_n t}}{\sqrt{1-\zeta^2}}\sin(\omega_d t + \theta) \quad (2.50)$$

The equation (2.50) is the time response of second order system when unit step input is given [for under dampled case]

where, $$\theta = \tan^{-1}\left[\frac{\sqrt{1-\zeta^2}}{\zeta}\right]$$

The response of underdamped second order system oscillates before setting to a final value. The oscillations depends on the value of damping ratio.

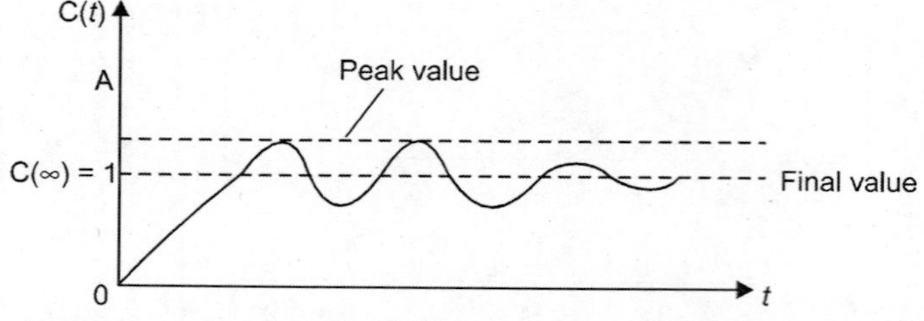

Fig. 2.10: Response of under damped second order system for unit step input.

Note: On constructing right angle triangle with ζ and $\sqrt{1-\zeta^2}$, we get,

$$\sin\theta = \sqrt{1-\zeta^2}$$

$$\cos\theta = \zeta$$

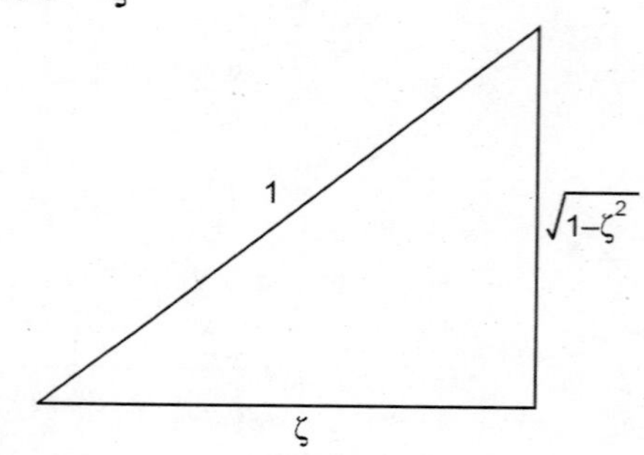

Fig. 2.11

$$\tan\theta = \frac{\sqrt{1-\zeta^2}}{\zeta}$$

$$\theta = \tan^{-1}\left[\frac{\sqrt{1-\zeta^2}}{\zeta}\right]$$

2.10.3 Time Response of Second Order System for Critically Damped Case When Input is Unit Step Input

The standard of closed loop transfer function of second order system is

$$\frac{C(s)}{R(s)} = \frac{\omega_n^2}{s^2 + 2\,s\omega_n s + \omega_n^2}$$

For critical damping $\zeta = 1$

$$\therefore \quad \frac{C(s)}{R(s)} = \frac{\omega_n^2}{s^2 + 2\zeta\omega_n s + \omega_n^2} = \frac{\omega_n^2}{(s+\omega_n)^2} \qquad (2.51)$$

when input is unit step, $r(t) = 1$ and $R(s) = 1/s$

∴ The response in s-domain,

$$C(s) = R(s)\frac{\omega_n^2}{(s+\omega_n)^2} = \frac{1}{s}\cdot\frac{\omega_n^2}{(s+\omega_n)^2}$$

$$C(s) = \frac{\omega_n^2}{s\,(s+\omega_n)^2} \tag{2.52}$$

By partial fraction, we can write,

$$C(s) = \frac{\omega_n^2}{s(s+\omega_n)^2} = \frac{A}{s} + \frac{B}{(s+\omega_n)^2} + \frac{C}{(s+\omega_n)} \tag{2.53}$$

$$A = s\,C(s)\big|_{s=0} = \left.\frac{\omega_n^2}{(s+\omega_n)^2}\right|_{s=0} = \frac{\omega_n^2}{\omega_n^2} = 1$$

$$A = 1$$

$$B = (s+\omega_n)^2\,C(s)\big|_{s=-\omega_n} = \left.\frac{-\omega_n^2}{s}\right|_{s=-\omega_n} = -\omega_n$$

$$B = -\,\omega_n$$

$$C = \frac{d}{ds}\Big[(s+\omega_n)^2\,C(s)\Big]_{s=-\omega_n} = \frac{d}{ds}\left[\frac{\omega_n^2}{s}\right]_{s=-\omega_n}$$

$$= \left.\frac{-\omega_n^2}{s^2}\right|_{s=-\omega} \frac{-\omega_n^2}{(-\omega_n)^2} = -1$$

$$C = -\,1$$

Substitute A, B and C values in equation (2.53)

$$C(s) = \frac{A}{s} + \frac{B}{(s+\omega_n)^2} + \frac{C}{(s+\omega_n)}$$

$$= \frac{1}{s} - \frac{\omega_n}{(s+\omega_n)^2} - \frac{1}{(s+\omega_n)} \tag{2.54}$$

The response in time domain, on taking inverse laplace transform,

$$C(t) = L^{-1}\,[C(s)] = L^{-1}\left[\frac{1}{s}\,\frac{-\omega_n}{(s+\omega_n)^2} - \frac{1}{(s+\omega_n)}\right]$$

$$C(t) = 1 - \omega_n\,t\,e^{-\omega_n t} - e^{-\omega_n t}$$

$$C(t) = 1 - e^{-\omega_n t}\,(1+\omega_n t)$$

Conclusion: The response of critically damped second order system have no oscillations.

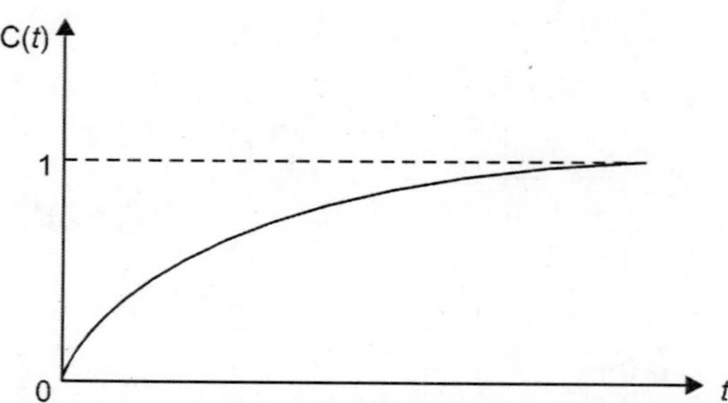

Fig. 2.12: Response of critically damped second order system for unit step input.

2.10.4 Response of Second Order System For Over Damped Case and When Input is Unit Step Input

The standard form of closed loop transfer function of second order system is

$$\frac{C(s)}{R(s)} = \frac{\omega_n^2}{s^2 + 2\zeta\omega_n s + \omega_n^2} \tag{2.55}$$

For overdamped system $\zeta > 1$. The roots of the denominator of transfer function are real and distinct. Let the roots of the denominator be s_a, s_b

$$s_a, s_b = -\zeta\omega_n \pm \omega_n\sqrt{\zeta^2 - 1} = -[\zeta\omega_n \pm \omega_n\sqrt{\zeta^2 - 1}] \tag{2.56}$$

Let, $s_1 = -s_a$ and $s_2 = -s_b$

$\therefore$
$$s_1 = \zeta\omega_n - \omega_n\sqrt{\zeta^2 - 1} \tag{2.57}$$

$$s_2 = \zeta\omega_n + \omega_n\sqrt{\zeta^2 - 1} \tag{2.58}$$

The closed loop transfer function can be written in terms of s_1 and s_2 as shown below,

$$\frac{C(s)}{R(s)} = \frac{\omega_n^2}{s^2 + 2\zeta\omega_n s + \omega_n^2} = \frac{\omega_n^2}{(s + s_1)(s + s_2)} \tag{2.59}$$

For unit step input $r(t) = 1$ and $R(s) = 1/s$

$\therefore$
$$C(s) = R(s) \cdot \frac{\omega_n^2}{(s + s_1)(s + s_2)} = \frac{\omega_n^2}{s(s + s_1)(s + s_2)} \tag{2.60}$$

By partial fraction expansion we can write,

$$C(s) = \frac{\omega_n^2}{s(s + s_1)(s + s_2)} = \frac{A}{s} + \frac{B}{(s + s_1)} + \frac{C}{(s + s_2)} \tag{2.61}$$

$$A = sC(s)\big|_{s=0} = s \cdot \frac{\omega_n^2}{s(s+s_1)(s+s_2)}\bigg|_{s=0} = \frac{\omega_n^2}{s_1\, s_2}$$

$$= \frac{\omega_n^2}{\left[\zeta\omega_n - \omega_n\sqrt{\zeta^2-1}\right]\left[\zeta\omega_n + \omega_n\sqrt{\zeta^2-1}\right]} = 1$$

$$= \frac{\omega_n^2}{\zeta^2\omega_n^2 - \omega_n^2(\zeta^2-1)} = \frac{\omega_n^2}{\omega_n^2} = 1$$

$$A = 1$$

$$B = (s+s_1)\,C(s)\big|_{s=-s_1} = \frac{\omega_n^2}{s\,(s+s_2)}\bigg|_{s=-s_1}$$

$$= \frac{\omega_n^2}{-s_1\,(-s_1+s_2)}$$

$$= \frac{-\omega_n^2}{s_1[-\zeta\omega_n + \omega_n\sqrt{s^2-1} + \zeta\omega_n + \omega_n\sqrt{\zeta^2-1}]}$$

$$= \frac{-\omega_n^2}{\left[2\,\omega_n\sqrt{\zeta^2-1}\right]s_1} = \frac{\omega_n}{2\sqrt{\zeta^2-1}} \times \frac{1}{s_1}$$

$$B = \frac{-\omega_n}{2\sqrt{\zeta^2-1}} \cdot \frac{1}{s_1}$$

$$C = C(s)\,(s+s_2)\big|_{s=-s_2} = \frac{\omega_n^2}{s\,(s+s_1)}\bigg|_{s=s_2}$$

$$C = \frac{\omega_n^2}{-s_2\,(-s_2+s_1)}$$

$$C = \frac{\omega_n^2}{-s_2\,[-\zeta\omega_n - \omega_n\sqrt{\zeta^2-1} + \zeta\omega_n - \omega_n\sqrt{\zeta^2-1}]}$$

$$C = \frac{\omega_n}{2\sqrt{\zeta^2-1}} \cdot \frac{1}{s_2}$$

Substitute *A*, *B* and *C* values in equation (2.61)

$$\therefore \quad = \frac{1}{s} - \frac{\omega_n}{2\sqrt{\zeta^2-1}} \cdot \frac{1}{s_1}\,\frac{1}{(s+s_1)} + \frac{\omega_n}{2\sqrt{\zeta^2-1}} \cdot \frac{1}{s_2}\,\frac{1}{(s+s_2)} \qquad (2.62)$$

on taking inverse Laplace transform we get,

The response in time domain,

$$c(t) = 1 - \frac{\omega_n}{2\sqrt{\zeta^2 - 1}} \cdot \frac{1}{s_1} e^{-s_1 t} + \frac{\omega_n}{2\sqrt{\zeta^2 - 1}} \cdot \frac{1}{s_2} e^{-s_2 t}$$

$$c(t) = 1 - \frac{\omega_n}{2\sqrt{\zeta^2 - 1}} \left[\frac{e^{-s_1 t}}{s_1} - \frac{e^{-s_2 t}}{s_2} \right] \tag{2.63}$$

where $s_1 = \zeta\omega_n - \omega_n\sqrt{\zeta^2 - 1}$

and $s_2 = \zeta\omega_n + \omega_n\sqrt{\zeta^2 - 1}$

The equation (2.63) is the time response of second order system when input is unit step input under overdamped.

The response of overdamped second order system has no oscillations but it takes longer time for the response to reach the final steady state value.

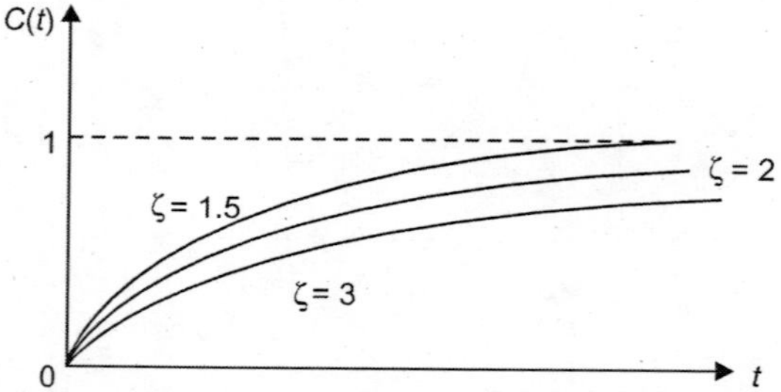

Fig. 2.13: Response of overdamped second order system for unit step input.

2.10.5 Location of Roots of Characteristic Equation and Time Response of Second Order System

$$\frac{C(s)}{R(s)} = \frac{\omega_n^2}{s^2 + 2\zeta\omega_n s + \omega_n^2}$$

$$s^2 + 2\zeta\,\omega_n\,s + \omega_n^2 = 0 \tag{2.64}$$

The roots of equation of (2.10..1)

$$s_1 = -\zeta\omega_n + j\omega_n\sqrt{1 - \zeta^2}$$

$$s_2 = -\zeta\omega_n + j\omega_n\sqrt{1 - \zeta^2}$$

The real part of the roots ($-s\omega_n$) represents the damping and imaginary part ($\omega_n\sqrt{1 - \zeta^2}$) represents the damped frequency.

From Fig. (2.14)

1. ω_n is the distance of root from origin

2. $\cos\theta = \dfrac{\zeta\omega_n}{\omega_n} = \zeta$ i.e. $\theta = \cos^{-1}\zeta$

   (when the roots are in Let-half of s-plane)

3. ω_d is the imaginary part of the root and is known as damped frequency (or) conditional frequency.

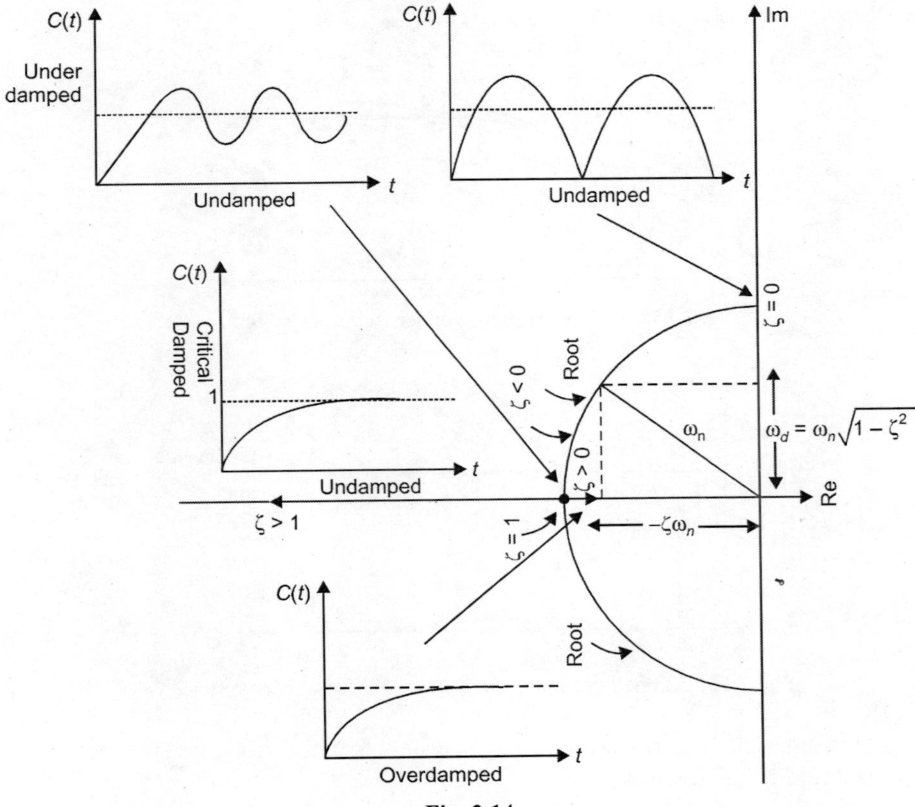

Fig. 2.14

2.10.6 Time Domain Specification

The desired performance characteristics of control systems are specified in terms of time domain specifications. Systems with energy storage elements cannot respond instantaneously and will exhibit transient responses, whenever they are subjected to inputs (or) disturbance.

The desired performance characteristics of a system of any order may be specified in terms of the transient response to a unit step input signal. The response of a second order system for unit-step input with various values of damping ratio is shown below

The transient response of a system to a unit step input depends on the initial conditions. Therefore to compare the time response of various systems it is necessary to start with standard initial conditions. The most practical standard is to start with the system at rest and so output and all time derivatives before $t = 0$ will be zero. The transient response of a practical control system often exhibits damped oscillation before reaching steady state. A typical damped oscillatory response of a system is shown below Fig. (2.10.6.2).

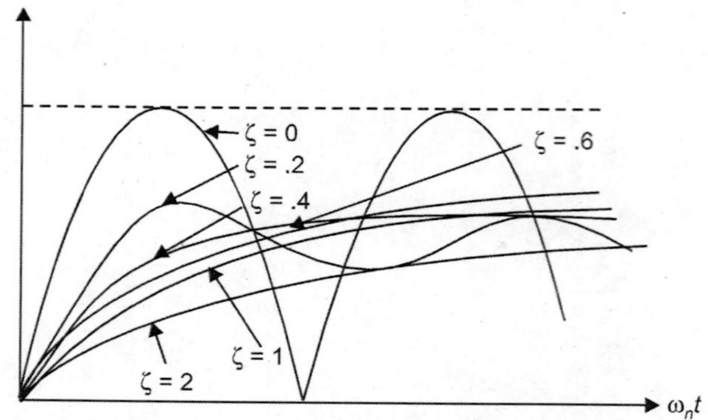

Fig. 2.15: Unit step response of second order system.

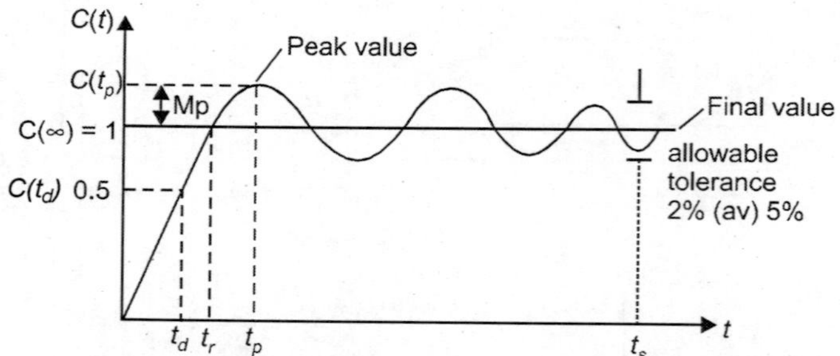

Fig. 2.16: Damped oscillatory response of second order system.

The transient response characteristics of a control system to a unit step input is specified in terms of the following time domain specifications.

1. Delay time, t_d
2. Rise time, t_r
3. Peak time, t_p
4. Maximum overshoot, M_p
5. Settling time, $t(s)$

The time domain specifications are defined as follows:

1. **Delay time (t_d):** It is the time taken for the response to reach 50% of the final value, for the very first time.
2. **Rise time (t_r):** It is the time taken for the response to raise from 0 to 100% for the very first time. For underdamped system, the rise time is calculated from 0 to 100%. But for overdamped system it is the time taken by the response to raise from 10% to 90%. For critically damped system, it is the time taken for response to raise from 5% to 95%.

3. **Peak time (t_p):** It is the time taken for the response to reach peak value for the very first time. (or) it is the time taken for the response to reach the peak overshoot, M_p.
4. **Peak overshoot (M_p):** It is defined as the ratio of the maximum peak value measured from peak value to the final value.

$$\text{Let final value} = C(\infty)$$

$$\text{Maximum value} = C(t_p)$$

$$\text{Peak overshoot, } M_p = \frac{C(t_p) - C(\infty)}{C(\infty)} \tag{2.65}$$

$$\%\text{ Peak overshoot, }\%\ M_p = \frac{C(t_p) - C(\infty)}{C(\infty)} \times 100 \tag{2.66}$$

5. **Settling time (t_s):** It is defined as the time taken by the response to reach and stay within a specified error. It is usually expressed as % of final values. The usual tolerable error is 2% or 5% of the final value.

Expression for Time Domain Specifications

Rise time (t_r): Response of second order system for underdamped case is given by,

$$C(t) = 1 - \frac{e^{-\zeta\omega_n t}}{\sqrt{1-\zeta^2}} \cdot \sin(\omega_d t + \theta)$$

at $t = t_r$,

$$C(t) = C(t_r) = 1$$

$$\therefore \quad C(t_r) = 1 - \frac{e^{-\zeta\omega_n \cdot t_r}}{\sqrt{1-\zeta^2}} \cdot \sin(\omega_d t_r + \theta) = 1$$

$$\therefore \quad = \frac{-e^{-\zeta\omega_n t_r}}{\sqrt{1-\zeta^2}} \cdot \sin(\omega_d t_r + \theta) = 0$$

Since $-e^{-\zeta\omega_n t_r} \neq 0$, the term, $\sin(\omega_d t_r + \theta) = 0$

$\sin\phi = 0$, when $\phi = 0, \pi, 2\pi, 3\pi$

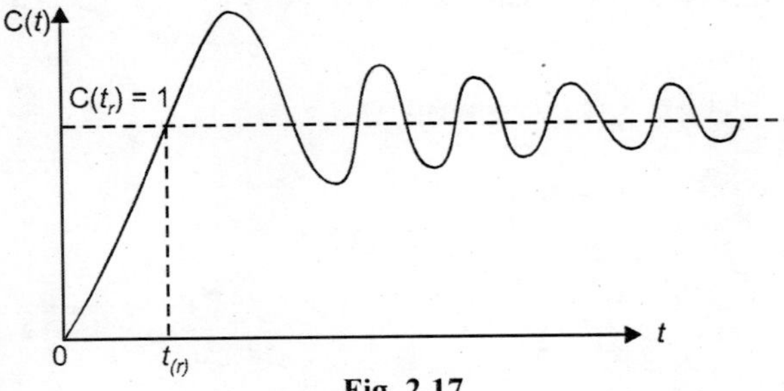

Fig. 2.17

$$\therefore \qquad \omega_d t_r + \theta = \pi$$

$$\omega_d t_r = \pi - \theta$$

$$\therefore \qquad \text{Rise time, } t_r = \frac{\pi - \theta}{\omega_d}$$

Here, $$\theta = \tan^{-1}\frac{\sqrt{1-\zeta^2}}{\zeta} \quad \text{(refer note 1)} \tag{1}$$

and damped frequency of oscillation, $\omega_d = \omega_n\sqrt{1-\zeta^2}$

$$\text{Rise time } t_r = \frac{\pi - \tan^{-1}\dfrac{\sqrt{1-\zeta^2}}{\zeta}}{\omega_n\sqrt{1-\zeta^2}} \tag{2.67}$$

Note: 1. On constructing right angle triangle with ζ and $\sqrt{1-\zeta^2}$, we get

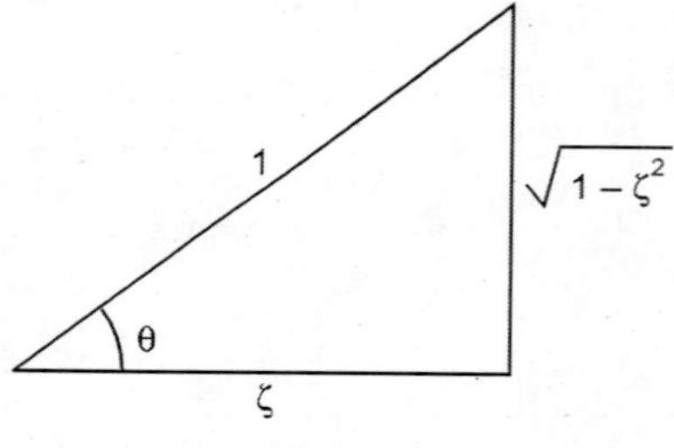

Fig. 2.18

Note: 2. $\tan^{-1}\dfrac{\sqrt{1-\zeta^2}}{\zeta}$ should be measured in radians.

Peak time (t_p): To find the expression for peak time, t_p, differentiate $C(t)$ with respect to t and equate to 0.

i.e. $$\left.\frac{d}{dt}C(t)\right|_{t=t_p} = 0$$

The unit step response of second order system is given by

$$C(t) = 1 - \frac{e^{-\zeta\omega_n t}}{\sqrt{1-\zeta^2}}\sin(\omega_d t + \theta)$$

Differentiating $C(t)$ with respect to t.

$$\frac{d}{dt}C(t) = \frac{-e^{\zeta\omega_n t}}{\sqrt{1-\zeta^2}}(-\zeta\omega_n)\sin(\omega_d t + \theta) + \left(\frac{-e^{-\zeta\omega_n t}}{\sqrt{1-\zeta^2}}\right)\cos(\omega_d t + \theta)\,\omega_d$$

Put, $$\omega_d = \omega_n \sqrt{1-\zeta^2}$$

$$\therefore \quad \frac{d}{dt}c(t) = \frac{e^{-\zeta\omega_n t}}{\sqrt{1-\zeta^2}}(\zeta\omega_n)\sin(\omega_d t + \theta) - \frac{\omega_n\sqrt{1-\zeta^2}}{\sqrt{1-\zeta^2}}e^{-\zeta\omega_n t}\cos(\omega_d t + \theta)$$

$$= \frac{\omega_n \cdot e^{-\zeta\omega_n t}}{\sqrt{1-\zeta^2}}\left[\zeta \sin(\omega_d t + \theta) - \sqrt{1-\zeta^2}\cos(\omega_d t + \theta)\right]$$

$$= \frac{\omega_n}{\sqrt{1-\zeta^2}} \cdot e^{-\zeta\omega_n t}\left[\cos\theta \sin(\omega_d t + \theta) - \sin\theta \cos(\omega_d t + \theta)\right]$$

$$= \frac{\omega_n}{\sqrt{1-\zeta^2}} \cdot e^{-\zeta\omega_n t}\left[\sin(\omega_d t + \theta)\cos\theta - \cos(\omega_d t + \theta)\sin\theta\right]$$

$$= \frac{\omega_n}{\sqrt{1-\zeta^2}} \cdot e^{-\zeta\omega_n t}\left[\sin\{(\omega_d t + \theta) - \theta\}\right]$$

$$= \frac{\omega_n}{\sqrt{1-\zeta^2}} \cdot e^{-\zeta\omega_n t} \cdot \sin(\omega_d \cdot t)$$

at $t = t_p$, $\frac{d}{dt}C(t) = 0$

$$\therefore \quad \left.\frac{d}{dt}C(t)\right|_{t=t_p} = \frac{\omega_n}{\sqrt{1-\zeta^2}} \cdot e^{-\zeta\omega_n t_p}\sin(\omega_d t_p) = 0$$

$$\therefore \quad \frac{\omega_n}{\sqrt{1-\zeta^2}} \cdot e^{-\zeta\omega_n t_p} \cdot \sin(\omega_d t_p) = 0$$

since, $0^{-\zeta\omega_n t_p} \neq 0$, the term, $\sin(\omega_d t_p) = 0$ when $\omega_d t_p = 0, \pi, 2\pi, 3\pi$

$$\therefore \quad \omega_d t_p = \pi$$

$\therefore$ Peak time, $$t_p = \pi/\omega_d \tag{2.68}$$

The damped frequency of oscillation , $\omega_d = \omega_n \sqrt{1-\zeta^2}$

Peak time, $$t_p = \frac{\pi}{\omega_n\sqrt{1-\zeta^2}} \tag{2.69}$$

Note 3: On constructing right angle triangle with ζ and $\sqrt{1-\zeta^2}$, we get

$$\sin\theta = \sqrt{1-\zeta^2}$$

$$\cos\theta = \zeta$$

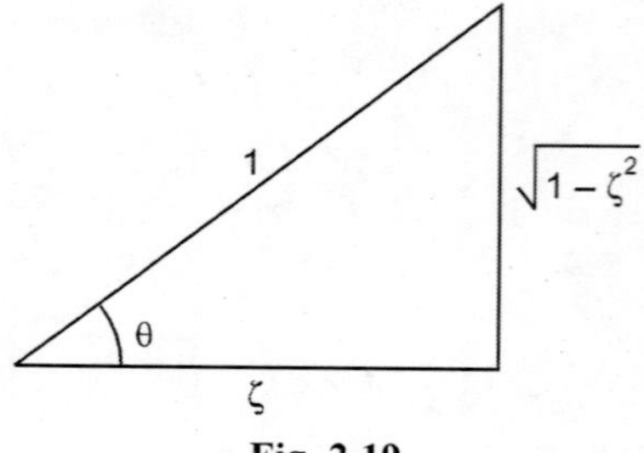

Fig. 2.19

Peak Overshoot (M_p)

$$\% \text{ Peak overshoot, } \% M_p = \frac{C(t_p) - C(\infty)}{C(\infty)} \times 100$$

where $C(t_p)$ = Peak response at $t = t_p$

The out step response of second order system is given by

$$C(t) = 1 - \frac{e^{-\zeta\omega_n t}}{\sqrt{1-\zeta^2}} \sin(\omega_d t + \theta)$$

$$\text{at } t = \infty, C(t) = C(\infty) = 1 - \frac{e^{-\infty}}{\sqrt{1-\zeta^2}} \sin(\omega_d t + \theta) = 1 - 0 = 1 \qquad (2.70)$$

$$\text{at } t = t_p = C(t) = C(t_p) = 1 - \frac{e^{-\zeta\omega_n t_p}}{\sqrt{1-\zeta^2}} \sin(\omega_d t_p + \theta)$$

put $t_p = \pi/\omega_d$

$$\therefore \qquad C(t_p) = 1 - \frac{e^{-\zeta\omega_n \cdot \frac{\pi}{\omega_d}}}{\sqrt{1-\zeta^2}} \sin\left(\omega_d \frac{\pi}{\omega_d} + \theta\right)$$

The damped frequency of oscillation,

$$\omega_d = \omega_n\sqrt{1-\zeta^2}$$

$$C(t_p) = 1 - \frac{e^{-\zeta\omega_n \cdot \frac{\pi}{\omega_n\sqrt{1-\zeta^2}}}}{\sqrt{1-\zeta^2}} \sin(\pi + \theta)$$

$$C(t_p) = 1 + \frac{e^{\frac{-\zeta\pi}{\sqrt{1-\zeta^2}}}}{\sqrt{1-\zeta^2}} \sin\theta \qquad [\text{since } (\pi + \theta) = -\sin\theta]$$

$$C(t_p) = 1 + \frac{e^{\frac{-\zeta\pi}{\sqrt{1-\zeta^2}}}}{\sqrt{1-\zeta^2}} \cdot \sqrt{1-\zeta^2} \qquad \text{[refer note (4)]}$$

$$C(t_p) = 1 + e^{\frac{-\zeta\pi}{\sqrt{1-\zeta^2}}} \tag{2.71}$$

$$\% \, M_p = \frac{C(t_p) - C(\infty)}{C(\infty)} \times 100$$

$$\% \, M_p = \frac{1 + e^{\frac{-\zeta\pi}{\sqrt{1-\zeta^2}}} - 1}{1} \times 100$$

Percentage peak overshoot $\% \, M_p = e^{\frac{-\zeta\pi}{\sqrt{1-\zeta^2}}} \times 100$ (2.72)

Note 4: On constructing right angle triangle with ζ and $\sqrt{1-\zeta^2}$, are get

$$\sin\theta = \sqrt{1-\zeta^2}$$

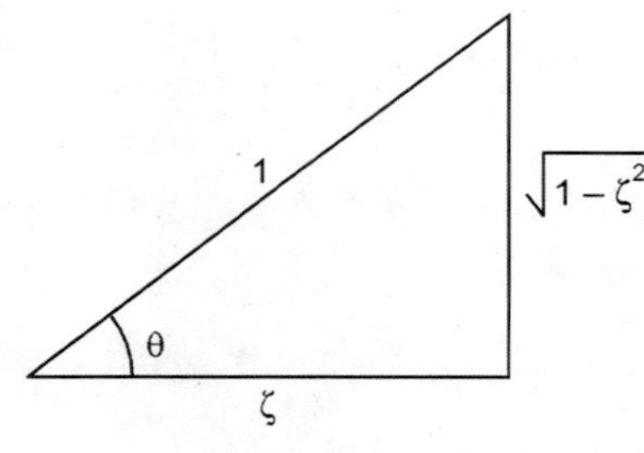

Fig. 2.20

Settling Time (t_s):

The response of second order system has two components they are,

1. Decaying exponential component, $\frac{e^{-\zeta\omega_n t}}{\sqrt{1-\zeta^2}}$
2. Sinusoidal component, $\sin(\omega_d t + \theta)$.

In this decaying exponential term reduces the oscillations produced by sinusoidal component. Hence the settling time is decided by the exponential component. The settling time can be found out by equating exponential component to percentage toterance errors.

For 2% toterance error band, at $t = t_s$,

$$\frac{e^{-\zeta\omega_n t_s}}{\sqrt{1-\zeta^2}} = 0.02$$

For least values of ζ, $e^{-\zeta\omega_n t_s} = 0.02$. On taking natural logarithm we get,

$$-\zeta\omega_n t_s = l_n\,(0.02)$$

$$-\zeta\omega_n t_s = -4$$

$\therefore$ Settling time, $$t_s = \frac{4}{\zeta\omega_n} = 4T;\ \text{for 2\% error.} \tag{2.73}$$

where, $$T = \frac{1}{\zeta\omega_n} = \text{Time constant of the system} \tag{2.74}$$

For 5% error, $e^{-\zeta\omega_n t_s} = 0.05$

on taking natural logarithm we get,

$$-\zeta\,\omega_n t_s = l_n\,(0.05)$$

$$-\zeta\,\omega_n t_s = -3$$

$\therefore$ Settling time, $$t_s = \frac{3}{\zeta\omega_n} = 3T;\ \text{for 5\% errors} \tag{2.75}$$

In general for a specified percentage error settling time,

$$t(s) = \frac{l_n(\%\ \text{error})}{\zeta\omega_n} \tag{2.76}$$

Note: The delay time can be found just dividing the rise time by the value of two. Because we know that the delay time is half the value of rise time (refer fig. 2.16).

2.10.7 Time Response of Second Order System with Unit Ramp Input

The closed loop transfer function of first order system,

$$\frac{C(s)}{R(s)} = \frac{\omega_n^2}{s^2 + 2\zeta\omega_n s + \omega_n^2}$$

It is input is unit step, then,

$$R(s) = 1/s^2$$

$\therefore$ The response in s-domain,

$$C(s) = \frac{R(s)\cdot\omega_n^2}{s^2 + 2\zeta\omega_n s + \omega_n^2} = \frac{\omega_n^2}{s^2(s^2 + 2\zeta\omega_n s + \omega_n^2)}$$

By partial fraction,

$$\frac{\omega_n^2}{s^2\,(s^2 + 2\zeta\omega_n s + \omega_n^2)} = \frac{A}{s} + \frac{B}{s^2} + \frac{Cs + D}{s^2 + 2\zeta\omega_n s + \omega_n^2} \tag{2.77}$$

$$\omega_n^2 = As(s^2 + 2\zeta\omega_n s + \omega_n^2) + B\,(s^2 + 2\zeta\omega_{n\,s} + \omega_n^2) + (Cs + D)\,s^2$$

Equating the coefficient of s^3, s^2, s and constant terms.

$$A + C = 0 \tag{2.78}$$

$$2\zeta\omega_n A + B + D = 0 \tag{2.79}$$

$$A\omega_n^2 + 2\zeta\omega_n B = 0 \tag{2.80}$$

$$B\omega_n^2 = \omega_n^2 \tag{2.81}$$

From equations (2.78) to (2.81)

$$A = -\frac{2\zeta}{\omega_n}; \quad C = \frac{2\zeta}{\omega_n}$$

$$B = 1; \quad D = -1 + 4\zeta^2$$

Substitute all these values in equation (2.77)

$$C(s) = \frac{-2\zeta}{\omega_n s} + \frac{1}{s^2} + \frac{\frac{2\zeta}{\omega_n}\cdot s + (4\zeta^2 - 1)}{s^2 + 2\zeta\omega_n s + \omega_n^2}$$

$$= \frac{1}{s^2} - \frac{2\zeta}{\omega_n s} + \frac{2\zeta}{\omega_n}\cdot\frac{(s + \zeta\omega_n - \zeta\omega_n)}{(s + \zeta\omega_n)^2 + \omega_n^2(1 - \zeta^2)} + \frac{4\zeta^2 - 1}{(s + \zeta\omega_n)^2 + \omega_n(1 - \zeta^2)}$$

$$= \frac{1}{s^2} - \frac{2\zeta}{\omega_n s} + \frac{2\zeta}{\omega_n}\cdot\frac{(s + \zeta\omega_n)}{(s+\zeta\omega_n)^2 + \omega_n^2(1 - \zeta^2)} - \frac{2\zeta}{\omega_n}\cdot\frac{\zeta\omega_n}{(s + \zeta\omega_n)^2 + \omega_n^2(1 - \zeta^2)} + \frac{4\zeta^2 - 1}{(s + \zeta\omega_n)^2 + \omega_n(1 - \zeta^2)}$$

$$C(s) = \frac{1}{s^2} - \frac{2\zeta}{\omega_n s} + \frac{2\zeta}{\omega_n}\cdot\frac{(s + \zeta\omega_n)}{(s + \zeta\omega_n)^2 + \omega_n^2(1 - \zeta^2)} + \frac{2\zeta^2 - 1}{(2 + \zeta\omega_n)^2 + \omega_n^2(1 - \zeta^2)}$$

We know,

$$\omega_d^2 = \omega_n^2(1 - \zeta^2)$$

$$C(s) = \frac{1}{s^2} - \frac{2\zeta}{\omega_n s} + \frac{2\zeta}{\omega_n}\cdot\frac{(s + \zeta\omega_n)}{(s + \zeta\omega_n)^2 + \omega_d^2} + \frac{2\zeta^2 - 1}{(s + \zeta\omega_n)^2 + \omega_d^2}\cdot\frac{\omega_d}{\omega_d}$$

$$C(s) = \frac{1}{s^2} - \frac{2\zeta}{\omega_n s} + \frac{2\zeta}{\omega_n}\cdot\frac{(s + \zeta\omega_n)}{(s + \zeta\omega_n)^2 + \omega_d^2} + \frac{2\zeta^2 - 1}{\omega_d}\cdot\frac{\omega_d}{(s + \zeta\omega_n)^2 + \omega_d^2} \tag{2.82}$$

On taking inverse Laplace transform of equation (2.82)

$$C(t) = t - \frac{2\zeta}{\omega_n} + \frac{2\zeta}{\omega_n}\cdot e^{-\zeta\omega_n t}\cos\omega_d t + \frac{2\zeta^2 - 1}{\omega_d}e^{-\zeta\omega_n t}\sin\omega_d t$$

$$= t - \frac{2\zeta}{\omega_n} + \frac{2\zeta\sqrt{1-\zeta^2}}{\omega_n\sqrt{1-\zeta^2}}e^{-\zeta\omega_n t}\cos\omega_d t + \frac{2\zeta^2 - 1}{\omega_n\sqrt{1-\zeta^2}}e^{-\zeta\omega_n t}\sin\omega_d t$$

CHAPTER 2

$$= t - \frac{2\zeta}{\omega_n} + \frac{e^{-\zeta\omega_n t}}{\omega_n\sqrt{1-\zeta^2}}\left[2\zeta\sqrt{1-\zeta^2}\cos\omega_d t + (2\zeta^2 - 1)\sin\omega_d t\right]$$

Put, $2\zeta^2 - 1 = \cos\phi$

$2\zeta\sqrt{1-\zeta^2} = \sin\theta$

$$C(t) = t - \frac{2\zeta}{\omega_n} + \frac{e^{-\zeta\omega_n t}}{\omega_n\sqrt{1-\zeta^2}}[\sin\phi\cos\omega_d t + \cos\phi\sin\omega_d t]$$

$$= t - \frac{2\zeta}{\omega_n} + \frac{e^{-\zeta\omega_n t}}{\omega_n\sqrt{1-\zeta^2}}\sin(\omega_d t + \phi)$$

$$C(t) = t - \frac{2\zeta}{\omega_n} + \frac{e^{-\zeta\omega_n t}}{\omega_n\sqrt{1-\zeta}}\sin\left[\left(\omega_n\sqrt{1-\zeta^2}\right)t + \phi\right] \quad (2.83)$$

where, $\phi = \cos^{-1}(2\zeta^2 - 1)$

or $\phi = \tan^{-1}\dfrac{2\zeta\sqrt{1-\zeta^2}}{2\zeta^2 - 1}$

The equation (2.83) is the time response of second order system under unit ramp input.

2.10.8 Time Response of Second Order System with Unit Impulse Input

Since, input is unit impulse $R(s) = 1$

$$C(s) = \frac{\omega_n^2}{s^2 + 2\zeta\omega_n s + \omega_n^2} \quad (2.84)$$

Break the equation (2.84) by partial fraction.

$$\frac{\omega_n^2}{s^2 + 2\zeta\omega_n s + \omega_n^2} = \frac{A}{s - s_1} + \frac{B}{s - s_2} \quad (2.84a)$$

The roots of the characteristics equation is,

$$s_1 = -\zeta\omega_n + j\omega_n\sqrt{1-\zeta^2}$$

$$s_2 = -\zeta\omega_n - j\omega_n\sqrt{1-\zeta^2}$$

$$\omega_d = \omega_n\sqrt{1-\zeta^2}$$

$$s_1 = -\zeta\omega_n + j\omega_d$$

$$s_2 = -\zeta\omega_n - j\omega_d$$

From Eq. (2.84a)

$$A = \left.\frac{\omega_n^2}{s - s_2}\right|_{s=s_1}, = \frac{\omega_n^2}{-\zeta\omega_n + j\omega_d + \zeta\omega_n + j\omega_d} = \frac{\omega_n^2}{2\,j\omega_d}$$

$$B = \frac{\omega_n^2}{-j\,2\omega_d}$$

Substitute A and B in (2.84a)

$$C(s) = \frac{\omega_n^2}{s^2 + 2\zeta\omega_n s + \omega_n^2} = \frac{\omega_n^2}{j\,2\omega_d}\cdot\frac{1}{s - s_1} - \frac{\omega_n^2}{j\,2\omega_d}\cdot\frac{1}{s - s_2} \tag{2.85}$$

On taking inverse Laplace transform for equation (2.85)

$$L^{-1}[C(s)] = C(t) = \frac{\omega_n^2}{2\,j\omega_d}[e^{s_1 t} - e^{s_2 t}]$$

$$= \frac{\omega_n^2}{j\,2\omega_d}[e^{(-\zeta\omega_n + j\omega_d)t} - e^{(-\zeta\omega_n - j\omega_d)t}]$$

$$= \frac{\omega_n^2}{j\,2\omega_d}[e^{-\zeta\omega_n t}\cdot e^{j\omega_d t} - e^{-\zeta\omega_n t}\cdot e^{-j\omega_d t}]$$

$$= \frac{\omega_n^2\cdot e^{-\zeta\omega_n t}}{j\,2\omega_d}[e^{j\omega_d t} - e^{-j\omega_d t}]$$

$$= \frac{\omega_n^2\cdot e^{-\zeta\omega_n t}}{j\,2\omega_d}[\cos\omega_d t + j\sin\omega_d t - \cos\omega_d t + j\sin\omega_d t]$$

$$= \frac{\omega_n^2\cdot e^{-\zeta\omega_n t}}{j\,2\omega_d}[j\,2\sin\omega_d t] = \frac{\omega_n^2\cdot e^{-\zeta\omega_n t}}{\omega_d}\cdot\sin\omega_d t \tag{2.86}$$

Put ω_d in eq. (2.86)

$$C(t) = \frac{\omega_n^2\cdot e^{-\zeta\omega_n t}}{\omega_n\sqrt{1-\zeta^2}}\sin\left(\omega_n\sqrt{1-\zeta^2}\right)t$$

$$C(t) = \frac{\omega_n}{\sqrt{1-\zeta^2}}\cdot e^{-\zeta\omega_n t}\cdot\sin\left(\omega_n\sqrt{1-\zeta^2}\right)t \tag{2.87}$$

(i) $\zeta < 1$:

$$C(t) = \frac{\omega_n}{\sqrt{1-\zeta^2}}\cdot e^{-\zeta\omega_n t}\sin\left(\omega_n\sqrt{1-\zeta^2}\right)t$$

(ii) $\zeta = 1$:

$$C(t) = \underset{\zeta \to 1}{Lt} \frac{\omega_n}{\sqrt{1-\zeta^2}} e^{-\zeta\omega_n t} \sin\left(\omega_n \sqrt{1-\zeta^2}\right) t$$

$$C(t) = t \cdot e^{-\omega_n t} \cdot \omega_n^2$$

(iii) $\zeta > 1$

Equation (2.84) can be written as

$$C(s) = \frac{\omega_n^2}{(s+\zeta\omega_n)^2 - \omega_n^2 (\zeta^2 - 1)} \quad (2.88)$$

Apply partial fraction and put $\omega_d^2 = \omega_n^2 (\zeta^2 - 1)$

$$\frac{\omega_n^2}{(s+s\omega_n)^2 - \omega_d^2} = \frac{\omega_n^2}{(s+\zeta\omega_n+\omega_d)(s+\zeta\omega_n-\omega_d)}$$

$$= \frac{A}{(s+\zeta\omega_n+\omega_d)} + \frac{B}{(s+\zeta\omega_n-\omega_d)} \quad (2.89)$$

From equation (2.88)

$$A = \left.\frac{\omega_n^2}{s+\zeta\omega_n-\omega_d}\right|_{s=-\zeta\omega_n-\omega_d}$$

$$A = \frac{-\omega_n}{2\sqrt{\zeta^2-1}}$$

Similarly, $$B = \frac{-\omega_n}{2\sqrt{\zeta^2-1}}$$

Substitute the values in eqn. (2.88)

$$C(s) = \frac{\omega_n^2}{(s+\zeta\omega_n)^2 - \omega_d^2}$$

$$= \frac{-\omega_n}{2\sqrt{\zeta^2-1}} \cdot \frac{1}{s+\zeta\omega_n+\omega_d} + \frac{\omega_n}{2\sqrt{\zeta^2-1}} \cdot \frac{1}{s+\zeta\omega_n-\omega_d}$$

Take inverse Laplace,

$$L^{-1}[C(s)] = C(t) = \frac{\omega_n}{2\sqrt{\zeta^2-1}}\left[-e^{-[\zeta\omega_n+\omega_d]t} + e^{-(\zeta\omega_n-\omega_d)t}\right]$$

$$C(t) = \frac{\omega_n}{2\sqrt{\zeta^2-1}}\left[e^{-\left(\zeta-\sqrt{\zeta^2-1}\right)\omega_n t} - e^{-\left(\zeta+\sqrt{\zeta^2-}\right)\omega_n t}\right] \quad (2.90)$$

The equation (2.90) in the time response of second order system under unit impulse input.

2.10.9 Time Response of Higher Order Control System

The input, output and transfer function are given by

$$C(s) = R(s) \cdot T(s)$$

where,

$C(s) \rightarrow$ Output of the system
$R(s) \rightarrow$ Input of the system
$T(s) \rightarrow$ Transfer function of the system

$$T(s) = \frac{K\,(s^m + a_1 s^{m-1} + \ldots a_m)}{(s^n + b_1\, s^{n-1} + \ldots b_n)}$$

$$C(s) = R(s) \cdot \frac{K\,(s^m + a_1\, s^{m-1} + \ldots + a_m)}{(s^n + b_1\, s^{n-1} + \ldots b_n)}$$

Apply partial fraction,

$$C(s) = \frac{K_1}{s + P_1} + \frac{K_2}{(s + P_2)} + \ldots \frac{K_n}{(s + P_n)}$$

For time response take inverse Laplace of each part of above equation

$$L^{-1}\, C(s) = L^{-1}\left(\frac{K_1}{s + P_1}\right) + L^{-1}\left(\frac{K_2}{s + P_2}\right) + \ldots L^{-1}\left(\frac{K_n}{s + P_n}\right)$$

$$C(t) = K_1\, e^{-P_1 t} + K_2 e^{-P_2 t} + \ldots + K_n e^{-P_n t} \quad (2.91)$$

The equation (2.91) is the time response of higher order system.

SOME WORKED EXAMPLES

EXAMPLE 2.1

Obtain the response of unity feedback system whose open loop transfer function is $G(s) = \dfrac{4}{s\,(s + 5)}$ and the input is unit step.

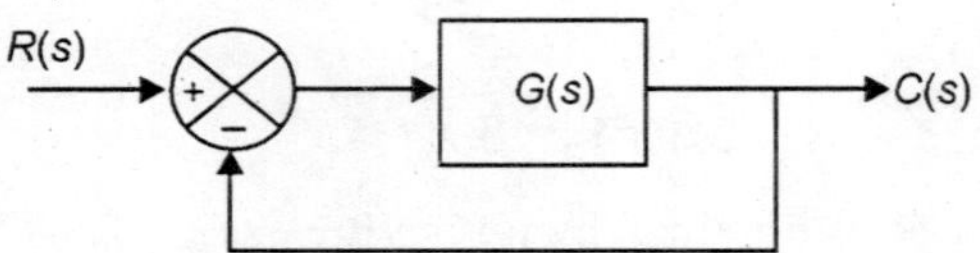

Fig. Ex. 2.1: Closed loop system.

CHAPTER 2

Solution: The closed loop system is shown in Fig. (Ex. 2.1)

The closed loop transfer function,

$$\frac{C(s)}{R(s)} = \frac{G(s)}{1+G(s)}$$

$$\frac{C(s)}{R(s)} = \frac{4/s(s+5)}{1+\dfrac{4}{s(s+5)}} = \frac{4/s(s+5)}{\dfrac{s(s+5)+4}{s(s+5)}} = \frac{4}{s(s+5)+4}$$

$$\frac{C(s)}{R(s)} = \frac{4}{s^2+5s+4} = \frac{4}{(s+4)\,(s+1)}$$

The response in s-domain

$$C(s) = R(s)\cdot\frac{4}{(s+1)\,(s+4)}$$

Since the input is unit step, $R(s) = 1/s$

$$\therefore \qquad C(s) = \frac{4}{s(s+1)\,(s+4)}$$

By Partial fraction expansion, we can write,

$$C(s) = \frac{4}{s(s+1)\,(s+4)} = \frac{A}{s} + \frac{B}{(s+1)} + \frac{C}{(s+4)}$$

$$A = C(s)\big|_{s=0} = \left.\frac{4}{(s+1)\,(s+4)}\right|_{s=0} = \frac{4}{1\times 4} = 1 \Rightarrow A = 1$$

$$B = C(s)\,(s+1)\big|_{s=-1} = \frac{4}{s(s+4)} = \frac{4}{-1\,(-1+4)} = \frac{-4}{3} \Rightarrow B = -4/3$$

$$C = C(s)\,(s+4)\big|_{s=-4} = \left.\frac{4}{s(s+1)}\right|_{s=-4}$$

$$= \frac{4}{-4(-4+1)} = \frac{1}{3} \Rightarrow C = 1/3$$

Substitute A, B and C values in $C(s)$,

$$C(s) = \frac{1}{s} - \frac{4}{3(s+1)} + \frac{1}{3(s+4)}$$

The time domain response $C(t)$ is obtained by taking inverse Laplace transform of $C(s)$.

The response in time domain,

$$C(t) = L^{-1}[C(s)] = L^{-1}\left[\frac{1}{s} - \frac{4}{3}\frac{1}{s+1} + \frac{1}{3}\frac{1}{s+4}\right]$$

$$= 1 - \frac{4}{3}e^{-t} + \frac{1}{3}e^{-4t} = 1 - \frac{1}{3}[4e^{-t} - e^{-4t}]$$

Results: The response of unity feedback system is:

$$C(t) = 1 - \frac{1}{3}[4e^{-t} - e^{-4t}]$$

EXAMPLE 2.2

A positional control system with velocity feedback is shown in Fig. (Ex. 2.2) what is the the response of the system for unit step input.

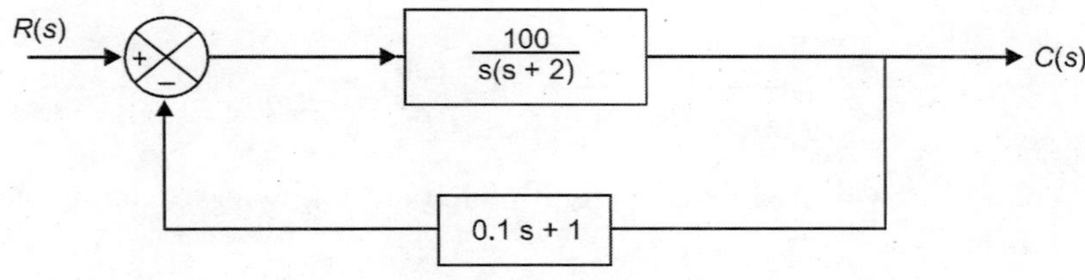

Fig. Ex. 2.2

Solution: The closed loop transfer function,

$$\frac{C(s)}{R(s)} = \frac{G(s)}{1 + G(s)\,H(s)}$$

Given that $G(s) = \dfrac{100}{s(s+2)}$ and $H(s) = .1s + 1$

$$= \frac{\dfrac{100}{s\,(s+2)}}{1 + \left[\dfrac{100}{s(s+2)}\right][0.1s+1]} = \frac{100/s(s+2)}{\dfrac{s(s+2) + 100\,(.1\,s+1)}{s(s+2)}}$$

$$= \frac{100}{s^2 + 2s + 10s + 100} = \frac{100}{s^2 + 12s + 100}$$

Here, $(s^2 + 12s + 100)$ is the characteristics polynomial. The roots of the characteristic polynomial are

$$s_1, s_2 = \frac{-12 \pm \sqrt{144 - 400}}{2} = \frac{-12 \pm j\,16}{2} = -6 \pm j\,8$$

The roots are complex conjugate. The system is underdamped and so the response of the system will have damped oscillations.

The response in s-domain

$$C(s) = R(s)\frac{100}{s^2 + 12s + 100}$$

since inut is unit step, $R(s) = 1/s$

$$\therefore \quad C(s) = \frac{1}{s}\cdot\frac{100}{s^2 + 12s + 100} = \frac{100}{s(s^2 + 12s + 100)}$$

By Partial fraction expansion we can write,

$$C(s) = \frac{100}{s(s^2 + 12s + 100)} = \frac{A}{s} + \frac{Bs + C}{s^2 + (12s + 100)}$$

The residue A is obtained by multiplying $C(s)$ by s and letting $s = 0$.

$$A = C(s)\big|_{s=0} = \left.\frac{100}{s^2 + 12s + 100}\right|_{s=0} = \frac{100}{100} = 1 \Rightarrow A = 1$$

The residue B and C are evaluated by cross multiplying the following equation and equating the coefficients of like power of s.

$$\frac{100}{s(s^2 + 12s + 100)} = \frac{A}{s} + \frac{Bs + C}{s^2 + 12s + 100}$$

$$100 = A(s^2 + 12s + 100) + (Bs + C)\,s$$

$$100 = As^2 + 12As + 100\,A + Bs^2 + Cs$$

on equating the coefficients of s^2 we get,

$$A + B = 0$$

$$\therefore \quad B = -A = -1$$

$$A = 1$$

on equating the coefficients of s we get,

$$0 = 12A + C \quad \therefore C = -12A = -12$$

$$C = -12$$

Substitute A, B and C values is equation (1)

$$\therefore \quad C(s) = \frac{1}{s} + \frac{-s - 12}{s^2 + 12s + 100}$$

$$= \frac{1}{s} - \frac{s + 12}{s^2 + 12s + 36 + 64} = \frac{1}{s} - \frac{s + 6 + 6}{(s + 6)^2 + 64}$$

$$= \frac{1}{s} - \frac{(s+6)}{(s+6)^2 + 64} - \frac{6}{(s+6)^2 + 64}$$

$$C(s) = \frac{1}{s} - \frac{(s+6)}{(s+6)^2 + 8^2} - \frac{6}{8} \cdot \frac{8}{(s+6)^2 + 8^2}$$

The time domain response is obtained by taking inverse Laplace transform of $C(s)$.

The response in time domain

$$C(t) = L^{-1}\,[C(s)] = L^{-1}\left[\frac{1}{s} - \frac{s+6}{(s+6)^2 + 8^2} - \frac{6}{8} \cdot \frac{8}{(s+6)^2 + 8^2}\right]$$

$$C(t) = 1 - e^{-6t}\cos t - \frac{6}{8} e^{-6t} \sin 8t$$

$$= 1 - e^{-6t}\left[\frac{6}{8} \sin 8t + \cos 8t\right]$$

The result can be converted to another standard form by constructing right angle triangle with ζ and $\sqrt{1-\zeta^2}$. The damping ratio ζ is evaluated by comparing the closed loop transfer function of the system with standard form of second order transfer function.

$$\frac{C(s)}{R(s)} = \frac{\omega_n^2}{s^2 + 2\zeta\omega_n s + \omega_n^2}$$

$$= \frac{100}{s^2 + 12s + 100}$$

on companing we get,

$$\omega_n^2 = 100$$

$$\boxed{\omega_n = 100}$$

$$2\zeta\omega_n = 12$$

$$\zeta = \frac{12}{\omega_n} = \frac{12}{2 \times 10} = .6 \Rightarrow \boxed{\zeta = .6}.$$

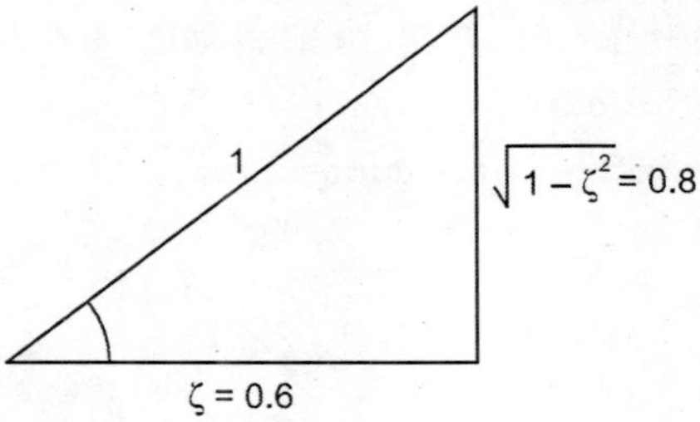

1

$\sqrt{1-\zeta^2} = 0.8$

$\zeta = 0.6$

constructing right angle triangle with ζ and $\sqrt{1-\zeta^2}$ we get,

$$\sin\theta = 0.8;\ \cos\theta = 0.6$$

$$\tan\theta = \frac{0.8}{0.6}$$

$$\theta = \tan^{-1}\left[\frac{0.8}{0.6}\right] = 53° = .925 \text{ rad}$$

$$\theta = .925 \text{ rad}$$

The response in time domain

$$C(t) = 1 - e^{-6t}\left[\frac{6}{8}\sin 8t + \cos 8t\right]$$

$$= 1 - e^{-6t}\,\frac{10}{8}\,e^{-6t}\left[\frac{6}{8}\sin 8t + \frac{8}{10}\cos 8t\right]$$

$$= 1 - \frac{10}{8}e^{-6t}\left[\sin 8t \times 0.6 + \cos 8t \times 0.8\right]$$

$$= 1 - 1.25\,e^{-6t}\left[\sin 8t + \cos\theta + \cos 8t \sin\theta\right]$$

$$= 1 - 1.25\,e^{-6t}\left[\sin 8t + \theta\right]$$

$$C(t) = 1 - 1.25e^{-6t}\sin\left[8t + .925\right]$$

Result: The response in time domain,

$$C(t) = 1 - e^{-6t}\left[\frac{6}{8}\sin 8t + \cos 8t\right]$$

$$C(t) = 1 - 1.25e^{-6t}\sin(8t + .925)$$

EXAMPLE 2.3

Measurements conducted on a servomechanism shows the system response to be $C(t) = 1 + 0.2.\ e^{-60t} - 1.2\ e^{-10t}$ when subject to a unit step input. obtain an expression for closed loop transfer function. determine the undamped natural frequancy and damping ration.

Solution: Given that $C(t) = 1 + 0.2\ e^{-60t} - 1.2\ e^{-10t}$

on taking Laplace transform of $c(t)$ we get,

$$L[C(t)] = C(s) = \frac{1}{s} + 0.2\frac{1}{(s+60)} - 1.2\frac{1}{(s+10)}$$

$$= \frac{(s+60)(s+10) + 0.2\,s\,(s+10) - 1.2\,s\,(s+60)}{s(s+60)(s+10)}$$

$$= \frac{s^2 + 70s + 600 + 0.2\,s^2 + 2s - 1.2s^2 - 72s}{s\,(s+60)(s+10)}$$

$$= \frac{600}{s(s+60)(s+10)}$$

$$C(s) = \frac{600}{s(s+60)(s+10)} = \frac{1}{s}\,\frac{600}{(s+60)(s+10)}$$

Since input is unit step, $R(s) = 1/s$

$$C(s) = R(s)\frac{600}{s^2 + 70s + 600}$$

The closed loop transfer function of the system,

$$\frac{C(s)}{R(s)} = \frac{600}{s^2 + 70s + 600}$$

The damping ratio and natural frequency of oscillation can be estimated by comparing the system transfer function with standard form of second order transfer function.

$$\frac{C(s)}{R(s)} = \frac{\omega_n^2}{s^2 + 2\zeta\omega_n s + \omega_n^2} = \frac{600}{s^2 + 70s + 600}$$

Compare L.H.S and R.H.S

$$\omega_n^2 = 600$$

$$\omega_n = \sqrt{600} = 24.49 \text{ rad/sec}$$

$$2\zeta\,\omega_n = 70$$

$$\zeta = \frac{70}{2\omega_n} = \frac{70}{2 \times 24.49} = 1.43.$$

Result:

- The closed loop transfer function of the system

$$\frac{C(s)}{R(s)} = \frac{600}{s^2 + 70s + 600}$$

- Natural frequency of oscillation, $\omega_n = 24.49$ rad/sec.
- Damping ratio, $\zeta = 1.43$ (over damped).

EXAMPLE 2.4

The unity feedback system is characterized by an open loop transfer function $G(s) = \dfrac{K}{s(s+10)}$.

Determine the gain K, so that the system will have a damping ratio of 0.5 for this value of K. Determine settling time, peak overshoot and time to peak overshoot for a unit step input.

Solution: This is the unity feedback system.

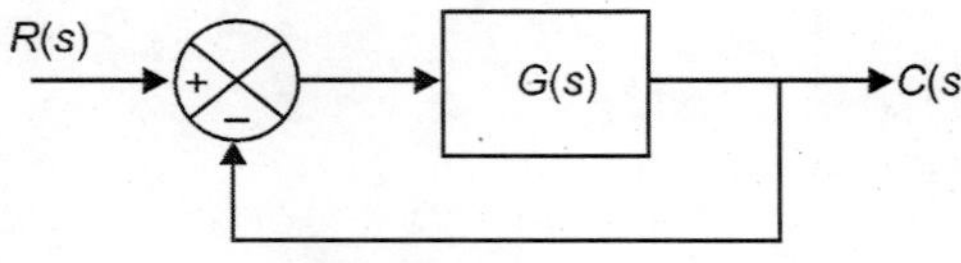

Fig. Ex. 2.4

The closed loop transfer function,

$$\frac{C(s)}{R(s)} = \frac{G(s)}{1+G(s)} \qquad [\because H(s)=1]$$

Given that, $$G(s) = \frac{K}{s(s+10)}.$$

So,

$$\frac{C(s)}{R(s)} = \frac{K/s(s+10)}{1+K/s(s+10)} = \frac{K}{s(s+10)+K} = \frac{K}{s^2+10s+K}$$

The value of K can be evaluated by comparing the system transfer function with standard form of second order transfer function.

$$\frac{C(s)}{R(s)} = \frac{\omega_n^2}{s^2+2\zeta\omega_n s+\omega_n^2} = \frac{K}{s^2+10s+K}$$

Compare the co-efficient of ω_n in both RHS and LHS

$\omega_n^2 = K$ $\qquad$ $2\zeta\omega_n = 10$

$\omega_n = \sqrt{K}$ $\qquad$ $\omega_n = 10/\text{rad/sec}$

Now $\zeta = .5$ and $\omega_n = \sqrt{K}$

$$\text{so, } 2 \times .5 \times \sqrt{K} = 10$$

$$\sqrt{K} = 10, \qquad \therefore \text{ gain, } K = 100$$

Percentage peak overshoot, $\% M_p = e^{-\zeta\pi/\sqrt{1-\zeta^2}} \times 100$

$$= e^{-0.5\pi/\sqrt{1-(0.5)^2}} \times 100$$

$$= 0.163 \times 100$$

$$\% M_p = 16.3\%$$

peak time, $$t_p = \pi/\omega_d = \frac{\pi}{\omega_n\sqrt{1-\zeta^2}}$$

$$= \frac{\pi}{10\sqrt{1-(0.5)^2}} = 0.363 \text{ sec}$$

Result: The value of gain, $K = 100$

Percentage peak overshoot $\% M_p = 16.3\%$

Peak time $t_p = 0.363$ sec.

EXAMPLE 2.5

The open loop transfer function of a unity feedback system is given by $G(s) = K/s\,(sT + 1)$ where K and T are positive constant. By what factor should the amplifier gain K be reduced, so that the peak overshoot of unit step response of the system is reduced step response of the system is reduced from 75% to 25%.

Solution: Unit feedback system is,

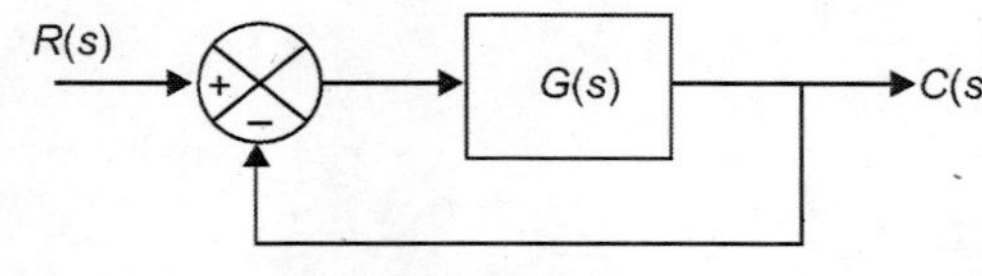

Fig. Ex. 2.5

The closed loop transfer function,

$$\frac{C(s)}{R(s)} = \frac{G(s)}{1+G(s)} \qquad [\because H(s) = 1]$$

Given that, $G(s) = K/s(sT + 1)$

$$\frac{C(s)}{R(s)} = \frac{K/s(sT+1)}{1+K/s(sT+1)} = \frac{K}{s(sT+1)+K} = \frac{K}{s^2T+s+K} = \frac{K/T}{s^2+\frac{1}{T}s+\frac{K}{T}}$$

Expression for ζ and ω_n can be obtained by comparing the transfer function with the standard form of second order transfer function.

$$\frac{C(s)}{R(s)} = \frac{\omega_n^2}{s^2+2\zeta\omega_n s+\omega_n^2} = \frac{K/T}{s^2+\frac{1}{T}\cdot s+\frac{K}{T}}$$

Compare the co-efficient of ω_n in RHS and LHS

$$\omega_n^2 = K/T$$

$$\omega_n = \sqrt{K/T}$$

$$2\zeta\omega_n = 1/T$$

$$\zeta = \frac{1}{2\omega_n T}$$

$$\zeta = \frac{1}{2\sqrt{\frac{K}{T}}\cdot T} = \frac{1}{2\sqrt{KT}}$$

The peak overshoot, M_p is reduced by increasing the damping ratio ζ. The damping ratio is increased by reducing the gain K.

when $M_p = 0.75$, Let $\zeta = \zeta_1$ and $K = K_1$

when $M_p = 0.25$, Let $\zeta = \zeta_2$ and $K = K_2$

Peak overshoot, $M_p = e^{-\zeta\pi/\sqrt{1-\zeta^2}}$ taking logarithm on both sides,

$$\ln M_p = \frac{-\zeta\pi}{\sqrt{1-\zeta^2}}$$

on squarring, we get,

$$(ln\, M_p)^2 = \frac{\zeta^2\,\pi^2}{1-\zeta^2}$$

on cross multiplication we get,

$$(1-\zeta^2)\,(ln\, M_p)^2 = \zeta^2\,\pi^2$$

$$(ln\, M_p)^2 - \zeta^2\,(ln\, M_p)^2 = \zeta^2\,\pi^2$$

$$(ln\, M_p)^2 = \zeta\,\pi^2 + \zeta^2\,(ln\, M_p)^2$$

$$(ln\, M_p)^2 = \zeta^2[\pi^2 + (\ln M_P)^2] \quad (1)$$

$$\therefore \quad \zeta^2 = \frac{(ln\, M_p)^2}{\pi^2 + (ln\, M_p)^2}$$

But

$$\zeta = \frac{1}{2\sqrt{KT}}$$

$$\zeta^2 = \frac{1}{4KT} \quad (2)$$

On equating equation (1) and (2)

$$\frac{1}{4KT} = \frac{(ln\, M_p)^2}{\pi^2 + (ln\, M_p)^2}$$

$$\frac{1}{K} = \frac{4T\,(ln\, M_p)^2}{\pi^2 + (ln\, M_p)^2}$$

$$K = \frac{\pi^2 + (ln\, M_p)^2}{4T\,(ln\, M_p)^2}$$

when, $K = K_1$, $M_p = 0.75$

$$K_1 = \frac{\pi^2 + (ln\, 0.75)^2}{4T\,(ln\, 0.75)^2} = \frac{9.952}{0.331T} = \frac{30.06}{T}$$

when, $K = K_2$, $M_p = 0.25$

$$K_2 = \frac{\pi^2 + (ln\, 0.25)^2}{4T\,(ln\, 0.25)^2} = \frac{11.79}{7.68T} = \frac{1.53}{T}$$

$$\frac{K_1}{K_2} = \frac{(1/T)\,30.06}{(1/T)\,1.53} = 19.6$$

$$K_1 = 19.6\,K_2 \text{ (or) } K_2 = \frac{1}{19.6}K_1$$

To reduce peak overshoot from 0.75 to 0.25, K should be reduced by 19.6 times. (approximately 20 times).

Results: The value of gain, K should be reduced approximately 20 times to reduce peak overshoot from 0.75 to 0.25.

EXAMPLE 2.6

A positional control system with velocity feedback is shown. What is the response $C(t)$ to the unit step input. Giventhat $\zeta = 0.5$. Also calculate, rise time, peak time, maximum overshoot and settling time.

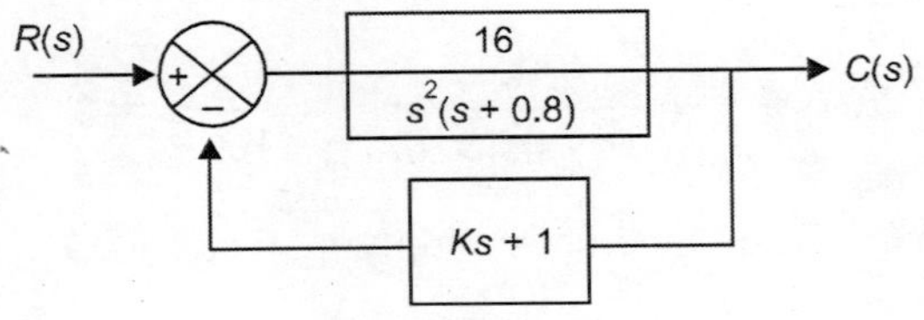

Fig. Ex. 2.6

Solution: The closed loop transfer function,

$$\frac{C(s)}{R(s)} = \frac{G(s)}{1 + G(s)\,H(s)}$$

Given that, $G(s) = 16/s\,(s + 0.8)$ and

$H(s) = Ks + 1$

$$\frac{C(s)}{R(s)} = \frac{16/s(s+0.8)}{1 + \dfrac{16}{s(s+0.8)}(Ks+1)} = \frac{16}{s^2 + (0.8 + 16K)\,s + 16}$$

The values of K and ω_n are obtained by comparing the system transfer function with standard form of second order transfer function.

$$\frac{C(s)}{R(s)} = \frac{\omega_n^2}{s^2 + 2\zeta\omega_n s + \omega_n^2} = \frac{16}{s^2 + (0.8 + 16K)\,s + 16}$$

Compare LHS and RHS

$\omega_n^2 = 16$ $\qquad 0.8 + 16\,K = 2\,\zeta\,\omega_n$

$\omega_n = 4$ rad/sec $\qquad K = \dfrac{2\zeta\omega_n - 0.8}{16} = \dfrac{2 \times 0.5 \times 4 - 0.8}{16} = 0.2$

$$\frac{C(s)}{R(s)} = \frac{16}{s^2 + (0.8 + 16 \times 0.2)\,s + 16} = \frac{16}{s^2 + 4s + 16} \quad 3$$

Given that the damping ratio, $\zeta = 0.5$. Hence the system is underdamped and so the response of the system will have damped oscillations. The roots of characteristics polynomial will be complex conjugate.

The response in s-domain, $C(s) = R(s)\dfrac{16}{s^2+4s+16}$

For unit step input, $R(s) = 1/s$

$$C(s) = \frac{1}{s}\cdot\frac{16}{s^2+4s+16} = \frac{16}{s(s^2+4s+16)}$$

apply partial fraction,

$$C(s) = \frac{16}{s(s^2+4s+16)} = \frac{A}{s} + \frac{Bs+C}{s^2+4s+16}$$

The residue A is obtained by multiplying $C(s)$ by s and letting $s = 0$

$$A = C(s)\cdot s\Big|_{s=0} = \frac{16}{s^2+4s+16}\Bigg|_{s=0} = \frac{16}{16} = 1 \Rightarrow A = 1$$

The residues B and C are evaluated by cross multiplying the following equation and equating the co-efficients of like power of s.

$$\frac{16}{s(s^2+4s+16)} = \frac{A}{s} + \frac{Bs+C}{s^2+4s+16}.$$

on cross multiplication we get,

$$16 = A(s^2+4s+16) + (Bs+C)s$$

$$16 = As^2 + 4As + 16A + Bs^2 + Cs.$$

Equate the co-efficients of s^2 terms, s terms and constant terms, we get

$O = A + B$ $\qquad\qquad O = 4A + C$

$\therefore\quad B = -A = -1$ $\qquad\qquad \therefore C = -4A = -4$

$\boxed{A = 1}$

$$C(s) = \frac{1}{s} + \frac{-s-4}{s^2+4s+16} = \frac{1}{s} - \frac{s+4}{s^2+4s+4+12}$$

$$= \frac{1}{s} - \frac{s+2+2}{(s+2)^2+12} = \frac{1}{s} - \frac{s+2}{(s+2)^2+12} - \frac{2}{\sqrt{12}}\frac{\sqrt{12}}{(s+2)^2+12}$$

Apply inverse Laplace transform of $C(s)$ to get time domain response,

$$C(t) = L^{-1}[C(s)] = L^{-1}\left[\frac{1}{s} - \frac{s+2}{(s+2)^2+12} - \frac{2}{\sqrt{12}}\frac{\sqrt{12}}{(s+2)^2+12}\right]$$

$$= 1 - e^{-2t}\cos\sqrt{12}t - \frac{2}{2\sqrt{3}}\cdot e^{-2t}\cdot\sin\sqrt{12}t$$

$$= 1 - e^{-2t}\left[\frac{1}{\sqrt{3}}\sin\left(\sqrt{12}t\right) + \cos\left(\sqrt{12}t\right)\right]$$

to get the result, use right angle triangle with ζ and $\sqrt{1-\zeta^2}$,

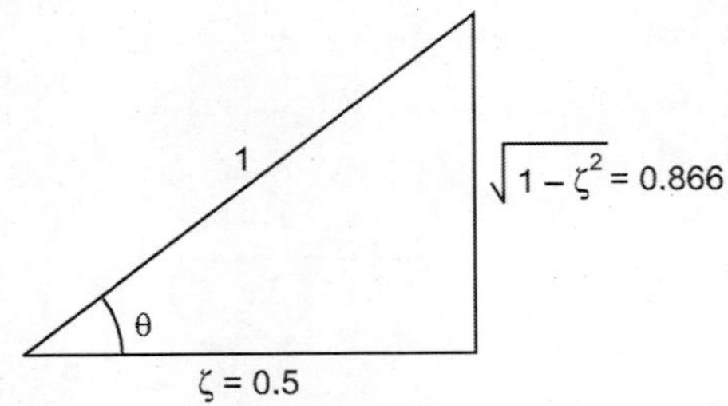

Fig. Ex. 2.6.1

$$\sin\theta = 0.866 = \frac{\sqrt{3}}{2}$$

$$\cos\theta = 0.5 = \frac{1}{2}$$

$$\tan\theta = 1.732$$

$$\theta = \tan^{-1} - 60° = 1.047 \text{ rad}$$

The response in time domain is,

$$C(t) = 1 - e^{-2t}\left[\frac{1}{\sqrt{3}} \times 2 \times \sin\sqrt{12}t \times \frac{1}{2} + \frac{2}{\sqrt{3}} \times \cos\sqrt{12}t \times \frac{\sqrt{3}}{2}\right]$$

$$= \left(1 - e^{-2t}\right)\left(\frac{2}{\sqrt{3}}\right)\left[\sin\sqrt{12}t\cos\theta + \cos\theta\sqrt{12}t\sin\theta\right]$$

$$= 1 - \frac{2}{\sqrt{3}}e^{-2t}\left[\sin\left(\sqrt{12}t + \theta\right)\right]$$

$$C(t) = 1 - \frac{2}{\sqrt{3}}e^{-2t}\left[\sin\left(\sqrt{12}t + 1.047\right)\right]$$

Damped frequency of oscillation

$$\omega_d = \omega_n\sqrt{1-\zeta^2} = 4\sqrt{1-(0.5)^2} = 3.464 \text{ rad/sec}$$

$$\therefore \quad \text{Rise time, } t_r = \frac{\pi - \theta}{\omega_d} = \frac{\pi - 1.047}{3.464} = 0.6046 \text{ sec}$$

$$\text{Peak time, } t_p = \frac{\pi}{\omega_d} = \frac{\pi}{3.464} = 0.907 \text{ sec}$$

$$\% \text{ Maximum overshoot, } \%M_p = e^{\frac{-\zeta\pi}{\sqrt{1-\zeta^2}}} \times 100$$

$$= e^{\frac{-0.5 \times \pi}{\sqrt{1-(0.5)^2}}} \times 100$$

$$= 0.163 \times 100 = 16.3\%$$

Settling time, $T_s = 3T$ for 5% error

$= 4T$ for 2% error

$$\text{Time constant, } T = \frac{1}{\zeta\omega_n} = \frac{1}{0.5 \times 4} = 0.5 \text{ sec}$$

For 5% error, settling time, $t_s = 3T = 3 \times 0.5 = 1.5$ sec

For 2% error, settling time, $t_s = 4T = 4 \times 0.5$

$= 2$ sec

Result: The time domain response,

$$C(t) = 1 - e^{-2t}\left[\frac{1}{\sqrt{3}} \sin\left(\sqrt{12}t\right) + \cos\left(\sqrt{12}t\right)\right]$$

or

$$C(t) = 1 - \frac{2}{\sqrt{3}} e^{-2t}\left[\sin\left(\sqrt{12}t + 1.047\right)\right]$$

Rise time, $t_r = 0.6046$ sec

Peak time $t_p = 0.907$ sec

Percentage maximum overshoot, $\% M_p = 16.3\%$

Settling time, $t_s = 1.5$ sec for 5% error

$= 2$ sec for 2% error

EXAMPLE 2.7

A unity feedback control system is characterized by the following open loop transfer function $G(s) = (0.4s + 1)/s(s + 0.6)$. Determine its transient response for unit step input and sketch response. Evaluate the maximum overshoot and the corresponding peak time.

Solution: The closed loop transfer function, $\dfrac{C(s)}{R(s)} = \dfrac{G(s)}{1 + G(s)\,H(s)}$

Given that, $G(s) = (0.4s + 1)/s(s + 0.6)$

For unity feedback system, $H(s) = 1$

$$\frac{C(s)}{R(s)} = \frac{G(s)}{1 + G(s)} = \frac{\dfrac{0.4s+1}{s(s+0.6)}}{1 + \dfrac{0.4s+1}{s(s+0.6)}} = \frac{0.4s+1}{s(s+0.6)\,0.4s+1}$$

$$= \frac{0.4s+1}{s^2 + 0.6s + 0.4s + 1} = \frac{0.4s+1}{s^2 + s + 1}$$

The s-domain response, $C(s)= R(s)\cdot\dfrac{0.4s+1}{s^2+s+1}$.

For step input, $R(s)=\dfrac{1}{s}$

$$C(s)=\frac{1}{s}\cdot\frac{0.4s+1}{s^2+s+1}=\frac{0.4s+1}{s(s^2+s+1)}$$

Apply partial fraction, $C(s)=\dfrac{0.4s+1}{s(s^2+s+1)}\dfrac{A}{s}+\dfrac{Bs+C}{s^2+s+1}$

multiply $C(s)$ by s and letting $s = 0$,

$$A=C(s)\cdot\big|_{s=0}=\left.\frac{0.4s+1}{s^2+s+1}\right|_{s=0}=1$$

The residues B and C are solved by cross multiplying the following equation and equating the coefficients of like power of s.

$$\frac{0.4s+1}{s\,(s^2+s+1)}=\frac{A}{s}+\frac{Bs+C}{s^2+s+1}$$

on cross multiplication we get

$$0.4s + 1 = A(s^2 + s + 1) + (Bs + C)s$$

$$0.4s + 1 = As^2 + As + A + Bs^2 + Cs.$$

on equating coefficients of s^2,

$$O = A + B$$

$$B = -A = -1$$

on equating coefficients of s,

$$0.4 = A + C$$

$$C = 0.4 - A = -0.6$$

$$C(s)=\frac{1}{s}+\frac{-s-0.6}{s^2+s+1}=\frac{1}{s}-\frac{s+0.6}{s^2+s+0.25+0.75}$$

$$=\frac{1}{s}-\frac{s+0.5+0.1}{(s+0.5)^2+0.75}=\frac{1}{s}-\frac{s+0.5}{(s+0.5)^2+0.75}-\frac{0.1}{\sqrt{0.75}}\frac{\sqrt{0.75}}{(s+0.5)^2+0.75}$$

take inverse Laplace transform of $C(s)$ to get time domain response,

$$C(t)=L^{-1}[C(s)]=L^{-1}\left[\frac{1}{s}-\frac{s+0.5}{(s+0.5)^2+0.75}-\frac{0.1}{\sqrt{0.75}}\cdot\frac{\sqrt{0.75}}{(s+0.5)^2+0.75}\right]$$

$$=1-e^{-0.5t}\cdot\cos\sqrt{0.75}t-\frac{0.1}{\sqrt{0.75}}e^{-0.5t}\cdot\sin\sqrt{.75}t$$

$$=1-e^{-0.5t}\left[.1155\sin\left(\sqrt{.75}t\right)+\cos\left(\sqrt{.75}t\right)\right]$$

The transient response is the part of the output which vanishes t tends to infinity. Here as t tends to infinity the exponential component $e^{0.5t}$ tends to zero. Hence the transient response is given by the damped sinusoidal component.

The transient response of $C(t) = e^{-0.5t}\left[0.1155 \sin\left(\sqrt{0.75}t + \cos\sqrt{0.75}t\right)\right]$

The value of ζ and ω_n can be estimated by comparing the characteristics equation of the system transfer function with standard form of second order characteristics equation.

$$s^2 + 2\,\zeta\,\omega_n\,s + \omega_n^2 = s^2 + s + 1$$

on comparing we get,

$$\omega_n^2 = 1 \qquad\qquad 2\,\zeta\omega_n = 1$$

$$\omega_n = 1 \text{ rad/sec} \qquad\qquad \zeta = \frac{1}{2\omega_n} = \frac{1}{2} = 0.5$$

Maximum overshoot, $M_p = e^{\frac{-\zeta\pi}{\sqrt{1-\zeta^2}}}$

$$= e^{\frac{-0.5\,\pi}{\sqrt{1-(0.5)^2}}} = 0.163$$

% Maximum overshoot, % $M_p = M_p \times 100 = 0.163 \times 100 = 16.3\%$

Peak time $\quad t_p = \dfrac{\pi}{\omega_d} = \dfrac{\pi}{\omega_n\sqrt{1-\zeta^2}} = \dfrac{\pi}{1 \times \sqrt{1-(0.5)^2}} = 3.628$ sec

The response of the system underdamped, it is shown below,

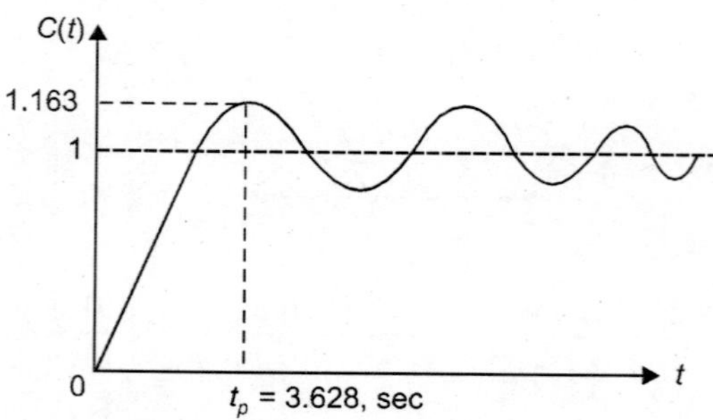

Fig. Ex. 2.7: Response of underdamped system.

Result: Transient response of the system = $e^{-0.5t}\left[0.1155 \sin\left(\sqrt{0.75}t\right) + \cos\left(\sqrt{0.75}t\right)\right]$

% Maximum peak overshoot = 16.3%

Peak time = 3.628 sec.

EXAMPLE 2.8

A unit feedback control system has an amplifier with gain $K_A = 10$ and gain ratio, $G(s) = 1/s\,(s+2)$ in the feed forward path. A derivative feedback, $H(s) = sK_0$ is introduced as a loop $G(s)$. Determine derivative feedback constant, K_0 so that the system damping factor is 0.6.

Solution: The given system can be represented by the block diagram shown in Fig. (Ex. 2.8).

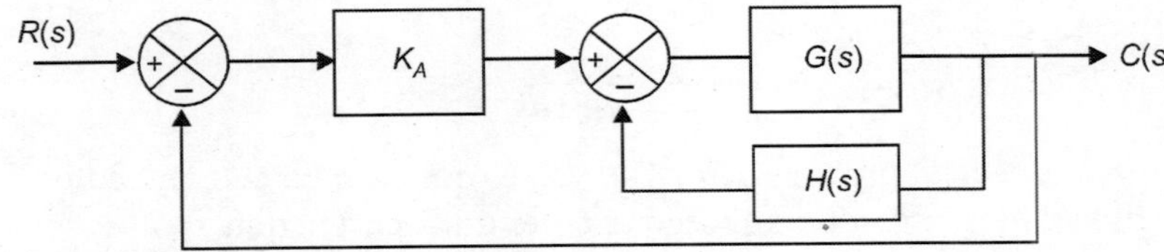

Fig. Ex. 2.8

Here, $K_A = 10$; $G(s) = \dfrac{1}{s(s+2)}$ and $H(s) = s \cdot K_0$

The closed loop transfer function of the system can be obtained by block diagram reduction techniques.

Step 1: Reducing the inner feedback loop.

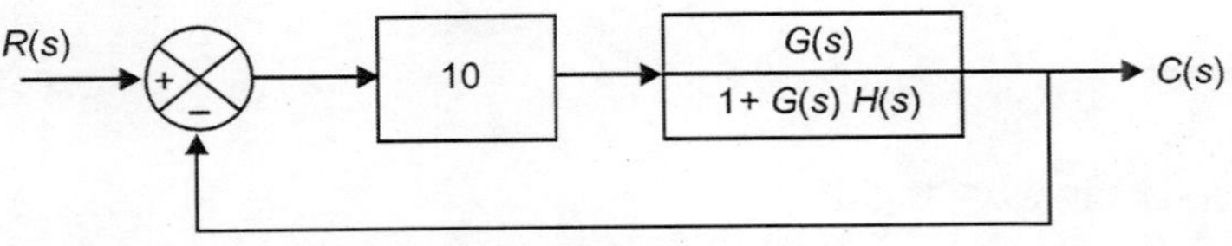

Fig. Ex. 2.8.1

$$\frac{G(s)}{1+G(s)\,H(s)} = \frac{1/s(s+2)}{1+\dfrac{1}{s(s+2)}\cdot sK_0} = \frac{1}{s(s+2)+sK_0}$$

$$= \frac{1}{s^2+2s+sK_0} = \frac{1}{s^2+s(2+K_0)}$$

Fig. Ex. 2.8.2

Step 2: Combining blocks in cascade (Rule 1).

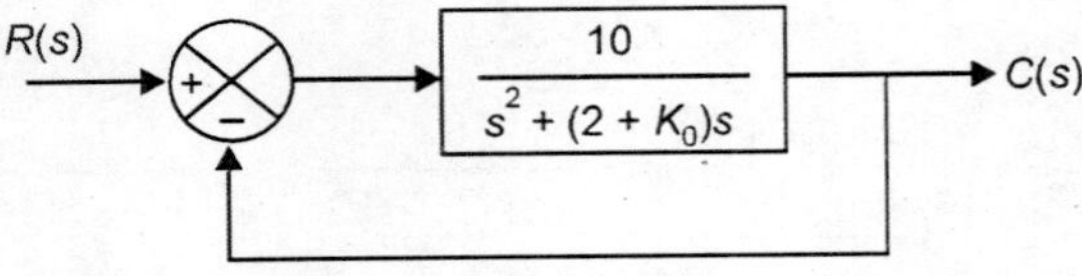

Fig. Ex. 2.8.3

Step 3: Reducing the unity feedback path (Rule 10),

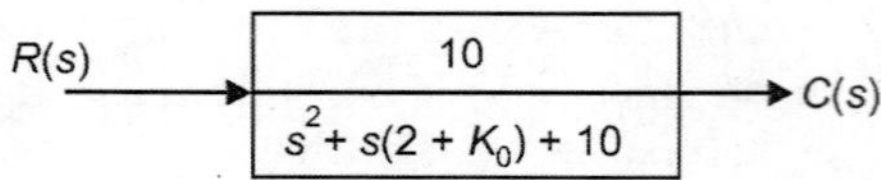

Fig. Ex. 2.8.4

The closed loop transfer function is,

$$\frac{C(s)}{R(s)} = \frac{10}{s^2 + s(2 + K_0) + 10}$$

The given system is a second order system. The value of K_0 can be determined by comparing the system transfer function with standardorm of second order transfer function.

$$\frac{C(s)}{R(s)} = \frac{\omega_n^2}{s^2 + 2\zeta\omega_n s + \omega_n^2} = \frac{10}{s^2 + s(2 + K_0) + 10}$$

on comparing we get,

$$\omega_n^2 = 10 \qquad 2 + K_0 = 2\zeta\omega_n$$

$$\omega_n = \sqrt{10} \qquad K_0 = 2\zeta\omega_n - 2$$

$$= 3.162 \text{ rad/sec} \qquad = (2 \times 0.6 \times 3.162) - 2 = 1.7944 \Rightarrow K_0 = 1.7944$$

Result: The value of constant, $K_0 = 1.7944$

EXAMPLE 2.9

A unity feedback control system has an open loop transfer function, $G(s) = 10/s\,(s + 2)$. Find the rise time, percentage overshoot, peak time and settling time for a step input of 12 units.

Solution: The unity feedback system is shown below.

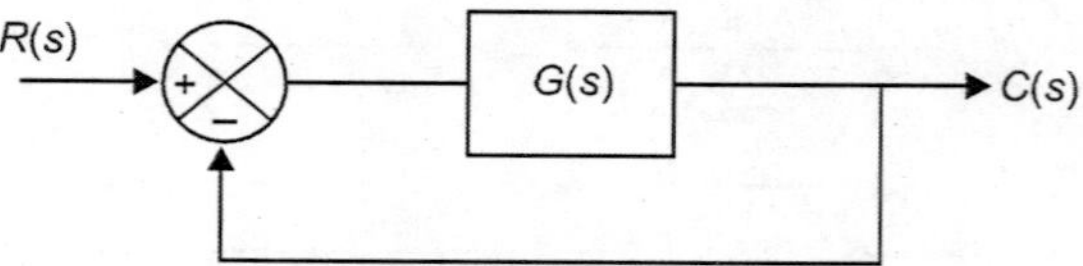

Fig. Ex. 2.9: Unity feedback system.

The closed loop transfer function,

$$\frac{C(s)}{R(s)} = \frac{G(s)}{1 + G(s)}$$

Given that, $G(s) = 10/s(s + 2)$

$$\frac{C(s)}{R(s)} = \frac{10/s(s+2)}{1 + \dfrac{10}{s(s+2)}} = \frac{10}{s(s+2) + 10} = \frac{10}{s^2 + 2s + 10}$$

The values of damping ratio ζ and natural frequency of oscillation ω_n are obtained by comparing the system function with standard form of second order transfer function.

$$\frac{C(s)}{R(s)} = \frac{\omega_n^2}{s^2 + 2\zeta\omega_n s + \omega_n^2} = \frac{10}{s^2 + 2s + 10}$$

On comparing we get,

$$\omega_n^2 = 10 \qquad 2\zeta\omega_n = 2$$

$$\omega_n = \sqrt{10} \qquad \zeta = \frac{2}{2\omega_n} = \frac{1}{3.162} = 0.316$$

$$\omega_n = 3.162 \text{ rad/sec}$$

$$\theta = \tan^{-1}\frac{\sqrt{1-\zeta^2}}{\zeta} = \tan^{-1}\frac{\sqrt{1-(0.316)^2}}{0.316} = 1.249 \text{ rad}$$

$$\omega_d = \omega_n\sqrt{1-\zeta^2} = 3.162\sqrt{1-(0.316)^2} = 3 \text{ rad/sec}$$

Rise time, $t_r = \dfrac{\pi - \theta}{\omega_d} = \dfrac{\pi - 1.249}{3} = 0.63 \text{ sec}$

Percentage peak over short $= e^{-\zeta\pi/\sqrt{1-\zeta^2}} \times 100$

$$= e^{\frac{-0.316\pi}{\sqrt{1-(0.316)^2}}} \times 100$$

$$= 0.3512 \times 100 = 35.12\%$$

Peak overshoot, M_p $= \dfrac{35.12}{100} \times 12 \text{ units}$

$$= 4.2144 \text{ units}$$

Peak time $t_p = \dfrac{\pi}{\omega_d} = \dfrac{\pi}{3} = 1.047 \text{ sec}$

Time constant, $T = \dfrac{1}{\zeta\omega_n} = \dfrac{1}{0.316 \times 3.162} = 1 \text{ sec}$

For 5% error, settling time, $t_s = 3T = 3$ sec

For 2% error, settling time, $t_s = 4T = 4$ sec

Result: Rise time, $t_r = 0.63$ sec

Percentage overshoot, % $M_p = 35.12\%$

Peak overshoot = 4.2144 units (for a input of 12 units)

CHAPTER 2

Peak time, $t_p = 1.047$ sec

Settling time, $t_s = 3$ sec for 5% error

$= 4$ sec for 2% error

EXAMPLE 2.10

A closed loop servo is represented by the differential equation $\frac{d^2c}{dt^2} + 8\frac{dc}{dt} = 64\, e$.

where c is the displacement of the output shaft, r is the displacement of the input and $e = r - c$. Determine undamped natural frequency, damping ratio and perdentage maximum overshoot for unit step unit.

Solution:

$$\frac{d^2c}{dt^2} + 8\frac{dc}{dt} = 64\, e \qquad (2.10.1)$$

and

$$e = r - c \qquad (2.10.2)$$

Put $e = r - c$ in eq. (2.10.1)

$$\frac{d^2c}{dt^2} + 8\frac{dc}{dt} = 64\,(r - c) \qquad (2.10.3)$$

Let

$$L[c] = C(s)$$

and

$$L[r] = R(s)$$

take Laplace transform for eq. (2.10.3)

$$s^2C(s) + 8s\, C(s) + 64\,[R(s) - C(s)]$$

$$s^2\, C(s) + 8s\, C(s) + 64\, C(s) = 64\, R(s)$$

$$(s^2 + 8s + 64)\, C(s) = 64\, R(s)$$

$$\frac{C(s)}{R(s)} = \frac{64}{s^2 + 8s + 64}$$

The ratio $C(s)/R(s)$ is the closed loop transfer function of the system. On comparing the system transfer function with standard form of second order transfer function, we can estimate the values of ζ and ω_n.

$$\frac{C(s)}{R(s)} = \frac{\omega_n^2}{s^2 + 2\zeta\omega_n s + \omega_n^2} = \frac{64}{s^2 + 8s + 64}$$

on comparing, we get

$$\omega_n^2 = 64 \qquad 2\,\zeta\omega_n = 8$$

$$\omega_n = 8 \text{ rad/sec} \qquad \zeta = \frac{8}{2\,\omega_n} = \frac{8}{2 \times 8} = 0.5$$

Percentage peak overshoot, $\% M_p = e^{\frac{-\zeta\pi}{\sqrt{1-(0.5)^2}}} \times 100 = 16.3\%$

Result: Undamped natural frequency of oscillation

$$\omega_n = 8 \text{ rad/sec}$$

Damping ratio, $\zeta = 0.5$

Percentage peak overshoot, % $M_p = 16.3\%$

2.10.10 Type Number of Control System

The type number is specified for loop transfer function $G(s) \cdot H(s)$. The number of poles of the loop transfer function using at the origin decides the type number of the system. In general if N is the number of poles at the origin then the type number is N.

The loop transfer function can be expressed as a ratio of two polynomials in s,

$$G(s) \cdot H(s) = K\frac{P(s)}{Q(s)} = \frac{K(s+z_1)(s+z_2)(s+z_3)}{s^n (s+P_1)(s+P_2)(s+P_3)} \tag{2.92}$$

where $Z_1, Z_2, Z_3 \rightarrow$ are zeros of transfer function.

$P_1, P_2, P_3 \rightarrow$ are poles of transfer function

$K \rightarrow$ is constant

$n \rightarrow$ is number of poles at the origin

The values of N in the denominator polynomial decides the type number of the system.

If $n = 0$, then the system is type – 0 system

If $n = 1$, then the system is type –1 system

If $n = 2$, then the system is type –2 system

If $n = 3$, then the system is type –3 system etc.,

EXAMPLE 2.11

A second order system has s transfer function given by $G(s) = 25/s^2 + 8s + 25$ Find the value of damping ratio.

Solution: $R(s) = 1/s$ (for unit step input)

$$G(s) = \frac{C(s)}{R(s)} \Rightarrow C(s) = G(s)\,R(s)$$

$$C(s) = \frac{25}{s(s^2+8s+25)} \tag{2.11.1}$$

Compare the eq. (2.11.1) with,

$$s^2 + 2\,\zeta\,\omega_n\,s + \omega_n^2 = 0$$

$$2\,\zeta\omega_n = 8$$

$$\omega_n = \sqrt{25}$$

$$= 5 \text{ rad/sec}$$

$$\zeta = .8$$

Result: The value of damping ratio is .8.

EXAMPLE 2.12

The block diagram of a feedback system is shown, below, Fig. (2.12)

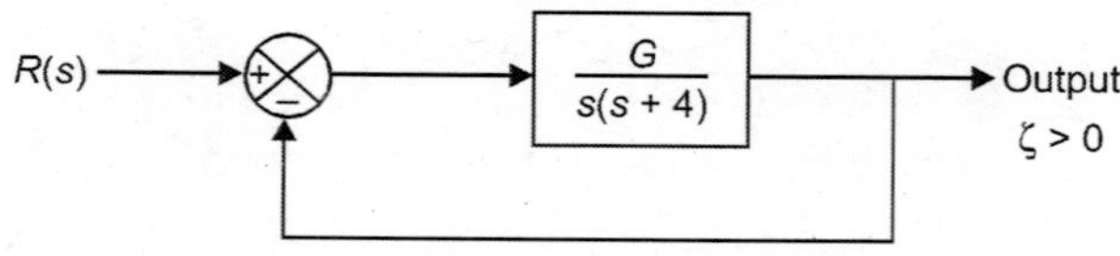

Fig. Ex. 2.12

(a) Find closed loop transfer function

(b) Find the minimum value of G for which the step response of the system would exhibit an overshoot as shown below (Fig. 2.12.1).

(c) For G equal to twice this minimum value, find the time period 'T' indicated in Fig. (2.12.1).

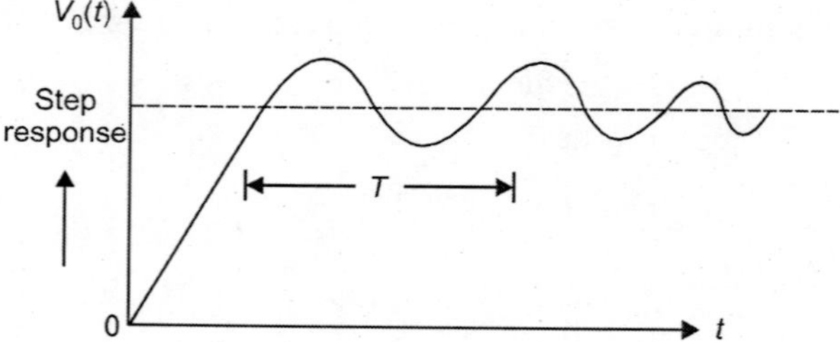

Fig. Ex. 2.12.1

Solution:

(a) Closed loop transfer function.

$$\frac{C(s)}{R(s)} = \frac{G/s(s+4)}{1+\dfrac{G}{s(s+4)}}$$

$$\frac{C(s)}{R(s)} = \frac{G/s(s+4)}{\dfrac{s(s+4)+G}{s(s+4)}} = \frac{G}{s(s+4)+G}$$

$$\frac{C(s)}{R(s)} = \frac{G}{s^2+4s+G}$$

(b) Characteristic equation, $s^2 + 4s + G = 0$. Compare the characteristics equation $s^2 + 4s + G = 0$ with the equation of $s^2 + 2\zeta\omega_n s + \omega_n^2$

so we get,

$$s^2 + 4s + G = s^2\, 2\,\zeta\omega_n s + \omega_n^2$$

compare the coefficient of s^2, s and constant in both RHS and LHS

$$2\zeta\omega_n = 4$$

$$\omega_n^2 = G \rightarrow \omega_n = \sqrt{G}$$

Let we assume $\zeta = .5$

$$2\,\zeta\omega_n = 4$$

$$2\zeta\sqrt{G} = 4$$

$$2 \times .5 \times \sqrt{G} = 4$$

$$\sqrt{G} = \frac{3}{2 \times .5} \Rightarrow \sqrt{G} = 3$$

$$G = 9$$

(c) $$G^1 = 2 \times 9 = 18$$

G = Twice the minimum value

$$\omega_n = \sqrt{9} = 3 \text{ rad/sec}$$

$$\zeta = 3/\,2\omega_n$$

$$= 3/2 \times 3.53 = 0.424$$

$$\omega_d = \omega_n\sqrt{1-\zeta^2}$$

$$= \omega_n\sqrt{1-\zeta^2}$$

$$= 3.53\sqrt{1-(0.424)^2} = 3.197$$

$$\frac{2\pi}{T} = \omega_d$$

$$\frac{2\pi}{T} = 3.197$$

$$T = 1.96 \text{ sec}$$

Result:

- The closed loop transfer function

$$\frac{C(s)}{R(s)} = \frac{G}{s^2 + 3s + G}$$

- Minimum value of G when the step response of the system would exhibit an overshoot $G = 9$
- $T = 1.96$ sec when $G' = 2G$.

EXAMPLE 2.13

The open loop transfer function of a servo system with feedback is a given by $G(s) = 10/(s+2)(s+5)$ Determine the damping ratio, undamped natural frequency of oscillation. What is the percentage overshoot of the response to a unit step input.

Solution: Given that, $G(s) = 10/(s+2)(s+5)$

$$H(s) = 1$$

$$T(s) = \frac{C(s)}{R(s)} = \frac{G(s)}{1+G(s)}$$

$$= \frac{\dfrac{10}{(s+2)(s+5)}}{1+\dfrac{10}{(s+2)(s+5)}}$$

$$T(s) = \frac{10}{(s+2)(s+5)+10}$$

The characteristic eqn. is,

$$(s+2)(s+5)+10 = 0$$

$$s^2 + 7s + 20 = 0 \quad (2.13.1)$$

Compare the eq. (2.14.1) with,

$$s^2 + 2\,\zeta\omega_n s + \omega_n^2 = 0$$

$$s^2 + 2\,\zeta\omega_n\, s + \omega_n^2 = s^2 + 7s + 20$$

compare the coefficients of s^2, s and constant from both the RHS and LHS

$$\omega_n^2 = 20$$

$$\omega_n = \sqrt{20}$$

$$= 4.472$$

$$\omega_n = 4.472 \text{ rad/sec}$$

and

$$2\zeta\omega_n = 7$$

$$2 \times \zeta \times 4.472 = 7$$

$$\zeta = 7/2 \times 4.472$$

$$\zeta = .7826$$

Percentage maximum overshoot,

$$\% M_p = e^{-\frac{\pi\zeta}{\sqrt{1-\zeta^2}}}$$

$$\% M_p = e^{-\frac{\pi \times 0.7826}{\sqrt{1-(0.7826)}}} \times 100$$

$$\% M_p = 1.92\%$$

Result: Damping ratio, $\zeta = 0.7826$. Natural frequency $\omega_n = 4.472$ rad/sec

Percentage maximum overshoot, $\% M_p = 1.92$ %

EXAMPLE 2.14

For a RLC series circuit, the value of $R = 1\text{k}\Omega$, $L = 10\text{mH}$ and $C = 0.01\ \mu\text{F}$ and the supply voltage $V = 6\text{V}$. DC supply. For this circuit, findout the undamped natural frequency and damping ratio of the circuit assume initial conditions are zero.

Solution:

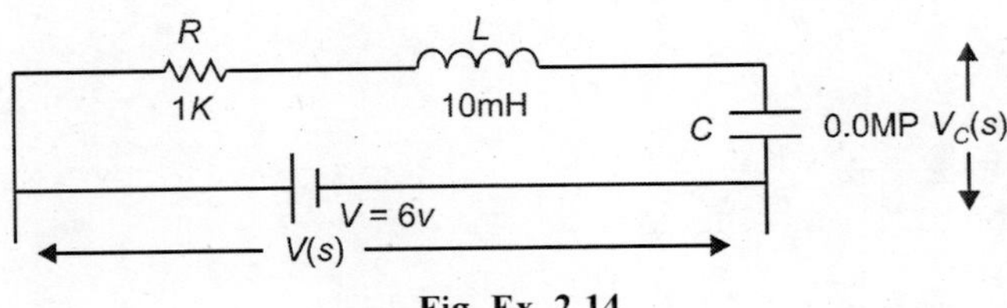

Fig. Ex. 2.14

For the above circuit first determine transfer function.

$$\text{Transfer function } C(s) = \frac{O/P}{I/P}$$

$$= \frac{O/P}{I/P} = \frac{V_C(s)}{V(s)}$$

$$V_c = \frac{1}{C}\int i\,dt$$

$$V_c(s) = \frac{I(s)}{Cs} \qquad (2.14.1)$$

$$V = R_i + \frac{Ldi}{dt} + \frac{1}{C}\int i\,dt$$

$$V(s) = RI(s) + Ls\, I(s) + \frac{I(s)}{Cs}$$

$$V(s) = I(s)\left[R + Ls + \frac{1}{Cs}\right]$$

$$V(s) = \left[\frac{RCs + LCs^2 + 1}{Cs}\right] I(s) \qquad (2.14.2)$$

Now divide equation (2.14.1) by (2.14.2)

$$\frac{V_c(s)}{V(s)} = \frac{I(s)/Cs}{I(s)\left[\frac{RCs + LCs^2 + 1}{Cs}\right]}$$

$$\frac{V_c(s)}{V(s)} = \frac{1}{RCs + LCs^2 + 1}$$

$$T(s) = \frac{\frac{1}{LC}}{\frac{RCs}{LC} + s^2 + \frac{1}{LC}}$$

Compare the equation (2.14.3) with
Characteristic equation of second order system

$$s^2 + 2\,\zeta\omega_n s + \omega_n^2 = s^2 + \frac{R}{L}s + \frac{1}{LC}$$

$$\omega_n^2 = \frac{1}{LC}$$

$$\omega_n = \frac{1}{\sqrt{LC}} \qquad (2.14.4)$$

$$2\zeta\omega_n = \frac{R}{L}$$

$$\zeta = \frac{R}{L2\omega_n}$$

where,

$$\omega_n = \frac{1}{\sqrt{LC}}$$

$$\zeta = \frac{R}{L \times 2 \times \frac{1}{\sqrt{LC}}}$$

$$= \frac{R\sqrt{LC}}{2L} = \frac{R\sqrt{LC}}{2\sqrt{L}\sqrt{L}}$$

$$\zeta = \frac{R}{2}\sqrt{\frac{C}{L}} \tag{2.14.5}$$

(i) $$\omega_n = \frac{1}{\sqrt{LC}} = 10^5 \text{ rad/sec} \qquad \left[\frac{1}{\sqrt{10 \times 10^{-3}} \times -01 \times 10^{-1}}\right]$$

$$\zeta = \frac{1000}{2}\sqrt{\frac{0.01 \times 10^{-6}}{10 \times 10^{-3}}} \qquad [\because R = 1000\ \Omega]$$

$$\zeta = 0.51$$

Result:

1. Damping ratio $\zeta = .5$
2. Natural frequency $\omega_n = 10^5$ rad/sec

EXAMPLE 2.15

Determine the value of K and H of the closed loop system Fig. (Ex. 2.15). So that the maximum overshoot in unit step response is 25% and the peaim is 2 sec. Assume $J = 1$ kg-m^2.

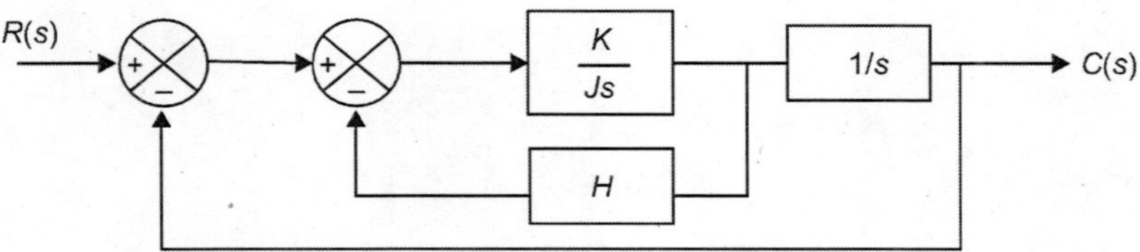

Fig. Ex. 2.15

Solution:
$$\frac{C(s)}{R(s)} = \frac{K}{Js^2 + KHs + K}$$

Put $J = 1$ (It is given in the problem)

$$\frac{C(s)}{R(s)} = \frac{K}{s^2 + KHs + K}$$

Characteristics eq. is $s^2 + KHs + K$.

Compare characteristics eq. with $s^2 + 2\zeta\omega_n s + \omega_n^2 = 0$

$$s^2 + 2\zeta\omega_n s + \omega_n^2 = s^2 + KHs + K$$

Compare the s term coefficient in both the sides in RHS and LHS.

$$\omega_n = \sqrt{K} \text{ and } 2\,\zeta\omega_n = KH \qquad (2.15.1)$$

Given that $M_p = 0.25;$ $\qquad [\therefore \%\, M_p = 25\%]$

$$M_p = e^{\frac{-\pi\zeta}{\sqrt{1-\zeta^2}}} = 0.25 \qquad (2.15.2)$$

Solve the eq. (2.15.2) to get the value of ζ. By taking log in both sides, we get,

$$\zeta = 0.404$$

Peak time

$$t_p = \frac{\pi}{\omega_n\sqrt{1-\zeta^2}}$$

$$\omega_n\sqrt{1-\zeta^2} = \frac{\pi}{t_p}$$

We know, $t_p = 2$[from problem it is given]

$$\omega_n\sqrt{1-\zeta^2} = \frac{\pi}{2}$$

Now substitute ζ value,

$$\omega_n\sqrt{1-(0.404)^2} = \frac{\pi}{2}$$

$$\omega_n = 1.717 \text{ rad/sec}$$

From (2.15.1)

$$\omega_n = \sqrt{K}$$

and

$$2\zeta\omega_n = \text{KH}$$

$$\omega_n = \sqrt{K}$$

$$K = \omega_n^2$$

$$= (1.73)^2 = 2.95 \text{ Nm.}$$

and

$$2\,\zeta\omega_n = KH$$

$$H = \frac{2\,\zeta\omega_n}{K} = \frac{2 \times 0.404 \times 1.717}{2.95} = .47$$

$$H = .47$$

Result:

$$K = 2.95 \text{ Nm}$$

$$H = 0.47$$

EXAMPLE 2.16

A second order control system is represented by a transfer function given below.

$$\frac{\theta_0(s)}{T(s)} = \frac{1}{Js^2 + fs + k}$$

where, $\theta(s)$ is proportional output and T is applied to the system and test results are given below.

(a) $\% M_p = 10\%$

(b) $t_p = 1$ sec

(c) The steady state error of the output is 0.5 radian. Determine the value of J, K and f.

Solution: Given that, $T(s) = 10/s$

$$\frac{\theta_0(s)}{T(s)} = \frac{1}{Js^2 + fs + K} \tag{2.16.1}$$

Put $T(s) = 10/s$ in eq. (2.18.1)

$$\theta_0(s) = \frac{10}{s\,(Js^2 + fs + K)}$$

$$\text{steady state value} = \lim_{s\to 0} s \cdot \theta_0(s)$$

$$0.5 = \lim_{s\to 0} s \cdot \frac{10}{s\,(Js^2 + fs + K)}$$

$$0.5 = \frac{10}{K}$$

$$K = 20$$

$$\% M_p = 10\%$$

$$M_p = 0.1$$

$$M_p = e^{-\pi\zeta/\sqrt{1-\zeta^2}} \times 100$$

$$0.1 = e^{-\pi\zeta/\sqrt{1-\zeta^2}}$$

$$\zeta = .59$$

$$t_p = 1 \text{ sec} \qquad \text{(Given)}$$

we know that, $$t_p = \frac{\pi}{\omega_n\sqrt{1-\zeta^2}}$$

Put $t_p = 1$, then we get $$1 = \frac{\pi}{\omega_n\sqrt{1-\zeta^2}}$$

$$1 = \frac{\pi}{\omega_n \sqrt{1-(.59)^2}}$$

$$\omega_n = 3.89 \text{ rad/sec}$$

From (2.16.1) eq. characteristic eq. is

$$Js^2 + fs + K = 0$$

$$s^2 + \frac{f}{J}s + \frac{K}{J} = 0 \quad (2.16.2)$$

Compare eq. (2.16.1) with $s^2 + 2\,\zeta\omega_n s + \omega_n^2 = 0$

$$\omega_n^2 = K/J \qquad 2\,\zeta\omega_n = \frac{f}{T}$$

$$J = \frac{K}{\omega_n^2} = 20/(3.89)^2 \qquad f = 2\,\zeta\omega_n J = 2 \times .59 \times 3.89 \times 1.32$$

$$J = 1.32 \qquad f = 6.06$$

Result:

$$J = 1.32$$
$$K = 20$$
$$f = 6.06$$

EXAMPLE 2.17

A feedback system is described by the following transfer function $G(s) = \frac{12}{s^2 + 4s + 16}$; $H(s) = Ks$.

The damping factor of the system is 0.6. Determine the overshoot of the system and the value of K.

Solution:

$$\frac{C(s)}{R(s)} = \frac{G(s)}{1 + G(s)\,H(s)}$$

$$\frac{C(s)}{R(s)} = \frac{12/s^2 + 4s + 16}{1 + \frac{12}{s^2 + 4s + 16} \cdot Ks} = \frac{\frac{12}{s^2 + 4s + 16}}{\frac{(s^2 + 4s + 16) + 12\,Ks}{(s^2 + 4s + 16)}}$$

$$= \frac{12}{s^2 + (4 + 16K)\,s + 16} \qquad (2.17.1)$$

Characteristics eq. is $s^2 + (4 + 16K)s + 16 = 0$

Compare the characteristics eq. with $s^2 + 2\,\zeta\,\omega_n\,s + \omega_n^2 = 0$

$$s^2 + (4 + 16K)s + 16 = s^2 + 2\,\zeta\omega_n\,s + \omega_n^2$$

Compare the s^2, s and constant terms in both the RHS and LHS

$$\omega_n^2 = 16 \qquad 2\,\zeta\omega_n = 4 + 16K$$

$$\omega_n = 4 \text{ rad/sec} \qquad 2 \times 0.8 \times 4 = 4 + 16K$$

$$K = .15$$

$$\frac{C(s)}{R(s)} = \frac{16}{s^2 + (4 + 16 \times .15)\, s + 16}$$

$$\% M_p = e^{-\frac{\pi\zeta}{\sqrt{1-\zeta^2}}} \times 100$$

$$= e^{-\frac{\pi\, 0.8}{\sqrt{1-(0.8)^2}}} \times 100$$

$$\% M_p = 1.5\%$$

Result:

$$K = 0.15$$

$$M_p = 1.5\%$$

CHAPTER 2

EXAMPLE 2.18

A system having a forward path transfer function $G(s) = 20/s(s + 2)$ and unity feedback. Determine the value of undamped natural frequency, damping ratio. If tachometer is introduced, the feedback path transfer function becomes $(1 + Ks)$. What should be the value of K to obtain damping ratio of 0.6. Also calculate the percentage peak overshoot, first undershoot, t_p and settling time within the 2% of final value.

Solution: The characteristic equation with unity feedback is,

$$1 + G(s) = 0 \quad \text{and} \quad H(s) = 1$$

$$G(s) = 20/s(s + 2)$$

So,

$$1 + \frac{20}{s\,(s+2)} = 0$$

$$1 + \frac{20}{(s^2 + 2s)} = 0$$

$$s^2 + 2s + 20 = 0 \qquad (2.18.1)$$

Compare the Eq. (2.18.1) with $s^2 + 2\,\zeta\omega_{n\,s} + \omega_n^2 = 0$

$$\omega_n^2 = 20 \qquad 2\,\zeta\omega_n = 2$$

$$\omega_n = 4.47 \text{ rad/sec} \qquad 2 \times \zeta \times 4.47 = 2 \Rightarrow 8.94\ \ \zeta = 2$$

$$\zeta = 2/8.94$$

$$\zeta = .22$$

Consider

$$H(s) = 1 + Ks$$

Here,

$$1 + G(s)\,H(s) = 0 \quad (2.18.2)$$

Put $G(s) = 20/s(s + 2)$ and $H(s) = 1 + Ks$ in Eq. (2.18.2)

$$1 + \frac{20}{s(s+2)}(1 + Ks) = 0$$

$$s^2 + s(2 + 20K) + 20 = 0 \quad (2.18.3)$$

Compare (2.18.3) with $s^2 + 2\zeta\omega_n s + \omega_n^2 = 0$

$$s^2 + (2 + 20K)\,s + 20 = s^2 + 2\,\zeta\omega_n s + \omega_n^2$$

Compare LHS and RHS

$\omega_n^2 = 20$	$2\,\zeta\omega_n = 2 + 20K$
$\omega_n = 4.47$ rad/sec	$2 \times 0.6\ \times 4.47 = 2 + 20K \Rightarrow 5.36 = 2 + 20K$
	$\Rightarrow 20K = 3.364 \Rightarrow K = .168$

Percentage maximum overshoot $\%\,M_p = e^{\frac{-\pi\zeta}{\sqrt{1-\zeta^2}}} \times 100$

when $\zeta = .22$,

$$\%\,M_p = e^{\frac{-\pi(0.22)}{\sqrt{1-(0.22)^2}}} \times 100$$

$$\%\,M_p = 49.3\%$$

when $\zeta = 0.6$,

$$\%\,M_p = e^{\frac{-\pi(0.6)}{\sqrt{1-(0.6)^2}}} \times 100$$

$$\%\,M_p = 9.48\%$$

$$\text{First undershoot} = e^{\frac{-2\pi}{\sqrt{1-\zeta^2}}}$$

$$= e^{\frac{-2\pi\,(0.6)}{\sqrt{1-(0.6)^2}}} = 0.009$$

Finding Peak Time:

when, $\zeta = .22$

$$t_p = \frac{\pi}{\omega_n\sqrt{1-\zeta^2}} = \frac{\pi}{4.47\sqrt{1-(.22)^2}}$$

$$t_p = .720 \text{ sec}$$

when $\zeta = 0.6$

$$t_p = \frac{\pi}{4\sqrt{1-(0.6)^2}} = 0.98 \text{ sec}$$

$$t_p = 0.98 \text{ sec}$$

Finding Settling Time t_s:

$$t_s = \frac{4}{\zeta\omega_n}$$

when $\zeta = .22$

$$t_s = \frac{4}{.22 \times 4.47} = \frac{4}{.983} = 4 \text{ sec}$$

when $\zeta = 0.6$

$$t_s = \frac{4}{4.47 \times 0.6} = \frac{4}{2.682} = 1.49$$

$$t_s = 1.49 \text{ sec}$$

Results:

1. The value of K when $\zeta = 0.6$ $K = .168$
2. Undamped natural frequency $\omega_n = 4.47$ rad/sec
3. Damping ratio $\zeta = .22$
4. % Peak overshoot % $M_p = 9.48\%$
5. First undershoot = .009
6. Peak time = .72 sec
7. Settling time within 2% of final value.
   (i) when $\zeta = .22$ $t_s = 4$ sec
   (ii) when $\zeta = 0.6$ $t_s = 1.49$ sec

2.10.11 Effect of Feedback on Time Constant of a control System

Consider the open loop transfer function as shown in below Fig. (2.10.11.1).

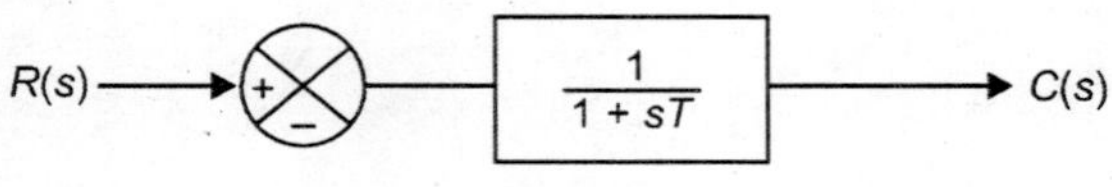

Fig. 2.10.11.1

$$\frac{C(s)}{R(s)} = \frac{1}{1+sT} \tag{2.93}$$

Let $R(s) = \frac{1}{s}$.

Put $R(s) = \frac{1}{s}$ in Eq. (2.93)

$$C(s) = \frac{R(s)}{1+sT}$$

$$C(s) = \frac{1}{1+sT} \cdot \frac{1}{s} \tag{2.94}$$

apply inverse Laplace for Eq. (2.94)

$$L^{-1}[C(s)] = L^{-1}\left[\frac{1}{s\,(1+sT)}\right]$$

$$C(t) = 1 - e^{-t/T} \tag{2.95}$$

In Eq. (2.95) T is the constant.

Now consider the closed loop system as shown in below Fig. (2.94)

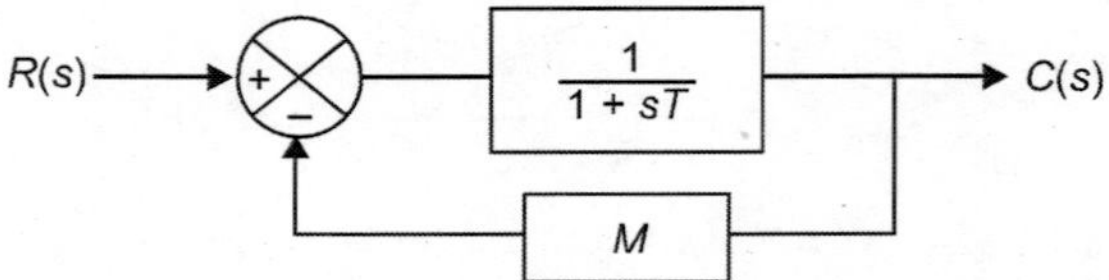

Fig. 2.10.11.2

If $H(s) = M$, then the overall transfer function is,

$$\frac{C(s)}{R(s)} = \frac{1/1+sT}{1+\dfrac{1}{1+sT}\cdot M}$$

$$\frac{C(s)}{R(s)} = \frac{1/1+sT}{\dfrac{1+sT+M}{1+sT}} = \frac{1}{1+sT+M}$$

$$\frac{C(s)}{R(s)} = \frac{1/T}{s+\left(\dfrac{1+M}{T}\right)} \tag{2.96}$$

apply inverse Laplace for Eq. (2.10.11.4)

$$L^{-1}[C(s)] = \frac{1/T}{s+\left(\dfrac{1+M}{T}\right)} = C(t)$$

$$C(t) = \frac{1}{1+M}\left[1 - e^{\frac{-t}{T/1+M}}\right] \tag{2.97}$$

From the Eq. (2.97) time constant is,

Time constant = $T/1 + M$

Conclusion: The time constant of closed loop system is less than the time constant of open loop system.

Less time constant means response is faster. Therefore, feedback improves the time response of the system.

2.11 STEADY STATE ERROR

The steady state error is the value of error signal $e(t)$, when t tends to infinity the steady state error is a measure of system accuracy. These errors arise from the nature of inputs, type of system and from non-linearity of system components. The steady state performance of a stable control system is generally judged by its steady state error to step, ramp and parabolic inputs.

Consider a closed loop system shown in Fig. (2.11.1).

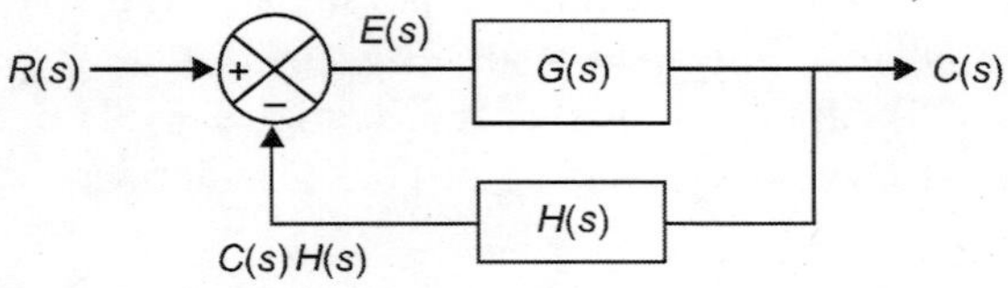

Fig. 2.11.1

From Fig. (2.11.1),

The error signal, $$E(s) = R(s) - C(s)\,H(s) \tag{2.98}$$

The output signal, $$C(s) = E(s)\,G(s) \tag{2.99}$$

Substitute the Eq. (2.99) in Eq. (2.98)

$$E(s) = R(s) - [E(s)\,G(s)] \cdot H(s)$$

$$E(s) + E(s)\,G(s)\,H(s) = R(s)$$

$$E(s)\,[1 + G(s)\,H(s)] = R(s)$$

$$E(s) = \frac{R(s)}{1 + G(s)\,H(s)} \tag{2.100}$$

Let $e(t)$ = error signal in time domain

$$e(t) = L^{-1}\,[E(s)] = L^{-1}\left[\frac{R(s)}{1 + G(s)\,H(s)}\right] \tag{2.101}$$

Let e_{ss} = steady state error

The steady state error is defined as the value of $C(t)$ when t tends to infinity.

$$e_{ss} = \lim_{t \to \infty} e(t) \tag{2.102}$$

The final Value theorem states that,

$$F(s) = L[f(t)] \text{ then,}$$

$$\lim_{t \to \infty} f(t) = \lim_{s \to 0} s \cdot F(s) \tag{2.103}$$

Using final value theorem,

The steady state error,

$$e_{ss} = \lim_{t\to\infty} e(t) = \lim_{s\to 0} s \cdot E(s)$$

$$= \lim_{s\to 0} \frac{s \cdot R(s)}{1+G(s)\,H(s)} \qquad (2.104)$$

2.11.1 Static Error Constants

When a control system is excited with standard input signal, the steady state error may be zero, constant or infinity. The value of steady state error depends on the type number and the input signal. Type-0 system will have a steady state error when the input is step signal. Type-1 system will have a constant steady state error when the input is ramp signal (or) velocity signal. Type-2 system will have a constant steady state error when the input is parabolic signal or acceleration signal. For the three cases mentioned above the steady state error is associated with one of the constants defined as follows.

Positional error constant, $K_p = \lim_{s\to 0} G(s)\cdot H(s)$ (2.105)

Velocity error constant, $K_V = \lim_{s\to 0} s\cdot G(s)\cdot H(s)$ (2.106)

Acceleration error constant, $K_a = \lim_{s\to 0} s^2\, G(s)\cdot H(s)$ (2.107)

The K_p, K_v and K_a are in general called static error constants.

2.11.2 Steady State Error when the Input is Step Signal

Steady state error, $e_{ss} = \lim_{s\to 0} \frac{sR(s)}{1+G(s)+H(s)}$ when the input is unit step, $R(s) = 1/s$

$$e_{ss} = \lim_{s\to 0} \frac{s\cdot 1/s}{1+G(s)\,H(s)}$$

$$= \lim_{s\to 0} \frac{1}{1+G(s)\,H(s)} = \frac{1}{1+\lim_{s\to 0} G(s)\,H(s)} = \frac{1}{1+K_P} \qquad (2.108)$$

where, $K_P = \lim_{s\to 0} G(s)\cdot H(s)$. The constant K_P is called positional error constant.

Type-0 System:

$$K_P = \lim_{s\to 0} G(s)\,H(s) = \lim_{s\to 0} K \frac{(s+Z_1)(s+Z_2)(s+Z_3)\ldots}{(s+P_1)(s+P_2)(s+P_3)\ldots}$$

$$= K\frac{Z_1\cdot Z_2\cdot Z_3}{P_1\cdot P_2\cdot P_3} = \text{constant}$$

$$e_{ss} = \frac{1}{1+K_P} = \text{constant}$$

In type-0 system when the input is unit step there will be constant steady state error.

Type-1 System

$$K_P = \lim_{s\to 0} G(s)\cdot H(s) = \lim_{s\to 0} K\frac{(s+Z_1)(s+Z_2)(s+Z_3)\ldots}{s(s+P_1)(s+P_2)(s+P_3)\ldots} = \infty$$

$$e_{ss} = \frac{1}{1+K_P} = \frac{1}{1+\infty} = 0$$

In system with type number 1 and above, for unit step input the value of K_P is infinity and so the steady state error is zero.

2.11.3 Steady State Error when the Input is Unit Ramp Signal

Steady state error, $e_{ss} = \lim_{s\to 0}\dfrac{sR(s)}{1+G(s)\,H(s)}$

when the input is unit ramp, $R(s) = 1/s^2$

$$e_{ss} = \lim_{s\to 0}\frac{s\cdot\frac{1}{s^2}}{1+G(s)\,H(s)} = \lim_{s\to 0}\frac{1}{s+sG(s)\,H(s)}$$

$$= \lim_{s\to 0}\frac{1}{sG(S)\,H(s)} = \frac{1}{K_V}$$

where, $K_V = \lim_{s\to 0} s\cdot G(s)\cdot H(s)$ The constant K_V is called velocity error constant.

Type-0 system:

$$K_V = \lim_{s\to 0} sG(s)\,H(S)\ \lim_{s\to 0} sK\frac{(s+Z_1)(s+Z_2)(s+Z_3)\ldots}{(s+P_1)(s+P_2)(s+P_3)\ldots} = 0 \Rightarrow K_V = 0$$

$$e_{ss} = \frac{1}{K_V} = \frac{1}{0} = \infty$$

In type-0 systems when the input is unit ramp, the steady state error is infinity.

Type-1 system:

$$K_V = \lim_{s\to 0} s\cdot G(s)\cdot H(s) = \lim_{s\to 0} sK\frac{(s+Z_1)(s+Z_2)(s+Z_3)\ldots}{s\,(s+P_1)(s+P_2)(s+P_3)\ldots}$$

$$= K\frac{Z_1\cdot Z_2\cdot Z_3\ldots}{P_1\cdot P_2\cdot P_3\ldots} = \text{constant}$$

$e_{ss} = 1/K_V = \text{constant}$

In type-1 system when the input is unit ramp there will be a constant steady state error.

Type-2 system:

$$K_V = \lim_{s\to 0} s\cdot G(s)\,H(s) = \lim_{s\to 0} sK\,\frac{(s+Z_1)(s+Z_2)(s+Z_3)\ldots}{s^2(s+P_1)(s+P_2)(s+P_3)\ldots} = \infty$$

$$e_{ss} = \frac{1}{K_V} = \frac{1}{\infty} = 0$$

In system with type number 2 and above, for unit ramp input, the value of K_V is infinity so the steady state error is zero.

2.11.4 Steady State Error when the Input is Unit Parabolic Signal

Steady state error, $e_{ss} = \lim_{s\to 0} \dfrac{sR(s)}{1+G(s)\,H(s)}$

when the input is unit parabola, $R(s) = 1/s^3$

$$e_{ss} = \lim_{s\to 0}\frac{s\cdot\frac{1}{s^3}}{1+G(s)\,H(s)} = \lim_{s\to 0}\frac{1}{s^2+s^2\,G(s)\,H(s)}$$

$$= \frac{1}{\lim_{s\to 0} s^2\,G(s)\,H(s)} - \frac{1}{K_a}$$

where, $K_a = \lim_{s\to 0} s^2\,G(s)\cdot H(s)$. The constant K_a is called acceleration error constant.

Type- 0 system:

$$K_a = \lim_{s\to 0} s^2\cdot G(s)\cdot H(s)$$

$$= \lim_{s\to 0} s^2K\,\frac{(s+Z_1)(s+Z_2)(s+Z_3)\ldots}{(s+P_1)(s+P_2)(s+P_3)\ldots} = 0 \Rightarrow K_a = 0$$

$$e_{ss} = \frac{1}{K_a} = \frac{1}{0} = \infty$$

In type- 0 systems for unit parabolic input, the steady state error is infinity.

Type-1 system:

$$K_a = \lim_{s\to 0} s^2\,G(s)\,H(s) = \lim_{s\to 0} s^2K\,\frac{(s+Z_1)(s+Z_2)(s+Z_3)\ldots}{s\,(s+P_1)(s+P_2)(s+P_3)\ldots} = 0 \Rightarrow K_a = 0$$

$$e_{ss} = \frac{1}{K_a} = \frac{1}{0} = \infty$$

Hence in type-1 systems for unit parabolic input the steady state error is infinity.

Type-2 systems:

$$K_a = \lim_{s\to 0} s^2\, G(s)\, H(s)$$

$$= \lim_{s\to 0} s^2 K \frac{(s+Z_1)(s+Z_2)(s+Z_3)\ldots}{s^2\,(s+P_1)(s+P_2)(s+P_3)\ldots}$$

$$K_a = K \cdot \frac{Z_1 \cdot Z_2 \cdot Z_3 \ldots}{P_1 \cdot P_2 \cdot P_3 \ldots} = \text{constant}$$

$$e_{ss} = \frac{1}{K_a} = \text{constant}$$

Hence in type-2 system when the input is unit parabolic signal there will be a constant steady state error.

Type-3 system:

$$K_a = \lim_{s\to 0} s^2\, G(s)\, H(s)$$

$$= \lim_{s\to 0} s^2 K \frac{(s+Z_1)(s+Z_2)(s+Z_3)\ldots}{s^3\,(s+P_1)(s+P_2)(s+P_3)\ldots}$$

$$e_{ss} = \frac{1}{K_a} = \frac{1}{\infty} = 0$$

In systems with type number 3 and above for unit parabolic input the value of K_a is infinity and so the steady state error is zero.

Table 2.2: The steady state error for various types of inputs

Type number	Steady state error when the input signal is (e_{ss})		
	Unit step	**Unit ramp**	**Unit parabolic**
0	$1/1 + K_P$	∞	∞
1	0	$1/K_V$	∞
2	0	0	$1/K_a$
3	0	0	0

2.11.5 Generalized Error Coefficient

The drawback in static error coefficient is that it does not show the variation of error with time and input should be a standard input. The generalized error coefficients gives the steady state error as a function of time. Also using the generalized error coefficients, the steady state error can be found for any type of input.

The error signal in s-domain, $E(s)$ can be expressed a product of two s-domain function.

$$E(s) = \frac{R(s)}{1+G(s)\,H(s)} = \frac{1}{1+G(s)\,H(s)} \cdot R(s) = F(s) \cdot R(s) \tag{2.109}$$

where, $F(s) = \dfrac{1}{1 + G(s)\,H(s)}$

Let

$$e(t) = L^{-1}\,[E(s)] \quad \text{(error signal in time domain)}$$
$$f(t) = L^{-1}\,[F(s)]$$
$$r(t) = L^{-1}[R(s)] \quad \text{(input signal in time domain)}$$

The convolution theorem states that the inverse Laplace transform of the product of two s-domain functions is equal to convolution of their time domain function.

i.e.,

$$L^{-}\,[F(s)\,R(s)] = \int_{-\infty}^{\infty} f(T)\cdot r(t-T)\cdot dT$$

Hence by convolution theorem the error function in time domain can be written as,

$$e(t) = \int_{-\infty}^{\infty} f(T)\cdot r\,(t-T)\cdot dT \tag{2.110}$$

where $T \rightarrow$ is a dummy variable. It is assumed that the input signal starts only at $t = 0$ and does not exist before $t = 0$. Hence the limit of integral can be changed as 0 to ∞.

$$e(t) = \int_{0}^{\infty} f(T)\,r(t-T)\,dT \tag{2.111}$$

Using Taylor's series expansion the signal $r(t - T)$ can be expressed as,

$$r(t-T) = r(t) - T\dot{r}\,(t) + \frac{T^2}{2!}\ddot{r}(t) - \frac{T^3}{3!}\dddot{r}\,(t) + \ldots + (-)^n\,\frac{T^n}{n!}\,r^n(t) \tag{2.112}$$

where, $r'(t)$ = 1st derivative of $r(t)$.

$r''(t)$ = 2nd derivative of $r(t)$.

and $r^n(t)$ = nth derivative of $r(t)$

On substituting the Taylor's series expansion of $r(t - T)$, the error $e(t)$ can be written as,

$$e(t) = \int_{0}^{\infty} f(T)\left[r(T) - Tr'(t) + \frac{T^2}{2!}r''\,(t) - \frac{T^3}{3!}r'''(t) + \ldots + (-1)^n\,\frac{T^n}{n!}r^{n'}\,r^n t \ldots\right] \times dT$$

$$e(t) = \int_{0}^{\infty} f(T)\,r(r)\;dT - \int_{0}^{\infty} f(T)\,Tr'(t)\,dT + \int_{0}^{\infty} f(T)\frac{T^2}{2!}r''(t)\cdot dT -$$

$$\int_{0}^{\infty} f(T)\frac{T^3}{3!}r'''\,(t)\cdot dT + \ldots + \int_{0}^{\infty} f(T)\,(-)^n\,\frac{T^n}{n!}r^{n'}\;r^n(t)\,dT \ldots \infty$$

where $r(t), r'(t), r''(t), \ldots\ r^{n'}(t)$ are constants. After integration with respect to T, the error signal can be written as,

$$e(t) = r(t)\int_0^\infty f(T)\,dT - r'(t)\int_0^\infty Tf(T)\,dT + \frac{r''(t)}{2!}\int_0^\infty T^2\, f(T)\cdot dT -$$

$$= \frac{r'''(t)}{3!}\int_0^\infty T^3\, f(T)\,dT + \ldots (-1)^n \frac{r^{n'}(t)}{n!}\int T^n\, f(T)\cdot dT \ldots$$

Let $C_0 = +\int_0^\infty f(T)\cdot dT$

$$C_1 = \int_0^\infty T\cdot f(T)\cdot dT$$

$$C_2 = +\int_0^\infty T^2\, f(T)\cdot dT$$

$$C_3 = -\int_0^\infty T^3\, f(T)\cdot dT$$

$$\vdots$$

$$C_n = (-1)^n \int_0^\infty T^n\, f(T)\cdot dT \tag{2.113}$$

$$e(t) = r(t)\,C_0 + r'(t)\,C_1 + r''(t)\frac{C_2}{2!} + r'''(t)\frac{C_3}{3!} + \ldots + r^{n'}C(t)\frac{C_n}{n} + \ldots$$

$$= C_0\, r(t) + C_1 r'(t) + \frac{C_2}{2!}r''(t) + \frac{C_3}{3!}r'''(t) + \ldots + \frac{C_n}{n!}r^{n'}r(t)\ldots \tag{2.114}$$

The coefficients $C_0, C_1, C_2 \ldots C_n$ are called the generalized error coefficients (or) dynamic error coefficients.

The steady state error e_{ss} is obtained by taking limit $r \to \infty$ on $e(t)$,

$$\therefore \qquad e_{ss} = \lim_{t\to\infty} e(t)$$

2.11.6 Evaluation of Generalized Error Coefficients (or) Dynamic Error Coefficients

The generalized error coefficients is given by,

$$C_n = (-)^n \int_0^\infty T^n f(T)\,dT$$

where, $F(s) = \dfrac{1}{1 + G(s)\,H(s)}$

we know that $L\,[f(t)] = F(s)$,

By the definition of Laplace transform

$$F(s) = \int_0^\infty f(T) \cdot e^{-sT} \cdot dT \tag{2.115}$$

take $\lim\limits_{s \to 0}$ on both sides for Eq. (2.115)

$$\lim_{s \to 0} F(s) = \lim_{s \to 0} \int_0^\infty f(T) \cdot e^{-sT} \cdot dT$$

$$= \int_0^\infty f(T) \lim_{s \to 0} e^{-sT} \cdot dT = \int_0^\infty f(T) \cdot dT = C_0$$

$$C_0 = \lim_{s \to 0} F(s) \tag{2.116}$$

on differentiating Eq. (2.115) with respect to s,

$$\frac{d}{ds} F(s) = \frac{d}{ds} \int_0^\infty f(T) \cdot e^{-sT} \cdot dT$$

$$= \int_0^\infty f(T) \cdot \frac{d}{ds} (e^{-sT}) \cdot dT = \int_0^\infty f(T) (-T) e^{-sT} \cdot dT$$

$$\frac{d}{ds} F(s) = -\int_0^\infty T\, f(T) \cdot e^{-sT} \cdot dT \tag{2.117}$$

take $\lim\limits_{s \to 0}$ on both sides,

$$\lim_{s \to 0} \frac{d}{ds} F(s) = \lim_{s \to 0} - \int_0^\infty T\, f(T) \cdot e^{-sT} \cdot dT$$

$$= -\int_0^\infty T\, f(T) \lim_{s \to 0} e^{-sT}\, dT = -\int_0^\infty T\, f(t)\, dT = C_1$$

$$C_1 = \lim_{s \to 0} \frac{d}{ds} F(s) \tag{2.118}$$

differentiate Eq. (2.117), w.r. to s

$$\frac{d}{ds}\left[\frac{d}{ds} F(s)\right] = \frac{d}{ds}\left[-\int_0^\infty T\, f(T) \cdot e^{-sT}\, dT\right]$$

$$\frac{d^2}{ds^2} F(s) = \left[-\int_0^\infty T\, f(T) \frac{d}{ds} (e^{-sT})\, dT\right] = -\int_0^\infty T\, f(T) (-T)\, e^{-sT} \cdot dT$$

$$\frac{d^2}{ds^2} F(s) = \int_0^{\infty} T^2 f(T) e^{-sT} \cdot dT \tag{2.119}$$

Apply the limit $s \to 0$

$$\lim_{s\to 0} \frac{d^2}{ds^2} F(s) = \lim_{s\to 0} \int_0^{\infty} T^2 f(T) \cdot e^{-sT} \cdot dT$$

$$\int_0^{\infty} T^2 f(T) \lim_{s\to 0} e^{-sT} \cdot dT = \int_0^{\infty} T^2 f(T) \cdot dT = C_2$$

$$C_2 = \lim_{s\to 0} \frac{d^2}{ds^2} F(s) \tag{2.120}$$

Similarly

$$C_n = \lim_{s\to 0} \frac{d^n}{ds^n} F(s) \tag{2.121}$$

2.11.7 Correlation between Static and Dynamic Error Coefficients

The values of dynamic error coefficients can be used to calculate static error coefficients. The following expressions shows the relationship between them.

$$C_0 = 1/1 + K_p \tag{2.122}$$

$$C_1 = 1/K_V \tag{2.123}$$

$$C_2 = 1/K_a \tag{2.124}$$

Proof:

$$C_0 = \lim_{s\to 0} F(s) = \lim_{s\to 0} \frac{1}{1+G(s)H(s)}$$

$$= \frac{1}{1+\lim_{s\to 0} G(s)H(s)} = \frac{1}{1+K_P}$$

2.11.8 Alternate Method for Generalized Error Coefficients

The error signal is s-domain, $E(s) = \dfrac{R(s)}{1+G(s)H(s)}$

$$\frac{E(s)}{R(s)} = \frac{1}{1+G(s)H(s)} \tag{2.125}$$

The (2.125) equation can be expressed as a power series of s as shown in Eq. (2.126).

$$\frac{E(s)}{R(s)} = \frac{1}{1+G(s)H(s)} = C_0 + C_1 s + \frac{C_2}{2!}s^2 + \frac{C_3}{3!}s^3 + \ldots \tag{2.126}$$

$$E(s) = C_0\ R(s) + C_1 s\ R(s) + \frac{C_2}{2!} s^2\ R(s) + \frac{C_3}{3!} s^3\ R(s) + \ldots \tag{2.127}$$

Take inverse Laplace transform of Eq. (2.127)

$$e(t) = C_0\ r(t) + C_1\ r'(t) + \frac{C_2}{2!} r''(t) + \frac{C_3}{3!} r'''\ (t) + \ldots \tag{2.128}$$

The Eq. (2.128) is same as that of Eq. (2.127). The coefficients C_0, C_1, $C_2 \ldots$ are called generalized error coefficients (or) dynamic error coefficients. The steady State error e_{ss} is obtained by taking limit $t \to \infty$ on $e(t)$.

$$e_{ss} = \lim_{t \to \infty} e(t)$$

EXAMPLE 2.20

For a unity feedback control system the open loop transfer function $G(s) = 10\ (s + 2)/s^2\ (s + 1)$. Find (a) the position, velocity and acceleration error constants, (b) the steady state error when the input is $R(s) = \frac{3}{s} - \frac{2}{s^2} + \frac{1}{3s^3}$.

Solution:

(a) *Static error constants*

For a unity feedback system, $H(s) = 1$

Position error constant,

$$K_P = \lim_{s \to 0} G(s)\ H(s) = \lim_{s \to 0} G(s)$$

$$= \lim_{s \to 0} \frac{10\ (s+2)}{s^2\ (s+1)} = \infty$$

Velocity error constant,

$$K_V = \lim_{s \to 0} s \cdot G(s)\ H(s)$$

$$= \lim_{s \to 0} s \cdot G(s)$$

$$= \lim_{s \to 0} s \cdot \frac{10\ (s+2)}{s^2\ (s+2)} = \infty$$

Acceleration error constant,

$$K_a = \lim_{s \to 0} s^2 \cdot G(s)\ H(s)$$

$$= \lim_{s \to 0} s^2 \cdot G(s)$$

$$= \lim_{s \to 0} s^2 \frac{10\ (s+2)}{s^2\ (s+1)} = \frac{10 \times 2}{1} = 20.$$

(b) *Steady state error*

Ist Method: The steady error for non-standard input is obtained using generalized error series,

The error signal, $e(t) = r(t)\,C_0 + r'(t)\,C_1 + r''(t)\dfrac{C_2}{2!} + \ldots + r^{n'}(t)\dfrac{C_n}{n!} + \ldots$

Given that, $$R(s) = \frac{3}{s} - \frac{2}{s^2} + \frac{1}{3s^3}$$

Input signal in time domain, $r(t) = L^{-1}[R(s)]$

$$= L^{-1}\left[\frac{3}{s} - \frac{2}{s^2} + \frac{1}{3s^2}\right] = \frac{1}{3} - 2t + \frac{1t^2}{3 \times 2!}$$

$$= \frac{1}{3} - 2t + \frac{t^2}{6}$$

$$r'(t) = \frac{d}{dt}r(t) = -2 + \frac{1}{6}2t = -2 + \frac{t}{3}$$

$$r''(t) = \frac{d^2}{dt^2}r(t) = \frac{d}{dt}r'(t) = \frac{1}{3}$$

$$r'''(t) = \frac{d^3}{dt^3}r(t) = \frac{d}{dt}r''(t) = 0$$

The derivatives of $r(t)$ is zero after second derivative. Hence we have to evaluate only three constants C_0, C_1 and C_2.

The generalized error constants are given by $C_0 = \lim\limits_{S\to 0} F(s)$; $C_1 = \lim\limits_{S\to 0}\dfrac{d}{ds}F(s)$; $C_2 = \lim\limits_{S\to 0}\dfrac{d^2}{ds^2}F(s)$.

$$F(s) = \frac{1}{1 + G(s)\,H(s)} = \frac{1}{1 + G(s)} = \frac{1}{1 + \dfrac{10(s+2)}{s^2\,(s+1)}}$$

$$= \frac{s^2\,(s+1)}{s^2\,(s+1) + 10\,(s+2)} = \frac{s^3 + s^2}{s^3 + s^2\;10s + 20}$$

$$C_0 = \lim_{s\to 0} F(s) = \lim_{s\to 0}\left[\frac{s^3 + s^2}{s^3 + s^2 + 10s + 20}\right] = 0$$

$$C_1 = \lim_{s\to 0}\frac{d}{ds}F(s) = \lim_{s\to 0}\frac{d}{ds}\left[\frac{s^3 + s^2}{s^3 + s^2 + 10s + 20}\right]$$

$$= \lim_{s\to 0}\left[\frac{(s^3 + s^2 + 10s + 20)\,(3s^2 + 2s) - (s^3 + s^2)\,(3s^2 + 2s + 10)}{(s^3 + s^2 + 10s + 20)^2}\right]$$

CHAPTER 2

$$= \lim_{s\to 0}\left[\frac{20s^3 + 70s^2 + 40s}{(s^3 + s^2 + 10s + 20)^2}\right] = 0$$

$$C_2 = \lim_{s\to 0}\frac{d^2}{ds^2}F(s) = \lim_{s\to 0}\frac{d}{ds}\left[\frac{d}{ds}F(s)\right]$$

$$= \lim_{s\to 0}\frac{d}{ds}\left[\frac{(20s^3 + 70s^2 + 40s)}{(s^3 + s^2 + 10s + 20)^2}\right]$$

$$= \lim_{s\to 0}\left[\frac{(s^3 + s^2 + 10s + 20)^2\,(60s^2 + 140s + 40) - (20s^3 + 70s^2 + 40s)\,2\times(s^3 + s^2 + 10s + 20)\,(3s^2 + 2s + 10)}{(s^3 + s^2 + 10s + 20)^4}\right]$$

$$= \frac{20^2 \times 40}{20^4} = \frac{1}{10}$$

Error signal, $e(t) = r(t)\,C_0 + r'(t)\,C_1 + r''(t)\frac{C_2}{2!}$

$$= \left(\frac{1}{3} - 2t + \frac{t^2}{6}\right)\times 0 + \left(-2 + \frac{t}{3}\right)\times 0 + \frac{1}{3}\times\frac{1}{10}\times\frac{1}{2!} = \frac{1}{60}$$

Steady state error,

$$e_{ss} = \lim_{t\to\infty} e(t) = \lim_{t\to\infty}\frac{1}{60} = \frac{1}{60}$$

IInd Method: The error signal in a s-domain, $E(s) = \dfrac{R(s)}{1 + G(s)\,H(s)}$

Given that $R(s) = \dfrac{3}{s} - \dfrac{2}{s^2} + \dfrac{1}{3s^3}$

$$G(s) = \frac{10\,(s+2)}{s^2\,(s+1)} \text{ and } H(s) = 1$$

$$E(s) = \frac{\frac{3}{s} - \frac{2}{s^2} + \frac{1}{3s^3}}{1 + \frac{10\,(s+2)}{s^2\,(s+1)}} = \frac{\frac{3}{s} - \frac{2}{s^2} + \frac{1}{3s^3}}{\frac{s^2\,(s+1) + 10\,(s+2)}{s^2\,(s+1)}}$$

$$= \frac{3}{s}\left[\frac{s^2\,(s+1)}{s^2\,(s+1) + 10\,(s+2)}\right] - \frac{2}{s^2}\left[\frac{s^2\,(s+1)}{s^2\,(s+1) + 10\,(s+2)}\right] + \frac{1}{3s^3}\left[\frac{s^2\,(s+1)}{s^2\,(s+1) + 10\,(s+2)}\right]$$

The steady state error e_{ss} can be obtained from final value theorem.

Steady state error, $e_{ss} = \underset{t\to\infty}{\text{Lim}}\ e(t) = \underset{s\to\infty}{\text{Lim}}\ s \cdot E(s)$

$$e_{ss} = \underset{t\to\infty}{\text{Lim}}\ s\left\{\frac{3}{5}\left[\frac{s^2(s+1)}{s^2(s+1)+10(s+2)}\right] - \frac{2}{s^2}\left[\frac{s^2(s+1)}{s^2(s+1)10(s+2)}\right] + \right.$$

$$\left. + \frac{1}{3s^3}\left[\frac{s^2(s+1)}{s^2(s+1)10(s+2)}\right]\right\}$$

$$= \underset{s\to 0}{\text{Lim}}\left\{\frac{3s^2(s+1)}{s^2(s+1)+10(s+2)} - \frac{2s(s+1)}{s^2(s+1)+10(s+2)} + \frac{(s+1)}{3s^2(s+1)+30(s+2)}\right\}$$

$$= 0 - 0 + \frac{1}{60} = \frac{1}{60}$$

$$e_{ss} = \frac{1}{60}$$

IIIrd Method:

Error signal in s-domain, $E(s) = \dfrac{R(s)}{1+G(s)H(s)}$

$$\frac{E(s)}{R(s)} = \frac{1}{1+G(s)H(s)},$$

Given that $G(s) = \dfrac{10(s+2)}{s^2(s+1)}$ and $H(s) = 1$

$$\frac{E(s)}{R(s)} = \frac{1}{1+\dfrac{10(s+2)}{s^2(s+1)}} = \frac{s^2(s+1)}{s^2(s+1)+10(s+2)} = \frac{s^3+s^2}{s^3+s^2+10s+20}$$

$$= \frac{s^2+s^3}{20+10s+s^2+s^3} = \frac{s^2}{20} + \frac{s^3}{40} + \ldots$$

$$E(s) = R(s)\left[\frac{s^2}{20} + \frac{s^3}{40} + \ldots\right]$$

$$= \frac{1}{20}s^2 R(s) + \frac{1}{40}s^3 R(s) + \ldots$$

On taking inverse Laplace transform,

$$e(t) = \frac{1}{20}r''(t) + \frac{1}{40}r'''(t) + \ldots$$

CHAPTER 2

Given that $$R(s) = \frac{3}{s} - \frac{2}{s^2} + \frac{1}{3s^3}$$

$$r(t) = L^{-1}[R(s)] = L^{-1}\left[\frac{3}{s} - \frac{2}{s^2} + \frac{1}{3s^3}\right]$$

$$= \frac{1}{3} - 2t + \frac{1t^2}{3 \times 2!} = \frac{1}{3} - 2t + \frac{t^2}{6}$$

$$r'(t) = \frac{d}{dt^2} r(t) = -2 + \frac{1}{6} 2t = -2 + \frac{t}{3}$$

$$r''(t) = \frac{d^2}{dt^3} r(t) = \frac{d}{dt} r''(t) = \frac{1}{3}$$

$$r'''(t) = \frac{d^3}{dt^3} r(t) = \frac{d}{dt} r\text{I}(t) = 0$$

Error signal in time domain, $e(t) = \frac{1}{20} \cdot \frac{d}{dt} r''(t)$

$$= \frac{1}{20}\left(\frac{1}{3}\right) = \frac{1}{60}$$

Steady state error, $$e_{ss} = \lim_{t \to \infty} e(t) = \lim_{t \to \infty} \frac{1}{60} = \frac{1}{60}$$

Result:

(a) Position error constant, $K_P = \infty$

Velocity error constant, $K_V = \infty$

Acceleration error constant, $K_a = 20$

(b) Steady state error when

$$R(s) = \frac{3}{s} - \frac{2}{s^2} + \frac{1}{3s^3} \text{ is } 1/60$$

i.e., $$e_{ss} = \frac{1}{60}$$

EXAMPLE 2.21

For servomechanisms with open loop transfer function given below explain what type of input signal give rise to a constant steady state error and calculate their values.

(i) $G(s) = \dfrac{20\,(s + 2)}{s\,(s + 1)\,(s + 3)}$

(ii) $G(s) = \dfrac{10}{(s+2)(s+3)}$

(iii) $G(s) = \dfrac{10}{s^2 (s+1)(s+2)}$

Solution:

(i) $G(s) = \dfrac{20(s+2)}{s(s+1)(s+3)}$

For unity feedback system, $\therefore$ $H(s) = 1$.

The open loop system has a pole at origin. Hence it is a Type-1 system. In system with type number-1, the velocity (ramp) input will give a constant steady state error.

The steady state error with unit velocity input, $e_{ss} = \dfrac{1}{K_V}$.

Velocity error constant, $K_V = \lim_{s\to 0} s \cdot G(s) \cdot H(s)$

$$= \lim_{s\to 0} s \cdot G(s)$$

$$= \lim_{s\to 0} s \frac{20(s+2)}{s(s+1)(s+3)} = \frac{20 \times 2}{1 \times 3} = \frac{40}{3}.$$

Steady state error, $e_{ss} = 1/K_V = \dfrac{3}{40} = 0.075$.

(ii) $G(s) = 10/(s+2)(s+3)$.

For unity feedback system, $H(s) = 1$

The open loop system has no pole at origin. Hence it is a type-0 system. In systems with type number-0, the step input will give a constant steady state error.

The steady state error with unit step input, $e_{ss} = 1/1 + K_P$ position error constant,

$$K_P = \lim_{s\to 0} G(s)\,H(s) = \lim_{s\to 0} G(s)$$

$$= \lim_{s\to 0} \frac{10}{(s+2)(s+3)} = \frac{10}{2 \times 3} = \frac{5}{3}$$

Steady state error, $e_{ss} = 1/1 + K_P = \dfrac{1}{1+5/3} = \dfrac{3}{3+5} = 0.375$

(iii) $G(s) = \dfrac{10}{s^2 (s+1)(s+2)}$

For unity feedback system, $H(s) = 1$. The open loop system has two poles at origin. Hence it is a Type-2 system. In systems with type number-2, the acceleration (Parabolic) input will give a constant steady state error.

The steady state error with unit acceleration in put,

Acceleration error constant, $K_a = \lim_{s\to 0} s^2 \cdot G(s) \cdot H(s)$

$$= \lim_{s\to 0} s^2\, G(s) = \lim_{s\to 0} s^2 \cdot \frac{10}{s^2\,(s+1)\,(s+2)} = \frac{10}{2\times 1} = 5$$

Steady state error, $e_{ss} = 1/K_a = 1/5 = 0.2$

Results: Steady state error in system

(i) With unit velocity input = 0.075

(ii) With unit step input = 0.375

(iii) With unit acceleration input = 0.2.

EXAMPLE 2.22

The open loop transfer function of a servo system with unity feedback is $G(s) = 10/s(0.1\,s + 1)$. Evaluate the static error constants of the system. Obtain the steady state error of the system. When subjected to an input given by the polynomial $r(t) = a_0 + a_1 t + \frac{a_2}{2} \cdot t^2$.

Solution: Finding static error constant

For unity feedback system, $H(s) = 1$.

loop transfer function, $G(s)\, H(s) = G(s)$

The static error constants are K_P, K_V and K_a

Position error constant, $K_P = \lim_{s\to 0} G(s) = \lim_{s\to 0} \frac{10}{s(0.1s+1)} = \infty$

Velocity error constant, $K_V = \lim_{s\to 0} s \cdot G(s) = s \cdot \frac{10}{s\,(0.1s+1)} = 10$

Acceleration error constant, $K_a = \lim_{s\to 0} s^2 \cdot G(s) = \lim_{s\to 0} s^2 \cdot \frac{10}{s\,(0\cdot 1s+1)} = 0$

Finding Steady State Error

Ist Method: The steady state error for non-standard input is obtained using generalized error series, given below.

The error signal, $e(t) = r(t)\,C_0 + \dot{r}(t)\,C_1 + \ddot{r}(t)\frac{C_2}{2!} + \ldots + r^n(t)\frac{C_n}{n!} + \ldots$

Given that, $$r(t) = a_0 + a_1 t + \frac{a_2}{2} t^2$$

$\therefore$ $$\dot{r}(t) = a_1 + a_2 t$$

$$\ddot{r}(t) = a_2$$
$$\dddot{r}(t) = 0$$

The derivatives of $r(t)$ is zero after second derivatives. Hence we have to evaluate only three constants C_0, C_1 and C_2.

The generalized error constants are given by,

$$C_0 = \lim_{s\to 0} F(s);\ C_1 = \lim_{s\to 0} \frac{d}{ds} F(s);\ C_2 = \lim_{s\to 0} \frac{d^2}{ds^2} F(s).$$

$$F(s) = \frac{1}{1+G(s)\,H(s)} = \frac{1}{1+G(s)} = \frac{1}{1+\dfrac{10}{s\,(0.1s+1)}} = \frac{s\,(0.1s+1)}{s\,(0.1s+1)+10}$$

$$F(s) = \frac{0.1s^2+s}{0.1s^2+s+10}$$

$$C_0 = \lim_{s\to 0} F(s) = \lim_{s\to 0} \frac{0.1s^2+s}{0.1s^2+s+10} = 0$$

$$C_1 = \lim_{s\to 0} \frac{d}{ds} F(s) = \lim_{s\to 0} \frac{d}{ds}\left[\frac{0.1s^2+s}{0.1s^2+s+10}\right]$$

$$= \lim_{s\to 0}\left[\frac{(0.1s^2+s+10)\,(0.2s+1)-(0.1s^2+s)\,(0.25+1)}{(0.1s^2+s+10)^2}\right]$$

$$= \lim_{s\to 0} \frac{2s+10}{(0.1s^2+s+10)} = \frac{10}{10^2} = 0.1$$

$$C_2 = \lim_{s\to 0} \frac{d^2}{ds^2} F(s) = \lim_{s\to 0} \frac{d}{ds}\left[\frac{d}{ds} F(s)\right] = \lim_{s\to 0} \frac{d}{ds}\left[\frac{2s+10}{(0.1s^2+s+10^2)}\right]$$

$$= \lim_{s\to 0}\left[\frac{(0.1s^2+s+10)^2\,(2)-(2s+10)\,(2)\,(0.1s^2+s+10)\,(0.2s+1)}{(0.1s^2+s+10)^4}\right]$$

$$= \frac{10^2\times 2-10\times 2\times 10\times 1}{10^4} = 0.$$

Error signal, $e(t) = r(t)\,C_0 + \dot{r}(t)\,C_1 + \ddot{r}(t)\dfrac{C_2}{2!}$

$= \dot{r}(t)\,C_1$ ($\therefore$ C_0 and C_2 are zero)

$= (a_1 + a_2\,t)\,0.1$

Steady state error is obtained by letting t to ∞

$\therefore$ Steady state error, $e_{ss} = \lim_{t\to\infty} e(t) = \lim_{t\to\infty}\left[(a_1+a_2t)\,0.1\right] = \infty$

IInd Method:

The error signal in s-domain, $E(s) = \dfrac{R(s)}{1 + G(s)\,H(s)}$

Given that

$$r(t) = a_0 + a_1 t + \frac{a_2}{2} t^2$$

$$G(s) = 10/s(0.1S + 1)$$

$$H(s) = 1$$

On taking Laplace transform of $r(t)$ we get $R(s)$.

$$R(s) = \frac{a_0}{s} + \frac{a_1}{s^2} + \frac{a_2}{2} \cdot \frac{2!}{s^3} = \frac{a_0}{s} + \frac{a_1}{s^2} + \frac{a_2}{s^3}$$

$$E(s) = \frac{R(s)}{1 + G(s)\,H(s)} = \frac{\dfrac{a_0}{s} + \dfrac{a_1}{s^2} + \dfrac{a_2}{s^3}}{1 + \dfrac{10}{s(0.1s + 1)}} = \frac{\dfrac{a_0}{s} + \dfrac{a_1}{s^2} + \dfrac{a_2}{s^3}}{\dfrac{s\,(0.1s + 1) + 10}{s(0.1s + 1)}}$$

$$= \frac{a_0}{s}\left[\frac{s(0.1s + 1)}{s(0.1s + 1) + 10}\right] + \frac{a_1}{s^2}\left[\frac{s(0.1s + 1)}{s(0.1s + 1) + 10}\right] + \frac{a_2}{s^3}\left[\frac{s(0.1s + 1)}{s(0.1s + 1) + 10}\right]$$

The steady state error e_{ss} can be obtained from find value theorem,

Steady state error, $e_{ss} = \lim_{t \to \infty} e(t) \lim_{s \to 0} s \cdot E(s)$

$$e_{ss} = \lim_{s \to 0} s \left\{ \frac{a_0}{s}\left[\frac{s\,(0.1s + 1)}{s\,(0.1s + 1) + 10}\right] + \frac{a_1}{s^2}\left[\frac{s\,(0.1s + 1)}{s(0.1s + 1) + 10}\right] + \left[\frac{s(0.1s + 1)}{s(0.1s + 1) + 10}\right]\frac{a_2}{s^3}\right.$$

$$= 0 + \frac{a_1}{10} + \infty = \infty$$

$$e_{ss} = \infty$$

IIIrd Method:

Error signal in s-domain,

$$E(s) = \frac{R(s)}{1 + G(s)\,H(s)}$$

$$\therefore \qquad \frac{E(s)}{R(s)} = \frac{1}{1 + G(s)\,H(s)}$$

Given that $\quad G(s) = \dfrac{10}{s\,(0.1s + 1)}$ and $H(s) = 1$

$$\therefore \qquad \frac{E(s)}{R(s)} = \frac{1}{1 + \dfrac{10}{s\,(0.1s + 1)}} = \frac{s\,(0.1s + 1)}{s\,(0.1s + 1) + 10}$$

$$\frac{E(s)}{R(s)} = \frac{.1s^2 + s}{0.1s^2 + s + 10} = \frac{s + 0.1s^2}{10 + s + 0.1s^2}$$

$$= \frac{s}{10} - \frac{s^3}{1000} + \ldots$$

$$\therefore \quad E(s) = \frac{s}{10} R(s) - \frac{s^3}{1000} R(s) + \ldots$$

$$\begin{array}{l}
\qquad\qquad\qquad \frac{s}{10} - \frac{s^3}{1000} \\
10s + s + 0.1s^2 \overline{\big)\, s + 0.1s^2} \\
\qquad\qquad\qquad s + \frac{s^2}{10} + \frac{s^3}{100} \\
\qquad\qquad\qquad \overline{\qquad\quad \frac{-s^3}{100}} \\
\qquad\qquad\qquad\quad -\frac{s^3}{100} - \frac{s^4}{1000} - \frac{s^5}{1000} \\
\qquad\qquad\qquad\qquad \overline{\quad \frac{s^4}{1000} + \frac{s^5}{1000}}
\end{array}$$

On taking inverse Laplace transform,

$$e(t) = \frac{1}{10}\dot{r} - \frac{1}{1000}\dddot{r}\,(t) + \ldots$$

Given that $r(t) = a_0 + a_1 t + \frac{a_2}{2} t^2$

$$\therefore \quad \dot{r} = \frac{d}{dt} r(t) = a_1 + a_2 t$$

$$\ddot{r}\,(t) = \frac{d}{dt}\dot{r}(t) = a_2$$

$$\dddot{r}\,(t) = \frac{d}{dt}\ddot{r}(t) = 0$$

$\therefore$ Error signal in time domain,

$$e(t) = \frac{1}{10}\dot{r}(t) = \frac{1}{10}(a_1 + a_2 t)$$

Steady state error, $\quad e_{ss} = \underset{t\to\infty}{\text{Lt}}\ e(t) = \underset{t\to\infty}{\text{Lt}} \frac{1}{10}(a_1 + a_2 t) = \infty$

Results:

1. Position error constant, $K_P = 0$

2. Velocity error constant, $K_V = 10$
3. Acceleration error constant, $K_a = 0$
4. Steady state error is ∞ i.e., $e_{ss} = \infty$

$$\left(\text{when input } r(t) = a_0 + a_1 + \frac{a_2 t^2}{2}\right),$$

$$e_{ss} = \infty$$

EXAMPLE 2.23

A unity feedback system with a closed loop transfer function $\frac{C(s)}{R(s)} = \frac{K_s + b}{s^2 + as + b}$. Determine the open loop transfer function $G(s)$. Show that the steady state error with unit ramp input is given by $\left(\frac{a-K}{b}\right)$.

Solution: For unity feedback system, $H(s) = 1$

The closed loop transfer function,

$$\frac{C(s)}{R(s)} = \frac{G(s)}{1+G(s)\,H(s)} = \frac{G(s)}{1+G(s)}$$

Let, $$M(s) = \frac{C(s)}{R(s)} = \frac{G(s)}{1+G(s)}$$

on cross multiplication we get,

$$G(s) = M(s)\,[1 + G(s)] = M(s) + M(s)\,G(s)$$

$$G(s) - M(s)\,G(s) = M(s)$$

$$G(s)\,[1 - M(s)] = M(s)$$

∴ $$G(s) = \frac{M(s)}{1 - M(s)}$$

Given that $$M(s) = \frac{K_s + b}{s^2 + as + b}$$

∴ Open loop transfer function,

$$G(s) = \frac{\dfrac{Ks + b}{s^2 + as + b}}{1 - \dfrac{Ks + b}{s^2 + as + b}} = \frac{Ks + b}{(s^2 + as + b) - (Ks + b)}$$

$$= \frac{Ks + b}{s^2 + as + b - Ks - b} = \frac{Ks + b}{s^2 + (a - K)\,s} = \frac{Ks + b}{s[s + (a - K)]}$$

Velocity error constant,

$$K_v = \underset{s\to 0}{\text{Lt}}\ s\cdot G(s)\,H(s) = \underset{s\to 0}{\text{Lt}}\ s\cdot G(s)$$

$$= \underset{s\to 0}{\text{Lt}}\ s\cdot \frac{Ks+b}{s\,[s+(a-K)]} = \frac{b}{a-K}$$

Steady state error with velocity input,

$$e_{ss} = \frac{1}{K_V} = \frac{a-K}{b}$$

Results: Open loop transfer function,

$$G(s) = \frac{Ks+b}{s[s+(a-K)]}$$

Steady state error with velocity input,

$$e_{ss} = \left[\frac{a-K}{b}\right]$$

EXAMPLE 2.24

A unity feedback system has the forward transfer function $G(s) = \dfrac{K_1\,(2s+1)}{s(5s+1)\,(1+s)^2}$. The input $r(t) = 1 + 6t$ is applied to the systems. Determine the minimum value of K_1 if the steady state error is to be less than 0.1.

Solution: Given that input $r(t) = 1 + 6t$.

On taking Laplace transform of $r(t)$ we get $R(s)$

$$\therefore \qquad R(s) = L[r\,(t)] = l[1+6t] = \frac{1}{s} + \frac{6}{s^2}$$

The error signal in s-domain

$$E(s) = \frac{R(s)}{1+G(s)\,H(s)}$$

$$\therefore \qquad E(s) = \frac{\frac{1}{s}+\frac{6}{s^2}}{1+\frac{K_1\,(2s+1)}{S\,(5s+1)\,(1+s)^2}} \qquad \text{(Here } H(s) = 1\text{)}$$

$$= \frac{\frac{1}{s}+\frac{6}{s^2}}{\frac{s(5s+1)\,(1+s)^2 + K_1(2s+1)}{s(5s+1)\,(1+s)^2}}$$

$$= \frac{1}{s}\left[\frac{3\,(5s+1)\,(1+s)^2}{s(5s+1)\,(1+s)^2+K_1(2s+1)}\right]+\frac{6}{s^2}\left[\frac{s\,(5s+1)\,(1+s)^2}{s(5s+1)\,(1+s^2)+K_1(2s+1)}\right]$$

The steady state error e_{ss} can be obtained from final value theorem.

$$e_{ss} = \underset{t\to\infty}{\text{Lt}}\; e(t) = \underset{s\to 0}{\text{Lt}}\; s\cdot E(s)$$

$$= \underset{s\to 0}{\text{Lt}}\; s\left\{\frac{1}{s}\left[\frac{s\,(5s+1)\,(1+s)^2}{s\,(5s+1)\,(1+s)^2+K_1\,(2s+1)}\right]+\frac{6}{s^2}\left[\frac{s\,(5s+1)\,(1+s)^2}{s\,(5s+1)\,(1+s)^2+K_1\,(2s+1)}\right]\right\}$$

$$= \underset{s\to 0}{\text{Lt}}\left\{\frac{s\,(5s+1)\,(1+s)^2}{s\,(5s+1)\,(1+s)^2+K_1\,(2s+1)}+\frac{6\,(5s+1)\,(1+s)^2}{s\,(5s+1)\,(1+s)^2+K_1\,(2s+1)}\right\}$$

$$= 0+\frac{6}{K_1}=\frac{6}{K_1}$$

Given that $e_{ss} < 0.1$

$$\therefore \qquad 0.1 = \frac{6}{K_1} \Rightarrow K_1 = \frac{6}{0.1} \Rightarrow K_1 = 60$$

Results: For steady state error $e_{ss} < 0.1$, the value of K_1 should be greater than 60.

EXAMPLE 2.25

The open loop transfer function of unity feedback system is given by

$$G(s) = \frac{50}{(1+0.1s)\,(s+10)}$$

Determine the static error coefficients K_P, K_v, and K_a.

Solution:

$$K_P = \underset{s\to 0}{\text{Lt}}\; G(s)\,H(s) = \underset{s\to 0}{\text{Lt}}\;\frac{50}{(1+0.1s)\,(s+10)} = 5$$

$$K_P = 5$$

$$K_v = \underset{s\to 0}{\text{Lt}}\; s\cdot\frac{50}{(1+0.1s)\,(s+10)} = 0$$

$$K_v = 0$$

$$K_a = \underset{s\to 0}{\text{Lt}}\; s^2\cdot\frac{50}{(1+0.1s)\,(s+10)} = 0$$

$$K_a = 0$$

Result: 1. $K_P = 5$,

2. $K_v = 0$.

EXAMPLE 2.26

The block diagram of an electronic pacemaker is given in Fig. (Ex. 2.26). Determine the steady state error for unit ramp input when $K = 400$ Also, determine the value of K for which the steady state error to a unit ramp will be 0.02

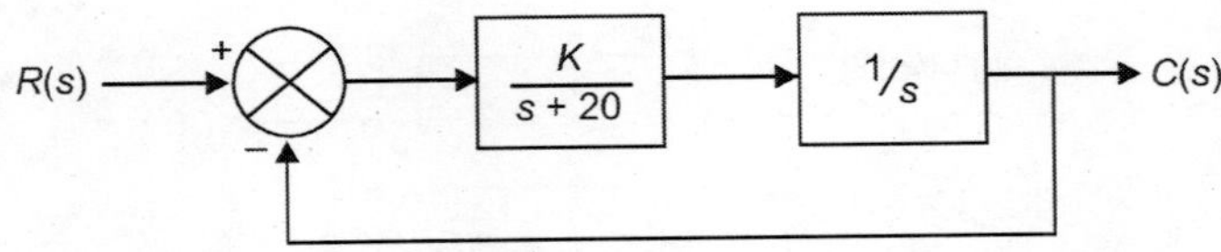

Fig. Ex. 2.26

Solution:

$$K = 400$$

$$R(s) = 1/s^2$$

$$H(s) = 1 \text{ (for unity feedback system)}$$

$$G(s)\,H(s) = K/s\,(s + 20)$$

steady state error is given by,

$$e_{ss} = \underset{s \to 0}{\text{Lt}}\ s \cdot \frac{1}{s^2} \frac{1}{1 + \dfrac{K}{s(s+20)}}$$

$$= \underset{s \to 0}{\text{Lt}} \frac{s + 20}{s(s+20) + 400} = 0.05$$

where,

$$K = 400 \text{ (given)}$$

$$R(s) = 1/s^2 \text{ (for unit ramp)}$$

Now for second case,

$$e_{ss} = .02 \text{ (given)}$$

$$e_{ss} = \underset{s \to 0}{\text{Lt}}\ s \cdot 1/s^2 \cdot \frac{1}{1 + \dfrac{K}{s(s+20)}}$$

$$0.02 = \underset{s \to 0}{\text{Lt}} \frac{s + 20}{s(s+20) + K}$$

$$= \frac{0 + 20}{0(0 + 20) + K} = \frac{20}{K}$$

$$.02 = \frac{20}{K} \Rightarrow K = 1000$$

Result:

(i) e_{ss} when $K = 400$ is .05

(ii) Value of K when $e_{ss} = .02$ is 1000.

EXAMPLE 2.27

A servomechanism is designed to keep a radar antenna pointed at a flying aeroplane. If the aeroplane is flying with a velocity of 600 km/hr. at a range of 2 km. and the maximum tracking error is to be within 1°, determine the required velocity error coefficient

Solution:

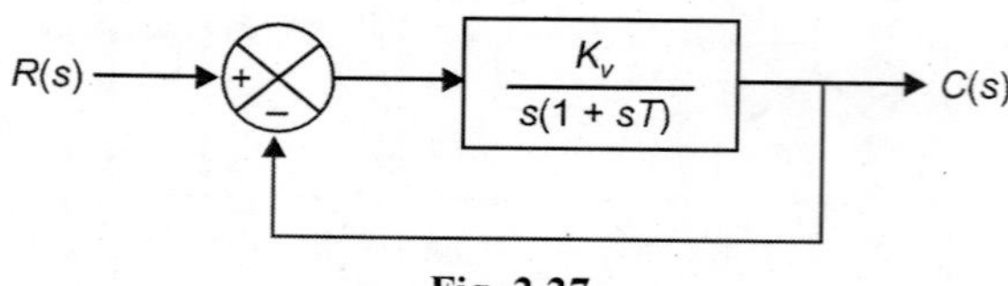

Fig. 2.27

Velocity = 600 km/hr.

Angular velocity $\omega = \dfrac{600}{2} = 300$ rad/hr.

$$= \frac{300}{3600} \text{ rad/sec } = 0.083 \text{ rad/sec}$$

$$R(s) = \frac{0.083}{s^2}$$

$$e_{ss} = \underset{s\to 0}{\text{Lt}}\; s \cdot R(s) \cdot \frac{1}{1 + G(s)\,H(s)}$$

$$= \underset{s\to 0}{\text{Lt}}\; \frac{0.083}{s^2} \cdot \frac{1}{1 + \dfrac{K_v}{s(1+sT)}}$$

$$e_{ss} = \underset{s\to 0}{\text{Lt}}\; \frac{0.083}{s} \cdot \frac{s(1+sT)}{s(1+sT) + K_v} = \frac{0.083}{K_v}$$

$$.1 = \frac{0.083}{K_v} \Rightarrow K_v = \frac{.083}{.1} = 83\%$$

Result: The required velocity error coefficient $K_v = 83\%$.

2.12 EFFECT OF P, PI AND PID CONTROLLER

2.12.1 Introduction

The automatic controller determines the value of controlled variable and compare the actual value to the desired value (reference input), then determines the deviation and produces a control signal that will reduce the deviation to zero (or) to a smallest possible value. The method by which the automatic controller produces the control signal is called control action.

The control action may operate through either mechanical, hydraulic, pneumatic (or) electromechanical means i.e. controllers can be electrical, hydraulic, pneumatic electromechanical or electronics types.

2.12.2 Elements of Industrial Automatic Controller

Figure (2.21) is a block diagram of an industrial controller. The controller consists of an error detector and amplifier. The measuring element is a device which converts the output variable to another suitable variable such as displacement, pressure or electrical signals, which can be used for comparing the output to the reference input signal. The set point of a controller must be converted to reference input of the same units as feedback signal from the measuring element. From Fig. (2.21).

$$e = r - b$$

i.e., deviation is the difference between controlled variable and set point (reference input).

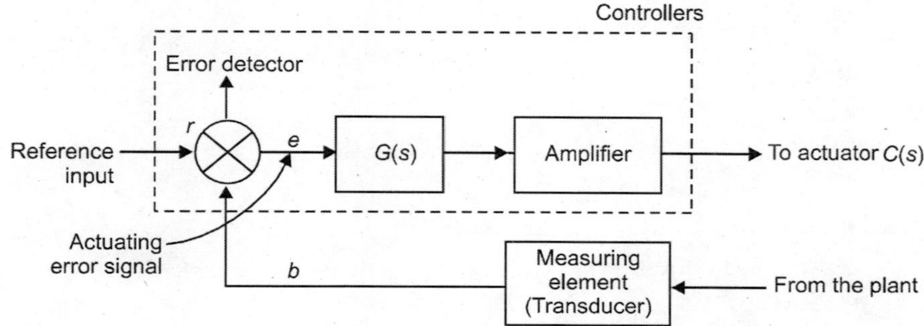

Fig. 2.21: Block diagram of industrial automatic controller.

2.12.3 Classification of Controllers

Controllers are classified depending upon the type of controlling action used. Therefore, they can be classified as

1. Two position or ON–OFF controllers
2. Proportional controllers
3. Integral controllers
4. Proportional-plus integral controllers
5. Proportional-plus derivative controllers
6. Proportional plus integral-plus derivative controllers.

They can also be classified according to the power source used for actuating mechanisms, such as electrical, electronics pneumatic and hydraulic controllers. Hydraulic controllers are used for controlling heavy loads, pneumatic controllers are suitable for shop floor applications.

The selection of a particular types of controller depends on the nature of plant, operating conditions such as safety, cost, availability, accuracy, weight and size.

2.12.4 Effect of Proportional Controller

The proportional controller produces an output signal which is proportional to error signal. The transferfunction of proportional controller is K_p. The term K_p is called the gain of the controller. Hence the proportional controller amplifies the error signal and increases the loop gain of the system. The following points are noted,

1. Steady state tracking accuracy.
2. Disturbance signal rejection.
3. Relative stability.

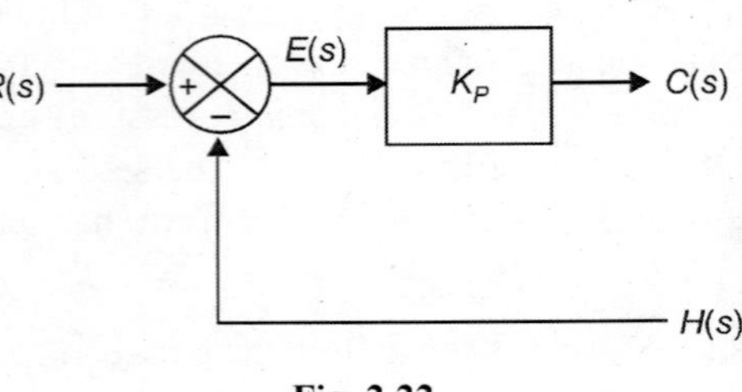

Fig. 2.22

In addition to increase in loop gain it decreases the sensitivity of the system to parameter variations. The drawback in proportional control action is that it produces a constant steady state error.

In a controller with proportional control action, there is a continuous linear relation between the output of the controller C and actuating error signal e(deviation).

Mathematically

$$C(t) = K_P\, e(t)$$

in terms of Laplace transform,

$$C(s) = K_P\, E(s)$$

$$K_P = \frac{C(s)}{E(s)} \tag{2.129}$$

2.12.5 Effect of Integral Controller [Re set Controller]

In a controller with integral control action, the output of the controller is changed at a rate which is proportional to the actuating error signal $e(t)$.

Mathematically,

$$\frac{d}{dt}C(t) = K_i e\,(t) \Rightarrow C(t) = K_i \int e(t)$$

where $K_i \rightarrow$ Integral constant

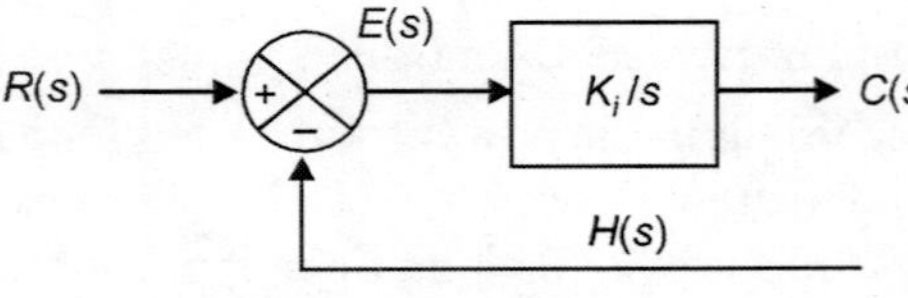

Fig. 2.23

$$C(t) = K_i \int e(t) \tag{2.130}$$

Laplace transform of equation (2.130)

$$C(s) = \frac{K_i}{s} E(s)$$

$$\frac{C(s)}{E(s)} = \frac{K_i}{s} \qquad (2.131)$$

Note: $T_i = 1/K_i$

Equation (2.131) is the transfer function of integral controller. The block diagram of integral controller is shown in Fig. (2.23). The inverse of *Ki* is called integral time T_i and defined as the time of change of output caused by a unit change of actuating error signal.

2.12.6 Effect of Derivative Controller [Refer Controller]

In a controller with derivative control action, the output of the controller depends on the rate of change of actuating error signal $e(t)$.

Mathematically,

$$C(t) = K_d \frac{d}{dt} e(t) \qquad (2.132)$$

where,

$K_d \rightarrow$ Derivative gain constant.

Laplace transform of Eq. (2.132)

$$C(s) = K_d \cdot s \cdot E(s)$$

$$\frac{C(s)}{E(s)} = K_d \cdot s \qquad (2.133)$$

Equation (2.133) is the transfer function of the controller. The block diagram is shown in Fig. (2.24).

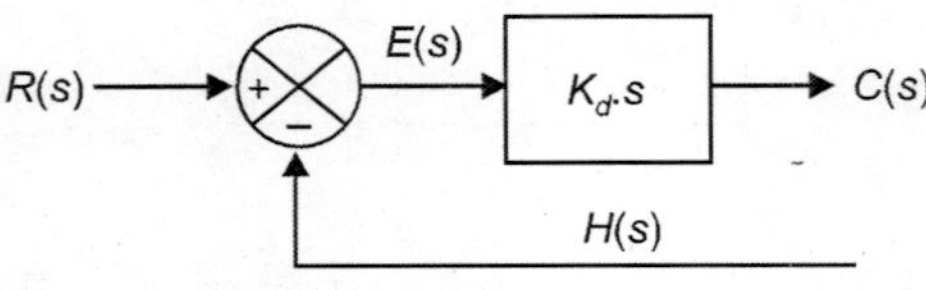

Fig. 2.24

From equation (2.132) it is clear that when the error is zero or constant, the output of the controller will be zero. Therefore this type of controller cannot be used alone. For this type of controller the gain should be small.

2.12.7 Effect of Proportional-Plus-Integral Controller [PI Controller]

The proportional plus integral controller (PI) produces an output signal consisting of two terms-one proportional to error signal and the other proportional to the integral of error signal.

$$C(t) = K_p\, e(t) + K_p K_i \int_0^t e(t)\, dt \qquad \textbf{Note :}\ T_i = \frac{1}{K_i} \text{ for } I \text{ controller}$$

$$C(t) = K_p\, e(t) + K_p \frac{1}{T_i} \int_0^t e(t)\, dt \qquad (2.134)$$

Laplace transform of Eq. (2.134)

$$C(s) = K_p\, E(s) + \frac{K_p}{sT_i} E(s)$$

$$C(s) = E(s)\left[1 + \frac{1}{sT_i}\right] K_p$$

$$\frac{C(s)}{E(s)} = \left[1 + \frac{1}{sT_i}\right] K_p \qquad (2.135)$$

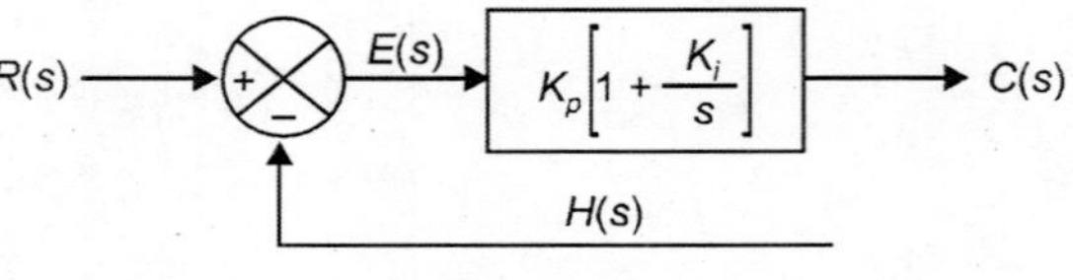

Fig. 2.25

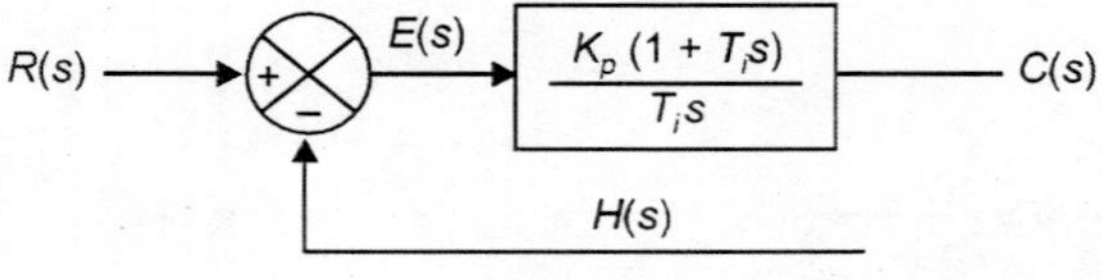

Fig. 2.26

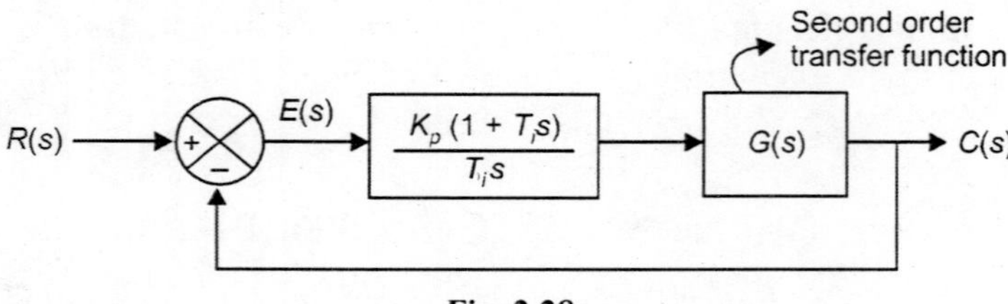

Fig. 2.27

Let this PI controller is connected with one second order system. Under this condition let we see the effect of PI controller on the system.

Fig. 2.28

Let $$G(s) = \frac{\omega_n^2}{s}[s + 2\zeta\omega_n]$$

Now the closed loop transfer function is given by

$$\frac{C(s)}{R(s)} = \frac{K_p\left[\frac{1+T_i s}{T_i s}\right]\left[\frac{\omega_n^2}{s[s+2\zeta\omega_n]}\right]}{1+K_p\left[\frac{1+T_i s}{T_i s}\right]\left[\frac{\omega_n^2}{s[s+2\zeta\omega_n]}\right]}$$

$$= \frac{K_p\omega_n^2(1+T_i s)}{s^2T_i\,(s+2\zeta\omega_n)+K_p\omega_n^2\,(1+T_i s)}$$

$$= \frac{K_p\omega_n^2\,(1+T_i s)}{T_i s^3 + 2\zeta\omega_n T_i s^2 + K_p\,\omega_n^2\,T_i s + K_p\omega_n^2}$$

$$= \frac{(K_p/T_i)\omega_n^2\,(1+T_i s)}{s^3 + 2\zeta\omega_n s^2 + K_p\omega_n^2 + \left(\frac{K_p}{T_i}\right)\omega_n^2}$$

$$\frac{C(s)}{R(s)} = \frac{K_i\,\omega_n^2(1+T_i s)}{s^3 + 2\zeta\omega_n s^2 + K_p\omega_n^2 s + K_i\,.\,\omega_n^2} \tag{2.136}$$

where, $K_i = K_p/T_i$ ***Note:*** $K_i = K_i/T_i$ for PI controller

From the closed loop transferfunction (Eq. 2.136) it is observed that the PI controller introduces a zero in the system and increases the order by one also the type number of open loop system is increased by one. The increase in type number results in reducing the steady state error.

The increase in order of the system results in a less stable system than the original one.

2.12.8 EFFECT OF PROPORTIONAL-PLUS-DERIVATIVE CONTROL ACTION (PD)

The proportional plus derivative controller produces an output signal consisting of two terms one proportional to error signal and the other proportional to the derivative of error signal.

$$C(t) = K_p\,e(t) + K_pT_d\,\frac{d}{dt}\,e(t) \tag{2.137}$$

Laplace transform of Eq. (2.137)

$$C(s) = K_p\,E(s) + K_p\,T_d\,s\cdot E(s)$$

$$= E(s)\,[1 + s\,T_d]\,K_p$$

$$\frac{C(s)}{E(s)} = K_p(1 + sT_d)$$

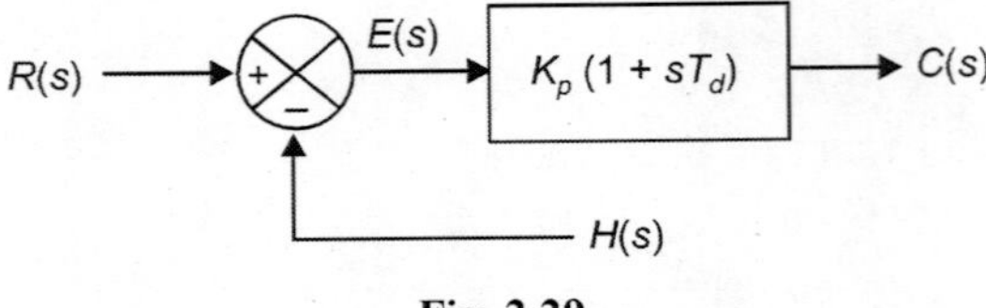

Fig. 2.29

T_d is known as derivative time constant. Derivative time is defined as the time interval by which the rate action advances the effect of the proportional control action.

Let we see the effect of PD controller on the second order system.

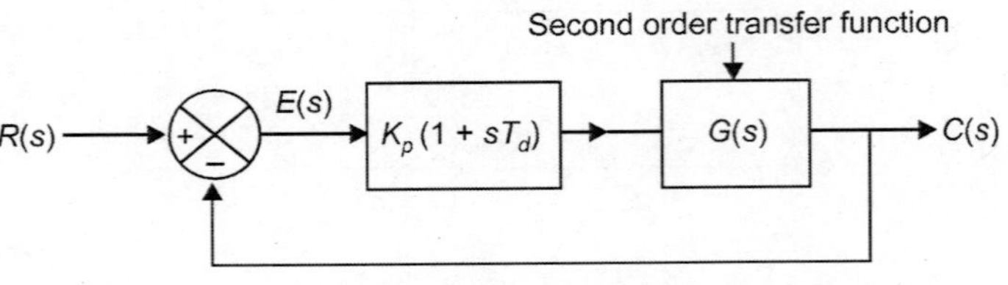

Fig. 2.30

$$\frac{C(s)}{R(s)} = \frac{K_p\,(1+T_d\cdot s)\cdot\dfrac{\omega_n^2}{s\,(s+2\zeta\omega_n)}}{1+K_p\,(1+T_d\cdot s)\cdot\dfrac{\omega_n^2}{s\,(s+2\zeta\omega_n)}}$$

$$= \frac{K_p\omega_n^2\,(1+T_d s)}{s\,(s+2\zeta\omega_n)+K_p\,\omega_n^2\,(1+T_d s)}$$

$$= \frac{K_p\,\omega_n^2\,(1+T_d s)}{s^2+2\zeta\omega_n\cdot s+K_p\cdot\omega_n^2+K_p\omega_n^2\,T_d s}$$

$$= \frac{K_p\,\omega_n^2\,(1+T_d\cdot s)}{s^2+(2\zeta\omega_n+K_p\,\omega_n^2\,T_d)s+K_p\omega_n^2}$$

$$\frac{C(s)}{R(s)} = \frac{\omega_n^2(K_p+K_d\cdot s)}{s^2+(2\zeta\omega_n+K_d\omega_n^2)\,s+K_p\omega_n^2}$$

where,

$$K_d = (K_p\cdot T_d) \qquad (2.138)$$

From the closed loop transfer function it is observed that the PD controller introduces a zero in the system and increase the damping ratio. The addition of the zero may increase the peak overshoot and reduce the rise time. But the effect of increased damping ultimately reduces the peak overshoot. It has faster response and improves the bandwidth.

2.12.9 Effect of Proportional-Plus Integral-Plus Derivative Controller (PID Controller)

The combination of proportional, integral and derivative control action is called PID control action and the controller is called three action controller (or) PID controller.

Mathematically,

$$C(t) = K_p\, e(t) + K_p \frac{1}{T_i} \int e(t)\, dt + K_p\, T_d \frac{d}{dt} e(t) \tag{2.139}$$

Laplace transform of Eq. (2.139)

$$C(s) = K_p\, E(s) + \frac{K_p}{sT_i} E(s) + K_p \cdot T_d \cdot sE(s)$$

$$\frac{C(s)}{E(s)} = K_P \left(1 + \frac{1}{sT_i} + sT_d\right) \tag{2.140}$$

K_p is the proportional gain, T_i is the integral time and T_d is the derivative time.

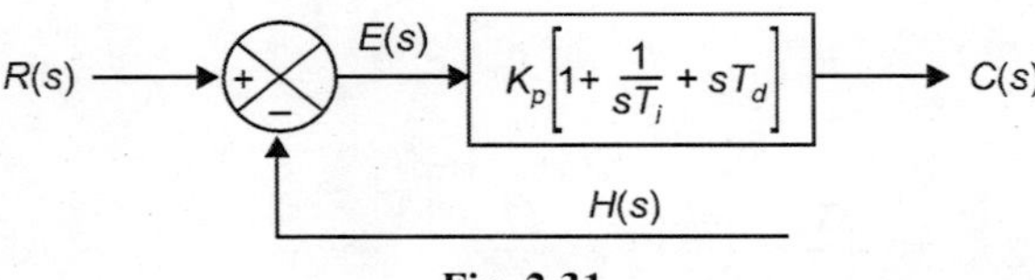

Fig. 2.31

Let this PID controller is connected with second order system. Under this condition let we see the effect of PID controller on the system.

Let $$G(s) = \frac{\omega_n^2}{s\,(s + 2\zeta\omega_n)}$$

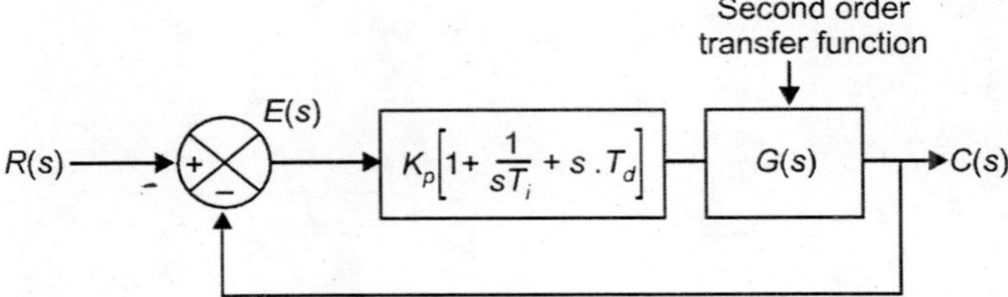

Fig. 2.12.9.2

Note: The transfer function of the Fig. (2.32) can be derived as same as previous cases.

After determining the closed loop transfer function. The following points are noted.

- The proportional controller stabilizes the gain but produces a steady state error.
- The integral controller reduces or eliminates the steady state error.
- The derivative controller reduces the rate of change of error. The combined effect of the three cannot be judged from the parameters K_p, K_i and K_d.

CHAPTER 2

EXAMPLE 2.28

A closed loop control system with unity feed back is shown in Fig. (Ex. 2.28.1 and 2.28.2). By using derivative control the damping ratio is to be made 0.7. Determine the value of delay time (T_d), also determine the rise time without derivative control and with derivative control. The input to the system is unit step.

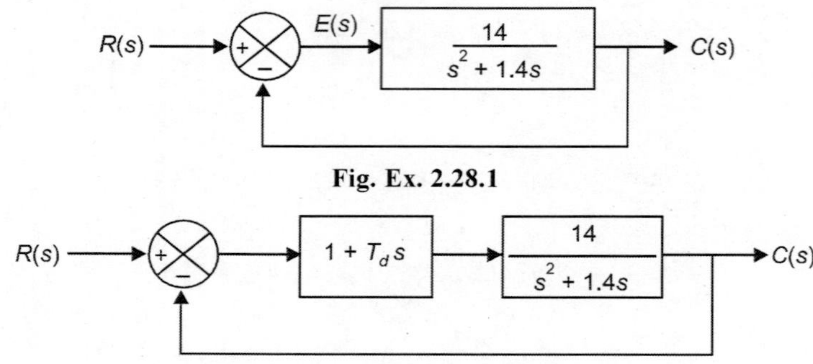

Fig. Ex. 2.28.1

Fig. Ex. 2.28.2

Solution: The closed loop transfer function of the system,

From Fig. (Ex. 2.28.1)
$$\frac{C(s)}{R(s)} = \frac{\dfrac{14}{s^2 + 1.4s}}{1 + \dfrac{14}{s^2 + 1.4s}} = \frac{14}{s^2 + 1.4s + 14}$$

The characteristic equation $s^2 + 1.4s + 14 = 0$

Compare with $s^2 + 2\zeta\omega_n s + \omega_n^2 = 0$

$$2\zeta\omega_n = 1.4$$

$$\omega_n^2 = 14 \Rightarrow \omega_n = 3.74 \text{ rad/sec}$$

$$2\zeta \times 3.74 = 1.4$$

$$\therefore \quad \zeta = .187$$

$$\zeta' = \zeta + \frac{\omega_n T_d}{2} = .187 + \frac{3.74\,(T_d)}{2}$$

$$.7 = .187 + \frac{3.74\,(T_d)}{2}$$

$$\therefore \quad T_d = .274 \text{ sec}$$

After substituting the value of Td in fig. 2.28.2 the figure becomes.

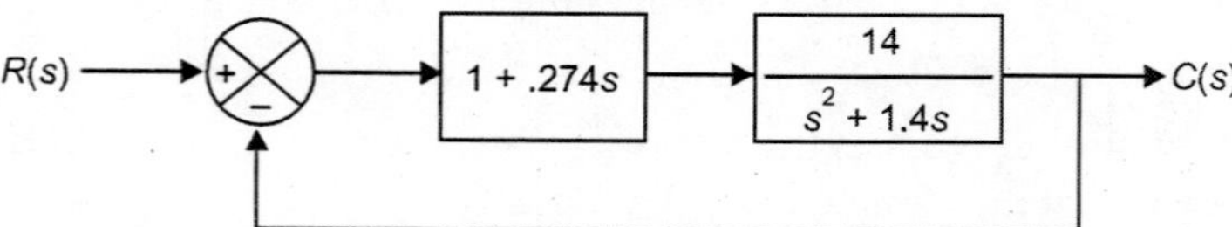

Fig. Ex. 2.28.3

Rise time $t_r = \dfrac{\pi - \tan^{-1}\dfrac{\sqrt{1-\zeta^2}}{\zeta}}{\omega_n\sqrt{1-\zeta^2}}$

$$= \frac{\pi - \tan^{-1}\dfrac{\sqrt{1-(.187)^2}}{.187}}{3.74\sqrt{1-(.187)^2}} = \frac{1.757}{3.67}$$

$$t_r = .478 \text{ sec}$$

Peak time, $t_p = \dfrac{\pi}{\omega_n\sqrt{1-\zeta^2}} = \dfrac{\pi}{3.74\sqrt{1-(.187)^2}} = .854 \text{ sec}$

$$t_p = .854 \text{ sec}$$

Percentage maximum ovrshoot

$$M_p = e^{\frac{-\zeta\pi}{e\sqrt{1-\zeta^2}}} \times 100$$

$$= e^{-\frac{\pi(.187)}{\sqrt{1-(.187)^2}}} \times 100$$

$$M_p = 55\%$$

The overall transferfunction with derivative control

For Fig. (2.28.3), $\dfrac{C(s)}{R(s)} = \dfrac{14\,(1+.274s)}{s^2+1.4s+14+3.836s}$

Note: Find transfer function of Fig. (2.28.2) using the formula $G/1+GH$.

$$\frac{C(s)}{R(s)} = \frac{14\,(1+.274s)}{s^2+5.236s+14}$$

Input $R(s) = 1/s$ (for step input)

$$C(s) = \frac{1}{s}\cdot\frac{14\,(1+.274s)}{s^2+5.236s+14} = \frac{1}{s}\left[\frac{14+3.836s}{s^2+5.236s+14}\right]$$

$$= \frac{1}{s}\cdot\frac{14}{s^2+5.236s+14} + \frac{1}{s}\cdot\frac{3.836s}{s^2+5.236s+14}$$

$$C(s) = \frac{A}{s} + \frac{Bs+C}{s^2+5.236s+14} + \frac{3.836s}{s^2+5.23s+14}$$

Now calculate the coefficients

$$14\,(1+.274s) = A\,(s^2+5.236s+14) + (B+C)s + 3.836s$$

Equating the coefficient of s^2, s and constant terms

$$A + B = 0$$

$$5.236A + C + 3.836 = 3.836$$

$$14A = 14 \Rightarrow A = 1 \Rightarrow B = -1$$

$$C = -5.236$$

Substitute the A, B and C values in $C(s)$

$$C(s) = \frac{1}{s} - \frac{s + 5.236}{s^2 + 5.236s + 14} + \frac{3.836}{s^2 + 5.236s + 14}$$

$$= \frac{1}{s} - \frac{s + 5.236}{(s + 2.618)^2 + (14 - 2.618)^2} + \frac{3.836}{(s + 2.618)^2 + (14 - 2.678)^2}$$

$$= \frac{1}{s} - \frac{s + 5.236}{(s + 2.618)^2 + (2.673)^2} + \frac{3.836}{(s + 2.618)^2 + (2.673)^2}$$

$$C(s) = \frac{1}{s} - \frac{+(s + 2.618 + 2.618)}{(s + 2.618)^2 + (2.673)^2} + \frac{3.386}{(s + 2.618)^2 + (2.673)^2}$$

$$= \frac{1}{s} - \frac{s + 2.618}{(s + 2.618)^2 + (2.673)^2} + \frac{1.218}{(s + 2.618)^2 + (2.673)^2}$$

$$C(s) = \frac{1}{s} - \frac{s + 2.618}{(s + 2.618)^2 + (2.673)^2} + \frac{2.673}{(5 + 2.618)^2 + (2.673)^2}\left(\frac{1.218}{2.673}\right)$$

on taking inverse Laplace transform

$$C(t) = 1 - e^{-2.618t} \cos 2.673t + .455\, e^{-2.618t} \sin 2.673t$$

For rise time $\quad C(t) = 1$

$$1 = 1 - e^{-2.618t} \cos 2.673t + 0.455\, e^{-2.618t} \sin 2.673t \qquad (2.28.1)$$

$$0.455 \sin 2.673t = \cos 2.673t$$

$$\tan 2.673t = 2.19$$

$$t = .427$$

$$t_r = .427 \text{ sec}$$

Result:

(i) Without derivative control

$t_r = .478$ sec

(ii) With Derivative control

$t_r = .427$ sec

(iii) $T_d = .274$ sec

EXAMPLE 2.29

Fig. (Ex. 2.29) shows a system employing proportional plus error control. Determine the value of error rate factor K_e so that the damping ratio is 0.5. Determine the value of settling time, maximum overshoot and steady state error (for unit ramp input) with and without error rate control.

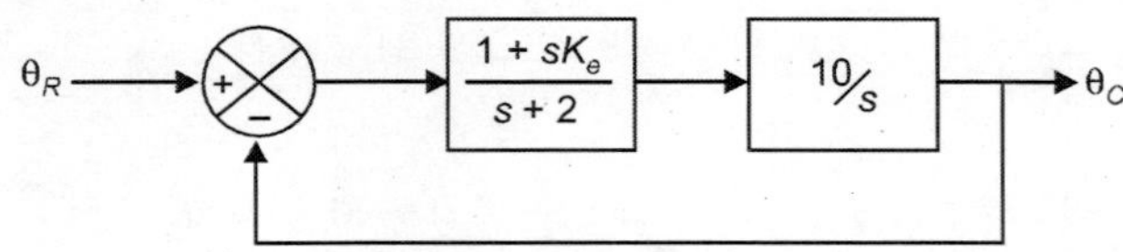

Fig. Ex. 2.29

Solution: Without error rate control i.e. $K_e = 0$ if $K_e = 0 \Rightarrow \left(\frac{1}{s+2}\right)\left(\frac{10}{s}\right) = \frac{10}{s(s+2)} = G(s)$

The overall transfer function $\frac{C(s)}{R(s)}$ will be $G(s)/1 + G(s)\,H(s)$

i.e.,
$$\frac{C(s)}{R(s)} = \frac{\frac{10}{s(s+2)}}{1+\frac{10}{s(s+2)}} = \frac{10}{s^2+2s+10}$$

Characteristic equation $s^2 + 2s + 10 = 0$

Compare without $s^2 + 2\zeta\omega_n s + \omega_n^2 = 0$

$$\omega_n^2 = 10$$

$$\therefore \quad \omega_n = 3.16 \text{ rad/sec}$$

$$2\zeta\omega_n = 2$$

$$2 \times \zeta \times 3.16 = 2$$

$$\zeta = .316$$

Maximum overshoot
$$M_p = e^{-\pi\zeta/\sqrt{1-\zeta^2}} \times 100$$

$$= e^{\frac{(-\pi \times .316)}{\sqrt{1-(.316)^2}}} \times 100$$

$$\% M_p = 35.14\%$$

Settling time
$$t_s = \frac{4}{\zeta\omega_n} = \frac{4}{0.316 \times 3.16} = 4 \text{ sec}$$

steady state error

$$e_{ss} = \frac{2\zeta}{\omega_n} = \frac{2 \times .316}{3.16}$$

$$e_{ss} = 0.2$$

With error control

The overall transferfunction of Fig. (2.29)

$$\frac{C(s)}{R(s)} = \frac{\frac{10\,(1+sK_e)}{s(s+2)}}{1+\frac{10\,(1+sK_e)}{s(s+2)}} = \frac{10\,(1+sK_e)}{s(s+2)+10\,(1+sK_e)}$$

characteristic equation $s(s+2)+10\,(1+sK_e)=0 \Rightarrow (s^2+2s)+10+10s\,Ke=0 \Rightarrow s^2+s(10\,Ke+2)+10$

compare with $\quad s^2+2\zeta\omega_n\,s+\omega_n^2=0$

$$\omega_n^2 = 10$$

$\therefore \quad \omega_n = 3.16 \text{ rad/sec} \Rightarrow 2\zeta\omega_n = 2+10\,K_e$

= 2 Ч.5 Ч3.16 = 2 + 10 K_e

$\therefore \quad K_e = .116$

The characteristic equation with $Ke = .116$ is

$$s^2+3.16s+10=0$$

Compare with $\quad s^2+2\zeta\omega_n s+\omega_n^2=0$

$$2\zeta\omega_n = 3.16$$

$$2\zeta(3.16) = 3.16$$

$$2\zeta = 1$$

$$\zeta = 1/2 = .5$$

Maximum overshoot, $\quad M_p = e^{-\pi\zeta/\sqrt{1-\zeta^2}} \times 100$

$$M_p = e^{\frac{-\pi\,(0.5)}{\sqrt{1-(0.5)^2}}} \times 100 = 16.3\%$$

$$\%M_p = 16.3\%$$

settling time

$$t_s = \frac{4}{\zeta\omega_n} = \frac{4}{0.5\times 3.16} = 2.53 \text{ sec}$$

$$t_s = 2.53 \text{ sec}$$

Steady state error $\quad e_{ss} = \underset{s\to 0}{\text{Lt}}\; sE(s)$

$$E(s) = \frac{1}{1+G(s)}\cdot R(s) \qquad [\because H(s)=1]$$

$$e_{ss} = \underset{s \to 0}{\text{Lt}} \; s \cdot \frac{1}{s^2} \cdot \frac{s(s+2)}{10 + 1.16s + s(s+2)} \quad [\text{Œ } R(s) = 1/s^2 \text{ for unit ramp i/p}]$$

$$e_{ss} = 0.2 \text{ rad}$$

Results:

(i) Without error control

(i) $\%M_p = 35.14\%$

(ii) $t_s = 4$ sec

(iii) $e_{ss} = 0.2$ rad

(ii) With error control

(i) $M_p = 16.3\%$

(ii) $t_s = 2.53$ sec

(iii) $e_{ss} = 0.2$ rad

EXAMPLE 2.30

A feedback system employing output-rate damping shown in Fig. (Ex. 2.30)

(a) In the absence of derivative feedback ($K_0 = 0$). (a) Determine the damping factor and natural frequency of the system. What is the steady state error resulting from unit ramp state error resulting from unit ramp input.

(b) Determine the derivative feedback constant K_0, which will increase the damping factor of the system to 0.6. What is the steady state error resulting from unit ramp input with this settling of the derivative feedback constant?

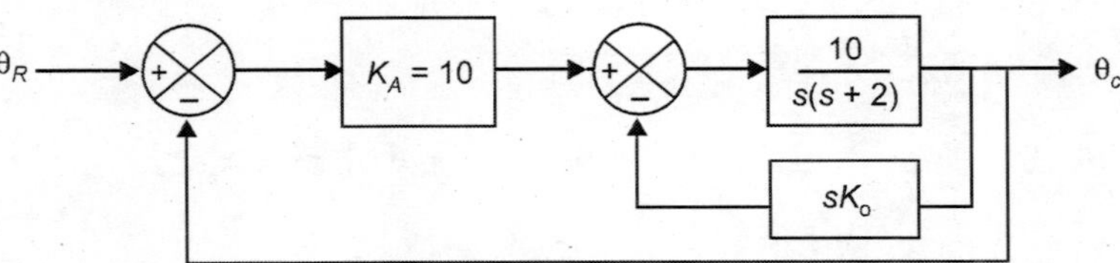

Fig. Ex. 2.30

Solution:

(a) Without derivative feedback

(a) When $K_0 = 0$

$$G(s) = \frac{10}{s(s+2)} = \frac{\frac{10}{s(s+2)}}{1 + \frac{10}{s(s+2)}} \qquad [\because H(s) = 1]$$

$$= \frac{\frac{10}{s(s+2)}}{\frac{s(s+2)+10}{s(s+2)}} = \frac{10}{s(s+2)+10}$$

$$= \frac{10}{s^2+2s+10}$$

Compare characteristic equation $s^2 + 2s + 10 = 0$ with $s^2 + 2\zeta\omega_n s + \omega_n^2 = 0$

$$\omega_n^2 = 0 \quad \omega_n = 3.16 \text{ rad/sec}$$

$$2\zeta\omega_n = 2$$

$$2 \times \zeta \times 3.16 = 2$$

$$\zeta = .316$$

Steady state error e_{ss} $\quad e_{ss} = \frac{2\zeta}{\omega_n} = \frac{2 \times .316}{3.16}$

$$e_{ss} = 0.2 \text{ rad}$$

(b) *With derivative feedback*

The overall transferfunction of Fig. (2.30),

$$\frac{C(s)}{R(s)} = \frac{G(s)}{1+G(s)\,H(s)} = \frac{\frac{10}{s(s+2)}}{1+\frac{10}{s(s+2)}(sK_0)}$$

$$\frac{C(s)}{R(s)} = \frac{10}{s^2+(2+K_0)\,s+10}$$

Characteristic equation $s^2 + (2 + K_0)\, s + 10$ compare with $s^2 + 2\zeta\omega_n s + \omega_n^2$

$$\omega_n^2 = 10$$

$\therefore$

$$\omega_n = 3.16 \text{ rad/sec}$$

$$2\zeta\omega_n = 2 + K_0$$

$$2 \times 0.6 \times 3.16 = 2 + K_0$$

$$K_0 = 1.792$$

Steady state error (for unit ramp input)

$$e_{ss} = \underset{s\to 0}{\text{Lt}}\ s \cdot E(s)$$

$$E(s) = \frac{R(s)}{1+G(s)\,H(s)}$$

[$\because R(s) = 1/s^2$ for unit ramp input]

$$e_{ss} = \underset{s\to 0}{\text{Lt}}\ s\frac{1}{s^2}\frac{s(s+2)+1.8s}{s(s+2)+1.8s+10}$$

$$e_{ss} = \underset{s\to 0}{\text{Lt}}\ \frac{s+2+1.8}{10} = 0.38$$

$$e_{ss} = 0.38 \text{ rad}$$

Results:

(i) Without derivative feedback

  (i) $\zeta = 0.316$

  (ii) $\omega_n = 3.16$ rad/sec

  (iii) $e_{ss} = 0.2$ rad

(ii) With derivative feedback

  (i) $K_0 = 1.792$

  (ii) $\omega_n = 3.16$ rad/sec

  (iii) $e_{ss} = 0.38$ rad

2.13 EFFECT OF ADDING POLES AND ZEROS TO THE SYSTEM

(i) Addition of a Pole:

Consider the transfer function of a closed loop system.

$$\frac{C(s)}{R(s)} = M(s) = \frac{\omega_n^2}{(s^2 + 2\zeta\omega_n s + \omega_n^2)} \tag{2.41}$$

Now a pole whose time constant is T_P is added to the system of Eq. (2.41). The closed loop transfer function becomes,

$$M(s) = \frac{\omega_n^2}{(s^2 + 2\zeta\omega_n s + \omega_n^2)(1 + s\,T_p)} \tag{2.42}$$

For various values of T_p the transient response for the step input is calculated and the graphs are shown in Fig. (2.33) for $\omega_n = 2$ rad/sec

From Fig. (2.33) it is evident that the addition of pole to the original second order system ($T_p = 0$), increases the rise time but decreases the over shoot. As the pole $s = \frac{-1}{T_p}$ is moved towards the origin, the number of oscillations decreases and the system becomes none and more sluggish. Hence a compromise has to be made in the selection of T_P so that the system response is very fast at the same time over shoot is not excessive.

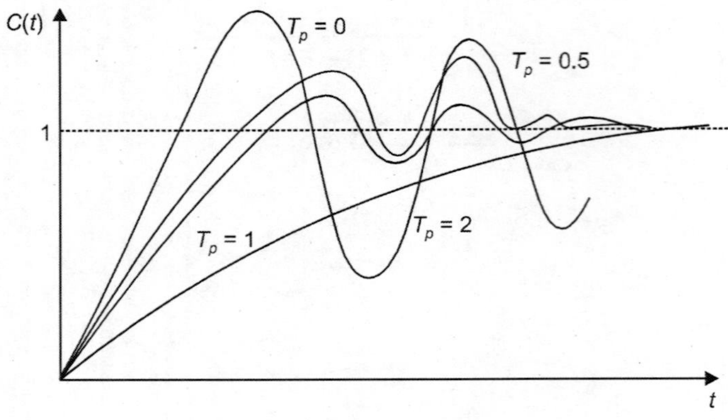

Fig. 2.33

(ii) Addition of a Zero:

Let $(1 + sT_z)$ be added to a second order system described by equation (2.41). Then equation (2.41) is rewritten in the following form with the addition of this zero.

$$\frac{C(s)}{R(s)} = M(s) = \frac{\omega_n^2 (1 + sT_z)}{(s^2 + 2\zeta\omega_n s + \omega_n^2)} \tag{2.43}$$

As before the values of $\omega_n = 2$ rad/sec and $\zeta = 0.6$ are chosen and the transient response for unit step input is calculated for various values of T_z The curves and plotted and shown in Fig. (2.13.2)

From Fig. (2.34) it is evident that the addition of a zero decreases the rise time and increases of oscillations. Even though the fastness of response is desirable, the increase in the amount of over-shoot and the number of oscillations per second are not desirable. A compromise has to be made between the two.

Conclusion:

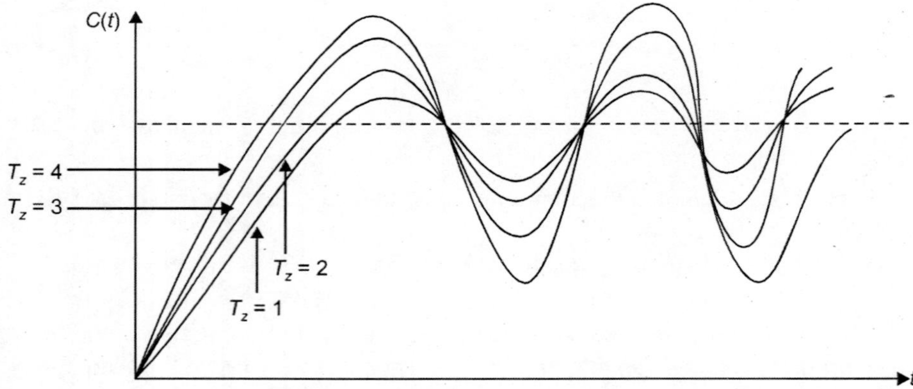

Fig. 2.34

(i) Adding a pole to the system increases the rise time

(ii) Addition of pole to the system decreases the over-shoot.

(iii) As the pole $s = \frac{-1}{+P}$ moved towards the origin, then the number of oscillations decreases.

(iv) Addition of zero to the system, decreases the rise time.

(v) Addition of zero to the system increases the overshoot and number of oscillations.

Note: Increase in the amount of overshoot and the number of oscillations are not desirable.

QUESTIONS AND ANSWERS

Q2.1 What is time Response?

Ans. The time response is the output of the closed loop system as a function of time. It is denoted by $C(t)$. It is given by inverse Laplace of the product of input and transfer function of the system.

The closed loop transfer function,

$$\frac{C(s)}{R(s)} = \frac{G(s)}{1+G(s)\,H(s)}$$

Response is s-domain,

$$C(s) = \frac{R(s)\,G(s)}{1+G(s)\,H(s)}$$

Response in time domain

$$C(t) = L^{-1}[C(s)] = L^{-1}\left[\frac{R(s)\,G(s)}{1+G(s)\,H(s)}\right]$$

Q2.2 What is transient and steady state response?

Ans. The transient response is the response of the system when the input changes from one state to another. The response of the system $t \rightarrow 0$ is called steady state response.

Q2.3 What is the importance of test signals?

Ans. The test signals resembles, the characteristics of actual input signals. The test signals are used to predetermine the performance of the system. If the response of a system is satisfactory for a test signal, then the system will be suitable for practical applications.

Q2.4 Name the test signals used in control system.

Ans. The commonly used test input signals in control system are impulse, step, ramp, acceleration and sinusoidal signals.

Q2.5 Define step signal.

Ans. The step signal is a signal whose values changes from 0 to A and remains constant at A for $t > 0$. The mathematical representation of step signal is

$$r(t) = Au(t)$$

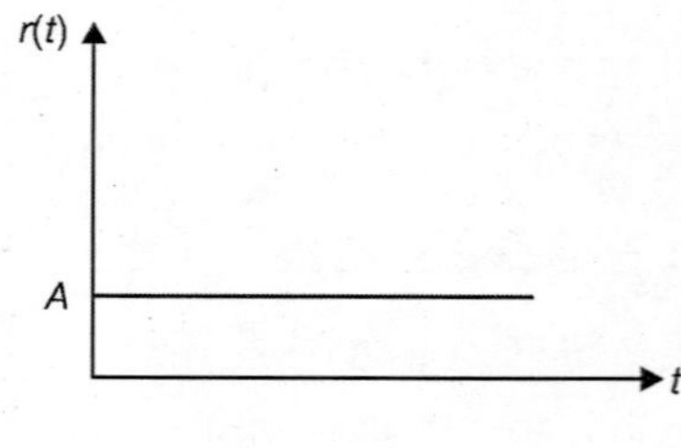

where, $u(t) = 1, t \geq 0$

$u(t) = 0, t < 0$

Q2.6 Define ramp signal.

Ans.

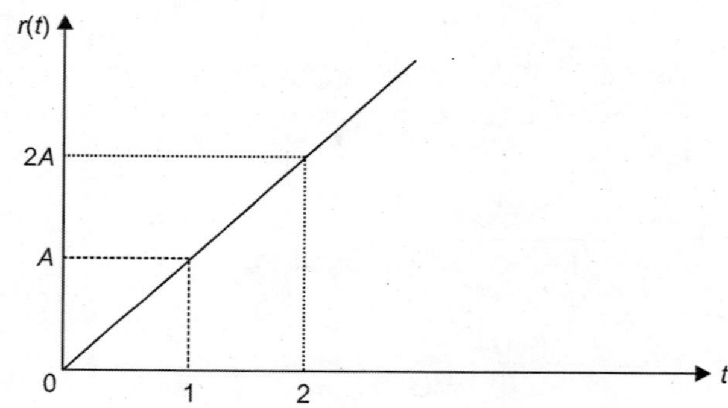

A ramp signal is a signal whose value increases linearly with time from an initial value of zero at $t = 0$. It is mathematically represented as

$$r(t) = At, \text{ when } t \geq 0$$

$$r(t) = 0; \text{ when } t < 0$$

Q2.7 Define parabolic signal

Ans.

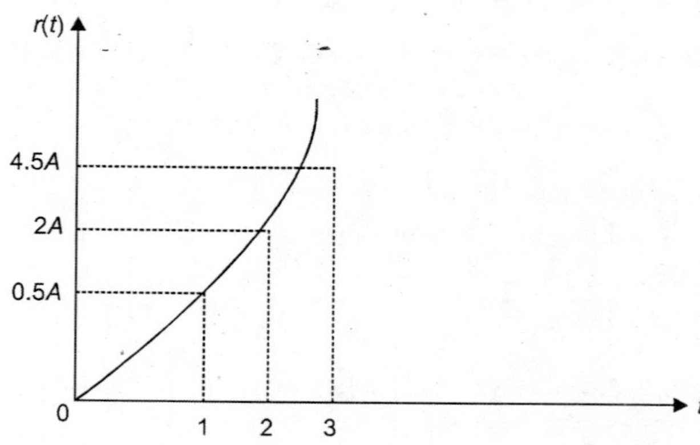

It is a signal in which the instantaneous values varies as square of the time from an initial value of zero at $t = 0$. It is mathematically represented as

$$r(t) = \frac{At^2}{2}, t \geq 0$$
$$r(t) = 0, t < 0$$

Q2.8 What is an impulse signal?

Ans. A signal which is available for very short duration is called impulse signal. Ideal impulse signal is a unit impulse signal which is defined as a signal having zero values at all time except at $t =$ 0; At
$t = 0$ the magnitude becomes infinite. It is denoted by $\delta(t)$ and mathematically expressed as

$$\delta(t) = 0 \text{ for } t \neq 0 \text{ and } \underset{t_1 \to 0}{\text{Lt}} \int_{-t_1}^{+t_1} \delta(t)\, dt = 1$$

Q2.9 What is weighting function?

Ans. The impulse response of system is called weighting function. It is given by inverse Laplace transform of system transfer function.

Q2.10 Define Pole?

Ans. The pole of a function $F(s)$ is the value at which the function $F(s)$ becomes infinite, where $F(s)$ is a function of complex variable s.

Q2.11 Define zero?

Ans. The zero of a function $F(s)$ is the value at which the function $F(s)$ becomes zero, where $F(s)$ is a function of complex variable s.

Q2.12 What is the order of a system?

Ans. The order of the system is given by the order of the differential equation governing the system. It is also given by the maximum power of s in the denominator polynomial of transfer function. The maximum power of s also gives the number of poles of the system and so the order of the system also given by number of poles of the transfer function.

Q2.13 Define damping ratio?

Ans. The damping ratio is defined as the ratio of the actual damping to critical damping.

Q2.14 Give the expression for damping ratio of mechanical and electrical system?

Ans. The damping ratio of second order mechanical translational system, $\zeta = \frac{B}{2\sqrt{MK}}$.

The damping ratio of second order mechanical rotational system, $\zeta = \frac{B}{2\sqrt{JK}}$.

The damping ratio of second order electrical system, $\zeta = \dfrac{R}{2\sqrt{\dfrac{L}{C}}}$

Q2.15 How the second order system is classified?

Ans. Depending on the value of damping? Depending on the value of damping, the system can be classified inot the following four Cases,

Case 1: Undamped system, $\zeta = 0$

Case 2: Underdamped system, $0 < \zeta < 1$

Case 3: Critically damped system, $\zeta = 1$

Case 4: Over damped system, $\zeta > 1$.

Q2.16 What will be the nature of response of a second order system with different types of damping?

Ans. For undamped system the response is oscillatory.

For underdamped system the response is damped oscillatory.

For critically damped system the response is exponentially rising.

For overdamped system the response is exponentially rising but the rise time will be very large.

Q2.17 Sketch the response of a second order underdamped system.

Ans.

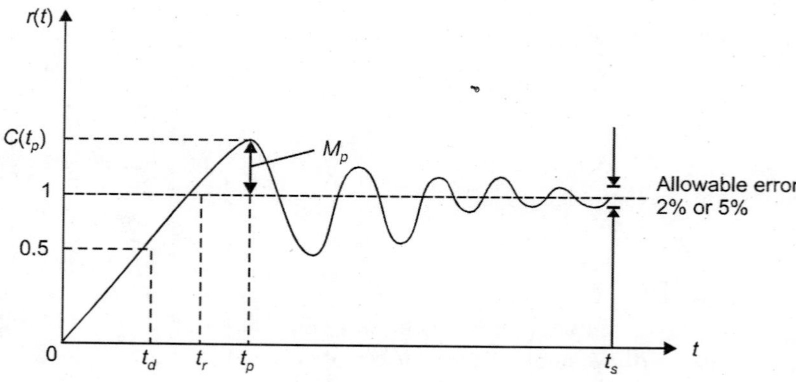

Q2.18 What is damped frequency of oscillation?

Ans. In underdamped system the response is damped oscillatory. The frequency of damped oscillation is given by $\omega_d = \omega_n \sqrt{1-\zeta^2}$

Q2.19 Give the expression for natural frequency of oscillations of electrical and mechanical system.

Ans. The natural frequency of oscillation of second order mechanical translational system

$$\omega_n = \sqrt{\frac{K}{M}}$$

The natural frequency of oscillation of second order mechanical rotational system

$$\omega_n = \sqrt{\frac{K}{J}}$$

The natural frequency of oscillation of second order electrical system

$$\omega_n = \frac{1}{\sqrt{LC}}$$

Q2.20 The closed loop transfer function of second order system is

$$\frac{C(s)}{R(s)} = \frac{10}{s^2 + 6s + 10}$$

What is the type of damping in the system?

Ans. Let us compare the given transfer function with the standard form of second order transfer function

$$\therefore \quad \frac{C(s)}{R(s)} = \frac{\omega_n^2}{s^2 + 2\zeta\omega_n s + \omega_n^2} = \frac{10}{s^2 + 6s + 10}$$

$$\therefore \quad \omega_n^2 = 10 \qquad 2\zeta\omega_n = 6$$

$$\omega_n = \sqrt{10} \qquad \zeta = 0.95$$

Since $\zeta < 1$ the system is underdamped.

Q2.21 The closed loop transfer function of a second order system is given by $\frac{200}{s^2 + 20s + 200}$. Determine the damping ratio and natural frequency of oscillation.

Ans. Let us compare the given transfer function to standard form of second order transfer function.

$$\frac{C(s)}{R(s)} = \frac{\omega_n^2}{s^2 + 2\zeta\omega_n s + \omega_n^2} = \frac{200}{s^2 + 20s + 200}$$

$$\therefore \quad \omega_n^2 = 200$$

$$\omega_n = \sqrt{200} = 14.14 \text{ rad/sec}$$

$$2\zeta\omega_n = 200$$

$$\zeta = .707$$

Result: Damping ratio = .707

Natural frequency ω_n = 14.14 rad/sec

Q2.22 List the time domain specifications

Ans. The time domain specifications are

(i) Delay time, (ii) Rise time, (iii) Peak time, (iv) Maximum peak overshoot, (v) Settling time.

Q2.23 Define delay time.

Ans. It is the time taken for response to reach 50% of the final value for the very first time.

Q2.24 Define rise time.

Ans. It is the time taken for response to meet 100% of the final value, for the very first time.

Q2.25 Define peak time.

Ans. It is the time taken for the response to reach the peak value for the very first time (or) It is the time taken for the response to reach peak overshoot, M_P.

Q2.26 Define peak overshoot.

Ans. It is defined as the ratio of the maximum peak value measured from final value to final value. Let final value = $C(\infty)$, maximum value = $C(t_p)$

$$\text{Peak overshoot, } M_p = \frac{C(t_p) - C(\infty)}{C(\infty)}$$

Q2.27 Define settling time.

Ans. It is defined as the time taken by the response to reach and stay within a specified error and the error is usually specified as % of final value. The usual tolerable error in 2% or 5% of the final value.

Q2.28 What is type of a system what is its significance?

Ans. The type number is given by number of poles of loop transfer function at the origin. The type number of the system decides the steady state error.

Q2.29 Distinguish between type and order of a system.

Ans.
1. Type number is specified for loop transfer function but order can be specified for any transfer function. (open loop or closed loop transfer function).
2. The type number is given by number of poles of loop transfer function lying at origin of s-plane but the order is given by the number of poles of transfer function.

Q2.30 For the system with following transfer function, determine type and orderof the system.

$$G(s)\,H(s) = \frac{K}{s(s+1)\,(s^2+6s+8)}$$

Ans. Type –1, order: 4

Q2.31 What is steady state error?

Ans. The steady state error is the value of error signal $e(t)$, when t tends to infinity. The steady state error is a measure of system accuracy. The errors arise from the nature of inputs, type of system and from non-linearity of system components.

Q2.32 What are static error constants?

Ans. The K_p, K_v and K_a are called static error constants. These constants are associated with steady state error in a particular type of system and for a standard input.

Q2.33 Define positional error constant?

Ans. The positional error constant

$$K_p = \lim_{s\to 0} G(s)\, H(s)$$

The steady state error in type – 0 system when the input is unit step is given by $1/1 + K_p$

Q2.34 Define velocity error constant?

Ans. The velocity error constant

$$K_v = \lim_{s\to 0} s \cdot G(s)\, H(s)$$

The steady state erin type – 1 system for unit ramp input is given by $1/K_v$.

Q2.35 Define acceleration error constant?

Ans. The acceleration error constant

$$K_a = \lim_{s\to 0} s^2\, G(s) \cdot H(s)$$

The steady state error in type–2 system for unit parabolic input is given by $1/K_a$.

Q2.36 A unity feedback system has a open loop transfer function of $G(s) = \dfrac{10}{(s+1)(s+2)}$. Determine the steady state error for unit step input?

Ans. The steady state error for unit step input,

$$e_{ss} = \frac{1}{1+K_p}$$

where $K_p = \lim_{s\to 0} G(s) \cdot H(s)$

for unity feedback system $H(s) = 1$

$$K_p = \lim_{s\to 0} G(s) = \lim_{s\to 0} \frac{10}{(s+1)(s+2)} = \frac{10}{3} = 3.33$$

$$e_{ss} = \frac{1}{1+K_p} = \frac{1}{1+3.33} = \frac{1}{4.33}$$

Q2.37 What are generalized error coefficients. They are the coefficients of generalized error series. The generalized error series is given by,

$$e(t) = C_0 r(t) + C_1 r'(t) + \frac{C_2}{2!} r''(t) + \frac{C_3}{3!} r'''(t) + \ldots + \frac{C_n}{n!} r^{n'}(t) \ldots$$

Ans. The coefficients C_0, C_1, C_2, … C_n are called generalized error coefficients (or) dynamic error coefficients.

The nth coefficient, $C_n = \lim_{s\to 0} \frac{d^n}{ds^n} F(s)$,

where $F(s) = \frac{1}{1+G(s)\,H(s)}$.

Q2.38 Give the relation between generalized and static error coefficients?

Ans. The following expression shows the relation between generalized and static error coefficient,

$$C_0 = \frac{1}{1+K_p},\; C_1 = \frac{1}{K_v},\; c_2 = \frac{1}{K_a}$$

Q2.39 Mention the advantage of generalized error constants over static error constants?

Ans. Generalized error series gives error signal as a function of time.

Q2.40 What is the effect on system performance when a proportional controller is introduced in a system?

Ans. The proportional controller improves the steady-state tracking accuracy, disturbance signal rejection and relative stability of the system. It also increases the loop gain of the system which results in reducing the sensitivity of the system to parameter variations.

Q2.41 What is the disadvantage in proportional controller?

Ans. The disadvantage of proportional controller is that it produces a constant steady state error.

Q2.42 What is the effect of PI controller on the system performance?

Ans. The PI controller increases the order of the system by one, which results in reducing the steady state error. But the system becomes less stable than the original system.

Q2.43 What is the effectof PD controller on the system performance?

Ans. The effect of PD controller is to increase the damping ratio of the system and the peak overshoot is reduced.

Q2.44 Why derivative controller is not used in control systems?

Ans. The derivative controller produces a control action based on rate of change of error signal and it does not produce corrective measures for any constant error. Hence derivative controller is not used in control systems.

EXERCISES

Q.1 What is the unit-step response of the following system.

Ans. $C(t) = -1.4s\ 5e^{-8.87t} + .145e^{-1.13t} + 1$

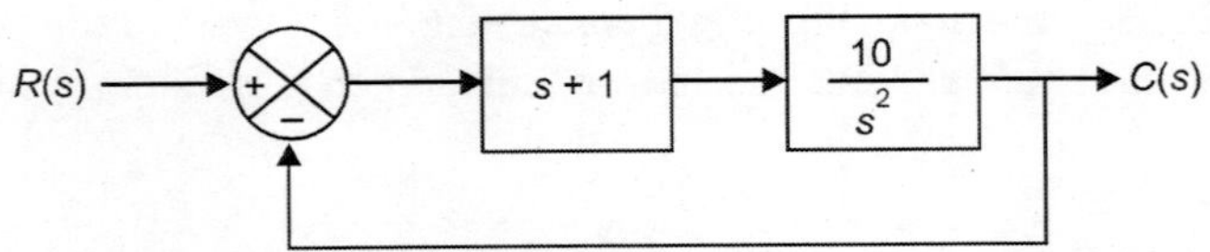

Q.2 Obtain the unit-step response of a unity feedback system whose open-loop transfer function is

$$G(s) = \frac{5(s+20)}{s(s+4.59)(s^2+3.41s+16.35)}$$

Ans. $C(t) = 1 + \frac{3}{8}e^{-t}\cos 3t - \frac{17}{24}e^{-t}\sin 3t - \frac{11}{8}e^{-3t}\cos t - \frac{13}{8}e^{-3t}\sin t$

Q.3 The open loop transfer function of an unity feedback control system is given by $G(s) = 100/s(s+2)(s+5)$. For unit step input, find time response of the closed loop system and determine % overshoot and the rise time.

Ans. $C(t) = \left[1 - 0.186\, e^{-7.45t} - 0.88\, e^{.225t}\cos(3.65t - 22°)\right]$

As t tends to infinity, $C(t)$ tends to infinity and so the system is unstable. Therefore % overshot and rise time are not defined.

Q.4 A servomechanism has its moment of inertia $J = 10 \times 10^{-6}$ kg – m^2, retarding friction, $B = 400 \times 10^{-6}$ $N - m$/(rad/sec) and elasticity coefficient, $K = 0.004$ N – m/ rad. Find the natural frequency and damping factor of the system.

Ans. Natural frequency, $\omega_n = 20$ rad/sec

Damping factor, $\zeta = 1$

Q.5 For a second order system whose open loop transfer function $G(s) = 4/s(s+2)$, determine the maximum overshoot, the time to reach the maximum overshoot when a step displacement of 18° is given to the system. Find the rise time, time constant and the settling time for an error of 7%.

Ans. Maximum overshoot = 0.16, when input is 18°, $M_p = 2.88°$

Peak time, $t_p = 1.181$ sec

Time constant, $T = 1$ sec

Rise time, $t_r = 1.21$ sec

Settling time for 7% error = 2.66 sec

Q.6 Consider the unity feedback closed loop system where the forward transfer functino is $G(s) = 25/s(s+5)$. Obtain the rise time, peak time, maximum overshoot and the settling time when the system is subjected to a unit-step input.

Ans. Rise time, $t_r = 0.55$ sec

Peak time, $t_p = 0.785$ sec

% peak overshoot, $M_p = 9.5\%$

Settling time, $t_s = 1.33$ sec (for 2% error)

= 1 sec (for 5% error)

Q.7 Consider the system given below (Fig. 7) where $\zeta = 0.6$ and $\omega_n = 0.5$ rad/sec. Determine the rise time, peak time, maximum overshoot and settling time, when the system is subjected to a unit step input.

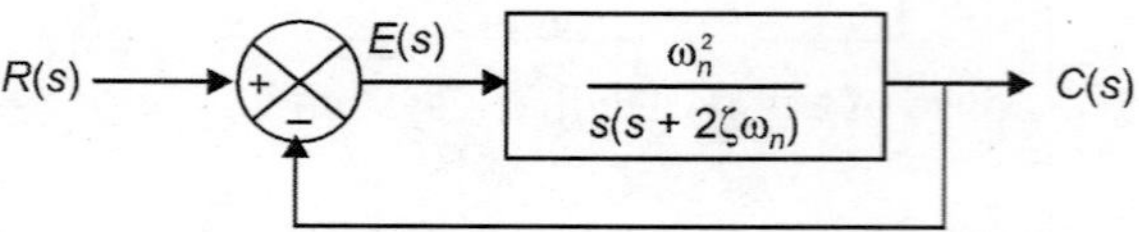

Ans. Rise time, $t_r = 0.55$ sec

Maximum overshoot, $M_p = 0.095$

Peak time, $t_p = 0.785$ sec

Settling time, $t_s = 1$ sec (for 5% criterian)

Q.8 For the system (given below), determine the values of K and K_h. So that the maximum overshoot in the unit step response is 0.2 and the peak time is 1 sec with these values of K and K_h, obtain rise time and settling time.

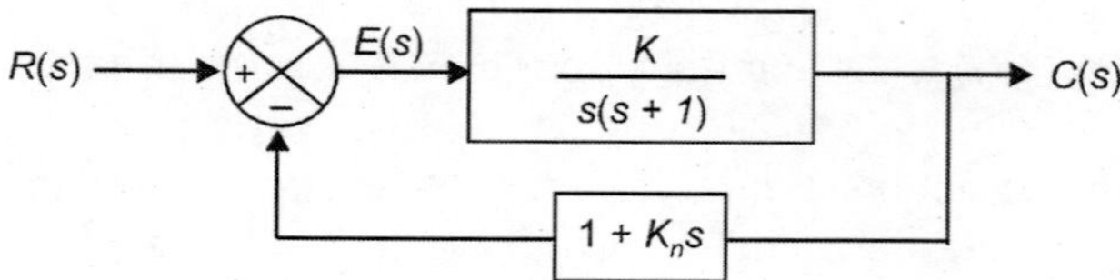

Ans. $K = 12.5$

$K_h = 0.178$

Rise time, $t_r = 0.65$ sec

Settling time, $t_s = 2.48$ sec (for 2% error)

$t_s = 1.86$ sec (for 5% error)

Q.9 The system (given below) subjected to a unit-step input. Determine the values of K and T, where the maximum overshoot of the system is 2.5 4% corresponding to $\zeta = 0.4$.

Ans. $K = 1.42$

$T = 1.09$

Q.10 Determine the values of K and T of the closed-loop system shown below. So that the maximum overshoot in unit-step response is 25% and the peak time is 2 sec. Answer that $J = 1$ kg – m^2.

Ans. $K = 2.95$ N – m

$T = 0.471$ sec

Q.11 A unity-feedback system is characterized by the open-loop transferfunction, $G(s) = 1/s(0.5s + 1)(0.2s + 1)$.

(a) Determine the steady-state errors to unit-step, unit-ramp and unit parabolic inputs.

(b) Determine rise time, peak time, peak overshoot and settling time of the unit-step response of the system.

Ans. (a) e_{ss} (unit step) = 0

e_{ss} (unit ramp) = 1

e_{ss} (unit parabola) = ∞

(b) Rise time, t_r = 1.91 sec

Peak time, t_p = 2.79 sec

Peak overshoot, M_p = 0.1265

Settling time, t_s = 5.4 sec

Q.12 For a system whose $G(s) = 10/s(s + 1)(s + 2)$. Find the steady state error when it is subjected to the input, $r(t) = 1 + 2t + 1.5t^2$.

Ans. The total steady state error is ∞.

Q.13 A unity feedback system has $G(s) = 1/s(1 + s)$. The input to the system is described by $r(t) = 4 + 6t + 2t^3$. Find the generalized error coefficients and steady state error.

Ans. $c_0 = 0 \quad c_1 = 1 \quad e_{ss} = \infty$

$c_2 = 0 \quad c_3 = -6$

Q.14 A unity feedback system has the forward path transfer function $G(s) = 10/(s + 1)$. Find the steady state error and the generalized error coefficient for $r(t) = t$.

Ans. $C_0 = 1/11 \quad e_{ss} = \infty$

$C_1 = 10/121$

Q.15 Find out the position, velocity and acceleration error coefficients for the following unity feedback systems having forward loop transfer function $G(s)$ as

(a) $100/(1 + 0.5s)(1 + 2s)$

(b) $K/s(1.01s)(1 + s)$

(c) $K/s^2(s^2 + 8s + 100)$

(d) $K(1 + s)(1 + 2s)/s^2(s^2 + 4s + 20)$

Ans. $K_p = 100$, $K_v = 0$ and $K_a = 0$ for the system $100/(1 + 0.5s)(1 + 2s)$

$K_p = \infty$, $K_v = K$ and $K_a = 0$ for the system $K/s(1 + 0.1s)(1 + s)$

$K_p = \infty$, $K_v = \infty$ and $K_a = K/100$ for the system $K/s^2(s^2 + 8s + 100)$

$K_p = \infty$, $K_v = \infty$ and $K_a = K/100$ for the system $K(1 + s)(1 + 2s)/s^2(s^2 + 4s + 20)$

Q.16 The open loop transferfunction of a unity feedback control system is $G(s) = 9(s + 1)$, using the generalized error series determine the error signal and steady state error of the system when the system is excited by (i) $r(t) = 2$, (ii) $r(t) = t$, (iii) $r(t) = 3t^2$ (iv) $r(t) = 1 + 2t + 3t^2/2$

Ans. (i) $e(t) = 0.2$; $e_{ss} = 0.2$

(ii) $e(t) = 0.1t + 0.09$; $e_{ss} = \infty$

(iii) $e(t) = 0.15t^2 + 0.24$; $e_{ss} = \infty$

(iv) $e(t) = 0.15t^2 + 0.77t + 0.226$; $e_{ss} = \infty$

Q.17 For unity feedback system having open loop transfer function as $G(s) = K(s + 2)/s^2(s^2 + 7s + 12)$. Determine (i) type of system, (ii) error constants K_p, K_v, and K_a, (ii) steady state error for parabolic input.

CHAPTER 2

Ans. (i) It is type 2 system

(ii) $K_p = \infty$

$K_v = \infty$

$K_a = K/6$

(iii) $e_{ss} = 6/K$

Q.18 Determine the values of K and T of the closed loop system. So that the maximum overshoot in unit step response is 25% and the peak time is 2 sec.

Ans. $K = 2.95$ N-m

$T = .471$ sec

SOME WORKED EXAMPLES

EXAMPLE 1

When the system shown in Fig. 1 in subjected to a unit step input, the system output responds such that it attains maximum peak overshoot of 25.4%. Determine the value of K and T. Assume $\zeta = .4$ $t_p = 3$.

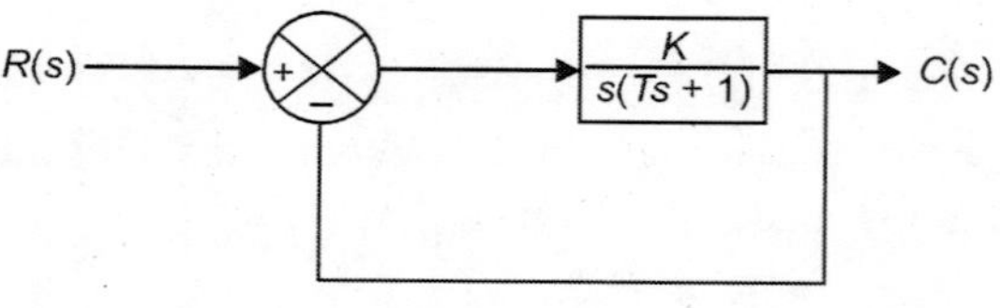

Fig. 1

Solution: We know

$$t_p = \frac{\pi}{\omega_d} = \frac{\pi}{\omega_n\sqrt{1-\zeta^2}}$$

$$\therefore \quad \frac{\pi}{\omega_n\sqrt{1-\zeta^2}} = 3$$

$$\frac{\pi}{\omega_n\sqrt{1-(.4)^2}} = 3$$

$$\therefore \quad \omega_n = \frac{\pi}{3\sqrt{1-(.4)^2}}$$

$$\boxed{\omega_n = 1.14}$$

From block diagram,

$$\frac{C(s)}{R(s)} = \frac{K}{Ts^2 + s + K}$$

This can be rewritten as (Divide by T in both numerator and denominator)

$$\frac{C(s)}{R(s)} = \frac{K/T}{s^2 + s/T + K/T}$$

Compare $s^2 + s/T + K/T$ with standard characteristic equation (i.e., $s^2 + 2\zeta\omega_n^s + \omega_n^2$)

$$\therefore \quad \omega_n^2 = K/T$$

$$\omega_n = \sqrt{K/T}$$

$$2\zeta\omega_n = 1/T$$

$$\therefore \quad T = \frac{1}{2\zeta\omega_n} = \frac{1}{2 \times .4 \times 1.14} = 1.09$$

$$\boxed{T = 1.09}$$

$$K = \omega_n^2 T \Rightarrow (1.14)^2 (1.09)$$

$$\boxed{K = 1.42}$$

Result: The value of K is 1.42 and T is 1.09.

EXAMPLE 2

Determine the values, K and k of the closed loop system shown in Fig. (2) so that the maximum overshoot in unit step response is 25% and the peak time is 2 sec. Assume that $J = 1$ kg-m^2.

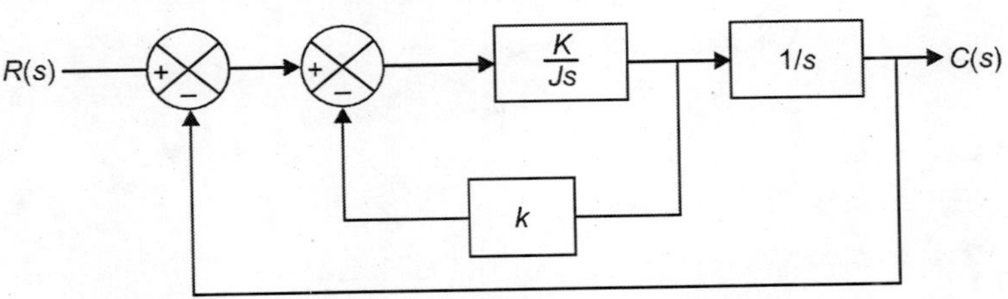

Fig. 2

Solution: The closed loop transfer function can be derived from Fig. (2), using the lock diagram reduction technique,

$$\frac{C(s)}{R(s)} = \frac{K}{Js^2 + Kks + K} \tag{1}$$

From given data, $J = 1$ kg-m^2, so substitute this value in equation (1),

$$\therefore \quad \frac{C(s)}{R(s)} = \frac{K}{s^2 + Kks + K}$$

Compare the characteristic equation $s^2 + Kks + K$ with standard characteristic equation of second order system $(s^2 + 2\zeta\omega_n + \omega_n^2)$

so,
$$\omega_n^2 = K \Rightarrow \omega_n = \sqrt{K}$$

$$2\zeta\omega_n = Kk \Rightarrow k = \frac{2\zeta\omega_n}{K}$$

we know,
$$\underset{\text{(Peak overshoot)}}{M_P} = e^{-\zeta\pi/\sqrt{1-\zeta^2}}$$

$$M_P = 25\% \text{ (given)}$$

$\therefore$
$$.25 = e^{-\zeta\pi/\sqrt{1-\zeta^2}}$$

from that,
$$\zeta\pi/\sqrt{1-\zeta^2} = 1.386$$

$\therefore$
$$\boxed{\zeta = .404}$$

we know t_p is 2 sec which is given is the problem

$$t_p = \frac{\pi}{\omega_d} = 2$$

$\therefore$
$$\boxed{\omega_d = 1.57}$$

$$\omega_n = \frac{\omega_d}{\sqrt{1-\zeta^2}} = \frac{1.57}{\sqrt{1-.404^2}} = 1.72$$

$$\boxed{\omega_n = 1.72}$$

$\therefore$
$$K = \omega_n^2 = 2.95 \text{ N-m}$$

and
$$k = \frac{2\zeta\omega_n}{K} = \frac{2 \times .404 \times 1.72}{2.95} = .471 \text{ sec}$$

Result: The value of K is 2.96 N-m and the value of k is 4.71 sec.

EXAMPLE 3

The Fig. (3) shows a mechanical vibratory system. When step input is applied to the system, (where step input value = 2) the mass oscillates such that it has M_P of 9.5 and t_p = 2 sec. Determine M, B and K of the system. The displacement x is measured from the equilibrium position. (The final value is 0.1 and $\zeta = 0.6$).

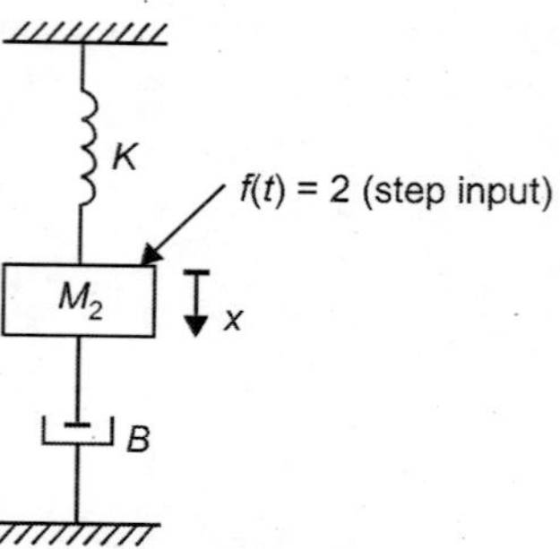

Fig. 3

Solution: From the Fig. (3), we can derive the transfer function of the system,

$$f(t) = M\frac{d^2x}{dt^2} + B\frac{dx}{dt} + Kx$$

on taking Laplace transform

$$F(s) = X(s)[Ms^2 + Bs + K]$$

$$\therefore \quad \frac{X(s)}{F(s)} = \frac{1}{Ms^2 + Bs + K}$$

The input is ie $r(t) = 2$

$$\therefore \quad R(s) = 2/s$$

Hence, $$X(s) = \frac{2}{s(Ms^2 + Bs + K)} \tag{1}$$

The steady state value of x is,

$$x(\infty) = \underset{s \to 0}{\text{Lt}}\ s\,X(s) = \underset{s \to 0}{\text{Lt}}\ s \cdot \frac{2}{s(Ms^2 + Bs + K)}$$

$$x(\infty) = \frac{2}{K} = 0.1 \text{ (Final value)}$$

$$\therefore \quad K = 20$$

Maximum peak over shoot is 9.5 and $\zeta = .6$ (given)

we know, $$t_p = \frac{\pi}{\omega_d} = \frac{\pi}{\omega_n\sqrt{1-\zeta^2}} = \frac{\pi}{.8\omega_n}$$

$$t_p = 2 \text{ sec [given]}$$

$$\therefore \quad 2 = \frac{\pi}{.8\omega_n} \Rightarrow \omega_n = 1.96$$

$$\boxed{\omega_n = 1.96 \text{ rad/sec}}$$

The equation (1) can be written as by dividing by M in both numerator and denominator,

$$X(s) = \frac{2/M}{s(s^2 + B/Ms + K/M)} \tag{2}$$

Compare $s^2 + \frac{B}{M}s + \frac{K}{M}$ with $s^2 + 2\zeta\omega_n s + \omega_n^2$ we get,

$$\omega_n^2 = \frac{K}{M} = \frac{20}{M}$$

$$M = \frac{20}{\omega_n^2} = \frac{20}{1.96^2} = 5.2$$

$$2\zeta\omega_n = B/M$$

$$\therefore \quad B = M \text{ Ч } 2 \text{ Ч } \zeta \text{ Ч } \omega_n = 2 \text{ Ч } .6 \text{ Ч } 1.96 \text{ Ч } 5.2$$

$$\boxed{B = 12.2}$$

Result: The value of

$$M = 5.2$$

$$B = 12.2$$

$$K = 20$$

EXAMPLE 4

For unity feedback system having open loop transfer function as $G(s) = K(s + 2)/s^2 (s^2 + 7s + 12)$. Determine (i) Type of system (iii) error constants K_P, K_v and K_a. (iii) Steady state error for step input.

Solution:

(i) Type of the system is 2

(ii) $K_P = \underset{s \to 0}{\text{Lt}}\, G(s)\, H(s) = \underset{s \to 0}{\text{Lt}} \dfrac{K(s+2)}{s^2(s^2 + 7s + 12)}$

$\boxed{K_P = \infty}$

$K_V = \underset{s \to 0}{\text{Lt}}\, s \cdot G(s)\, H(s) = \underset{s \to 0}{\text{Lt}}\, s \cdot \dfrac{K(s+2)}{s^2(s^2 + 7s + 12)}$

$\boxed{K_V = \infty}$

$K_a = \underset{s \to 0}{\text{Lt}}\, s^2\, G(s) H(s) = s^2 \cdot \dfrac{K(s+2)}{s^2(s^2 + 7s + 12)}$

$\boxed{K_a = K/6}$

(iii) The steady state error for unit step input

$$e_{ss} = \frac{1}{1 + K_p} = \frac{1}{1 + \infty} = 0$$

$$e_{ss} = 0 \quad \text{for unit step input}$$

Result:

(i) Type of the system is 2

(ii) $K_p = \infty$, $K_v = \infty$, $K_a = K/6$

(iii) $e_{ss} = 0$ [For unit step].

EXAMPLE 5

Find out K_P, K_V and K_a for the unity feedback system $G(s) = 100/(1 + .5s)(1 + 2s)$.

Solution: The open loop transfer function is

$$G(s) = \frac{100}{(1 + .5s)(1 + 2s)}$$

$$K_p = \underset{s \to 0}{\text{Lt}}\ G(s)\,H(s) = \underset{s \to 0}{Lt}\ \frac{100}{(1+.5s)\,(1+2s)}$$

$$\boxed{K_P = 100}$$

$$K_v = \underset{s \to 0}{Lt}\ s \cdot G(s)\,H(s)\ \underset{s \to 0}{Lt}\ s \cdot \frac{100}{(1+5s)\,(1+250}$$

$$\therefore\ \boxed{K_V = 0}$$

$$K_a = \underset{s \to 0}{Lt}\ s^2 G(s)\,H(s) = \underset{s \to 0}{2t}\ s^2 - \frac{100}{(1+.5s)\,(1+2s)}$$

$$\therefore\ \boxed{K_a = 0}$$

Result: $K_P = 100$, $K_v = 0$ and $K_a = 0$

EXAMPLE 6

Consider the unity feedback closed loop system where the transfer function is $G(s) = 25/s\,(s+5)$. Obtain t_p when subjected to unit step input.

Solution: The open loop transfer function is

$$\frac{C(s)}{R(s)} = \frac{\dfrac{25}{s(s+5)}}{1+\dfrac{25}{s(s+5)}} = \frac{25}{s(s+5)+25}$$

$$\frac{C(s)}{R(s)} = \frac{25}{s^2+5s+25} \qquad (1)$$

Comparing the equation (1) with standard second order system characteristic equation we get,

$$\omega_n^2 = 25 \Rightarrow \boxed{\omega_n = 5}$$

$$2\zeta\omega_n = 5$$

$$2\ ч\ \zeta(5) = 5$$

$$\zeta = \frac{5}{10} = .5$$

we know,

$$t_p = \frac{\pi}{\omega_n\sqrt{1-\zeta^2}}$$

$$t_p = \frac{\pi}{5\sqrt{1-.5^2}}$$

$$\boxed{t_p = .785}\ \text{sec}$$

Result: $t_p = .785$ sec, when subjected to unit step input.

EXAMPLE 7

Consider the system given in Fig. (4) where $\zeta = .6$ and $\omega_n = .5$ rad/sec. Determine the rise time and peak time when the system is subjected to a unit step input.

Solution: We know,

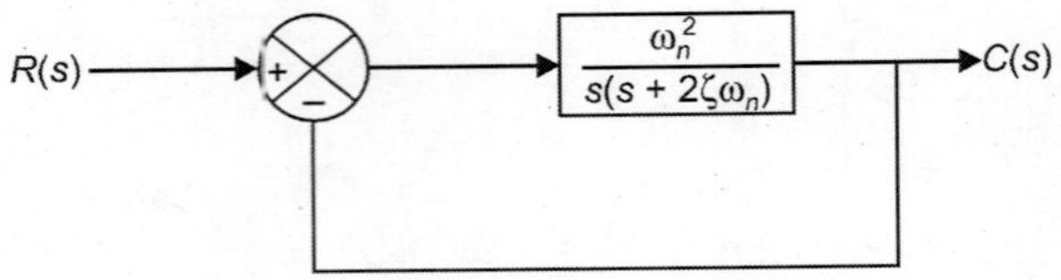

$$t_r = \frac{\pi - \tan^{-1}\sqrt{\frac{1-\zeta^2}{\zeta}}}{\omega_n\sqrt{1-\zeta^2}}$$

$$t_r = \frac{\pi - \tan^{-1}\sqrt{\frac{1-.6^2}{.6}}}{.5\sqrt{1-.6^2}} = .55 \text{ sec}$$

$$\boxed{t_r = .55 \text{ sec}}$$

$$t_p = \frac{\pi}{\omega_d} = \frac{\pi}{\omega_n\sqrt{1-\zeta^2}} - \frac{\pi}{\omega_n\sqrt{1-.6^2}}$$

$$\boxed{t_p = .785} \text{ sec}$$

Fig. 4

Result:

$t_r = .55$ sec

$t_p = .785$ sec.

3

Control System Components

3.1 INTRODUCTION

From the above Fig. (3.1), it is understood that without amplifier, Error detector, Actuator and feedback system, the whole system becomes open loop system. So from this we can define what is control system component. The control system component can be of either, controlling device (or) feedback device. The basic control system components are mentioned in Fig. (3.1), i.e. Error detector, Actuator, Amplifier and feedback system. Now we will see the functions and necessity of those components is detail.

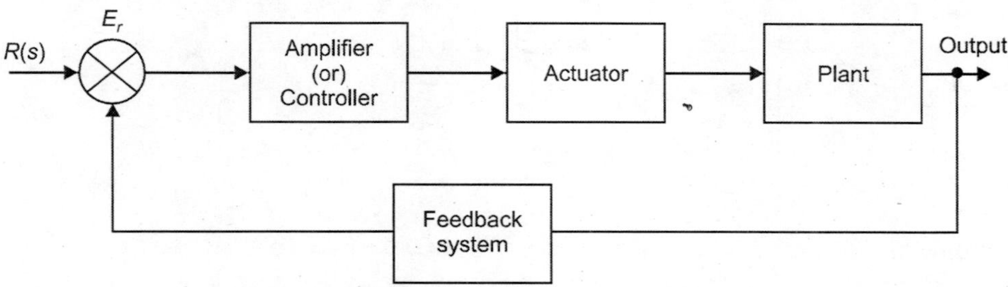

Fig. 3.1: Block diagram of closed loop control system.

1. **Error detector:** It is to compare the actual signal with reference signal . (i.e.,) it is used to measure the amount of deviation from the desired value.
2. **Amplifier:** It is the next stage of error detector. Generally the error signal will be weak signal and so it has to be amplified and then modified for better control action. In most of the system the controller amplifies the error signal.

   Examples: P, PI, PD and PID controllers.
3. **Actuator:** It is next stage of controller (or) Amplifier which converts the output of controller to the required form of energy that can be acceptable for the plant.
4. **Feedback system:** The feedback system samples the output to produce a feedback signal which is proportional to output. It also convert the output variable into another suitable variable such as displacement, pressure (or) voltage. So that it can be used to compare with the reference input.

   Examples: Transducer, Tachogenerators etc.,

3.2 POTENTIOMETERS

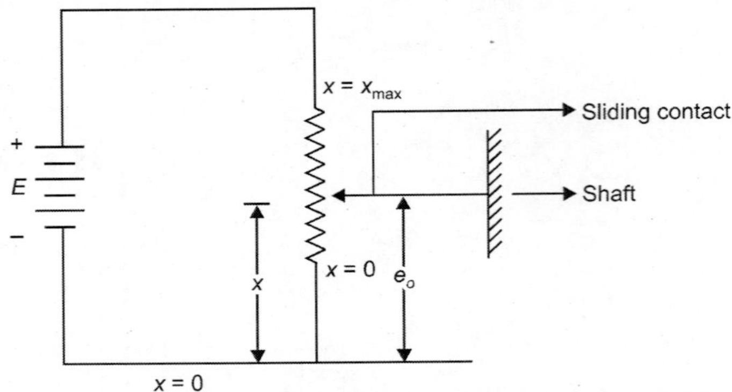

Fig. 3.2: Linear displacement potentiometer.

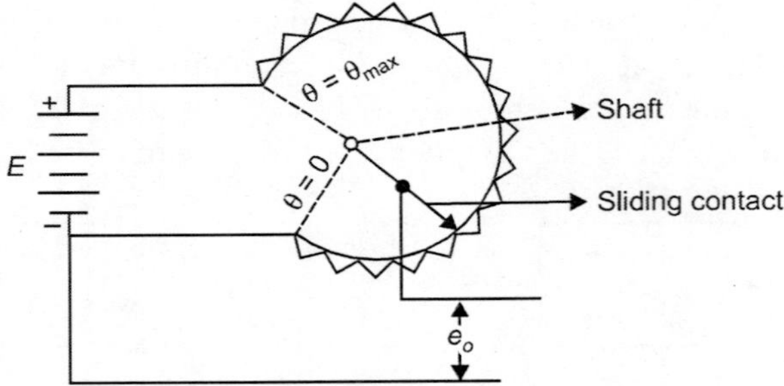

Fig. 3.3: Angular displacement potentiometer.

A potentiometer is a device which can be used to convert a liner (or) angular displacement into a voltage. A potentiometer has variable resistance whose value varies according to the linear/angular displacement of the sliding contact (moving contact).

Construction and Working

The resistance element is made up of wire on a former (or) by depositing a conducting material on a plastic base. It has one input shaft in which the sliding contact is attached. Sliding contact will move when displacement is applied to the shaft. The output voltage is obtained according to the position of the sliding contact.

The potentiometer can be excited by dc (or) ac voltage. The output is measured at sliding contact with respect to reference.

Note: The AC potentiometer will have inductive effect in addition to resistance which leads to difficulty in balancing the potentiometers used as error detectors.

Output equation

Ref. Fig. (3.2) and (3.3)

(i) Linear displacement potentiometer:

x = Displacement of the sliding contact from reference ($x = 0$)

x_{max} = Maximum displacement of sliding contact.

E = Excitation voltage [AC (or) DC]

e_o = Output voltage

If $x = 0$, then $e_o = 0$

If $x = x_{max}$, then $e_o = E$

$\therefore$ For displacement x, the output voltage,

$$e_o = \frac{E}{x_{max}} \cdot x$$

$$\boxed{e_o = K_P x}$$

where, $$K_P = \frac{E}{x_{max}}$$

= sensitivity of the potentiometer in volts/mm

(ii) Angular displacement potentiometer:

θ = Angular displacement of sliding contact from reference ($\theta = 0$)

θ_{max} = Maximum displacement of sliding contact

E = Excitation voltage

e_o = Output voltage

If $\theta = 0$, then $e_o = 0$

If $\theta = \theta_{max}$, then $e_o = E$

$\therefore$ For a displacement θ_1 the output voltage, $\dfrac{E}{\theta_{max}}\theta = K_P\theta$

where $$K_P = \frac{E}{\theta_{max}}$$

= sensitivity of the potentiometer in volts/deg.

Applications of potentiometers

1. It can be used as a feedback system.
2. It can be used as a error detector.

Potentiometers as a error detector

The Fig.(3.4) shows how the potentiometers are used as a error detector. When two ideal potential meters are connected across same supply then it act as a error detector.

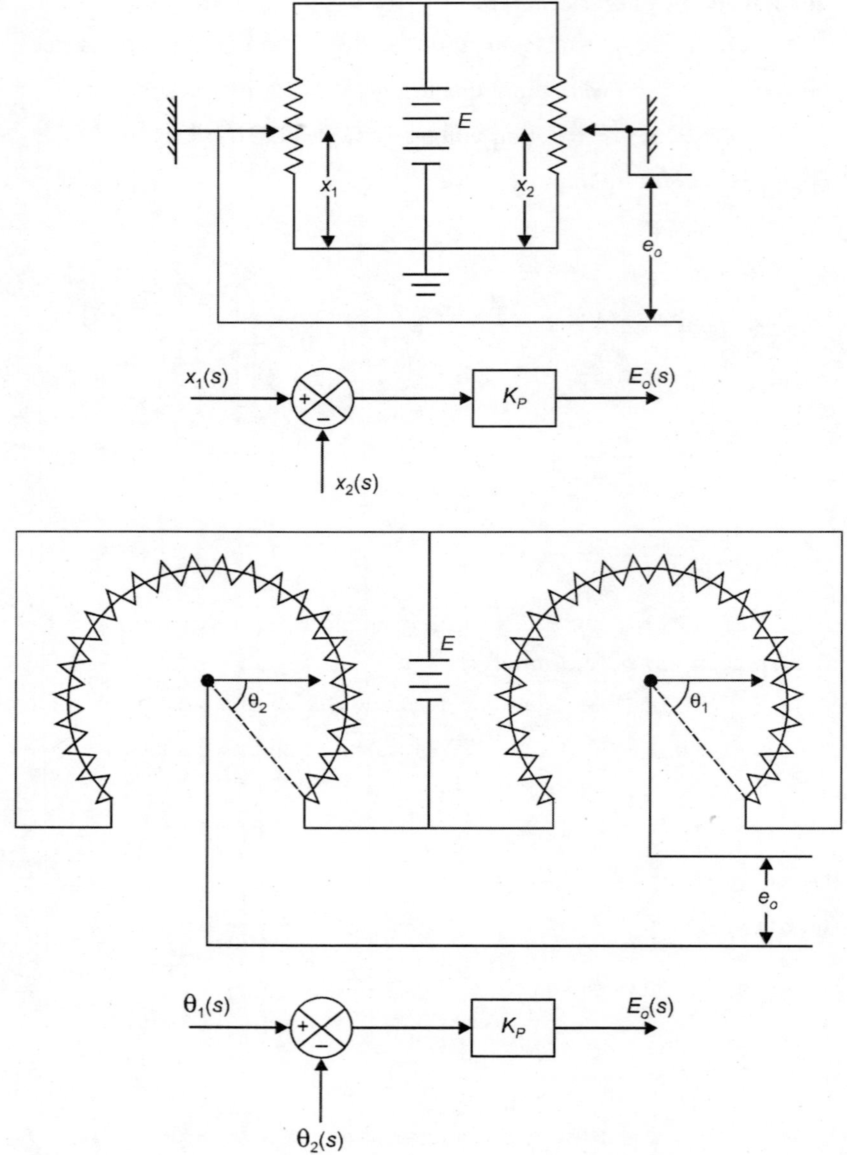

Fig. 3.4: Potentiometers as error detector.

3.3 SYNCHROS

Synchro is a device which works on the principle of rotating transformer. A synchro system formed by interconnection of the devices called synchro transmitter and synchro receiver. This also called a synchro pair.

(i) Synchro transmitter

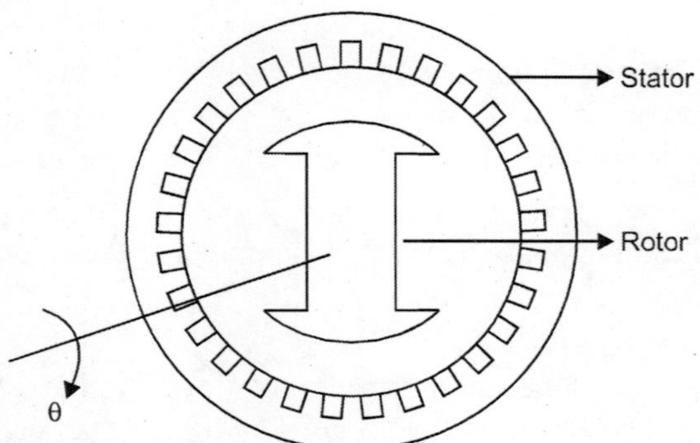

Fig. 3.5: Constructional features of synchro transmitter.

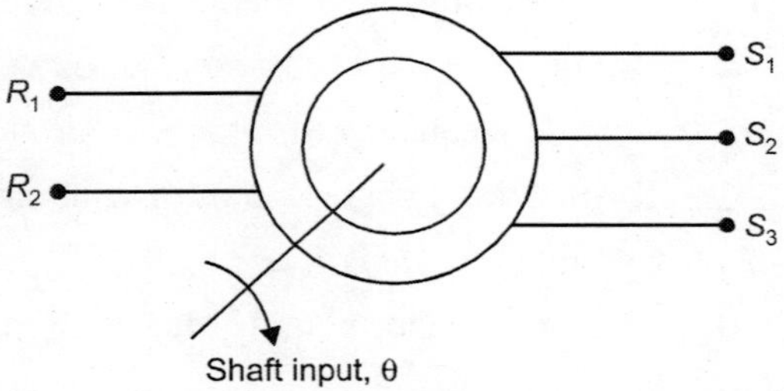

Fig. 3.6: Schematic symbol of synchro transmitter.

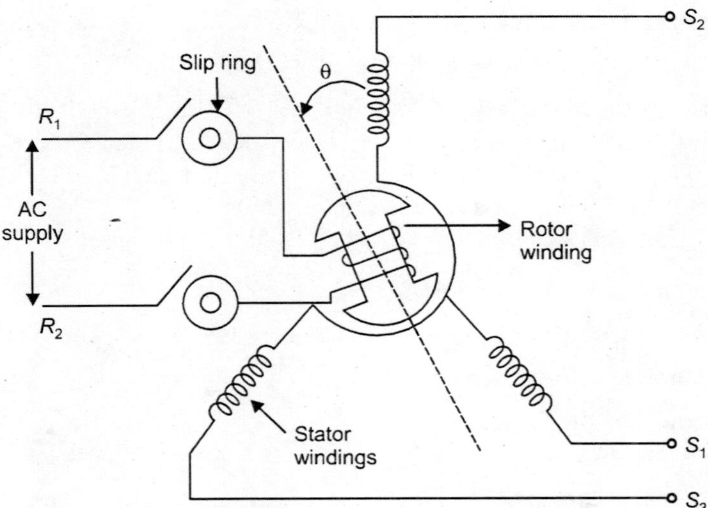

Fig. 3.7: Electrical circuit of synchro transmitter.

Construction

The two major parts of synchro transmitter are stator and rotor. The stator is identical to the stator of three phase alternator. It is made of laminated silicon steel and slotted as the inner periphery to accommodate a balanced three phase winding. The stator winding is concentric type with the axis of three coils (120є apart. The stator winding is star connected.

The rotor is of dumbbell construction with a single winding. The ends of the rotor winding are terminated on two sliprings. A single phase ac voltage is applied to the rotor through the sliprings.

Working principle

When the rotor is excited by ac voltage, the rotor current flows, and a magnetic field is produced. The rotor magnetic field induces an emf in the stator coils by transformer action. The effective voltage induced in any stator coil depends upon the angular position of the coil's axis with respect to rotor axis.

Let, e_r = Instantaneous value of ac voltage applied to rotor

e_{s1}, e_{s2}, e_{s3} = Instantaneous value of emf induced in stator coils s_1, s_2, s_3

E_r = Maximum value of rotor excitation voltage

ω = Angular frequency of rotor excitation voltage

K_t = Turns ratio of stator and rotor windings

K_c = Coupling coefficient

θ = Angular displacement of rotor with respect to reference

The instantaneous value of excitation voltage

$$e_r = E_r \sin \omega t$$

The magnitude of induced emfs are proportional to the turns ratio and coupling coefficient. The turns ratio, K is a constant, but coupling coefficient, Kc is a function of rotor angle position.

$\therefore$ Induced emf is stator coil = $K_t\, K_c\, E_r \sin \omega t$

Let e_{s2} be reference vector when $\theta = 0$, the flux linkage [refer Fig. (3.7)] of coil s_2 is maximum and when θ = 90є, the flux linkage of coil s_2 is zero. Hence the flux linkage of coil s_2 is function of $\cos \theta$ (ie, $K_c = K_t \cos \theta$ for coil s_2). The flux linkage of coil s_3 will be maximum after a rotation of 120є in anticlockwise direction and that of s_1 after a rotation of 240є

$\therefore$ Coupling coefficient

$$K_c \text{ for coil } s_2 = K_1 \cos \theta \tag{3.1}$$

Coupling coefficient,

$$K_c \text{ for coil} - s_3 = K_1 \cos (\theta - 120є) \tag{3.2}$$

Coupling coefficient,

$$K_c \text{ for coil} - s_1 = K_1 \cos (\theta - 240є) \tag{3.3}$$

Hence the emfs of stator coils with respect to neutral can be expressed as follows:

$$e_{s2} = K_t K_1 \cos \theta\, E_r \sin \omega t = K\, E_r \cos \theta \sin \omega t \tag{3.4}$$

$$e_{s3} = K_t K_1 \cos (\theta - 120є) E_r \sin \omega t = K\, E_r \cos (\theta - 120є) \sin \omega t \tag{3.5}$$

$$e_{s1} = K_t K_1 \cos(1 - 240є)\, E_r \sin \omega t = K\, E_r \cos(\theta - 240є) \sin \omega t \qquad (3.6)$$

where, $K = K_t K_1$

with reference to Fig. (6.3.4), by KVL the coil to coil emf can be written as

$$e_{s1}\, s_2 = e_{s1} - e_{s2} = \sqrt{3}\, K\, E_r \sin(\theta + 240є) \sin \omega t \qquad (3.7)$$

$$e_{s2}\, s_3 = e_{s2} - e_{s3} = \sqrt{3}\, K\, E_r \sin(\theta + 120є) \sin \omega t \qquad (3.8)$$

$$e_{s3}\, s_1 = e_{s3} - e_{s1} = \sqrt{3}\, K\, E_r \sin\theta \sin \omega t \qquad (3.9)$$

From equation (3.4), we can say that maximum emf is induced in coil s_2 but from equation (3.9) it is observed that the coil voltage $e_{s3}\, s_1$ is zero.

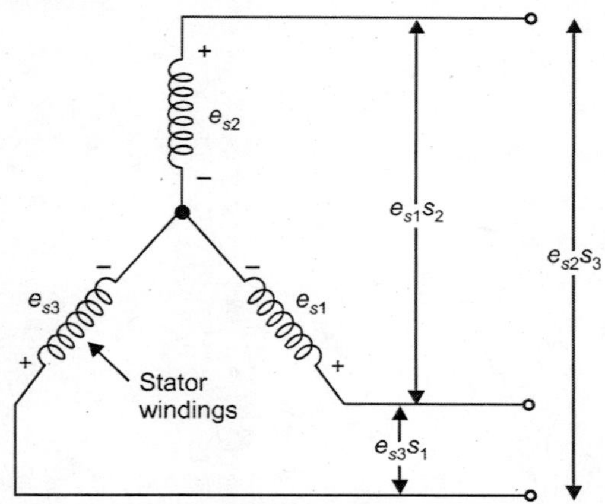

Fig. 3.8: Induced emf in stator coils.

(ii) Synchro receiver

Construction

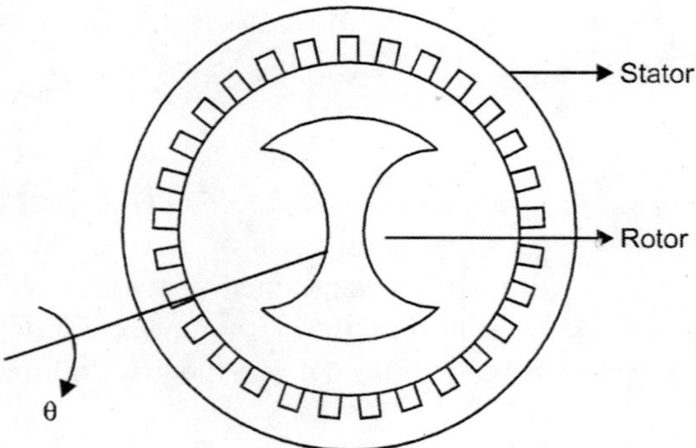

Fig. 3.9: Constructional features of synchro receiver.

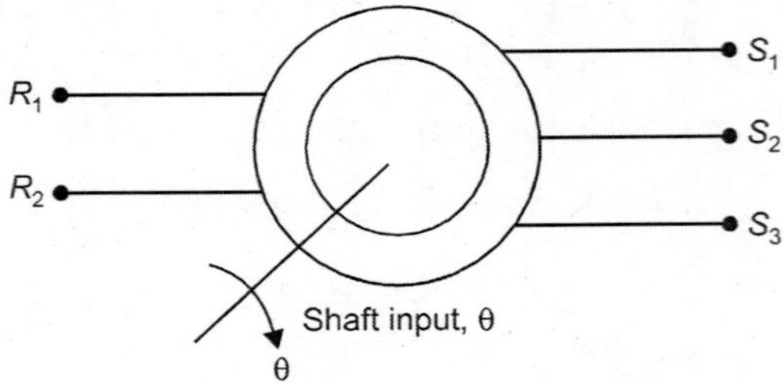

Fig. 3.10: Schematic symbol of a synchro receiver.

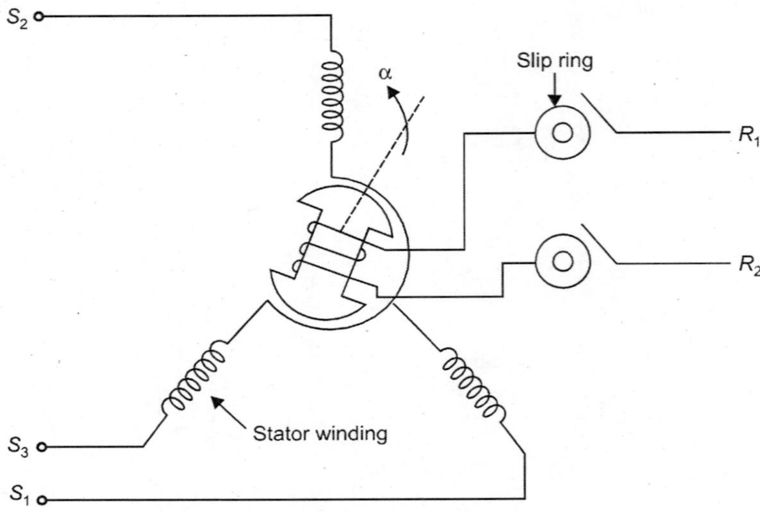

Fig. 3.11: Electrical circuit of synchro receiver.

The constructional features of synchro control transformer is similar to that of synchro transmitter, except the shape of rotor. The rotor of the control transformer in made cylindrical so that the air gap is practically uniform. This feature of the control transformer minimizes the changes in the rotor impedance with the rotation of the shaft. The construction features electrical circuit and a schematic symbol of control transformer shown in Fig. (3.9), (3.10) and (3.11).

Working

The output of the synchro transmitter act as input to synchro receiver. That is the generated emf of the synchro transmitter is applied as input to the stator coils of control transformer. The rotor shaft is connected to the load whose position has to be maintained at the desired value. Depending on the current position of the rotor and the applied emf on the stator, an emf is induced on the rotor winding. This emf can be measured and used to drive a motor so that the position of the load is corrected.

Applications

This can be used as a error detector and applicable to correct the position of motor.

Synchro as Error Detector

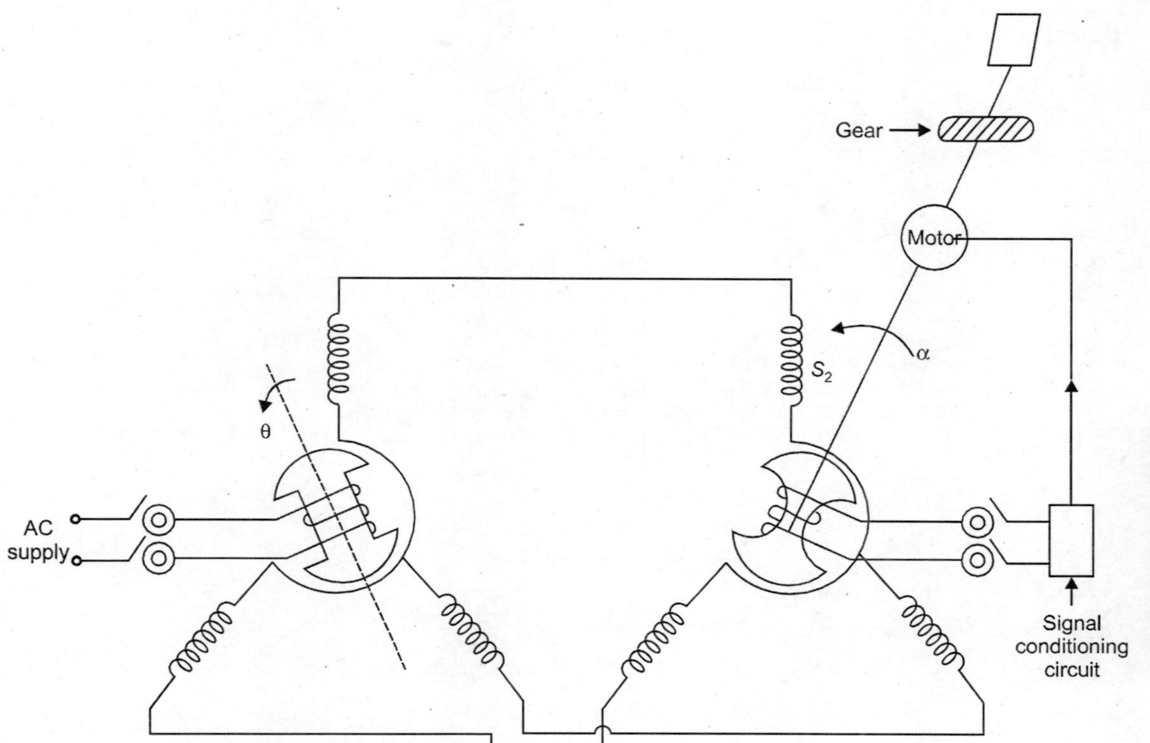

Fig. 3.12: Synchro as error detector.

* Synchro error detector is formed by interconnecting synchro transmitter and receiver. The angular position of the transmitter is the reference input and the rotor is excited by ac supply with frequency ω.

The synchro receiver rotor is connected to a servo motor, whose position has to be changed (or) controlled.

The transmitter rotor will be in electrical zero position and the receiver rotor will be in null position and angular separation of both rotor axis in aligned position is 90є.

When the rotor of transmitter is excited, the flux is set-up and emfs are induced in stator coils. These induced emfs are impressed on the stator coils of control transformer (receiver). The current in the stator coils set up flux in receiver. Due to the similarity in the magnetic construction, the flux patterns produced in the two synchros will be the same if all losses are neglected.

Let the rotor of transmitter rotates through an angle of from its electrical zero position. Now the rotor of the receiver will rotate in the same direction through an angle α from its null position. The net angular separation of the two rotor is equal to (90 – θ + α) and the voltage induced in the receiver rotor is proportional to the cosine of this angle. The error voltage is amplified and used to drive a servomotor. The motor drives the shaft of the synchro receiver until it comes to a new position at which the error voltage is zero.

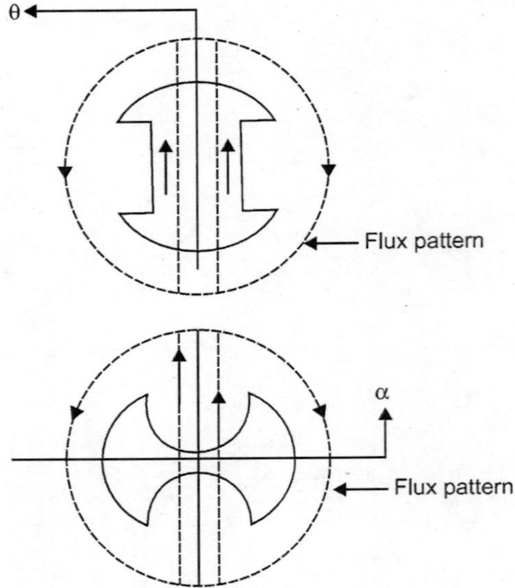

Fig. 3.13: Rotor positions and flux pattern.

∴ voltage across slip rings of control transformer (modulated error voltage) }

$$e_m = K' E_r \cos(90 - \theta + \alpha) \sin \omega t$$
$$= K' E_r \cos[90(\theta - \alpha)] \sin \omega t$$
$$= K' E_r \sin(\theta - \alpha) \sin \omega t$$

where K' is a proportional constant.

Let $\phi(t) = \theta - \alpha$

For small values of $\phi(t)$, $\sin(\theta - \alpha) = \sin \phi(t) \approx \phi(t)$

∴ $$e_m = K_1 E_r \phi(f) \sin \omega t \quad (3.10)$$

from equation (3.10) we can say that the output voltage of the synchro error deter is modulated signal with carrier frequency ω.

The magnitude of the modulated carrier wave is proportional to $\phi(t)$ and the phase reversals of the modulated wave depend on the sign of $\phi(t)$. The signal conditioning circuit demodulates the voltage available across slip rings and develops a demodulated and amplified error voltage to drive the motor.

The demodulated error voltage, $e = K_s \phi(t)$ (3.11)

where, K_s = sensitivity of the synchro error detector in volts/deg

on taking Laplace transform of equation (6.3.11), we get,

$$E(s) = K_s \phi(s)$$

∴ $$\frac{E(s)}{\phi(s)} = K_s \quad (3.12)$$

The equation (3.12) is the transfer function of the synchro error detector.

3.4 CONTROLLERS

A controller is a device which is introduced in the system to modify the error signal and to produce a control signal. The controller modifies the transient response of the system.

Depending upon control action, the controllers are classified as follows,

1. Two position (or) on-off controllers.
2. Proportional controllers (P)
3. Integral controllers (I)
4. Proportional-plus-integral controllers (PI)
5. Proportional-plus-integral-plus derivative controllers (PID)
6. Proportional-plus-derivative controllers (PD)

On-off (or) two position controller

The on-off (or) two position controller has only two fixed positions. They are either on or off. The on-off control system is very simple in construction and hence less expensive. The on off control action is provided by a relay. Relay is a device which has NO and NC (normally open and normally closed) contacts, whose operations are controlled by relay coil. When realy coil is excited, the relay operates and the contacts change their positions.

Let the output signal from the controller be $u(t)$ and the actuating error signal be $e(t)$. In this controller, $u(t)$ remains at either a maximum (or) minimum value.

$$u(t) = u_1 \text{ for } e(t) > 0$$
$$u(t) = u_2 \text{ for } e(t) < 0 \quad (3.13)$$

From Fig. (3.14),

$$E(s) = L[e(t)]$$
$$U(s) = L[u(t)]$$

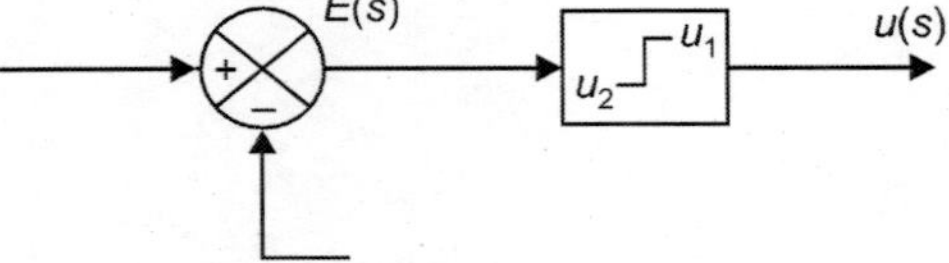

Fig. 3.14: Block diagram of on-off controller.

Proportional Controller (*P*-Controller).

It is device that produces a control signal, $u(t)$ which is proportional to the input error signal, $e(t)$.

In *P*-controller, $u(t) \propto e(t)$

$$\therefore \quad u(t) = K_P\, e(t) \quad (3.14)$$

where, K_P = Proportional gain

on taking laplace transform of equation (6.4.2)

$$U(s) = K_P\, E(s) \quad (3.15)$$

$$\therefore \quad \frac{U(s)}{E(s)} = K_P \quad (3.16)$$

The equation (3.15) gives the output of the P-controller for the input $E(s)$ and equation (6.4.2) is the transfer function of the P-controller.

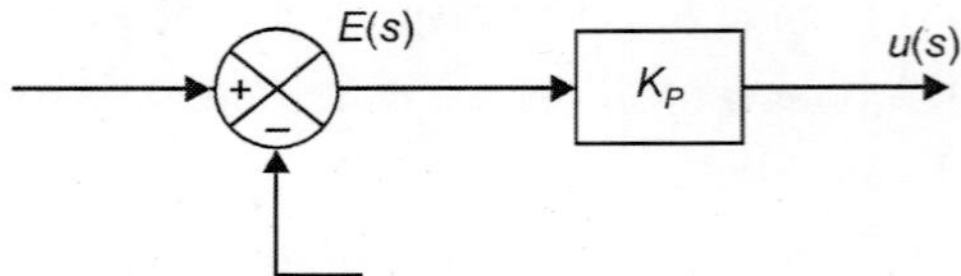

Fig. 3.15: Block diagram of P-controller.

From equation (3.15) it is understood that the proportional controller amplifies the error signal by an amount K_P. Introduction of the controller on the system increases the loop gain by an amount K_P.

***From P-controller following points are noted.**

- It increases the loop gain.
- It increases steady state tracking accuracy.
- It increases signal rejection and relative stability.
- System becomes less sensitive to parameter variations.
- Increment of gain to very large values may lead to instability of the system.
- The drawback in P-controller is that it leads to a constant steady state error.

Example of *P*-controller

The equivalent circuit for the Fig. (3.16) can be draw as shown in Fig. (3.18).

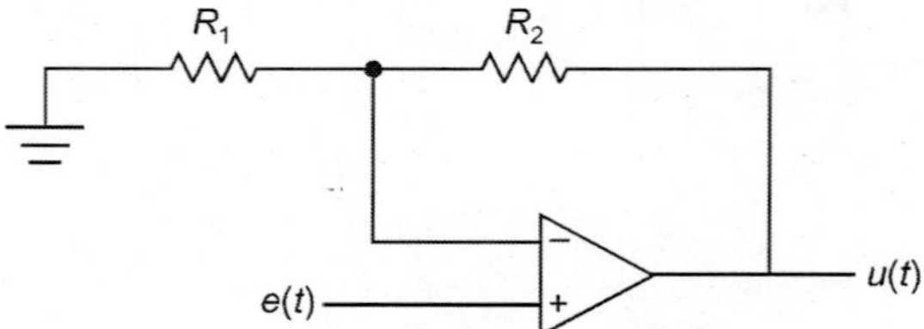

Fig.3.16: OP-amp-P controller using non-inverting amplifier.

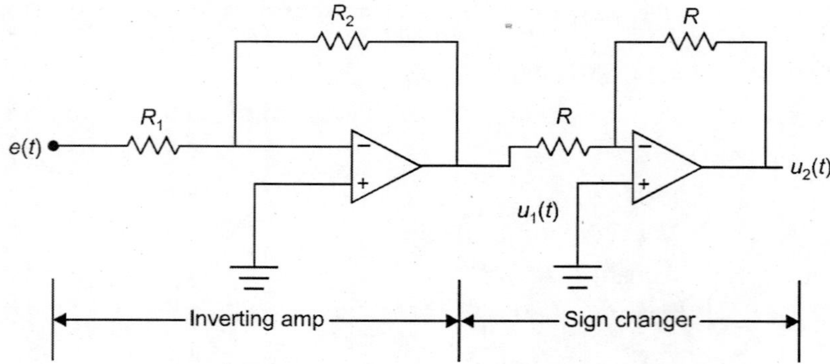

Fig. 3.17: OP-amp P controller using inverting amplifier.

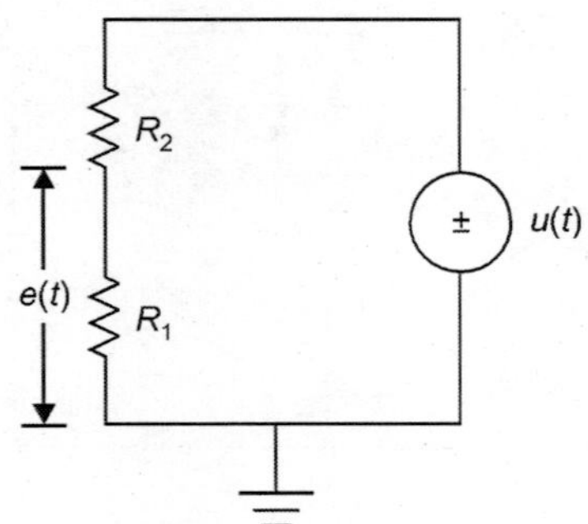

Fig. 3.18: Equivalent circuit of *P*-controller [non-inverting Amp].

From Fig. (3.18),

By voltage division rule,

$$e(t) = \frac{R_1}{R_1 + R_2} u(t) \quad (3.17)$$

$$\therefore \quad u(t) = \frac{R_1 + R_2}{R_1} e(t) \quad (3.18)$$

on taking laplace transform of equation (6.4.6), we get

$$U(s) = \frac{R_1 + R_2}{R_1} E(s)$$

$$\therefore \quad \frac{U(s)}{E(s)} = \frac{R_1 + R_2}{R_1} \quad (3.19)$$

By comparing equations (3.19) and (3.16)

$$K_P = \frac{R_1 + R_2}{R_1} \quad (3.20)$$

* Therefore by adjusting the values of R_1 and R_2 the value of gain, K_P can be varied.

The equivalent circuit for Fig. (3.17),

From Fig. (3.19),

$$e(t) = I_1 R_1$$

$$\therefore \quad I_1 = \frac{e(t)}{R_1} \quad (3.21)$$

$$u_1(t) = -I_1 R_2 \quad (3.22)$$

substitute for I_n from Eq. (3.21) in equation (3.22)

$$u_1(t) = \frac{-e(t)}{R_1} \cdot R_2 \quad (3.23)$$

CHAPTER

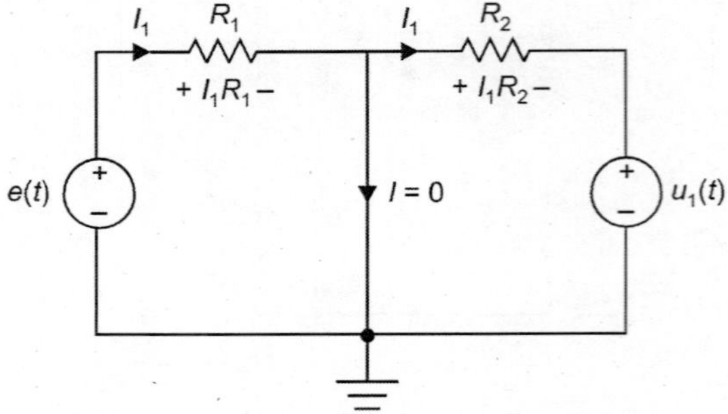

Fig. 3.19: Equivalent circuit of controller.

From Fig. (3.20),

$$u_2(t) = -I_2R$$

$$\therefore \quad I_2 = \frac{u_2(t)}{R} \tag{3.24}$$

$$u_1(t) = I_2R \tag{3.25}$$

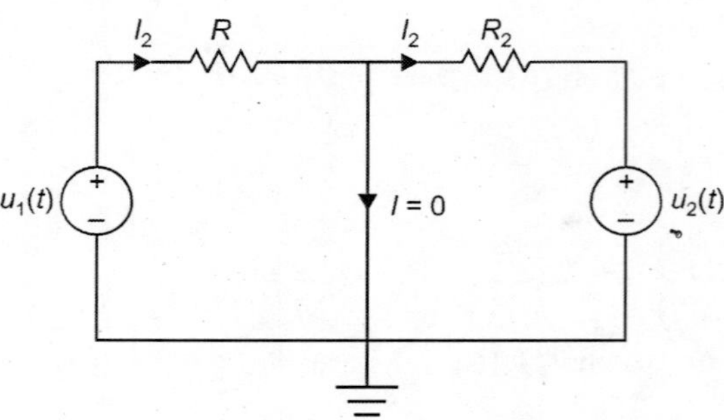

Fig. 3.20: Equivalent circuit of sign changer (Inverting).

Substitute I_2 from equation (3.24) in equation (3.25) we get

$$-u_2(t) = \frac{-e(t)}{R_1} \cdot R_2$$

$$u_2(t) = \frac{R_2}{R_1} e(t) \tag{3.26}$$

on taking laplace transform of equation (3.26) we get,

$$U_2(s) = \frac{R_2}{R_1} E(s)$$

$$\therefore \quad \frac{U_2(s)}{E(s)} = \frac{R_2}{R_1} \tag{3.27}$$

on comparing the equation (3.27) with equation (3.16), we get

Proportional gain, $$K_p = \frac{R_2}{R_1} \tag{3.28}$$

Conclusion: Therefore by adjusting the values of R_1 and R_2 the value of gain K_P can be varied.

Integral Controller (*I*-controller)

It is a device which produce a control signal $u(t)$ which is proportional to integral of the input error signal, $e(t)$.

In *I*-controller, $u(t) \propto \int e(t)\,dt$

$$\therefore \quad u(t) = K_i \int e(t)\,dt \tag{3.29}$$

where, K_i = Integral gain (or) constant

on taking laplace transform of equation (3.29) with zero initial conditions we get

$$U(s) = K_i \frac{E(s)}{s} \tag{3.30}$$

$$\therefore \quad \frac{U(s)}{E(s)} = \frac{i}{s} \tag{3.31}$$

The equation (3.31) is called transfer function of the *I*-controller.

The block diagram of *I*-controller is shown in Fig. (3.21).

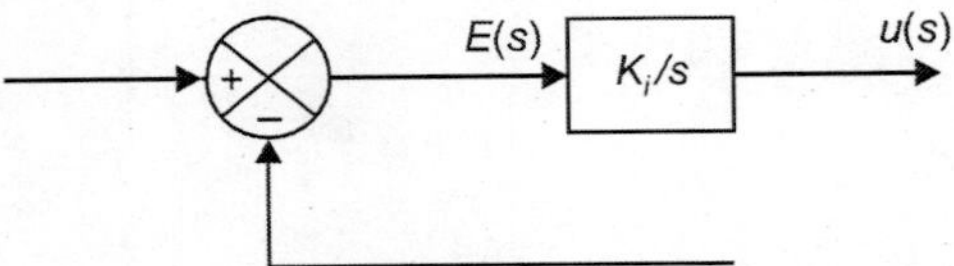

Fig. 3.21: Block diagram of an integral controller.

From *I*-controller the following points are noted.

- *I*-controller removes (or) reduces the steady state error without need for manual reset.
- It may lead to oscillatory response of increasing (or) decreasing amplitude which is undesirable and the system may become unstable.

Example of I-controller

The Fig. (3.22) shows the circuit diagram of *i*-controller using OP-amp.

The Equivalent circuit diagram for Fig. (3.22) can be drawn as

*** Assumption mode**

1. The voltages of both inputs are equal.
2. The input current is zero.

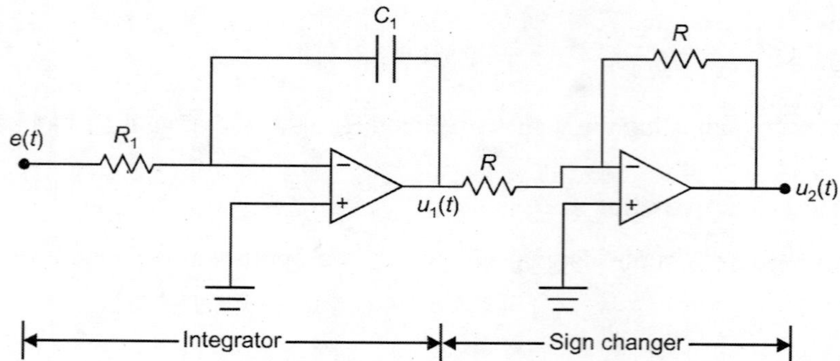

Fig. 3.22: Circuit diagram of *I*-controller.

From Fig. (3.23),

$$e(t) = i_1 R_1,$$

$$\therefore \quad i_1 = \frac{e(t)}{R_1} \tag{3.32}$$

$$u_1(t) = \frac{1}{C_1}\int i_1 \, dt \tag{3.33}$$

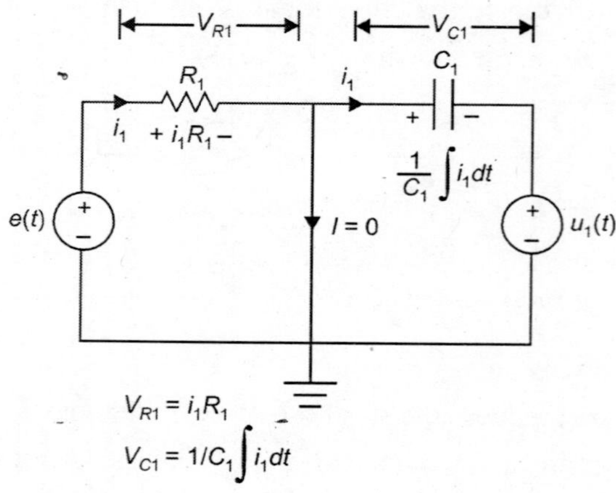

Fig. 3.23: Eqt circuit of integrator.

Substitute i_1 from equation (3.32) in equation (3.33)

$$\therefore \quad u_1(t) = \frac{-1}{C_1}\int \frac{e(t)}{R_1} dt$$

$$u_1(t) = -\frac{1}{R_1 C_1}\int e(t) dt \tag{3.34}$$

From Fig. (3.24)

$$u_2(t) = -I_2R$$

$$\therefore \quad I_2 = \frac{-u_2(t)}{R} \tag{3.35}$$

$$\therefore \quad u_1(t) = I_2R \tag{3.36}$$

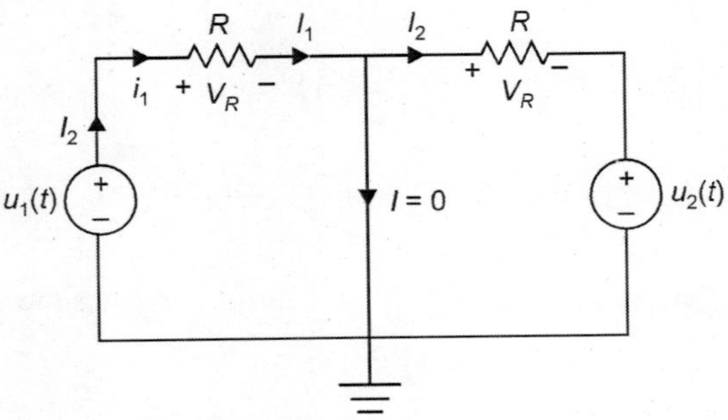

Fig. 3.24: Eqt circuit of sign changer.

Substitute for I_2 from equation (3.35) in equation (3.36)

$$\therefore \quad u_1(t) = \frac{-u_2(t)}{R}R = -u_2(t) \tag{3.37}$$

on equating the equations (3.37) and (3.34) we get,

$$-u_2(t) = \frac{-1}{R_1C_1}\int e(t)\,dt$$

$$\therefore \quad u_2(t) = \frac{-1}{R_1C_1}\int e(t)\,dt \tag{3.38}$$

on taking laplace transform of equation (3.38) with zero intial conditions, we get

$$U_2(s) = \frac{-1}{R_1C_1}\frac{E(s)}{s} \tag{3.39}$$

$$\frac{U_2(s)}{E(s)} = \frac{1}{R_1C_1}\frac{1}{s} \tag{3.40}$$

The equation (3.40) is called as transer function of *I*-controller.

On comparing the equations (3.40) with (3.31) we get

Integral gain, $$K_i = \frac{1}{R_1C_1} \tag{3.41}$$

Conclusion: *Therefore by adjusting the values of R_1 and C_1 the value of gain K_i can be varied.

CHAPTER 3

Proportional Plus Integral Controller (*PI*-controller)

The PI controller produces an output signal of two terms-one is proportional to error signal and the other is proportional to the integral of error-signal.

In PI controller,

$$u(t) \propto \left[e(t) + \int e(t)\, dt \right]$$

$$u(t) = K_P e(t) + \frac{K_P}{T_i} \int e(t)\, dt \qquad (3.42)$$

where, $K_P \rightarrow$ proportional gain constant

$T_i \rightarrow$ Integral time

on taking laplace transform of equation (3.42) with zero intial conditions we get,

$$U(s) = K_P E(s) + \frac{K_P}{T_i} \frac{E(s)}{s} \qquad (3.43)$$

$$\therefore \quad \frac{U(s)}{E(s)} = K_P \left[1 + \frac{1}{T_i s} \right] \qquad (3.44)$$

The equation (3.44) is the transfer function of the PI controller.

The block diagram of PI controller shown in Fig. (3.25).

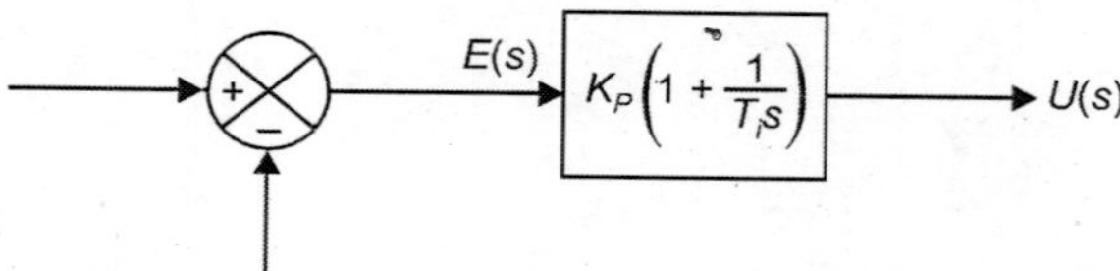

Fig. 3.25: Block diagram of PI controller.

From PI controller the following prints are noted.

1. The advantages of both *P* and *I* controller are combined.
2. *P* controller increases the loop gain.
3. *I* controller reduces steady state error.
4. The change in value of K_P affects both the proportional and integral parts of control action.

Example of PI controller

The equivalent circuit diagram for Fig. (3.26) is shown below,

Assumptions mode

1. The voltages at both inputs one equal.
2. The input current is zero.

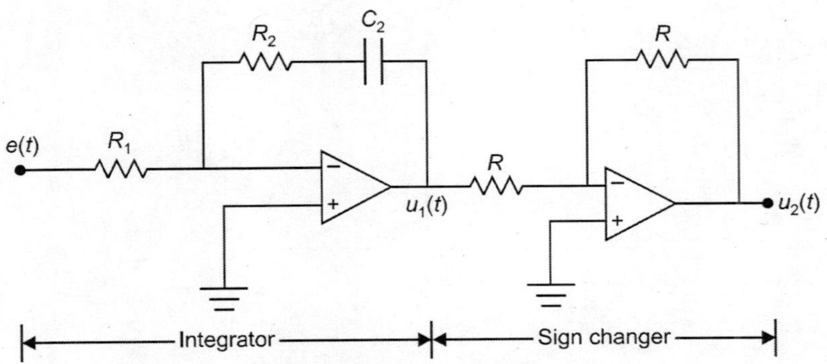

Fig. 3.26: Circuit diagram of PI controller.

From Fig. 3.27,

$$e(t) = i_1 R_1$$

$$\therefore \quad i_1 = \frac{e(t)}{R_1} \tag{3.45}$$

$$u_1(t) = -i_2 R_2 - \frac{1}{C_2}\int i_1 dt \tag{3.46}$$

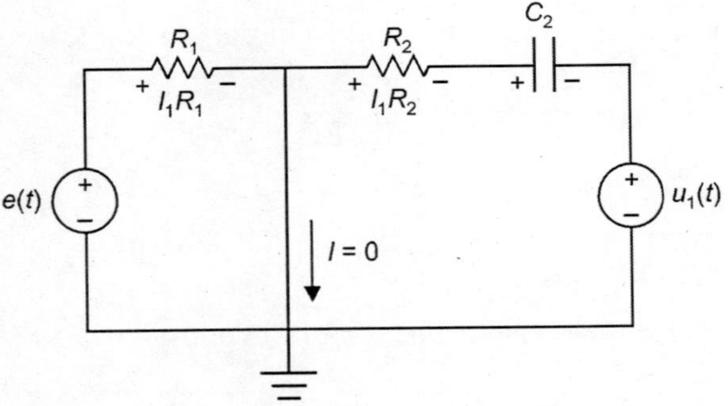

Fig. 3.27: Eqt circuit of Integrator.

Substitute i_1 from equation (3.45) in equation (3.46), we get,

$$\therefore \quad u_1(t) = -\frac{e(t)}{R_1} R_2 - \frac{1}{C_2}\int \frac{e(t)}{R_1} dt \tag{3.47}$$

From Fig. (3.28), we get

$$u_2(t) = -I_2 R,$$

$$\therefore \quad I_2 = \frac{-u_2(t)}{R} \tag{3.48}$$

$$u_1(t) = I_2 R \tag{3.49}$$

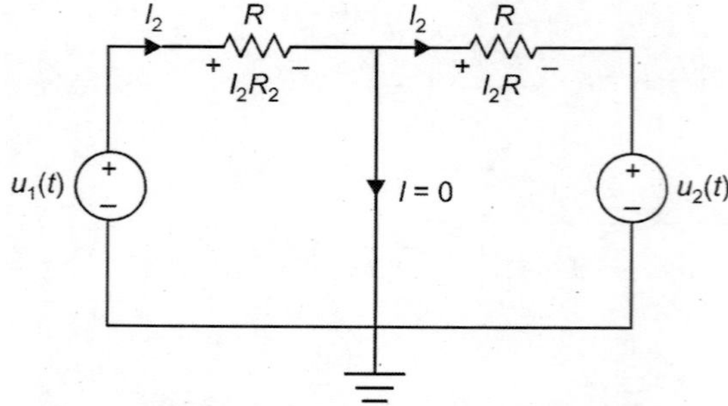

Fig. 3.28: Equivalent circuit of sign changer.

Substitute I_2 from equation (3.48) in equation (3.49), we get,

$$\therefore \qquad u_1(t) = \frac{-u_2(t)}{R} R = -u_2(t) \qquad (3.50)$$

on equating the equations (3.50) and (3.47) we get,

$$-u_2(t) = \frac{-e(t)}{R_1} R_2 - \frac{1}{C_2} \int \frac{e(t)}{R_1} dt$$

$$\therefore \qquad u_2(t) = \frac{R_2}{R_1} e(t) + \frac{1}{R_1 C_2} \int e(t)\, dt \qquad (3.51)$$

on taking laplace transform of equation (3.51) with zero initial junctions, we get.

$$U_2(s) = \frac{R_2}{R_1} E(s) + \frac{1}{R_1 C_2 \dfrac{E(s)}{s}}$$

$$\frac{U_2(s)}{E(s)} = \frac{R_2}{R_1}\left[1 + \frac{1}{R_2 C_2 s}\right] \qquad (3.52)$$

The equation (3.52) is called the transfer function of PI controller,

where $\qquad K_P = \dfrac{R_2}{R_1} \Rightarrow$ Proportional gain

$T_i = R_2 C_2 \Rightarrow$ Integral time constant

Conclusion: By varying the values of R_1 and R_2 the values of K_P and T_i can be adjusted.

Proportional, Plus Integral Plus Derivative Controller (PID Controller)

The PID controller produce an output signal consisting of three terms-one proportional to error signal, another are proportional to integral of error signal and the third one proportional to derivative of error signal.

In PID-controller,

$$u(t) \propto \left[e(t) + \int e(t) + \frac{d}{dt} e(t) \right]$$

$$\therefore \quad u(t) = K_P e(t) + \frac{K_P}{T_i} \int e(t)\, dt + K_P T_d \frac{d}{dt} e(t) \tag{3.53}$$

where,

K_P = Proportional gain

T_i = Integral time constant

T_d = Derivative time constant

on taking laplace transform of equation (3.53)

$$U(s) = K_P\, E(s) + \frac{K_P}{T_i} \frac{E(s)}{s} + K_P T_d\, E(s) \cdot s \tag{3.54}$$

$$\therefore \quad \frac{U(s)}{E(s)} = K_P \left[1 + \frac{1}{T_i s} + T_{d \cdot s}\right] \tag{3.55}$$

Eq. (3.55) in the transfer function of PID controller.

The block diagram of PID controller shown in Fig. (3.29).

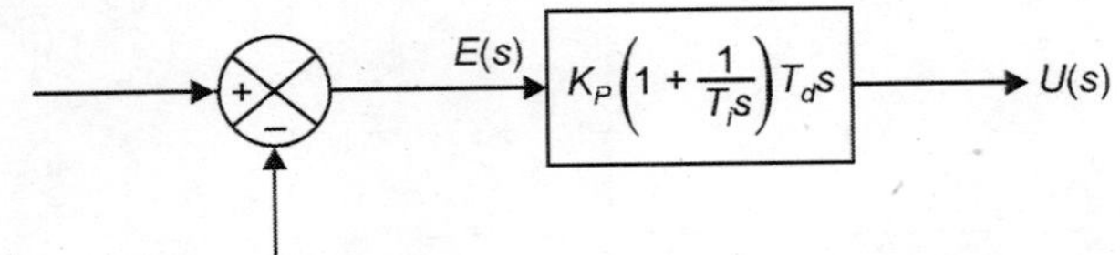

Fig. 3.29: Block diagram of PID controller.

1. It is having the advantages of P, I and D controllers.
2. P increases gain, I reduces steady state error and D reduces the rate of change of error.

Example of PID controller

The circuit diagram of PID controller is shown in below Fig. (3.30).

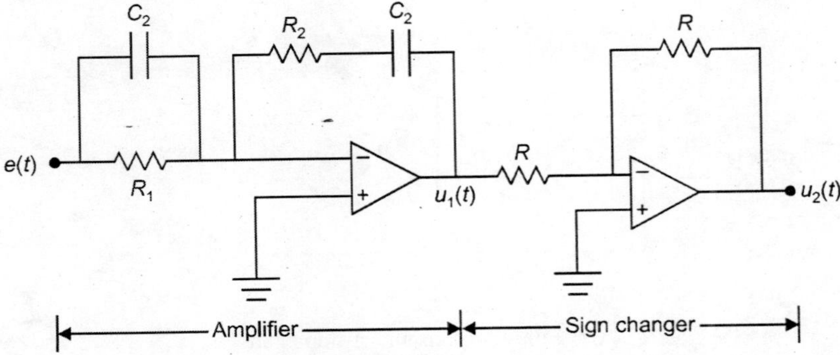

Fig. 3.30: Circuit diagram of PID controller.

The equivalent circuit of the Fig. (3.30) can be drawn as shown in Fig. (3.31)

From Fig. (3.31), we get

CHAPTER 3

$$i_1 = \frac{e(t)}{R_1} + C_1 \frac{de(t)}{dt} \tag{3.56}$$

on taking laplace transform of equation (3.56) with zero intial conditions we get,

$$I_1(s) = \frac{1}{R_1} E(s) + C_1 s\ E(s)$$

$$I_1(s) = \left[\frac{1}{R_1} + C_1 s\right] E(s) \tag{3.57}$$

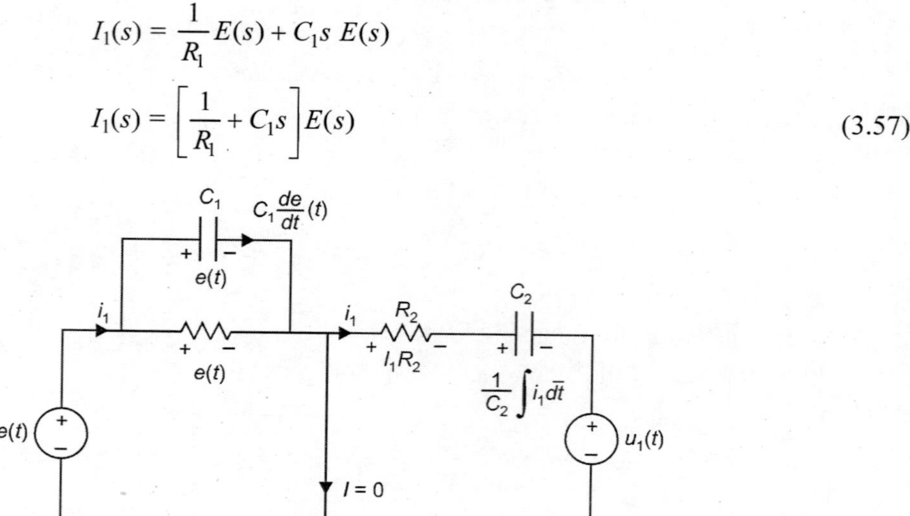

Fig. 3.31: Equivalent circuit of amplifier.

From Fig. (3.32), we get,

$$i_1 R_2 + \frac{1}{C_2}\int i_1 dt = -u_1(t) \tag{3.58}$$

Fig. 3.32: Equivalent circuit of sign changer.

on taking laplace transform of equation (3.58), with zero intial condition, we get

$$I_1(s) R_2 + \frac{1}{C_2}\frac{I_1(s)}{s} = -U_1(s)$$

$$\therefore \quad I_1(s)\left[R_2 + \frac{1}{C_2 s}\right] = -U_1(s) \tag{3.59}$$

Substitute for $I_1(s)$ from equation (3.57) in equation (3.59)

$$\therefore \quad \left[\frac{1}{R_1} + C_1 s\right] E(s)\left[C_2 + \frac{1}{C_2 s}\right] = -U_1(s) \tag{3.60}$$

From Fig. (6.4.19)

$$u_2(t) = -I_2 R,$$

$$K_P\left[\frac{T_i + T_d}{T_i} + \frac{1}{T_i S} + T_d S\right].$$

$$\therefore \quad I_2 = \frac{-u_2(t)}{R} \tag{3.61}$$

$$u_1(t) = \frac{-u_2(t)}{R} R = -\,u_2(t) \tag{3.62}$$

on taking laplace transform of equation (3.62) we get,

$$+U_1(s) = -U_2(s) \tag{3.63}$$

From equations (3.60) and (3.63) we get,

$$U_2(s) = \left[\frac{R_2}{C_1} + \frac{C_1}{C_2} + \frac{1}{R_1 C_2 s} + R_2 C_1 s\right] E(s)$$

$$\therefore \quad \frac{U(s)}{E(s)} = \left[\frac{R_2 C_2 + R_1 C_1}{R_1 C_2} + \frac{1}{R_1 C_2 s} + R_2 C_1 s\right]$$

$$= \frac{R_2}{R_1}\left[\frac{R_2 C_2 + R_1 C_1}{R_2 C_2} + \frac{1}{R_2 C_2 s} + R_1 C_1 s\right] \tag{3.64}$$

The equation (3.64) is the transfer function of PID controller.

From above derivations,

Proportional gain constant, $= K_p\left[\frac{T_i + T_d}{T_i} + \frac{1}{T_i S} + T_d S\right]$

$$K_P = \frac{R_2}{R_1}$$

Derivative time constant,

$$T_d = R_1 C_1$$

Integral time constant,

$$T_i = R_2 C_2$$

Also, $$\frac{R_1 C_1 + R_2 C_2}{R_2 C_1} = 1$$

Conclusion: By varying the values of R_1 and R_2 the value of K_P, T_d and T_i are adjusted.

3.5 SERVO MOTORS

The motors which is used in automatic control systems are called servomotors and the system which is used to control the position of an object is called servomechanism.

It is used to convert electrical signal in to an angular displacement. They can operate either by continuous supply (or) by step by step supply depending upon their construction.

Depending upon their supply it can be classified in to two types.

1. DC servomotor
2. AC servomotor

Generally both AC and DC servo motors are having following features.

1. Steady state stability
2. Linear relationship in both electrical and mechanical characteristics.
3. Fast response because of low inertia.
4. Wide range of speed control.

(i) DC Servomotors

The DC servomotors are generally used for large power applications such as machine tools and robotics.

It can be classified in to many types mainly it can be classified into two types

1. Permanent magnet motors
2. Electromagnetic field motors

1. Permanent magnet motors

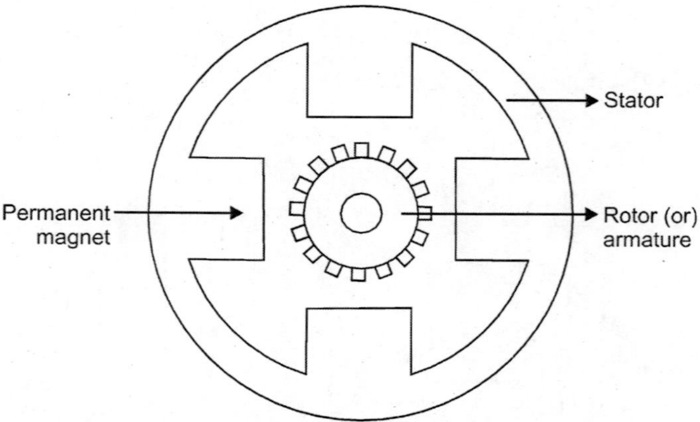

Fig. 3.33: Constructional feature of PM motor supply.

Construction

In this type, the field winding consist of permanent magnets. The rotor employs special type of constructions to reduce the weight and so the inertia of the rotating system also reduces. Armature is of cylindrical with small diameter and longer axial length to get high speed. It can be of two types either disc type armature (or) hollow type armature.

Working

Principle: Whenever the current carrying conductor is placed in the magnetic field it experiences mechanical force. This servomotor receive the electrical signal by controlling devices as soon as it get signal, the rotor start rotates. the input to the motor is directly proportional to error signal.

Advantage

1. Simple, more reliable, [Œ No field power supply].
2. Higher efficiency [Œ No field loss]
3. Linear mechanical characteristics.
4. Higher power output.

Disadvantages

1. Magnets get deteriorate with time.
2. Magnets get demagnetised by large transient currents.

2. Electromagnetic field motors

Construction

This type of motors are similar to conventional dc motors by construction but it has some special features when compared to ordinary dc motor.

1. The number of slots and commutator segments are more to improve commutation.
2. Compoles and compensating windings are provided to eliminate sparking.
3. The diameter to length ratio of motor is small to reduce inertia.
4. Oversize shafts are employed to withstand the high torque stress.
5. Eddy currents are reduced by complete lamination and by using low loss steel.

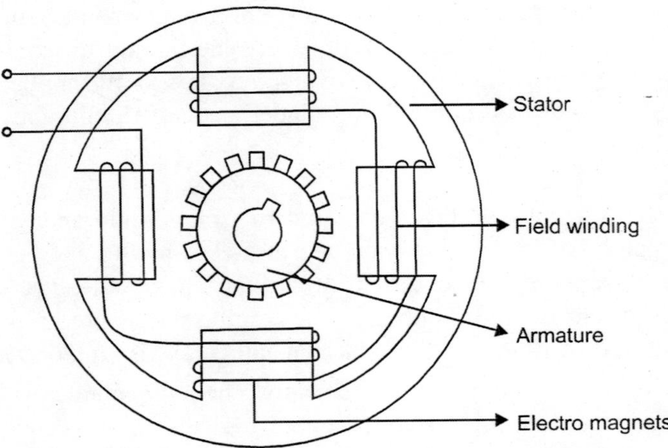

Fig. 3.34: Constructional feature of electromagnetic field motors.

Working

The principle of this type of motor is also same as previous type but the only difference is that, the magnetic fields produced in case of PM type motor is without separate power supply but in this case separate power supply is needed. [working is same as PM type motor].

Advantages

1. Speed can be varied.
2. Improved commutation process.
3. High speed application.

Disadvantages

1. Field losses due to field excitation
2. Comparatively less efficienty.
3. Less power output due to power loss in field winding.

(ii) AC Servomotor

An ac servomotor is basically a two phase induction motor. Only small difference between AC servo motor and induction motor. The excitation voltage applied to stator windings should be displaced by 90є and the rotor is built with high resistance to obtain linear mechanical characteristics.

Construction

The AC servomotor is basically a two phase induction motor with some special design features. The stator consists of two pole-pairs (A–B and C–D) mounted on the inner periphery of the stator, such that their axis are at an angle of 90є in space. Each pole pair carries a winding. One winding is called reference winding and the other is called control winding. The supply used to drive the motor is single phase and so a phase capacitor is connected to one of phase to produce a phase difference of 90є.

The rotor construction is usually squirrel cage or drag-cup type. The squirrel cage rotor is made of laminations. The rotor bars are placed on the slots and short circuited at both ends by end rings. The diameter of the rotor is kept small in order to reduce inertia and to obtain good accelerating characteristics. The Drug-cup construction is employed for very low inertia applications. In this type of construction the rotor will be in the form of hollow cylinder made of aluminium.

Working

The stator windings are excited by voltages of equal rms magnitude and 90є phase difference. This results in exciting currents i_1 and i_2 that are phase displaced by 90є and have equal rms values. These currents give rise to rotating magnetic field of constant magnitude. The direction of rotation depends on the phase relationship of the two currents. The exciting current produces a clockwise rotating magnetic field and phase shift of 180є. It will produce an anti clockwise rotating magnetic field.

The rotating magnetic field sweeps over the conductors. The rotor conductors experience a change in flux and so voltages are induced in rotor conductors. This voltage circulates current in the short circuited rotor conductors and the currents create rotor flux. Due to the interaction of stator and rotor flux, a mechanical force is developed in the rotor and so the rotor starts moving in the same direction as that of rotating magnetic field.

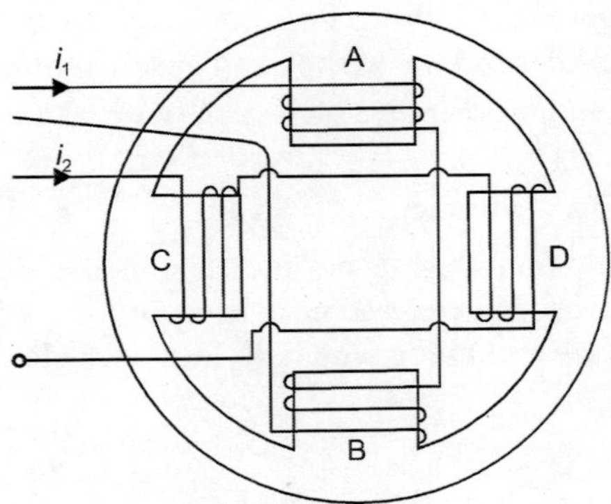

Fig. 3.35: Stator of AC servomotor.

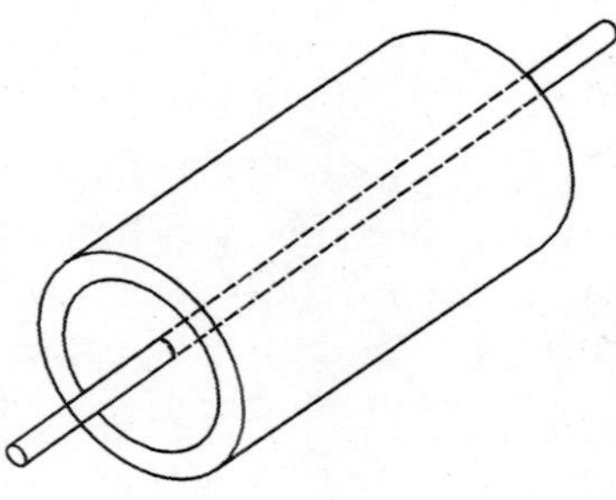

Fig. 3.36: Rotor of AC servomotor.

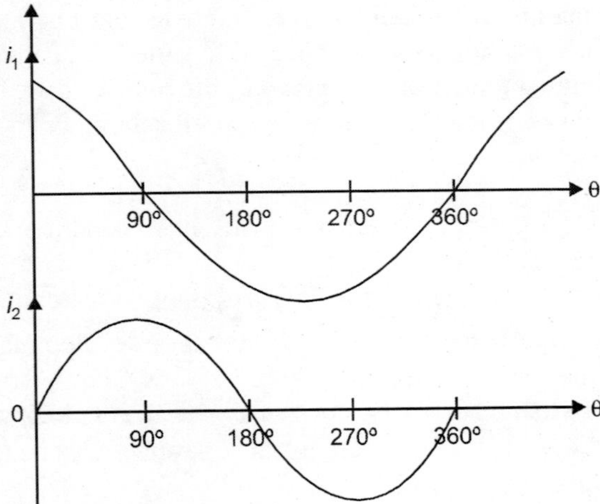

Fig. 3.37: Exciting currents of AC servomotor.

Advantages

1. Fast response
2. Steady state stability
3. Accurate position control
4. Easy construction.

Disadvantages

1. Lower power output compared to DC servomotor.
2. Linearity characteristics are not achieved easily as DC servomotor.
3. Starting torque is low as compared to DC servomotor.

3.6 STEPPER MOTOR

A stepper motor transforms electrical pulses into equal increments of rotary shaft motion called steps. They work in conduction with electronic switching devices. It has a wound stator and a non-excited rotor. They are classified as variable reluctance, permanent magnet (or) hybrid, depending on the type of rotor. They are also classified as 2-phase, 3 phase (or) 4-phase depending on the number of windings on the stator. The number of teeth or poles on the rotor and the number of poles on the stator determine the size of the step called step angle.

step angle = 360e/number of steps per revolution.

Motors are available with step rates of 200 steps per revolution. Further increase in step rate is limited by mechanical and physical constraints. This limitation has been overcome by elecytronic methods of reducing the step size. Half stepping and micro stepping are technique whcih divides each step into two half steps (or) *n* number of microsteps.

(i) Permanent Magnet Rotor Stepper Motor

Construction

The stator of this type of stepper motor has salient poles carrying control windings. Each pole carries a control winding. A pair of control windings are connected in series and called a phase. It is wound for any number of phases. (Maximum number of phase is four) the rotor is made is the form of a permanent magnet spider cast integral (or) assembled of number of permanent magnets.

Working

Consider the stepper motor having 4-pole stator with two phase windings. Let the rotor be made of permanent magnet with 2 poles. The stator poles are marked A, B, C and D and they are excited with pulses supplied by switching devices. Each control pulse applied to the control windings of stator.

In first excitation is applied to A and C [ref. Fig. (3.38)], they develop magnetic polarity in upward direction, at this point let assume rotor is in vertical position. If now A and C are switched OFF and B and D are excited then the rotor turn to the horizontal position by 90є in anticlock wise direction. Then again for next step another 90є and vice-versa. The direction can be reversed by changing the current direction.

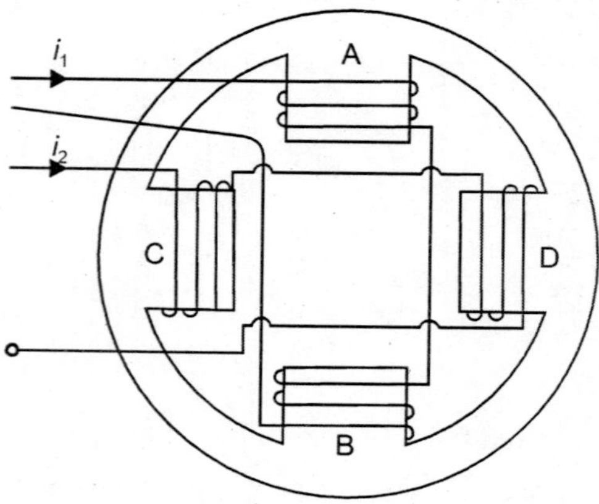

Fig. 3.38: Stator of permanent magnet rotor stepper motor.

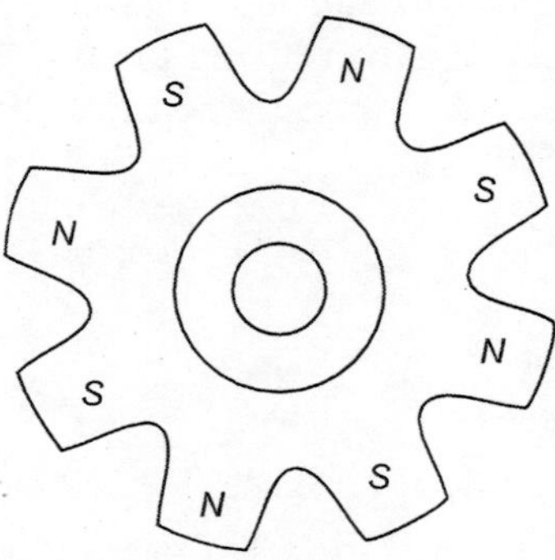

Fig. 3.39: Cast integral type rotor.

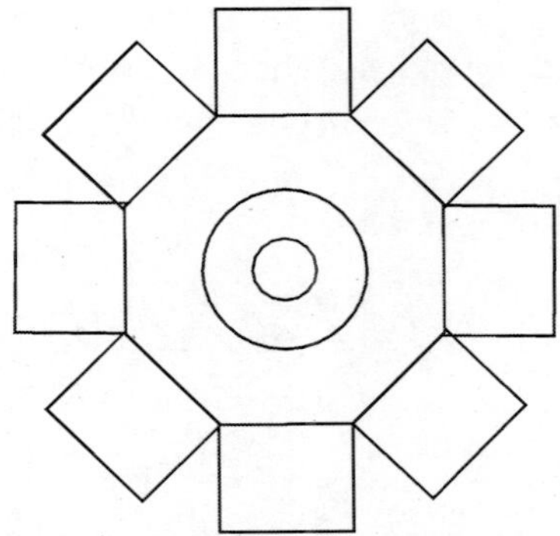

Fig. 3.40: Assembled separated magnets.

Variable Reluctance Stepper Motor

Construction

It consist of stator and rotor. The rotor is made up of soft-iron with the number of teeth, the stator also has teeth in addition to a number of wound poles. The rotor construction is giving appearance of a gear. The main constructional feature is that this type of motor has variable reluctance in their magnetic field.

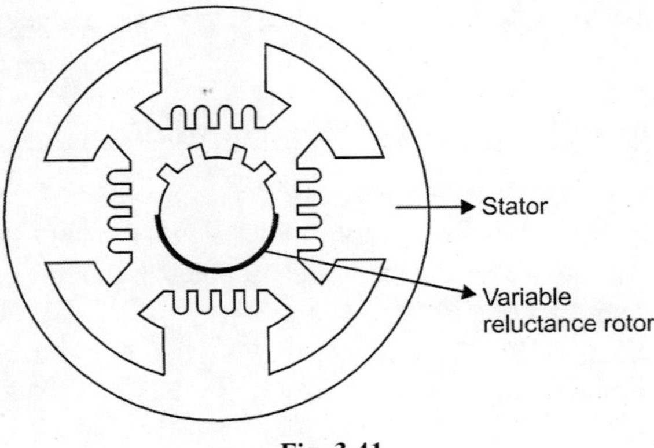

Fig. 3.41

Working

When the supply is applied to coil in the stator, then the magnetic flux is generated in the stator coils. The flux in the stator coils causes teeth in the rotor to line up with teeth in the stator. When the current is switched to the next winding, the rotor moves a distance of one step angle, and a new set of teeth line up. By this way rotor gives the rotation according to the electric pulses.

Hybrid Stepper Motors

Construction

The construction of hybrid steppers motor is same as variable reluctance stepper motor. The magnets are replaced by permanent magnet. The rotor has same construction as variable reluctance stepper motor. The magnet is usually in the rotor in this type of motor. The stator has eight poles with two center-tapped windings. When the centre taps are used, the motor is considered to be a four phase motor. When the centre taps are removed or left open circuited, the motor is considered a two phase motor.

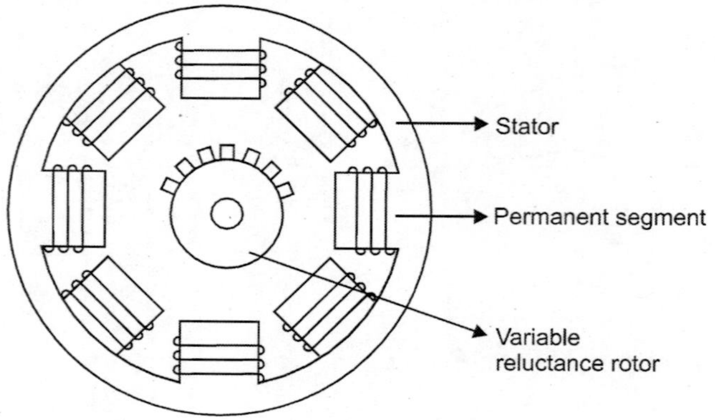

Fig. 3.42

Working

When electrical pulses are given to the rotor due to the interaction between the permanent magnet flux and rotor flux there will produce torque which makes the rotor to rotate. This type of motor have high torque, high inertia and small step angles. A typical hybrid stepper has 50 teeth in the rotor and 50 teeth in the stator.

Important types of Stepping Operation of Stepper Motors

Full stepping

Full step = 360є/No. of rotor poles Ч No. of stator pairs

Let a stepper motor having 4 poles in the stator and 10 poles in the rotor then,

Full step angle = 360є/10 Ч 2

Full step angle =18є

Let we see how the excitation of stator poles are done.

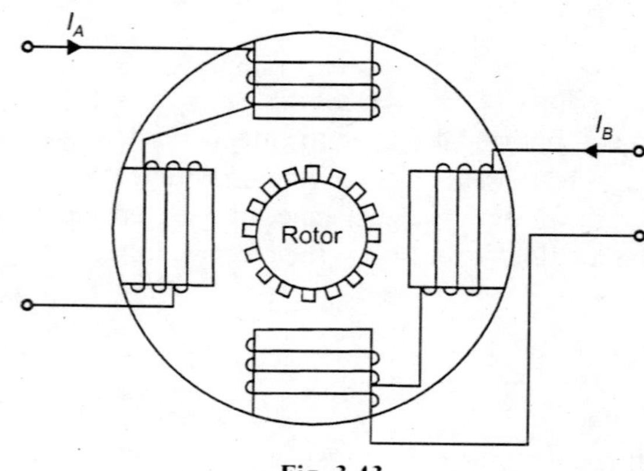

Fig. 3.43

Step	θ(deg)	I_A	I_B
0	0є	1	0
1	18є	0	1
2	36є	–1	0
3	54є	0	–1
4	72є	1	0
5	90є	0	1

Half step operation

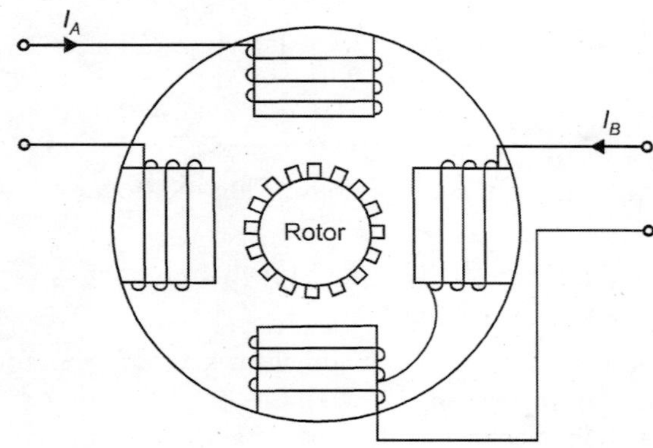

Fig. 3.44

Step	θ(deg)	I_A	I_B
0	0	1	0
1	9	.707	.707
2	18	0	1
3	27	–.707	.707
4	36	–1	0
5	45	–.707	–.707
6	54	0	–1
7	63	.707	.707
8	72	1	0

Half steps are accomplished by applying partial current to both phase windings to position the rotor halfway between two full step position.

Micro stepping

Micro stepping simply extends the half stepping technique to more than one mid position by using different values of current in each phase. The microstep sizes that are most commonly used are 1/10, 1/16, 1/32 and 1/125 of a full step. It provides finer resolution.

The value of current which is used to provide micro stepping can be written as,

$$I_A = \cos\frac{(90n)}{s} I_m$$

$$I_B = \sin\frac{(90n)}{s} I_m$$

where, $I_m \rightarrow$ Maximum value of phase current
$I_A \rightarrow$ Current in phase A
$I_B \rightarrow$ Current in phase B
$n \rightarrow n^{\text{th}}$ microstep from start position
$s \rightarrow$ Number of microsteps in a full step

3.7 MAGNETIC AMPLIFIER

A magnetic amplifier is a device which is used for controlling the flow of power to a load by means of saturating the magnetic core.

The magnetic material which are specially suited for this purpose is rectangular hysterisits loop with a sharp saturation and low hysterisis loss.

The construction is same as transformer. The main part is saturable reactor. The saturable reactor is consist of saturable core carrying minimum of two windings one of this windings is fed with ac and this is called the gate winding. The excitation ia adjusted such that flux swing in the core lies between knees of the B-H curve. When DC is applied to the second winding which is called the control winding. The core will saturate during half cycle due to additional magnetisation.

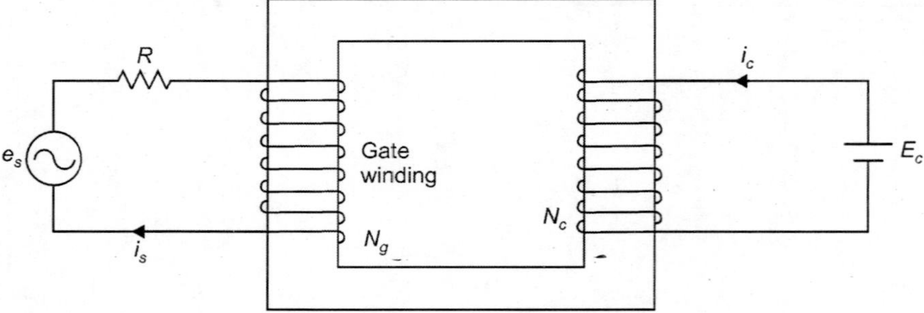

Fig. 3.45: Magnetic amplifier.

when no dc is applied to control winding, the core flux variation is within saturation limit. Under this condition the flux swing is very large and so gate winding offers large inductance and current drawn from ac source is only the magnetising current.

Let, dc supply is given to control winding then current i_c is passed through control winning, then the operating point is suddenly shifted. Under this condition, the change in flux is very small compared to the case when i_c is zero. Hence the inductance of gate winding reduces to a very small value and now entire supply voltage appears across the load resistance.

With higher dc control current the core saturates earlier in the positive half cycle, thus delivering more current to the load. By varying the control current the load current can be controlled. Thus with an expenses of very small power in the control side, large power in the ac side is controlled in the saturable reactor.

The main drawback is that the output is delivered during one half only, in the other half cycle, the cone being unsaturated. To overcome these difficulties, a windings are so connected that no ac voltage is induced in the control windings. The gate windings can be connected either in parallel (or) in series.

According to this, the magnetic amplifiers are classified as

1. Series magnetic amplifier.
2. Parallel magnetic amplifier.

3.8 MODULATORS AND DEMODULATORS

(i) Modulators

The device which converts dc signal to ac signal is called a modulator. The modulator consists of a chopper, input transformer, a vibrating contact and a vibrating coil. Inside the chopper the reed is vibrated by means of the vibrating coil at carrier (or) line frequency. The dc signal is applied across the terminals. When the reed is in the up position current flows in the upper half of the centre tapped input transformer primary winding. In the next half cycle the reed is in the down position and current flows in the down position and current flows in the lower half of the transformer primary winding. The input transformer's secondary winding first senses a flow of current up the primary winding, then a current down the primary windings. This action produces ac voltage across the secondary winding.

(ii) Demodulators

This is the reverse process of modulators. The device which converts ac signal to a dc signal is called demodulators same modulator can be used as Demodulator.

The Fig. (3.46) can be used as both modulator and demodulator.

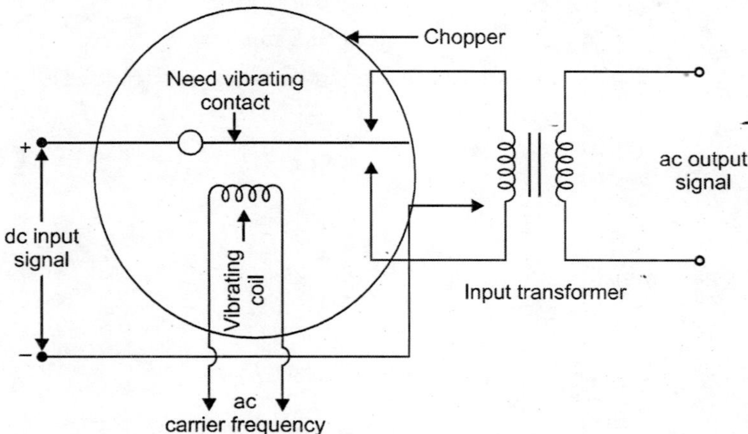

Fig. 3.46: Modulator and demodulator.

3.9 TACHO GENERATOR

Definition

It is an electromechanical device which produces an output voltage proportional to its shaft speed. It can be employed as an analogue speed indicator, velocity feedback device or signal integrator. There are two types.

1. DC tacho generator
2. AC tacho generator

1. **DC tacho generator:** A dc techno generator is a small dc generator with linear characteristics. It consist of stator with a permanent magnet field, a rotating armature circuit and a commutator with brush assembly. The rotor is a connected to ths shaft whose speed has to be measured.

   The output voltage of the generator is proportional to the angular velocity of the shaft. The polarity of the output voltage is dependent on the rotation of the shaft. The schematic diagram of DC tachogenerator is shown in Fig. (3.47)

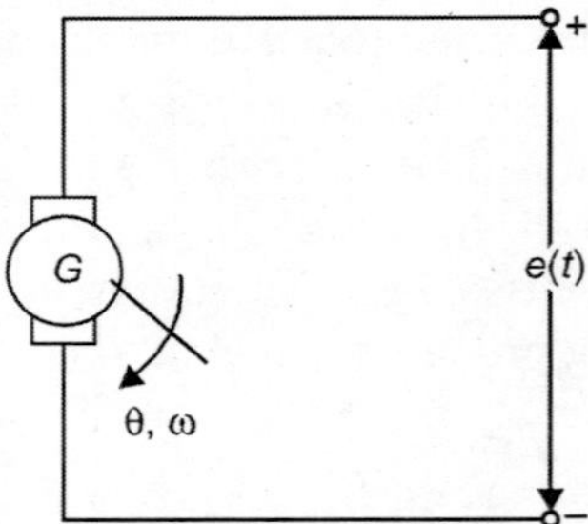

Fig. 3.47: Schematic diagram of DC tacho generator.

Let

$$e(t) = \text{Output voltage}$$

$$\theta = \text{Angular displacement, rad}$$

$$\omega = d\theta/dt = \text{Angular speed, rad/sec}$$

$$K_t = \text{Sensitivity of tachogenerator volts/(rad/sec)}$$

$$e(t) \propto \frac{d\theta}{dt}$$

$$\therefore \quad e(t) = K_t \frac{d\theta}{dt}$$

on taking Laplace transform

$$E(s) = K_E\, s\, \theta(s)$$

$$\boxed{\therefore \frac{E(s)}{\theta(s)} = s\, K_E} \qquad (3.65)$$

The equation (3.65) is the transfer function of dc tachogenerator.

2. **AC tachogenerator:** The ac tachogenerator resembles a two phase induction motor. It consist of two stator windings displaced by 90є. It has squirrel cage rotor. One stator winding is called as control winding in which an emf is induced when the rotor rotates by induction principle and the other winding is called reference winding and it is excitred by a sinusoidal voltage of frequency, ω_r. when the rotor is stationary, no emf is induced in the control winding (or) output winding and the therefore the output voltage is zero. When the rotor rotates, a voltage at the reference frequency is induced in the control winding (or) output winding. The magnitude of the output voltage is directly proportional to the rotational speed.

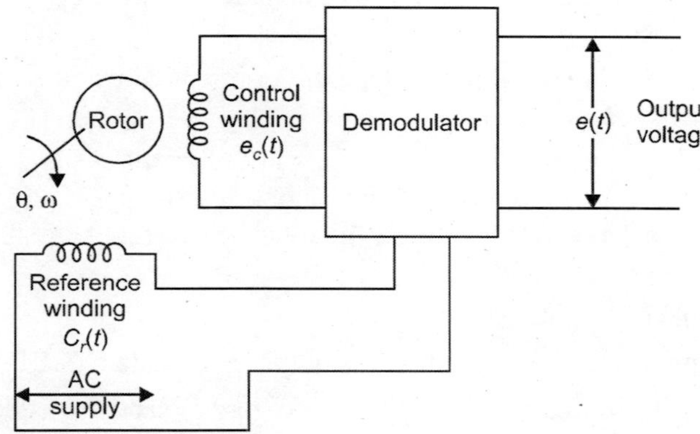

Fig. 3.48: Schematic diagram of AC tacho generator.

Output voltage,

$$e(t) \propto \frac{d\theta}{dt}$$

$$\therefore \quad e(t) = K_t \frac{d\theta}{dt}$$

$$\therefore \quad \frac{E(s)}{\theta(s)} = s\,K_t \tag{3.66}$$

The equation (3.66) is the transfer function of AC tachogenerator.

3.10 CONCEPT OF STABILITY

The term stability refers to the stable working condition of a control system. Every working system is designed to be stable. In a stable system the response (or) output is predictable, finite and stable for a given input (or) changes in input (or) changes in system parameters.

The different definitions of the stability are the following

1. A system is stable if its output is bounded (finite) for any bounded (finite) input.
2. A system is asymptotically stable if in the absence of input, the output tends towards zero (or to the equilibrium state) irrespective of initial conditions.

3. A system is stable if for a bounded disturbing input signal the output vanishes ultimately as t approaches infinity.
4. A system is unstable if for a bounded disturbing input signal the output is of infinite amplitude (or) oscillatory.
5. For a bounded input signal, if the output has constant amplitude oscillations then the system may be stable (or) unstable under some limited constraints such a system is called limitedly stable.
6. If a system output is stable for all variations of its parameters then the system is called absolutely stable system.
7. If a system output is stable for a limited range of variations of its parameters, then the system is called conditionally stable system.

The concept of stability is very important to analyse and design the system. A system is said to be stable if its response cannot be made to increase indefinitely by the application of a bounded input excitation. If the output approaches towards infinite value for sufficiently large time, the system is said to be unstable.

A Linear Time Invariant (LTI) system is stable if

1. The system is excited by a bounded input, the output is bounded. (BIBO stability criteria)
2. In the absence of the input, the output tends towards zero (the equilibrium state of the system)

This is known as asymptotic stable.

Consider the transfer function.

$$\frac{C(s)}{R(s)} = \frac{a_m s^m + a_{m-1} s^{m-1} + \ldots a_1 s + a_0}{b_n s^n + b_{n-1} s^{n-1} + \ldots b_1 s + b_0} \tag{3.67}$$

The output is given by

$$C(t) = \int_0^\infty g(\tau)\, r(t-\tau)\, d\tau$$

where $g(\tau) = L^{-1}\, G(s)$ = Impulse response of the system. So, a system is said to be stable if the impulse response approaches zero for sufficiently large time. If the impulse response approaches infinity for sufficiently large time, the system is said to be unstable. If the impulse response approaches a constant value for sufficiently large time, the system is said to be marginally stable.

3.11 RESPONSE OF A SYSTEM

Let the closed loop transfer function (or overall transfer function)

$$\frac{C(s)}{R(s)} = M(s)$$

∴ The response (or) output in s-domain.

$$C(s) = M(s)\ R(s)$$

Let,

$$C(t) = L^{-1}\ [C(s)] \qquad \text{(Response in time domain)}$$

$$r(t) = L^{-1}\ [R(s)] \qquad \text{(Input in time domain)}$$

For an impulse input,

$$r(t) = \delta(t);$$

$$R(s) = L[\delta(t)] = 1$$

The impulse response of the system

$$L^{-1}\ [C(s)] = L^{-1}\ [M(S)\ R(s)]$$

$$= L^{-1}\ M(S) = m(t) \qquad (3.68)$$

Hence the impulse response of a system is the inverse Laplace transform of the system transfer function. The advantage of impulse response is that the knowledge of impulse response can be used to determine the output of a system for any arbitrary input.

The output is s-domain is the product of transfer function and input is s-domain. The convolution theorem states that the inverse Laplace transform of the product of two s-domain functions is equal to the convolution of their time domain functions.

i.e.,
$$L^{-1}\ [M(s)\mathrm{R}(s)] = \int_{-\infty}^{+\infty} m(\tau) r(t-\tau) d\tau$$

∴ The response of a system in time domain

$$C(t) = L^{-1}\ [M(s)\ R(s)]$$

$$= \int_{-\infty}^{+\infty} m(\tau)\, r(t-\tau)\, d\tau \qquad (3.69)$$

From the equation (3.2.2) it can be concluded that the response for any arbitrary input is given by convolution of impulse response and the input.

3.12 BOUNDED-INPUT BOUNDED OUTPUT STABILITY (BIBO)

A linear relaxed system is said to have BIBO stability if every bounded (finite) input results in a bounded (finite) output. A test for BIBO stability can be obtained from convolution theorem.

The response of a system $C(t)$ for any input is given by convolution of the input and impulse response.

The response in time domain,

$$C(t) = \int_0^{\infty} m(\tau)\, r(t-\tau)\, d\tau \qquad (3.70)$$

Note: A relaxed system is one in which the initial conditions are zero. Hence the limits of integration is from 0 to ∞.

If the input $r(t)$ is bounded then there exist a constant A, such that $|r(t)| \le A_1 < \infty$. The condition for bounded output for this bounded input condition can be derived as follows.

On taking the absolute value on both sides of equation (3.70), we get

$$|C(t)| = \left| \int_0^{\infty} m(\tau) \cdot r(t-\tau)\, d\tau \right| \qquad (3.71)$$

Since the absolute value of an integral is not greater then the integral of the absolute value of the integrand the equation (3.71) can be written as,

$$|C(t)| \le \int_0^{\infty} |\, m(\tau)\, r(t-\tau)\, d\tau \,|$$

$$\le \int_0^{\infty} |m(\tau)\, r(t-\tau)\, d\tau$$

$$\le \int_0^{\infty} |m(\tau)\, A_1\, d\tau \qquad [\because \text{For bounded input } |r\,(t-\tau) < A_1\,]$$

$$\le A_1 \int_0^{\infty} |\, m(\tau)\,|\, d\tau \qquad (3.72)$$

It the output $C(t)$ is bounded then there exists a constant A_Q

Such that $\qquad |\, C(t)\,| \le A_2 < \infty$

$$\therefore \qquad A_1 \int_0^{\infty} |\, m(\tau)\,|\, d\tau \le A_2 < \infty \qquad (3.73)$$

The above condition is satisfied if

$$\int_0^{\infty} |\, m(\tau)\,|\, d\tau < \infty$$

Hence the bounded output,

$$\int_0^{\infty} |\, m(\tau)\,|\, d\tau < \infty$$

therefore we can conclude that a system with impulse response $m(\tau)$ is BIBO stable if and only if the impulse response is absolutely integrable i.e.,

$$\int_0^\infty |m(\tau)|\, d\tau \text{ is finite}$$

This means that area under the absolute value curve of the impulse response $m(\tau)$ evaluated from $t = 0$ to $t = \infty$ must be finite.

3.13 EFFECT OF LOCATION OF POLES ON STABILITY

The closed loop transfer function, $M(s)$ can be expressed as a ratio of two polynomials in s. The denominator polynomial of closed loop transfer function is called characteristic equation. The roots of characteristic equation are poles of closed loop transfer function.

For BIBO stability the integral of impulse response should be finite. Which implies that the impulse response should be finite as t tends to infinity. (The impulse response is the inverse Laplace transform of the transfer function). This requirement for stability can be linked to the location of roots of characteristic equation in the s-plane.

The closed loop transfer function $M(s)$ can be expressed as ratio of two polynomials in s as shown in equation (3.4.1)

$$M(s) = \frac{b_0 s^m + b_1 s^{m-1} + b_2 s^{m-2} + \ldots + b_{m-1} s + b_m}{a_0 s^n + a_1 s^{n-1} + a_2 s^{n-2} + \ldots a_{n-1} s + a_n} \tag{3.74}$$

$$= \frac{(s + Z_1)(s + Z_2)(s + Z_3) + \ldots (s + Z_m)}{(s + P_1)(s + P_2)(s + P_3) + \ldots (s + P_n)} \tag{3.75}$$

The roots of numerator polynomial $Z_1, Z_2 \ldots Z_n$ are zeros. The roots of denominator polynomial P_1, P_2, ... P_n are poles. The denominator polynomial is the characteristic equation and so the poles are roots of characteristic equation.

By partial fraction expansion we can write,

$$M(s) = \frac{A_1}{s + P_1} + \frac{A_2}{s + P_2} + \frac{A_3}{s + P_3} + \ldots \frac{A_n}{s + P_n} \tag{3.76}$$

The poles, $P_1, P_2, P_3 \ldots P_m$ may be at origin or lying on imaginary axis or lying on right or left half s-plane. The impulse response is given by inverse Laplace transform of $M(s)$. The inverse Laplace transform of each term of $M(s)$ depends on the location of poles. The impulse response of various terms of $M(s)$ are shown in Table (3.4.1).

From Table (3.4.1) the following conclusions are drawn based on the location of roots of characteristic equation.

1. If all the roots of characteristic equation have negative real parts (i.e. lying on left half s-plane)

then the impulse response is bounded (i.e., it decreases to zero as t tends to ∞). Hence $\int_0^\infty m(\tau)\,d\tau$ is finite and the system is bounded-input bounded output stable.

2. If any roots of the characteristic equation has a positive real part (i.e., lying on left half s-plane) then impulse response is unbounded (i.e. increases to ∞ as t tends to ∞). Hence $\int_0^\infty m(\tau)\,d\tau$ is infinite and so system is unstable.

Table 3.4.1

Transfer function and location of roots on s–plane	Impulse response (inverse Laplace transform and) the sketch of time domain function)
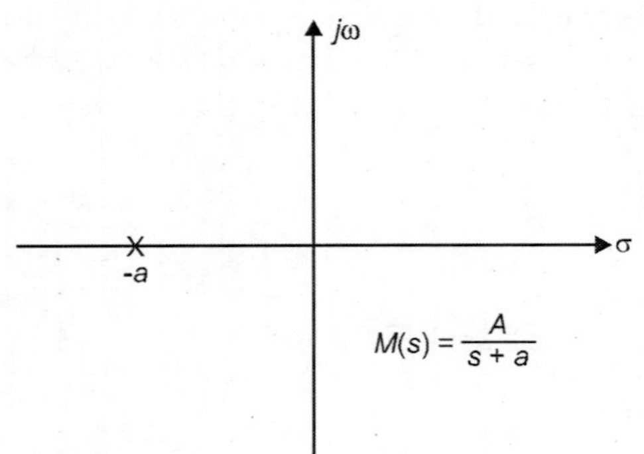 Root on negative real axis	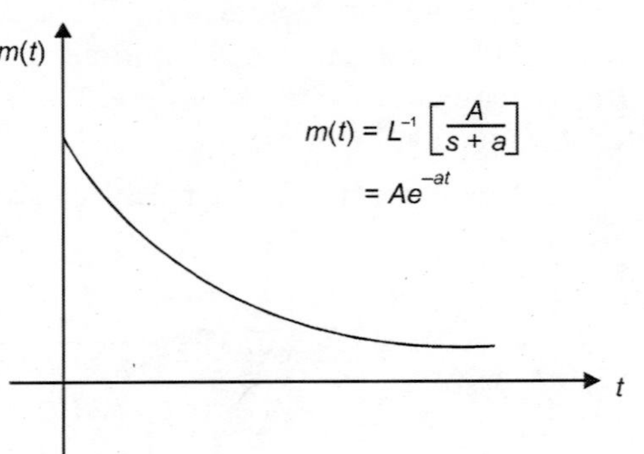 Response is exponentially decaying
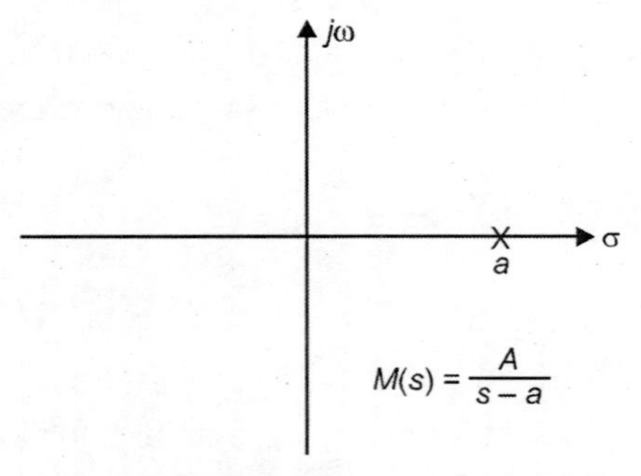 Root on positive real axis	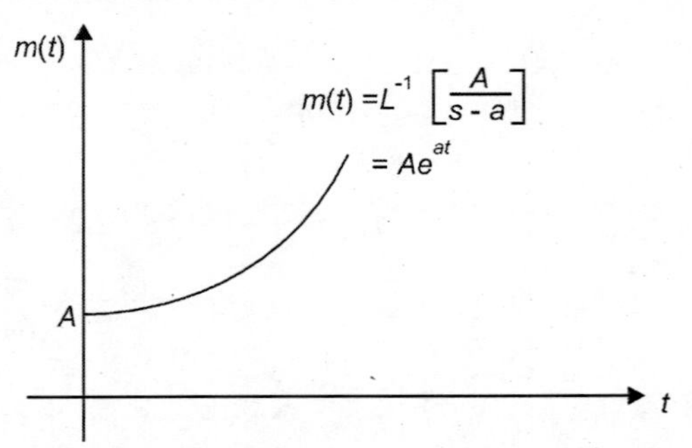 Response is exponentially increasing
$M(s)\dfrac{A}{s+a+jb}+\dfrac{A}{s+a-jb}$	$m(t)-L^{-1}\left[\dfrac{A}{s+a+jb}+\dfrac{A}{s+a-jb}\right]$ $= Ae^{-(a+jb)t}+A^{-(a-jb)t}$ $= 2Ae^{-at}\cos bt = 2\,Ae^{-at}\sin(bt+90°)$

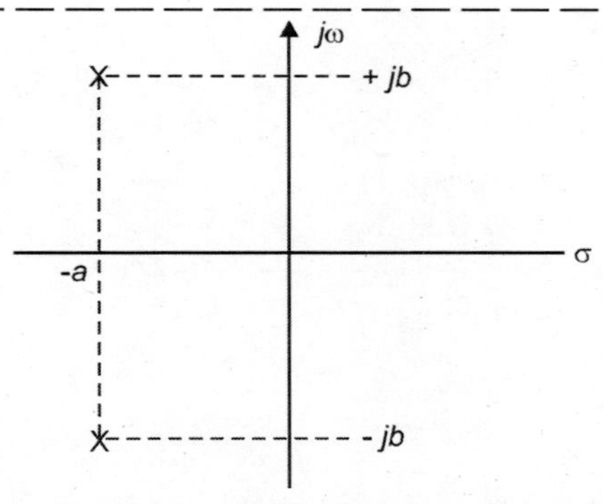 Complex conjugate roots on left half S-plane	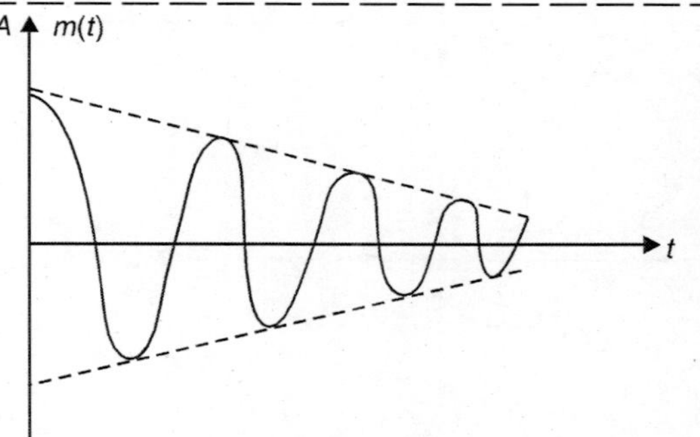 The response is damped sinusoidal (i.e. damped oscillatory)
$M(s) = \frac{A}{s-a+jb} + \frac{A}{s-a-jb}$ 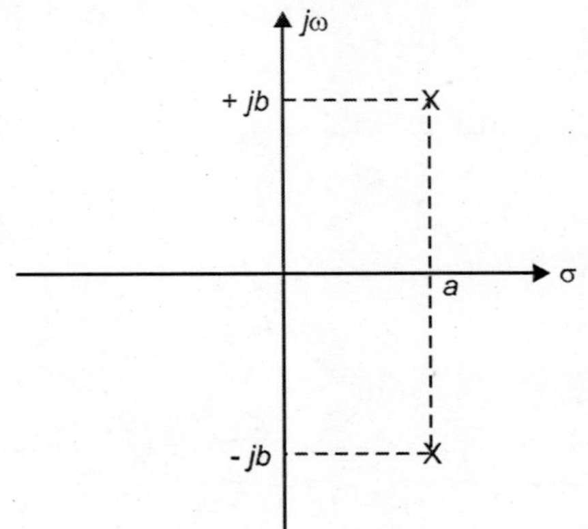 Complex conjugate roots on right half s-plane	$m(t) = L^{-1}\left[\frac{A}{s-a+jb} + \frac{A}{s-a-jb}\right]$ $= Ae^{-(a+jb)t} + Ae^{-(a-jb)t}$ $= 2\,Ae^{at}\cos bt$ $= 2\,Ae^{at}\sin(bt + 90°)$ 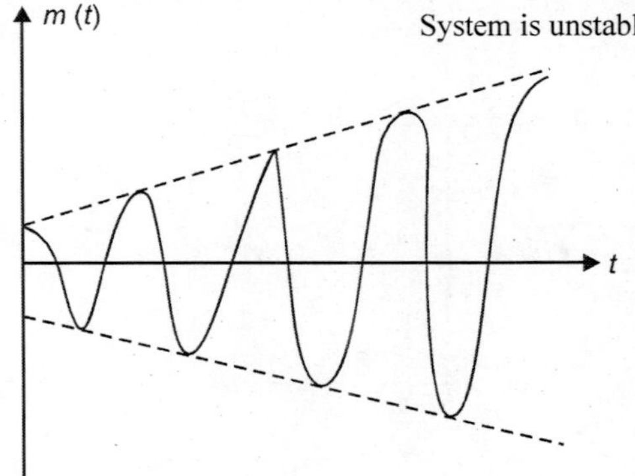 The response is exponentially increasing sinusoidal (i.e. amplitude of oscillations exponentially increases with time)
$M(s) = \frac{A}{s+jb} + \frac{A}{s-jb}$	$m(t) = L^{-1}\left[\frac{A}{s+jb} + \frac{A}{(s-jb)}\right]$ $= Ae^{-jb} + Ae^{+jb}$ $= 2A\cos bt$ $= 2A\sin(bt + 90°)$

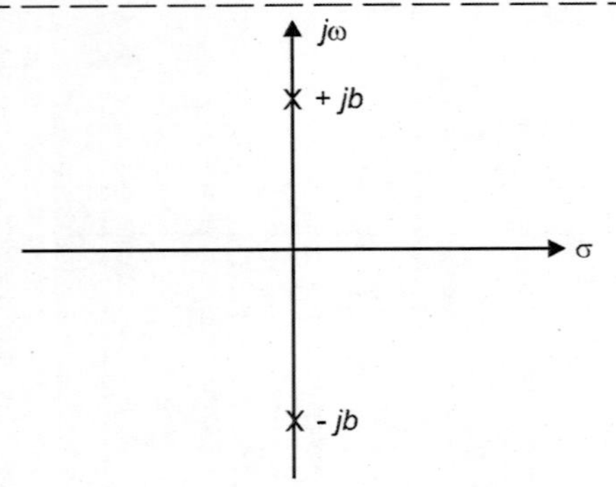

Single pair of roots on imaginary axis.

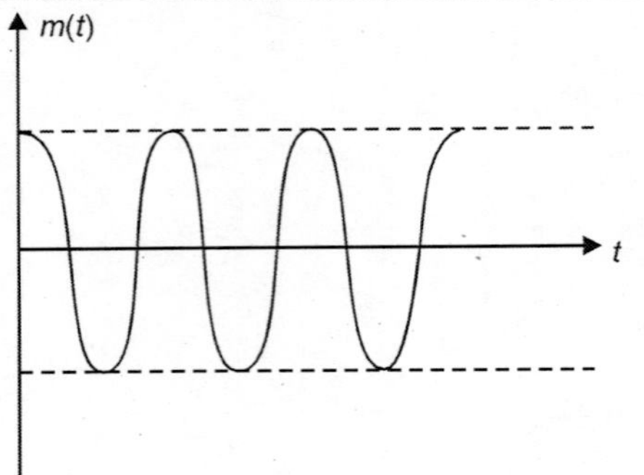

The response is oscillatory

$$M(s) = \frac{A}{(s+jb)^2} + \frac{A}{(s-jb)^2}$$

$$M(t) = L^{-1}\left[\frac{A}{(s+jb)^2} + \frac{A}{(s-jb)^2}\right]$$

$$= At\,e^{-jb} + At\,e^{jb}$$

$$= 2\,At\cos bt = 2At\sin(bt + 90°)$$

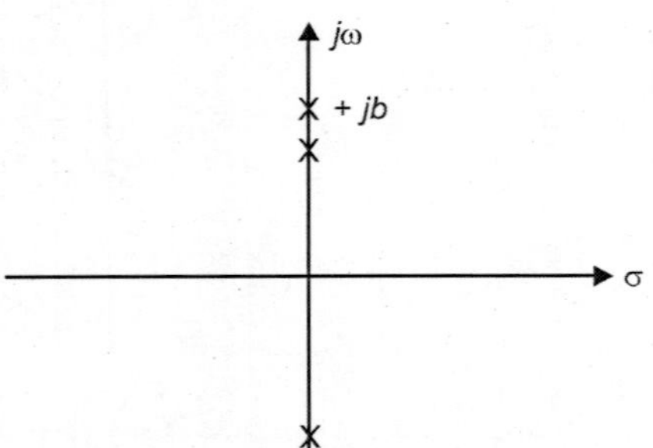

Double pair of roots on imaginary axis

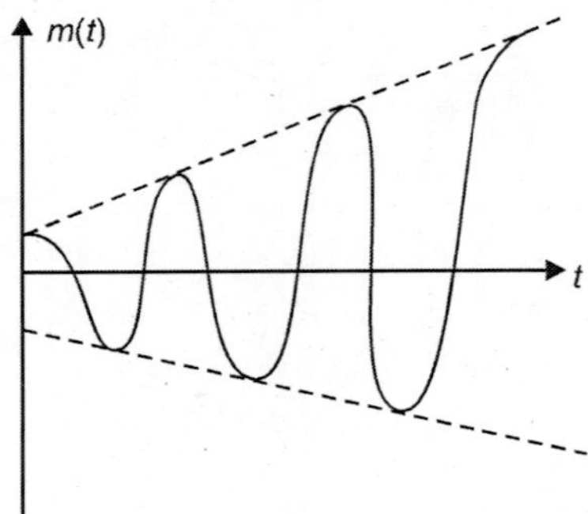

Response is Linearly increasing sinusoidal (i.e. amplitude of oscillations linearly increases with time)

$$M(s) = \frac{A}{s}$$

$$m(t) = L^{-1}[A/s] = A$$

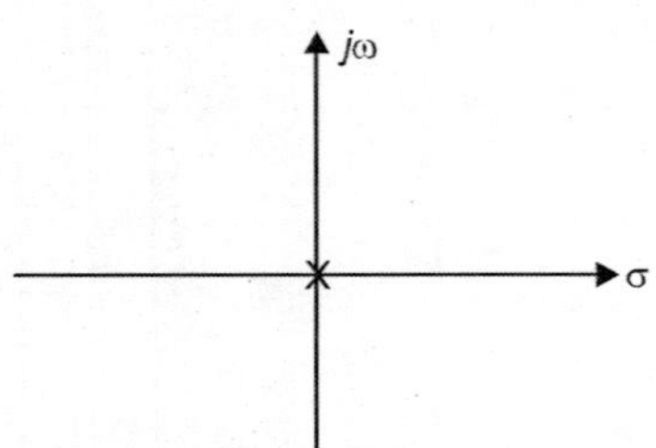

Single root at origin

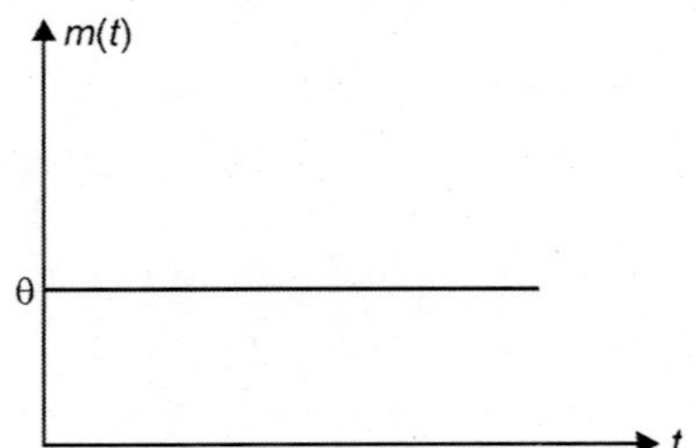

The response is constant

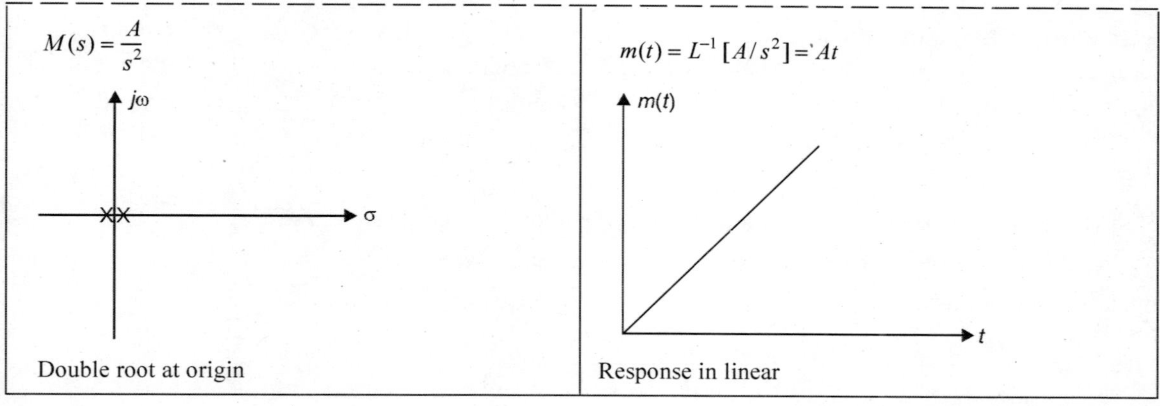

Double root at origin

Response in linear

3. If the characteristic equation has repeated roots on the imaginary axis then impulse response is unbounded (i.e., it increases to ∞ as t tends to ∞). Hence $\int_0^\infty m(\tau)\, d\tau$ is finite and so system is unstable.
4. If one (or) more non repeated roots of the characteristic equation are lying on the imaginary axis, then impulse response is bounded (i.e., it has constant amplitude oscillations) but $\int_0^\infty m(\tau)\, d\tau$ is infinite aid so the system is unstable.
5. If the characteristic equation has single root at origin then the impulse response is bounded (i.e., it has constant amplitude) but $\int_0^\infty m(\tau)\, d\tau$ is infinite and so the system is unstable.
6. If the characteristic equation has repeated roots at origin then the impulse response is unbounded (i.e. it linearly increases to infinity as t tends to ∞) and so the system is unstable.
7. In system with one (or) more repeated roots on imaginary axis (or) with single root at origin, the output is bounded for bounded inputs except for the inputs having poles matching the system poles. These cases may be treated as acceptable (or) non-acceptable. Hence when the system has non repeated poles on imaginary axis (or) single pole at origin, it is referred as limitedly (or) marginally stable system.

In summary the following three points may be stated regarding the stability of the system depending on the location of roots of characteristic equation.

(i) If all the roots of characteristic equation has negative real parts, then the system is stable.

(ii) If any roots of the characteristic equation has a positive real part (or) if there is a repeated root on the imaginary axis then the system is unstable.

(iii) If the condition (i) is satisfied except for the presence of one (or) more non repeated roots on the imaginary axis, then the system is limitedly (or) marginally stable.

In order to ascertain the stability of a system, it is necessary to determine if any of the roots of the characteristic equation lie in the right half s-plane. The characteristic equation is given by the denominator polynomial of closed loop transfer function, [Eq. (3.75)]

Consider the nth order characteristic equation shown below.

$$a_0 s^n + a_1 s^{n-1} + a_2 s^{n-2} + \dots a_{n-1} s + a_n = 0 \quad (3.77)$$

Let the roots of nth order characteristic equation [equation (3.4.3)] be $s = r, r_2 \dots r_n$

These roots are functions of the coefficients $a_0, a_1, a_2 \dots a_{n-1}, a_n$

Consider a second order polynomial,

$$a_0 s^2 + a_1 s + a_2 = a_0 \left(s^2 \frac{a_1}{a_0} s + \frac{a_2}{a_0} \right)$$

$$= a_0 (s - r_1)(s - r_2)$$

$$= a_0 s^2 - a_0 (r_1 + r_2) s + a_0 r_1 r_2 \quad (3.78)$$

Consider a third order polynomial

$$a_0 s^3 + a_1 s^2 + a_2 s + a_3 = a_0 \left[s^3 + \frac{a_1}{a_0} s^2 + \frac{a_2}{a_0} s + \frac{a_3}{a_0} \right]$$

$$= a_0 (s - r_1)(s - r_2)(s - r_3)$$

$$= a_0 s^3 - a_0 (r_1 + r_2 + r_3) s^2 + a_0 (r_1 r_2 + r_1 r_3 + r_2 r_3) s - a_0 r_1 r_2 r_3 \quad (3.79)$$

on extending this expansion to the nth order polynomial, we get,

$$a_0 s^n + a_1 s^{n-1} + \dots a_{n-1} s + a_n = a_0 s^n - a_0 \text{ (sum of all the roots) } s^{n-1} +$$
$$a_0 \text{ (sum of the products of the roots taken 2 at a time) } s^{n-2} -$$
$$a_0 \text{ (sum of the products of the roots taken 3 at a time) } s^{n-3} + \dots +$$
$$a_0 (-1)^n \text{ (product of all the } n \text{ roots)} \quad (3.80)$$

If the roots of a polynomial are real and in the left half s-plane, then all r_1 in equations (3.78) and (3.79) are real and negative. Therefore all polynomial coefficients are positive. This characteristic also applies to the general case of equation (3.80). If at least one root is in the right half of s-plane then some of the coefficients will be negative. Also, it can be observed that if all the roots are in the left half plane, no coefficient can be zero.

Since the characteristic polynomial coefficients are real, the complex roots should occur as conjugate pairs.

From equation (3.80) when polynomial coefficients are formed, the imaginary parts of roots products of roots will cancel.

Therefore, if all roots occur in the left half plane, (whether it is complex (or) real) then all coefficients of the general polynomial of equation (3.80) will be positive. Presence of a negative coefficient implies that there is atleast one root in the right half plane.

A zero coefficient indicates presence of complex–conjugate roots on the imaginary axis and /or one or more roots in the right half plane.

In summary the following conclusions can made about the coefficients of characteristic polynomial.

1. If all the coefficients are positive and if no coefficient is zero then all the roots are in the left half plane.
2. If any coefficient a_i is equal to zero then some of the roots may be one the imaginary axis (or) on the right half plane.
3. If any coefficient a_i is negative then atleast one root is in the right half plane.

It can be concluded that the absence (or) negativeness of any of the coefficients of a characteristic polynomial indicates that the system is either unstable (or) at most marginally stable. Thus the necessary condition for stability of the system is that all the coefficients of its characteristic polynomial be positive. If any coefficient is zero/negative, we can immediately say that the system is unstable.

When all the coefficients are positive, the system is not necessarily stable, because there may be roots in the right half plane and/or on the imaginary axis. In order for all the roots to have negative real parts, it is necessary but not sufficient that all of the coefficients of characteristic equation be positive. [For example, consider the characteristic polynomial with all positive coefficients $s^3 + s^2 + 2s + 8 = 0$. The characteristic polynomial can be written as

$$(s^3 + s^2 + 2s + 8) = (s + 2)\left(s - \frac{1}{2} - j\frac{\sqrt{15}}{2}\right)\left(s - \frac{1}{2} + j\frac{\sqrt{15}}{2}\right) = 0$$

The coefficients of the polynomial are all positive but two roots are on right half plane and so the system is unstable.

3.14 ROUTH HURWITZ CRITERION

The closed loop transfer function of a system is a ratio of two polynomials in s. The denominator polynomial of closed loop transfer function is called the characteristic equation of the system.

The roots of the characteristic equation of a stable system should lie on the left of s-plane. Hence the roots should have negative real parts. The Routh–Hurwitz stability criterion is an analytical procedure for determining whether all the roots of a polynomial have negative real parts (or) not.

The first step in analysing the stability of a system is to examine its characteristic equation. The necessary condition for stability is that all the coefficients of the polynomial be positive. If some of the coefficients are zero (or) negative it can be concluded that the system is not stable.

When all the coefficients are positive, the system is not necessarily stable Even though the coefficients are positive, some of the roots may lie on the right half of s-plane (or) on the imaginary axis. In order for all the roots to have negative real parts, it is necessary but not sufficient that all of the coeffcients of the characteristic equation be positive. If all the coeffcients of the characteristic equation are positive, then the system may be stable and one should proceed further no examine the sufficient conditions of stability.

The Hurwitz criterion is in terms of determinants and Routh criterion is in terms of array formulation.

The Routh stability criterion is based on ordering the coefficients of the characteristic equation,

$$a_0 s^n + a_1 s^{n-1} + a_2 s^{n-2} + \ldots + a_{n-1} s + a_n = 0$$

where $a_0 > 0$ Routh array as shown in below.

$$
\begin{array}{llllll}
s^n & : & a_0 & a_2 & a_4 & a_6 & a_8 & \cdots \\
s^{n-1} & : & a_1 & a_3 & a_5 & as_7 & a_9 & \cdots \\
s^{n-2} & : & b_0 & b_1 & b_2 & b_3 & b_4 & \cdots \\
s^{n-1} & : & \cdots\cdots & & & & & \\
\vdots & & \vdots & \vdots & \vdots & \vdots & \vdots & \\
s^1 & : & . & . & . & . & . & \\
s^0 & : & . & . & . & . & . &
\end{array}
$$

Definiton of routh stability criterion **"The necessary and sufficient condition for stability is that all of the elements in the first column of the routh array be positive. If this condition is not met, the system is unstable and the number of sign changes in the elements of the first column of the Rough array corresponds to the number of roots of the characteristic equation in the Right half of the *s*-plane"**.

Note: If the order of sign column elements is +, +, –, + and +. Then + to – is considered as one sign change and – to + as another sign change.

3.14.1 Construction of Routh Array

Let the characteristic polynomial be

$$a_0 s^n + a_1 s^{n-1} + a_2 s^{n-2} + \ldots a_{n-1} s^1 + a_n s^0$$

The coefficients of the polynomial are in two rows as shown below

$$
\begin{array}{l}
s^n : a_0 \;\; a_2 \;\; a_4 \;\; a_6 \ldots \\
s^{n-1} : a_1 \;\; a_3 \;\; a_5 \;\; a_7 \ldots
\end{array}
$$

If n is even then s^n row is formed by coefficients of even order terms and s^{n-1} row is formed by coefficients of odd order terms (i.e. coefficients of odd power of s).

If n is odd, then s^n row is formed by coefficients of even order terms (i.e., coefficients of even powers of s).

The other rows of routh array upto s^0 row can be formed by the following procedure. Each row of routh array is constructed by using the elements of previous two rows.

Consider two consecutive rows of routh array as shown below.

$$
\begin{array}{l}
s^{n-x} : x_0 \;\; x_1 \;\; x_2 \;\; x_3 \;\; x_4 \ldots \\
s^{n-x-1} : y_0 \;\; y_1 \;\; y_2 \;\; y_3 \;\; y_4 \ldots
\end{array}
$$

Let the next row be, $s^{n-x-2} : z_0 \;\; z_1 \;\; z_2 \;\; z_3 \ldots$

The elements of s^{n-x-2} row are given by,

$$z_0 = (-1)\begin{vmatrix} x_0 & x_1 \\ y_0 & y_1 \end{vmatrix} = \frac{y_0 x_1 - y_1 x_0}{y_0}$$

$$z_1 = (-1)\begin{vmatrix} x_0 & x_2 \\ y_0 & y_2 \end{vmatrix} = \frac{y_0 x_2 - x_0 y_2}{y_0}$$

$$z_2 = (-1)\begin{vmatrix} x_0 & x_3 \\ y_0 & y_3 \end{vmatrix} = \frac{y_0 x_3 - y_3 x_0}{y_0}$$

The elements z_0, z_1, z_2, z_3 ... computed until an element equals to zero for all possible computations as shown above.

In the process of constructing Routh array the missing terms are considered as zeros. Also, all the elements of any row can be multiplied or divided by a positive constant for simplification.

For constructing Routh array the following three cases are important.

Case I: Normal Routh array (non-zero elements in the first column of routh array)

Case II: A row of all zeros

Case III: First element of a row is zero but some or other elements are not zero but some or other elements are not zero.

Case I: Normal routh array

In this case there is no difficulty in forming Routh array. The routh array can be constructed as explained above. The sign changes are not noted to find the number of roots lying on the right half of s-plane and the stability of the system can be estimated.

In this case if there is no sign change in the first column of routh array then all the roots are lying on left half of s-plane and the system is stable.

If there is sign change in the first column of routh array, then the system is unstable and the number of roots lying on the right half of *s*-plane is equal to number of sign changes the remaining roots are lying on the left half of *s*-plane.

Case II: A row of all zeros

An all zero row indicates the existence of an even polynomial as a factor of the given characteristic equation.

In an even polynomial the exponents of s are even integers (or) zero only. This even polynomial factor is called the auxiliary polynomial. The coefficients of the auxiliary polynomial will always be the elements of the row directly above the row of zeros in the array.

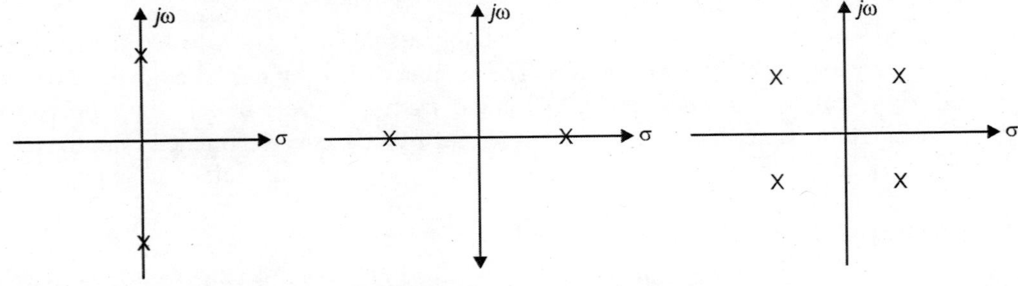

Fig. 3.49: The roots of an even polynomial.

Method 1

1. Determine the auxiliary polynomial, $A(s)$
2. Differentiate the auxiliary polynomial with respect to s, get d $A(s)$ ds.
3. The row of zeros is replaced with coefficients of $dA(s)/ds$.
4. Continue the construction of the array in the usual manner (as that of case–I) and the array is interpreted as follows.
   (a) If there are sign changes in the first column of routh array then the system is unstable. The number of roots lying on right half of s-plane is equal to number of sign charges. The number of roots on imaginary axis can be estimated from the roots of auxiliary polynomial. The remaining roots are lying on the left half of s-plane.
   (b) If there are sign changes in the first column of routh array then the all zeros row indicate the existence of purely imaginary roots and so the system is limitedly or marginally stable. The roots of auxiliary equation lies on imaginary axis and the remaining roots lies on left half of s-plane.

Method 2:

1. Determine the auxiliary polynomial, $A(s)$.
2. Divide the characteristic equation by auxiliary polynomial.
3. Construct routh array using the coefficients of quotient polynomial.
4. The array is interpreted as follows.
   (a) If there are sign changes in the first column of routh array of quotient polynomial then the system is unstable. The number of roots of quotient polynomial lying on right half of s-plane is given by number of sign changes in first column of routh array

   The roots of auxiliary polynomial are directly calculated to find whether they are purely imaginary or purely real or complex.

   The total number of roots on right half of s-plane is given by the sum of number of sign changes and the number of roots of auxiliary polynomial with positive real part. The number of roots on imaginary axis can be estimated from the roots of auxiliary polynomial. The remaining roots are lying on the left half of s-plane.

   (b) If there is no sign change in the first column of routh array of quotient polynomial then the system is limitedly (or) marginally stable. Since there is no sign change all the roots of quotient polynomial are lying on the left half of s-plane.

   The roots of auxiliary polynomial are directly calculated to find whether they are purely imaginary (or) purely real or complex. The number of roots lying on imaginary axis and on the right half of s-plane can be estimated from the roots of auxiliary polynomial. The remaining roots are lying on the left half of s-plane can be estimated from the roots of auxiliary polynomial. The remaining roots are lying on the left half of s-plane.

Case III: First element of a row is zero

While constructing routh array, if a zero is encountered as first element of a row then all the elements of the next row will be infinite. To overcome this problem let 0 – ∈ and complete the construction of array in the usual way (as that of case)

Finally let $\in \to 0$ and determine the values of the elements of the array which are functions of ∈. The resultant array is interpreted as follows.

Note: If all the elements of a row are zeros then the solution is attempted by considering the polynomial as case II polynomial. Even if there is a single element zero one's row. It is considered as a row of all zeros.

(a) If there is no sign change in first column of routh array and if there is no row with all zeros, then all the roots are lying on left half of *s*-plane and the system is stable.

(b) if there are sign changes in first column of routh array and there is no row with all zeros, then some of the roots are lying on the right half of s-plane is equal to number of sign changes and the remaining roots are lying on the left half of s-plane.

(c) If there is a row of all zeros after leting $\in \to 0$, then there is a possibility of roots on imaginary axis. Determine the auxiliary polynomial and divide the characteristic equation by auxiliary polynomial to eliminate the imaginary roots. The routh array is constructed using the coefficients of quotient polynomial and divide the characteristic equation by auxiliary polynomial to eliminate the imaginary roots. The routh array is constructed using the coefficients of quotient polynomial and the characteristic equation is interpreted as explained in method-2 of Case II: polynomial

Statement of Routh–Hurwitz Criterion: The system is stable if and only if all the elements in the first column have the same algebraic sign. If all elements are not of the same sign then the number of sign changes of the elements in first column equals the number of roots of the characteristic equation in the right of the *s*-plane.

CHAPTER 3

EXAMPLE 3.1

Using Routh criterion, determine the stability of the system represented by the characteristic equation, $s^4 + 8s^3 + 18s^2 + 16s + 5 = 0$. Comment on the location of the roots of characteristic equation.

Solution: The characteristic equation of the system is $s^4 + 8s^3 + 18s^2 + 16s + 5 = 0$. The given characteristic equation is 4th order equation and so it has 4 roots. Since the highest power of *s* is even number, from the first row of routh array using the coefficients of even powers of *s* and from the second row using the coefficients of odd powers of *s*.

$$\begin{array}{llll} s^4: & 1 & 18 & 5 & \ldots\text{Row–1} \\ s^3: & 8 & 16 & & \ldots\text{Row–2} \end{array}$$

The elements of s^3 row can be divided by 8 to simplify the computations.

$$\begin{array}{lllll} s^4: & 1 & 18 & 5 & \ldots\text{ Row 1} \\ s^3: & 1 & 2 & & \ldots\text{ Row 2} \\ s^2: & 16 & 5 & & \ldots\text{ Row 3} \\ s^1: & 1.7 & & & \ldots\text{ Row 4} \\ s^0: & 5 & & & \ldots\text{ Row 5} \end{array}$$

Column – 1

$s^2 : \dfrac{1\times 18 - 2\times 1}{1}$ $\qquad \dfrac{1\times 5 - 0\times 1}{1}$

$s^2 : 16 \qquad 5$

$s^1 : \dfrac{16\times 2 - 5\times 1}{16}$

$s^1 : 1.6875 \;1\; 1.7$

$s^0 : 1.7\times 5 - 0\times 16/1.7$

$s^0 : 5$

on examining the elements of first column of routh array it is observed that all the elements are positive and there is no sign change. Hence all the roots are lying on the left half of s-plane and system is stable.

Result

1. Stable system
2. All the four roots are lying on the left of s-plane.

EXAMPLE 3.2

Construct Routh array and determine the stability of the system whose characteristic equation is $s^6 + 2s^5 + 8s^4 + 12s^3 + 20s^2 + 16s + 16 = 0$. Also determine the number of roots lying on right half of s-plane, left half of s-plane and on imaginary axis.

Solution: The characteristic equation of the system is $s^6 + 2s^5 + 8s^4 + 12s^3 + 20s^2 + 16s + 16 = 0$. The given characteristic polynomial is 6th order equation and so it has 6 roots. Since the highest power of s is even number, form the first row of routh array using the coefficients of even powers of s and form the second row using the coefficient of odd powers of s.

s^6 :	1	8	20	16	...Row 1
s^5 :	2	12	16		...Row 2

The elements s^5 can be divided by 2 to simplify the calculations.

s^6 :	1	8	20	16	... Row 1
s^5 :	1	6	8		... Row 2
s^4 :	1	6	8		... Row 3
s^3 :	0	0			... Row 4
s^3 :	1	3			... Row 5
s^2 :	3	8			... Row 6
s^1 :	.33				... Row 7
s^0 :	8				... Row 8

Column – 1

s^4 :	$\dfrac{1\times 8-6\times 1}{1}=2$	$\dfrac{1\times 20-8\times 1}{1}=12$	$\dfrac{1\times 16-0\times 1}{1}=16$
Divided by 2			
s^4 :	1	6	8
s^3 :	$\dfrac{1\times 6-6\times 1}{1}$	$\dfrac{1\times 8-8\times 1}{1}$	
s^3 :	0	0	

The auxiliary equation is

$$A = s^4 + 6s^2 + 8 \quad \text{(This is } s^4 \text{ row)}$$

On differentiating A with respect to s we get

$$\frac{dA}{dS} = 4s^3 + 12s$$

The coefficients of $\frac{dA}{dS}$ are used to form s^3 row. $\left(\text{Form } s^3 \text{ row again using } \frac{dA}{dS}\right)$

s^3 :	4	12
Divide by 4		
s^3 :	1	3
s^2 :	$\dfrac{1\times 6-3\times 1}{1}$	$\dfrac{1\times 8-0\times 1}{1}$
s^2 :	3	8
s^1 :	$\dfrac{3\times 3-8\times 1}{3}$	
s^1:	0.33	
s^0 :	$\dfrac{0.33\times 8-0\times 3}{0.33}$	
s^0 :	8	

On examining the elements of 1st column of routh array it is observed that there is no sign change the row with all zeros indicate the possibility of roots on imaginary axis. Hence the system is limitedly (or) marginally stable.

The auxiliary polynomial is

$$s^4 + 6s^2 + 8 = 0$$

Let $\qquad s^2 = x$

$$\therefore \qquad x^2 + 6x + 8 = 0$$

The roots of quadratic are,

$$x = \frac{-6 \pm \sqrt{6^2 - 4\times 8}}{2} = -3 \pm 1 = -2 \text{ (or)} - 4.$$

The roots of auxiliary polynomial is,

$$s = \pm\sqrt{x} = \pm\sqrt{-2} \quad \text{and} \quad \pm\sqrt{-4}$$
$$= +j\sqrt{2}, -j\sqrt{2}, +j2 \quad \text{and} \quad -j2$$

The roots of auxiliary polynomial are also roots of characteristic equation. Hence 4 roots are lying on imaginary axis and the remaining two roots are lying on the left half of s-plane.

Result:

1. The system is limitedly (or) marginally stable.
2. Four roots are lying on imaginary axis and the remaining two roots are lying on the left half of s-plane.

EXAMPLE 3.3

Construct Routh array and determine the stability of the system represented by the characteristic equation $s^5 + s^4 + 2s^3 + 2s^2 + 3s + 5 = 0$. comment on the location of the roots of characteristic equation.

Solution: The characteristic equation of the system is

$$s^5 + s^4 + 2s^3 + 2s^2 + 3s + 5 = 0$$

The given characteristic polynomial is 5th order equation and so it has to roots. Since the highest power of s is odd number, form the first row of routh array using the coefficients of odd powers of s and form the second row using the coefficients of even powers of s.

s^5 :	1	2	3	... Row 1
s^4 :	1	2	5	... Row 2
s^3 :	$\in$	-2		... Row 3
s^2 :	$\dfrac{2\in+2}{\in}$	5		... Row 4
s^1 :	$-\left(\dfrac{5\in^2+4\in+4}{2\in+2}\right)$			... Row 5
s^0 :	5			... Row 6

On letting $\in \to 0$, we get

	Column – 1			
s^5 :	1	2	3	... Row 1
s^4 :	1	2	5	... Row 2
s^3 :	0	-2		... Row 3
s^2 :	∞	5		... Row 4
s^1 :	-2			... Row 5
s^0 :	5			... Row 6

On observing the elements of first column of routh array, it is found that there are two sign changes. Hence two roots are lying on right half of s-plane and the system is unstable. The remaining three roots are lying on the left half of s-plane.

s^3 :	$\frac{1\times 2 - 2\times 1}{1}$	$\frac{1\times 3 - 5\times 1}{1}$
s^3 :	0	-2
Replace 0 by $\in$		
s^3 :	$\in$	-2
s^2 :	$\frac{\in\times 2 - (-2\times 1)}{\in}$	$\frac{\in\times 5 - 0\times 1}{\in}$
s^2 :	$\frac{2\in + 2}{\in}$	5
s^1 :	$\frac{\frac{2\in+2}{\in}\times(-2)-(5\times\in)}{2\frac{\in+2}{\in}}$	
s^1 :	$\frac{-(5\in^2+4\in+4)}{2\in+2}$	
s^0 :	$\frac{\frac{-(5\in^2+4\in+4)}{2\in+2}\times 5 - 0\times\frac{2\in+2}{\in}}{\frac{-(5\in^2+4\in+4)}{2\in+2}}$	
s^0 :	5	

Result:

1. The system is unstable.
2. Two roots are lying on the right of s-plane and three roots are lying on the left half of s-plane.

EXAMPLE 3.4

By routh stability criterion determine the stability of the system represented by the characteristic equation $9s^5 - 20s^4 + 10s^3 - s^2 - 9s - 10 = 0$. Comment on the location of roots of characteristic equation.

Solution: The characteristic polynomial of the system is $9s^5 - 20s^4 + 10s^3 - s^2 - 9s - 10 = 0$ characteristic polynomial, it is found that some of the coefficients are negative and so some roots will lie on the right of s-plane. Hence the system is unstable. The routh array can be constructed to find the number of roots lying on right half of s-plane.

$$9s^5 - 20s^4 + 10s^3 - s^2 - 9s - 10 = 0$$

CHAPTER 3

The given characteristic polynomial is 5th order equation and so it has 5 roots. Since the highest power of s is odd number, form the first row of routh array using the coefficients of odd powers of s and form the second row using the coefficients of even powers of s.

$$\begin{array}{lcccl} s^5: & 9 & 10 & -9 & \ldots \text{Row 1} \\ s^4: & -20 & -1 & -10 & \ldots \text{Row 2} \\ s^3: & 9.55 & -13.5 & & \ldots \text{Row 3} \\ s^2: & -29.3 & -10 & & \ldots \text{Row 4} \\ s^1: & -16.8 & & & \ldots \text{Row 5} \\ s^0: & -10 & & & \ldots \text{Row6} \end{array}$$

Column – 1

$s^3:$	$\dfrac{-20 \times 10 - (-1 \times 9)}{-20}$	$\dfrac{-20 \times (-9) - (-10) \times 9}{-20}$
$s^3:$	9.55	– 13.5
$s^2:$	$\dfrac{9.55 \times (-1) - (-13.5) \times (-20)}{9.55}$	$\dfrac{9.55 \times (-10)}{9.55}$
$s^2:$	– 29.3	– 10
$s^1:$	$\dfrac{-29.3 \times (-13.5) - (-10) \times 9.55}{-29.3}$	
$s^1:$	– 16.8	
$s^0:$	$\dfrac{-16.8 \times (-10)}{-16.8}$	
$s^0:$	– 10	

By examining the elements of 1st column of routh array it is observed that there are three sign changes and so three roots are lying on the right half of s-plane and the remaining two roots are lying on the left half of s-plane.

Result:

1. The system is unstable
2. Three roots are lying on the right half and two roots are lying on the left half of s-plane.

EXAMPLE 3.5

The characteristic polynomial of a system is $s^7 + 9s^6 + 24s^5 + 24s^4 + 24s^3 + 24s^2 + 23s + 15 = 0$. Determine the location of roots on s-plane and hence the stability of the system.

Solution: The characteristic equation is $s^7 + 9s^6 + 24s^5 + 24s^4 + 24s^3 + 24s^2 + 23s + 15 = 0$. The given characteristic polynomial is 7th order equation and so it has 7 roots. Since the highest power of s is odd number, form the first row of array using the coefficients of odd powers of s and form the second row using the coefficients of even power of s as shown below.

s^7 :	1	24	24	23	... Row 1
s^6 :	9	24	24	15	... Row 2

Divide s^6 row by 3 to simplify the computations.

	Column – 1				
s^7 :	1	24	24	23	... Row 1
s^6 :	3	8	8	5	... Row 2
s^5 :	1	1	1		... Row 3
s^4 :	1	1	1		... Row 4
s^3 :	0	0			... Row 5
s^3 :	2	1			... Row 5
s^2 :	0.5	1			... Row 6
s^1 :	– 3				... Row 7
s^0 :	1				... Row 8

Column – 1

s^5 : $\dfrac{3 \times 24 - 8 \times 1}{3}$ $\quad \dfrac{3 \times 24 - 8 \times 1}{3}$ $\quad \dfrac{3 \times 23 - 5 \times 1}{3}$

s^5 : 21.33 21.33 21.33

Divide by 21.33

s^5 : 1 1 1

s^4 : $\dfrac{1 \times 8 - 1 \times 3}{1}$ $\quad \dfrac{1 \times 8 - 1 \times 3}{1}$ $\quad \dfrac{1 \times 5 - 0 \times 3}{1}$

s^4 : 5 5 5

Divide by 5

s^4 : 1 1 1

s^3 : $\dfrac{1 \times 1 - 1 \times 1}{1}$ $\quad \dfrac{1 \times 1 - 1 \times 1}{1}$

s^3 : 0 0

The auxiliary polynomial is

$$A = s^4 + s^2 + 1$$

Differentiate A with respect to s

$$\frac{dA}{dS} = 4s^3 + 2s$$

CHAPTER 3

Divided by 2		
s^3 :	4	2
s^3 :	2	1
s^2 :	$\frac{2\times1-1\times1}{2}$	$\frac{2\times1-0\times1}{2}$
s^2 :	0.5	1
s^1 :	$\frac{0.5\times1-1\times2}{0.5}$	
s^1 :	-3	
s^0 :	$\frac{-3\times1}{-3}$	
s^0 :	1	

On examining the first column elements of routh array it is found that there are two sign changes. Hence two roots are lying on the right half of s-plane and so the system is unstable the row of all zeros indicate the possibility of roots an imaginary axis. This can be tested by evaluating the roots of auxiliary polynomial.

The auxiliary equation is $s^4 + s^2 + 1 = 0$.

Put $s^2 = x$ in the auxiliary equation

$$\therefore \quad s^4 + s^2 + 1 = x^2 + x + 1 = 0$$

The roots of quadratic are,

$$x = \frac{-\pm\sqrt{1-4}}{2} = -\frac{1}{2} \pm j\frac{\sqrt{3}}{2}$$

$$= 1\angle120° \text{ or } 1\angle-120°$$

But

$$s^2 = x \therefore s = \pm\sqrt{x} = \pm\sqrt{1\angle120°} \text{ (or) } \pm\sqrt{1\angle-120°}$$

$$= \pm\sqrt{1\angle120°/2} \text{ (or) } \pm\sqrt{1\angle-120°/2}$$

$$= \pm\angle60° \text{ (or) } \pm-60°$$

$$= \pm(0.5 + j0.866) \text{ (or) } \pm(0.5 - j0.866)$$

The roots of auxiliary polynomial are lying on the right half of s-plane and the remaining two on the left half of s-plane the roots of auxiliary equation one also the roots of characteristic polynomial. The two roots lying on the right half of s-plane are indicated by two sign changes in the first column of routh array. The remaining five roots are lying on the left half of s-plane. No roots are lying on imaginary axis. So the system in unstable.

Result:

1. The system is unstable.
2. Two roots are lying on right half of s-plane and five roots are lying on left half of s-plane.

EXAMPLE 3.6

The characteristic polynomial of a system is $s^7 + 5s^6 + 9s^4 + 4s^3 + 20s^2 + 36s + 36 = 0$. Determine the location of roots on the s-plane and hence the stability of the system.

Solution: The characteristic equation is

$$s^7 + 5s^6 + 9s^5 + 9s^4 + 4s^3 + 20s^2 + 36s + 36 = 0$$

The given characteristic polynomial is 7th order equation and so it has 7 roots. Since the highest power of is odd number, form the first row of array using the coefficients of odd powers of s and form the second row of array using the coefficients of even powers of s as shown below.

s^7 :	1	9	4	36	... Row 1
s^6 :	5	9	20	36	... Row 2

Divide s^6 row by 5 to simplify the computations.

s^7 :	1	9	4	36	... Row 1
s^6 :	1	1.8	4	7.2	... Row 2
s^5 :	1	0	4		... Row 3
s^4 :	1	0	4		... Row 4
s^3 :	0	0			... Row 5

The row of all zeros indicate the existence of even polynomial, which is also the auxiliary polynomial. the auxiliary polynomial is $s^4 + 4 = 0$. Divide the characteristic equation by auxiliary equation to get the quotient polynomial.

The characteristic equation can be expressed as a product of quotient polynomial and auxiliary equation.

$$\therefore \quad s^7 + 5s^6 + 9s^5 + 9s^4 + 4s^3 + 20s^2 + 36s + 36 = (s^4 + 4)\,(s^3 + 5s^2 + 9s + 9) = 0$$

The routh array is constructed for quotient polynomial as shown below.

s^3 :	1	9
s^2 :	5	9
s^1 :	7.2	
s^0 :	9	

Column – 1

s^5 :	$\dfrac{1\times 9 - 1.8\times 1}{1}$	$\dfrac{1\times 4 - 4\times 1}{1}$	$\dfrac{1\times 36 - 7.2\times 1}{1}$
s^5 :	7.2	0	28.8
Divide by 7.2			
s^5 :	1	0	4
s^4 :	$\dfrac{1\times 1.8 - 0\times 1}{1}$	$\dfrac{1\times 4 - 4\times 1}{1}$	$\dfrac{1\times 7.2 - 0\times 1}{1}$
s^4 :	1.8	0	7.2

CHAPTER 3

Divide by 1.8			
s^4 :	1.8	0	7.2
s^4 :	1	0	4
s^3 :	$\frac{1\times 0-0\times 1}{1}$	$\frac{1\times 4-4\times 1}{1}$	
s^3 :	0	0	
s^1 :	$\frac{5\times 9-9\times 1}{5}$		
s^1 :	7.2		
s^0 :	$\frac{7.2\times 9-0\times 5}{7.2}$		
s^0 :	9		

There is no sign change in the elements of first column of routh array of quotient polynomial. Hence all the roots of quotient polynomial are lying on the left half of s-plane.

To determine the stability the roots of auxiliary polynomial should be evaluated.

The auxiliary polynomial is

$$s^4 + 4 = 0$$

Put $s^2 = x$ in the auxiliary equation.

$$\therefore \quad s^4 + 4 = x^2 + 4 = 0$$

$$\therefore \quad x^2 = -4$$

$$x = \pm\sqrt{-4} = \pm\, j2 - 2\angle 90\epsilon\,(\text{or})\; 2\angle -90\epsilon$$

But, $s = \pm\sqrt{x}$,

$$= \pm\sqrt{2\angle 90°}\ (\text{or})\ \pm\sqrt{2\angle -90°}$$

$$= \pm\sqrt{2}\angle 90°/2\ (\text{or})\ \pm\sqrt{2}\angle -90°/2$$

$$= \pm\sqrt{2}\angle 45°\ (\text{or})\ \pm\sqrt{2}\angle -45°$$

$$= \pm(1 + j1)\ (\text{or}) \pm (1 - j1)$$

The roots of auxiliary equation are complex and has quadrantal symmetry. Two roots of auxiliary equation are lying on the right half of s-plane and the other two on the left half of s-plane.

The roots of characteristic equation are given by roots of quotient polynomial and auxiliary polynomial. Hence we can conclude that two roots of characteristic equation are lying on the right half of s-plane and so the system is unstable. The remaining five roots are lying on the left half of s-plane

---

Result:

1. The system is unstable.

2. Two roots of s-plane and five roots are lying on the left of s-plane.

$$
\begin{array}{r|l}
 & s^3 + 5s^2 + 9s + 9 \\
\hline
s^2 + 4 & s^7 + 5s^6 + 9s^5 + 9s^4 + 4s^3 + 20s^2 + 36s + 36 \\
 & s^7 \qquad\qquad\qquad + 4s^3 \\
\hline
 & 5s^6 + 9s^5 + 9s^4 \qquad + 20s^2 + 36s + 36 \\
 & 5s^6 \qquad\qquad\qquad + 20s^2 \\
\hline
 & 9s^5 + 9s^4 \qquad + 36s + 36 \\
 & 9s^5 \qquad\qquad + 36s \\
\hline
 & 9s^4 \qquad + 36 \\
 & 9s^4 \qquad + 36 \\
\hline
 & 0 \\
\hline
\end{array}
$$

EXAMPLE 3.7

Use the routh stability criterion to determine the location of roots on the s-plane and hence the stability for the system represented by the characteristic equation $s^5 + 4s^4 + 8s^3 + 8s^2 + 7s + 4 = 0$.

Solution: The characteristic equation of the system is $s^5 + 4s^4 + 8s^3 + 8s^2 + 7s + 4 = 0$. The given characteristic polynomial is 5th order equation and so it has 5 roots. Since the highest power of s is odd number, form the first row of routh array using the coefficients of odd powers of s and form the second row using the coefficients of even power of s.

s^5 :	1	8	7	... Row 1
s^4 :	4	8	4	... Row 2

Divide s^4 row by 4 to simplify the calculation

	Column – 1			
s^5 :	1	8	7	... Row 1
s^4 :	1	2	1	... Row 2
s^3 :	1	1		... Row 3
s^2 :	1	1		... Row 4
s^1 :	$\in$			... Row 5
s^0 :	1			... Row 6

when $\in \to 0$, there is no sign change in the first column of routh array. But we have a row of all zeros (s^1 row (or) row 5) and so there is a possibility of roots on imaginary axis. This can be found from the roots of auxiliary polynomial. Here the auxiliary polynomial is given by s^2 row.

$$s = \pm\sqrt{-1} = \pm\, j1$$

s^3 :	$\dfrac{1\times 8 - 2\times 1}{1}$	$\dfrac{1\times 7 - 1\times 1}{1}$
s^3 :	6	6

Divide by 6		
s^3 :	1	1
s^2 :	$\frac{1\times2-1\times1}{1}$	$\frac{1\times1-0\times1}{1}$
s^2 :	1	1
s^1 :	$\frac{1\times1-1\times1}{1}$	
s^1 :	1	
s^0 :	$\frac{\in\times1-0\times1}{\in}$	
s^0 :	1	

The roots of auxiliary polynomial are $+j1, j1$, lying on imaginary axis. The roots of auxiliary polynomial are also roots of characteristic equation are lying on imaginary axis and so the system is limitedly (or) marginally stable. The remaining three roots of characteristic equation are lying on the left half of s-plane.

Results:

1. The system is limitedly (or) marginally stable.
2. The two roots are lying on imaginary axis and three roots are lying on left of s-plane.

EXAMPLE 3.8

Use the routh stability criterion to determine the location of roots on the s-plane and hence the stability for the system represented by the characteristic equation $s^6 + s^5 + 3s^4 + 3s^3 + 3s^2 + 2s + 1 = 0$.

Solution: The characteristic polynomial of the system is

$$s^6 + s^5 + 3s^4 + 3s^3 + 3s^2 + 2s + 1 = 0$$

The given characteristic polynomial is 6th order equation and so it has 6 roots. Since the highest power of s is even number, form the first row of routh array using the coefficients of even powers of s and form the second row using the coefficients of odd powers of s as shown below.

s^4 :	$\frac{1\times3-3\times1}{1}$	$\frac{1\times3-2\times1}{1}$	$\frac{1\times1-0\times1}{1}$
s^4 :	0	1	1
Let $0 \to \in$			
s^4 :	$\in$	1	1
s^3:	$\frac{\in\times3-1\times1}{\in}$	$\frac{\in\times2-1\times1}{\in}$	

$$s^3: \quad \frac{3\in-1}{\in} \qquad \frac{2\in-1}{\in}$$

$$s^2: \quad \frac{\dfrac{3\in-1}{\in}-\dfrac{2\in-1}{\in}\times\in}{\dfrac{3\in-1}{\in}} \qquad \frac{\dfrac{3\in-1}{\in}\times 1-0\times\in}{\dfrac{3\in-1}{\in}}$$

$$s^2: \quad -2\in^2+4\in-1/3\in-1 \qquad 1$$

$$s^1: \quad \frac{\dfrac{-2\in^2+4\in-1}{3\in-1}\times\dfrac{2\in-1}{\in}-\dfrac{3\in-1}{\in}\times 1}{\dfrac{-2\in^2+4\in-1}{3\in-1}}$$

$$s^1: \quad \frac{(-2\in^2+4\in-1)(2\in-1)-(3\in-1)(3\in-1)}{\in(-2\in^2+4\in-1)}$$

$$s^1: \quad \frac{-4\in^2+\in^2}{s(-2\in^2+4\in-1)}=\frac{4\in^2-\in}{2\in^2-4\in+1}$$

$$s^0: \quad \frac{\dfrac{4\in^2-\in}{2\in^2-4\in+1}\times 1-0\times\dfrac{-2\in^2+4\in-1}{3\in-1}}{(4\in^2-\in)/(2\in^2-4\in+1)}$$

$$s^0: \quad 1$$

s^6 :	1	3	3	1	... Row 1
s^5 ;	1	3	2		... Row 2
s^4:	$\in$	1	1		... Row 3
s^3 :	$\frac{3\in-1}{\in}$	$\frac{2\in-1}{\in}$			... Row 4
s^2:	$\frac{-2\in^2+4\in-1}{3\in-1}$	1			... Row 5
s^1 :	$\frac{4\in^2-\in}{2\in^2-4\in+1}$				... Row 6
s^0 :	1				... Row 7

On letting $\in \to 0$, we get

s^6 :	1	3	3	1	...Row 1
s^5 :	1	3	2		...Row 2
s^4 :	0	1	1		...Row 3

	Column 1		
s^3 :	$-\infty$	$-\infty$	...Row 4
s^2 :	1	1	...Row 5
s^1 :	0		...Row 6
s^0 :	1		...Row 7

Since there is a row of all zeros (s^1 row) there is a possibility of roots on imaginary axis. The auxiliary polynomial is $s^2 + 1 = 0$.

The roots of auxiliary polynomial are

$$s = \pm\sqrt{-1} = \pm\, j1$$

The roots of auxiliary polynomial are also roots of characteristic equation. Hence roots are lying on imaginary axis. Therefore divide the characteristic polynomial by auxiliary equation and construct the routh array for quotient polynomial to find the roots lying on right half of s-plane.

The characteristic polynomial can be expressed as a product of auxiliary polynomial and quotient polynomial.

$$\therefore \quad s^6 + s^5 + 3s^4 + 3s^3 + 3s^2 + 2s + 1$$
$$= (s^2 + 1)\,(s^4 + s^3 + 2s^2 + 2s + 1) = 0$$

The routh array for quotient polynomial is constructed as shown below.

s^4 :			1	... Row 1
s^3 :	1	**2**		... Row 2
s^2 :	$\in$	1		... Row 3
s^1 :	$\dfrac{2\in-1}{\in}$			... Row 4
s^0 :	1			... Row 5

$$\begin{array}{r|l} & s^4 + s^3 + 2s^2 + 1 \\ \hline s^2 + 1 & s^6 + s^5 + 3s^4 + 3s^3 + 3s^2 + 2s + 1 \\ - & s^6 \quad + s^4 \\ \hline & s^5 + 2s^4 + 3s^3 + 3s^2 + 3s^2 + 2s + 1 \\ & s^5 \quad + s^3 \\ \hline & 2s^4 + 2s^3 + 3s^2 + + 2s + 1 \\ & 2s^4 \quad + 2s^2 \\ \hline & 2s^3 + s^2 + 2s + 1 \\ & 2s^3 \quad + 2s \\ \hline & s^2 + 1 \\ & s^2 + 1 \\ \hline & 0 \end{array}$$

s^2 :	$\frac{1\times 2-2\times 1}{1}$	$\frac{1\times 1-0\times 1}{1}$
s^2 :	0	
Let $0 \to \in$		
s^2 :	$\in$	1
s^1 :	$\frac{\in\times 2-1\times 1}{\in}$	
s^1 :	$\frac{2\in-1}{\in}$	
s^0 :	$\frac{\frac{2\in-1}{\in}\times 1-0\times\in}{(2\in-1)/\in}$	
s^0 :	1	

On letting $\in \to 0$ we get,

s^4 :	1	2	1	...Row 1
s^3 :	1	2		...Row 2
s^2 :	0	1		...Row 3
s^1 :	$-\infty$			...Row 4
s^0 :	1			...Row 5

Column 1

On examining the first column of routh array of quotient polynomial, we found that there are two sign changes. Hence two roots are lying on imaginary axis, two roots are lying on right half of s-plane and the remaining two roots are lying on left half of s-plane. Hence the system is unstable.

Result:

1. The system is unstable.
2. The two roots are lying on imaginary axis, two roots are lying on right half s-plane and two roots are lying on left half s-plane.

EXAMPLE 3.9

Determine the range of K for stability of unity feedback system whose open loop transfer function is $G(s) = K/s(s+1)(s+2)$.

Solution: The closed loop transfer function,

$$\frac{C(s)}{R(s)} = \frac{G(s)}{1+G(s)}$$

$$= \frac{K / s(s+1)(s+2)}{1 + \dfrac{K}{s(s+1)(s+2)}}$$

$$= \frac{K}{s(s+1)(s+2) + K}$$

The characteristic equation is

$$s(s+1)(s+2) + K = 0$$

$$s(s^2 + 3s + 2) + K = 0$$

$$s^3 + 3s^2 + 2s + K = 0$$

The routh array is constructed as shown below.

The highest power of s in the characteristic polynomial is odd number. Hence form the first row using the coefficients of odd powers of s and form the second row using the coefficients of even powers of s.

	Column 1	
s^3 :	1	2
s^2 :	3	K
s^1 :	$\frac{6-K}{3}$	
s^0 :	K	

Column 1

s^1 :	$\frac{2 \times 2 - K \times 1}{3}$
s^1 :	$\frac{6-K}{3}$
s^0 :	$\frac{\frac{6-K}{3} \times K - 0 \times 3}{(6-K)/3}$
s^0 :	K

For the system to be stable there should not be any sign change in the elements of first column. Hence choose the value of K so that the first column elements are positive.

Form s^0 row, for the system to be stable,

$$K > 0$$

Form s^1 row, for the system to be stable,

$$\frac{6-K}{3} > 0$$

For $\frac{6-K}{3} > 0$, the value of K should be less than 6.

∴ The range of K for the system to be stable is $0 < K < 6$.

Result: The value of K is in the range $0 < K < 6$ for the system to be stable.

EXAMPLE 3.10

The open loop transfer function of a unity feedback control system is given by

$$G(s) = \frac{K}{(s+2)(s+4)(s^2+6s+25)}$$

By applying the routh criterion, discuss the stability of the closed-loop system as a function of K. Determine the value of K which will cause sustained oscillations in the closed loop system. What are the corresponding oscillating frequencies?

Solution: The closed loop transfer function

$$\frac{C(s)}{R(s)} = \frac{G(s)}{1+G(s)} = \frac{\dfrac{K}{(s+2)(s+4)(s^2+6s+25)}}{1+\dfrac{K}{(s+2)(s+4)(s^2+6s+2s)}}$$

$$= \frac{K}{(s+2)(s+4)(s^2+6s+25)+K}$$

The characteristic equation is given by the denominator polynomial of closed loop transfer function. The characteristic equation is

$$(s+2)(s+4)(s^2+6s+25)+K = 0$$

$$(s^2+6s+8)(s^2+6s+25)+K = 0$$

$$s^4+12s^3+69s^2+198s+200+K = 0$$

The routh array is constructed as shown below. The highest power of s is the characteristic equation is even number. Hence form the first row using the coefficients of even powers of s and form the second new using the coefficients of odd powers of s.

s^4 :	1	69	$200+K$	...Row 1
s^3 :	12	198		...Row 2

Divide s^3 row by 12 to simplify the calculations.

s^4 :	1	69	$200+K$	...Row 1
s^3 :	1	16.5		...Row 2
s^2 :	52.5	$200+K$		...Row 3
s^1 :	$\frac{666.25-K}{52.5}$			...Row 4
s^0 :	$200+K$			...Row 5
	Column – 1			

CHAPTER 3

s^2 :	$\frac{1 \times 69 - 16.5 \times 1}{1}$	$\frac{1 \times (200 + K)}{1}$
s^2 :	52.5	$200 + K$
s^1 :	$\frac{52.5 \times 16.5 - (200 + K) \times 1}{52.5}$	
s^1 :	$\frac{666.25 - K}{52.5}$	
s^0 :	$\frac{666.25 - K}{52.5} \times (200 + K)$	
s^0 :	$200 + K$	

For the system to be stable there should not be any sign change in the elements of first column. Hence choose the value of K so that the first column elements are positive.

From s^1 row, for the system to be stable, $(666.25 - K) > 0$

Since $(666.25 - K) > 0$, K should be less than 666.25.

From s^0 row, for the system to stable, $(200 + K) > 0$

Since $(200 + K) > 0$, K should be greater than – 200, but practical value of K-starts from 0. Hence K should be greater than 0.

∴ The range of K for the system to be stable is $0 < K < 666.25$.

When $K = 666.25$ the s^1 row becomes zero, which indicates the possibility of roots on imaginary axis. A system will oscillate if it has roots on imaginary axis and no roots on right half of s-plane.

When $K = 666.25$, the coefficients of auxiliary equation are given by the s^2 row.

∴ The auxiliary equation is

$$52.5\, s^2 + 200 + K = 0$$

$$52.5\, s^2 + 200 + 666.25 = 0$$

$$s^2 = \frac{-200 - 666.25}{52.5} = -16.5$$

$$s = \pm j\, 4.06$$

When $K = 666.25$, the system has roots on imaginary axis and so it oscillates. Frequency of oscillation is given by the value of root on imaginary axis.

∴ The frequency of oscillation,

$$\omega = 4.06 \text{ rad/sec}$$

Result:

1. The range of K for stability is $0 < K < 666.25$.
2. The system oscillates when $K = 666.25$.
3. The frequency of oscillation, $\omega = 4.06$ rad/sec (when $K = 666.25$).

EXAMPLE 3.11

The open loop transfer function of a unity feedback system is given by $G(s) = \frac{K(s+1)}{s^3 + as^2 + 2s + 1}$. Determine the value of K and a so that the system oscillates at a frequency of 2 rad/sec.

Solution: The closed loop transfer function

$$\frac{C(s)}{R(s)} = \frac{G(s)}{1+G(s)}$$

$$= \frac{\frac{K(s+1)}{s^3 + as^2 + 2s + 1}}{1 + \frac{K(s+1)}{s^3 + as^2 + 2s + 1}} = \frac{K(s+1)}{s^3 + as^2 + 2S + 1 + K+s(+1)}$$

∴ The characteristic equation is

$$s^3 + as^2 + 2s + 1 + K(s + 1) = 0$$
$$s^3 + as^2 + 2s + 1 + Ks + K = 0$$
$$s^3 + as^2 + (2 + K)s + 1 + K = 0$$

The routh array of characteristic polynomial is constructed as shown below. The maximum power of s is odd, hence the first row of routh array is formed using coefficients of odd powers of s and the second row of routh array is formed using coefficients of even powers of s.

If the elements of s row are all zeros then there exist an even polynomial (for auxiliary polynomial). If the roots of the auxiliary polynomial are purely imaginary then the roots are lying on imaginary axis and the system oscillates. The frequency of oscillation is the root of auxiliary polynomial.

From routh array , From s row.

$$\frac{a(2+K)-(1+K)}{a} = 0$$
$$a(2 + K) - (1 + K) = 0$$
$$2a + Ka - 1 - K = 0$$
$$2a - 1 + K(a - 1) = 0$$
$$\text{Put } K = 4a - 1$$

$$\therefore 2a - 1 + (4a - 1)(a - 1) = 0$$
$$2a - 1 + 4a^2 - 4a - a + 1 = 0$$
$$4a^2 - 3a = 0 \text{ (or) } a(4a - 3) = 0$$
$$\text{since a} \neq 0,\ 4a - 3 = 0$$
$$a = 3/4$$
$$\text{when } a = (3/4),\ K = 4a - 1$$
$$K = 2$$

Routh array

s^3 :	1	$2 + K$
s^2 :	a	$1 + K$
s^1 :	$\frac{a(2+K)-(1+K)}{a}$	
s^0 :	$1 + K$	

The auxiliary polynomial is

$$as^2 + (1 + K) = 0$$

$$as^2 = -(1 + K)$$

$$s = \pm j\sqrt{\frac{1+K}{a}}$$

Given that $s = \pm j2$

$$\therefore \quad \sqrt{\frac{1+K}{a}} = 2$$

$$\frac{1+K}{a} = 4$$

$$K = 4a = 1$$

Result: When the system oscillates at a frequency of 2 rad/sec, $K = 2$ and $a = 3/4$.

EXAMPLE 3.2

A feedback system has open loop transfer function of $G(s) = Ke^{-s}/s(s^2 + 5s + 9)$ determine the maximum value of K for stability of closed loop system.

Solution: Generally control systems have very low band width which implies that it has very low frequency range of operation. Hence for low frequency ranges the term e^{-s} can be replaced by $(1 - s)$ [i.e. e^{-sT} Ä $(1 - sT)$]

$$\therefore \quad G(s) = \frac{Ke^{-s}}{s(s^2 + 5s + 9)} \simeq \frac{K(1-s)}{s(s^2 + 5s + 9)}$$

The closed loop transfer function

$$\frac{C(s)}{R(s)} = \frac{G(s)}{1 + G(s)} = \frac{\dfrac{K(1-s)}{s(s^2 + 5s + 9)}}{1 + \dfrac{K(1-s)}{s(s^2 + 5s + 9)}}$$

$$= \frac{K(1-s)}{s(s^2 + 5s + 9) + K(1-s)}$$

The characteristic equation is given by the denominator polynomial of closed loop transfer function.

$\therefore$ The characteristic equatino is

$$s(s^2 + 5s + 9)\,(K\,(1 - s)) = 0$$

$$s(s^2 + 5s + 9) + K(1 - s) = s^3 + 5s^2 + 9s + K - Ks = 0$$

$$\therefore \quad s^3 + 5s^2 + (9 - K)\,s + K = 0$$

The routh array of characteristic polynomial is constructed as shown.

The maximum power of s in the characteristic polynomial is odd, hence from the first row of routh array using coefficients of odd powers of s and second row of routh array using coefficients of even powers of s.

$$\begin{array}{lcc}
s^3: & 1 & 9-K \\
s^2: & 5 & K \\
s^1: & 9-1.2K & \\
s^0: & K & \\
s^0: & K & \\
s^1: & \dfrac{5\times(9-K)-K\times 1}{5} & \\
s^1: & 45-5K-K15 & \\
s^1: & \dfrac{45-6K}{5}\ \ddot{A}-1.2K & \\
s^0: & \dfrac{(9-1.2K)\times K}{(19-1.2K)} & \\
s^0: & K &
\end{array}$$

From s^1 row, for stability of the system, $(-1.2K) > 0$.

If $(9-1.2K) > 0$ then $1.2K < 9$

$$\therefore \qquad K < \frac{9}{1.2} = 7.5$$

From s^0 row, for stability of the system, $K > 0$.

Finally we can conclude that for stability of the system k is in the range of $0 < K < 7.5$

Result: For stability of the system K is in the range of $0 < K < 7.5$.

3.15 ROOT LOCUS

Root locus is a simple technique which is used to analyse the systems. The root locus technique is a powerful tool for adjusting the location of closed loop poles to achieve the desired system performance by varying one (or) more system parameters.

Consider the open loop transfer function of system,

$$G(s) = K/s(s+P_1)(s+P_2)$$

The closed loop transfer function of the system with unity feedback is given by

$$\frac{C(s)}{R(s)} = \frac{G(s)}{1+G(s)}$$

CHAPTER 3

$$= \frac{\dfrac{K}{s(s+p_1)(s+p_2)}}{1+\dfrac{K}{s(s+p_1)(s+p_2)}}$$

$$= \frac{K}{s(s+p_1)(s+p_2)+K}$$

The denominator polynomial of $C(s)/R(s)$ is the characteristic equation of the system. The characteristic equation is given by

$$s(s+p_1)(s+p_2)+K=0$$

The roots of characteristic equation is a function of open loop gain K. When the gain K is varied from 0 to ∞, the roots of characteristic equation will take different values. When $K = 0$, the roots are given by open loop poles. When $K \to \infty$, the roots will take the value of open loop zeroes.

The path taken by the roots of characteristic equation when open loop gain K is varied. from 0 to ∞ are called root loci.

Note: In general the roots of characteristic equation can be varied by varying any other system parameter other than gain.

In general the closded loop transfer function of the system with multiple loops is obtained from the signal flow graph of the system using Mason's gain formula.

$$\frac{C(s)}{R(s)} = T(s) = \frac{1}{\Delta}\sum P_K \Delta_K$$

The determinant, Δ is the denominator polynomial of $C(s)/R(s)$. The characteristic equation of the system is given by, $\Delta = 0$. For the single loop system shown in Fig. (3.50).

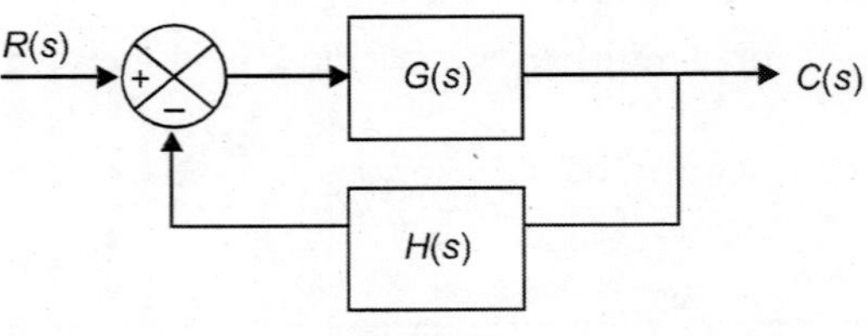

Fig. 3.50

$$\frac{C(s)}{R(s)} = \frac{G(s)}{1+G(s)H(s)}$$

The characteristic equation, is

$$1 + G(s)H(s) = 0$$

Let,
$$D(s) = G(s)H(s)$$

$$1 + G(s)H(s) = 1 + D(s) = 0$$

or
$$D(s) = -1 \qquad (3.81)$$

From equation (3.8.1) it can be concluded that the roots of the characteristic equation occur only for those values of s for which, $D(s) = -1$

The equation (3.8.1) can be converted to two Even s conditions given below,

$$| D(s) | = 1 \tag{3.82}$$

$$\angle D(s) = \pm 180° (2q + 1) \tag{3.83}$$

where $q = 0, 1, 2, 3...$

The equation (3.82) is called magnitude criterion and equation (3.83) is called angle criterion. The magnitude criterion states that $s = s_a$ will be a point on root Locus if for that value of s, $|D(s)| = |G(s)\, H(s)| = 1$. The angle criterion states that $s = s_a$ will be a point on root Locus if for that value of s, $\angle D(s) = \angle G(s)\, H(s)$ is equal to an odd multiply of 180°.

The function $D(s)$ can be expressed as a ratio of two polynomials in s as shown in below.

$$D(s) = G(s)\, H(s) = K \frac{(s + z_1)(s + z_2)(s + z_3)\ldots}{(s + p_1)(s + p_2)(s + p_3)\ldots} \tag{3.84}$$

$$\therefore \quad D(s) = K \frac{|s + z_1||s + z_2||s + z_3|\ldots}{|s + p_1||s + p_2||s + p_3|\ldots} = K \frac{\prod_{i=1}^{m} |s + z_i|}{\prod_{i=1}^{n} |s + p_i|}$$

where, m = Number of zeros

n = Number of poles

The magnitude criterion states that

$$|D(s)| = 1$$

$$\therefore \quad K = \frac{\sum_{i=1}^{m} |s + z_i|}{\sum_{i=1}^{n} |s + p_i|} = 1$$

or

$$K = \frac{\sum_{i=1}^{n} |s + p_i|}{\sum_{i=1}^{m} |s + z_i|} \tag{3.84}$$

The opne-loop gain K corresponding to a point $s = s_a$ on root Locus can be calculated using equation (3.84). If can be shown that $|s + p_i|$ is equal to the length of vector drawn from $s = p_i$ to $s = s_a$ and $|s + z_i|$ is equal to the length of vector drawn from $s = z_i$ to $s = s_a$. Hence the equation for K can be written as

$$K = \frac{\text{Product of length of vector from open loop poles to the point } s = s_a}{\text{Product of length of vectors from open loop zeros of the point } s = s_a}$$

From equation (3.83)

$$\angle D(s) = \angle(s+z_1) + \angle(s+z_2) + \angle(s+z_3) \ldots - \angle(s+p_1) - \angle(s+p_2) - \angle(s+p_3) \ldots$$

$$= \sum_{j=1}^{m} \angle(s+z_i) - \sum_{i=1}^{n} \angle(s+p_i)$$

The angle criterion states that

$$\angle D(s) = \pm\, 180° \; (2q+1)$$

$$\therefore \qquad \sum_{i=1}^{m} \angle(s+z_i) - \sum_{i=1}^{n} \angle(s+p_i) = \pm\, 180° \; (2q+1) \qquad (3.85)$$

The equation (3.85) can be used to check whether a point $s = s_a$ is a point on root Locus or not. It can be shown that $\angle(s + p_i)$ is equal to the angle of vector drawn from $s = p_i$ to $s = s_a$ and $\angle\, s + z_i$ is equal to the angle of vector drawn from $s = z_i$ to $s = s_a$. Hence equation (3.85) can be written as

$$\begin{bmatrix} \text{Sum of angles vectors from} \\ \text{open loop zeros to the point} \\ s = s_a \end{bmatrix} - \begin{bmatrix} \text{Sum of angles of vectors from} \\ \text{open loop poles to the point} \\ s = s_a \end{bmatrix}$$

$$= \pm\, 180° \; (2q+1)$$

Construction of Root Locus

The sketching of root locus is difficult task, so we always go for trial and error method. In this method, the poles and zeros of $G(s)\ H(s)$ are located on the s-plane on a graph sheet and a trial point $s = s_a$ is selected. Determine the angles of vectors drawn from poles and zeros to the trial point. From the angle criterion, determine the angle to be contributed by these vectors to make the trial point as a point on root Locus. Shift the trial point suitably so that the angle criterion is satisfied.

A number of points are determined using the above procedure. Join the points by a smooth curve which is the root Locus. The value of K for a particular root can be obtained from the magnitude criterion.

Since, the trial and error method is difficult, we have developed set of rules to reduce the difficulties involved in sketching root locus and to develop a quick approximate sketch. From the approximate sketch a more accurate root locus can be obtained.

Rules for Construction of Root Locus

Rule 1: The root locus is a symmetrial about the real axis.

Rule 2: Each Branch of the root locus originates from an open-loop pole corresponding to $K = 0$ and terminates at either on a finite open loop zero corresponding to $K = \infty$. The number of branches of the root locus terminating on infinity is equal to $(n - m)$,

where $\qquad m \rightarrow$ Number of zeros

$\qquad n \rightarrow$ Number of poles

Rule 3: Segments of the real axis having an odd number of real axis open-loop poles plus zeros to their right side is a part of the root locus.

Rule 4: Asymptotes making angles with the real axis is given by,

$$\phi_A = \frac{\pm 180° [2q+1]}{(n-m)}; \; q = 0, 1, 2, \ldots (n-m)$$

Rules 5: The point of intersection of the asymptotes with the real axis is at $s = \sigma_A$

where $$\sigma_A = \frac{\text{Sum of poles} - \text{Sum of zeros}}{n-m}$$

Rule 6: The breakaway and break in points of the root locus are determined from the roots of the equation $dK/ds = 0$. If r numbers of branches of root locus meet at a point, then they away at angle of $\pm 180°/r$.

Rule 7: The angle of departure from a complex open-loop pole is given by

$$\phi_p = \pm 180° (2q+1) + \phi; \; q = 0, 1, 2, \ldots$$

where ϕ is the net angle contribution at the pole by all other open loop poles and zeros.

Rule 8: The intersection of root Locus branches with the imaginary axis can be determined by use of the routh criterion, or by letting $s = j\omega$ in the characteristic equation and equating the real part and imaginary part to zero, to solve for ω and K. the value of ω is the intersection point on imaginary axis and K is the value of gain at the intersection point.

Rule 9: The open-loop gain K at any point $s = s_a$ on the root locus is given by

$$K = \frac{\text{Product of vector lengths from open loop poles to the point } s_a}{\text{Product of vector lengths from open zeros to the point } s_a}$$

Note: The length of vector should be measured to scale. If there is no finite zero then the product of vector lengths from zeros is equal to 1.

Procedure for Constructing Root Locus

Step 1: Locate the poles and zeros of $G(s)\,H(s)$ on the s-plane. The root locus branch start from open loop poles and terminate at zeros.

Step 2: Determine the root locus on real axis.

Step 3: Determine the asymptotes of root locus branches and meeting point of asymptotes with in real axis.

Step 4: Find the break away and breaking points.

Step 5: Determine the angle of departure from the complex pole if there is a complex pole. Determine the angle of arrival at the complex zero, if there is a complex zero.

Step 6: Find the points where the root loci crosses the imaginary axis.

Step 7: The value of gain K at any point on the locus can be determined from magnitude condition. The magnitude conditon is, given by

$$\text{Gain } K \text{ at a point } (s = s_a) = \frac{\text{Product of length of vectors from poles to the point } (s = s_a)}{\text{Product of length of vectors from finite zeros to the point } (s = s_a)}$$

Note: When there is no finite zero, the denominator is taken as unity. The length of vectors should be measured to scale.

Brief Explanation for the Various Steps in the Procedure for Constructing Root Locus

Step 1: Location of Poles and zeros:

Draw the real and imaginary axis on an ordinary graph sheet and choose same scales both on real and imaginary axis. The poles are marked by cross "X" and zeros and marked by small circle "O". The number of root locus branches is equal to number of poles of open loop transfer function. The origin of a root locus is at a pole and the end is at a zero.

If n = Number of poles and

m = Number of finite zeros,

then n root locus branches ends at finite zeros. The remaining $(n - m)$ root locus branches will end at zeros at infinity.

Step 2: Root locus on real axis:

To decide the part of root locus on real axis, take a test point on real axis. If the total number of poles and zeros on the real axis to the right of the test point is odd number then the test point lies on the root locus. if it is even then the test point does not lie on the root locus.

Step 3: Angles of asymptotes and centroid:

If n = Number of poles and

m = Number of zeros,

then $(n - m)$ root locus branches will terminate at zeros at infinity. These root locus branches will go along an asymptotic path and meets the asymptotes at infinity. Hence number of asymptotes is equal to number of root locus branches going to infinity. The formula for angles of asymptotes and the centroid is,

$$\text{Angles of asymptotes} = \frac{\pm 180\,(2q + 1)}{n - m}$$

where, $q = 0, 1, 2, 3\ldots (n - m)$

$$\text{centroid} = \frac{\text{Sum of poles} - \text{Sum of zeros}}{n - m}$$

Step 4: Breakaway and break in points:

The breakaway (or) breaking points either lie on real axis (or) exists as complex conjugate pairs. If there is a root locus on real axis between two poles then there exists a breakaway point.

If there is root locus on real axis between two zeros then there exist a breakin point. If there is a root locus on real axis between pole and zero then there may be (or) may not be breakaway (or) breakin point.

Let the characteristic equation be in the form

$$B(s) + K\,A(s) = 0$$

$$\therefore \quad K = -\frac{B(s)}{A(s)}$$

The breakaway and breakin point is given by roots of the equation $\frac{dK}{ds} = 0$. The roots of $\frac{dK}{ds} = 0$

are actual breakaway (or) breakin point provided for this value of root the gain K should be positive and real.

Step 5: Angle of departure and angle of arrival

$$\text{Angle of departure (from a complex pole } A) = 180° - \begin{bmatrix}\text{Sum of angles of vector to} \\ \text{the complex pole A from} \\ \text{other poles}\end{bmatrix} + \begin{bmatrix}\text{Sum of angles of vectors} \\ \text{to the complex pole A from} \\ \text{zeros}\end{bmatrix}$$

If poles are complex then they exist only as conjugate pair. Consider the two complex conjugate roles A and A* shown in Fig. (3.51).

$$\text{Angle of departure at pole } A = 180° - (\theta_1 + \theta_3 + \theta_5) + (\theta_2 + \theta_4)$$

$$\text{Angle of departure at pole } A^* = [\text{Angle of departure at pole } A]$$

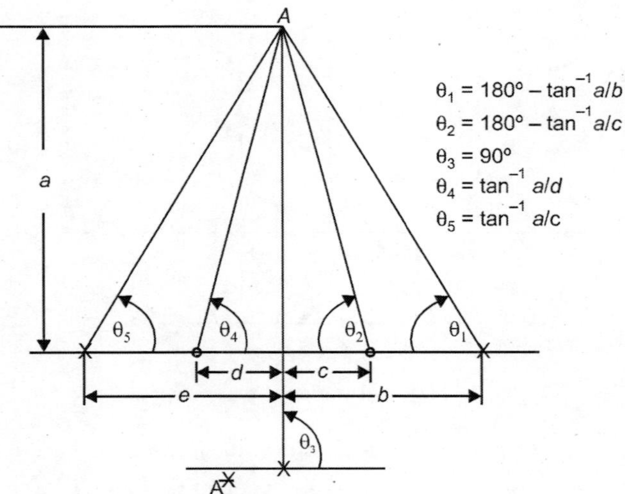

Fig. 3.51: Calculation of angles of departure.

Note: The angles can be calculated as shown in Fig. (3.51) or they can be measured using protractor.

$$\text{Angle of arrival at a complex zero } A = 180° - \begin{bmatrix}\text{Sum of angles of vectors} \\ \text{to the complex zero A from} \\ \text{all other zeros}\end{bmatrix} + \begin{bmatrix}\text{Sum of angles of vectors} \\ \text{to the complex zero } A \text{ from} \\ \text{poles}\end{bmatrix}$$

CHAPTER 3

From Fig. (3.52)

$$\theta_1 = -180° - \tan^{-1}(a/c)$$

$$\theta_2 = 180° - \tan^{-1}(a/b)$$

$$\theta_3 = 90°$$

$$\theta_4 = \tan^{-1}(a/d)$$

$$\theta_5 = \tan^{-1}(a/e)$$

If zeros are complex then they exist only as conjugate pairs. Consider the two complex conjugate zeros A and A^* as shown in Fig. (3.52)

Angle of arrival at zero $A = 180° - (\theta_1 + \theta_3) + (\theta_2 + \theta_4 + \theta_5)$

Angle of arrival at zero $A^* = -$ [Angle of arrive at zero A]

Note: The angles can be calculated as shown in Fig. (3.52) or they can be measured using protractor.

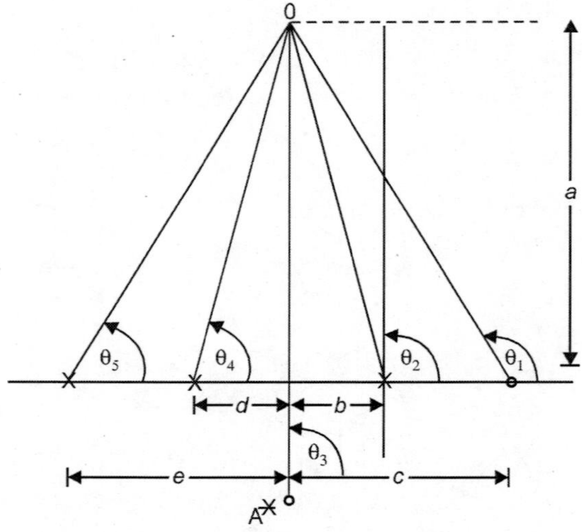

Fig. 3.52: Calculation of angle of arrival.

Step 6: Point of intersection of root locus with imaginary axis:

The point where the root loci intersects the imaginary axis can be found by following three methods.

1. By Routh Hurwitz array.
2. By trial and error approach.
3. Letting $s = j\omega$ in the characteristic equation and separate the real part and imaginary part. Two equations are obtained-one by equating real part to zero and the other by equating imaginary part to zero. Solve the two equations for ω and K. The value of ω gives the point where the root locus crosses imaginary axis. The value of K gives the value of gain K at this crossing point. Also this value of K is the limiting value of K for stability of the system.

Step 7: Test points and root locus:

Choose a test point. Using a protractor roughly estimate the angles of vectors drawn to this point and adjust the point to satisfy angle criterion. Repet the procedure for few more test points. Sketch the root locus from the knowledge of typical sketches and the informations obtained in step 1 through 6.

Determination of Open Loop Gain for a Specified Damping of the Dominant Roots

The dominant pole is a pair of complex conjugate pole which decides the transient response of the system. In higher order systems the dominant poles are given by the poles which are very close to origin, provide all other poles are lying far away from the dominant poles. The poles which are far away from the origin will have less effect on the transient response of the system.

The transfer function of higher order systems can be approximated to a second order transfer function, whose standard form of closed loop transfer function is

$$\frac{C(s)}{R(s)} = \frac{\omega_n^2}{s^2 + 2\zeta\omega_n s + \omega_n^2}$$

The dominant poles (*sd* and *sd**) are given by the roots of quadratic factor,

$$(s^2 + 2\zeta\omega_n s + \omega_n^2) = 0$$

$$\therefore \quad s_d = \frac{-2\zeta\,\omega_n \pm \sqrt{4\zeta^2\omega_n{}^2 - 4\omega_n{}^2}}{2}$$

$$= -\zeta\omega_n \pm j\omega_n\sqrt{1-\zeta^2}$$

The dominant pole can be plotted on the *s*-plane as shown in Fig. (3.53)

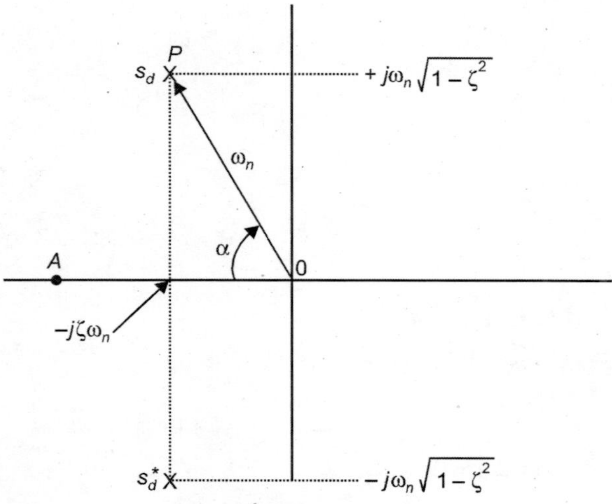

Fig. 3.53: Dominant pole, s_d.

In the right angle triangle, *OAP*

$$\cos\alpha = \frac{\zeta\omega_n}{\omega_n} = \zeta$$

$$\therefore \qquad \alpha = \cos^{-1} \zeta$$

To fix a dominant pole on root locus, draw a line at an angle of $\cos^{-1} \zeta$ with respect to negative real axis. The meeting point of this line with root locus will give the location of dominant pole. The value of K corresponding to dominant pole can be obtained from magnitude condition.

The gain K corresponding to dominant pole,

$$s_d = \frac{\text{Product of length of vectors from open loop poles to dominant pole}}{\text{Product of length of vectors from open loop zeros to dominant pole}}$$

Advantages of Root Locus

- The root locus technique is an important tool in designing control systems with desired performance characteristics. The desired performance of the system can be achieved by adjusting the location of its closed-loop poles in the s-plane by varying one (or) more system parameters.
- The root locus can be plotted in the s-plane by varying a system parameter (usually gain, K) over the complete range of values. The roots corresponding to a particular value of the system parameter can then be located on the locus (or) the value of the parameter for a desired root location an be determined from the locus.
- The root locus technique is also used for stability analysis. Using root locus the range of values of K, for a stable system can be determined. It is also easier to study the relative stability of the system from the knowledge of location of closed loop poles. The dominant roots are used to estimate the damping ratio and natural frequency of oscillation of the system. From ζ and ω_n the time domain specifications can be calculated.

3.15.1 General Layout of Rod locus:

EXAMPLE 3.13

A unity feedback control system has an open loop transfer function $G(s) = \frac{K}{s(s^2 + 4s + 13)}$. Sketch the root locus.

Solution:

Step 1: To locate poles and zeros:

The poles of open loop transfer function are the roots of the equation, $s(s^2 + 4s + 13) = 0$

The roots of the quadratic are,

$$s = -4 \pm \frac{\sqrt{4^2 - 4 \times 13}}{2} = -2 \pm j3$$

$\therefore$ The poles are $0, -2 + j3$ and $-2 - j3$

The poles are marked by X (cross) as shown in Fig. (Ex. 3.13.1)

Step 2: To find the root locus on real axis:

There is only one pole on real axis at the origin. Hence if we choose any test point on the negative real axis then to the right of that point the total number of real poles and zeros is one, which is an odd number. Hence the entire negative real axis will be part of root locus. The root locus on real axis shown as a bold line in Fig. (Ex. 3.13.1).

Note: For the given transfer function one root locus branch will start at the pole at the origin and travel through the negative real axis to meet the zero at infinity.

Step 3: To find angles of asymptotes and centroid:

Since there are s poles, the number of root locus branches are three. There is no finite zero. Hence all the three root locus branches ends at zeros at infinity. The number of asymptotes required are three.

$$\text{Angles of asymptotes} = \frac{\pm 180° \, (2q+1)}{n-m}$$

Here $n = 3$, and $m = 0$

$$\text{If } q = 0, \text{ Angles} = \frac{\pm 180°}{3} = \pm 60°$$

$$\text{If } q = 1, \text{ Angles} = \frac{180° \times 3}{3} = \pm 180°$$

$$\text{If } q = 2, \text{ Angles} = \frac{\pm 180° \times 5}{3} = \pm 300° = \pm 60°$$

$$\text{If } q = 3, \text{ Angles} = \frac{\pm 180° \times 7}{3} = \pm 420° = \pm 60°$$

Since ±60є is repeating, the angle of asymptotes = ±60є.

$$\text{Centroid} = \frac{\text{Sum of poles} - \text{Sum of zeros}}{n-m}$$

$$= \frac{0 - 2 + j3 - 2 - j3 - 0}{3} = \frac{-4}{3} = -1.33$$

The centroid is marked on real axis and from the centroid the angles of asymptotes are marked using a protractor. The asymptotes are drawn as dotted lines as shown in Fig. (Ex. 3.13.1)

Step 4: To find the break away and breakin points:

The closed loop transfer function,

$$\frac{C(s)}{R(s)} = \frac{G(s)}{1+G(s)} \qquad [\because H(s) = 1]$$

$$\frac{C(s)}{R(s)} = \frac{K}{s(s^2+4s+13)+K}$$

The characteristic equation is

$$s\,(s^2 + 4s + 13) + K = 0$$

$$s^3 + 4s^2 + 13s + K = 0$$

$$\therefore \qquad K = -s^3 - 4s^2 - 13s$$

on differentiating the equation of K w.r.to s we get

$$\frac{dK}{ds} = -(3s^2 + 8s + 13)$$

Put

$$\frac{dK}{ds} = 0$$

$$\therefore \quad -(3s^2 + 8s + 13) = 0$$

$$3s^2 + 8s + 13 = 0$$

$$\therefore \quad s = \frac{-8 \pm \sqrt{8^2 - 4 \times 13 \times 3}}{2 \times 3}$$

$$= -1.33 \pm j\,1.6$$

check for K

when $s = (-1.33 + j\,1.6)$, the value of K is given by

$$K = -(s^3 + 4s^2 + 13s)$$

$$= -[-1.33 + j(1.6)^3 + 4(-1.33 + j1.6)^2 + 13(-1.33 + j1.6)]$$

$$\neq \text{Positive and real}$$

when $s = -1.33 - j1.6$. The value of K is not equal to real and positive.

Since the value of K for $s = (-1.33 \pm j1.6)$ are not real and positive, the points are not on actual break away (or) break in points. The root locus has neither breakaway nor breakin point.

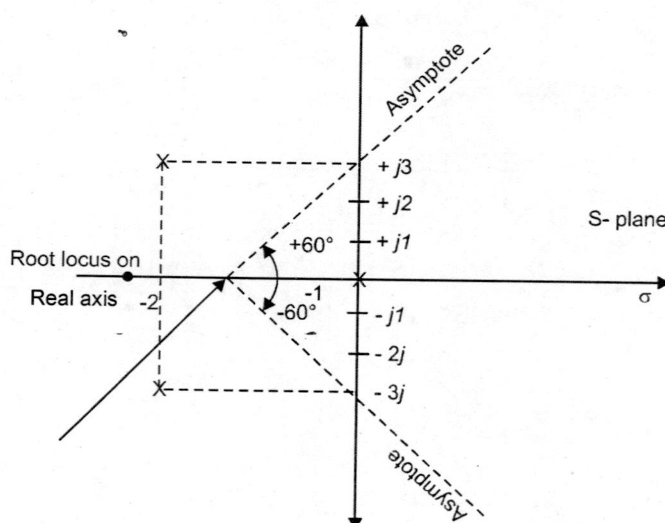

Fig. Ex. 3.13.1: Figure showing the asymptotes, root locus on real axis and location of poles and centroid.

Step 5: To find angle of departure

Let us consider the complex pole A as shown in Fig. (Ex. 3.13.1). Draw vectors from all other poles to the Pole A as shown in Fig. (Ex. 3.13.1). Let the angles of these vectors be θ_1 and θ_2

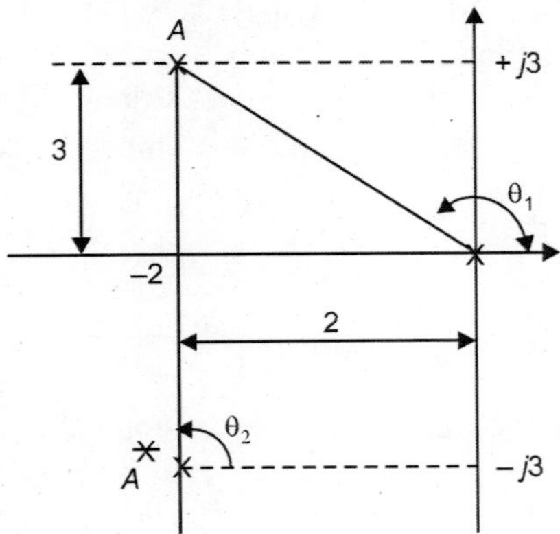

Fig. Ex. 3.13.1

Here, $\theta_1 = 180° - \tan^{-1}(3/2) = 123.7°$

$\theta_2 = 90°$

Angle of departure from the complex pole A

$$= 180° - (\theta_1 + \theta_2)$$

$$= 180° - (123.7° + 90°) = -33.7°$$

The angle of departure at complex pole A^* is negative of the angle of departure at complex pole A.

$\therefore$ Angle of departure at pole $A^* = +33.7$

Make the angles of departure at complex pole using protractor.

Step 6: To find the crossing point on imaginary axis:

The characteristic equation is given by $s^3 + 4s^2 + 13s + K = 0$

Put $s = j\omega$

$$(j\omega)^3 + 4(j\omega)^2 + 13(j\omega) + K = 0$$

$$-j\omega^3 - 4\omega^2 + 13\,j\omega + K = 0$$

on equating imaginary part to zero,

we get, $-\omega^3 + 13\omega = 0$

$$-\omega^3 = -13\omega$$

$$\omega = \pm\sqrt{13}$$

$$\omega = \pm 3.6$$

on equating real part to zero, we get,

$$-4\omega^2 + K = 0$$

$$K = 4\omega^2$$

$$K = 4(13) = 52$$

$$K = 52$$

The crossing point of root locus is $\pm j3.6$. The value of the K at this crossing point is $K = 52$.

The complete root locus sketch is shown in Figure (Ex. 3.13.3). The root locus has three branches one branch starts at the pole at origin and travel through negative real axis to meet the zero at infinity. The other two root locus branches starts at complex poles (along the angle of departure) crosses the imaginary axis at $\pm j3.6$ and travel parallel to asymptotes to meet the zeros at infinity.

EXAMPLE 3.14

Sketch the root locus of the system whose open loop transfer function is $G(s) = \dfrac{K}{s(s+2)(s+4)}$.

Find the value of K so that the damping ratio of the closed loop system is 0.5.

Solution:

Step 1: To locate poles and zeros

The poles of open loop transfer function are the roots of the equation, $s(s + 2)\ (s + 4) = 0$.

$\therefore$ The poles are, $s = 0, -2, -4$.

The poles one marked by X (cross) as shown in Figure (Ex. 3.14.1)

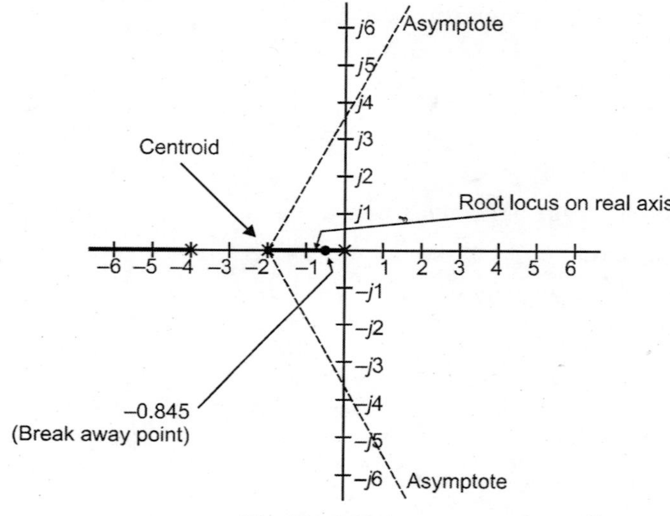

Fig. Ex. 3.14.1

Step 2: To find root locus on real axis

There are 3 poles on the real axis. Choose test point on real axis between $s = 0$ and $s = -2$. To the right of this point the total number of real poles and zeros is one, which is an odd number. Hence the real axis between $s = 0$ and $s = -2$ will be part of root locus. Chose a test point on real axis between $s = -2$ and $s = -4$. to the right of this point the total number of real poles and zeros is two which is an even number. Hence the real axis between $s = -2$ and $s = -4$ will not be a part of root locus.

Choose a test point on real axis to the lest of $s = -4$. To the right of this point, the total number of real poles and zeros is three, which is an odd number. Hence the entire negative real axis from $s = -4$ to $-\infty$ will be a part of root locus.

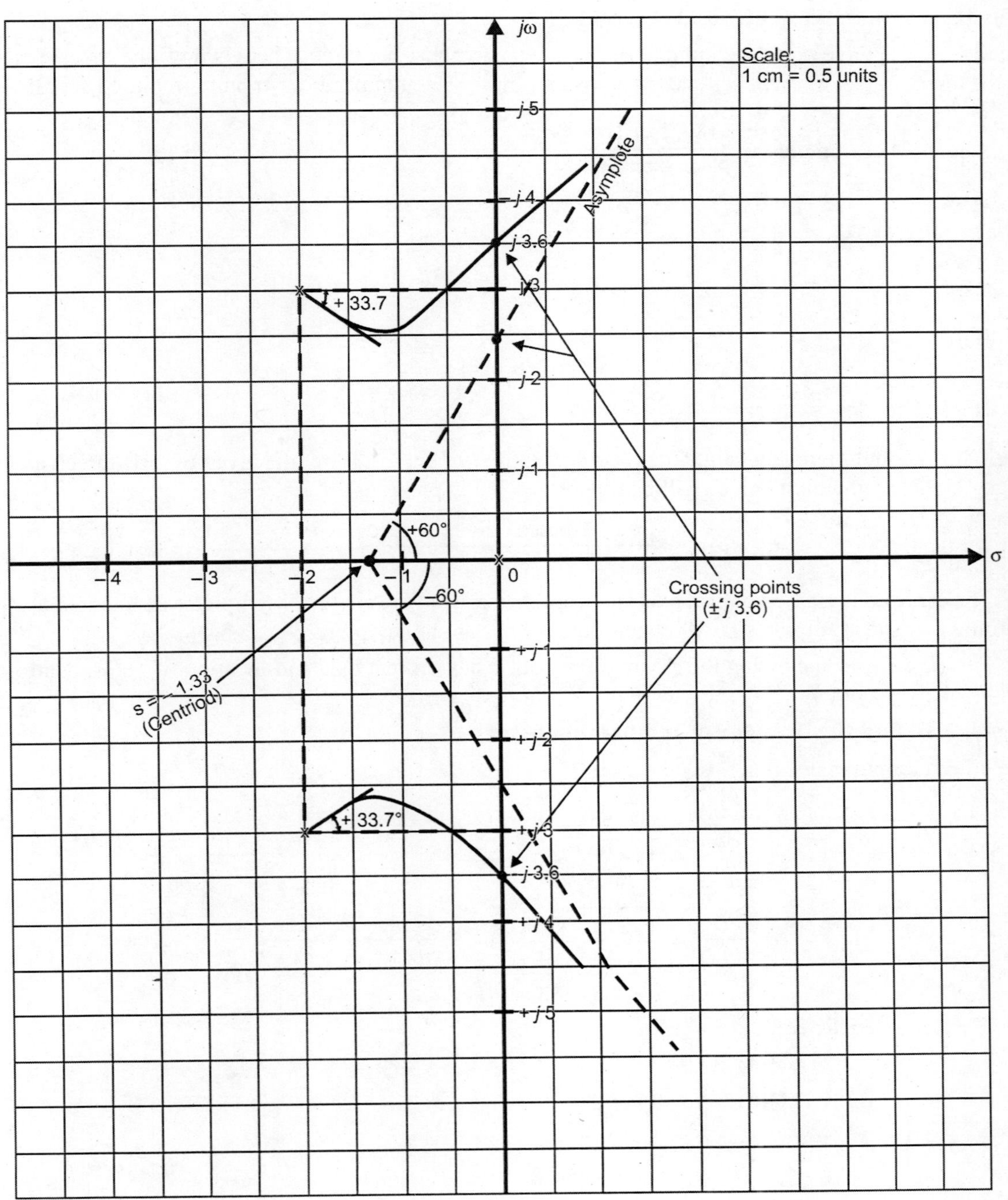

Fig. Ex. 3.13.3: Root locus of $1 + G(s) = 1 + \dfrac{K}{s(s^2 + 4s + 13)}$.

Step 3: To find asymptotes and centroid

Since there are three poles the number of root locus branches are three. There is no finite zero. Hence all the three root locus branches ends at zeros at infinity. The number of asymptotes required are three.

$$\text{Angles of asymptotes} = \pm\frac{180(2q+1)}{n-m}$$

Hence $n = 3$ and $m = 0$

$$\text{If } q = 0, \text{ angle} = \pm\frac{180}{3} = \pm 60$$

$$\text{If } q = 1, \text{ angles} = \pm\frac{180\times 3}{3} = \pm 180°$$

$$\text{If } q = 2, \text{ angles,} = \pm\frac{180\times 5}{3} = \pm 300° = \mp 60°$$

Note: It is enough if you calculate the required number of angles. Here it is given by first three values of angles. The remaining values will be repetitions of the previous values.

$$\text{Centroid} = \frac{\text{Sum of poles} - \text{Sum of zeros}}{n-m} = \frac{0-2-4-0}{3} = -2.$$

The centroid is marked on real axis and from the centroid the angle of asymptotes are marked using a protractor. The asymptotes are drawn as dotted lines as shown in Fig. (Ex. 3.14.1).

Fig. (Ex. 3.14.1) showing the asymptotes, root locus on real axis and location of poles, centroid and breakaway points.

Step 4: To find the breakaway and breakin point

The closed loop transfer function,

$$\frac{C(s)}{R(s)} = \frac{G(s)}{1+G(s)} \qquad [\text{Œ}H(s) = 1]$$

$$\therefore \quad \frac{C(s)}{R(s)} = \frac{\dfrac{K}{s(s+2)(s+4)}}{1+\dfrac{K}{s(s+2)(s+4)}} = \frac{K}{s(s+2)(s+4)+K}$$

The characteristic equation is given by

$$s(s+2)\,(s+4) + K = 0$$

$$s(s^2 + 6s + 8) + K = 0$$

$$s^3 + 6s^2 + 8s + K = 0$$

$$\therefore \quad K = -s^3 - 6s^2 - 8s$$

On differentiating the equation of K with respect to s we get,

$$\frac{dK}{ds} = -(3s^2 + 12s + 8)$$

Put $$\frac{dK}{ds} = 0$$

∴ $$-(3s^2 + 12s + 8s) = 0$$

$$(3s^2 + 12s + 8s) = 0$$

$$s = \frac{-12 \pm \sqrt{12^2 - 4 \times 3 \times 8}}{2 \times 3} = -0.815 \text{ (or)} - 3.154$$

Check for K

when $s = -0.845$, the value of K is given by

$$K = [-(-0.845)^3 + 6(-0.0845)^2 + 8(-0.845)] = 3.08$$

Since K is positive and real for $s = -0.845$, this point is actual breakaway point,

when $s = -3.154$, the value of K is given by

$$K = [(-3.154)^3 + 6(-3.154)^2 + 8(-3.154)] = -3.08$$

Since K is negative for $s = -3.154$, this is not a actual breakaway point. The breakaway point is marked on the negative real axis as shown in Fig. (Ex. 3.14.1).

Step 5: To find angle of departure

Since there are no complex pole or zero, we need not find angle of departure or arrival.

Step 6: To find the crossing point of imaginary axis

The characteristic equation is given by

$$s^3 + 6s^2 + 8s + K = 0$$

Put $$s = j\omega$$

$$(j\omega)^3 + 6(j\omega)^2 + 8(j\omega) + K = 0$$

$$-j\omega^3 - 6\omega^2 + j8\omega + K = 0$$

Equating imaginary part of zero

$$-j\omega^3 + j8\omega = 0$$

$$-j\omega^3 = j8\omega$$

$$\omega^2 = 8$$

$$\omega = \pm\sqrt{8} = \pm 2.8$$

$$\omega = \pm 2.8$$

Equating real part to zero

$$-6\omega^2 + K = 0$$

$$K = 6\omega^2 = 6 \text{ Ч } 8 = 48$$

$$K = 48$$

The crossing point of root locus ± 2.8. The value of K corresponding to this point is $K = 48$. (This is two limiting value of K for the stability of the system.

The complete root locus sketch is shown in Fig. (3.14.2). The root locus has three branches. One branch starts at the pole at meet the zero at infinity. The other two root locus branches starts at

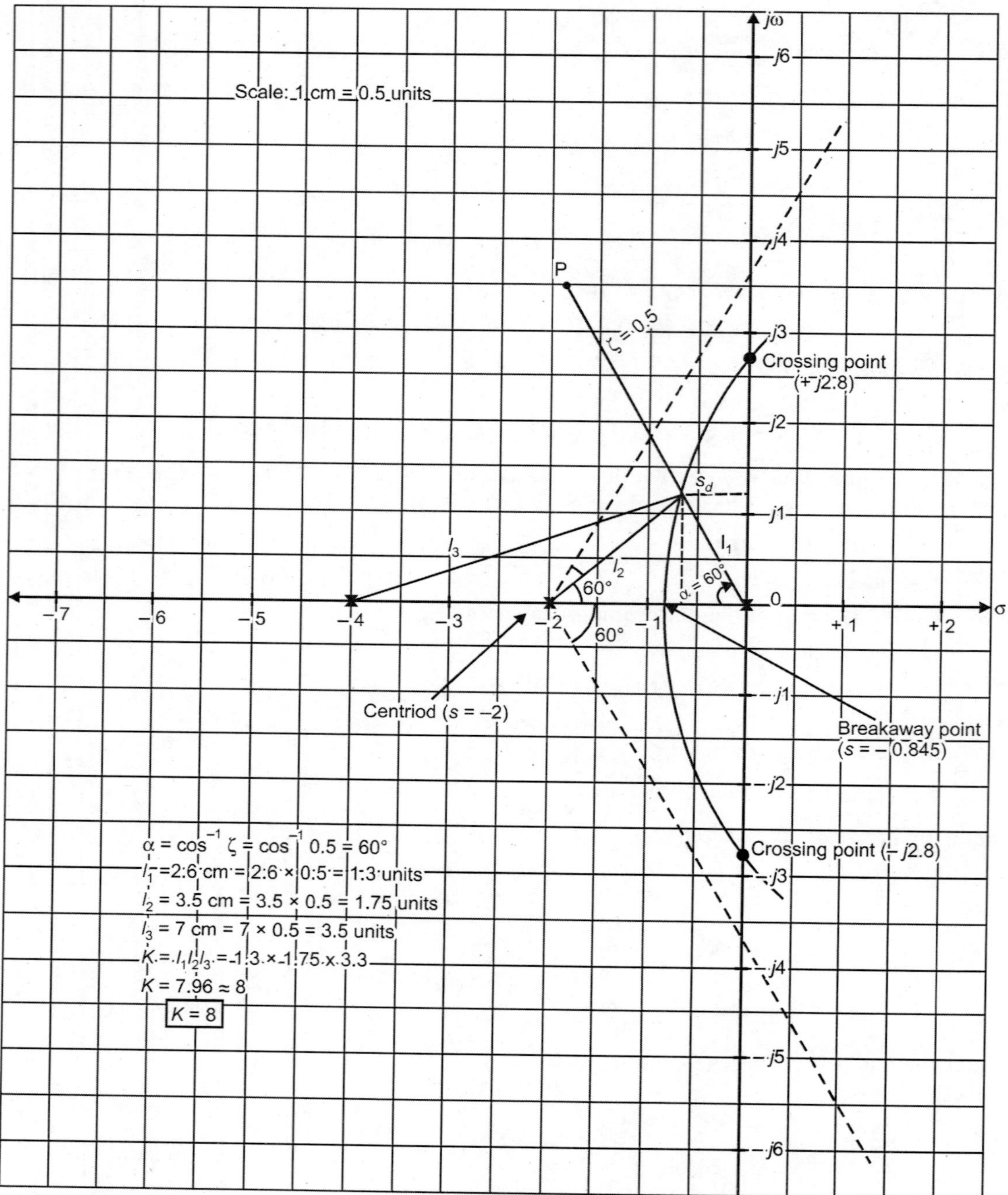

Fig. Ex. 3.14.2: Root locus of $1 + G(s) = 1 + \dfrac{1}{s(s+2)(s+4)}$.

$s = 0$ and $s = -2$ and travel through negative real axis, breakaway from real axis at $s = -0.845$, then crosses imaginary axis at $s = \pm j2.8$ and travel parallel to asymptotes to meet the zeros at infinity.

To find the value of K corresponding to $\zeta = 0.5$

Given that $\zeta = 0.5$

Let $\alpha = \cos^{-1} \zeta = \cos^{-1} 0.5 = 60°$.

Draw a loine OP, such that the angle between line OP and negative real axis is 60° ($\alpha = 60°$) as shown in Fig. (Ex. 3.14.2). The meeting point of the line OP and root locus gives the dominant pole, *sd*.

The value of K corresponding to the point, $s = s_d$

$$K = \frac{\text{Product of length of vector from all poles to the points, } s = s_d}{\text{Product of length of vectos from all zeros to the point, } s = s_d}$$

$$= \frac{l_1 . l_2 . l_3}{1}$$

K = 1.3 Ч 1.75 Ч 3.5 = 7.96 ≈ 8

Result: $K = 8$ at $\zeta = 0.5$.

EXAMPLE 3.15

The open loop transfer function of a unity feedback system is given by $G(s) = \dfrac{K(s+9)}{s(s^2+4s+11)}$. Sketch the root locus of the system.

Solution:

Step 1: To locate poles and zeros

The poles of open loop transfer function are the root locus of the equation

$$s(s^2 + 4s + 11) = 0$$

The roots of the quadratic are,

$$s = \frac{-4 \pm \sqrt{4^2 - 4 \times 11}}{2} = -2 \pm j\,2.64$$

∴ The poles are lying at s = $0, -2 + j2.64, -2 - j2.64$

The zeros are lying at $s = -9$ and infinity

The poles are marked by *X*(cross) and zeros by "0" (circle) as shown in Fig. (Ex. 3.15.1).

Step 2: To find root locus on real axis

One pole and one zero lines on real axis choose a test point to the left of $s = 0$, then to the right of this point, the total number of poles and zeros is one which is an odd numbers. Hence the portion of real axis from $s = 0$ to $s = -9$ will be a part of root locus. If we choose a test point of the left of $s = -9$ then to the right of this point, the total number of pole and zeros is two, which is an even number. Hence the real axis from $s = -9$ to $-\infty$ will not be a part of root locus. The root locus on real axis is shown as bold line in Fig. (Ex. 3.15.1).

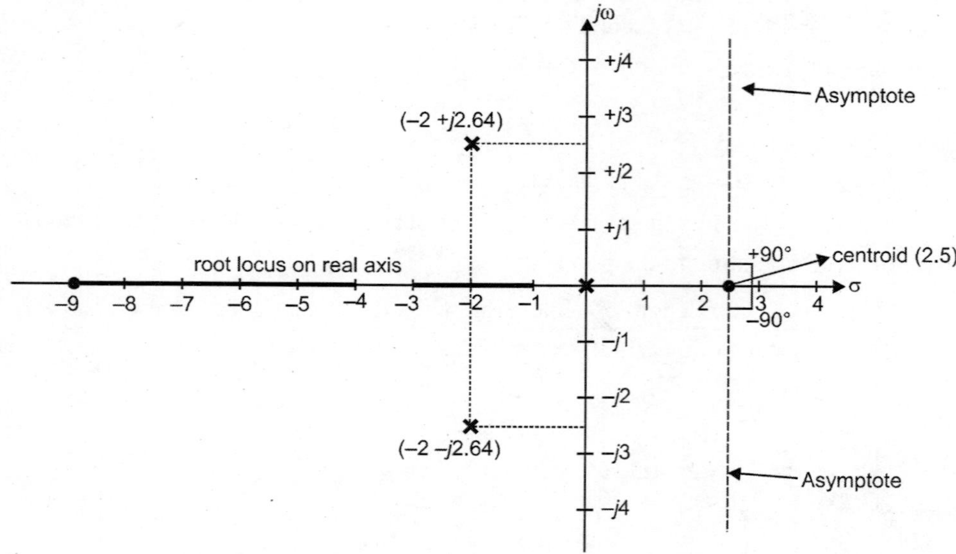

Fig. Ex. 3.15.1: Showing the asymptotes root locus on real axis and location of poles, zeros and centroid.

Step 3: To find angles of asymptotes and centroid

Since there are 3 poles, the number of root locus branches are three one root locus branch start at the pole at origin and travel along negative real axis to meet the zero at $s = -9$. The other tow root locus branches meet the zeros at infinity, The number of asymptotes required are 2.

$$\text{Angle of asymptotes} = \pm\frac{180°(2q+1)}{n-m}$$

Here $n = 3$ and $m = 1$

$$\text{If } q = 0, \text{ angles} = \pm\frac{180}{2} = \pm 90°$$

$$\text{If } q = 2, \text{ angles} = \pm\frac{180\times 3}{2} = \mp 90°$$

$$\text{If } q = 3, \text{ angles} = \pm\frac{180\times 5}{2} = \pm 90°$$

Note: It is enough if you calculate the required number of angles. Here it is given by first two value of angles. The remaining values will be repetitions of the previous values.

$$\text{Centroid} = \frac{\text{Sum of poles} - \text{sum of zeros}}{n-m}$$

$$= 0 - 2 + j2.64 - 2 - j\,2.64 - (-9)/2 = 2.5$$

The centroid is marked and from the centroid, the angles of asymptotes are makrked using a protractor. The asymptotes are drawn as dotted lines as shown in Fig. (Ex. 3.15.1).

Step 4: To find the breakaway and breakin points

From the location of poles and zeros and from the knowledge of typical sketches of root locus, it can be concluded that there is no possibility of breakaway (or) breakin points.

Step 5: To find the angle of departure

Let us consider the complex pole A as shown in Fig. (Ex. 3.15.2). Draw vectos from all other poles and zero to the pole A. Let the angles of these vectors be θ_1, θ_2 and θ_3.

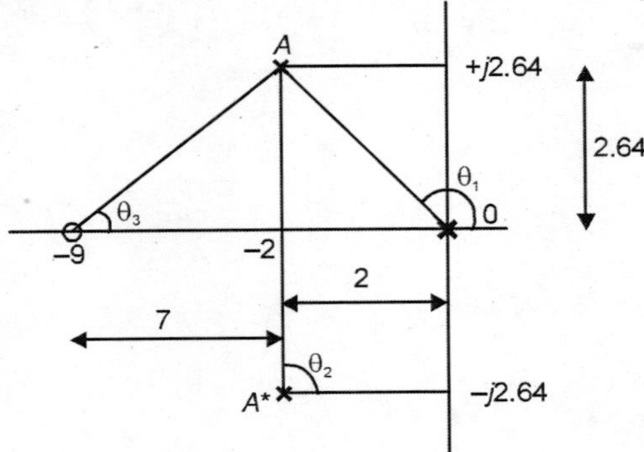

Fig. Ex. 3.15.2

Here $$\theta_1 = 180° - \tan^{-1}\frac{2.64}{2} = 127.1\epsilon$$

$$\theta_2 = 90°$$

$$\theta_3 = \tan^{-1}\frac{2.64}{7} = 20.7\epsilon$$

Angle of departure from the complex pole A

$$= 180° - (\theta_1 + \theta_2 + \theta_3)$$

$$= 180° - (127.1 + 90) + 20.7$$

$$= -16.4°$$

The angle of departure at the complex pole A^* is negative of the angle of departure at complex pole A,

$\therefore$ Angle of departure at pole $A^* = -(-16.4) = +16.4°$

Mark the angles of departure at complex poles using protractor.

Step 6: To find the crossing point of imaginary axis

The closed loop transfer fucntion,

$$\frac{C(s)}{R(s)} = \frac{G(s)}{1+G(s)}$$

$$= \frac{\dfrac{K(s+9)}{s(s^2+4s+11)}}{1+\dfrac{K(s+9)}{s(s^2+4s+11)}}$$

$$= \frac{K(s+9)}{s(s^2+4s+11)+K(s+9)}$$

The characteistic equation is the denominator polynomial of $\dfrac{C(s)}{R(s)}$

∴ The characteristic equations,

$$s(s^2 + 4s + 11) + K(s + 9) = 0$$

$$(s^3 + 4s^2 + 11s) + Ks + 9K = 0$$

Put $s = j\omega$

$$(j\omega)^3 + 4(j\omega)^2 + 11(j\omega) + K(j\omega) + 9K = 0$$

$$-j\omega^3 - 4\omega^2 + j11\omega + 9K = 0$$

on equating imaginary part to zero,

$$-j\omega^3 + j\,11\,\omega + jK\omega = 0$$

$$-j\omega^3 = -j\,11\,\omega - jK\omega$$

$$\omega^2 = 11 + K$$

Put $K = 8.8$

∴ $\omega^2 = 11 + 8.8 = 19.8$

$$\omega = \pm\sqrt{19.8}$$

$$\omega = \pm\, 4.4$$

on equating real part to zero

$$-4\omega^2 + 9K = 0$$

$$9K = 4\omega^2$$

Put $\omega^2 = 11 + K$

∴ $9K = 4(11 + K)$

$$9K = 44 + 4K$$

$$9K - 4K = 44$$

$$5K = 44$$

$$K = \frac{44}{5} = 8.8$$

$$K = 8.8$$

The crossing point of root locus is $\pm j\,4.4$. The value of K at this crossing point is $K = 8.8$ (This is limiting value of K for the stabiltiy of the system).

The complete root locus sketch is shown in Fig. (Ex. 3.15.3). The root locus has three branches one branch starts at pole at origin and travel through neagtive real axis to meet the zero at $s = -9$. The other two root locus branches starts at complex poles (along the angle of departure) crosses the imaginary axis at $\pm j\,4.4$ and travel parallel t asymptotes to meet the zeros at infinity.

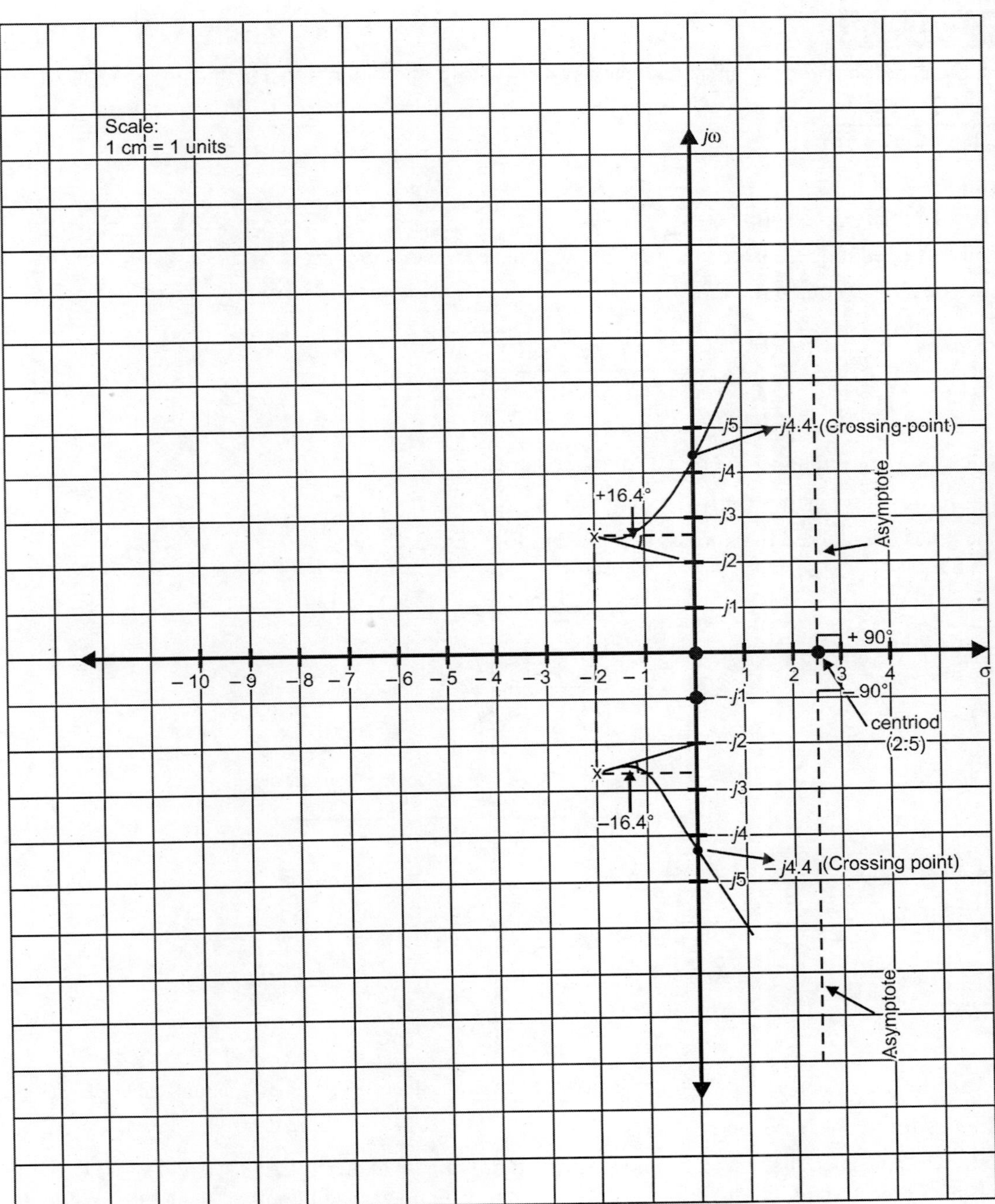

Fig. Ex. 3.15.3: Root locus sketch of $1 + G(s) = 1 + \dfrac{K(s+9)}{s(s^2+4s+11)}$.

EXAMPLE 3.16

Sketch the root locus for the unity feedback system whose open loop transfer funciton is $G(s)\,H(s) = \dfrac{K}{s(s+4)(s^2+4s+20)}$.

Solution:

Step 1: To locate poles and zeros

The poles of open loop transfer function are the roots of the equation, $s(s + 4)\,(s^2 + 4s + 20) = 0$.

The roots of the quadratic are,

$$s = \frac{-4 \pm \sqrt{4^2 - 4 \times 1 \times 20}}{2}$$

$$= -2 \pm j4$$

$\therefore$ The poles are lying at $s = 0, -4, -2 + j4$ and $-2 - j4$.

The zeros are lying at infinity.

The poles are marked by x(cross) as shown in Fig. (Ex. 3.16.1).

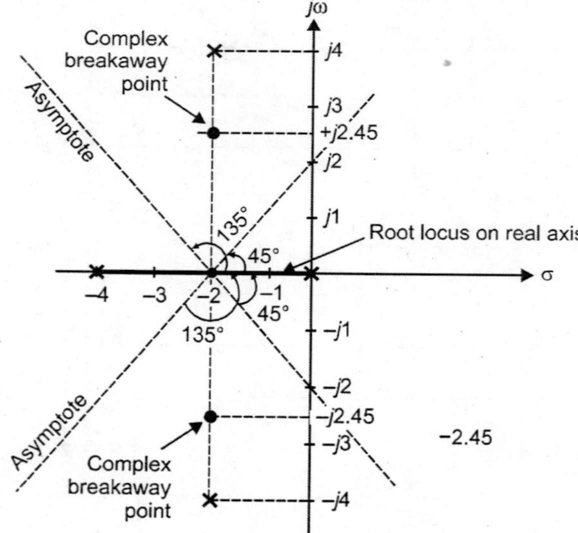

Fig. Ex. 3.16.1: Showing the asymptotes, root locus on real axis and location of poles, centroid and breakaway.

Step 2: To find root locus on real axis

There are two polesa on the real axis. Choose a test point on real axis between $s = 0$ and $s = -4$. To the right of this point, the total number of real poles is one which is an odd number. hence the real axdis between $s = 0$ and $s = -4$ will be a part of root locus. Choose a test point to the left of $s = -4$, now to the right of this test point the total number of poles and zeros is two which is even number. Hence the real axis from $s = -4$ to $s = -\infty$ will not be a part of root locus. The root locus on real axis is shown as a bold line in Fig. (Ex. 3.16.1).

Step 3: To find angles of asymptotes and centroid

Since there are four poles, the number of root locus branches are four. There is no finite zero. Hence all the four root locus branchse ends at zeros at infinity. Hence the number of asymptotes required is four.

$$\text{Angle of asymptotes} = \frac{\pm 180(2q+1)}{n-m}$$

Here $n = 4$ and $m = 0$

If $q = 0$, angles $= \pm\dfrac{180}{4} = \pm 45°$

If $q = 1$, angles $= \pm\dfrac{180\times 3}{4} = \pm 135°$

If $q = 2$, angles $= \pm\dfrac{180\times 5}{4} = \pm 225° = \mp 135°$

If $q = 3$, angles $= \pm\dfrac{180\times 7}{4} = \pm 315° = \mp 45°$

Note: It is enoguh if you calculate the required number of angles. Here it is given by first four value of angles. The remaing values will be repetitions of the previous values.

$$\text{Centroid} = \frac{\text{Sum of poles} - \text{Sum of zeros}}{(n-m)}$$

$$= \frac{0-4-2+j4-2-j4-0}{4-0} = -\frac{8}{4} = -2.$$

The centroid is marked on real axis and from the centroid, the angles of asymptotes are marked using a protractor. The asymptotes are drawn as dotted lines as shwon in Fig. (Ex. 3.16.1).

Step 4: To find the breakaway and breakin point

The closed loop transfer function, $\dfrac{C(s)}{R(s)} = \dfrac{G(s)}{1+G(s)}$

$$\therefore \quad \frac{C(s)}{R(s)} = \frac{\dfrac{K}{s(s+4)(s^2+4s+20)}}{1+\dfrac{K}{s(s+2)(s^2+4s+20)}} = \frac{K}{s(s+4)(s^2+4s+20)+K}$$

The charactertistic equation is given by

$$s(s+4)(s^2+4s+20) + K = 0$$

$$\therefore \quad K = -s(s+4)(s^2+4s+20) = -(s^2+4s)(s^2+4s+20)$$

$$K = -(s^4 + 8s^3 + 36s^2 + 80s)$$

CHAPTER 3

on differentiating the equation of K with respect to s we get,

$$\frac{dK}{ds} = -(4s^3 + 24s^2 + 72s + 80)$$

Put $$\frac{dK}{ds} = 0$$

$\therefore$ $$-(4s^3 + 24s^2 + 72s + 80) = 0$$

$$4s^3 + 24s^2 + 72s + 80 = 0$$

on dividing by 4 we get,

$$s^3 + 6s^2 + 18s + 20 = 0$$

The equation $s^3 + 6s^2 + 18s + 20 = 0$ will have atleast one real root. By trial and error, the real root is found to be $s = -2$ **(Refer Lin's method)**

The polynomial $(s^3 + 6s^2 + 18s + 20) = 0$ can be expressed as

$$s^3 + 6s^2 + 18s + 20 = (s + 2)(s^2 + 4s + 10) = 0$$

The root of the quadratic $(s^2 + 4s + 10)$ are given by,

$$s = \frac{-4 \pm \sqrt{4^2 - 4 \times 10}}{2} = -2.45$$

Check for K

When $s = -2$, $$K = -(s^4 + 8s^3 + 36s^2 + 80s)$$

$$= -[(-2)^4 + 8(-2)^3 + 36(-2)^2 + 80(-2)]$$

$$= -[-64] = 64$$

When $s = -2 \pm j\,2.45$,

$$K = -[s^4 + 8s^3 + 36s^2 + 80s]$$

$$= -[3.16\angle\pm129°)^4 + 8(3.16\angle\pm129°)^3 + 36(3.16\angle\pm129°)^2$$

$$+80(3.16\angle\pm129°)$$

$$= -[99.7\angle\pm156° + 252.4\angle\pm27° + 359.5\angle\pm258° + 252.8\angle\pm129°]$$

For postive value of angles,

$$K = -[-91 + j40° + 225 + j115 - 75 - j351 - 159 + j196]$$

$$= -[-100] = 100$$

For negative values of angles,

$$K = -[-91 + j140 + 225 + j115 - 75 - j351 - 159 - j196]$$

$$= -[100] = 100$$

For all the roots of the equation , the value of K is positive and real. Hence all the three roots are actual breakaway points. The breakaway points are shown in Fig. (Ex. 3.16.1).

Step 5: To find angle of departure

Let us consider the complex pole A shown is Fig. (Ex. 3.16.2). Draw vectos from all other poles to the pole A as shwon is Fig. (Ex. 3.16.2). Let angles of these vector be θ_1, θ_2 and θ_3.

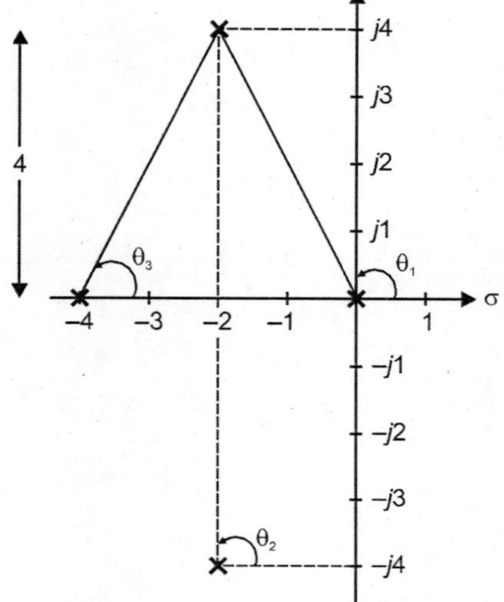

Fig. Ex. 3.16.2

Hence,

$$\theta_1 = 180° - \tan^{-1}\frac{4}{2} = 117°$$

$$\theta_2 = 90°$$

$$\theta_3 = \tan^{-1}\frac{4}{2} = 63°$$

Angle of departure from complex pole A,

$$= 180° - (\theta_1 + \theta_2 + \theta_3)$$

$$= 180° - (117 + 90 + 63) = -90°$$

The angle of departure at complex pole A^* is negative of the angle of departure at complex pole A.

∴ Angle of departure from complex pole $A^* = +90°$ mark the angle of departure at complex poles using protactor.

Step 6: To find the crossing point on imaginary axis

The characteristic equation is given by,

$$s^4 + 8s^3 + 36s^2 + 80s + K = 0$$

Put $s = j\omega$,

$$(j\omega)^4 + 8(j\omega)^3 + 36(j\omega)^2 + 80j\omega + K = 0$$

$$\omega^4 - j\omega^3 - 36(\omega)^2 + j80\omega + K = 0$$

on equating imaginary part of zero,

$$-j8\omega^3 + j80\omega = 0$$

$$-j8\omega^3 = -j80\omega$$

$$\omega^2 = 10$$

$$\omega = \pm\sqrt{10} = \pm 3.2$$

on equating real part of zero,

$$\omega^4 - 36\omega^2 + K = 0$$

$$K = -\omega^4 + 36\omega^2$$

Put $\qquad \omega^2 = 10$

$\therefore \qquad K = -(10)^2 + (36 \text{ Ч } 10) = 260.$

The crosing point of root locus is ± 3.2. The value of K at this crossing point is $K = 260$. (This is the limiting value of K for stability).

The complete root locus is sketched as shown in Fig. (Ex. 3.16.3). The root locus has four branches. All the root locus branches goes to infinity along the asymptotic lines to meet the zero at infinity.

To find real root of $\dfrac{dK}{ds} = 0$ by Lin's method,

The first trial divisor is choosen as the last two terms of the polynomial.

Ist Trial:

$$\text{Trial divisor} = 18s + 20 = s + \frac{20}{18} = s + 1.11$$

$$\begin{array}{r|l} & s^2 + 4.89s + 12.57 \\ \hline s + 1.11 & s^3 + 6s^2 + 18s + 20 \\ & s^3 + 1.11s^2 \\ \hline & 4.89s^2 + 18s \\ & 4.89s^2 + 5.43s \\ \hline & 12.57s + 20 \quad \leftarrow \text{Next Trial Divisor} \\ & 12.57s + 13.95 \\ \hline & \underline{6.05} \end{array}$$

IInd Trial:

$$\text{Trial divisor} = 12.57\,s + 20 = s + \frac{20}{12.57} = s + 1.59$$

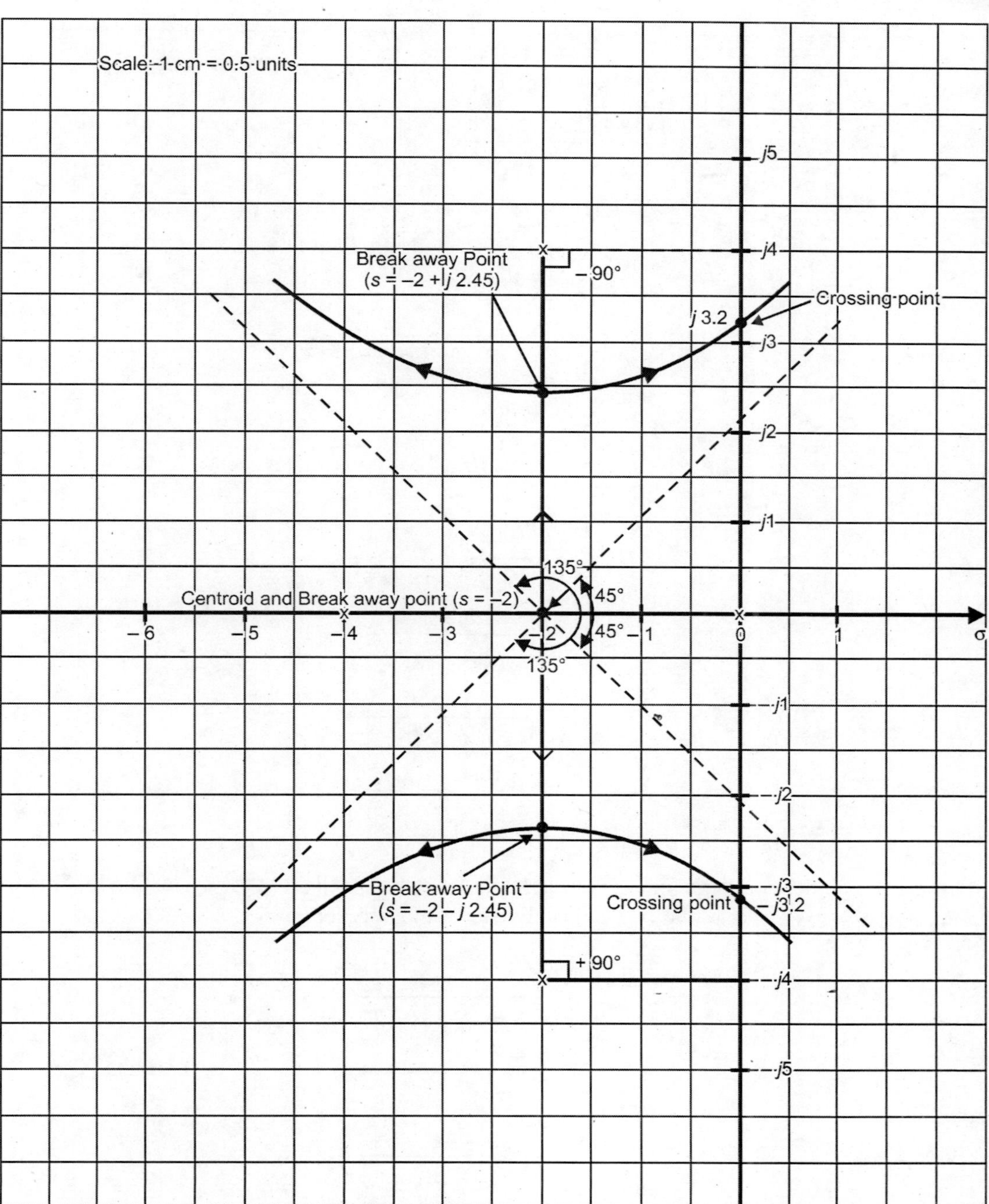

Fig. Ex. 3.16.3: Root locus sketch of $1 + G(s) = 1 + \dfrac{K}{s(s+4)(s^2+4s+20)}$.

$$\begin{array}{r|l} & s^2 + 4.41s + 11 \\ \hline s + 1.59 & s^3 + 6s^2 + 18s + 20 \\ & s^3 + 1.59s^2 \\ \hline & 4.41s^2 + 18s \\ & 4.41s^2 + 7s \\ \hline & 11s + 20 \quad \leftarrow \text{Next Trial Divisor} \\ & 11s + 17.49 \\ \hline & 2.51 \end{array}$$

IIIrd Trial:

$$\text{Trial divisor} = 11s + 20 = s + \frac{20}{11} = s + 1.82$$

$$\begin{array}{r|l} & s^2 + 4.18s + 10.4 \\ \hline s + 1.89 & s^3 + 6s^2 + 18s + 20 \\ & s^3 + 1.82s^2 \\ \hline & 4.18s^2 + 18s + 20 \\ & 4.18s^2 + 7.6s \\ \hline & 10.4s + 20 \\ & 10.4s + 18.9 \\ \hline & 1.1 \end{array}$$

Since the remainder converge for every trial, let us approximate the root to $s = -2$. On dividing the polynomial by $s + 2$, we found that $(s + 2)$ is a divisor of the polynomial.

$$\begin{array}{r|l} & s^2 + 4s + 10 \\ \hline s + 2 & s^3 + 6s^2 + 18s + 20 \\ & s^3 + 2s^2 \\ \hline & 4s^2 + 18s \\ & 4s^2 + 8s \\ \hline & 10s + 20 \\ & 10s + 20 \\ \hline & 0 \end{array}$$

The real root is to be $s = -2$.

EXAMPLE 3.17

Sketch the root locus for the unity feedback system whose open loop transfer function is $G(s)H(s) = \frac{K(s+1.5)}{s(s+1)(s+5)}$.

Solution:

Step 1: To locate poles and zeros

The poles of open loop transfer function are the roots of the equation, $s(s + 1)(s + 5) = 0$ and the zeros are the roots of the equation $(s + 1.5) = 0$.

The poles are lying at $s = 0, -1, -5$.

The zeros and lying at $s = -1.5$ and infinity. The poles are marked by X (cross) and zeros by "0" (circle) as shown in Fig. (Ex. 3.17.1).

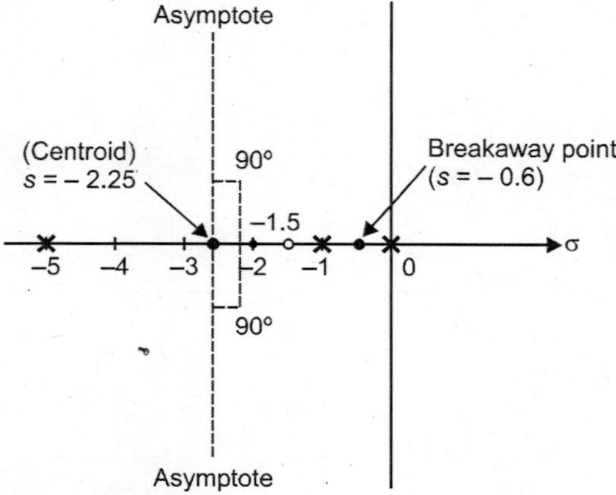

Fig. Ex. 3.17.1: Showing the asymptote, root locus on real axis and location of poles, zeros, centroid and breakaway point.

Step 2: To find root locus on real axis

The segment of real axis between $s = 0$ and $s = -1$ and the segment of real axdis between $s = -1.5$ and $s = -5$ will be part of root locus. Before if we choose a test point in this segment to the right of this point we have odd number of real poles and zeros. The root locus on real axis are shwon as bold lines is Fig. (Ex. 3.17.1).

Step 3: To find angles of asymptotes and centriod

Since there are three poles, the number of root locus branches are three. There is no finite zero, so one root locus branch will end at finite zero. The other two branches will meet the zero at infinity. Hence the number of asymptotes required is two.

$$\text{Angle of asymptotes} = \pm\frac{180(2q+1)}{n-m}$$

Here $n = 3$ and $m = 1$,

CHAPTER 3

If $q = 0$, angles $= \pm\frac{180}{2} = \pm 90$

$$\text{Centroid} = \frac{\text{Sum of poles} - \text{Sum of zeros}}{n-m}$$

$$= \frac{0-1-5-(-1.5)}{2} = -2.25$$

The centroid is marked on real axis and from the centroid the angles of asymptotes are marked using a protractor. The asymptotes are drawn as dotted lines as shown in Fig. (Ex. 3.17.1).

Step 4: To find the breakaway and breakin points

The closed loop transfer function,

$$\frac{C(s)}{R(s)} = \frac{G(s)}{1+G(s)}$$

$$= \frac{\frac{K(s+1.5)}{s(s+1)(s+5)}}{1+\frac{K(s+1.5)}{s(s+1)(s+5)}}$$

$$= \frac{K(s+1.5)}{s(s+1)(s+5)+K(s+1.5)}$$

The characteristic equation is, $s(s + 1)(s + 5) + K(s + 1.5) = 0$

$$\therefore \quad K = \frac{-(s+1)(s+5)}{s+1.5} = \frac{-s(s^2+6s+5)}{s+1.5}$$

$$= \frac{-(s^3+6s^2+5s)}{s+1.5}$$

on differentiating K with respect to s we get,

$$\frac{dK}{ds} = \frac{-(3s^2+12s+5)(s+1.5)-[s^3+6s^2+5s](1)}{(s+1.5)^2}$$

$$= \frac{-3s^2-4.5s^2-12s^2-18s-5s-7.5+s^3+6s^2+9s}{(s+1.5)^2}$$

$$= \frac{-2s^3-10.5s^2-18s-7.5}{(s+1.5)^2} = \frac{-2(s^3+5.25s^2+9s}{(5+1.5)^2}$$

For $\frac{dK}{ds} = 0$, the numerator should be zero.

$$\therefore \quad s^3 + 5.25s^2 + 9s + 3.75 = 0$$

The third order polynomial will have one real root. The real root of the above polynomial can be determined by Lin's method (Refer Lin's Method)

The polynomial cane bexpressed $s^3 + 5.25s^2 + 98 + 3.75 = (s + 0.6)(s^2 + 4.65s + 6.2) = 0$.

The roots of the quadratic $(s^2 + 4.65s + 6.2)$ are $\dfrac{s = -4.65 \pm \sqrt{(4.65)^2 - 4 \times 6.2}}{2}$ =-2.3 1 j 0.89.

Check for K

When $s = -0.6$. $K = -(s^3 + 6s^2 + 5s)/s + 1.5$

$$= \frac{-\left[(-0.6)^3 + 6(-0.6)^2 + 5(-0.6)\right]}{-0.6 + 1.5} = 1.17$$

For $s = -0.6$, the value of K is positve and real and so it is actual breakaway point. It can be shown that for $s = -2.3 \pm j\,0.86$ the value of K is not positve and real and so they cannot be breakaway point. The actual breakaway point is shown in Fig. (Ex. 3.17.1)

To find the real root of $s^3 + 5.25\,s^2 + 3.75 = 0$ by Lin's method.

The least two terms of the polynomial are choosen as I[st] trial divisor.

I[st] trial:

I[st] trial divisor = $9s + 3.75$

$s + 3.75/9$

$s = s + 0.42$

$$\begin{array}{r|l} & s^2 + 4.83s + 6.97 \\ \hline s + 0.42 & s^3 + 5.25s^2 + 9s + 3.75 \\ & s^3 + 0.42s^2 \\ \hline & 4.83s^2 + 9s \\ & 4.83s^2 + 2.03s \\ \hline \text{II}^{nd}\text{ trial divisor} \longrightarrow & 6.97s + 3.75 \\ & 6.97s + 2.93 \\ \hline & 0.82 \end{array}$$

IInd trial:

IInd divisor = $6.97s + 3.75$

$= s + \dfrac{3.75}{6.97}$

$= s + .54$

$$\begin{array}{r|l} & s^2 + 9.713s + 6.46 \\ \hline s + .54 & s^3 + 5.25s^2 + 9s + 3.75 \\ & s^3 + .54s^2 \\ \hline & 4.71s^2 + 9s \\ & 4.71s^2 + 2.54s \\ \hline \text{III}^{rd}\text{ trial divisor} \longrightarrow & 6.46s + 3.75 \\ & 6.46s + 3.49 \\ \hline & .26 \end{array}$$

IIIrd trial:

IIIrd trial divisor = $6.46s + 3.75$

$= s + \dfrac{3.75}{6.46}$

IVth trial divisor = $6.3\,s + 3.75$

$= s + \dfrac{3.75}{6.3} = s + 0.6$

$$
\begin{array}{r|l}
 & s^2 + 4.67s + 6.3 \\
\hline
s + .58 & s^3 + 5.25s^2 + 9s + 3.75 \\
 & s^3 + .58s^2 \\
\hline
 & 4.67s^2 + 9s \\
 & 4.67s^2 + 2.7s \\
\hline
\text{IV}^{\text{th}} \text{ trial divisor} \longrightarrow & 6.3s + 3.75 \\
 & 6.3s + 3.65 \\
\hline
 & 0.1
\end{array}
\qquad
\begin{array}{r|l}
 & s^2 + 4.65s + 6.2 \\
\hline
s + 0.6 & s^3 + 5.25s^2 + 9s + 3.75 \\
 & s^3 + 0.6s^2 \\
\hline
 & 4.65s^2 + 9s \\
 & 4.65s^2 + 2.8s \\
\hline
 & 6.2s + 3.75 \\
 & 6.2s + 3.72 \\
\hline
 & .03
\end{array}
$$

On neglecting the small value of 0.03, one of the root of the polynomial is, $s = -0.6$

Step 5: To find angle of departure

Since there are no complex pole or zero we need not find angle of departure or arrival.

Step 6: To find crossing point of imaginary axis

The characteristic equation is

$$s(s + 1)(s + 5) + K(s + 1.5) = 0$$

$$s(s^2 + 6s + 5) + Ks + 1.5K = 0$$

$$s^3 + 6s^2 + 5s + Ks + 1.5K = 0$$

Put $s = j\omega$

$$(j\omega)^3 + 6(j\omega)^2 + 5(j\omega) + K(j\omega) + 1.5K = 0$$

$$-j\omega^3 - 6\omega^2 + j5\omega + jK\omega + 1.5K = 0$$

On equating imaginary part to zero we get,

$$-j5\omega^3 = jK\omega = 0$$

$$-j\omega^3 = -j5\omega - jK\omega$$

$$\omega^2 = 5 + K$$

On equating real part to zero we get,

$$-6\omega^2 + 1.5K = 0$$

Put $\omega^2 = 5 + K$

$$-6(5 + K) + 1.5K = 0$$

$$-30 - 4.5K = 0$$

$$K = \frac{30}{4.5} = -6.67$$

Since the value of K is negative, there is no crossing point on imaginary axis, or for any positive values of K the root locus will not cross imaginary axis.

The complete root locus sketch is shown in Fig.(Ex. 3.17.2). The root locus has three branches.

One branch starts at $s = -5$ and ends at finite zero at $s = -1.5$. The other two root locus starts at $s = 0$ and $s = -1$ and breakaway from real axis at $s = -0.6$, then travel parallel to asymptotes to meet the zeros get infinity.

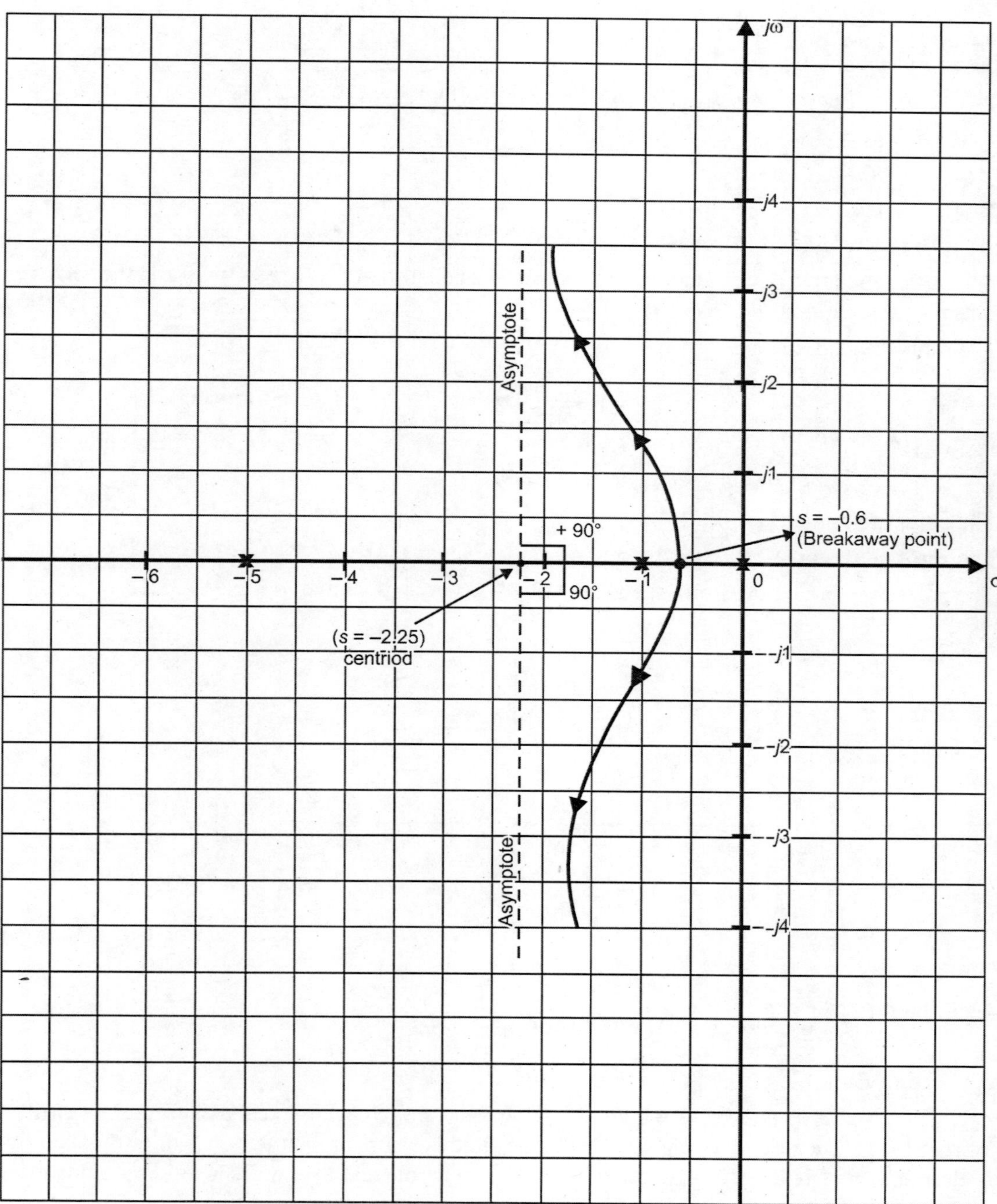

Fig. Ex. 3.17.2: Root locus sketch of $1 + G(s) = 1 + K(s + 1.5)/s(s + 1)(s + 5)$.

EXAMPLE 3.18

Sketch the root locus for the unity feedback system whose open loop transfer function is

$$G(s) = \frac{K(s^2 + 6s + 25)}{s(s+1)(s+2)}.$$

Solution:

Step 1: To locate poles and zeros

The poles of open loop transfer function are roots of the equation $s(s + 1)(s + 2) = 0$ and the zerros are the roots of the equation $(s^2 + 6s + 25) = 0$.

The roots of quadratic are,

$$s = \frac{-6 \pm \sqrt{6^2 - 4 \times 25}}{2}$$

$$= -3 \pm j4$$

The poles are lying at $s = 0, -1, -2$.

The zeros are lying at $s = -3 + j4, -3 - j4$.

The poles are marked by X(cross) and zeros by "o"(circle as shown in Fig. (Ex. 3.18.1).

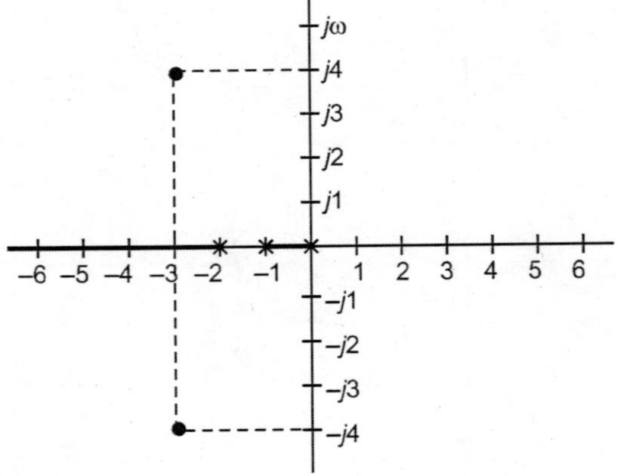

Fig. Ex. 3.18.1: Showing the root locus on real axis, location of poles and zeros.

Step 2: To find root locus on real axis

The segment of real axis between $s = 0$ and $s = -1$ and the entire negative real axis from $s = -2$ will be part of root locus. Because if we choose a test point in this segment then to the rigth of this point we have odd number of real poles and zeros. The root locus on real axis are shown as a bold line in Fig. (Ex. 3.18.1).

Step 3: To find angles of asymptotes and centroid

Since there are three poles the number of root locus branches are three. There are two finite zeros, so two root locus branch will end at finite zeros. The thrid root locus will meet the zero at finity by travelling througy negative real axis. Here the number of asymptotes is one and the angle fo asymptores is ± 180є.

Step 4: To find the breakways and breaking points:

The closed loop transfer function

$$\frac{C(s)}{R(s)} = \frac{G(s)}{1+G(s)} = \frac{\dfrac{K(s^2+6s+25)}{s(s+1)(s+2)}}{1+\dfrac{K(s^2+6s+25)}{s(s+1)(s+2)}}$$

$$= \frac{K(s^2+6s+25)}{s(s+1)(s+2)+K(s^2+6s+25)}$$

The characteristic equation is,

$s(s+1)(s+2) + K(s^2+6s+25) = 0$

$$\therefore \quad K = \frac{-s(s+1)(s+2)}{s^2+6s+25} = \frac{-s(s^2+3s+2)}{s^2+6s+25}$$

$$= \frac{-s^3-3s^2-2s}{s^2+6s+25}$$

On differentiating K with respect to s we get,

$$\frac{dK}{ds} = \frac{(-3s^2-6s-2)(s^2+6s+25)-(s^3-3s^2-2s)(2s+6)}{(s^2+6s+25)^2}$$

$$= \frac{-(s^4+12s^3+91s^2+150s+50)}{(s^2+6s+25)^2}$$

For $\dfrac{dK}{ds} = 0$, the numerator should be zero.

$\therefore \quad s^4 + 12s^3 + 91s^2 + 150s + 50 = 0$

The fourth order polynomial can be split into two quadratic equations. The two quadratic factors can be obtained by lin's method. (Refer Lin's method).

To find quadratic factors by Lin's method.

The first trial divisor can be last three terms.

Ist trial:

$$\text{Ist trial divisor} = 91s^2 + 150s + 50$$

$$= s^2 + \frac{150}{91}s + \frac{50}{91} = s^2 + 1.65s + .55$$

$$
\begin{array}{r|l}
 & s^2 + 10.35s + 73.37 \\
\hline
s^2 + 1.65s + .55 & s^4 + 12s^3 + 91s^2 + 150s + 50 \\
 & s^4 + 1.65s^3 + 0.55s^2 \\
\hline
 & 10.35s^3 + 90.45s^2 + 150s \\
 & 10.35s^3 + 17.08s^2 + 5.7s \\
\hline
\text{II}^{\text{nd}} \text{ trial divisor} \longrightarrow & 73.37s^2 + 144.3s + 50 \\
 & 73.37s^2 + 121.1s + 40.35 \\
\hline
 & 23.2s + 9.65
\end{array}
$$

IInd trial:

IInd trial divisor = $73.37s^2 + 144.4s + 50$

$$= s^2 + \frac{1443}{73.37}s + \frac{50}{73.37} = s^2 + 2s + .7$$

$$
\begin{array}{r|l}
 & s^2 + 10s + 70.3 \\
\hline
s^2 + 2s + .7 & s^4 + 12s^3 + 91s^2 + 150s + 50 \\
 & s^4 + 2s^3 + .7s^2 \\
\hline
 & 10s^3 + 90.3s^2 + 150s \\
 & 10s^3 + 20s^2 + 7s \\
\hline
 & 70.3s^3 + 143s + 50 \\
 & 73.3s^2 + 140.6s + 49.2 \\
\hline
 & 2.4s + 0.8
\end{array}
$$

On neglecting the small remainder we can write

$s^4 + 12s^3 + 91s^2 + 150s + 50$ Ë $(s^2 + 2s + .7)\,(s^2 + 10s + 70.3)$

The roots of the quadratic $(s^2 + 2s + .7)$ are

$$s = \frac{-2 \pm \sqrt{2^2 - 4 \times .7}}{2} = -0.45, -1.55$$

The roots of the quadratic $(s^2 + 10s + 70.3)$ are

$$s = \frac{-10 \pm \sqrt{10^2 - 4 \times 70.3}}{2} = -5 \pm j6.73$$

Here, $s = 1.55$ is not a point on root locus, hence it cannot be a breakaway point.

Check the other three values for actual breakway point

when $$s = 0.45,\ K = \frac{-s^3 - 3s^2 - 2s}{s^2 + 6s + 25}$$

$$= \frac{-(-4.5)^3 - 3(-.45)^2 - 2(-.45)}{(-.45)^2 + 6(-.45) + 25} = 0.017$$

For $s = -0.45$, the value of K is positive and real and so it is actual breakaway point. It can be shown that for $s = -5 \pm j6.73$ the value of K is not positive and real and so thay cannot be breakwasy points. The actual breakwasy point is shown in Fig. (Ex. 3.18.1).

Step 5: To find angle of arival.

Let us consider the complex zero. A shown in Fig. (Ex. 3.18.2). Draw vectors from all other poles to the A as shown in Fig. (Ex. 3.18.2). Let the angles of these vectors be θ_1, θ_2, θ_3 and θ_4.

Hence,

$$\theta_1 = 180\epsilon \tan^{-1}\frac{4}{3} = 126.9\epsilon$$

$$\theta_2 = 180\epsilon \tan^{-1}\frac{4}{2} = 116.6\epsilon$$

$$\theta_3 = 180\epsilon \tan^{-1}\frac{4}{1} = 104\epsilon$$

$$\theta_4 = 90\epsilon$$

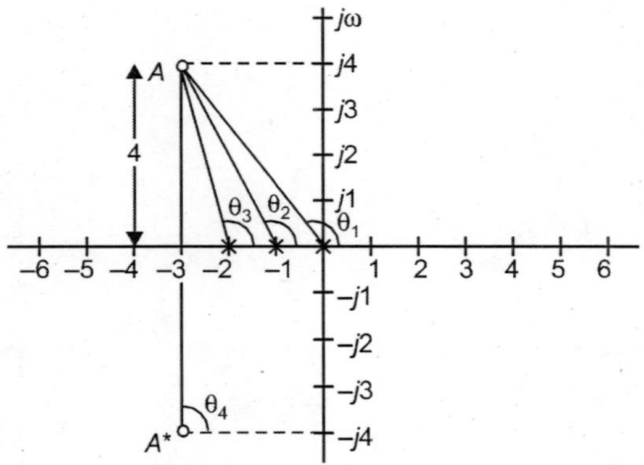

Fig. Ex. 3.18.2

Angle of arrival at complex zero A

$$= 180\epsilon - (\theta_4) + (\theta_1 + \theta_2 + \theta_3)$$

$$= 180\epsilon - 90\epsilon + 126.9\epsilon + 116.6\epsilon + 10.4\epsilon$$

$$= 437.5\epsilon = 77.5\epsilon$$

Angle of arrival at complex zero A^* is negative of the angle of arrival at complex zero A.

∴ Angle of arrival at complex zero $A^* = -77.5\epsilon$

Mark the angles of arrival at complex zeros using protractor.

Step 6: To find the crossing point on imaginary axis:

The characteristic equation is

$$s(s+1)(s+2) + K(s^2 + 6s + 25) = 0$$

$$s(s^2 + 3s + 2) + Ks^2 + 6Ks + 25K = 0$$

$$s^3 + 3s^2 + 2s + Ks^2 + 6Ks + 25K = 0$$

$$s^3 + (3 + K)s^2 + (2 + 6K)s + 25K = 0$$

Put $s = j\omega$

$$(j\omega)^3 + (3 + K)(j\omega)^2 + (2 + 6K)(j\omega) + 25K = 0$$

$$-j\omega^3 - (3 + K)\omega^2 + j(2 + 6K)\omega + 25K = 0$$

On equating imaginary part to zero

$$-j\omega^3 + j(2 + 6K)\omega = 0$$

$$-j\omega^3 = -j(2 + 6K)\omega$$

$$\omega^2 = (2 + 6K)$$

On equating real part to zero,

$$-(3 + K)\omega^2 + 25K = 0$$

Put $\omega^2 = 2 + 6K$

$$-(3 + K)(2 + 6K) + 25K = 0$$

$$-(6 + 18K + 2K + 6K^2) + 25K = 0$$

$$-6K^2 + 5K - 6 = 0$$

$$K = \frac{-5 \pm \sqrt{5^2 - 4 \times (-6)(-6)}}{2 \times (-6)}$$

$$K = 0.4 \pm j\,0.9$$

Since the value of K is not real and positive, there is no crossing point on imaginary axis, or for any positive values of K the root locus will not cross imaginary axis.

Step 7: To find points on root locus

Choose test points a, b, c, d on the s plane and adjust the test points to satisfy angle criterion. The test points are shown in Fig. (Ex. 3.18.3).

On the upper half of splane the root locus in sketched through the test points a, b, c and d. The root locus on the lower half of s-plane is the mirror image of the root locus on the upper half of splane.

The complete root locus sheten is shown in Fig. (Ex. 3.18.3). the root locus has three branches, one branch starts at $s = -2$ and goes to infinity along negative real axis. The other two root locus branches starts at $s = 0$ and $s = -1$ and breaks from real axis at $s = -0.45$, then meets the complex zeros.

EXAMPLE 3.19

Sketch the root locus for the unity feedback system whose open loop transfer function is $G(s) = K/s(s^2 + 6s + 10)$.

Solution:

Step 1: To locate poles and zeros

The poles of open loop transfer function are the roots of the equation $s(s^2 + 6s + 10) = 0$.

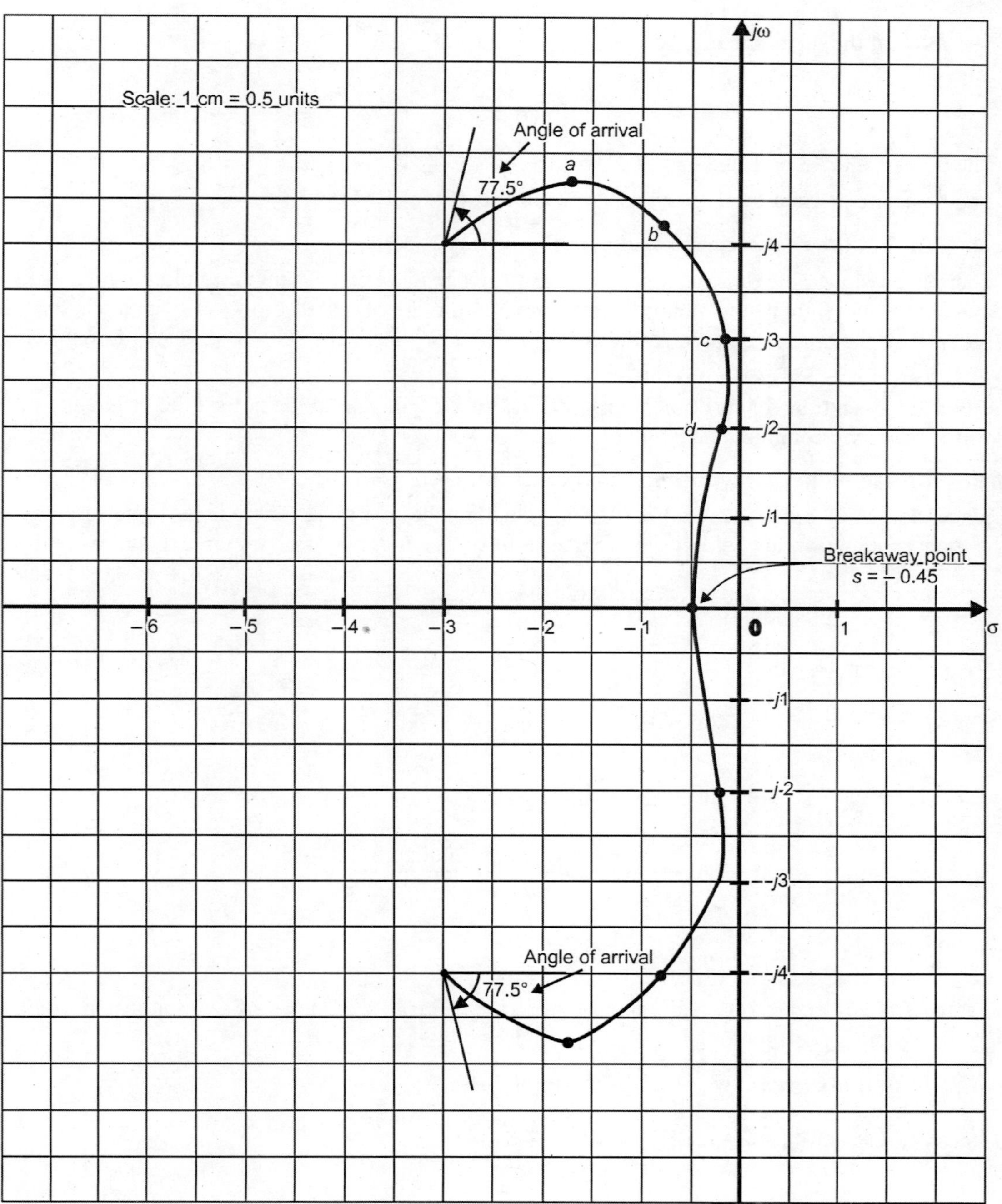

Fig. Ex. 3.18.3: Root locus sketch of $1 + G(s) = 1 + \dfrac{K(s^2 + 6s + 2s)}{s(s+1)(s+2)}$.

The roots of the quadratic are,

$$s = \frac{-6 \pm \sqrt{6^2 - 4 \times 10}}{2}$$

The poles are 0, $-3 + j1$ and $-3 - j1$.

The poles are marked by Ч(cross) as shown in Fig. (Ex. 3.19.1).

Step 2: To find the root locus on real axis.

There is only one pole on real axis at the origin. Hence if we choose any test point on the negative real axis then to the right of that point the total number of real poles and zero sis one, which is san odd number. Hence the entire negtaive real axis will be part of root locus. The root locus on real axis is shown as bold line is Fig. (Ex. 3.19.1).

Note: For the given transfer function are root locus branch will start at the pole at the origin and meet the zero at infinity through the negative real axis.

Step 3: To find angles of asymptotes and centroid:

Since there are 3 poles the number of root locus branches are three. There is no finite zero. hence all the three root locus branches ends at zeros at infinity. The number of asymptores required are three.

$$\text{Angle of asymptotes} = \frac{\pm 180є(2q+1)}{n-m}$$

Here, $n = 3$ and $n = 0$,

If $q = 0$, $\quad \text{Angles} = \pm\frac{180}{3} = \pm 60є$

If $q = 1$, $\quad \text{Angles} = \pm\frac{180 \times 3}{3} = \pm 180є$

$$\text{Centroid} = \frac{\text{Sum of poles} - \text{Sum of zeros}}{n-m}$$

$$\text{Centroid} = \frac{-3 + j1 - 3 - j1}{3} = -2$$

The centroid is marked on real axis and from the centroid the angles of asymptotes are marked using or protractor. The asymptotes are drawn as dotted lines as shown in Fig. (Ex. 3.19.1).

Step 4: To find the breakaway and breaking points

The closed loop transfer function, $\dfrac{C(s)}{R(s)} = \dfrac{G(s)}{1+G(s)}$

$$\therefore \quad \frac{C(s)}{R(s)} = \frac{\dfrac{K}{s(s^2+6s+10)}}{1+\dfrac{K}{s(s^2+6s+10)}} = \frac{K}{s(s^2+6s+10)+K}$$

The characteristic equation is $s(s^2 + 6s + 10) + K = 0$

$$\therefore \quad K = -s(s^2 + 6s + 10) = -s^3 - 6s^2 - 10s$$

on differentiating the equation of K with respect to s we get,

$$\frac{dK}{ds} = -3s^2 - 12s - 10$$

Put $\frac{dK}{ds} = 0$

$$-3s^2 - 12s - 10 = 0$$

$$3s + 12s + 10 = 0$$

$$\therefore \quad s = \frac{-12 \pm \sqrt{12^2 - 4 \times 3 \times 10}}{2 \times 3} = -1.18 \text{ (or)} - 2.82$$

Check for K

when, $s = -1.18$, $K = -s^3 - 6s^2 - 10s$

$$= -(-1.18)^3 - 6(-1.18)^2 - 10(-1.18) = 5.09$$

when, $s = -2.82$, $K = -s^3 - 6s^2 - 10s$

$$= -(-2.82)^3 - 6(-2.82)^2 - 10(-2.82) = 2.91$$

Since the values of K for $s = -1.18$ and -2.82 and positive and real, both the points are actual breakaway or break in points. It can be proved that $s = -2.82$ in breaking point and $s = -1.18$ is a breakaway points. The breakin and breakaway points are shown in Fig. (Ex. 3.19.1)

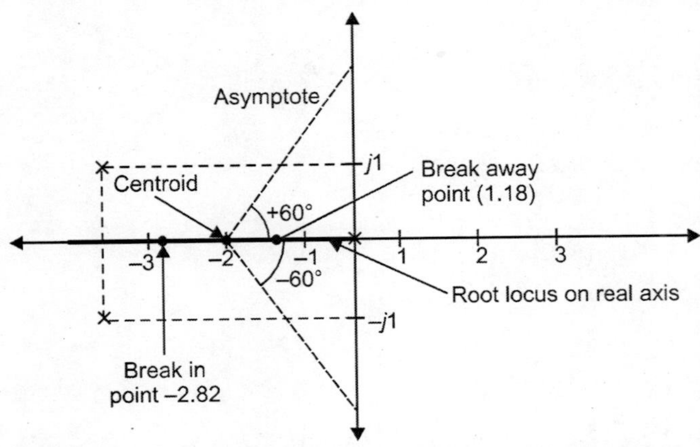

Fig. Ex. 3.19.1: Figure showing the asymptote, root locus on real axis and location of poles, centroid, breakin and breakaway points.

Step 5: To find the angle of departure

Consider the complex pole A shown in Fig. (Ex. 3.19.2) draw vectors from all other poles to the pole A as shown in Fig. (Ex. 3.19.2). Let the angle of these vectors be θ_1 and θ_2.

Here, $\theta_1 = -180\epsilon - \tan^{-1}(1/3) = 161.6\epsilon$

$$\theta_2 = 90\epsilon$$

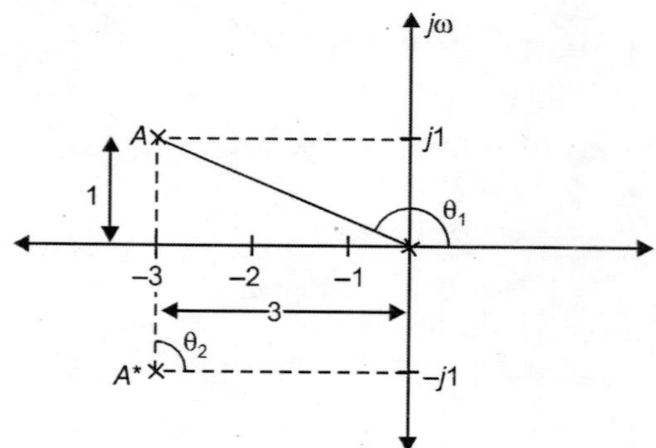

Fig. Ex. 3.19.2

Angle of departure from the complex pole A = 180є – ($\theta_1 + \theta_2$)

= 180є –(61.6є + 90є) = –71.6є ≈ –72є

The angle of departure at complex pole A^* is negative of the angle of departure at complex pole A.

∴ Angle of departure at pole $A^* = +72$

Mark the angles of departure at complex poles using protractors.

Step 6: To find the corssing point on imaginary axis.

The characteristics equationis given by

$$s\,(s^2 + 6s + 10) + K = 0$$

$$s^3 + 6s^2 + 10s + K = 0$$

Put $s = j\omega$

$$(j\omega)^3 + 6(j\omega)^2 + 10(j\omega) + K = 0$$

$$-j\omega^3 - 6\omega^2 + j\,10\omega + K = 0$$

On equating imaginary part to zero we get,

$$-\omega^3 + 10\omega = 0$$

$$\omega^3 = 10\omega$$

$$\omega^2 = 10$$

$$\omega = \pm\sqrt{10} = \pm 3.16 \approx \pm 3.2$$

On equating real part to zero we get,

$$-6\omega^2 + K = 0$$

$K = 6\omega^2 = 6$ Ч $10 = 60 \Rightarrow K = 60$

The root locus crossing imaginary axis at $\pm j32$ and the gain K corresponding to this point is 60. This is the limiting value of K for the stability of the system.

The complete root locus sketch is shown in Fig. (Ex. 3.19.3). The root locus has 3 branches. One branch starts at $s = 0$ and goes to infinity along negative real axis. The other two root locus branches starts at $s = -3 + j1$ and enter the real axis at $s = -2.82$ and then breakaway from real axis at $s = -1.18$. Finally they travel paralllel to asymptotes to meet the zeros at infinity.

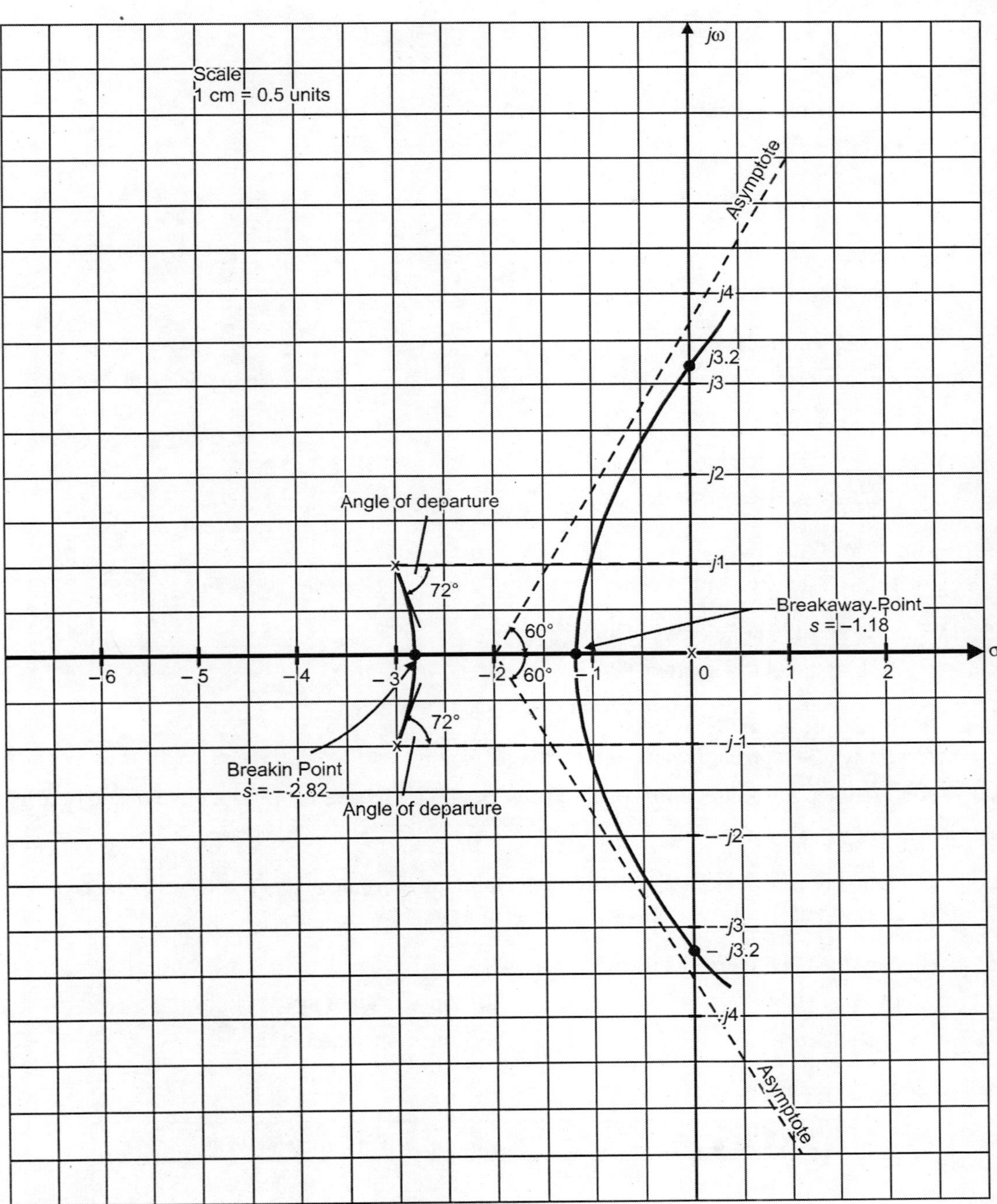

Fig. Ex. 3.19.3: Root 10 cm sketch of $= 1+G(s) = \dfrac{K}{s(s^2+6s+10)}$.

QUESTIONS AND ANSWERS

Q3.1 What are the basic components of an automatic control system?

Ans.
- Error detector
- Amplifier
- Actuator
- Plant
- Feedback system

Q3.2 What is a potentiometer?

Ans. A potentiometer is a device that can be used to convert linear or angular displacement into a voltage.

Q3.3 What are the applications of potentiometer?

Ans.
- It can be used as error detector.

Q3.4 What is syncro?

Ans. Synchro is a device used to convert an angular motion to an electrical signal.

Q3.5 What are all the applications of synchro?

Ans.
- Positional control system
- To transmit angular motion from one place to another.

Q3.6 Why derivative control is not employed in isolation?

Ans. A derivative control mode in isolation provides no corrective efforts for any constant errors.

Q3.7 What is servomotor?

Ans. The motor used in automatic control system or in servomechanism are called servomotors. Which converts electrical signal to angular motion.

Q3.8 What are the characteristics of servomotors?

Ans. 1. Steady state stability 2. Wide range of speed control 3. Fast response 4. Linear characteristics

Q3.9 Compare AC and DC servomotors.

Ans.

DC servomotor	AC servomotor
1. High power output	1. Relatively less power output
2. Linear characteristics	2. Non-liner
3. Fast response	3. Slower response
4. Large power applications	4. Low power applications

Q3.10 What is stepper motor?

Ans. A stepper motor is a device which transforms electrical pulses into equal increments of rotary shaft motion called steps.

Q3.11 What is full-step, half-step and micro-step?

Ans. In steeper motors the number of teeth or poles on the rotor and the number of poles on the stator determine the size of the step (or) step angle this basic size of step is called full-step

$$\text{Full step} = \frac{360\epsilon}{\text{Number of rotor poles} \times \text{No. of stator pole pairs}}$$

Q3.12 How the roots of characteristic equation are related to stability?

Ans. If the roots of characteristic equation has positive real part then the impulse response of the system is not bounded. Hence the system will be unstable. If the roots has negative real parts then the impulse response is bounded. Hence the system will be stable.

Q3.13 What is the necessary and sufficient conditions for stability?

Ans.

1. The necessary condition for stability is that all the coefficients of the characteristic polynomial be positive.
2. All the elements in the 1st volume of the routh array should be positive.

Q3.14 What is the relation between stability and coefficinet of characteristic polynomial?

Ans. If the coefficients of characteristic polynomial are negative or zero, then some of roots lies on right half of s-plane. Hence the system is unstable. If the coefficnets of characteristic polynomial are positive and if no coefficient is zero then there is a possibility of the system to be stable provided all the roots are lying on left half of s-plane.

Q3.15 What is routh stability criterion?

Ans. Routh criterion states that the necessary and sufficient conditions for stability is that all of the elements in the first column of the routh array be positive. If this condition is root met, the system is unstable and the number of sign changes is the elements of the first column of routh array corresponds to the number of roots of characteristic equation is the right half of the s plane.

Q3.16 What is auxiliary polynomial?

Ans. In the construction of routh array a row of all zero indicates the existence of an even polynomial as a factor of the given characteristic equation. In an even polynomial the exponents of s are integers or zero only. This even polynomial factor is called auxiliary polynomial. The coefficients of auxiliary polynomial are given by the elements of the row just above the row of all zeros.

Q3.17 What is limitedly stable system?

Ans. For a bounded input signal, if the output has constant amplitude oscillations then the system may be stable or unstable undersome constraints. Such a system is called limitedly stable system.

Q3.18 What is root locus?

Ans. The path taken be a root of characteristic equation where open loop gain K is varied from 0 to ∞ is called locus.

Q3.19 What are asymptotes and what is centroid?

Ans.
1. Asymptotes are straight lines which are parallel to root locus going to infinity and meet the root locus at infinity.
2. The meeting point of asymptotes with real axis called centroid.

Q3.20 What is breakaway and breaking points?

Ans.
- The point where the root locus breaks from the real axis to enter into the complex plane is called breakaway point.
- The point where the root locus enters the real axis from the complex plane is called break in point.

Q3.21 What is dominant pole?

Ans. The dominant pole is a pair of complex conjugate pole which decides transient response of the system.

In higher order systems the dominant poles are very close to origin and all other poles of the system are widely separated and so they have less effect on transient response of the system.

EXERCISES

Q.1 Using routh criterion determine the locations of the root of the following characteristic equations and commention the stability of the system..

(a) $3s^4 + 10s^3 + 5s^2 + 5s + 3 = 0$

(b) $s^5 + 2s^4 + 24s^3 + 48s^2 - 25s - 50 = 0$

(c) $2s^6 + 4s^5 + s^4 - 32s^3 + 51s^2 + 3s + 15 = 0$

(d) $s^6 + 13s^5 + 5s^4 + 9s^3 + 8s^2 + 6s + 4 = 0$

Q.2 The characteristic equations for certain feedback control systems are given below. In each case, determine the range of values of *K*, for which the system is stable.

(a) $s^4 + 3s^3 + 3s^2 + s + K = 0$

(b) $s^4 + s^3 + s^2 + s + K = 0$

Q.3 Open loop transfer functions of certain using feedback systems are given below. In case determine the location of closed loop poles in the *s* plane, using routh criterion. Comment on the stability of closed loop system.

(a) $G(s) = 200(1 + s)/s(1 + 0.1s)\ (1 + 0.2s)\ (1 + 0.5s)$

(b) $G(s) = 10/(s + 2)\ (s + 4)\ (s^2 + 6s + 25)$

Q.4 The open loop transfer functions of certain unity feedback systems are given below. Sketch the root locus of each system.

(a) $G(s) = K(s + 2)/(s + 3)^2\ (s^2 + 2s + 17)$

(b) $G(s)\ K(s + 4)/s(s + .5)\ (s + 2)$

(c) $G(s) = K/s(s + 3)\ (s^2 + s + 2)$

(d) $G(s) = K/s(s + 2)\ (s^2 + 2s + s)$

(e) $G(s) = K(s + 1)/s^2(s + 12)$

(f) $G(s) = K(s^2 + 1)/s(s + 2)$

Q.5 A unity feedback system has an open loop transfer function, $G(s) = K/s(s^2 + 8s + 32)$. Sketch the root locus and determine the dominant closed loop poles with $\zeta = .5$. Determine the value of *K* at this point.

4

Correlation

4.1 CORRELATION BETWEEN TIME AND FREQUENCY RESPONSE

The Correlation between time and frequency response has an explict form only for first and second order systems. The correlation for second order system is discussed here.

Consider the standard form of transfer function of second order system.

$$\frac{C(s)}{R(s)} = \frac{\omega_n^2}{s^2 + 2\zeta\omega_n s + \omega_n^2}$$

where,

ζ = Damping factor

ω_n = undamped natural frequency the sinusoidal transfer function of the system is obtained by letting $s = j\omega$,

$$\frac{C}{R}(j\omega) = T(j\omega) = \frac{\omega_n^2}{(j\omega)^2 + 2\zeta\omega_n (j\omega) + \omega_n^2}$$

$$= \frac{\omega_n^2}{-\omega^2 + j2\zeta\omega_n (j\omega) + \omega_n^2}$$

$$= \frac{\omega_n^2}{\omega_n^2\left[\dfrac{-\omega^2}{\omega_n^2} + j2\zeta\dfrac{\omega}{\omega_n} + 1\right]} = \frac{1}{1 - \left[\dfrac{\omega}{\omega_n}\right]^2 + j\,2\zeta\left[\dfrac{\omega}{\omega_n}\right]}$$

$$= \frac{1}{(1 - u^2) + j\,2\zeta\,u}$$

where, $u = \left[\dfrac{\omega}{\omega_n}\right]$ is the normalised frequency magnitude of closed loop system,

$$M = |\,T(j\omega)\,| = \frac{1}{\sqrt{(1 - u^2) + (2\zeta u)^2}} \tag{4.1}$$

Phase of closed loop system

$$\alpha = \angle T(j\omega) = -\tan^{-1}\left[\frac{2\zeta u}{1-u^2}\right] \tag{4.2}$$

The magnitude and phase angle characteristics for normalized frequency u, for certain values of ζ are shown in Fig. (4.1) and (4.2). The frequency at which M has a peak value is known as the reasonent frequency. The peak values of the magnitude in the reasonent peak M_r. At this frequncy the slope of the magnitude curve is zero.

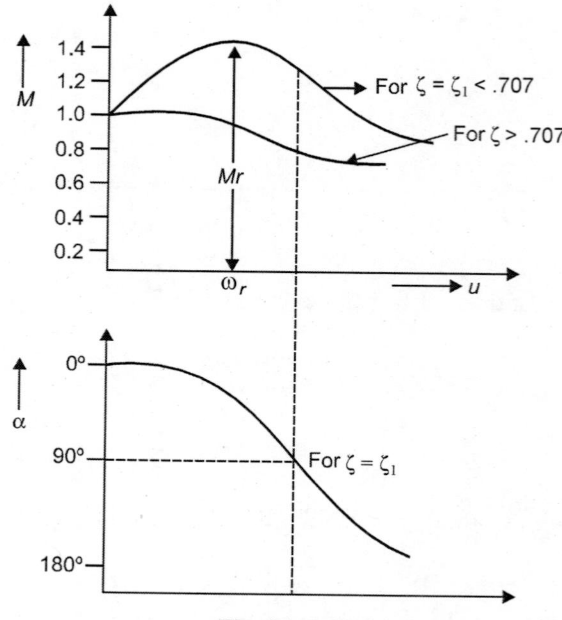

Fig. 4.1 and 4.2

Let ω_r be the reasonant frequency and $u_r = \left[\frac{\omega_r}{\omega_n}\right]$ be the normalised reasonent frequency.

The expression for reasonant frequency ω_r and the corresponding phase α_r can be obtained by substituting the expression for ω_r in the equation of M and α.

It can be shown that,

$$M_r = \frac{1}{2\zeta\sqrt{1-\zeta^2}}; \qquad M_r = \sqrt{1-2\zeta^2}$$

$$\alpha_r = -\tan^{-1}\left[\frac{\sqrt{1-2\zeta^2}}{\zeta}\right] \text{ and } \omega_r = \omega_n\sqrt{1-2\zeta^2}$$

when $\zeta = 0$

$$\omega_r = \omega_n\sqrt{1-2\zeta^2} = \omega_n$$

$$M_r = \frac{1}{2\zeta\sqrt{1-\zeta^2}} = \infty$$

Hence, it is clear that as s tends to zero, ω_r approaches ω_n and M_r approaches infinity.

For $0 < \zeta < \frac{1}{\sqrt{2}}$, the reasonent frequency always hasa value less than ω_n and the reasonent peak has a value greater then one.

For $\zeta < \frac{1}{\sqrt{2}}$, the condition $\left(\frac{dM}{dc}\right) = 0$, will not be satisfied for any real value of ω.

Hence when $\zeta < \frac{1}{\sqrt{2}}$, the magnitude decreases motrotonically from $M = 1$ at $u = 0$ with increasing u.

It follows that for $\zeta > \frac{1}{\sqrt{2}}$, there is no reasonent peak and the greatest value of M equals one.

The frequency at which M has a value of $\frac{1}{\sqrt{2}}$ is specialsignificance and is called the cut-off frequency ω_c. the signal frequencies above cut-off are greatly alternated on passing through a system.

For feedback control system, the range of frequencies over which $M \geq \frac{1}{\sqrt{2}}$ is defined as Bandwidth ω_b. Control system being low-pass filter, the bandwidth ω_b is equal to cut-off frequency ω_c.

In general the bandwidth of a control system indicates the noise-filtering characteristics of the system. Also, band width gives a measure of the transiont response.

The normalised bandwidth, $u_b = \left[\frac{\omega_b}{\omega_n}\right]$

$$u_b = \left[1 - 2\zeta^2 + \sqrt{2 - 4\zeta^2 + \zeta^4}\right]$$

u_b

from the equation of u_b it is clear that u_b is a function of ζ is shown in Fig. (4.3)

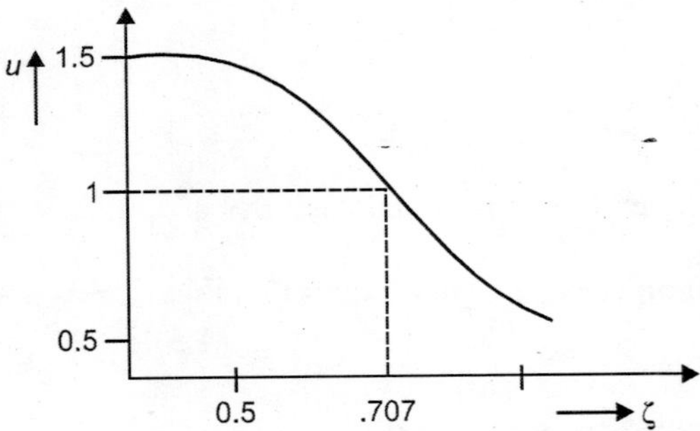

Fig. 4.3: Normalized bandwidth as a function of ζ.

The expression for the dumped frequency of oscillation ω_b and peak overshoot M_p of the step response for $0 \leq \zeta \leq 1$ are

Damped frequency, $\omega_d = \omega_n\sqrt{1-\zeta^2}$

$$\text{Peak overshoot, } M_p = \frac{-\zeta\pi}{\sqrt{1-\zeta^2}}$$

Comparison of the equation of M_r and M_p reveals that both are functions of only ζ.

The sketch of M_r and M_p for various value of ζ are shown in Fig. (4.4) and (4.5). The sketch reveals that a system with a given value of M_r must exhibit a corresponding value of M_p if subjected to a step input. For $\zeta > \frac{1}{\sqrt{2}}$ the reasonent peak M_r does not exist and the correlation breaks down. This is not a serious problem as for this range of ζ, the step response oscillations are well dampled and M_p is negligible.

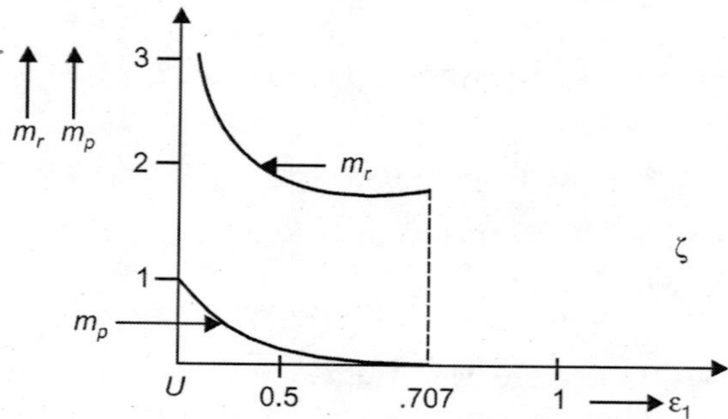

Fig. 4.4: M_r and M_p as a function ζ.

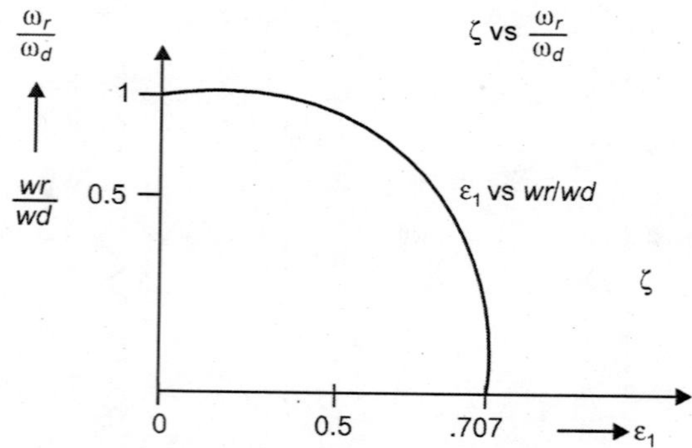

Fig. 4.5: ω_r/ω_d as a function of ζ.

The comparison of the equation of ω_r and ω_d reveals that there exists a definite correlation between them. The sketch of ω_r/ω_d with respect of ζ is shown in Fig. (4.5).

4.2 NYQUIST STABILITY CRITERION

Let take the closed loop transfer function

$$\frac{C(s)}{R(s)} = \frac{G(s)}{1+G(s)H(s)}$$

The characteristic equation of the system is given by the condition, $1 + G(s)\,H(s) = 0$

Let, $$F(s) = 1 + G(s)\,H(s)$$

The loop transfer function $G(s)\,H(s)$ can be written as

$$G(s)\,H(s) = \frac{K(s+z_1)(s+z_2)\dots(s+z_m)}{(s+p_1)(s+p_n)\dots(s+p_n)} \tag{4.3}$$

where $m \le n$

$$\therefore \quad F(s) = 1 + G(s)\,H(s) = 1 + \frac{K(s+z_1)\dots(s+z_m)}{(s+p_1)\dots(s+p_n)}$$

$$= \frac{(s+p_1)(s+p_2)\dots(s+p_n) + K(s+z_1)(s+z_2)\dots(s+z_m)}{(s+p_1)(s+p_2)\dots+(s+p_n)} \tag{4.4}$$

$$= \frac{(s+z_1')(s+z_2')\dots(s+z_n')}{(s+p_1)(s+p_2)\dots(s+p_n)} \tag{4.5}$$

where, $z_1', z_2' \dots z_n'$ are zeros of $F(s)$ which are obtained by combining the numerator and denominator polynomial of $G(s)\,H(s)$.

For the condition $F(s) = 0$, the numerator of $F(s)$ should be equal to zero

$$\therefore \quad (s+z_1')(s+z_2')\dots(s+z_n') = 0 \tag{4.6}$$

we can say that Eq. (Ex. 4.3) is the characteristic equation of the system. For the stability of the system the roots of the characteristic equation should not lie on the right half s-plane. The roots of characteristic equation are zeros of $F(s)$ and also they are poles of closed loop transfer function.

Hence we can conclude that for the stability of closed loop system the zeros of $F(s)$ should not lie on the right half S-plane.

Note: For unity feedback system

$$\frac{C(s)}{R(s)} = \frac{K(s+z_1)(s+z_2)\dots(s+z_m)}{(s+z_1')(s+z_2')\dots(s+z_n')}$$

Let us choose an arbitrary contour in the s-plane which encircles the right half zeros and poles of $F(s)$ [Eq. (4.4)]. The principle of argument states that (explained in section 4.3) the corresponding contour in $F(s)$-plane, N times in the anticlock wise direction.

where $$N = p - z \tag{4.7}$$

Here,

p = Number of poles of $F(s)$ lying on right half s-plane

z = Number of zeros of $F(s)$ lying on right half s-plane

For the stability of the system the roots of characteristic equation and zeros of $F(s)$ should not lie on the right half of s-plane. Hence for a stable system $z = 0$. Hence from Eq. (4.7) when $z = 0$, we get

when $$z = 0, \quad N = p \tag{4.8}$$

$$z \ne 0, \quad N \ne p \tag{4.9}$$

From Eq. (4.3) and (4.4) we can say that the poles of $F(s)$ are also poles of open loop transfer function, hence for the stability of the system, (with reference to Eq. (4.8) and (4.9) number of poles of loop transfer function lying on right of s-plane should be equal to counter clock wise encirclement of the origin of $F(s)$ plane. If this condition is not met the system is unstable.

The loop transfer function, $G(s)\,H(s)$ can be expressed as

$$G(s)\,H(s) = [1 + G(s)\,H(s)] - 1 = F(s) - 1 \tag{4.10}$$

From Eq. (4.10) it can be concluded that the $F(s)$-plane contour drawn with respect to origin is same as the $F(s)$-plane contour of $F(s) - 1$ drawn with respect to $(-1 + j0)$ as shown in Fig. (4.6).

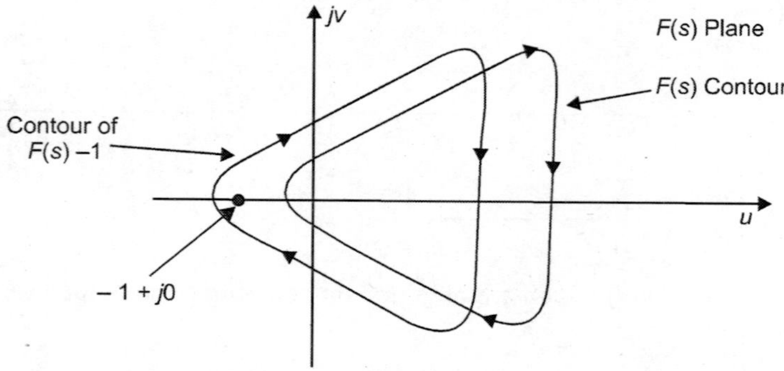

Fig. 4.6

Thus the encirclement of the origin of $F(s)$ plane by the $F(s)$ contour is equivalent to the encirclement of the pint $(-1 + j0)$ by the $G(s)\,H(s)$ contour. The Nyquist stability criterion have been proposed based on this concept.

In order to investigate the presence of poles of $G(s)\,H(s)$ on the right half s-plane a contour, C is chosen such that it encloses the entire right half s-plane as shown in Fig. (4.7) such contour C is called Nyquist contour.

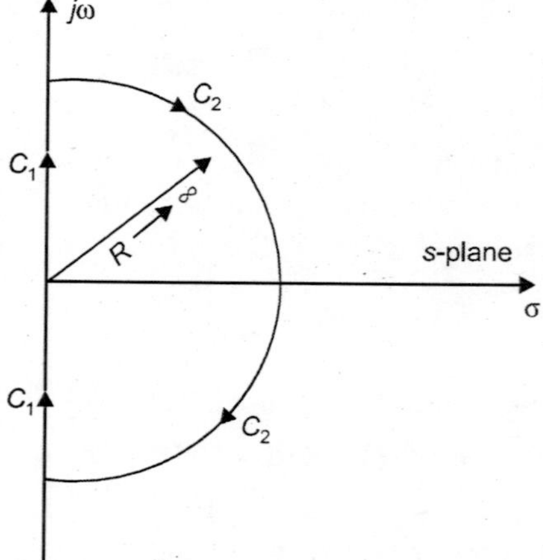

Fig. 4.7: Nyquist contour.

The nyquist contour is directed clockwise and comprise of two segments.

(i) An infinite line segment C_1 along the imaginary axis.

(ii) An arc, C_2 of ∞ radius.

along C_1, $s = j\omega$ with s varying form $-j\infty$ to $+j\infty$.

Along C_2, $s = \underset{K \to \infty}{\text{Lt}} \; Ke^{j\theta}$ with θ varying from $+\pi/2$ to $-\pi/2$.

Note: The s-plane is a complex plane. Any point on a complex plane can be expressed by the complex number in polar form, $Re^{j\theta}$, where R is the magnitude and θ is the argument (or phase).

Definition of Nyqist stability criterion:

The nyquist stability criterion can be stated follows. "If the $G(s)\,H(s)$ contour in the $G(s)\,H(s)$ plane corresponding to nyquist contour in the s-plane encircles the pint $(-1 + j0)$ in the anticlockwise direction as many times as the number of right half s-plane poles of $G(s)\,H(s)$, then the closed loop system is stable".

Important points to be noted.

1. There is no encirclement of $-1 + j0$ point. This implies that the system is stable if there are no poles of $G(s)\,H(s)$ in the right half s-plane. If there are poles on right half s-plane then the system is unstable.
2. An anticlockwise encirclement of $-1 + j0$ point. In this case the system is stable if the number of anticlockwise encirclements is same as the number of poles of $G(s)\,H(s)$ in the right half s-plane. If the number of encirclement is not equal to number of poles an right half *s-plane* then the system is unstable.
4. There is a clock-wise encirclement of the $(-1 + j0)$ point. In this case the system is always unstable.

Steps to Analyse the Stability using Nyquist Criterion

The following procedure can be followed to investigate the stability of closed loop system from the knowledge of open loop system, using Nyquist stability criterion.

1. Choose a nyquist contour as shown in Fig. (4.8) which encloses the entire right half s-plane except the singular points. The nyquist contour encloses all the right half s-plane poles and zeros of $G(s)\,H(s)$.

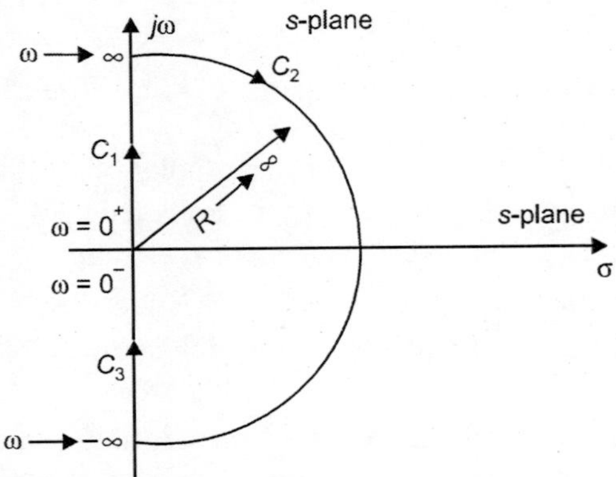

Fig. 4.8(a): Nyquist contour when there is no pole on imaginary.

2. The nyquist contour should be mapped to the $G(s)\ H(s)$ plane using the function $G(s)\ H(s)$ to determine the encirclement $-1 + j0$ point in the $G(s)\ H(s)$ plane. The nyquist contour of Fig. 4.8(b) can be divided into four sections C_1, C_2, C_3 and C_4. The mapping of the four sections in the $G(s)\ H(s)$ plane can be carried sectionwise and then combined together to get entire $G(s)\ H(s)$ contour.
3. In section C_1 the value of ω varies from 0 to ∞. The mapping of section C_1 is obtained by letting $s = j\omega$ in $G(s)\ H(s)$ and varying ω from 0 to +∞.

   i.e., $G(s)\,H(s)\big|_{\substack{s = j\omega \\ \omega = 0 \text{ to } \infty}} = G(j\omega)\,H(j\omega)\big|_{\omega = 0 \text{ to } \infty}$

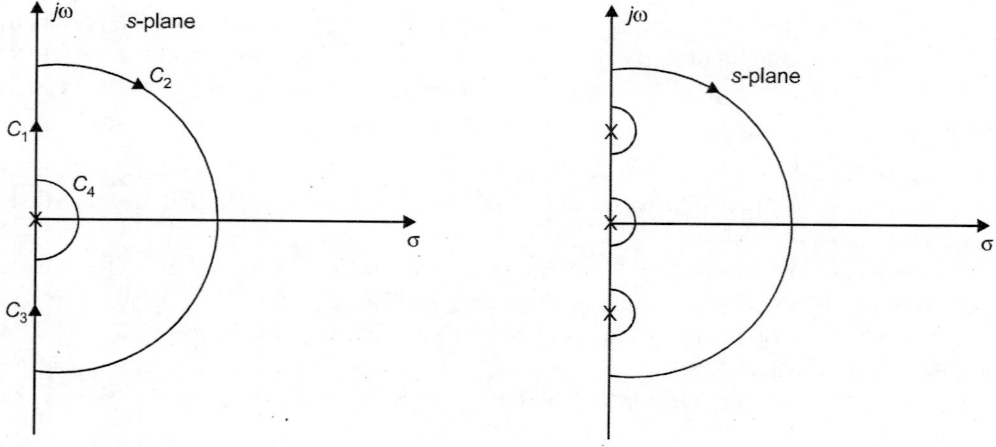

(b) Nyquist contour when there are poles at origin.

(c) Nyquist contour when there are poles on imaginary axis and at origin.

Fig. 4.8(b) and(c): Nyquist contour.

The locus of $G(j\omega)\ H(j\omega)$ as ω is varies from 0 to ∞ will be the $G(s)\ H(s)$ contour in $G(s)\ H(s)$ plane corresponding to section C_1 in s plane. This locus is the polar plot of $G(j\omega)\ H(j\omega)$. There are three ways of mapping this section of $G(s)\ H(s)$ contour they are,

(i) Calculate the values of $G(j\omega)\ H(j\omega)$ for various values of ω and sketch the actual locus of $G(j\omega)\ H(j\omega)$.

(ii) Separate the real part and imaginary part of $G(j\omega)\ H(j\omega)$. Equate the imaginary part to zero, to find the frequency at which the $G(j\omega)\ H(j\omega)$ locus crosses real axis.Substitute this frequency on real part and find the crossing point of the locus on real axis. Sketch the approximate locus of $G(j\omega)\ H(j\omega)$ form the knowledge of type number and order of the system (or) from the value of $G(j\omega)\ H(j\omega)$ at $\omega = 0$ and $\omega = \infty$.

(iii) Separate the magnitude and phase of $G(j\omega)\ H(j\omega)$ to -180 and solve for ω. This value of ω is the phase crossover frequency and the magnitude at this frequency is the crossing point on real axis. Sketch the approximate root locus.

4. The section C_2 of nyquist contour has a semicircle of infinite radius. Therefore, every point on section C_2 has infinite magnitude but the argument varies from $+\pi/2$, to $-\pi/2$. Hence the mapping

of section C_2 from s-plane to $G(s)\,H(s)$ plane can be obtained by letting $s = \underset{R \to \infty}{\text{Lt}}\; Re^{j\theta}$ in $G(s)$ $H(s)$ and varying θ from $+\pi/2$ to $-\pi/2$.

Consider the loop transfer function in time constant form and with y number of poles at origin, as shown here,

$$G(s)\,H(s) = \frac{K(1+sT_1)\,(1+sT_2)}{s^y\,(1+sT_a)\,(1+sT_b)}$$

Let $G(s)\,H(s)$ has n zeros and n poles including poles at orign. For practical systems, $m > n$ since $s \to Re^{j\theta}$ and $R \to \infty$, the term $(1 + sT)$ can be approximated to sT [i.e., $(1 + sT) \simeq sT$]

$$\therefore \qquad G(s)\,H(s) \simeq K \cdot \frac{sT_1 \cdot sT_2 \cdot sT_3 \ldots.}{s^y \cdot sT_a \cdot sT_b \cdot sT_c \ldots}$$

$$= K_1 \frac{s^m}{s^n} = \frac{K_1}{s^{n-m}}$$

$$G(s)\,H(s) \approx K_1/s^n - s^m$$

on letting,

$$s = \underset{R \to \infty}{\text{Lt}}\; Re^{j\theta}$$

we get

$$\frac{G(s)\,H(s)}{s = \underset{R\to\infty}{\text{Lt}}\; Re^{j\theta}} = \frac{K_1}{\underset{R\to\infty}{\text{Lt}}\; (Re^{j\theta})^{n-m}} = 0e^{-j\theta(n-m)}$$

when $\theta = \pi/2$,

$$G(s)\,H(s) = 0\,e^{-j\,\pi/2(n-m)}$$

when $\theta = -\pi/2$,

$$G(s)\,H(s) = 0\,e^{+j\,\pi/2(n-m)}$$

From the above two equations we can conclude that the section C_2 of nyquist contour in s-plane is mapped as circles/circular are around is mapped as circles /circular are around origin with radius tending to zero in the $G(s)\,H(s)$ plane.

5. In section C_3 the value of ω varies from $-\infty$ to 0. The mapping of section C_3 is obtained by letting $s = +j\omega$ in $G(s)\,H(s)$ and varying from $-\infty$ to 0.

i.e.,

$$G(s)\,H(s)\big|_{\substack{s=+j\omega \\ \omega=-\infty \text{ to } 0}} = G(j\omega)\,H(j\omega)\big|_{\omega=-\infty \text{ to } 0}$$

The locus of $G(j\omega)\,H(j\omega)$ as ω is varies form $-\infty$ to 0 will be the $G(s)\,H(s)$ contour in $G(s)\,H(s)$ plane corresponding section C_3 in s-plane. This locus is the inverse polar plot of $G(j\omega)\,H(j\omega)$. The inverse polar plot is given by the mirror image of polar plot with respect to real axis.

6. The section C_4 of nyquist contour has a semicircle of zero radius. Therefore every point on semicircle has zero magnitude but the argument varies from $-\pi/2$ to $+\pi/2$. Hence the mapping of section C_4 from s-plane to $G(s)\,H(s)$ plane can be obtained by letting $s = \underset{R \to 0}{\text{Lt}}\; Re^{j\theta}$ in $G(s)$ $H(s)$ and varying θ from $-\pi/2$ to $+\pi/2$.

Consider the loop transfer function in time constant from and with y number of poles at orign as shown below.

$$G(s)\,H(s) = \frac{K(1+sT_1)(1+sT_2)(1+sT_3)\ldots}{s^y\,(1+sT_a)(1+sT_b)(1+sT_c)\ldots}$$

Let $G(s)\,H(s)$ has m zeros and n poles including poles at origin. For practical systems $m > n$. Since $s \to Re^{j\theta}$ and $R \to 0$ the term $1 + sT$ can be approximated to 1 [i.e., $(1 + sT) \approx 1$].

$$\therefore \qquad G(s)\,H(s) \approx K\frac{1}{sy}$$

on letting $s = \underset{R\to 0}{\text{Lt}}\ Re^{j\theta}$ we get

$$G(s)\,H(s)\Big|_{s=\underset{R\to 0}{Lt}\ \text{Re}^{j\theta}} = \frac{K}{\underset{R\to 0}{Lt}\ (Re^{j\theta})^y} = \infty e^{-j\theta y}$$

when $\theta = -\pi/2$,

$$G(s)\,H(s) = \infty\ e^{j\pi/2y}$$

when $\theta = \pi/2$,

$$G(s)\,H(s) = \infty\ e^{-j\pi/2y}$$

From the above two equations we can conclude that the section of nyquist. Contour in s-plane is mapped as circles/circular are in $G(s)\,H(s)$ plane with origin as centre and infinite radius.

Note:

1. If there are no poles on the origin then the section C_4 of nyquist contour will be absent.
2. If there are poles on imaginary axis as shown below then the nyquist contour is divided into the following 8 sections and the mapping is performed section wise.

Section C_1: From $+jo^+$ to $+j\omega_1^-$

Section C_2: From $+j\omega_1^-$ to $+j\omega_1^+$

Section C_3: From $+j\omega_1^+$ to $+j\infty$

Section C_4: From $+j\infty$ to $-j\infty$

Section C_5: From $-j\infty$ to $-j\omega_1^-$

Section C_6: From $-j\omega_1^-$ to $-j\omega_1^+$

Section C_7: From $-j\omega_1^+$ to $-jo^-$

Section C_8: From $-jo^-$ to $+jo^+$

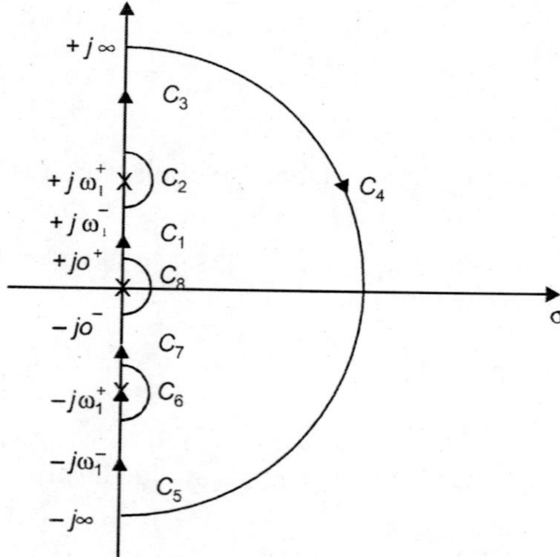

Fig. 4.9

4.3 MATHEMATICAL PRELIMINARIES FOR NYQUIST STABILITY CRITERION

Let
$$F(s) = \frac{(s - z_1)(s - z_2)\ldots(s - z_m)}{(s - p_1)(s - p_2)\ldots(s - p_n)} \tag{4.11}$$

The roots of numerator polynomial are zeros and the roots of denominator polynomial are poles.

The function has m number of zeros and n number of poles.

Let s be a complex variable represented by $s = \sigma + j\omega$ on the complex s-plane. The function $F(s)$ is also complex and may be defined as $F(s) = u + jv$ and represented on the complex $F(s)$-plane with coordinates u and v.

The Eq. (4.11) indicates that, the function $F(s)$ maps the point in the s-plane into the $F(s)$ plane.

Note: A function is analytic in the s-plane provide the function and all its derivatives exist. The points in the s-plane where the function (or it derivatives) does not exist are called singular points.

Since any number of points of analycity in the s-plane can be mapped into the s-plane which does not go through any singular point, there exists a corresponding contour in the $F(s)$-plane as shown in Fig. (4.10).

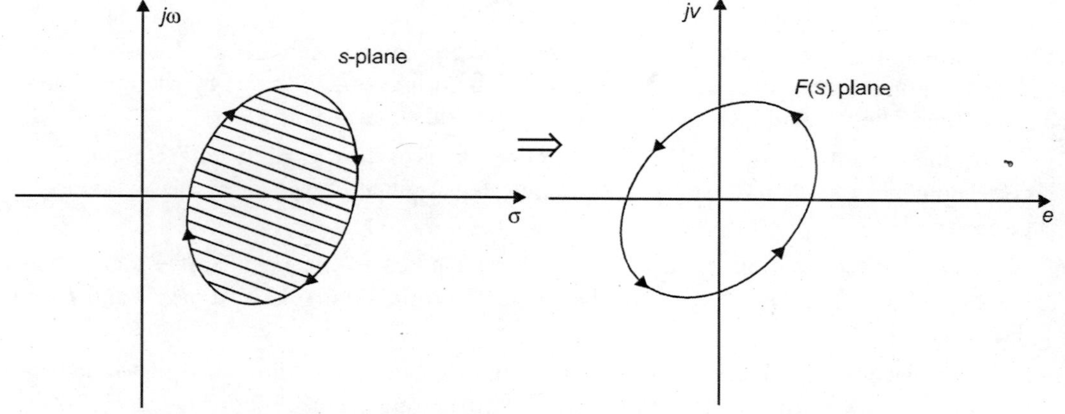

Fig. 4.10: An arbitrary contour in s-plane and its corresponding contour in $F(s)$ plane.

Note: For the development of nyquist criterion, the exact shape of the contour is not required but only the number of encirclements of the origin of the $F(s)$-plane is essential.

Concept of encircled and enclosed

It is important to distinguish between the concept of encircled and enclosed which are frequently used to apply nyquist criterion.

Encircled: A point is said to be encircled by a closed path if it is found inside the path with reference to Fig. (4.11a) the point A is encircled in the clockwise direction and the point B is not encircled.

Enclosed: Any point or region is said to be enclosed by a closed path, if it is found to the right of the path when the path is traversed in the prescribed direction. The shaded regions in Fig. (4.11b and c) are the region enclosed by the closed path with reference to Fig. (4.11c) the point A is not enclosed by closed path but point B is enclosed.

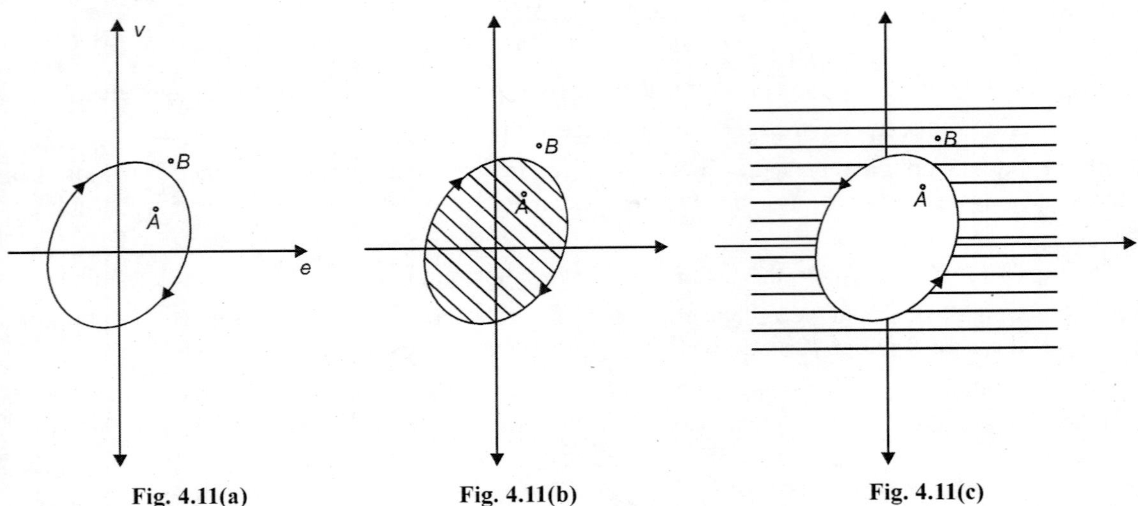

Fig. 4.11(a) Fig. 4.11(b) Fig. 4.11(c)

It can be proved that there exists a relationship between the enclosure of poles and zeros by the s-plane closed contour and number of encirclements of the origin of $F(s)$ plane by the corresponding $F(s)$-plane contour. The following points is the summary of this relationship.

1. If s-plane closed contour encloses z-number of zeros in the right half s-plane then the corresponding contour in $F(s)$ plane will encircle, the origin of $F(s)$ plane Z times in the clockwise direction.
2. If s-plane closed contour encloses P number of poles in the right half s-plane then the corresponding contour in $F(s)$ plane will encircle the origin of $F(s)$ plane p times in anticlockwise direction.
3. If the s plane closed contour encloses Z zeros and P poles in the right half s-plane and if $P > Z$, then the corresponding contour in $F(s)$-plane will encircle the origin of $F(s)$-plane $(P\text{-}Z)$ times in the anticlockwise direction.
4. If the s-plane closed contour enclosed Z zeros and P poles in the right half s-plane and if $P < Z$ then the corresponding contour in $F(s)$-plane will encircle the orign of $F(s)$-plane $(Z - P)$ times in the clockwise direction.
5. If the s-plane closed contour encloses Z zeros and P poles in the right half x-plane and if $P = Z$, then the corresponding contour in $F(s)$-plane will not encircle the origin of $F(s)$-plane.
6. If the s-plane closed contour does not enclose any pole (or) zero then the corresponding contour in $F(s)$-plane will not encircle the origin of $F(s)$-plane.

The table (4.9.1.1) shows examples of arbitrary s-plane contours and their corresponding $F(s)$-plane contours. Here zeros are marked by small circles (0) and poles by (X).

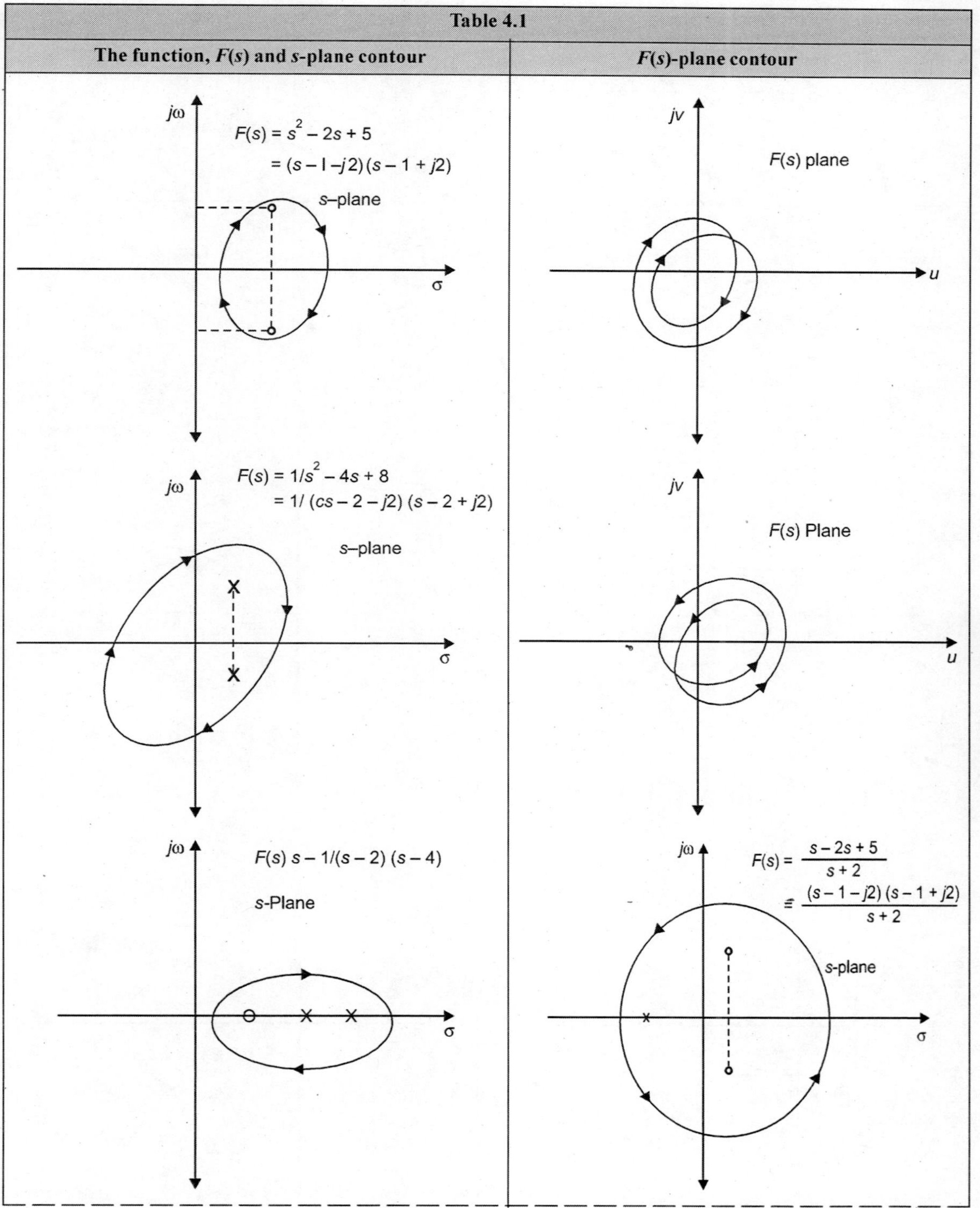
Table 4.1
The function, F(s) and s-plane contour
F(s)-plane contour
jω
F(s) = s² – 2s + 5
= (s – l –j2) (s – 1 + j2)
s–plane
σ
jv
F(s) plane
u
jω
F(s) = 1/s² – 4s + 8
= 1/ (cs – 2 – j2) (s – 2 + j2)
s–plane
σ
jv
F(s) Plane
u
jω
F(s) s – 1/(s – 2) (s – 4)
s-Plane
σ
jω
F(s) = s – 2s + 5 / s + 2
= (s – 1 – j2) (s – 1 + j2) / s + 2
s-plane
σ

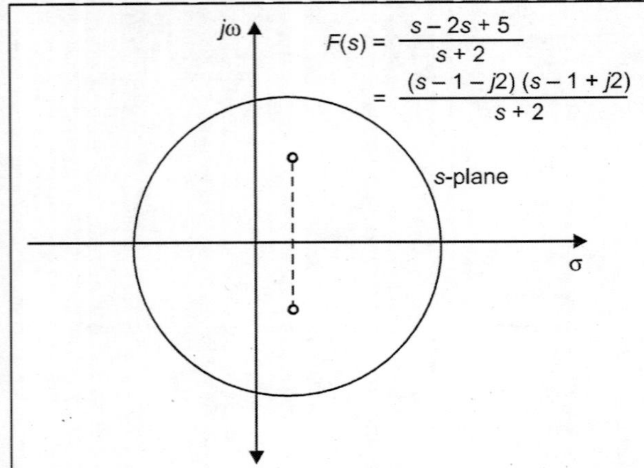
jω
F(s) = (s − 2s + 5)/(s + 2)
= (s − 1 − j2) (s − 1 + j2)/(s + 2)
s-plane
σ

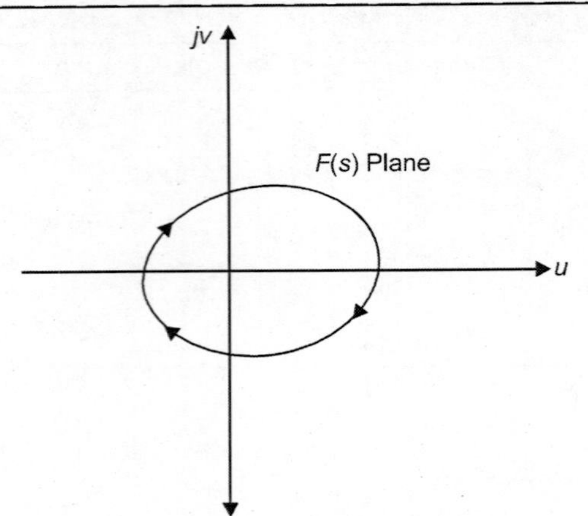
jv
F(s) Plane
u

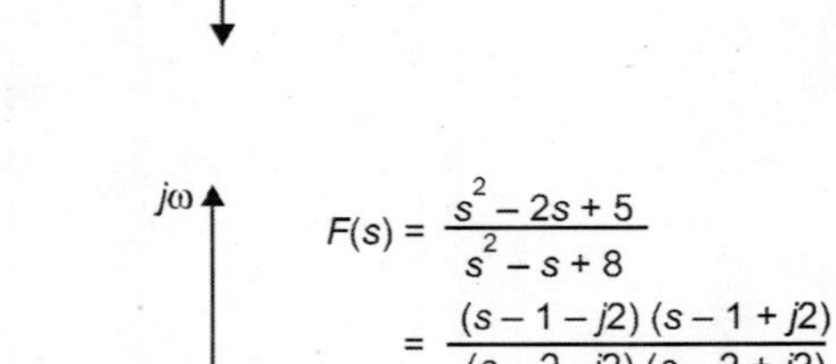
jω
F(s) = (s² − 2s + 5)/(s² − s + 8)
= (s − 1 − j2) (s − 1 + j2)/(s − 2 −j2) (s − 2 + j2)

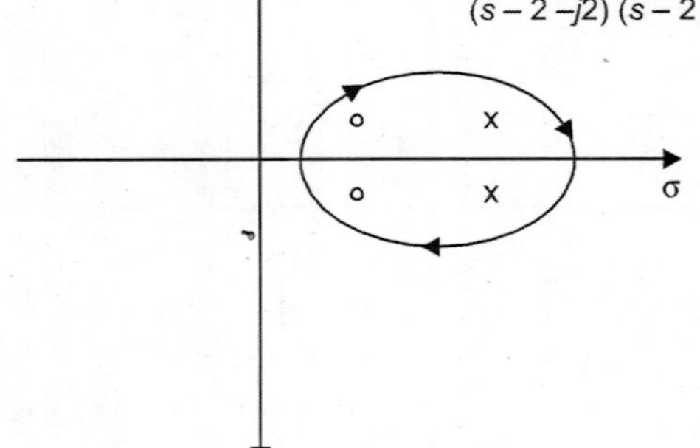
σ

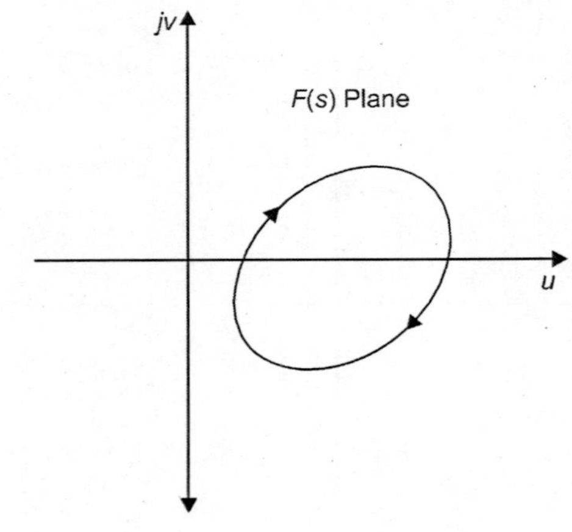
jv
F(s) Plane
u

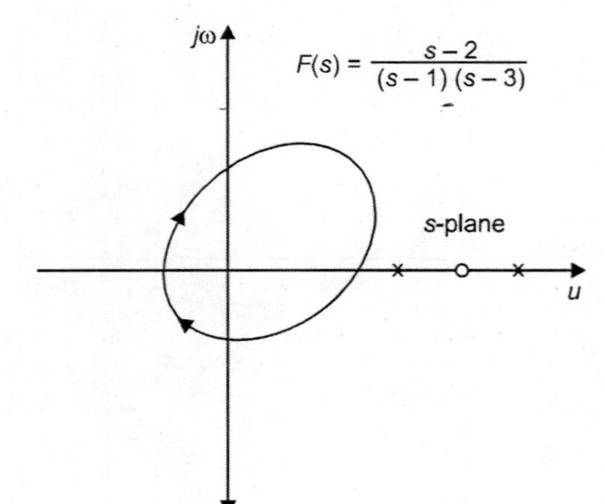
jω
F(s) = (s − 2)/(s − 1) (s − 3)
s-plane
u

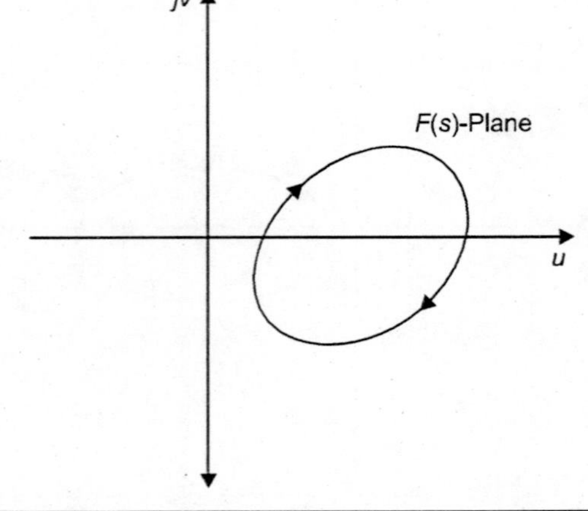
jv
F(s)-Plane
u

The relation between the enclosure of poles and zeros of $F(s)$ lying on the right half s-plane by the s-plane contour and the encirclements of the origin of $F(s)$-plane by the corresponding $F(s)$-plane contour is called principle of argument.

EXAMPLE 4.1

Draw the Nyquist plot for the system whose open loop transfer function is $G(s)\,H(s) = \dfrac{K}{s(s+2)\,(s+10)}$.

Determine the range of K for which closed loop system is stable.

Solution: Given that,

$$G(s)\,H(s) = \frac{K}{s(s+2)\,(s+10)} = \frac{K}{s \times 2\left[\frac{s}{2}+1\right] \times 10\left[\frac{s}{10}+1\right]} \quad (4.1.1)$$

$$= \frac{.05K}{s(1+.5s)\,(1+0.1s)} \Rightarrow Mag \ \frac{0.05k}{\omega\sqrt{1+(.5\omega)^2}\,\sqrt{1+(0.1\omega)^2}}$$

$$Angle = \frac{0.05}{90^\circ + \tan^{-1}.5\omega + \tan^{-1}0.1\omega} \quad or \quad -90^\circ - \tan^{-1}.5\omega - \tan^{-1}0.1\omega \quad (4.1.2)$$

The open loop transfer function has a pole at origin. Hence choose the Nyquist contour on s-plane enclosing the entire right half plane except the origin as shown in Fig. (Ex. 4.1.1).

The Nyquist contour has 4 sections C_1, C_2, C_3 and C_4. The mapping of each section's performed separately and the overall Nyquist plot is obtained by combining the individual sections.

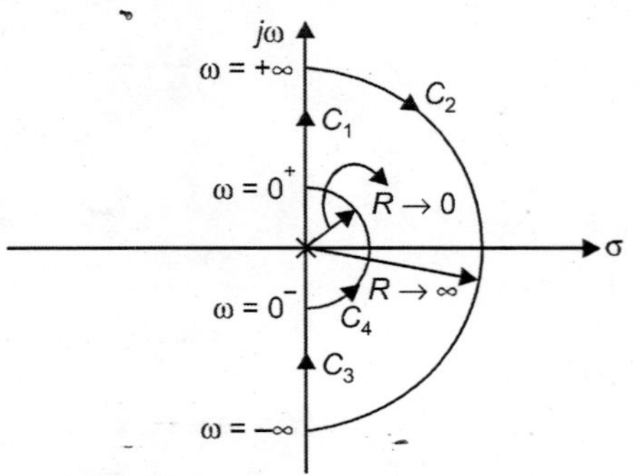

Fig. Ex. 4.1.1: Nyquist contour in s-plane.

Mapping of Section C_1

In section C_1, ω varies from, 0 to ∞. The mapping of section C_1 is given by the locus of $G(j\omega)$ as ω is varies from 0 to ∞. This locus is the polar plot of $G(j\omega)\,H(j\omega)$.

$$G(s)\,H(s) = 0.05K/s(1+.5s)\,(1+0.1s)$$

Let, $s = j\omega$,

$$\therefore \quad G(j\omega)\,H(j\omega) = 0.05K/j\omega\,(1+j0.5\omega)\,(1+j0.1\,\omega)$$

$$= .05K/{-0.6\omega^2} + j\omega(1-0.5\omega^2)$$

when the locus of $G(j\omega)\ H(j\omega)$ crosses real axis the imaginary term will be zero and the corresponding frequency is the phase crossover frequency, ω_{pc}.

$\therefore$ At $\omega = \omega_{pc}$, $\omega_{pc}\,(1 - 0.5\ \omega_{pc}^2) = 0$

$$1 - 0.5\,\omega_{pc}^2 = 0$$

$$\omega_{pc} = \sqrt{\frac{1}{.05}} = 4.472 \text{ rad/sec}$$

At $\omega = \omega_{pc} = 4.472$ rad/sec,

$$G(j\omega)\ H(j\omega) = .05K/{-0.6\omega^2} = \frac{-.05K}{0.6 \times (4.472)^2}$$

$$G(j\omega)\ H(j\omega) = -0.00417\ K$$

The open loop system is type-1 and third order system it is a minimum phase system with all poles. Hence the polar plot of $G(j\omega)\ H(j\omega)$ starts at –90° axis at infinity, crosses real axis at –0.0417 K and ends at origin in second quadrant. The section C_1 and its mapping is show in Fig. (Ex. 4.1.2) and (Ex. 4.1.3) are drawn accordingly to values which are obtained from equation (4.1.3) and (4.1.4).

Substitute $\omega = 0$ in equation (4.1.1) and (4.1.2), we get $\infty\ \angle{-90^\circ}$ (4.1.3)

Then substitute $\omega = \infty$ in equation (4.1.1) and (4.1.2), we get **0** $\angle{-270^\circ}$ (4.1.4)

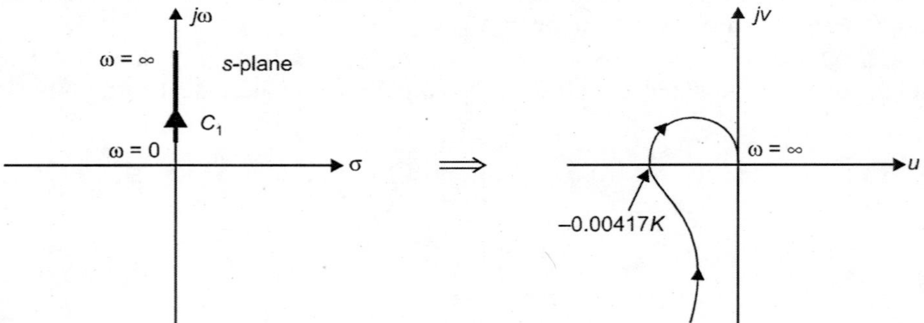

Fig. Ex. 4.1.2: Section C_1 in s-plane.

Fig. Ex. 4.1.3: Mapping of section C_1 in $G(s)\ H(s)$ s-plane.

Mapping of Section C_2

The mapping of section C_2 from s-plane to $G(s)\ H(s)$ s-plane is obtained by letting $s = \underset{R\to\infty}{Lt}\ Re^{j\theta}$ in $G(s)\ H(s)$ and varying θ from $+\pi/2$ to $-\pi/2$. Since $s \to Re^{j\theta}$ and $R \to \infty$, the $G(s)\ H(s)$ can approximated as shown below [i.e., $(1 + sT)$ 1 sT]

$$G(s)\ H(s) = .05K/s(1 + .5s)\ (1 + .1s)\ 1\ \frac{.05K}{s \times .5s \times 1.s} = \frac{K}{s^3}$$

Let, $s = \underset{R\to\infty}{Lt}\ Re^{j\theta}$

$$\therefore \quad G(s)\ H(s)\Big|_{s = \underset{R\to\infty}{Lt} Re^{j\theta}} = \frac{K}{s^3}\Bigg|_{s = \underset{R\to\infty}{Lt} Re^{j\theta}} = \frac{K}{\underset{R\to\infty}{Lt}\ (Re^{j\theta})^3} = 0e^{-j3\theta}$$

when $\theta = \pi/2$, $G(s)\ H(s) = oe^{-j3\pi/2}$

when $\theta = -\pi/2$, $G(s)\ H(s) = oe^{+j3\pi/2}$

From above two equations, we can say that section C_2 in s-plane is mapped as circular arc of zero radius around origin $G(s)$ $H(s)$-plane with argument varying from $-3\pi/2$ to $+3\pi/2$ as shown in Fig. (Ex. 4.1.4) and (Ex. 4.1.5).

Mapping of Section C_3

In section C_3, ω varies from $-\infty$ to 0. The mapping of section C_3 is given by the locus of $G(j\omega)$ $H(j\omega)$ as ω is varied from $-\infty$ to 0. This locus is the inverse polar plot of $G(j\omega)$ $H(j\omega)$. The inverse plot is given by the mirror image of polar plot with respect to real axis. This section is s-plane and its corresponding contour $G(s)$ $H(s)$ plane are shown in Fig. (Ex. 4.1.6) and (Ex. 4.1.7).

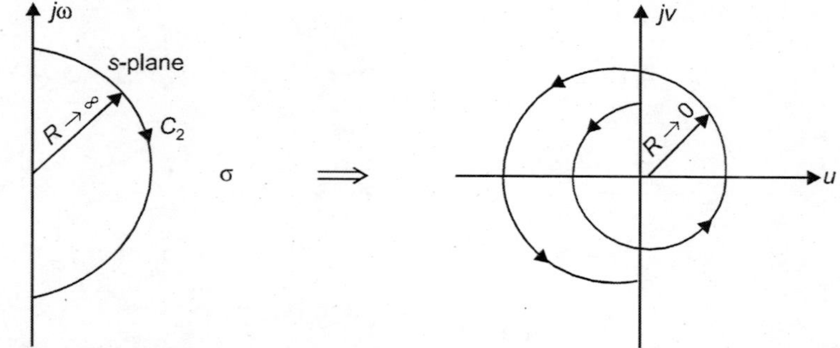

Fig.Ex. 4.1.4: Section C_2 in s-plane.

Fig. Ex. 4.1.5: Mapping of section C_2 in $G(s)$ $H(s)$ s-plane.

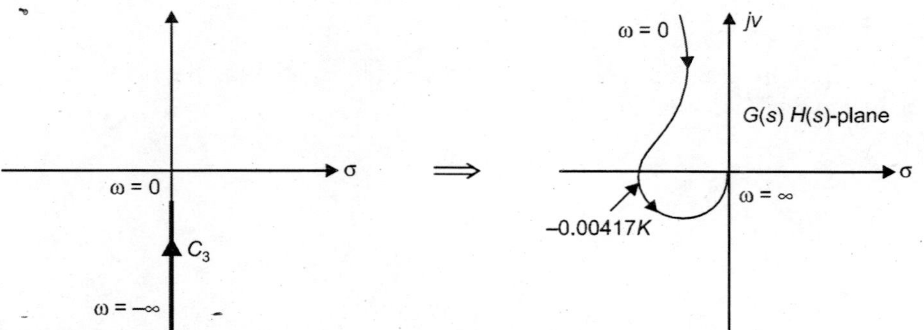

Fig. Ex. 4.1.6: Section C_3 in s-plane.

Fig. Ex. 4.1.7: Mapping of section C_3 in $G(s)$ $H(s)$ s-plane.

Mapping of Section C_4

The mapping of section C_4 from s-plane to $G(s)$ $H(s)$-plane is obtained by letting $s = \underset{R \to 0}{Lt}\ Re^{j\theta}$ in $G(s)$ H(s) and varying θ from $-\pi/2$ to $+\pi/2$. Since $s \to Re^{j\theta}$ and $R \to 0$ the $G(s)$ $H(s)$ can be approximated 1 + sT 1 1 [Œ 1 Á sT]

$$G(s)\,H(s) = \frac{.05K}{s(1+.5s)(1+.1s)} \approx \frac{.05K}{s \times 1 \times 1} = \frac{.05K}{s}$$

Let $$s = \underset{R \to 0}{Lt}\ Re^{j\theta}$$

$$\therefore \quad G(s)\,H(s)\Big|_{\underset{R\to 0}{Lt}\ Re^{j\theta}} = \left.\frac{.05K}{s}\right|_{s=\underset{R\to 0}{Lt}\ Re^{j\theta}} = \frac{.05K}{\underset{R\to 0}{Lt}\ (Re^{j\theta})} = \infty e^{-j\theta}$$

when $\theta = -\pi/2$, $\quad G(s)\,H(s) = \infty e^{+j\pi/2}$

when $\theta = +\pi/2$, $\quad G(s)\,H(s) = \infty e^{-j\pi/2}$

From above two equations, we can say that section C_4 in s-plane is mapped as circular arc of infinite radius with argument varying from $+\pi/2$ to $-\pi/2$ as shown in Fig. (Ex. 4.1.8) and (Ex. 4.1.9).

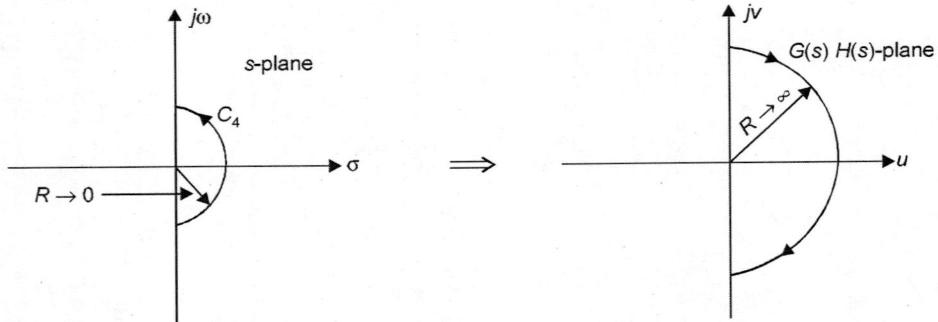

Fig. 4.1.8: Section C_4 in s-plane.

Fig. 4.1.9: Mapping of section C_4 in $G(s)\,H(s)$ s-plane.

Complete nyquist plot

The entire Nyquist plot in $G(s)\,H(s)$-plane can be obtained by combining the mappings of individual sections, as shown in Fig. (Ex. 4.1.10).

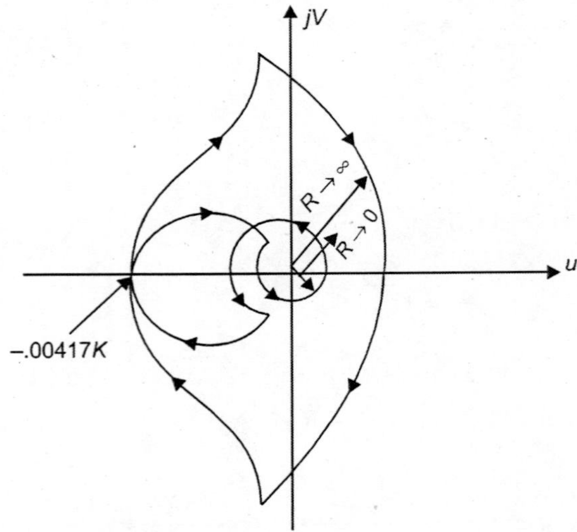

Fig. Ex. 4.1.10: Nyquist plot of $G(s)H(s) = K/s(s + 2)\,(s + 10)$.

Stability analysis

When, $-0.00417K = -1$, the contour passes through $(-1 + j0)$ point and corresponding value of K is the limiting value of K for stability.

$$\therefore \qquad \text{Limiting value of } K = \frac{1}{.00417} = 240$$

when K < 240

when K is less than 240, the contour crosses real axis at a point between 0 and -1. On travelling through Nyquist plot along the indicated direction it is found that the point $-1 + j0$ is not encircled. Also the open loop transfer function has no poles on the right half of s-plane. Therefore the closed loop system is stable.

when K > 240

when K is greater than 240, the contour crosses real axis at a point between -1 and $-\infty$. On travelling through Nyquist plot along the indicated direction it s found that the point $-1 + j0$ is encircled in clockwise direction two times. Therefore the closed loop system is unstable.

Result: The value of K for stability is $0 < K < 240$.

EXAMPLE 4.2

Construct the Nyquist plot for a system whose open loop transfer function is given by $G(s)\,H(s) = \frac{K(1+s)^2}{s^3}$. Find the range of K for stability.

Solution: Given that,

$$G(s)\,H(s) = \frac{K(1+s)^2}{s^3}$$

The open loop transfer function has three poles at origin. Hence choose the Nyquist contour on s-plane enclosing the entire right half plane except the origin as shown in Fig. (Ex. 4.2.1). The Nyquist contour has four sections C_1, C_2, C_3 and C_4. The mapping of each section is performed separately and the overall Nyquist plot is obtained by combining the individual sections.

Mapping of Section C_1

In section C_1, ω varies from, 0 to $+\infty$. The mapping of section C_1 is given by the locus of $G(j\omega)\,H(j\omega)$ as ω is varied from 0 to ∞. This locus is the polar plot of $G(j\omega)\,H(j\omega)$.

$$G(s)\,H(s) = \frac{K(1+s)^2}{s^3}$$

Let, $s = j\omega$,

$$\therefore \qquad G(j\omega)\,H(j\omega) = \frac{K(1+j\omega)^2}{(j\omega)^3} = \frac{K(1-\omega^2+2j\omega)}{-j\omega^3}$$

$$= \frac{K(1-\omega^2)}{-j\omega^3} + \frac{K2j\omega}{-j\omega^3} = \frac{-2K}{\omega^2} + j\frac{K(1-\omega^2)}{\omega^3}$$

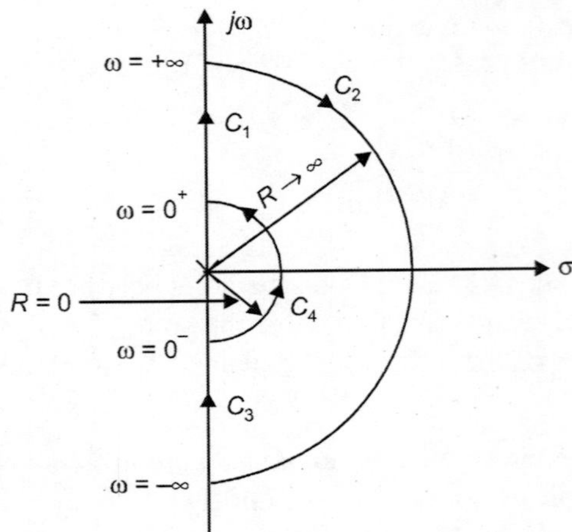

Fig. Ex. 4.2.1: Nyquist contour in *s*-plane.

when the locus of $G(j\omega)\ H(j\omega)$ crosses real axis the imaginary term will be zero and the corresponding frequency is the phase crossover frequency, ω_{pc}.

$\therefore$ At $\omega = \omega_{pc}, \omega_{pc}$ $K(1 - \omega_{pc}^2) = 0$

$\therefore$ $1 - \omega_{pc}^2 = 0$ $\qquad \omega_{pc} = 1$ rad/sec

At $\qquad \omega = \omega_{pc} = 1$ rad/sec,

$$G(j\omega)\ H(j\omega) = \frac{2K}{\omega^2} = \frac{-2K}{1^2} = -2K$$

$$G(j\omega)\ H(j\omega) = K(1 + j\omega)^2 / (j\omega)^3$$

$$= K\sqrt{1+\omega^2}\ \angle \tan^{-1}\omega \sqrt{1+\omega^2}\ \angle \tan^{-1}\omega \,/\, \omega^3\ \angle 270°$$

$$= \frac{K(1+\omega^2)}{\omega^3} \angle (2\tan^{-1}\omega - 270°)$$

As $\omega \to 0$, $\qquad G(j\omega)\ H(j\omega) \to \infty \angle -270°$

As $\omega \to \infty$, $\qquad G(j\omega)\ H(j\omega) \to 0 \angle -90°$

From above two equations we can say that the polar plots starts at –270° axis at infinity, crosses axis at $-2K$ and ends at origin in third quadrant. The section C_1 and its mapping are shown in Fig. (Ex. 4.2.2) and (Ex. 4.2.3).

Mapping of Section C_2

The mapping of section C_2 from *s*-plane to $G(s)\ H(s)$-plane is obtained by letting $s = \underset{R\to\infty}{Lt} Re^{j\theta}$ in $G(s)$ $H(s)$ and varying θ from $+\pi/2$ to $-\pi/2$. Since $s \to Re^{j\theta}$ and $R \to \infty$, the $G(s)\ H(s)$ can approximated as $(1 + sT)$ 1 $sT]$

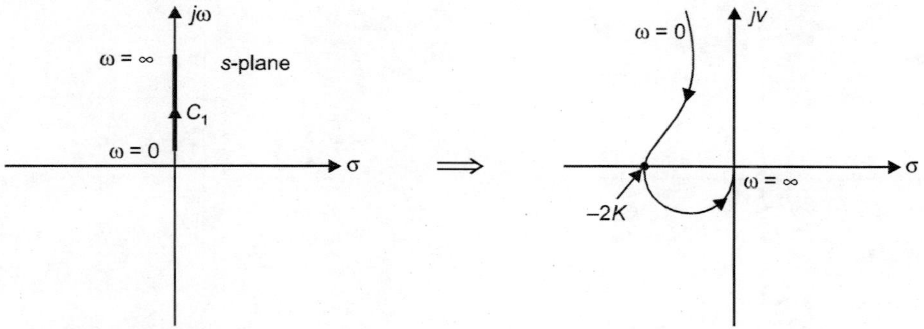

Fig. Ex. 4.2.2: Section C_1 in *s*-plane.

Fig. Ex. 4.2.3: Mapping of section C_1 in $G(s)\ H(s)$ *s*-plane.

$$G(s)\ H(s) = \frac{K(1+s)^2}{s^3} = \frac{Ks^2}{s^3} = K/s$$

Let, $$s = \underset{R\to\infty}{Lt}\ \mathrm{Re}^{j\theta}$$

$$\therefore \qquad G(s)\ H(s)\Big|_{s = \underset{R\to\theta}{Lt}\ \mathrm{Re}^{j\theta}} = \frac{K}{s}\Bigg|_{s = \underset{R\to\infty}{Lt}\ \mathrm{Re}^{j\theta}} = \frac{K}{\underset{R\to\infty}{Lt}\ \mathrm{Re}^{j\theta}} = oe^{-j\theta}$$

when $\theta = \pi/2$, $\qquad G(s)\ H(s) = oe^{-j\pi/2}$

when $\theta = -\pi/2$, $\qquad G(s)\ H(s) = oe^{j\pi/2}$

From above two equations, we can say that section C_2 in *s*-plane is mapped as circular arc of zero radius around origin $G(s)\ H(s)$-plane with argument varying from $-\pi/2$ to $+3\pi/2$ as shown in Fig. (Ex. 4.2.4) and (Ex. 4.2.5).

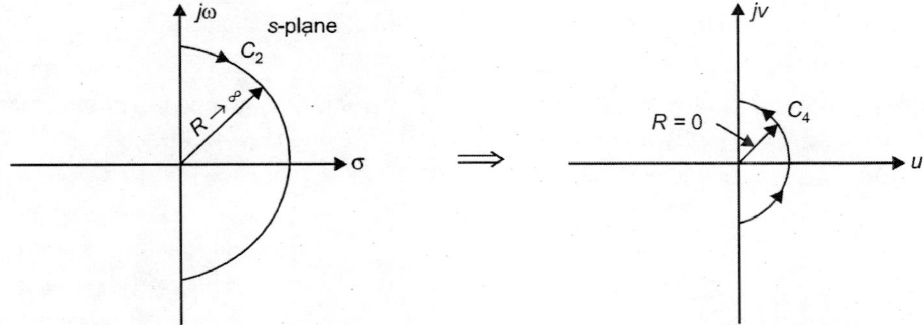

Fig. 4.2.4: Section C_2 in *s*-plane.

Fig. 4.2.5: Mapping of section C_2 in $G(s)\ H(s)$ *s*-plane.

Mapping of Section C_3

In section C_3, ω varies from –∞ to 0. The mapping of section C_3 is given by the locus of $G(j\omega)\ H(j\omega)$ as ω is varied from –∞ to 0. This locus is the inverse polar plot of $G(j\omega)\ H(j\omega)$.

The inverse plot is given by the mirror image of polar plot with respect to real axis. The section is C_3 in *s*-plane and its corresponding contour $G(s)\ H(s)$ plane are shown in Fig. (Ex. 4.2.6) and (Ex. 4.8.2.7).

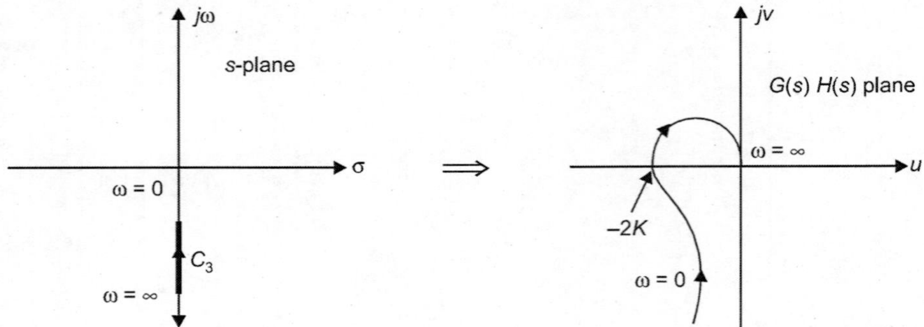

Fig. 4.8.2.6: Section C_3 in s-plane.

Fig. 4.8.2.7: Mapping of section C_3 in $G(s)$ $H(s)$ s-plane.

Mapping of Section C_4

The mapping of section C_4 from s-plane to $G(s)$ $H(s)$-plane is obtained by letting $s = \underset{R \to 0}{Lt} \text{Re}^{j\theta}$ in $G(s)$ H(s) and varying θ from $-\pi/2$ to $+\pi/2$. Since $s \to \text{Re}^{j\theta}$ and $R \to 0$ the $G(s)$ $H(s)$ can be approximated 1 + sT 1 1 [Œ sT À 1]

$$G(s)\,H(s) = \frac{K(1+s)^2}{s^3} \approx \frac{K \times 1}{s^3} = K / s^3$$

Let
$$s = \underset{R \to 0}{Lt} \text{Re}^{j\theta}$$

$$\therefore \quad G(s)\,H(s)\Big|_{\underset{R \to 0}{Lt} \text{Re}^{j\theta}} = \frac{K}{s^3}\Bigg|_{s = \underset{R \to 0}{Lt} \text{Re}^{+j\theta}} = \frac{K}{\underset{R \to 0}{Lt} (\text{Re}^{+j\theta})^3} \infty e^{-j3\theta}$$

when $\theta = -\pi/2$, $\quad G(s)\,H(s) = \infty e^{+j3\pi/2}$

when $\theta = \pi/2$, $\quad G(s)\,H(s) = \infty e^{-j3\pi/2}$

From above two equations, we can say that section C_4 in s-plane is mapped as circular arc of finite radius varying from $+3\pi/2$ to $-3\pi/2$ as shown in Fig. (Ex. 4.2.8) and (Ex. 4.2.9).

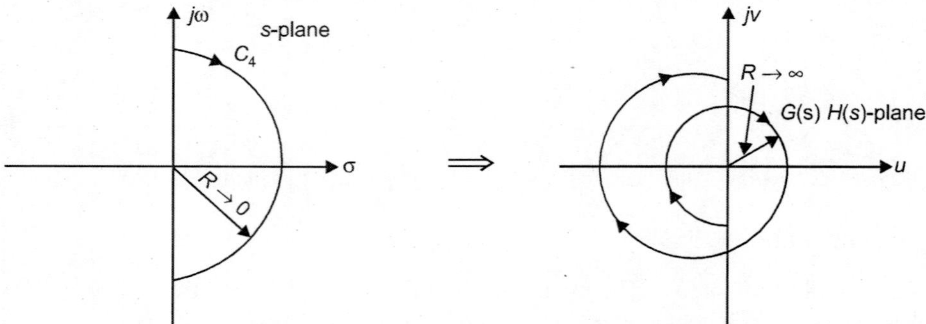

Fig. Ex. 4.2.8: Section C_4 in s-plane.

Fig. Ex. 4.2.9: Mapping of section C_4 in $G(s)$ $H(s)$ s-plane.

Complete plot

The complete Nyquist plot can be obtained by combining the Fig's. (Ex. 4.2.3, 4.2.5, 4.2.7) and (4.2.9).

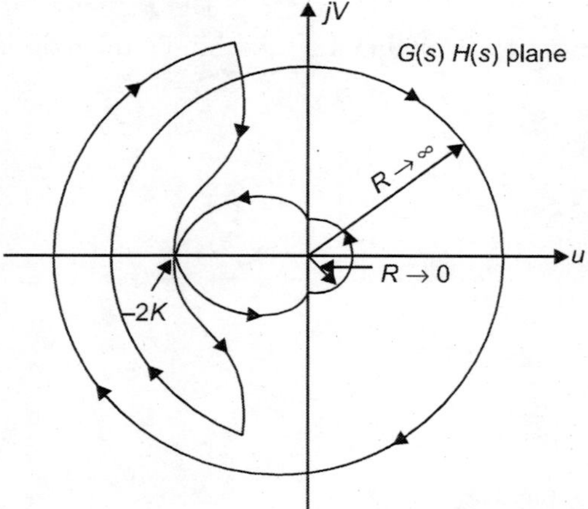

Fig. Ex. 4.2.10: Nyquist plot of $G(s)\,H(s) = \dfrac{K(1+s)^2}{s^3}$.

Stability analysis

When, $-2K = -1$, the contour passes through $(-1 + j0)$ point and corresponding value of K is the limiting value of K for stability.

$\therefore$ Limiting value of $K = 1/2 = .5$

when K < 0.5

when K is less than .5, the contour crosses real axis at a point between 0 and –1. On travelling through Nyquist plot along the indicated direction it is observed that the $-1 + j0$ point is encircled in clockwise direction one time therefore the system is unstable.

when K > .5

when $K > 0.5$, the contour crosses real axis at a point between –1 and $-\infty$. On travelling through Nyquist plot along the indicated direction it is observed that $(-1 + j0)$ point is encircled in both clockwise and anticloskwise direction one time. Hence net encirclement is zero. Also the open loop system has no poles at the right half of s-plane. Therefore the closed loop system is stable.

Result: The value is stable for the value of $K > 0.5$.

EXAMPLE 4.3

Determine the range of values of K for which the system is stable for the system with the open loop transfer function $G(s)\,H(s) = \dfrac{K(1+.5s)(1+s)}{(1+10s)(s-1)}$.

Solution: Given that,

$$G(s)\,H(s) = \frac{K(1+0.5s)(1+s)}{(1+10s)(s-1)}$$

The open loop transfer function does not have a pole at origin. Hence choose the Nyquist contour on s-plane enclosing the entire half plane as shown in Fig. (Ex. 4.3.1).

The Nyquist contour has 3 sections. C_1, C_2 and C_3. The mapping of each section is performed separately and the overall Nyquist plot is obtained by combining the individual sections.

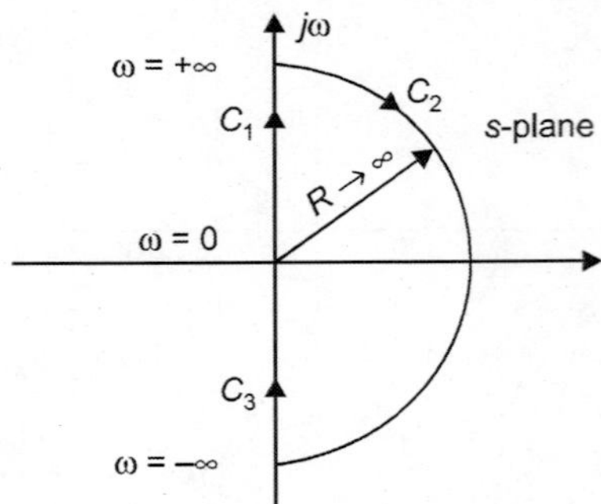

Fig. Ex. 4.3.1: Nyquist contour in s-plane.

Mapping of Section C_1

In section C_1, ω varies from 0 to +∞. The mapping of section C_1 is given by the locus of $G(j\omega)\ H(j\omega)$ as w is varied from 0 to ∞. This locus is the polar plot of $G(j\omega)\ H(j\omega)$

$$G(s)\ H(s) = \frac{K(1+.5s)\,(1+s)}{(1+10s)\,(s-1)}$$

Let $s = j\omega$

$$\therefore \quad G(j\omega)\ H(j\omega) = \frac{K(1+j0.5\omega)\,(1+j\omega)}{(1+j\,10\omega)\,(-1+j\omega)} = \frac{K(1+j1.5\omega-0.5\omega^2)}{-1-j9\omega-10\omega^2}$$

$$= \frac{K(1-.5\omega^2)+j1.5\omega K}{-(1+10\omega^2)-j9\omega}$$

$$G(j\omega)\ H(j\omega) = \frac{K(1-0.5^2\omega)+j1.5\omega K}{-(1+10\omega^2)-j9\omega} \times \frac{-1(1+10\omega^2)+j9\omega}{-(1+10\omega^2)+j9\omega}$$

$$= \frac{-K(1-0.5\omega^2)\,(1+10\omega^2)-13.5\omega^2K + j[9\omega K(1-0.5\omega^2)-1.5\omega K(1+10\omega^2)}{(1+10\omega^2)^2+(9\omega)^2}$$

when the$G(j\omega)\ H(j\omega)$ locus crosses real axis the imaginary term is zero and the corresponding frequency is the phase crossover frequency.

$\therefore$ At $\omega = \omega_{pc}$, $9\omega_{pc}\ K(1-0.5\,\omega_{pc}^2) - 1.5\ \omega_{pc}\ K\,(1+10\,\omega_{pc}^2) = 0$

$$9\omega_{pc}\ K\left(1-0.5\,\omega_{pc}^2\right) = 1.5\,\omega_{pc}K\left(1+10\omega_{pc}^2\right)$$

$$1 - .5\omega_{pc}^2 = .167 + 1.67\omega_{pc}^2$$

$$2.17\omega_{pc}^2 = .833$$

$$\omega_{pc} = \sqrt{\frac{.833}{2.17}}$$

$$\omega_{pc} = .62 \text{ rad/sec}$$

At $\omega = \omega_{pc} = .62$ rad/sec

$$G(j\omega)\,H(j\omega) = \frac{-K(1-.5\omega^2)\,(1+10\omega^2) - 13.5\omega^2 K}{(1+10\omega^2)\,(9\omega^2)}$$

$$= K\left[\frac{1-0.5\times.62^2\ (1+10\times.62^2)+13.5\times.62^2}{(1+10\times.62^2)^2+(9\times.62)^2}\right]$$

$$G(j\omega)\,H(j\omega) = -.167\,K$$

The $G(j\omega)\ H(j\omega)$ locus crosses real axis at a point –0.167.

The shape of $G(j\omega)\ H(j\omega)$ locus is determined by calculating the magnitude and phase of $G(j\omega)\ H(j\omega)$ for various values of ω.

$$G(j\omega)\,H(j\omega) = \frac{K(1+j.5\omega)\,(1+j\omega)}{(1+j10\omega)\,(-1+j\omega)}$$

$$= \frac{K\sqrt{1+(.5\omega)^2}\ \angle\tan^{-1}.5\omega\sqrt{1+\omega^2}\angle\tan^{-1}\omega}{\sqrt{1+(10\omega)^2}\ \angle\tan^{-1}10\omega\sqrt{1+\omega^2}\ \angle(180^\circ-\tan^{-1}\omega)}$$

$$= \frac{K\sqrt{1+.25\omega^2}}{\sqrt{1+100\omega^2}}\angle(\tan^{-1}.5\omega + 2\tan^{-1}\omega - \tan^{-1}10\omega - 180^\circ)$$

$$|G(j\omega)\,H(j\omega)| = K\sqrt{1+.25\omega^2}\Big/\sqrt{1+100\omega^2}$$

$$\angle G(j\omega)\,H(j\omega) = \tan^{-1}.5\omega + 2\tan^{-1}\omega - \tan^{-1}10\omega - 180^\circ$$

ω, rad/sec	0	0.1	0.5	1.5	2.0	5.0	∞
$\lvert G(j\omega)\ H(j\omega)\rvert$	K	$.707K$	$.202K$	$.083K$	$.07K$	$.05K$	0
$\angle G(j\omega)\ H(j\omega)$, deg	–180	–210	–191	–116	–95	–43	0

It is understood from the above table is that

1. The locus of $G(j\omega)\ H(j\omega)$ starts at $K\angle{-180^\circ}$ when $\omega = 0$ and travels in second quadrant,
2. Locus crosses axis at $-.167\ K$ and enters third quadrant.
3. Then the locus corsses negative imaginary axis and enters 4th quadrant.
4. Finally the locus ends at origin when $\omega = \infty$.

The section C_1 in s-plane end its corresponding mappings are shown in below Fig. (Ex. 4.3.2) and (Ex. 4.3.3).

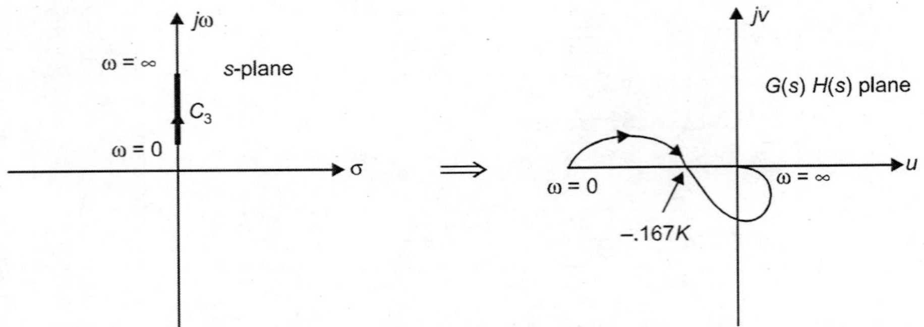

Fig. Ex. 4.3.2: Section C_1 in s-plane.

Fig. Ex. 4.3.3: Mapping of section C_1 in $G(s)\ H(s)$ s-plane.

Mapping of Section C_2

The mapping of section C_2 from s-plane to $G(s)\ H(s)$-plane is obtained by letting $s = \underset{R\to\infty}{Lt}\ Re^{j\theta}$ in $G(s)\ H(s)$ and varying θ from $-\pi/2$ to $+\pi/2$. Since $s \to Re^{j\theta}$ and $R \to \infty$, the $G(s)\ H(s)$ can be approximated as (1 + sT) 1 sT] [Œ sT Á 1]

$$G(s)\ H(s) = \frac{K(1+.5s)\,(1+s)}{(1+10s)\,(s-1)} \approx \frac{K.5s \times s}{10s \times s} = .05K$$

The approximate $G(s)\ H(s)$ is independent of s and so the contour of section C_2 in s-plane is mapped as a point at $.05K$ in $G(s)\ H(s)$ plane.

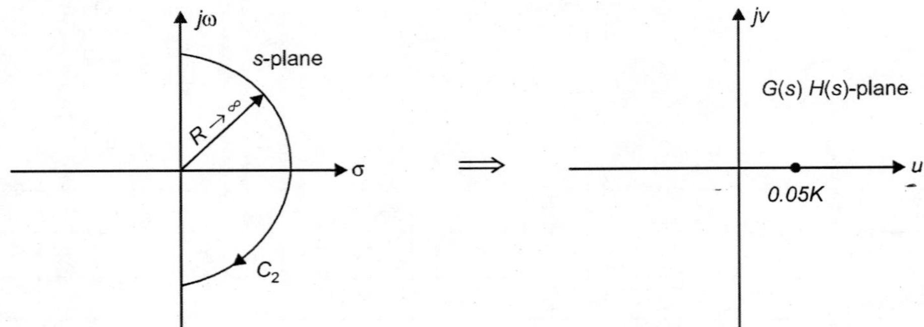

Fig. Ex. 4.3.4: Section C_2 in s-plane.

Fig. Ex. 4.3.5: Mapping of section C_2 in $G(s)\ H(s)$ s-plane.

Mapping of Section C_3

In section C_3, ω varies from $-\infty$ to 0. The mapping of section C_3 is given by the locus of $G(j\omega)\ H(j\omega)$ as ω is varied from $-\infty$ to 0. This locus is the inverse polar plot of $G(j\omega)\ H(j\omega)$.

The inverse plot is given by the mirror image of polar plot with respect to real axis. The section is C_3 in s-plane and its corresponding contour $G(s)\ H(s)$ plane are shown in below Fig. (Ex. 4.3.6) and (Ex. 4.3.7).

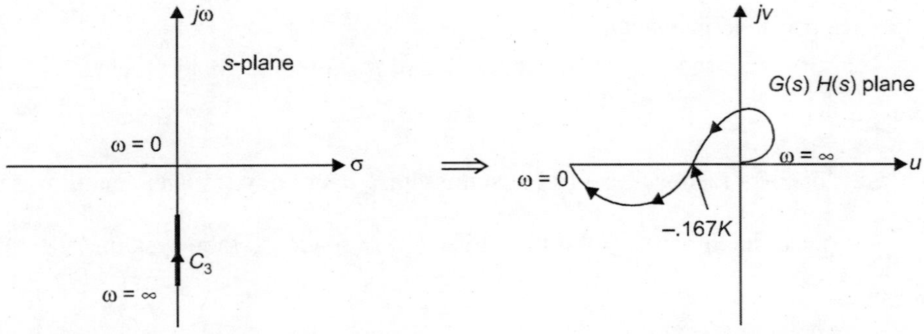

Fig. Ex. 4.3.6: Section C_3 in s-plane.

Fig. Ex. 4.3.7: Mapping of section C_3 in $G(s)\ H(s)$ s-plane.

Complete nyquist plot

The entire nyquist plot in $G(s)\ H(s)$ is obtained by combining Fig. (Ex. 4.3.3.), (Ex. 4.3.5) and (Ex. 4.3.7). The complete plot is shown in Fig. (Ex. 4.3.8).

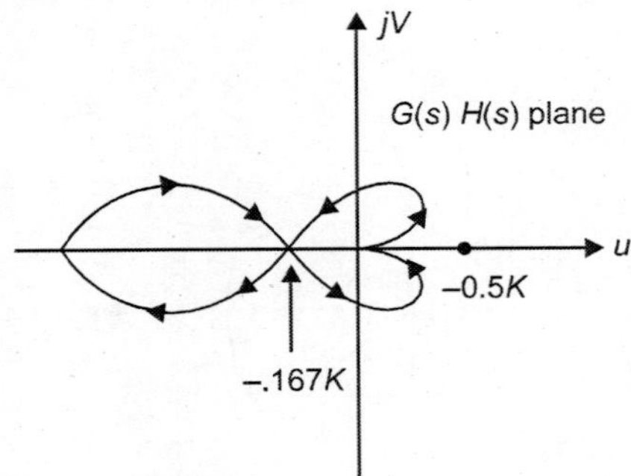

Fig. Ex. 4.3.8: The Nyquist plot of $G(s) = \dfrac{K(1+.5s)\,(1+s)}{(1+10s)\,(s-1)}$.

Stability analysis

When, $-0.167K = -1$, the contour passes through $(-1 + j0)$ point and this value of K is the limiting value for stability.

$$\text{The limiting value of } K = \frac{1}{.167} = 6$$

when K < 6

when $K < 6$, the locus crosses red axis between 0 and -1. On travelling through the locus it is observed that the $-1 + j0$ point is encircled clockwise and so the closed loop system is unstable.

when K > 6

when $K > 6$, the locus crosses real axis between -1 and $-\infty$. On travelling through the locus is observed

CHAPTER 4

that the $-1 + j0$ point is encircled anticlockwise and one time. Hence net encirclement is zero. Also the open loop system has one pole at the right half of s-plane. Hence the system is stable.

Result:

1. The open loop system is unstable.
2. For stability of the closed loop system the value of K should be greater them six ($K > 6$).

EXAMPLE 4.4

Construct Nyquist plot for a feedback control system whose open loop transfer function is given by, $G(s)\,H(s) = \dfrac{5}{s(1-s)}$. Commenton the stability of open loop and closed loop system.

Solution: Given that,

$$G(s)\,H(s) = \frac{5}{s(1-s)}$$

The open loop transfer function has a pole at origin. Hence choose the Nyquist contour on s-plane, enclosing the entire right half s-plane except the origin as shown in Fig. (Ex. 4.4.1).

The Nyquist contour has 4 sections, C_1, C_2, C_3 and C_4. The mapping of each section is performed separately and the overall plot is obtained by combining all individual sections.

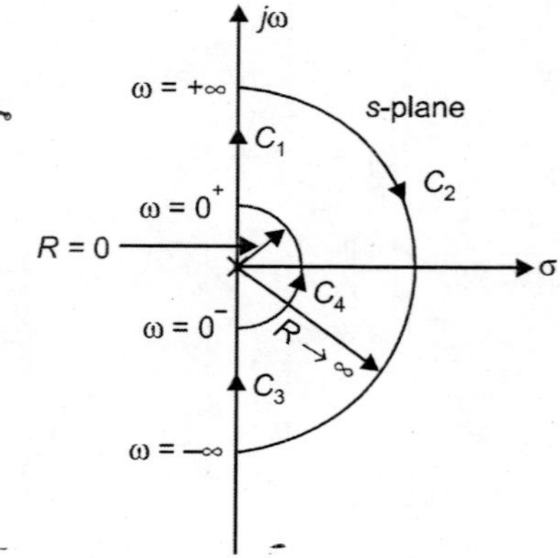

Fig. Ex. 4.4.1: Nyquist contour in s-plane.

Mapping of Section C_1

In section C_1, ω varies from 0 to $+\infty$. The mapping of section C_1 is given by the locus of $G(j\omega)\,H(j\omega)$ as ω is varied from 0 to ∞. This locus is the polar plot of $G(j\omega)\,H(j\omega)$

$$G(s)\,H(s) = 5/s\,(1-s)$$

Let, $s = j\omega$

$$\therefore \qquad G(j\omega)\,H(j\omega) = \frac{5}{j\omega(1-j\omega)} = \frac{5}{\omega\,\angle 90^\circ \sqrt{1+\omega^2}\,\angle -\tan^{-1}\omega}$$

$$= \frac{5}{\omega\sqrt{1+\omega^2}} \angle(-90^\circ + \tan^{-1}\omega)$$

$\therefore$
$$|G(j\omega)\,H(j\omega)| = 5/\omega\sqrt{1+\omega^2}$$
$$\angle G(j\omega)\,H(j\omega) = -90^\circ + \tan^{-1}\omega$$

The exact shape of $G(j\omega)\,H(j\omega)$ locus is determined by calculating the magnitude and phase of $G(j\omega)\,H(j\omega)$ for various values of ω.

ω, rad/sec	0	0.6	1.0	2.0	10.0	∞
$\lvert G(j\omega)\,H(j\omega)\rvert$	∞	7.15	3.53	1.12	.05	0
$\angle G(j\omega)\,H(j\omega)$, deg	–90	–59	–45	–26	–5	0

$G(j\omega)\,H(j\omega)$ locus in section C_1 starts at –90° axis at infinity for $\omega = 0$ and meets the origin along 0°axis when $\omega = \infty$. The mapping of section C_1 is shown in Fig. (Ex. 4.4.2) and (Ex. 4.4.3).

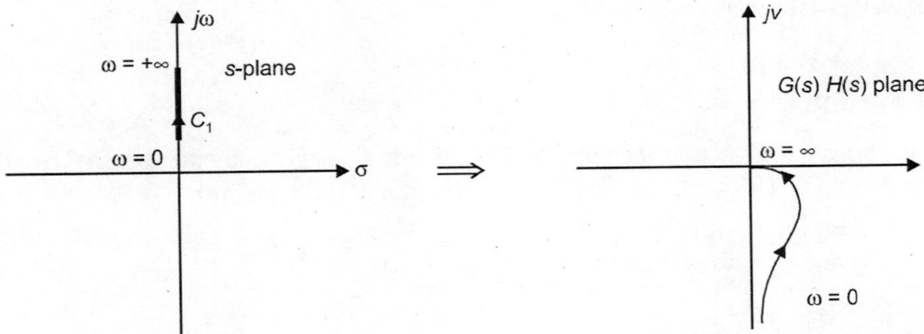

Fig. Ex. 4.4.2: Section C_1 in s-plane.

Fig. Ex. 4.4.3: Mapping of section C_1 in $G(s)\,H(s)$ plane.

Mapping of Section C_2

The mapping of section C_2 from s-plane to $G(s)\,H(s)$-plane is obtained by letting $s = \underset{R\to\infty}{Lt}\ Re^{j\theta}$ in $G(s)$ $H(s)$ and varying θ from $+\pi/2$ to $-\pi/2$. Since $s \to Re^{j\theta}$ and $R \to \infty$, the $G(s)\,H(s)$ can be approximated as $[(1-s)\ 1\ s]$

$\therefore$
$$G(s)\,H(s) = \frac{5}{s(1-s)} \approx \frac{5}{s(-s)} = \frac{5}{s^2 e^{j\pi}}$$

Let,
$$s = \underset{R\to\infty}{Lt}\ \text{Re}^{j\theta}$$

$\therefore$
$$G(s)\,H(s)\Big|_{s = \underset{R\to\theta}{Lt}\ \text{Re}^{j\theta}} = \frac{5}{\underset{R\to\infty}{Lt}\ (\text{Re}^{j\theta})^2\, e^{j\pi}} = oe^{-j(2\theta+\pi)}$$

when $\theta = \pi/2$, $G(s)\ H(s) = oe^{-j2\pi}$

when $\theta = -\pi/2$, $G(s)\ H(s) = oe^{jo}$

From above two equations, we can say that section C_2 in s-plane is mapped as circular arc of zero radius around origin in $G(s)\ H(s)$-plane with argument varying from -2π to $+0$ as shown in Fig. (Ex. 4.4.4) and (Ex. 4.4.5).

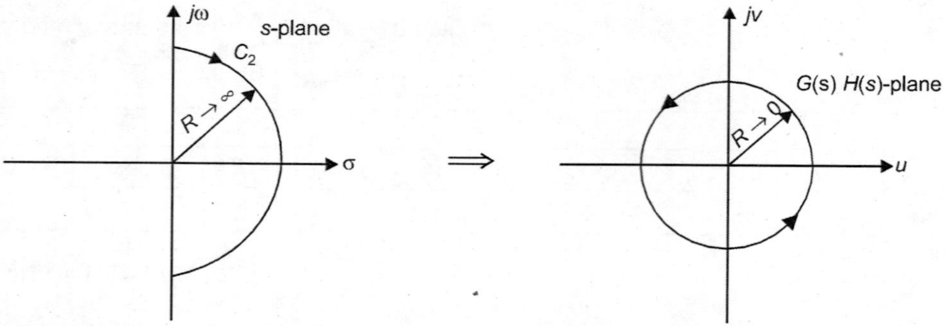

Fig. Ex. 4.4.4: Section C_2 in s-plane.

Fig. Ex. 4.4.5: Mapping of section C_2 in $G(s)\ H(s)$ s-plane.

Mapping of Section C_3

In section C_3, ω varies from $-\infty$ to 0. The mapping of section C_3 is given by the locus of $G(j\omega)\ H(j\omega)$ as ω is varied from $-\infty$ to 0. This locus is the inverse polar plot of $G(j\omega)\ H(j\omega)$.

The inverse polar plot is with respect to real axis. The section is C_3 in s-plane and its corresponding contour $G(s)\ H(s)$ plane are shown in below Fig. (Ex. 4.4.6) and (Ex. 4.4.7).

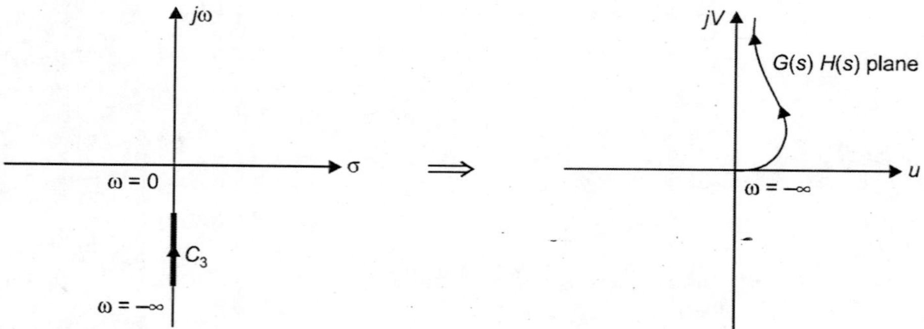

Fig. 4.4.6: Section C_3 in s-plane.

Fig. 4.4.7: Mapping of section C_3 in $G(s)\ H(s)$ s-plane.

Mapping of Section C_4

The mapping of section C_4 from s-plane to $G(s)\ H(s)$-plane is obtained by letting $s = \underset{R \to 0}{Lt}\ Re^{j\theta}$ in $G(s)$ $H(s)$ and varying θ from $-\pi/2$ to $+\pi/2$. Since $s \to Re^{j\theta}$ and $R \to 0$, $G(s)\ H(s)$ can be approximated 1 – s 1 1 [Œ s À 1]

$$\therefore \quad G(s)\,H(s) = \frac{5}{s(1-s)} \approx \frac{5}{s \times 1} = 5/s$$

Let $$s = \underset{R \to 0}{Lt}\ \mathrm{Re}^{j\theta}$$

$$\therefore \quad G(s)\,H(s)\big|_{s = \underset{R\to 0}{Lt}\ \mathrm{Re}^{j\theta}} = \frac{5}{\underset{R\to 0}{Lt}\ \mathrm{Re}^{j\theta}} = \infty e^{-j\theta}$$

when $\theta = -\pi/2$, $\quad G(s)\,H(s) = \infty e^{j\pi/2}$

when $\theta = +\pi/2$, $\quad G(s)\,H(s) = \infty e^{-j\pi/2}$

From above two equations, we can say that section C_4 in s-plane is mapped as circular arc of infinite radius with argument varying from $+\pi/2$ to $-\pi/2$ as shown in Fig. (Ex. 4.4.8) and (Ex. 4.4.9).

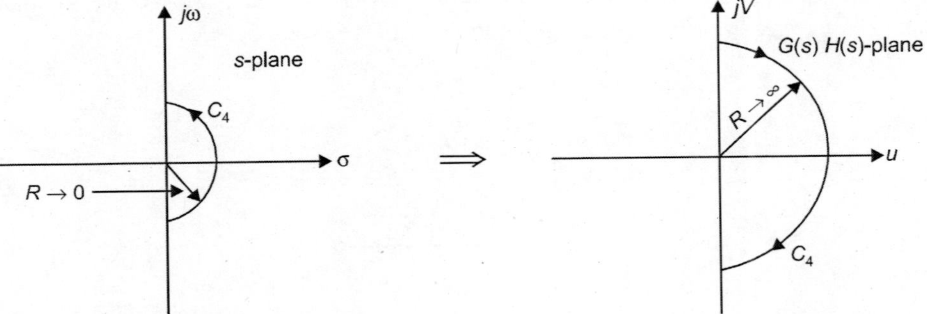

Fig. Ex. 4.4.8: Section C_4 in s-plane.

Fig. Ex. 4.4.9: Mapping of section C_4 in $G(s)\,H(s)$ s-plane.

Complete nyquist plot

The complete nyquist plot can be obtained by Fig. (Ex. 4.4.3, 4.4.5, 4.4.7) and (Ex. 4.4.9).

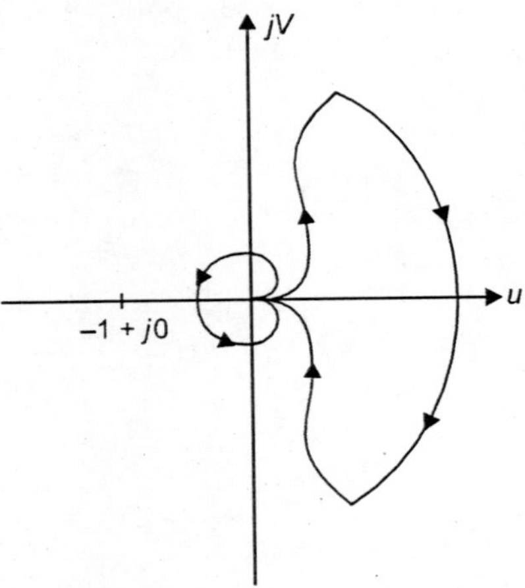

Fig. Ex. 4.4.10: Complete Nyquist plot of $G(s) = 5s(1-s)$.

Stability analysis

The nyquist contour in $G(s)\ H(s)$ place does not encircle the point $(-1 + j0)$ but the open loop transfer function has one pole on the right half s-plane. $\therefore$ The system is unstable.

Result: Both open and closed loop systems are unstable.

EXAMPLE 4.5

Determine the stability of closed loop system, whose open loop transfer function in $G(s)\ H(s) = (s+2)/(s+1)\ (s-1)$, use nyquist stability criterion. Comment on the stability of open-loop and closed loop system.

Solution: Given that

$$G(s)\ H(s) = (s+2)\big/(s+1)\ (s-1)$$

The open loop transfer function does not have a pole at origin. Hence choose the Nyquist contour on s-plane enclosing the entire right half place as shown in Fig. (Ex. 4.5.1).

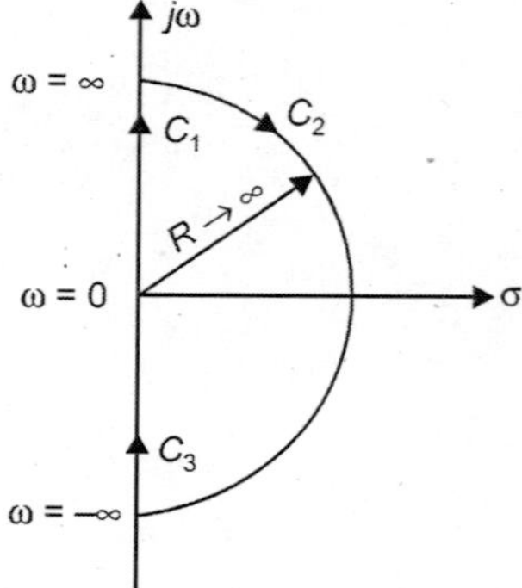

Fig. Ex. 4.5.1: Nyquist contour in s-plane.

The Nyquist contour has 3 sections C_1, C_2, C_3. The mapping of each section is performed separately and the overall Nyquist plot is obtained by combining them.

Mapping of Section C_1

In section C_1, ω varies from 0 to $+\infty$. The mapping of section C_1 is given by the locus of $G(j\omega)\ H(j\omega)$ as ω is varied from 0 to ∞. This locus is the polar plot of $G(j\omega)\ H(j\omega)$

$$G(s)\ H(s) = (s+2)\big/(s+1)\ (s-1) = \frac{2(1+.5s)}{(1+s)\ (-1+s)}$$

Let, $s = j\omega$

$$\therefore \qquad G(j\omega)\ H(j\omega) = \frac{2(1+j0.5\omega)}{(1+j\omega)\ (-1+j\omega)} = \frac{2\sqrt{1+.25\omega^2}\ \angle \tan^{-1} .5\omega}{\sqrt{1+\omega^2}\ \angle \tan^{-1}\omega \sqrt{1+\omega^2}\ \angle 180° - \tan^{-1}\omega}$$

$$= \frac{2\sqrt{1+2.5\omega^2}}{1+\omega^2} \angle(\tan^{-1}.5\omega - 180°)$$

$$\therefore \qquad |G(j\omega)\,H(j\omega)| = \frac{2\sqrt{1+.25\omega^2}}{1+\omega^2}$$

$$\angle G(j\omega)\,H(j\omega) = -180° + \tan^{-1}.5\omega$$

The exact shape of $G(j\omega)\,H(j\omega)$ locus as determined by calculating the magnitude and phase of $G(j\omega)\,H(j\omega)$ for various of ω values.

ω, rad/sec	0	0.4	1	2	10	∞
$\|G(j\omega)\,H(j\omega)\|$	2	1.76	1.12	.57	0.1	0
$\angle G(j\omega)\,H(j\omega)$, deg	–180	–168	–153	–135	–101	–90

From above values, we can say that $G(j\omega)\,H(j\omega)$ locus starts at two axis at a magnitude for –2 for ω = 0 and meets the origin along –90° axis when $\omega = \infty$.

The section C_1 in s-plane and its corresponding mapping in $G(s)\,H(s)$ plane is shown in Fig. (Ex. 4.5.2) and (4.5.3).

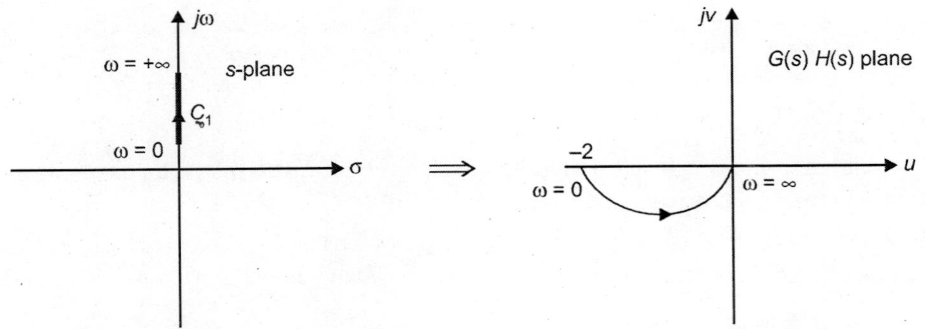

Fig. Ex. 4.5.2: Section C_1 in s-plane.

Fig. Ex. 4.5.3: Mapping of section C_1 in $G(s)\,H(s)$ plane.

Mapping of Section C_2

The mapping of section C_2 from s-plane to $G(s)\,H(s)$-plane is obtained by letting $s = \underset{R\to\infty}{Lt}\; Re^{j\theta}$ in $G(s)\,H(s)$ and varying θ from $+\pi/2$ to $-\pi/2$. Since $s \to Re^{j\theta}$ and $R \to \infty$, the $G(s)\,H(s)$ can be approximated as 1 + sT 1 sT

$$\therefore \qquad G(s)\,H(s) = \frac{2(1+.5s)}{(1+s)\,(1-s)} \approx \frac{2 \times .5s}{s \times s} = \frac{1}{s}$$

Let, $\qquad s = \underset{R\to\infty}{Lt}\; Re^{j\theta}$

$$\therefore \quad G(s)\,H(s)\Big|_{s=\underset{R\to\infty}{Lt}\,\mathrm{Re}^{j\theta}} = \frac{1}{\underset{R\to\infty}{Lt}\,\mathrm{Re}^{j\theta}} = oe^{-j\theta}$$

when $\theta = \pi/2$, $\quad G(s)\,H(s) = oe^{-j\pi/2}$

when $\theta = -\pi/2$, $\quad G(s)\,H(s) = oe^{j\pi/2}$

From above two equations, we can say that section C_2 in s-plane is mapped as circular arc of zero radius around origin in $G(s)\,H(s)$-plane with argument varying from $-\pi/2$ to $+\pi/2$ as shown in Fig. (Ex. 4.5.4) and (Ex. 4.5.5).

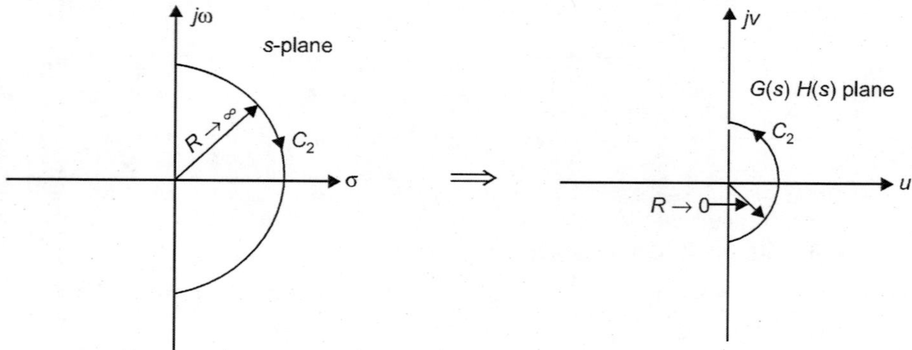

Fig. Ex. 4.5.4: Section C_2 in s-plane.

Fig. Ex. 4.5.5: Mapping of section C_2 in $G(s)\,H(s)$ s-plane.

Mapping of Section C_3

In section C_3, ω varies from -0 to 0. The mapping of section C_3 is given by the locus of $G(j\omega)\,H(j\omega)$ as ω is varied from $-\infty$ to 0. This locus is the inverse polar plot of $G(j\omega)\,H(j\omega)$.

The inverse polar plot is by the mirror image of polar plot with respect to real axis. The section is C_3 in s-plane and its corresponding contour $G(s)\,H(s)$ plane are shown in below Fig. (Ex. 4.5.6) and (Ex. 4.5.7).

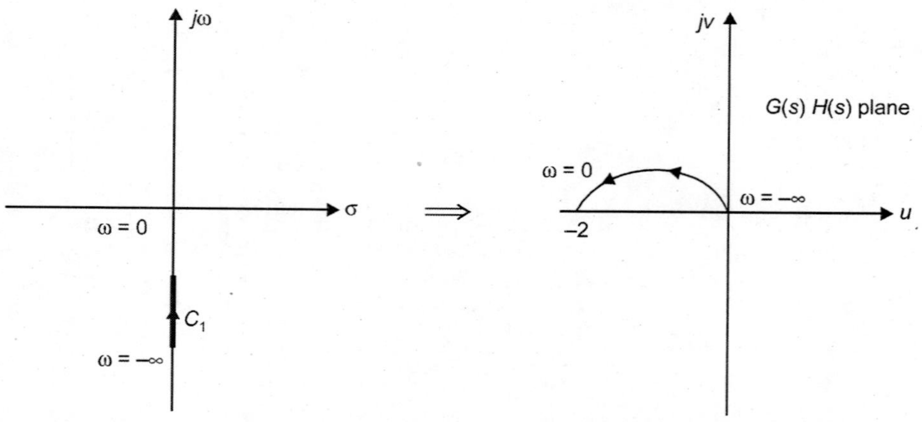

Fig. Ex. 4.5.6: Section C_3 in s-plane.

Fig. Ex. 4.5.7: Mapping of section C_3 in $G(s)\,H(s)$ s-plane.

Complete nyquist plot

The complete nyquist plot can be obtained by combining Fig. (Ex. 4.5.3), (4.5.5) and (4.5.7).

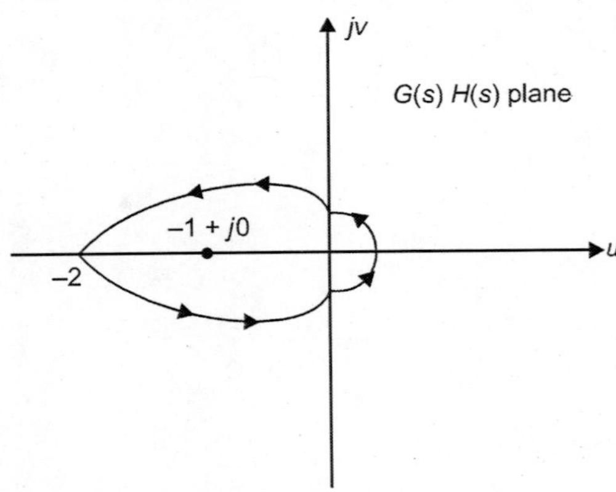

Fig. Ex. 4.5.8: Complete Nyquist plot of $G(s)\ H(s) = s + 2/(s + 1)\ (s - 1)$.

Stability analysis

On travelling through nyquist contour it is observed that $-1 + j0$ point is encircled is anticlockwise direction one time. Also the open loop transfer function has one pole at right half *s*-plane. Since the number of anticlockwise encirclement is equal to number of open loop poles on right half *s*-plane, the closed loop system is stable.

Result:

- Open loop system is unstable.
- Closed loop system is stable.

4.4 *M* AND *N* CIRCLES

The magnitude of closed loop transfer function with unity feedback can be shown to be value of M.These circles are called M-circles.

If the phase of closed loop transfr function with unity feedback is α; then it can be shown that tan α will be the form of circle for every value of α. These circles are called *N* circles.

The *M* and *N* circles are used to find the closed loop frequency response graphically from the open loop frequency response $G(j\omega)$ without calculating the magnitude and phase of the closed loop transfer function at each frequency.

The *M* and *N* circles are available at standard chart. The chart consists of *M* and *N* circles superimposed on ordinary graph sheet. Using ordinary graph the Locus of $G(j\omega)$ will cut the *M*-circles and *N*-circles at various points the intersection of $G(j\omega)$ Locus with *M* and *N* circles gives the magnitude and phase of the closed loop system at frequencies corresponding to the cutting point of $G(j\omega)$.

The *M* and α for various values of ω are tabulated. The magnitude and phase response of closed loop system are sketched on semilog graph sheet by taking ω on the logarithmic scale on *X*-axis.

M Circles

Consider the closed loop transfer function,

$$\frac{C(s)}{R(s)} = \frac{G(s)}{1+G(s)\,H(s)} \qquad (\because\ H(s)=1)$$

for unity feedback system,

$$\frac{C(s)}{R(s)} = \frac{G(s)}{1+G(s)}$$

Put $S = j\omega$,

$$\frac{C(j\omega)}{R(j\omega)} = \frac{G(j\omega)}{1+G(j\omega)} = \frac{G(j\omega)}{1+G(j\omega)}$$

Let,

$$G(j\omega) = X + jY$$

where X = real part of $G(j\omega)$

Y = Imaginary part of $G(j\omega)$

$$\therefore \qquad \frac{C(j\omega)}{R(j\omega)} = \frac{X+jY}{1+X+jY} = \frac{\sqrt{X^2+Y^2}\tan^{-1}\dfrac{Y}{X}}{\sqrt{(1+X^2)+Y^2}\tan^{-1}\dfrac{Y}{1+X}}$$

Let M = Magnitude of $\dfrac{C(j\omega)}{R(j\omega)}$

$$\therefore \qquad M = \frac{\sqrt{X^2+Y^2}}{\sqrt{(1+X^2)+Y^2}}$$

on squaring

$$M^2 = \frac{X^2+Y^2}{(1+X^2)+Y^2}$$

$$M^2\,[(1+X^2)+Y^2] = X^2+Y^2$$

$$M^2\,(1+X^2+2X+Y^2) = X^2+Y^2$$

$$M^2 + M^2X^2 + M^2\,2X + M^2Y^2 - X^2 - Y^2 = 0$$

$$X^2\,(M^2-1) + M^2\,2X + M^2 + Y^2\,(M^2-1) = 0 \qquad (4.12)$$

when $M = 1$, the Eq. (4.9.1) represents a straight line. Consider Eq. (4.12) when $M = 1$

$X^2\,(1-1) + 2X + 1 + Y^2\,(1-1) = 0$

$$\Rightarrow \qquad 2X + 1 = 0$$

$$\therefore \qquad X = -1/2$$

Hence when $M = 1$, the Eq. represents straight line passing through $X = -1/2$ and $Y = 0$

when $M = 1$, the equation (4.12) represents a family of circles. Consider Eq. (4.12)

$X^2\,(M^2-1) + M^2\,2X + M^2 + Y^2\,(M^2-1) = 0$

$$\div \text{ by } (M^2 - 1)$$

$$X^2 + \frac{M^2}{M^2-1} \cdot 2X + \frac{M^2}{M^2-1} \cdot Y^2 = 0$$

Add $\frac{M^2}{(M^2-1)^2}$ on both sides,

$$X^2 + \frac{M^2}{M^2-1} 2X + \frac{M^2}{M^2-1} + \frac{M^2}{(M^2-1)^2} + Y^2 = \frac{M^2}{(M^2-1)^2}$$

$$X^2 + \frac{M^2}{M^2-1} 2X + \frac{M^2(M^2-1)+M^2}{(M^2-1)^2} + Y^2 = \frac{M^2}{(M^2-1)^2}$$

$$X^2 + \frac{M^2}{M^2-1} \cdot 2X + \frac{M^4}{(M^2-1)^2} + Y^2 = \frac{M^2}{(M^2-1)^2}$$

$$\left[X + \frac{M^2}{M^2-1}\right]^2 + Y^2 = \frac{M^2}{[M^2-1]} \tag{4.13}$$

The equation of circle with centre at (X_1 Y_1) and radius r given by

$$(X - X_1)^2 + (Y - Y_1)^2 = r^2 \tag{4.14}$$

On companing Eqs. (4.13) and (4.14) it can be concluded that the Eq. (4.13) represents a family circles with centre at $(- M^2/M^2 - 1, 0)$ and with radius, $r = M/(M^2 - 1)$. The circles given by Eq. (4.12) are called M circles.

when $M = 0$

$$X_1 = \frac{-M^2}{M^2-1} = 0$$

$$Y_1 = 0$$

$$r = \frac{M}{M^2-1} = 0$$

Hence when $M = 0$ the magnitude circle becomes a point at (0, 0).

when $M = \infty$

$$X_1 = \frac{-M^2}{M^2-1} \approx \frac{-M^2}{M^2} = -1$$

$$Y_1 = 0$$

$$r = \frac{M}{M^2-1} \approx \frac{M}{M^2} = \frac{1}{M} = 0$$

Hence when $M = \infty$, the magnitude circle becomes a point at (– 1, 0).

From the above analysis it is clear that the magnitude of closed loop transfer function will be in the form of circles when $M \neq 1$ and when $M = 1$, the magnitude is a straight line passing through (–1/2, 0).

For value of M less them 1, the magnitude is a circle to the right of the straight line corresponding to $M = 1$. It is observed that the circles for $M < 1$ passes through (– 1/2, 0) and (0, 0) on the negative real axis. For decreasing values of M, the radius decreases and the circle, becomes a point at (0, 0) when $M = 0$

For values of $M > 1$, the magnitude is a circle to the left of the straight line corresponding to $M = 1$ It is observed that circle passes between the points (–1, 0) and (–1/2, 0) on the negative real axis. For increasing values of M the radius decreases and the circle becomes a point at (– 1, 0) when $M = \infty$ the family of M-circles are shown in Fig. (4.12)

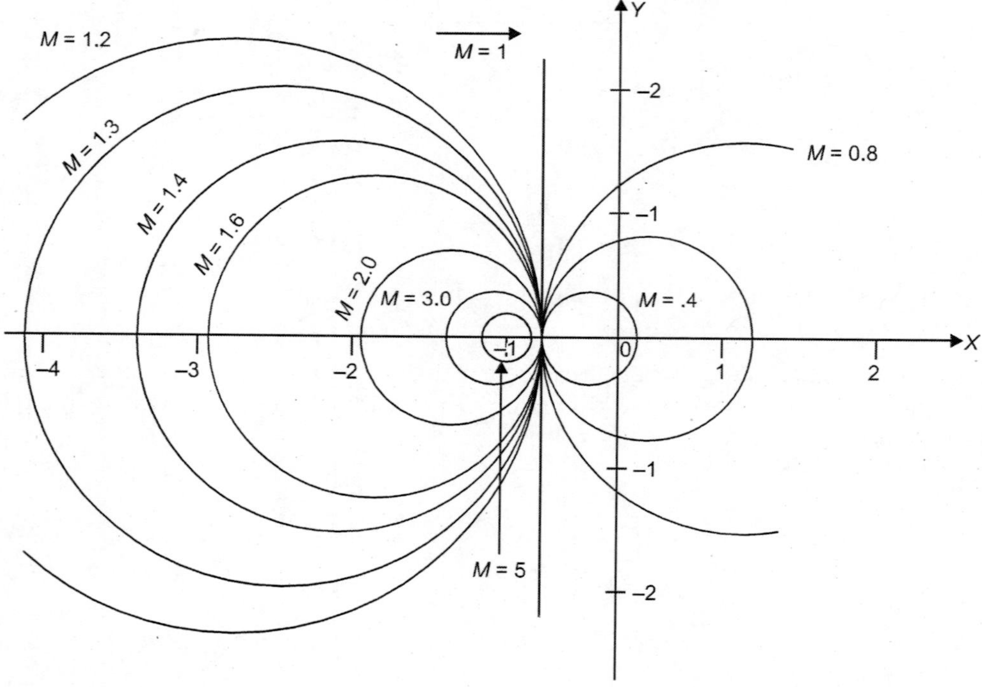

Fig. 4.12: *M* circles.

N Circles

Considered the closed loop transfer function,

$$\frac{C(s)}{R(s)} = \frac{G(s)}{1 + G(s)\,H(s)}$$

for unity feedback system, $\dfrac{C(s)}{R(s)} = \dfrac{G(s)}{1 + G(s)}$

Put $s = j\omega$, $$\frac{C(j\omega)}{R(j\omega)} = \frac{G(j\omega)}{1+G(j\omega)}$$

Let $$G(j\omega) = X + jY$$

where X = Real part of $G(j\omega)$

Y = Imaginary part of $G(j\omega)$

$$\therefore \quad \frac{G(j\omega)}{R(j\omega)} = \frac{X+jY}{1+X+jY} = \frac{\sqrt{X^2+Y^2}\,\tan^{-1}\left(\dfrac{Y}{X}\right)}{\sqrt{1+X^2+Y^2}\,\tan^{-1}\dfrac{Y}{1+x}}$$

Let $$\alpha = \text{Arg}\left[\frac{C(j\omega)}{R(j\omega)}\right]$$

$$\therefore \quad \alpha = \tan^{-1}\left[\frac{Y}{X}\right]\tan^{-1}\left(\frac{Y}{1+X}\right)$$

Let $$N = \tan\alpha$$

$$\therefore \quad N = \tan\left[\tan^{-1}\frac{Y}{X} - \tan^{-1}\frac{Y}{1+X}\right]$$

$$= \frac{\tan\left[\tan^{-1}\left(\dfrac{Y}{X}\right)\right] - \tan\left[\tan^{-1}\dfrac{Y}{1+X}\right]}{1+\tan\left[\tan^{-1}\dfrac{Y}{X}\right]\tan\left[\tan^{-1}\dfrac{Y}{1+X}\right]}$$

Note: $$\tan(A-B) = \frac{\tan A - \tan B}{\tan A \times \tan B}$$

$$= \frac{\dfrac{Y}{X} - \dfrac{Y}{1+X}}{1+\dfrac{Y}{X} - \dfrac{Y}{1+X}} = \frac{\dfrac{Y(1+X)-XY}{X(1+X)}}{\dfrac{X(1+X)+Y^2}{X(1+X)}} = \frac{Y+XY-XY}{X+X^2+Y^2}$$

$$\therefore \quad N = \frac{Y}{X+X^2+Y^2}$$

on rearranging we get,

$$X + X^2 + Y^2 = Y/N$$

$$X + X^2 + Y^2 - Y/N = 0$$

Add thet term $\dfrac{1}{4} + \left[\dfrac{1}{2N}\right]^2$ on both sides.

$$X + X^2 + Y^2 - \frac{Y}{N} + \frac{1}{4} - 1\left[\frac{1}{2N}\right]^2 = \frac{1}{4} + \left[\frac{1}{2N}\right]^2$$

$$\left[X^2 + \frac{1}{4} + X\right] + \left[Y^2 \frac{1}{(2N)^2} - \frac{Y}{N}\right] = \frac{1}{4} + \frac{1}{(2N)^2}$$

$$\left[X + \frac{1}{2}\right]^2 + \left[Y - \frac{1}{2N}\right]^2 = \frac{1}{4} + \frac{1}{(2N)^2} \qquad (4.15)$$

The equation of circle with centre at (X_1, Y_1) and radius r is

$$(X - X_1)^2 + (Y - Y_1)^2 = r^2 (4.11.5)$$

on comparing Eqs. (4.15) and (4.16), it can be concluded that the Eq. (4.15) represents a family of circle with centre at (–1/2, 1/2 N) and with radius.

Radius $\sqrt{\frac{1}{4} + \frac{1}{(2N)^2}}$ the circles given by the Eq. (4.12) are called N circles.

The equation of N circles is satisfied at two points (0, 0) and (–1, 0). Hence the N-circles passes through these two points for all values of α ($N = \tan \alpha$)

When $X = 0$ and $Y = 0$

$$\left[X + \frac{1}{2}\right]^2 + \left[Y - \frac{1}{2N}\right]^2 = \frac{1}{4} + \frac{1}{(2N)^2}$$

$$\left(\frac{1}{2}\right)^2 + \left(-\frac{1}{2N}\right)^2 = \frac{1}{4} + \frac{1}{(2N)^2}$$

$$\frac{1}{4} + \frac{1}{4N^2} = \frac{1}{4} + \frac{1}{4N^2}$$

Consider the equation of N circles when $X = -1$ and $Y = 0$.

$$\left(X + \frac{1}{2}\right)^2 + \left(Y - \frac{1}{2N}\right)^2 = \frac{1}{4} + \frac{1}{(2N)^2}$$

$$\left(-1 + \frac{1}{2}\right)^2 + \left(-\frac{1}{2N}\right)^2 = \frac{1}{4} + \frac{1}{(4N)^2}$$

$$\frac{1}{4} + \frac{1}{4N^2} = \frac{1}{4} + \frac{1}{4N^2}$$

The above analysis shows that the equation of N circle is satisfied at points (0, 0) and (– 1, 0). where $\alpha = 180°$ the circle becomes a straight line passing through real axis. It is also observed that the circle for $\alpha = \theta°$ above the real axis will be part of circle for $\propto = \theta°$ below the real axis, as shwon in Fig. (4.12), the family of N circles are shown in Fig. (4.13).

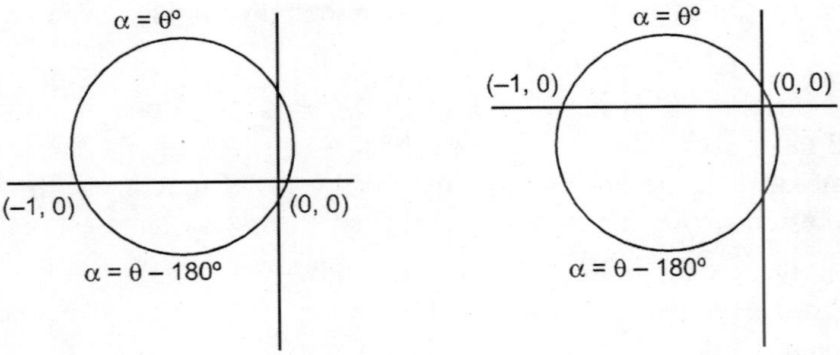

Fig. 4.12

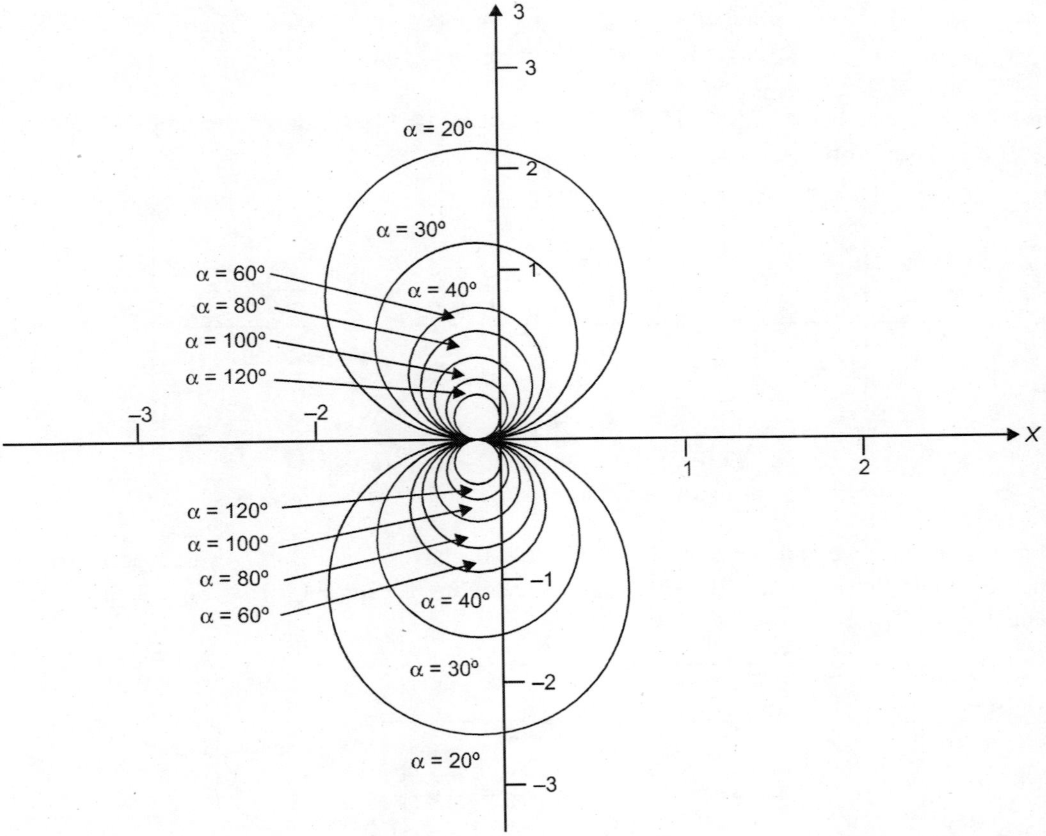

Fig. 4.13: Constant *N* circles.

4.5 ALL PASS AND MINIMUM PHASE SYSTEM

The system is said to be all-pass and minimum phase system when all poles and zeros in the left half of the *s*-plane.

To explain this concept let we consider one example.

For every pole in the left half plane, there is a zero in the mirror image position.

$$G(j\omega) = 1 - j\omega T/1+j\omega T \quad (4.17)$$

Whose pole-zero configuration is shown in Fig. (4.5). It has magnitude of unity at all frequencies and a phase angle ($-2 \tan^{-1} \omega T$) which varies form 0° to –180° as ω is increased from 0 to ∞. The property of unit magnitude at all frequencies applied to all transfer function with antisymmetric pole zero pattern. Physical systems with this property, are called all pass systems.

Let we consider another example, the transfer function has poles in the left half s-plane and zeros in both the left and right-half s-plane. Poles are not permitted to lie in the right-half s-plane because such a system would be unstable.

$$G_1(j\omega) = (1 - j\omega T)/(1 + j\omega T_1)\,(1 + j\omega T_2) \quad (4.18)$$

$$G_1(j\omega) = [(1 + j\omega T)/(1 + j\omega T_1)\,(1 + j\omega T_2)] \left[\frac{(1 - j\omega T)}{1 + j\omega T}\right] \quad (4.19)$$

The pole-zero pattern of Eq. (4.18) is shown in Fig. (4.14) similarly the pole-zero pattern of Eq. (4.19) is shown in Fig. (4.14c).

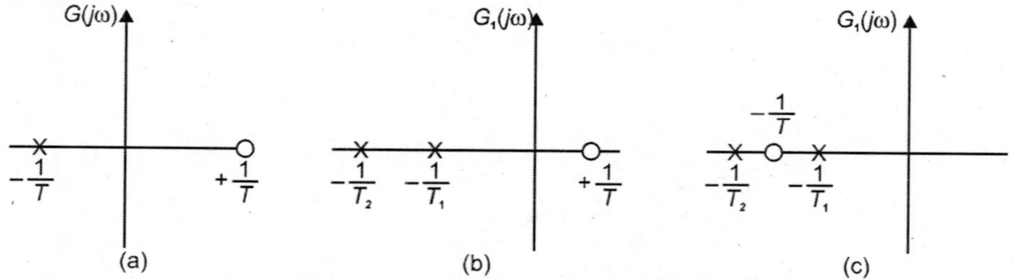

Fig. 4.14: Pole zero pattern for (a) All pass function (b) Nonminimum phase function (c) minimum phase function.

From Fig. (4.15), we can see that $G_1(j\omega)$ and $G_2(j\omega)$ have identical curves of magnitude versus frequency but their phases versus frequency curves are deferent with $G_2(j\omega)$ have in a smaller range of phase angle them $G_1(j\omega)$.

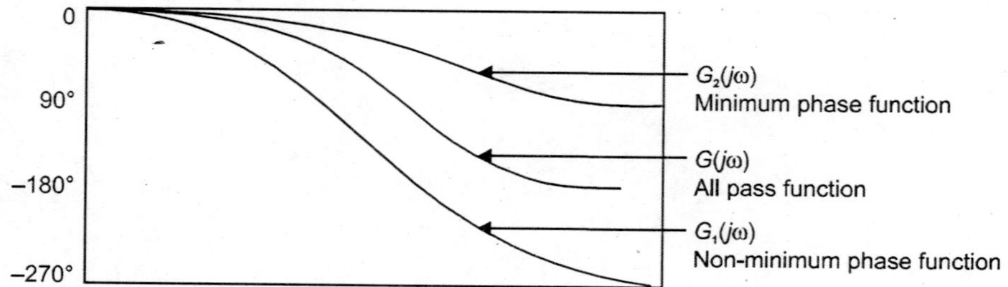

Fig. 4.15: Phase angle characteristics of minimum phase and nonminimum-phase functions.

It means that for $G_1(j\omega)$, there is no unique relationship between magnitude and phase 1 as it is always possible to alert the phase curve without affecting the associated magnitude curve by the addition of an all-pass transfer function.

A Transfer function which has one or more zeros in the right half s-plane is known as non minimum phase transfer function.

If a transfer function has any zeros in the right half *s*-plane, it is possible to extract them one by one by associating them with all - pass transfer function in a mannor shown in Fig. (4.10.2). Each time this is done, the magnitude curve remains unaltered but the phase log is reduced, until we are left with a function which contains no zeros in the right half s-plane. Such a transfer function has the least (minimum) phase angle range for a given magnitude curve and is called a minimum phase function.

4.6 FREQUENCY RESPONSE

The frequency response is the steady state response of a system when the input to the system is a sinusoidal signal. Consider a linear time invariant (LTI) system, H shown in Fig. (4.16). Let $x(t)$ be the input. The output is $y(t)$ is also a sinusoidal signal of the same frequency but magnitude and phase angle are not same.

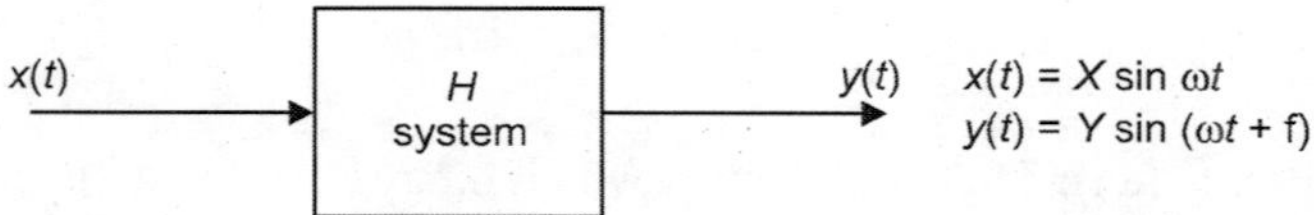

Fig. 4.16: LTI system.

The magnitude and phase relationship between the sinusoidal input and the steady state output of a system is termed as frequency response. In LTI systems the frequency respnse is independent of amplitude and phase of input. The freuqency response of a system is obtained by varying the frequency of the input signal by keeping the magnitude of the input signal at a constant value.

In they system transfer function $T(s)$, if s is replaced by $j\omega$, then the transfer function is called sinusoidal transfer function. The frequency response can be obtained by letting S by $j\omega$ i.e., obtained directly from $T(j\omega)$. $T(j\omega)$ is complex function of frequency.

The frequency response can be evaluated for both open loop system and closed loop system.

The advantages of frequency response analysis:

1. The absolute and relative stability of the closed loop system can be estimated from the knowledge of their open loop frequency response.
2. The practical testing of systems can be easily carried with available sinusoidal signal generators and precise measurement equipments.
3. The transfer function of complicated systems can be determined experimentally by frequency response tests.
4. The design and parameter adjustment of the open loop transfer function of a system for specified closed loop performance is carried out more easily in frequency domain.
5. When the system is designed by use of frequency analysis, the effects of noise disturbance and parameter variations are relatively easy to visualize and incorporate corrective measure.
6. The frequency response analysis and designs can be extended to certain non linear control system.

Frequency domain specifications

The performance and characteristics of a system in frequency domain are measured in terms of frequency domain specifications. The requirements of a system to be designed are usually specified in terms of these specifications.

The frequency domain specifications are:

1. Resonant peak, M_r
2. Resonant frequency, ω_r
3. Band width
4. Cut-off rate
5. Gain margin (Kg)
6. Phase margin (γ)

1. **Resonant peak (M_r):** The maximum value of the magnitude of closed loop transfer function is called the resonant peak. A large resonant peak corresponds to a large overshoot is transient response.
2. **Resonant frequency (ω_r):** The frequency at which the resonant peak occurs is called resonant frequency ω_r. This is related to the frequency of oscillation in the step response and thus its indicative of the speed of transient responses.
3. **Bandwidth (ω_b):** The bandwidth is the range of frequencies for which the system gain is more than –3 *db* the frequency at which the gain is –3 *db* is called cut of frequency. Bandwidth is usually defined for closed loop system and it transmits the signals whose frequencies are less than cut off frequency. The bandwidth is a measure of the ability of a feedback system to produced the input signal noise rejection characteristics and rise time. A large bandwidth corresponds to a small rise time (or) fast response.
4. **Cut off rate:** The slope of the log magnitude curve near the cut off frequency is called the cut off rate. The cut off rate indicates the ability of the system to distinguish the signal from noise.
5. **Gain margin, K_g:** The gain margin K_g is defined as the reciprocal of the magnitude of open loop transfer function at phase cross over frequency. The frequency at which the phase of open loop transfer function is 180° is called the phase cross – over frequency ω_{pc}.

$$\text{Gain margin, } K_g = \omega_{Pc} \frac{1}{|G(j\omega_{Pc})|}$$

The gain margin in *db* can be expressed as,

$$K_g \text{ in } db = 20 \log K_g = 20 \log \frac{1}{|G(j\omega_{Pc})|} = -20 \log |G(j\omega_{Pc})|$$

Note: $|G(j\omega_{Pc})|$ is the magnitude of $G(j\omega)$ at $\omega = \omega_{Pc}$. (Gain margin is *db* is defined as the negative of the *db* magnitude of $G(j\omega)$ at phase cross over frequency)

The gain margin indicated the amount by which the gain of the system can be increased with out affecting the stability of the system.

6. **Phase Margin (γ):** The phase margin γ, is that amount of additional pahse Lag at the gain cross over frequency required to bring the system to the verge of instability. The gain cross over frequency ω_{gc} is the frequency at which the magnitude of the open loop transfer function is unity (or is frequency at which the *db* magnitude is zero).

   The phase margin γ, is obtained by adding 180° to the phase angle φ of the open loop transfer function at the gain cross over frequency.

   Phase margin, $\gamma = 180° + \phi_{g_c}$, where,

$$\phi_{g_c} = \angle G(j\omega_{gc})$$

   Note: $\angle G(\omega_{gc})$ is the phase angle of $G(j\omega)$ at $\omega = \omega_{gc}$.

Estimation of Frequency Domain Specifications for a Second Order System

Resonant Peak (M_r)

Consider the closed loop transfer function of second order system

$$\frac{C(s)}{R(s)} = \frac{\omega_n^2}{s^2 + 2\zeta\omega_n s + \omega_n^2}$$

The sinusoidal transfer function $T(j\omega)$ is obtained by letting $s = j\omega$

$$\therefore \quad T(j\omega) = \frac{C(s)}{R(s)} = \frac{\omega_n^2}{(j\omega)^2 + 2\zeta\omega_n(j\omega) + \omega_n^2}$$

$$= \frac{\omega_n^2}{-\omega^2 + j2\zeta\omega_n\omega + \omega_n^2}$$

$$= \frac{\omega_n^2}{\omega_n^2\left[-\frac{\omega^2}{\omega_n^2} + j2\zeta\frac{\omega}{\omega_n} + 1\right]}$$

$$= \frac{1}{1 - \left(\frac{\omega}{\omega_n}\right)^2 + j2\zeta\left(\frac{\omega}{\omega_n}\right)}$$

Let, normalised frequency, $u = \left(\frac{\omega}{\omega_n}\right)$

$$\therefore \quad T(j\omega) = \frac{1}{\left(1 - u^2\right) + j2\zeta u} = \frac{1}{\sqrt{\left(1 - u^2\right)^2 + (2\zeta u)^2} \angle \tan^{-1}\frac{2\zeta u}{1 - u^2}}$$

Let, $M \rightarrow$ Magnitude of closed loop transfer function

$\alpha \rightarrow$ phase of closed loop transfer function

CHAPTER 4

$$M = |T(j\omega)| = \left[\frac{1}{\left(1-u^2\right)^2 + \left(2\zeta u\right)^2}\right]^{\frac{1}{2}} = \left[\left(1-u^2\right)^2 + 4\zeta^2 u^2\right]^{\frac{1}{2}}$$

$$\alpha = \angle T(j\omega) = -\tan^{-1}\left(\frac{2\zeta u}{1-u^2}\right)$$

The resonant peak is the maximum value of M. The condition for maximum value of M can be obtained by differentiating the Eq. of M with respect to u and letting $\frac{dM}{du} = 0$ when $u = M_r$, where,

$u_r = \frac{\omega_r}{\omega}$ = Normalised resonant frequency

on differentiating equation of M with respect to u we get,

$$\frac{dM}{du} = \frac{d}{du}\left[\left(1-u^2\right)^2 + 4\zeta^2 u^2\right]^{\frac{1}{2}}$$

$$= -\frac{1}{2}\left[\left(1-u^2\right)^2 + 4\zeta^2 u^2\right]^{\frac{3}{2}}\left[2(1-u^2)(-2u) + 8\zeta^2 u\right]$$

$$= \frac{-\left[-4u(1-u^2) + 8\zeta^2 u\right]}{2\left[(1-u^2)^2 + 4\zeta^2 u^2\right]^{3/2}} = \frac{4u(1-u^2) - 8\zeta^2 u}{2\left[(1-u^2)^2 + 4\zeta^2 u^2\right]^{3/2}}$$

Replace u by u_r in above equation

$$\frac{4u_r(1-u_r^2) - 8\zeta^2 u_r}{2\left[(1-u_r^2)^2 + 4\zeta^2 u_r^2\right]^{3/2}} = 0$$

Equating numerator to zero in above equation we get,

$$4u_r(1 - u_r^2) - 8\zeta^2 u_r = 0$$

$$4u_r - 4u_r^3 - 8\zeta^2 u_r = 0$$

$$4u_r^3 = 4u_r - 8\zeta^2 u_r$$

$$u_r^2 = 1 - 2\zeta^2$$

$$u_r = \sqrt{1-2\zeta^2}$$

The resonant peak occurs when $u_r = \sqrt{1-2\zeta^2}$ put this condition is equation for M and solve for M_r

$$\therefore \quad M_r = \left.\frac{1}{\left[\left(1-u^2\right)^2 + 4\zeta^2 u^2\right]^{1/2}}\right|_{u = u_r}$$

$$\therefore \qquad M_r = \frac{1}{\left[\left(1-u_r^2\right)^2 + 4\zeta^2 u_r^{\,2}\right]^{1/2}}$$

$$M_r = \frac{1}{\left[1-\left(1-2\zeta^2\right)\right]^2 + 4\zeta^2(1-2\zeta^2)^{1/2}}$$

$$M_r = \frac{1}{\left(4\zeta^4 + 4\zeta^2 - 8\zeta^4\right)^{1/2}} = \frac{1}{\left(4\zeta^2 - 4\zeta^4\right)^{1/2}} = \frac{1}{\left[4\zeta^2(1-\zeta^2)\right]^{1/2}}$$

$$= \frac{1}{2\zeta\sqrt{1+\zeta^2}}$$

$\therefore$ The resonant peak, $$M_r = \frac{1}{2\zeta\sqrt{1-\zeta^2}}$$

Resonant frequency (ω_r)

The normalised resonant frequency,

$$u_r = \frac{\omega_r}{\omega_n} = \sqrt{1-2\zeta^2}$$

The resonant frequency $\quad \omega_r = \omega_n\sqrt{1-2\zeta^2}$

Bandwidth (ω_b)

Let, normalized bandwidth, $u_b = \dfrac{\omega_b}{\omega_n}$

when $u = u_b$, the magnitude M, of the closed loop system is $\dfrac{1}{\sqrt{2}}(\text{or} - 3db)$.

Hence in the equation of M, put $u = u_b$ and equate to $\dfrac{1}{\sqrt{2}}$.

$$M = \frac{1}{\sqrt{2}} = \frac{1}{\left[(1-u_b^2)^2 + 4\zeta^2 u_b^2\right]^{1/2}}$$

on squaring and cross multiplying,

$$(1 - u_b^2)^2\, 4\zeta^2\, u_b^2 = 2$$

$$1 + u_b^4 + 2u_b^2 + 4\zeta^2\, u_b^2 = 2$$

$$u_b^4 - 2u_b^2\,(1 - 2\zeta^2) - 1 = 0$$

Let, $x = u_b^2$

$$\therefore \qquad x^2 - 2(1 - 2\zeta^2)x - 1 = 0$$

CHAPTER 4

$$\therefore \qquad x = 2(1-2\zeta^2) \pm \frac{\sqrt{4(1-2\zeta^2)^2+4}}{2}$$

By taking only the positive sign,

$$x = 1 = -2\zeta^2 + \sqrt{2-4\zeta^2+4\zeta^4}$$

But,
$$u_b = \sqrt{x} = 1\left[1-2\zeta^2+\sqrt{2-4\zeta^2+4\zeta^4}\right]^{1/2}$$

Also,
$$u_b = \frac{\omega_b}{\omega_n}$$

$$\therefore \quad \text{bandwidth, } \omega_b = \omega_n \times u_b = \omega_n\left[1-2\zeta^2+\sqrt{2-4\zeta^2+4\zeta^4}\right]^{1/2}$$

Phase margin (γ)

The open loop transfer function of second order system,

$$G(s) = \frac{\omega_n^2}{s(s+2\zeta\omega_n)}$$

The sinusoidal transfer function $G(j\omega)$ is obtained by letting $s = j\omega$

$$G(j\omega) = \frac{\omega_n^2}{j\omega(j\omega+2\zeta\omega_n)} = \frac{\omega_n^2}{-\omega^2+j2\zeta\omega\omega_n}$$

$$= \frac{\omega_n^2}{\omega_n^2\left[-\frac{\omega^2}{\omega_n^2}+j2\zeta\frac{\omega}{\omega_n}\right]}$$

Let normalised frequency, $u = \frac{\omega}{\omega_n}$ on substituting $u = \frac{\omega}{\omega_n}$ is above equation we get,

$$G(j\omega) = \frac{1}{u^2+j2\zeta u}$$

Magnitude of
$$G(j\omega) = |G(j\omega)| = \frac{1}{\sqrt{u^4+4\zeta^2u^2}}$$

Phase of
$$G(j\omega) = \angle G(j\omega) = -\tan^{-1}\frac{2\zeta u}{-u^2} = -\tan^{-1}\left(-\frac{2\zeta}{u}\right)$$

$$= \tan^{-1}\frac{2\zeta}{u}$$

At the gain cross-over frequency code, the magnitude of $G(j\omega)$ is unity

Let normalised gain cross over frequency,

$$u_{gc} = \frac{\omega_{gc}}{\omega_n}$$

substitute u by u_{gc} in the equation for magnitude of $G(j\omega)$ and equates to unity

$$\therefore \quad \text{At } u = u_{gc}, \qquad |G(j\omega)| = 1 = \frac{1}{\sqrt{u_{gc}^4 + 4\zeta^2 u_{gc}^2}}$$

$$\therefore \quad u_{gc}^4 + 4\zeta^2 u_{gc}^2 = 1$$

$$u_{gc}^4 + 4\zeta^2 u_{gc}^2 - 1 = 0$$

$$\text{Let} \quad x = u_{gc}^2$$

$$\therefore \quad x^2 + 4\zeta^2 x - 1 = 0$$

$$x = \frac{-4\zeta^2 \pm \sqrt{16\zeta^4 + 4}}{2}$$

$$x = -2\zeta^2 \pm \sqrt{4\zeta^4 + 1}$$

on taking only the +ve sign

$$u_{gc} = \sqrt{x} = \left[-2\zeta^2 + \sqrt{4\zeta^4 + 1}\right]^{1/2}$$

The phase margin, $\gamma = 180 + \angle G(j\omega)/\ \omega = \omega_{gc},\ u = u_{gc}$

Substuituting for $\angle G(j\omega)$ from equations phase of $G(j\omega)$ and phase marging (γ) we get,

$$\gamma = 180\epsilon + \tan^{-1}\frac{2\zeta}{u_{gc}} = 180\epsilon + \tan^{-1}\frac{2\zeta}{\left[-2\zeta^2 + \sqrt{4\zeta^4 + 1}\right]^{-1/2}}$$

Note: Gain margin of 2nd order system is infinite.

4.7 FREQUENCY RESPONSE PLOTS

Frequency response analysis of control system can be carried either mathematically (or) graphically. The various graphical techniques to Frequency response is

1. Bode plot
2. Polar plot (or Nyauist Plot)
3. Nichols Plot
4. *M* and *N* circles
5. Nichols Chart

Bode plot, polar plot and Nichols plot are usually drawn for open loops system. The M and N circle and Nichols chart are used to graphically determine the frequency response of unity feedback closed loop system from the knowledge of open loop response.

The Frequency response plots are used to determine the Frequency domain specifications, to study the stability of the system and to adjust the gain of the system to satisfy the designed specifications.

4.8 BODE PLOTS

A bode plot consists of two graphs one is magnitude plot of sinusoidal transfer function versus log ω. The other is a plot of the phase angle of a sinusoidal transfer function versus log ω.

The bode plot can be drawn for both open loop and closed loop transfer function usually the bode plot is drawn for open loop system The standard representation the logarithmic magnitude of open loop transfer function is $G(j\omega)$ is $20\log|G(j\omega)|$ where base of the logarithm is 10.

The unit used is this representation of the magnitude is the decibel usually abbreviated as *db*. The curves are drawn on semilog paper, using the log scale for frequency and the linear scale for either magnitude (or) phase angle (in degree).

The main advantage of the bode plot is that multiplication of magnitude can be converted into addition. Also a simple method for sketching an approximate log-magnitude curve is available.

Consider the open loop transfer function,

$$G(s) = K(1 + sT_1)/s(1 + sT_2)(1 + sT_3)$$

$$G(j\omega) = K(1 + j\omega T_1)/j\omega(1 + j\omega T_2)(1 + j\omega T_3)$$

$$= \frac{K\angle 0\epsilon\sqrt{1+\omega^2 T_1^2}\ \angle\tan^{-1}\omega T_1}{\omega\angle 90\epsilon\sqrt{1+\omega^2 T_2^2}\angle\tan^{-1}\omega T_2\sqrt{1+\omega^2 T_3^2}\ \angle\tan^{-1}\omega T_3}$$

The magnitude of $G(j\omega) = |G(j\omega)| = \dfrac{K\sqrt{1+\omega^2 T_1^2}}{\omega\sqrt{1+\omega^2 T_2^2}\sqrt{1+\omega^2 T_3^2}}$

The phase angle of the $G(j\omega) = \angle G(j\omega)$

$$= \tan^{-1}\omega T_1 - 90° - \tan^{-1}\omega T_2 - \tan^{-1}\omega T_3$$

The magnitude or $G(j\omega)$ can be expressed in decibels as shown below —

$|G(j\omega)|$ in $db = 20\log|G(j\omega)|$

$$= 20\log\left[\frac{K\sqrt{1+\omega^2 T_1^2}}{\omega\sqrt{1+\omega^2 T_2^2}\sqrt{1+\omega^2 T_3^2}}\right]$$

$$= 20\log\left[\frac{K}{\omega}\times\sqrt{1+\omega^2 T_1^2}\times\frac{1}{\sqrt{1+\omega^2 T_2^2}}\times\frac{1}{\sqrt{1+\omega^2 T_3^2}}\right]$$

$$= 20\log\frac{K}{\omega} \times 20\log\sqrt{1+\omega^2 T_1^2} + 20\log\frac{1}{\sqrt{1+\omega^2 T_2^2}} + 20\log\frac{1}{\sqrt{1+\omega^2 T_3^2}}$$

$$= 20\log\frac{K}{\omega} \times 20\log\sqrt{1+\omega^2 T_1^2} - 20\log\sqrt{1+\omega^2 T_2^2} - 20\log\sqrt{1+\omega^2 T_3^2} \qquad (4.20)$$

From the Eq. (4.11.1) it si clear that, when the magnitude is expressed is *db*, the multiplication is converted to addition. Hence in magnitude plot, the *db* magnitude of individual factors of $G(j\omega)$ can be added.

Basic factors of G(jω)

1. Constant gain, K
2. Integral factor, $\frac{K}{j\omega}$ or $\frac{K}{(j\omega)^n}$
3. Derivative factor, $K(j\omega)$ or $K(j\omega)^n$
4. Fist order factor in denominator, $\frac{1}{1+j\omega T}$ or $\frac{1}{(1+j\omega T)^m}$
5. First order factor in numerator, $(1 + j\omega T)$ or $(1 + j\omega T)^m$
6. Second order factor in denominator, $\left[\frac{1}{1+2\zeta(j\omega/\omega_n)+(j\omega/\omega_n)^2}\right]$
7. Quadratic factor in numerator, $\left[1+2\zeta\left(\frac{j\omega}{\omega_n}\right)+\left(\frac{j\omega}{\omega_n}\right)^2\right]$

Constant Gain, K

Let $G(s) = K$

$\therefore$ $G(j\omega) = K = K\angle 0°$

$A = |G(j\omega)|\ db = 20\log K$

$\phi = \angle G(j\omega) = 0°$

The magnitude plot for a constant gain K is a horizontal straight line at the magnitude of 20 log K *db*. The phase plot is straight line at 0°.

when $K > 1$, 20 log K is positive

when $0 < K < 1$, 20 log K is negative

when $K = 1$, 20 log K is zero

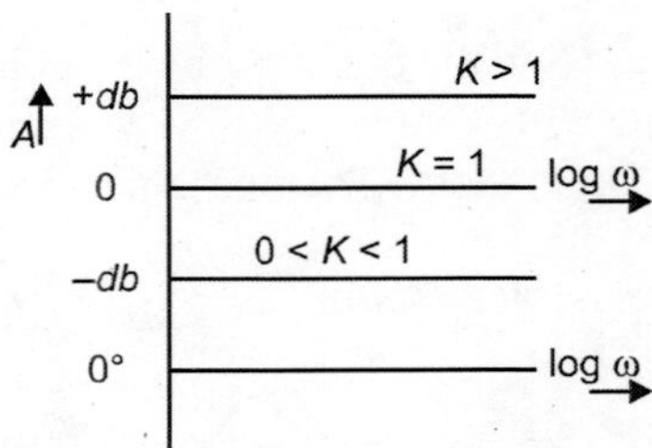

Fig. 4.17

Integral factor

Let
$$G(s) = \frac{K}{s}$$

$$G(j\omega) = \left(\frac{K}{j\omega}\right) = \frac{K}{\omega} \angle -90°$$

$$A = |G(j\omega)| \text{ in } db = 20 \log \left(\frac{K}{\omega}\right)$$

and
$$\phi = \angle G(j\omega) = -90°$$

when $\omega = 0.1\ K$, $A = 20 \log \left(\frac{1}{0.1}\right) = 20\ db$

when $\omega = K$, $A = 20 \log 1 = 0\ db$

when $\omega = 10\ K$, $A = 20 \log \frac{1}{10}\ db = -20\ db$

From the above analysis it is evident that the magnitude plot of the integral factor is a straight line with a slope of – 20 *db*/dec and passing through zero *db*, when $\omega = K$. Since the $\angle G(j\omega)$ is constant and independent of ω the phase plot is a straight line at – 90°.

When an integral factor has multiplicity of *n*, then

$$G(s) = \frac{K}{s^n}$$

$$G(j\omega) = \frac{K}{(j\omega)^n} = \frac{K}{\omega^n} \angle -90n\epsilon$$

$$A = |G(j\omega)| \text{ in } db = 20 \log \frac{K}{\omega^n}$$

$$= 20 \log \left[\frac{K^{1/n}}{\omega}\right]^n = 20n \log \left[\frac{K^{1/n}}{\omega}\right]$$

$$\phi = \angle G(j\omega) = -90n°$$

Now the magnitude plot of the integral factor in straight line with a slope of –20 *n db*/dec and passing through zero *db* with $\omega = K^{1/n}$. The phase plot is a straight line at – 90*n*°

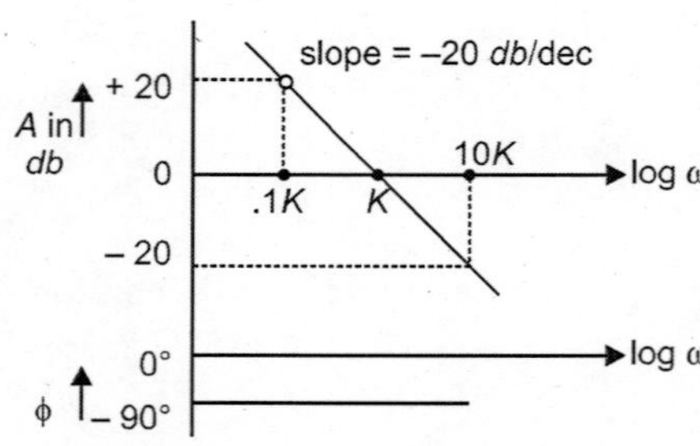

Fig. 4.18: Bode plot of integral factor $\left(\frac{k}{s}\right)$.

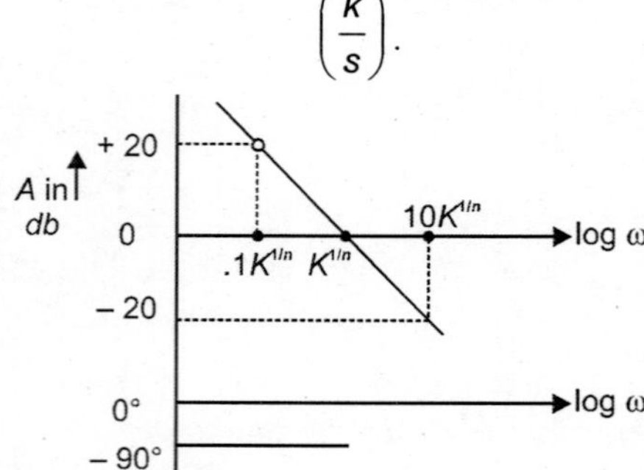

Fig. 4.19: Bode plot of integral factor $\frac{k}{(j\omega)^n}$

Derivative factor

Let
$$G(s) = Ks$$

$G(j\omega) = K(j\omega) = (K\omega) \angle 90°$

$A = |G(j\omega)|$ in $db = 20 \log (K\omega)$

and $\phi = \angle G(j\omega) = +90°$

when $\omega = 0.1/K, A = 20 \log (0.1) = -20\ db$

when $\omega = \dfrac{1}{K}, A = 20 \log 1 = 0\ db$

when $\omega = \dfrac{10}{K}, A = 20 \log 10 = +20\ db.$

From above analysis it is evident that the magnitude plot of the derivative factor is a straight line with a slope of + 20 *db*/dec and passing through zero *db* when $\omega = \dfrac{1}{K}$. Since the $\angle G(j\omega)$ is a constant and impedance of ω, the phase plot is a straight line at +90°.

When the derivative factor has multiplicity of *n* then

$$G(s) = Ks^n$$

$$\therefore \quad G(j\omega) = K(j\omega)^n = K\omega^n \angle 90n°$$

$A = |G(j\omega)|$ in $db = 20 \log (K\omega^n) = 20 \log (K^{1/n}\omega) = 20 \log (K^{1/n}\omega)$ and $\phi = \angle G(j\omega) = 90n°$

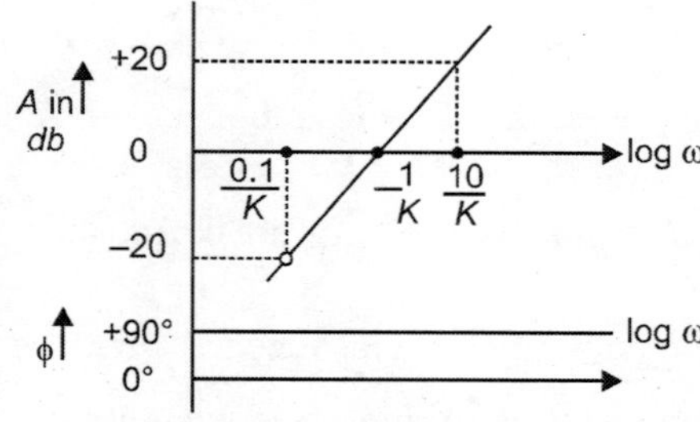

Fig. 4.20: Bode plot of derivative factor $K(j\omega)$.

Now the magnitude plot of the derivative factor is a straight line with a slope of + 20 *db*/dec and passing through zero *db* when $\omega = 1/K^{1/n}$. The phase plot is a straight line at $90n°$.

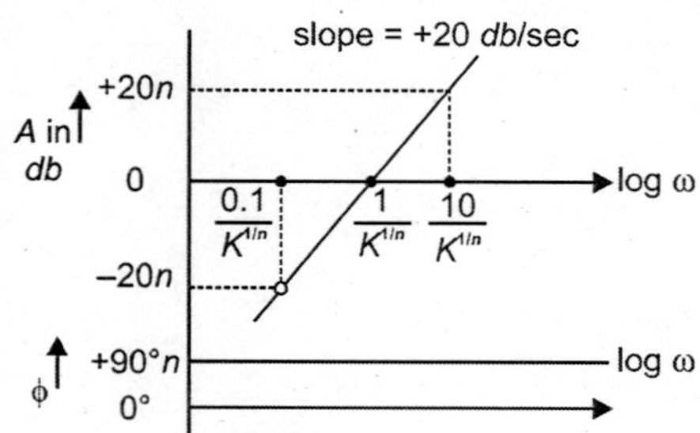

Fig. 4.21: Bode plot of derivative factor $K(j\omega)$.

First order factor in denominator

$$G(s) = \frac{1}{1 + sT}$$

CHAPTER 4

$$G(j\omega) = \frac{1}{1 + j\omega T} = \frac{1}{\sqrt{1+\omega^2 T^2}} \angle - \tan^{-1} \omega T$$

$$A = |G(j\omega)\text{in } db = 20\log \frac{1}{\sqrt{1+\omega^2 T^2}} = -20\log\sqrt{1+\omega^2 T^2}$$

At very low frequency $\omega T << 1$

$\therefore$ $$A = -20\log\sqrt{1+\omega^2 T^2} = -20\log 1 = 0$$

At very high frequencies $\omega T >> 1$

$\therefore$ $$A = -20\log\sqrt{1+\omega^2 T^2} \approx -20\sqrt{\omega^2 T^2} = -20\log \omega T$$

at $$\omega = \frac{1}{T}, A = -20 \log 1 = 0$$

at $$\omega = \frac{10}{T}, A = -20 \log 10 = -20\ db$$

The above analysis shows that the magnitude plot of the factor $\frac{1}{(1+j\omega T)}$ can be approximated by two straight lines, one is *s* straight line at *odb* for the frequency range $0 < \omega < \frac{1}{T}$ and the other to a straight line with slope – 20 *db*/dec for the frequency or break frequency. For the factor $\frac{1}{(1+j\omega T)}$ the frequency, $\omega = \frac{1}{T}$ is the corner frequency, ω_c. It divides the frequency response curve into two regions, a curve for low frequency region and a curve for high frequency region.

The actual magnitude at the corner frequency, $\omega_c = 1/T$ is

$$A = -20\log\sqrt{1+1} = -3\,db.$$

Hence by this approximation the loss in *db* at the corner frequency is – 3*db*. The phase plot is obtained by calculating the phase angle of $G(j\omega)$ for various values of ω

phase angle $$\phi = \angle G(j\omega) = -\tan^{-1} \omega T$$

At the corner frequency, $\omega = \omega_c = \frac{1}{T}, \phi = -\tan^{-1} \omega T = -\tan^{-1}(1) - 45°$

As $\omega \to 0$, $\phi \to 0$ and As $\omega \to \infty$, $\phi \to -90°$

The phase angle of the factor $\left[\frac{1}{(1+j\omega T}\right]$ varies form 0° to – 90° as ω is varied from zero to ∞.

The phase plot is a curve passing through – 45° at ω_c.

When the first order factor is the denominator has a multiplicity of *m* then,

$$G(s) = \frac{1}{(1+sT)^m}$$

and $$G(j\omega) = \frac{1}{(1+j\omega T)^m} = \frac{1}{\left[\sqrt{(1+j\omega^2 T^2)}\right]^m} \angle m \tan^{-1} \omega T$$

$$A = |G(j\omega)| \text{ in } db = 20 \log \frac{1}{\sqrt{(1+\omega^2 T^2)}^m} = -20 \log \sqrt{1+\omega^2 T^2}$$

$$\phi = \angle G(j\omega) = -m \tan^{-1} \omega T$$

Now the magnitude plot of the factor $\frac{1}{(1+j\omega T)^m}$ can be approximated by two straight lines, one is a straight line at zero *db* for the frequency range $0 < \omega < \frac{1}{T}$ and the other is a straight line with slope –20 m *db*/dec for the frequency range $\frac{1}{T} < \omega < \infty$. The corner frequency, $\omega_c = \frac{1}{T}$ and the loss in *db* at the corner frequency is – 3 *mdb*.

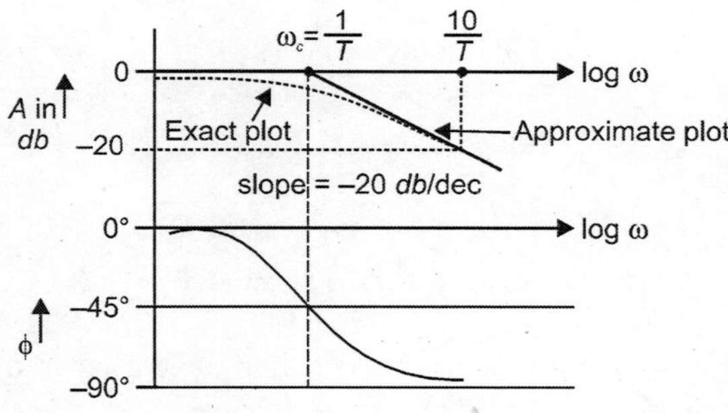

Fig. 4.22: Bode plot of the factor $\frac{1}{1+j\omega T}$.

The phase angle of the factor $\frac{1}{(1+j\omega T)^m}$ various from 0° to – 90° as ω is varied from zero to ∞. The phase plot is a curve through – 45 *m*° at ω_c.

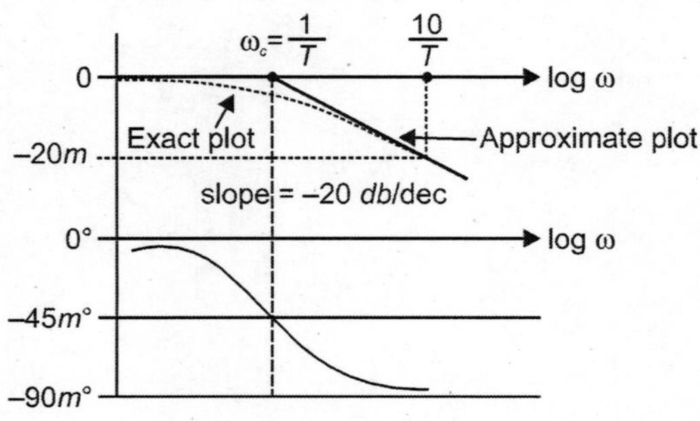

Fig. 4.23: Bode plot of the factor $\frac{1}{(1+j\omega)^m}$.

CHAPTER 4

First order factor in the numerator

$$G(s) = 1 + sT$$

$$G(j\omega) = 1 + j\omega T = \sqrt{1+\omega^2 T^2} \angle - \tan^{-1} \omega T$$

$$A = |G(j\omega) \text{ in } db = 20\log\sqrt{1+\omega^2 T^2} \text{ and } \phi = < G(j\omega) = \tan^{-1} \omega T$$

By analysis similar to that of previous section if can be shown that the magnitude plot of the factor $(1 + j\omega T)$ can be approximated by two staright lines, one is a straight line at zero *db* for the frequency range $0 < \omega < \frac{1}{T}$ and the other is a straight line with slope + 20 *db*/dec for the frequency range $\frac{1}{T} < \omega < \infty$. The two straight lines are asymptotes of the exact curve.

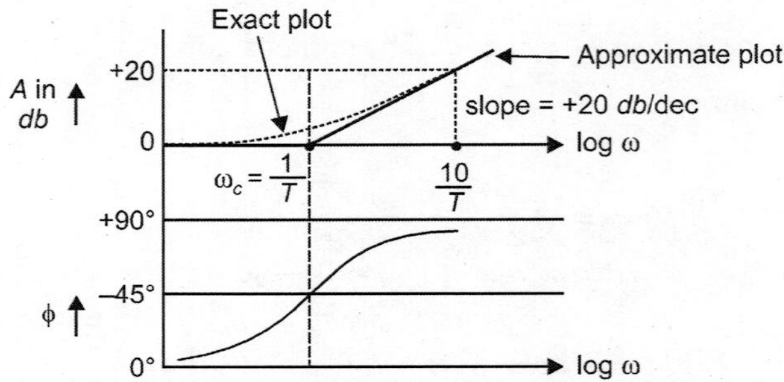

Fig. 4.24: Bode plot of the factor $(1 + j\omega T)$.

The frequency at which the two asymptotes meet is called the corner frequency or break frequency. For the factor $(1 + j\omega T)$, the frequency, $(\omega = 1/T)$ is corner frequency, ω_c. By this approximation the loss in *db* at the corner frequency is + 3 *db*. The phase angle of the factor $(1 + j\omega T)$ varies from zero to +90° as ω is varied from 0 to ∞. The phase plot in a curve passing through + 45° at ω_c.

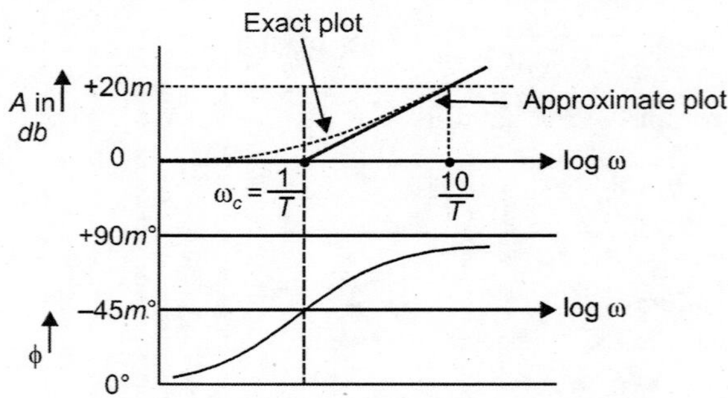

Fig. 4.25: Bode plot of the factor $(1 + j\omega T)^m$

When the first order factor in the numerator has multiplicity of m, then $G(s) = (1 + sT)^m$

$$G(j\omega) = (1 + sT)^m = \left(\sqrt{1+\omega^2 T^2}\right)^m \angle m \tan^{-1} \omega T$$

$$A = |G(j\omega)| \text{ in } db = 20\log\left(\sqrt{1+\omega^2T^2}\right)^m$$

$$= 20m \log\sqrt{1+\omega^2T^2}$$

$$\phi = \angle G(j\omega) = m \tan^{-1} \omega T$$

Now the magnitude plot of the factor $(1 + j\omega T)^m$ can eb approximated by two straight lines, one is a straight line at zero *db* for the frequency range $0 < \omega < \frac{1}{T}$ and the other in a straight line with a slope of + 20 *m db*/dec for the frequency $\frac{1}{T} < \omega < \infty$. The corner frequency $\omega_c = \frac{1}{T}$ and the loss in *db* at this comes frequency is + 3 *m db*.

The phase angle of the factor $(1 + j\omega T)^m$ varies from zero to + 90 $m°$ as ω is varied form zero intinity. The phase plot is a curve passing through + 45 $m°$ at ω_c.

Quadratic factor in the denominator

$$G(s) = \frac{1}{1+2\zeta\frac{s}{\omega_n}+\left[\frac{s}{\omega_n}\right]^2}$$

$$G(j\omega) = \frac{1}{1+j\frac{2\zeta\omega}{\omega_n}+\left[\frac{j\omega}{\omega_n}\right]^2} = \frac{1}{1-\left[\frac{\omega}{\omega_n}\right]^2+j2\zeta\frac{\omega}{\omega_n}}$$

$$= \frac{1}{\sqrt{\left[1-\frac{\omega^2}{\omega_n^2}\right]^2+4\zeta^2\frac{\omega^2}{\omega_n^2}}} \angle \tan^{-1}\frac{2\zeta\frac{\omega}{\omega_n}}{1-\frac{\omega^2}{\omega_n^2}}$$

Let $A = |G(j\omega)|$ in *db*

$$A = 20\log\frac{1}{\sqrt{\left(1-\frac{\omega^2}{\omega_n^2}\right)^2+4\zeta^2\frac{\omega^2}{\omega_n^2}}} = -20\log\sqrt{\left(1-\frac{\omega^2}{\omega_n^2}\right)^2+4\zeta^2\frac{\omega^2}{\omega_n^2}}$$

$$= -20\log\frac{1}{1-\frac{\omega^4}{\omega_n^4}-2\frac{\omega^2}{\omega_n^2}+4\zeta^2\frac{\omega^2}{\omega_n^2}} - 20\log\sqrt{1-\frac{\omega^2}{\omega_n^2}+(2-4\zeta^2)+\frac{\omega^4}{\omega_n^4}}$$

At very low frequencies when $\omega << \omega_n$ the magnitude is

$$A = -20\log\sqrt{1-\frac{\omega^2}{\omega_n^2}\left(2-4\zeta^2\right)+\frac{\omega^4}{\omega_n^4}} \approx -20\log 1 = 0$$

At very high frequencies when $\omega >> \omega_n$ the magnitude is

$$A = -20\log\sqrt{1-\frac{\omega^2}{\omega_n^2}\left(2-4\zeta^2\right)+\frac{\omega^4}{\omega_n^4}} \approx -20\log\sqrt{\frac{\omega^4}{\omega_n^4}} = -20\log\frac{\omega^2}{\omega_n^2}$$

$$\therefore \qquad A = -40\log\frac{\omega}{\omega_n}$$

At $\omega = \omega_n$, $A = -40 \log 1 = 0$ *db*

At $\omega = 10\,\omega_n$, $A = -40 \log 10 = -40$ *db*

From the above analysis it is evident that the magnitude plot of the quadratic factor in the denominator can be approximated by two straight line, one is a straight line at 0 *db* for the frequency range $0 < \omega < \omega_n$ and the other is a straight line with slope – 40 *db*/bec for the frequency range $\omega_n < \omega < \infty$. The two straight lines ar asymptotes of the exact curve. The frequency at which the two asymptotes meet is called the corner frequency. For the quadratic factor, the frequency ω_n is the corner frequency, ω_c.

The two asymptotes of the exact curve are independent of the damping ratio, ζ. In the exact magnitude plot resonant peak occurs near the corner frequency and the magnitude of resonant peak depends on ζ. Lower the value of ζ, larger will be resonent peak. Hence by this approximation the error at the corner frequency depends on damping ratio ζ. The phase plot is obtained by calculating the phase angle of $G(j\omega)$ for various value of ω.

$$\phi = \angle G(j\omega) = \tan^{-1}\left[\frac{2\zeta\dfrac{\omega}{\omega_n}}{1-\dfrac{\omega^2}{\omega_n^2}}\right]$$

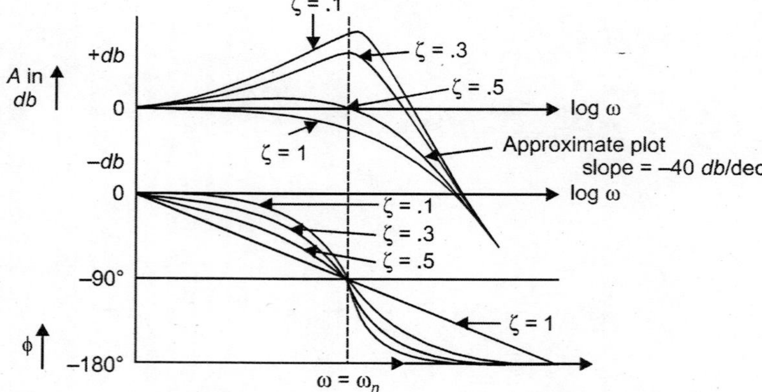

Fig. 4.26: Bode plot of qualratic factor in denominator.

At $\omega = \omega_n$, $\phi = \tan^{-1}\dfrac{2\zeta}{0} = \tan^{-1}\infty = -90°$

At $\omega = 0$, $\phi \to 0$ and as $\omega \to \infty$, $\phi = -180€$

The phase angle of the quadratic factor varies from 0 to – 180° as ω is varied from 0 to ∞. The phase plot is a curve passing through –90° at ω_c. At the corner frequency phase angle is – 90° and independent of ζ, but at all other frequency depends on ζ.

Quadratic factor in the numerator

$$G(s) = 1 + 2\zeta\left(\frac{s}{\omega_n}\right) + \left(\frac{s}{\omega_n}\right)^2$$

$$G(j\omega) = 1 + 2\zeta\frac{\omega}{\omega_n} + \left(\frac{j\omega}{\omega_n}\right)^2 = \sqrt{\left[1 - \left(\frac{\omega^2}{\omega_n^2}\right)\right]^2 + 4\zeta\frac{\omega^2}{\omega_n^2}}\angle\tan^{-1}\frac{2\zeta\dfrac{\omega}{\omega_n}}{\dfrac{1-\omega^2}{\omega_n^2}}$$

The magnitude plot of the quadratic factor is the numerator can be approximated by two straight lines, one is a straight line at 0 *db* for the frequency range $0 < \omega < \omega_n$ and the other is a straight line with slope + 40 *db*/dec for the frequency range $\omega_n < \omega < \infty$. The corner frequency is ω_n. Due to this approximation the error at the corner frequency depends on ζ.

The phase angle varies from 0 to + 180°, as ω is varied from 0 to ∞. At the corner frequency the phase angle is + 90° and independent of ζ, but at all other frequency it depends on ζ.

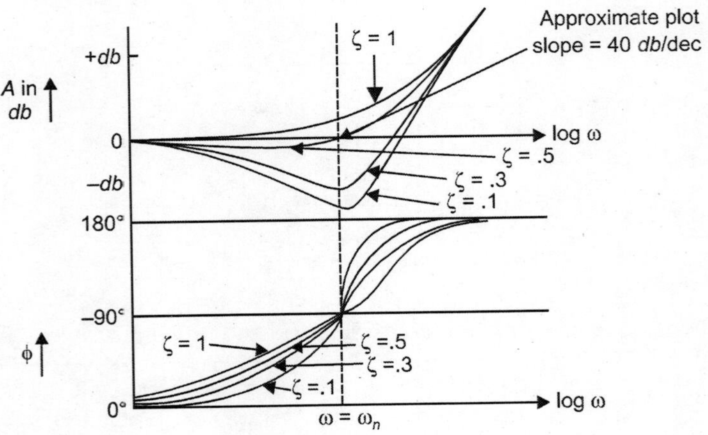

Fig. 4.27: Bode plot of quadratic factor in numerator.

Tite step by step procedure for plotting the magnitude plot is given below

Step 1: Convert the transfer function into bode form (or) time constant form. The bode form of the transfer function is

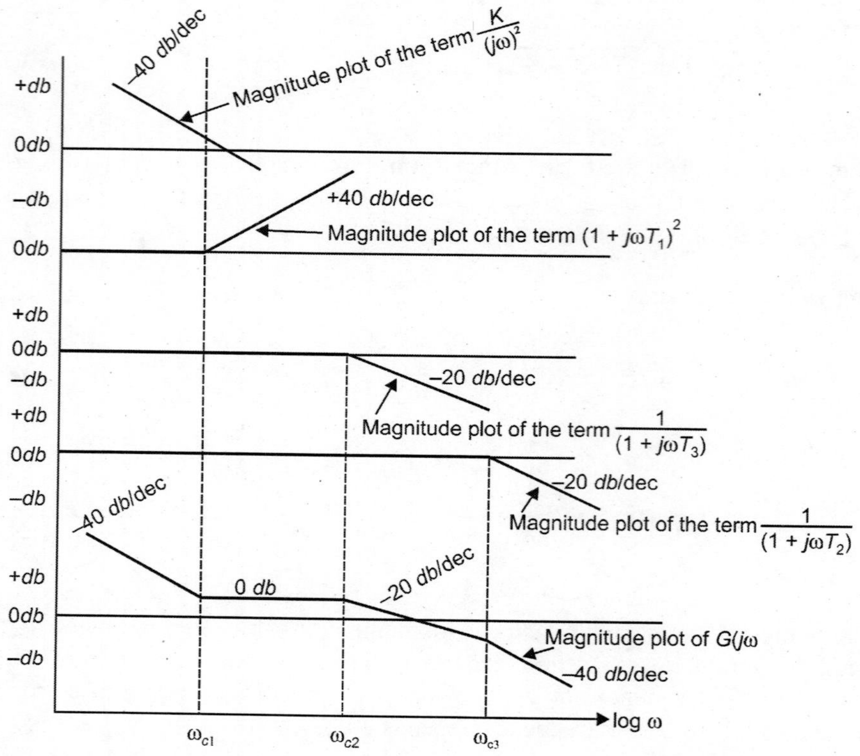

Fig. 4.28: Magnitude bode plot of $G(j\omega) = \dfrac{K(1 + j\omega T_1)^2}{(j\omega)^2 (1 + j\omega T_2)(1 + j\omega T_3)}$.

$$G(s) = \frac{K(1 + sT_1)}{\left(1 + \frac{s^2}{\omega_n^2} + 2\zeta\frac{s}{\omega_n}\right) s(1 + sT_2)}$$

$$G(j\omega) = \frac{K(1 + j\omega T_1)}{j\omega(1 + j\omega T_2)\left[1 - \frac{\omega^2}{\omega_n^2} + j2\zeta\frac{\omega}{\omega_n}\right]}$$

Step 2: List the corner frequencies in the increasing order and prepare a table as shown below :

Term	**Corner frequency rad/sec**	**Slope of each term *db*/dec**	**Change in slope from term to another term *db*/dec**

In the above table enter K or $\frac{K}{(j\omega)^n}$ or $K(j\omega)^n$ as the first term and the other terms in the increasing order of corner frequencies. Then enter the corner frequency, slope contributed by each term and change in slope at every corner frequency.

Step 3: Choose an arbitrary frequency ω_l which is lesser than the lowest corner frequency. Calculate the *db* magnitude of K or $\frac{K}{(j\omega)^n}$ or $K(j\omega)^n$ at use and at the lowest corner frequency.

Step 4: Then calculate the gain (*db* magnitude) at every corner frequency are by one by using the formula, for first two terms the formual will be,

A_1 (gain of first term) = 20 log |first term|

A_2 (gain of second term) = 20 log |first term|

A_3 (gain of third term) = change in slope from ω_{c1} to ω_{c2} Ч log

$$\left|\frac{\text{Present frequency}}{\text{Previous frequency}}\right| + \text{Previous gain } (A_2)$$

The same formula should be used till n^{th} term.

Step 5: Choose an arbitrary frequency ω_n. Which is greater than the highest corner frequency. Calculate the gain at ω_n by using the formula in step 4.

Step 6: In a semilog graph sheet mark the required range of frequency on *x*-axis (log scale) and the range of *db* magnitude on *y*-axis (ordinary scale).

Step 7: Mark all the points obtained in step 4 and 5 on the graph and join the points by straight lines. Mark the slope at every part of the graph.

Note: The magnitude plot obtained above is an approximate plot. If an exact plot is needed then appropriate correction should be made at every corner frequencies.

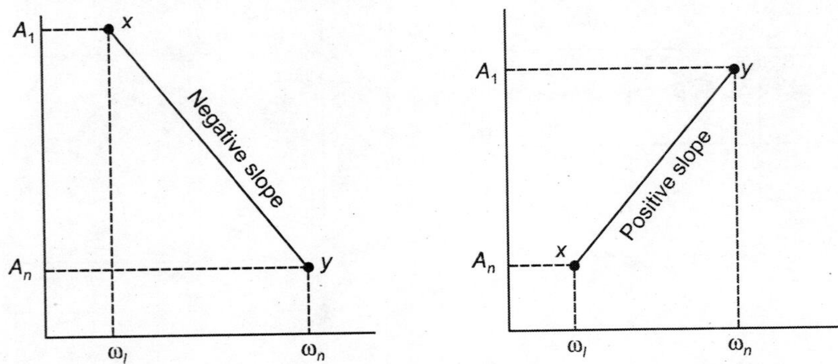

Fig. 4.29: Representation of magnitude plot for both +ve and –ve slopes.

Procedure for phase plot of bode plot

The phase point is an exact plot and no approximations are made while drawing the phase plot (i.e., plot can be drawn without using scale whereas magnitude plot should be drawn using scale. The exact phase angles of $G(j\omega)$ are computed for various values of ω and tabulated.

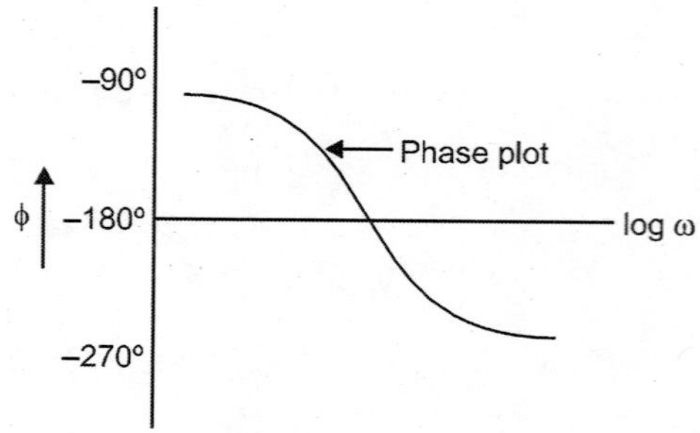

Fig. 4.30: Representation of phase plot.

The choice of frequencies are same as magnitude plot. The magnitude plot and phase plot are drawn in a single semilong sheet on a common frequency scale.

Take another y-axis in the graph where the magnitude plot is drawn and in this y-axis mark the range of phase angles. From the tabulated values of ω and phase angles, mark all the points on the graph. Join the points by a smooth curve.

Determination of Gain margin and phase margin from bode plot

The gain margin in *db* is given by the negative of *db* magnitude of $G(j\omega)$ at the phase cross-over frequency, ω_{pc}. The ω_{pc} is the frequency at which phase of $G(j\omega)$ is –180°. If the *db* magnitude of $G(j\omega)$ at ω_{pc} is negative then gain margin is positive and vice versa.

Let ϕ_{gc} be the phase angle of $G(j\omega)$ at gain cross over frequency ω_{gc}. The ω_{gc} is the frequency at which the *db* magnitude of $G(j\omega)$ is zero. Now the phase margin γ is given by, $\gamma = 180° + \phi_{gc}$. If ϕ_{gc} is less then –180° than phase margin is positive and vice versa. The positive and negative gain margins are shown in Fig. (4.31).

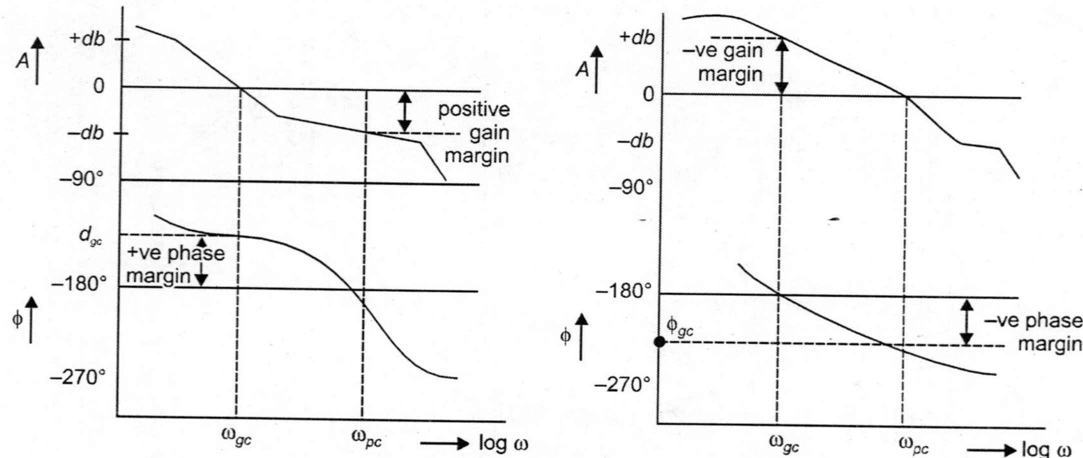

Fig. 4.31: Bode plot showing (i) Phase margin (ii) Gain margin (kg).

Gain adjustment using bode plot: In the open loop transfer function $G(j\omega)$ the constant K contributes only magnitude. Hence by charging the values of K the system can be adjusted to meet the desired specifications. The desired specification are gain margin, phase margin, ω_{pc} and ω_{gc}. In a system transfer function if the value of K required to be estimated to satisfy a desired specification then draw the bode plot of the system with $K = 1$. The constant K can add 20 log K to every point of the magnitude plot and due to this addition the magnitude plot will shift vertically, up or down. Hence shift the magnitude plot vertically up or down to meet the desired specification. Equate the vertical distance by which the magnitude plot is shifted to 20 log K and solve for K.

Let x = change in db (The required db on which the plot have to be shifted)

Now $20 \log K = x$; $\log K = \dfrac{x}{20}$ $\therefore K = 10^{\frac{x}{20}}$

EXAMPLE 4.6

Sketch bode plot for the following tarnster function and determine the system gain K for the gain cross over frequency to be 5 rad/sec $G(s) = \dfrac{Ks^2}{(1+0.2s)(1+.02s)}$.

Solution:

Step 1: Replace s by $j\omega$ in given transfer function

The sinusoidal transfer function $G(j\omega)$ is obtained by replacing s by $j\omega$ in the given s-domain transfer function

$$\therefore \quad G(j\omega) = \frac{K(j\omega)^2}{(1+0.2j\omega)(1+.02j\omega)}$$

Let $K = 1$,

$$\therefore \quad G(j\omega) = \frac{(j\omega)^2}{(1+0.2j\omega)(1+.02j\omega)}$$

Step 2: Find the required informations to draw magnitude plot

The corner frequencies are

$$\omega_{c1} = \frac{1}{0.2} = 5 \text{ rad/sec}$$

$$\omega_{c2} = \frac{1}{.02} = 50 \text{ rad/sec}$$

The various terms of $G(j\omega)$ are listed in Table Ex. 4.6.1 is the increasing order of their corner frequency. Also the table shows the slope contributed by each term and the change is slop at the corner frequency.

Choose a low frequency ωl such that $\omega l < \omega_{c1}$ and choose a high frequency ω_h such that $\omega_h > \omega_{c2}$

Let $\omega_l = .5$ rad/sec and $\omega_h = 100$ rad/sec

Let $A = |G(j\omega)|$ in db

Let us calculate A at ω_l, ω_{c1}, ω_{c2} and ω_h.

At $\omega = \omega_l$, $A_1 = 20 \log |(j\omega)^2| = 20 \log (\omega^2) = 20 \log (.5)^2 = -12\ db \Rightarrow A_1 = -12\ db$

At $\omega = \omega_{c1}$, $A_2 = 20 \log |(j\omega)^2| = 20 \log (\omega^2) = 20 \log (5)^2 = 28\ db \Rightarrow A_2 = 28\ db$

Table Ex. 4.6.1

Term	Corner frequency rad/sec	Slope of each term db/dec	Change in slope from one term to another term db/dec
$(j\omega)^2$	–	+40	
$\frac{1}{1+j0.2\omega}$	$\omega_{c1} = \frac{1}{.2} = 5$	– 20	+40 – 20 = 20
$\frac{1}{1+j.02\omega}$	$\omega_{c2} = \frac{1}{.02} = 50$	– 20	20 – 20 = 0

At $\omega = \omega_{c2}$. A_3 = [slope from one term to another term) Ч log $\frac{\text{Present frequency}}{\text{Previous frequency}}$ + Previous gain (A_2)

$$\text{i.e., } A_3 = \left[\text{slope from } \omega_{c1} \text{ to } \omega_{c2} \text{ Ч} \log \frac{\omega_{c2}}{\omega_{c1}}\right] + A_2 \text{ (at } \omega = \omega_{c1}) = 20 \log \frac{50}{5} + 28 = 48\,db$$

At $\omega = \omega_h$, A_4 = [slope from one term to another term) Ч log $\frac{\text{Present frequency}}{\text{Previous frequency}}$ + Previous gain (A_3)

$$\text{i.e., } A_4 = \left[\text{slope from } \omega_{c2} \text{ to } \omega_h \text{ Ч} \log \frac{\omega_h}{\omega_{c2}}\right] + A \text{ (at } \omega = \omega_{c2}) = 0 \times \log\frac{100}{50} + 48 = 48\ db.$$

Let the points *a*, *b*, *c* and d be the points corresponding to frequencies ω_l, ω_{c1}, ω_{c2} and ω_h respectively on the magnitude plot.

Phase plot

The phase angle of $G(j\omega)$ as a function of ω is given by $\phi = \angle G(j\omega) = 180° - \tan^{-1} 0.2\,\omega - \tan^{-1} .02\omega$.

The phase angle of $G(j\omega)$ are calculated for various value for ω and listed Table (Ex. 4.6.2)

Table Ex. 4.6.2

ω, rad/sec	$\phi = \angle G(j\omega)$ deg
0.5	173.7 ≈ 174
1	167.6 ≈ 168
5	129.3 ≈ 130
10	105.3 ≈ 106
50	50.7 ≈ 50
100	29.5 ≈ 30

when
$$\omega = .5,\ \phi = 180° - \tan^{-1} .2\,(.5) - \tan^{-1} .02\,(.5) \approx 174°$$
$$\omega = 1,\ \phi = 180° - \tan^{-1} .2\,(1) - \tan^{-1} .02\,(1) \approx 168°$$
$$\omega = 5,\ \phi = 180° - \tan^{-1} .2\,(5) - \tan^{-1} .02\,(1) \approx 130°$$
$$\omega = 10,\ \phi = 180° - \tan^{-1} .2\,(10) - \tan^{-1} .02\,(10) \approx 106$$
$$\omega = 50,\ \phi = 180° - \tan^{-1} .2\,(50) - \tan^{-1} .02\,(50) \approx 50$$
$$\omega = 100,\ \phi = 180° - \tan^{-1} .2\,(100) - \tan^{-1} .02\,(100) \approx 30$$

On the same semi log sheet choose scale on the *y*-axis on the right side of semilog sheet. Mark the calculated phase angle on the graph sheet. Join the points by smooth curve.

Calculation of K

We know that the gain cross over frequency 5 rad/sec (given). At $\omega = 5$ rad/sec the gain is 28 *db*. If gain cross over frequency is 5 rad/sec then at that frequency the *db* gain should be zero. Hence to every point of magnitude plot a *db* gain of –28 *db* should be added. The addition of –28 *db* shifts the plot downwards. The magnitude correction is independent of frequency. Hence the magnitude of –28 *db* is contributed by the term *K*. The value of *K* is calculated by equating 20 log *k* to –28 *db*.

$$20 \log K = -28\ db$$

$$\log K = -\frac{28}{20}$$

$$\therefore \quad K = 10^{-(28/20)} = .0398$$

The magnitude plot with $K = 1$ and .0398 and the phase plot are shown in Fig. (4.6.1)

Note: The frequency $\omega = 5$ rad/sec is a corner frequency. Hence in the exact plot the *db* gain at $\omega = 5$ rad/sec will be 3 *db* less than the approximate plot. Therefore for exact plot the 20 log *K* will contribute a gain of – 25 *db*.

$$\therefore \quad 20 \log K = -25\ db$$

$$\log K = -\frac{25}{20}$$

$$\therefore \quad K = 10^{-\frac{25}{20}} = .0562$$

$$K = .0562$$

Results: The value of *K* at 5 rad/sec is .0562.

$$\boxed{K = .0562}$$

4.9 RELATIONSHIP BETWEEN SYSTEM TYPE AND LOG MAGNITUDE CURVE

Consider the unity-feed back control system. The static position, velocity and acceleration error constant describe the low frequency behaviour of type 0, type 1, type 2 systems, respectively. For a given system,

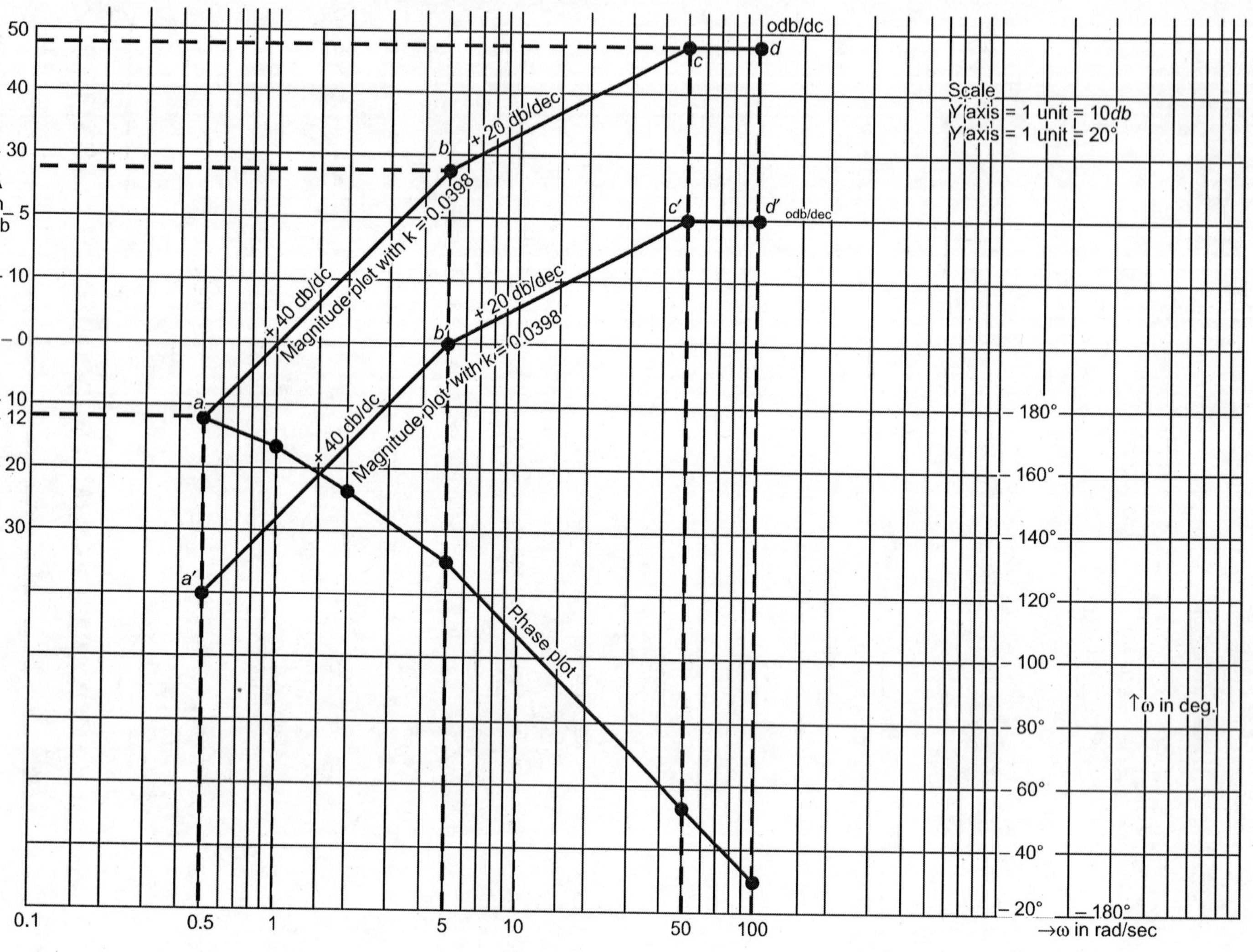

Fig. 4.6.1: Bode plot of transfer function $G(s0 = ks2/(1 + 0.2s)(1 + .2s)$.

only one of the static error constant is finite and significant. (The larger value of the finite static error constant, the higher the loop gain; as ω approaches zero).

The type of the system determines the slope of the log magnitude curve at low frequencies. Thus, information concerning the existence and magnitude of the steady state error of a control system to a given input can be determined from the observation of the low frequency region of the log-magnitude curve.

Determination of static position error constants: Consider the unity feedback control system shown in Fig. (4.32). Assume that the open - loop transfer function is given by Fig. (4.33).

$$G(s) = \frac{K(T_a s + 1)(T_b s + 1).....(T_m s + 1)}{s^n (T_1 s + 1)(T_2 s + 1).....(T_p s + 1)}$$

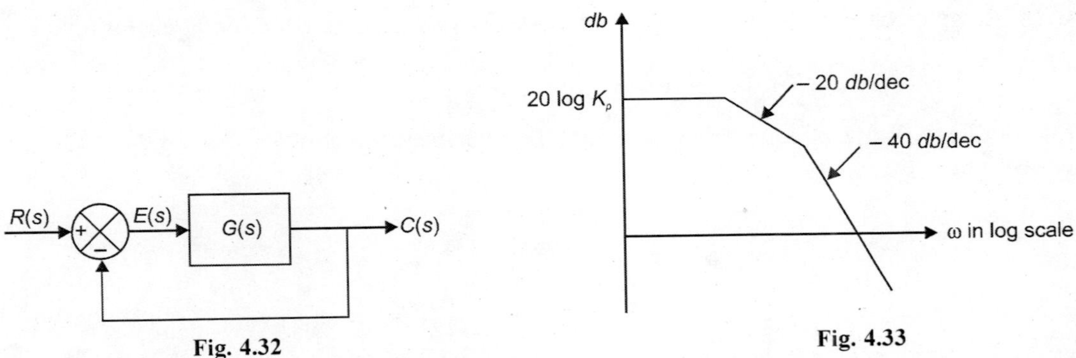

Fig. 4.32

Fig. 4.33

Fig. (4.33) shows an example of the log-magnitude plot of a type 0 system. In such a system the magnitude of $G(j\omega)$ equals K_p at low frequencies, or

$$\underset{\omega \to 0}{Lt}\ G(j\omega) = K = K_p$$

If follows that the low-frequency asymptotes is a horizontal line at 20 log K_p *db*.

Determination of static velocity error constants: Consider the unity feedback control system shown in Fig. (4.32) and (4.34) shows an example of the log magnitude plot of a type 1 system. The intersection of the initial – 20 *db*/dec segment with the line ω = 1 has the magnitude 20 log K_v. This can be explained as below.

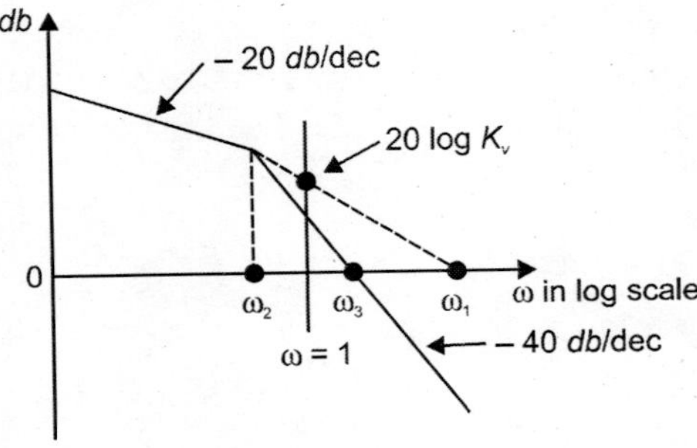

Fig. 4.34

CHAPTER 4

In a type 1 system, $G(j\omega) = \dfrac{K_v}{j\omega}$, for $\omega << 1$

thus $$20\log\left|\frac{K_v}{j\omega}\right|_{\omega=1} = 20 \log K_v$$

The intersection of the initial – 20 log *db*/dec segment with the 0 *dB* line has a frequency equal to K_v. To see this,

$$\left|\frac{K_v}{j\omega_1}\right|_{\omega=1} = 1 \text{ or } K_v = \omega_1$$

As an example consider the type system with unity feedback whose open loop transfer function is

$$G(s) = \frac{K}{s}\,(Js + F)$$

If we define the corner frequency to be ω_2 and the frequency at the intersection of the – 40 *db*/dec segment with 0*db* line to be ω_3, then

$$\omega_2 = \frac{F}{J},\ \omega_3^2 = \frac{K}{J}$$

Since $$\omega_1 = K_v = \frac{K}{F}$$

is follows that $$\omega_1\omega_2 = \omega_3^2$$

$$\frac{\omega_1}{\omega_3} = \frac{\omega_3}{\omega_2}$$

on the bode diagram

$$\log \omega_1 - \log \omega_3 = \log \omega_3 - \log \omega_2$$

Thus the ω_3 point is just midway between the ω_2 and ω_1 points. The damping ratio ζ of the system is then

$$\zeta = \frac{F}{2\sqrt{KJ}} = \frac{\omega_2}{2\omega_3}$$

Determination of Static Acceleration Error constants: Consider the unity feedback system shown Fig. (4.32) and Fig. (4.35) shows an example of the log magnitude plot of a type 2 system. The intersection of the initial – 40 *db*/dec segment with the $\omega = 1$ lines has the magnitude of 20 log K_a. Since at low frequencies

$$G(s) = \frac{K_a}{(j\omega)^2} \text{ for } \omega << 1$$

it follows that $$20\log\left|\frac{K_a}{(j\omega)^2}\right|_{\omega=1} = 20 \log K_a$$

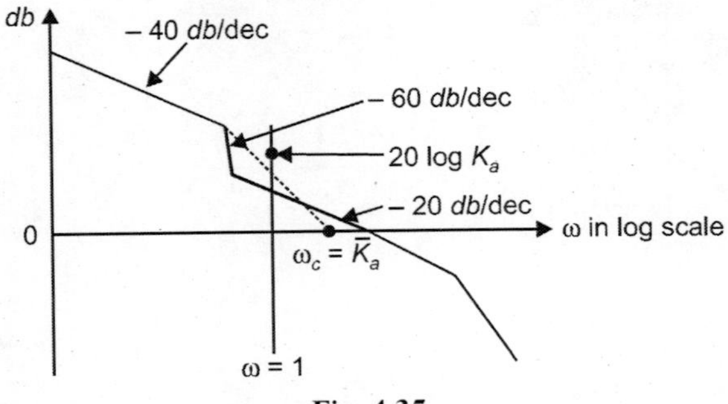

Fig. 4.35

The frequency ω_a at the intersection of the initial – 40 *db*/dec segment with the 0*db* line gives he square root of K_a. This can be seen from the following

$$20\log\left|\frac{K_a}{(j\omega)^2}\right| = 20\log 1 = 0$$

which yields $\omega_a = \sqrt{K_a}$

4.10 DETERMINATION OF TRANSFER FUNCTION FROM BODE MAGNITUDE PLOT

If one is able to draw the bode magnitude plot for given transfer function then it should not be difficult to get the transfer function from given if the bode magnitude plot. However the method described below is valid only for transfer function of minimum phase.

The following steps are followed to get the transfer function from bode plot.

Step 1: Identify low frequency response portion (start) of the plot. 0 *db*/dec. (horizontal plot) corresponds to *K*, +20 db/dec. corresponds to *Ks*, 40 *db*/dec. Corresponds to Ks^2 etc. and – 20 *db*/sec. Corresponds to $\frac{K}{s}$, – 40 *db*/dec. corresponds to $\frac{K}{s^2}$ etc.

Step 2: Whenever there is a change of slope, assign the corner frequencies T_1, T_2 etc.

Step 3: Change of slope (for eg) from say + 20 *db*/dec to 0 *db*/dec at corner frequency *T*, corresponds to a pole, from + 20 *db*/dec. to –20 *db*/dec. corresponds to two repeated poles, from 0 *db*/dec. to + 20 *db*/dec corresponds to a zero, form – 20 *db*/dec to + 20 *db*/dec. corresponds to repeated zeros. Thus the poles and zero and the time constant of the transfer function are determined.

(i.e.) Identify change of slope at every corner frequencies.

Step 4: From the given bode plot, obtain the decibel. The gain and the corner frequencies are obtained by writing appropriate equation and then the complete transfer function is obtained.

EXAMPLE 4.7

For the bode magnitude plot shown in Fig. (Ex. 4.7.1) obtain the transfer function.

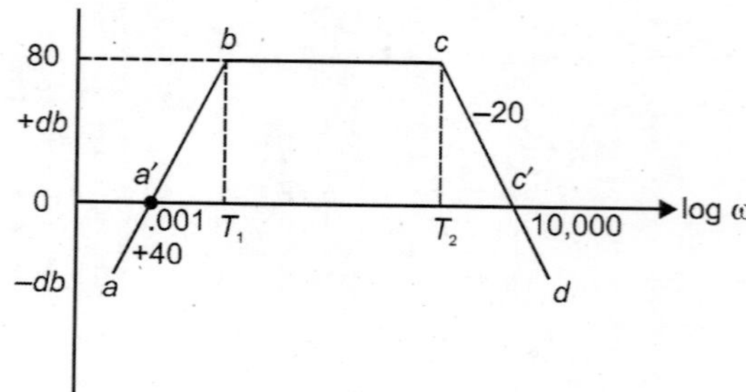

Fig. Ex. 4.7.1

Solution:

Step 1: The slope of the straight lines changes at point b and c. Let the corner frequencies corresponding to these points be T_1 and T_2. Since the low frequency lines $a\,a'b$ starts with $+40$ *db*/dec, the corresponding transfer function will be $G(s) = Ks^2$. At point b there should be two repeated poles with corner frequency T_1, to nulify the $+ 40$ *db*/dec lines. Hence

$$G_2(s) = \frac{1}{\left(1+\frac{s}{T_1}\right)^2}$$

At point c there should be another pole so that the straight line cd goes with $- 20$ *db*/dec.

Hence $$G_3(s) = \frac{1}{(1+sT_1)}$$

the complete transfer function is

$$G(s) = G_1(s)\,G_2(s)\,G_3(s) = \frac{Ks^2}{\left(1+\frac{s}{T_1}\right)^2\left(1+\frac{s}{T_2}\right)}$$

Now it is necessary to determine K_1, T_1 and T_2 from Fig. (4.7.1).

The straight line $aa'b$ is due to the function Ks^2. Hence

$$|G_1(j\omega)| = |K \cdot \omega^2| = 1 \text{ at } \omega = .001$$

$$K(.001)^2 = 1$$

$$\therefore \quad K = 10^6$$

Also from the triangle, $a'bT_1$

$$40 \log \frac{T_1}{.001} = 80$$

From which, $$T_1 = .1$$

$$20 \log \frac{10,000}{T_2} = 80$$

$$T_2 = 1$$

Hence $$G(s) = \frac{10^6 s^2}{(1+10s)^2(1+s)}$$

Result: The transfer function for the given bode magnitude plot is $\boxed{G(s) = \frac{10^6 s^2}{(1+10s)^2(1+s)}}$.

4.11 FREQUENCY RESPONSE SPECIFICATION

The gain margin

Consider the $G_1 + (j\omega)$ plot shown in Fig. (4.36) where the magnitude of $(G_1 + j\omega)$ is crosses the negative real axis is d, then $\frac{1}{d}$ is called the gain margin and it is – 20 log d is decibel. The frequency at which the gain is unity is called the gain cross–over frequency. The gain margin is defined as the reciprocal of the gain of open lop sinusoidal transfer function evaluated at the frequency ω_g at which the phase shift is 180°.

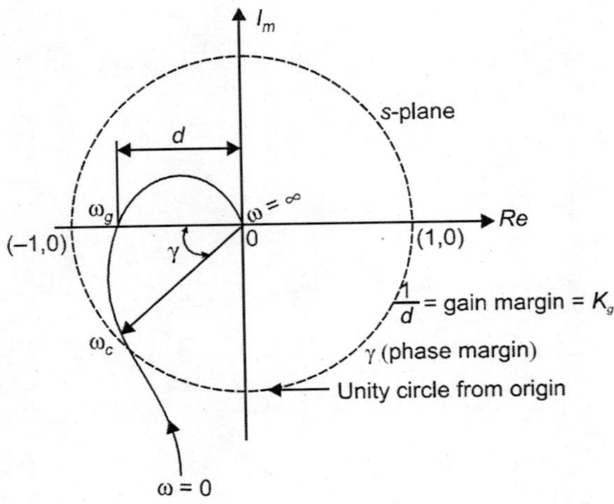

Fig. 4.36

The phase margin

The phase margin is defined as the angle through which $G_1 + (j\omega)$ plot is rotated in order that unity magnitude point on the locus passes through the critical point (–1, 0). In other words the phase margin ϕ_m as in Fig. (4.36) is – 180° plus the phase angle at unity gain.

Damping factor

There is an approximate relationship relating phase margin (γ) and damping factor ζ of a second order system. This is given by

$$\zeta = \frac{\pi}{360°}\gamma$$

Where γ is in degrees. For larger values of phase margin and gain margin, the stability is higher. For a maximum phase transfer function both gain margin and phase margin should be positive for the system to be stable. Negative phase margin indicates that the system is unstable.

Resonant peak

The resonant peak is the peak value of $M(j\omega) = \frac{C}{R}(j\omega)$ for a second order system.

$$M_P = \frac{1}{2\zeta\sqrt{1-\zeta_1^2}}$$

Resonant frequency

It is the frequency at which the resonance peak M_P occurs. For a second order system.

$$\omega_p = \omega_n\sqrt{1-2\zeta^2}$$

Bandwidth

The bandwidth is defined as the frequency at which the magnitude gain of the frequency response curve is reduced to $\frac{1}{\sqrt{2}} = .707$ of its low frequency gain.

Consider a second order system.

$$M(s) = \frac{\omega_n^2}{s^2 + 2\zeta\omega_n s + \omega_n^2}$$

$$|M(j\omega)| = \frac{\omega_n^2}{\left(\omega_n^2 - \omega^2\right) + 4\zeta^2\omega_n^2\omega^2}$$

Let ω_b be the frequency at which the gain is reduced to $\frac{1}{\sqrt{2}}$ times its low frequency value.

$$\frac{1}{\sqrt{2}} = \frac{\omega_n^2}{\sqrt{\left(\omega_n^2 - \omega_b^2\right) + 4\zeta^2\omega_b^2\omega_n^2}}$$

Solving the equation we get

$$\omega_b = \omega_n \left[1 - 2\zeta^2 + \sqrt{4\zeta^4 - 4\zeta^2 + 2} \right]^{1/2}$$

Hence the bandwidth of a second order system,

$$\text{B.W} = \omega_n \left[1 - 2\zeta^2 + \sqrt{4\zeta^4 - 4\zeta^2 + 2} \right]^{1/2}$$

Note: It is to be noted that for some system, large band width is necessary for the system to respond to a wide range of input signal. For certain other systems small band width is necessary so that it provides attenuation to the high frequency noise in the input signal. Large bandwidth corresponds to a small rise time and fasters of response. However large bandwidth requires high cost components.

EXAMPLE 4.8

For the bode magnitude plot shown in Fig. (Ex. 4.8) determine the transfer function.

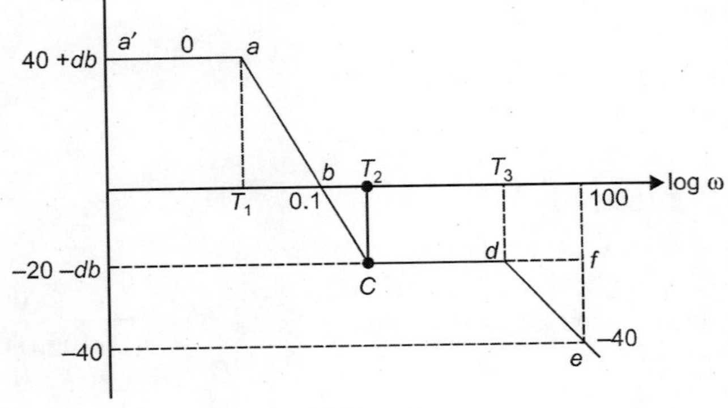

Fig. 4.8

CHAPTER 4

Solution: Let the corner frequencies where the slope changes be T_1, T_2 and T_3

For the horizontal line $a'a$, the transfer function is

1. $G_1(s) = K$

2. For the straight line abc, $G_2(s) = \dfrac{1}{\left(1 + \dfrac{s}{T_1}\right)}$

3. For the horizontal line cd, $G_3(s) = \left(1 + \dfrac{s}{T_2}\right)$

4. For the straight line de, $G_4(s) = \left(1 + \dfrac{s}{T_3}\right)^2$

5. Hence the resultant transfer function is $G(s) = \dfrac{K\left(1+\dfrac{s}{T_2}\right)}{\left(1+\dfrac{s}{T_1}\right)\left(1+\dfrac{s}{T_3}\right)^2}$

6. Now, K, T_1, T_2 and T_3 are to be determined. For the horizontal line $a'a$, only k is contributing db upto the point a.

   Hence $20 \log K = 40$ or $K = 100$.

7. From the triangle ab T_1,

   $$20 \log \frac{1}{T_1} = 40 \text{ or } T_1 = \frac{1}{100} = .001$$

8. From the triangle bc T_2,

   $$20 \log \frac{T_2}{.1} = 20 \text{ or } T_2 = 1$$

9. From the triangle def,

   $$40 \log \frac{100}{T_3} = \text{height } ef = 20$$

   $T_3 = 31.6$

10. Hence $G(s) = \dfrac{100(1+s)}{(1+1000s)(1+.032s)^2}$

Result: The transfer function of given magnitude bode plot is $G(s) = \dfrac{100(1+s)}{(1+1000s)(1+.032s)^2}$

EXAMPLE 4.9

For the log magnitude plot shown in Fig. (Ex. 4.9), obtain the transfer function.

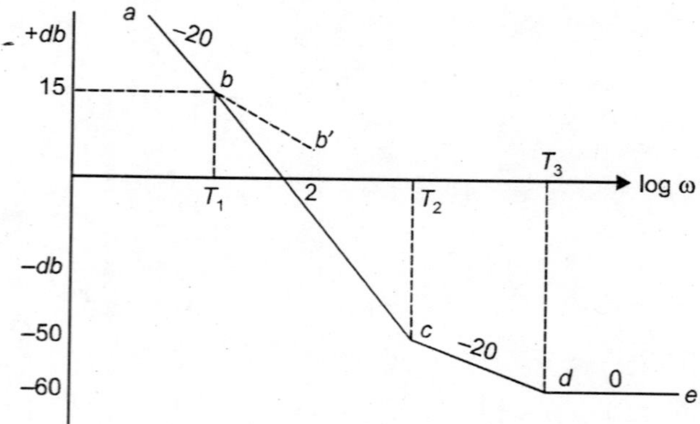

Fig. Ex. 4.9

Solution:

1. Let T_1, T_2, T_3 be the corner frequencies where the slopes change.
2. For the line *ab*, $G_1(s) = \dfrac{K}{s}$
3. For the line *bc*, $G_2(s) = \left(\dfrac{1}{1+\dfrac{s}{T_1}}\right)$
4. For the line *cd*, $G_3(s) = \left(1+\dfrac{s}{T_2}\right)$
5. For the line *de*, $G_4(s) = \left(1+\dfrac{s}{T_3}\right)$
6. The resultant transfer function is $G(s) = \dfrac{K\left(1+\dfrac{s}{T_2}\right)\left(1+\dfrac{s}{T_3}\right)}{s\left(1+\dfrac{s}{T_1}\right)}$
7. Now we have to determine K, T_1, T_2 and T_3

   From straight line *abb′*

   $\dfrac{K}{\omega} = 1$ at $\omega = 2$ hence $K = 2$.
8. At point *b*, the magnitude contribution is due to $\dfrac{K}{\omega}$.

   Hence at $\omega = T_1$, $20 \log \dfrac{K}{T_1} = 15$ or $T_1 = .36$.
9. From the straight line *bc*, $40 \log \dfrac{T_2}{T_1} = 65$ or $T_2 = 15$.
10. From the straight line *cd*, $20 \log \dfrac{T_3}{T_2} = (60-50) = 10$ or $T_3 = 47.4$.
11. Hence the transfer function is, $G(s) = \dfrac{2(1+.067s)(1+.021s)}{s(1+2.78s)}$.

Result: The transfer function of given bode magnitude is $G(s) = \dfrac{2(1+.067s)(1+.021s)}{s(1+2.78s)}$.

CHAPTER 4

EXAMPLE 4.10

Determine analytically, the gain margin and phase margin for the system with the following transfer function $G(s)\,H(s) = \dfrac{5}{s\left(1+\dfrac{s}{2}\right)\left(1+\dfrac{s}{6}\right)}$.

Solution:

$$GH(j\omega) = \frac{5}{\omega\sqrt{\left(1+\dfrac{\omega^2}{4}\right)\left(1+\dfrac{\omega^2}{36}\right)}} \cdot -90° - \tan^{-1}\frac{\omega}{2} - \tan^{-1}\frac{\omega}{6} \qquad (4.11.5)$$

Let ω_g be the frequency at which the polar plot crosses the –ve real axis. Then,

$$-90° - \tan^{-1}\frac{\omega_g}{2} - \tan^{-1}\frac{\omega_g}{6} = -180°$$

$$\frac{\tan^{-1}\omega_g\left(\dfrac{1}{2}+\dfrac{1}{6}\right)}{1-\dfrac{\omega_g^2}{12}} = +90°$$

$\dfrac{1-\omega_g^2}{12} = 0$ form which $\omega_g = 3.46$ rod/sec

Substituting the value of ω_g in Eq. (4.11.5) for magnitude

$$d = \frac{5}{3.46\sqrt{(1+3)\left(1+\dfrac{1}{3}\right)}} = \frac{5}{8}$$

Hence gain margin = $1/d = 8/5 = 1.6 = K_g$

Let ω_c be the frequency at unity gain. Then from Eq. (4.11.5)

$$\frac{5}{\omega_c\sqrt{\left(1+\dfrac{\omega_c^2}{4}\right)\left(1+\dfrac{\omega_c^2}{36}\right)}} = 1$$

Form which $\omega_c = 2.79$ rad/sec

The phase margin (γ) $= -90° - \tan^{-1}\dfrac{\omega_c}{2} - \tan^{-1}\dfrac{\omega_c}{6} + 180\epsilon$

$$= -90° - \tan^{-1}\frac{2.79}{2} - \tan^{-1}\frac{2.79}{6} + 180° = 10.36°.$$

Result: Phase margin (γ) = 10.36
Gain margin (K_g) = 1.6

EXAMPLE 4.11

Sketch the bode plot for the following transfer function determine phase margin and gain margin $G(s) = \dfrac{7.5(1+0.2s)}{s(s^2+16s+100)}$.

Solution: The sinusoidal transfer function $G(j\omega)$ is obtained replacing s by $j\omega$ in the given s-domain.

Given that $$G(s) = \frac{7.5(1+0.2s)}{s(s^2+16s+100)}$$

On comparing the quadratic factor of $G(s)$ with standard form of quadratic factor we can find ζ and ω_n.

$$\therefore \quad s^2 + 16s + 100 = s^2 + 2\zeta\,\omega_n\,s + \omega_n^2$$

On comparing we get, $\omega_n^2 = 100$

$$\therefore \quad \omega_n = 10$$

$$2\zeta\omega_n = 16$$

$$\therefore \quad \zeta = 16/2\,\omega_n$$

$$\zeta = .8$$

$$G(s) = \frac{75(1+0.2s)}{s\times 100\left(\frac{s^2}{100}+\frac{16s}{100}+1\right)} = \frac{.75(1+0.2s)}{s\left(0.01s^2+0.16s+1\right)}$$

$$\therefore \quad G(j\omega) = \frac{.75(1+0.2\,j\omega)}{j\omega\left(1+0.01(j\omega)^2+0.16\,j\omega\right)}$$

$$G(j\omega) = \frac{.75(1+j0.2\omega)}{j\omega\left(1-0.01\omega^2+j0.16\omega\right)}$$

Magnitude plot

The corner frequencies are $\omega_{c1} = \dfrac{1}{.2} = 5$ rad/sec and $\omega_{c2} = 10$ rad/sec.

The various terms of $G(j\omega)$ are listed in Table (Ex. 4.11.1) in the increasing order to their corner frequencies. Also the table shows the slope contributed by each term and the change in slope at the corner frequency.

Choose a low frequency ω_l such that $\omega_l < \omega_{c1}$ and choose a high frequency ω_h such that $\omega_h > \omega_{c2}$

Let $\omega_l = .5$ rad/sec and $\omega_h = 20$ rad/sec

Let $A = |G(j\omega)|$ in *db*, let us calculate A at ω_l, ω_{c1}, ω_{c2} and ω_h

At, $\omega = \omega_l$, $\quad A_1 = 20\log|\text{first term}|$

CHAPTER 4

Table Ex. 4.11.1

Term	Corner frequency rad/sec	Slope of each term *db*/dec	Change in slope from one term to another term *db*/dec
$\frac{.75}{j\omega}$	–	– 20	
$1 + j\,0.2\,\omega$	$\omega_{c1} = \frac{1}{.2} = 5$	+ 20	– 20 + 20 = 0
$\frac{1}{1-.01\omega^2 + j0.16\omega}$	$\omega_{c2} = \omega_n = 10$	– 40	0 – 40 = – 40

i.e., $A_1 = 20\log\left|\frac{.75}{j\omega}\right| = 20\log\frac{.75}{.5} = 3.5 = db$

At $\omega = \omega_{c1}$, $A_2 = 20 \log |\text{first term}|$

$$A = 20\log\left|\frac{.75}{j\omega}\right| = 20\log\frac{.75}{.5} = -16.5\,db$$

At $\omega = \omega_{c2}$, $A_3 = [\text{slope from one term to another term}) \text{ Ч} \log \left|\frac{\text{Present frequency}}{\text{Previous frequency}}\right| + \text{Previous gain } (A_2)$

i.e., $A_2 = \left[\text{slope from } \omega_{c1} \text{ to } \omega_{c2} \text{ Ч} \log \frac{\omega_{c2}}{\omega_{c1}}\right] + A_2 \,(\text{at } \omega = \omega_{c1})$

$$= 0 \times \log\frac{10}{5} + (-16.5) = -16.5\,db$$

At $\omega = \omega_h$, $A_4 = [\text{slope from one term to another term}) \text{ Ч} \log \left|\frac{\text{Present frequency}}{\text{Previous frequency}}\right| + \text{Previous gain } (A_3)$

i.e., $A_4 = \left[\text{slope from } \omega_{c2} \text{ to } \omega_h \text{ Ч} \log \frac{\omega_h}{\omega_{c2}}\right] + A_3 \,(\text{at } \omega = \omega_{c2})$

$$= -40 \times \log\frac{20}{10} + (-16.5) = -28.5\,db.$$

Let the points *a*, *b*, *c* and d be the points corresponding to frequencies ω_l, ω_{c1}, ω_{c2} and ω_h respectively on the magnitude plot.

Phase plot

The phase angle of $G(j\omega)$ as a function of ω is given by

$$\phi = \angle G(j\omega) = \tan^{-1} 0.2\omega - 90° - \tan^{-1} \frac{.16\omega}{1-.01\omega^2} \quad \text{for } \omega \le \omega_n$$

$$\phi = \angle G(j\omega) = \tan^{-1} 0.2\omega - 90° \left[\tan^{-1} \frac{.16\omega}{1-.01\omega^2} + 180° \right] \quad \text{for } \omega \le \omega_n$$

The phase angle of $G(j\omega)$ are calculated for various values of ω and listed in Table (Ex. 4.11.2).

Table Ex. 4.11.2

ω rad/sec	$\phi = \angle G(j\omega)$ deg
0.5	$-88.9 \approx -88$
1	$-87.9 \approx -88$
5	$-91.8 \approx -92$
10	$-116.6 \approx -116$
20	$-147.3 \approx 148$
50	$-167.3 \approx -168$
100	$-173.7 \approx -174$

Note: In quadratic factors the angles varies from 0° to 180 But the calculator calculates $\tan^{-1}$ only between 0° to 90°. Hence a correction factor of 180° should be added to the phase angle after corner frequency.

On the same semilog sheet choose a scale of 1 unit = 20° on the y-axis on the right side of semi log sheet. Mark the calculated phase angle on the graph sheet. Join the points by a smooth curve.

The magnitude plot and the phase plot are shown in Fig. (Ex. 4.11). From the Fig. (Ex. 4.11) we find the phase angle at gain cross over frequency $(\omega_{gc}) = \phi_{gc} = 88°$.

$\therefore$ Phase margin $= 180° + \phi_{gc} = 180є + \phi_{gc} = 180° - 88° = 92°$.

Here, K_g, Gain margin $= +\infty$

Note: The phase plot crosses – 180° only at infinite. The $|G(j\omega)|$ at infinity is $-\infty db$. Hence the gain margin is $+\infty$.

Result: Phaser margin, $\gamma = 92°$

Gain margin, $K_g = +\infty$

EXAMPLE 4.12

Given $G(s) = \dfrac{Ke^{-0.2s}}{s(s+2)(s+8)}$. Find K so that the system is stable under two conditions (a) gain margin equal to 6 db and (b) phase margin equal to 45°.

CHAPTER 4

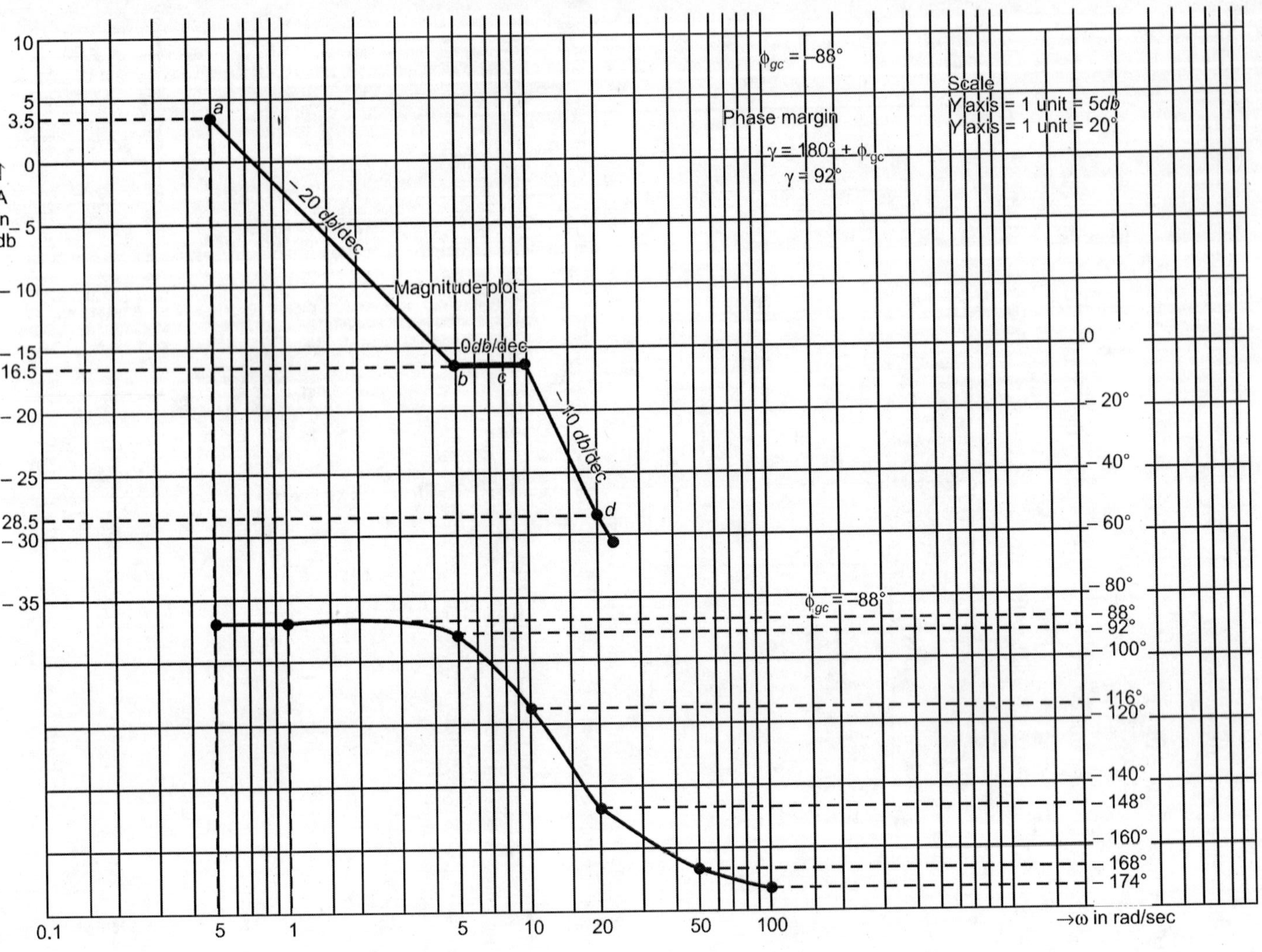

Fig. Ex. 4.11.1: Bode plot of transfer function $G(s) = \dfrac{75(1+0.2s)}{s(s^2+16s+100)}$.

Solution: Given that

$$G(s) = \frac{Ke^{-0.2s}}{s(s+2)(s+8)} = \frac{Ke^{-0.2s}}{s \times 2\left(1+\frac{s}{2}\right) \times \left(1+\frac{s}{8}\right)} = \frac{\frac{1}{16}Ke^{-.25}}{s(1+.5s)(1+.125s)}$$

$$= \frac{0.0625ke^{-0.2s}}{s(1+0.5s)(1+.125s)}$$

Let $K = 1$

The sinusoidal transfer function $G(j\omega)$ is obtained by replacing s by $j\omega$ in the gives domain transfer function.

$$G(j\omega) = \frac{0.0625\ Ke^{-0.2j\omega}}{j\omega(1+j0.5\omega)\ (1+j0.125\omega)}$$

Note: $|.0625\ e^{-j0.2\omega}| = .0625$

and $\angle e^{-j0.2\omega} = -0.2\ \omega$, radians

Magnitude plot

The corner frequencies are

$$\omega_{c1} = \frac{1}{0.5} = 2\text{ rad/sec} \text{ and } \omega_{c2} = \frac{1}{.125} = 8\text{ rad/sec}$$

The various terms of $G(j\omega)$ are listed in Table Ex. 4.12.1 in the increasing order of their corner frequencies. Also the table shows the slope contributed by each term and the change in slope at the corner frequency.

Table Ex. 4.12.1

Term	**Corner frequency rad/sec**	**Slope of each term *db*/dec**	**Change in slope from one term to another term *db*/sec**
$\frac{0.0625}{j\omega}$	–	– 20	
$\frac{1}{1+j0.5\omega}$	$\omega_{c1} = \frac{1}{.5} = 2$	– 20	– 20 – 20 = – 40
$\frac{1}{1+j0.125\omega}$	$\omega_{c2} = \frac{1}{.125} = 8$	– 20	– 40 – 20 = – 60

Choose a low frequency ω_l such that $\omega_l = \omega_{c1}$ and choose a high frequency ω_h such that $\omega_h = \omega_{c2}$.

Let $\omega_l = 0.5$ rad/sec and $\omega_h = 50$ rad/sec

Let $A = |G(j\omega)|$ in *db*.

Let us calculated A at ω_l, ω_{c1}, ω_{c2} and ω_h

$$\text{At } \omega = \omega_l, A_1 = 20\log\left|\frac{0.625}{j\omega}\right| = 20\log\frac{0.0625}{0.5} = -18\,db \;\; [A_1 = 20 \text{ log (first term)}]$$

$$\text{At } \omega = \omega_{c1}, A_2 = 20\log\left|\frac{0.625}{j\omega}\right| = 20\log\frac{0.0625}{2} = -30\,db \;\; [A_2 = 20 \text{ log |first term|}]$$

$$\text{At } \omega = \omega_{c2}, A_3 = [\text{slope from one term to another term}] \text{ Ч log} \left|\frac{\text{Present frequency}}{\text{Previous frequency}}\right| + \text{Previous gain}$$

$$A_3 = [\text{slope from } \omega_{c1} \text{ to } \omega_{c2} \text{ Ч log } \frac{\omega_{c2}}{\omega_{c1}}] + A_2 \text{ (at } \omega = \omega_{c1})$$

$$= -40 \text{ Ч log } \frac{8}{2} + (-30) = -54\ db.$$

$$\text{At } \omega = \omega_h, A_4 = [\text{slope from one term to another term}] \text{ Ч log} \left|\frac{\text{Present frequency}}{\text{Previous frequency}}\right| + \text{Previous gain}$$

i.e.,
$$A_4 = \left[\text{slope from } \omega_{c2} \text{ to } \omega_h \text{ Ч log } \frac{\omega_h}{\omega_{c2}}\right] + A_3 \text{ (at } \omega = \omega_{c2})$$

$$A_4 = -60 \text{ Ч log } \frac{50}{8} \text{ (54)} = -102\ db.$$

Let the points *a*, *b*, *c*, *d* be two points corresponding to frequencies ω_l, ω_{c1}, ω_{c2} and ω_h respectively on the magnitude plot.

Phase plot

The phase angle of $G(j\omega)$ as a function of ω is given by

$$\phi = -0.2\omega \times \frac{180}{\pi} - 90 - \tan^{-1} 0.5\omega - \tan^{-1} 0.125\omega$$

The phase angle of $G(j\omega)$ are calculated for various values of ω and listed in Table (Ex. 4.12.2)

Table Ex. 4.12.2

ω, rad/sec	φ = deg
0.01	– 90.4 ≈ – 90
0.1	– 94.7 ≈ – 94
0.5	– 113.3 ≈ – 114
1	– 134.4 ≈ – 134
214	– 171.9 ≈ – 172
3	– 201.2 ≈ 202
4	– 225.8 ≈ 226

On the same semilog sheet choose a scale on the y-axis on the right side of the semilog sheet. Mark the calculated phase angle on the graph sheet. Join the points by smooth curve.

The magnitude and phase plot are shown in Fig.. (4.12).

Calculate of K

Phase margin, $\gamma = 180 + \phi_{gc}$, where ϕ_{gc} is the phase of $G(j\omega)$ at $\omega = \omega_{gc}$.

where $\gamma = 45°$, $\phi_{gc} = \gamma - 180\epsilon = 45\epsilon - 180\epsilon = -135°$

With $K = 1$, the *db* gain at $\phi = -135°$ is -24 *db*.

This gain should be made zero to have to PM of 45°, hence to every point of magnitude plot a *db* gain of 24 *db* should be added. The corrected magnitude plot is obtained by shifting the plot with $K = 1$ by 24 *db* upwards. The magnitude correction is independent of frequency. hence the magnitude of 24 *db* is contributed by the term *K*. The value of *K* is calculated by equating 20 log be to 24 *db*.

$$\therefore \qquad 20 \log K = 24, K = 10^{24/10}, K = 15.84$$

with $K = 1$, the gain margin $= -(-.34) = 34$ *db*. But the required margin is 6 *db*. Hence to every point of magnitude plot a *db* gain of 28 *db* should be added. This addition of 28 *db* shifts the plot upwards. The magnitude correction is independent of frequency. Hence the magnitude of 28 *db* is contributed by the term *k*. The value of *k* is calculated by equating 20 log k to 28 *db*.

$$\therefore \qquad 20 \log K = 28, K = 10^{28/20}, K = 25.12$$

The magnitude plot with $K = 15.84$ and 25.12 are shown in Fig. (Ex. 4.12).

Result:

(a) $K = 15.84$

(b) $K = 25.12$

EXAMPLE 4.13

Plot the bode diagram for the following transfer function and obtain the gain and phase cross over frequencies $G(s) = 10/s\,(1 + 0.4s)\,(1 + 0.1s)$.

Solution: Given that $$G(s) = \frac{10}{s(1+0.4s)(1+0.1s)}$$

The sinusoidal transfer function of $G(j\omega)$ is obtained by replacing s by $j\omega$ in the given transfer function.

$$\therefore \qquad G(j\omega) = 10/j\omega\,(1 + 0.4\,j\omega)\,(1 + 0.1\,j\omega)$$

Magnitude Plot

The corner frequencies are,

$$\omega_{c1} = \frac{1}{0.4} = 2.5\,\text{rad/sec}$$

$$\omega_{c2} = \frac{1}{0.1} = 10\,\text{rad/sec}$$

The various terms of $G(j\omega)$ are listed in Table (Ex. 4.13.1) in the increasing order of their corner frequencies. Also the table shows the slope contributed by each term and the change in slope at the corner frequency.

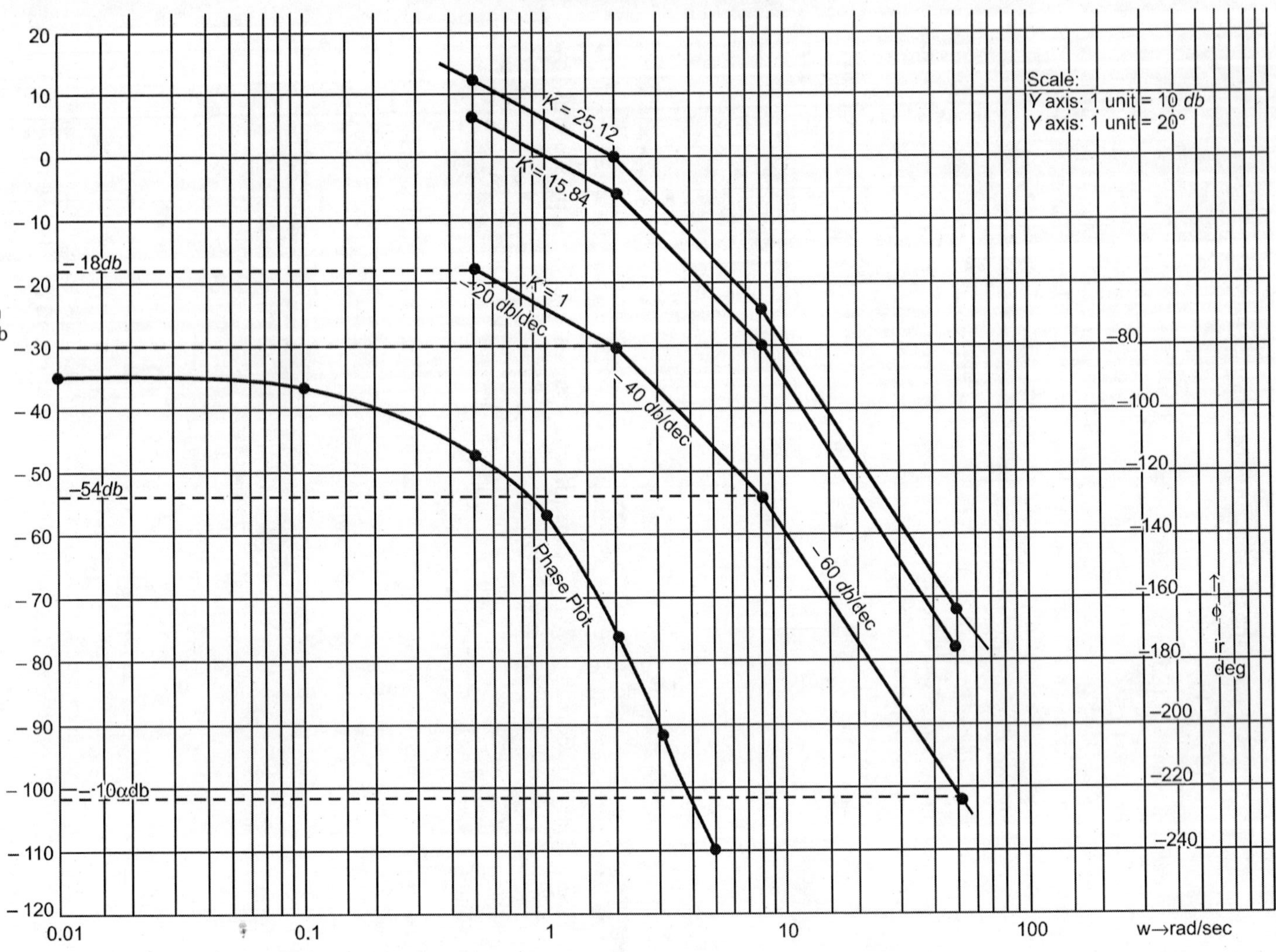

Fig. Ex. 4.12: $G(s) = 0.0625\ Ke^{-0.2s}/s(1 + 0.5s)\ (1 + 0.125s)$.

Table Ex. 4.13.1			
Term	**Corner frequency rad/sec**	**Slope of each term** *db*/**dec**	**Change in slope from one term to another term** *db*/**sec**
$\frac{10}{j\omega}$	–	– 20	
$\frac{1}{1+j0.4\omega}$	$\omega_{c_1} = \frac{1}{0.4} = 2.5$	– 20	– 20 – 20 = – 40
$\frac{1}{1+j0.1\omega}$	$\omega_{c2} = \frac{1}{0.1} = 10$	– 20	– 40 – 20 = – 60

Choose a low frequency ω_l such that $\omega_l < \omega_{c1}$ and choose a high frequency ω_n such that $\omega_h < \omega_{c1}$.
Let $\omega_l = 0.1$ rad/sec, and $\omega_h = 50$ rad/sec

Let $A = |G(j\omega)|$ in *db*

Let us calculate A at ω_l, ω_{c1}, ω_{c2} and ω_h

At $\omega = \omega_l$, $$A_1 = 20\log\left|\frac{10}{j\omega}\right| = 20\log\left|\frac{10}{0.1}\right| = 40\ db \quad [\text{Œ}A_1 = 20 \text{ log } |\text{first term})]$$

At $\omega = \omega_{c1}$ $$A = 20\log\left|\frac{10}{j\omega}\right| = 20\log\left|\frac{10}{2.5}\right| = 12\ db \quad [\text{Œ}A_2 = 20 \text{ log } |\text{first term})]$$

At $\omega = \omega_{c2}$, A_3 = [slope from one term to another term] Ч log $\left|\frac{\text{Present frequency}}{\text{Previous frequency}}\right|$ + Previous gain (A_2)

$$A_3 = [\text{slope from } \omega_{c1} \text{ to } \omega_{c2} \text{ Ч log } \left.\frac{\omega_{c2}}{\omega_{c1}}\right] + A_2 \text{ (at } \omega = \omega_{c1})$$

$$A_3 = -40 \times \log\frac{10}{2.5} + 12 = -12\ db$$

At $\omega = \omega_h$, A_4 = [slope from one term to another term] Ч log $\left|\frac{\text{Present frequency}}{\text{Previous frequency}}\right|$ + Previous gain (A_3)

$$A_4 = [\text{slope from } \omega_{c2} \text{ to } \omega_h \text{ Ч log } \left.\frac{\omega_h}{\omega_{c2}}\right] + A_3(\text{at } \omega = \omega_{c2})$$

$$= -60 \times \log\frac{50}{10} + (-12) = -54\ db$$

Let the points a,b,c, and d be the points corresponding to frequencies ω_l, ω_{c1}, ω_{c2} and ω_h respectively on the magnitude plot.

Phase plot

The phase angle of $G(j\omega)$ as a function of ω is given by

$$\phi = -90° - \tan^{-1} 0.4\,\omega - \tan^{-1} 0.1\,\omega$$

The phase angles of $G(j\omega)$ are calculated for various values of ω and listed in Table (Ex. 4.13.2)

Table Ex. 4.13.2

ω rad/sec	**φ, deg**
0.1	– 92.86 ≈ – 92
1	– 117.5 ≈ – 118
2.5	– 149 ≈ – 150
4	– 169.79 ≈ – 170
10	– 210.96 ≈ – 210
20	– 236.3 ≈ – 236

on the same semilog sheet choose a scale on the y-axis on the right side of semilog sheet. Mark the calculated phase angle on the graph sheet. Join the points by a smooth curve.

The magnitude and phase plots are shown in Fig. (Ex. 4.13) from the graph the gain and phase cross over frequencies are found to be 5 rad/sec.

Result: Gain cross-over frequency = 5 rad/sec

Phase cross-over frequency = 5 rad/sec.

EXAMPLE 4.14

For the following transfer function of raw bode plot and obtain gain cross-over frequency $G(s) = \dfrac{20}{s(1+3s)(1+4s)}$.

Solution: The sinusoidal transfer function of $G(j\omega)$ is obtained by replacing s by $j\omega$ in the given transfer function.

$$G(j\omega) = \frac{20}{j\omega(1+j3\omega)(1+j4\omega)}$$

Magnitude plot

The corner frequencies are $\omega_{c1} = \dfrac{1}{4} = .25$ rad/sec, $\omega_{c2} = \dfrac{1}{3} = .33$ rad/sec

The various terms of $G(j\omega)$ are listed in Table (Ex. 4.14.1) in the increasing order of their corner frequencies. Also the table shows the slope contributed by each term and change in slope at the corner frequency.

Choose a frequency ω_l such that $\omega_l < \omega_{c1}$ and choose a high frequency ω_h such that $\omega_h > \omega_{c2}$.

Let $\omega_l = .15$ rad/sec, and $\omega_h = 1$ rad/sec

Let $A = |G(j\omega)|$ in db

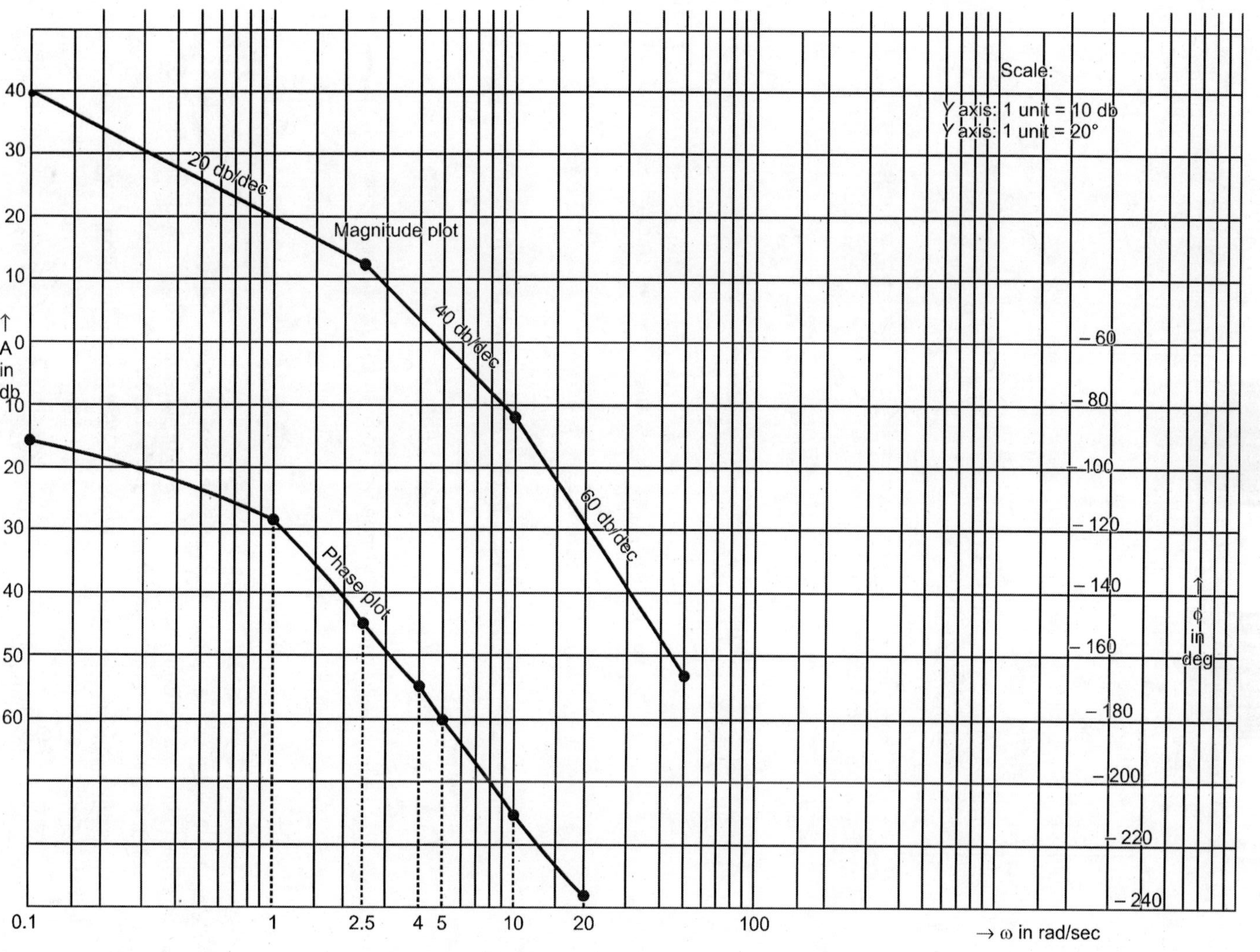

Fig. 4.13: Bode plot for $G(s) = 10/s(1 + 0.4s)\ (1 + 0.1s)$.

Table Ex. 4.14.1

Term	Corner frequency rad/sec	Slope of each term *db*/dec	Change in slope from one term to another term *db*/sec
$\frac{20}{j\omega}$	–	– 20	
$\frac{1}{1+j4\omega}$	$\omega_{c1} = \frac{1}{4} = .25$	– 20	– 20 – 20 = – 40
$\frac{1}{1+j3\omega}$	$\omega_{c2} = \frac{1}{3} = .33$	– 20	– 40 – 20 = – 60

Let us calculate A at ω_l, ω_{c1}, ω_{c2} and ω_h

At $\omega = \omega_l$, $$A_1 = |G(j\omega)| = 20 \log \left|\frac{20}{.15}\right| = 42.5\, db$$

At $\omega = \omega_{c1}$, $$A_2 = |G(j\omega)| = 20 \log \left|\frac{20}{.25}\right| = 38\, db$$

At $\omega = \omega_{c2}$, $$A_3 = \left[\text{slope from } \omega_{c1} \text{ to } \omega_{c2} \text{ Ч } \log \frac{\omega_{c2}}{\omega_{c1}}\right] + A \text{ (at } \omega = \omega_{c1})$$

$$= -40 \times \log \frac{.33}{.25} + 38 = 33\, db$$

At $\omega = \omega_h$, $$A_4 = [\text{slope from } \omega_{c2} \text{ to } \omega_h \text{ Ч } \log \frac{\omega_h}{\omega_{c2}}] + A \text{ (at } \omega = \omega_h)$$

$$= -60 \times \log \frac{1}{.33} + 33 = -14\, db$$

Let the points *a*, *b*, *c* and *d* be the points corresponding to frequencies ω_l, ω_{c1}, ω_{c2} and ω_h respective on the magnitude plot.

Phase plot

The phase angle of $G(j\omega)$, $\phi = -90 - \tan^{-1} 3\omega - \tan^{-1} 4\omega$

The phase angle of $G(j\omega)$ are calculated for various values of ω and listed in Table (Ex. 4.14.2) on the same semilog sheet choose a scale on the *y*-axis on the right side of semilog sheet. Mark the calculated phase angle on the graph sheet. Join the points by a smooth curve.

The magnitude and phase plots are shown in Fig. (Ex. 4.14). From the graph thegain cross over frequency is found to be

$$\omega_{gc} = 1.1 \text{ rad/sec}$$

Result: Gain cross over frequency = 1.1 rad/sec.

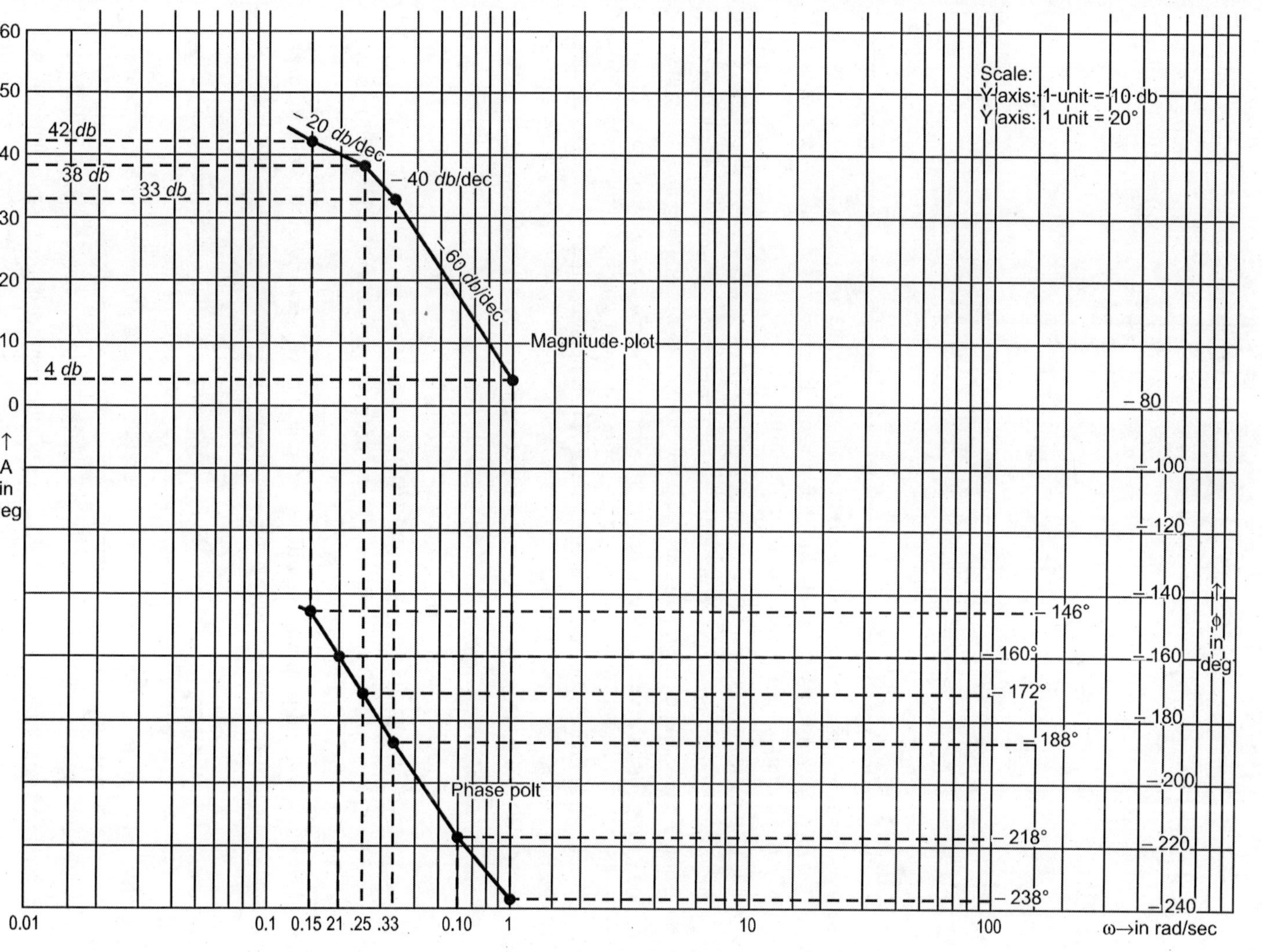

Fig. Ex. 4.14: Bode plot for $20s(1 + 3s)(1 + 4s)$.

CHAPTER 4

Table Ex. 4.14.2	
ω rad/sec	**φ, deg**
.15	–146є
.2	–160є
.25	–172є
.33	–188є
.5	–210є
.6	–218є
1	–238є

4.12 POLAR PLOT

The polar plot of a sinusoidal transfer function $G(j\omega)$ is a plot of the magnitude of $G(j\omega)$ versus the phase angle of $G(j\omega)$ on polar coordinates as ω is varied from zero to intinity. Thus the polar plot is the locus of vectors $|G(j\omega)| \angle G(j\omega)$ as ω is varied from zero to infinity.

The polar plot is also called as Nyquist plot.

The polar graph sheet has concentric circles and radial lines. The circles represent the magnitude and radial lines represent the phase angle. Each point on the polar graph has a magnitude and phase angle.

The magnitude of a point is given by the value of the circle passing through that point and the phase angle is given by the radial line passing through that point. In polar graph sheet a positive phase angle is measured in anticlockwise from the reference axis (0) and negative angle is measured clock wise from the reference axis (0).

Alternatively, if $G(j\omega)$ can be expressed in rectangular coordinates as,

$$G(j\omega) = G_R(j\omega) + G_I(j\omega)$$

where, $G_R(j\omega)$ = Real part of $G(j\omega)$

and $G_I(j\omega)$ = Imaginary part of $G(j\omega)$,

Then the polar plot can be plotted in ordinary graph sheet between $G_R(j\omega)$ and $G_1(j\omega)$ as ω is varied from 0 to ∞. This is also called as inverse polar plot.

To plot the polar plot, first compute the magnitude and phase of $G(j\omega)$ for various value of ω and tabulate them. Usually the choice of frequencies are corner frequencies and frequencies around corner frequencies.

Choose proper scale for the magnitude circles. Fix all the points on polar graph sheet and join the points by smooth curve. Write the frequency corresponding to each point of the plot.

To plot the polar plot on imaginary graph sheet, Compute the magnitude and phase for various values of ω. For inverse polar plot convert the polar coordinates to rectangular coordinates using $P \rightarrow R$ conversion (polar to rectangular) in the calculator. Sketch the polar plot using rectangular coordinates.

For minimum phase transfer function with only poles, the type number of the system determines at what quadrant the polar plot starts and the order of the system determines at what quadrant the polar plot ends.

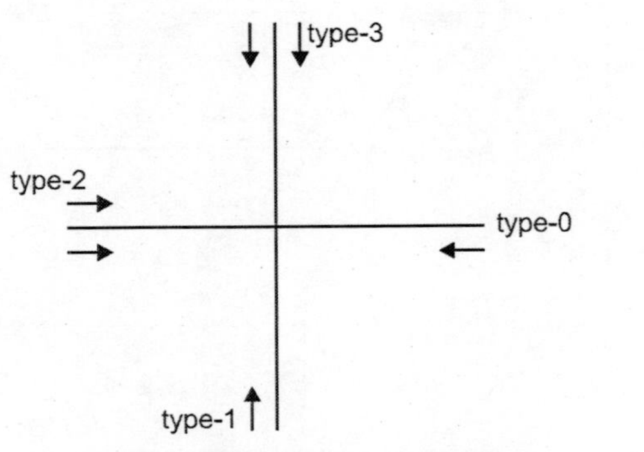

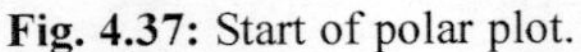
Fig. 4.37: Start of polar plot.

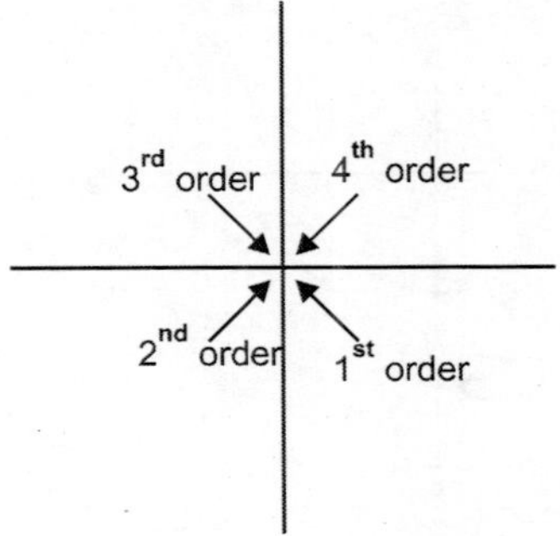

Fig. 4.38: End of polar plot.

Sketches of polar plot for different type and order of the system

Type: 0, Order: 1	*Type: 1, Order: 2*
$G(s) = \dfrac{1}{1+sT}$ $G(s) = \dfrac{1}{1+j\omega T}$	$G(s) = \dfrac{1}{s(1+sT)}$ $G(j\omega) = \dfrac{1}{j\omega(1+j\omega T)}$
Type: 0, Order: 2	*Type: 0, Order: 3*
$G(s) = \dfrac{1}{(1+sT_1)(1+sT_2)}$ $G(j\omega) = \dfrac{1}{(1+j\omega T_1)(1+j\omega T_2)}$	$G(s) = \dfrac{1}{(1+sT_1)(1+sT_2)(1+sT_3)}$ $G(j\omega) = \dfrac{1}{(1+j\omega T_1)(1+j\omega T_2)(1+j\omega T_3)}$

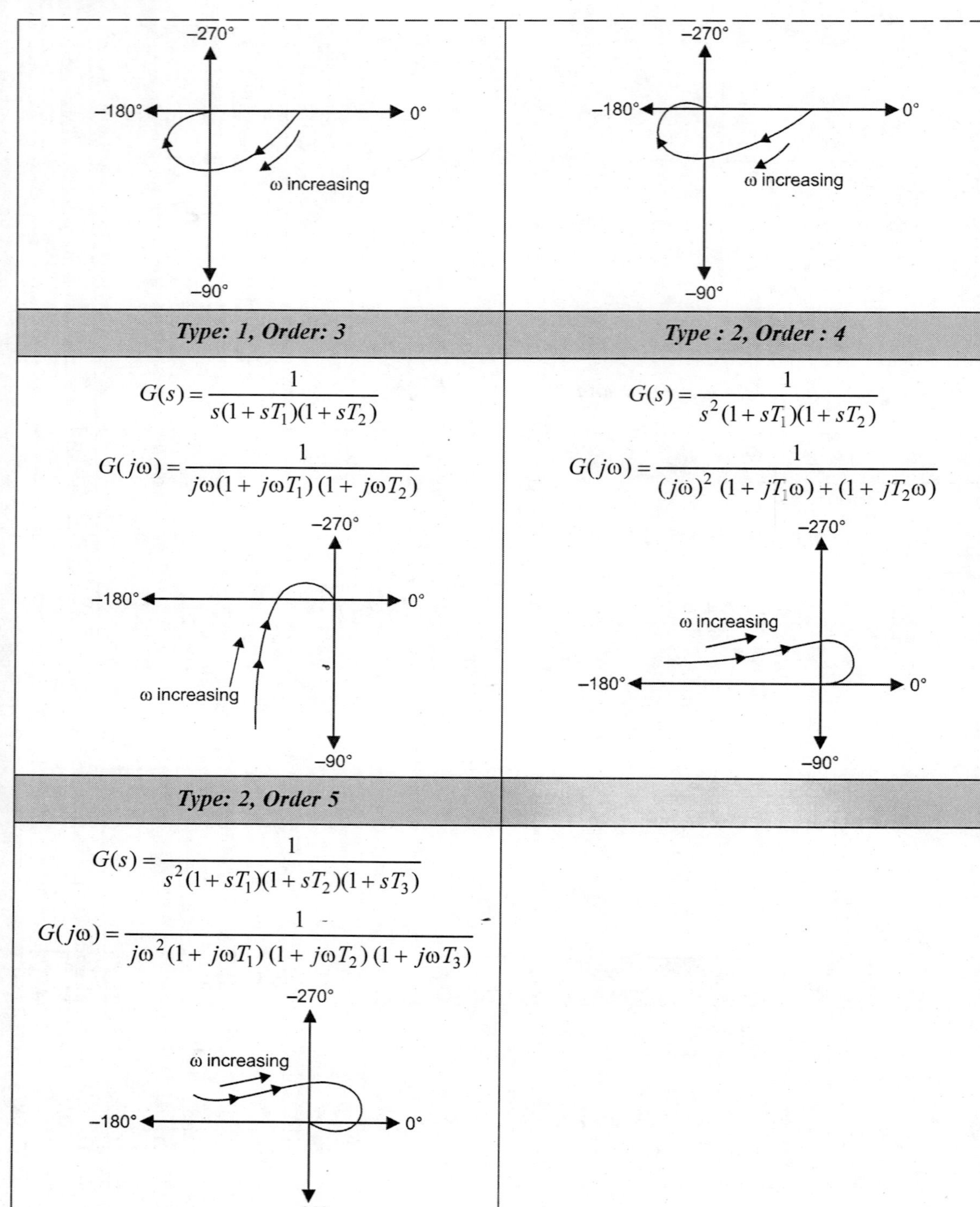

Type: 1, Order: 3	Type : 2, Order : 4
$G(s)=\frac{1}{s(1+sT_1)(1+sT_2)}$ $G(j\omega)=\frac{1}{j\omega(1+j\omega T_1)(1+j\omega T_2)}$	$G(s)=\frac{1}{s^2(1+sT_1)(1+sT_2)}$ $G(j\omega)=\frac{1}{(j\omega)^2\,(1+jT_1\omega)+(1+jT_2\omega)}$

Type: 2, Order 5	
$G(s)=\frac{1}{s^2(1+sT_1)(1+sT_2)(1+sT_3)}$ $G(j\omega)=\frac{1}{j\omega^2(1+j\omega T_1)(1+j\omega T_2)(1+j\omega T_3)}$	

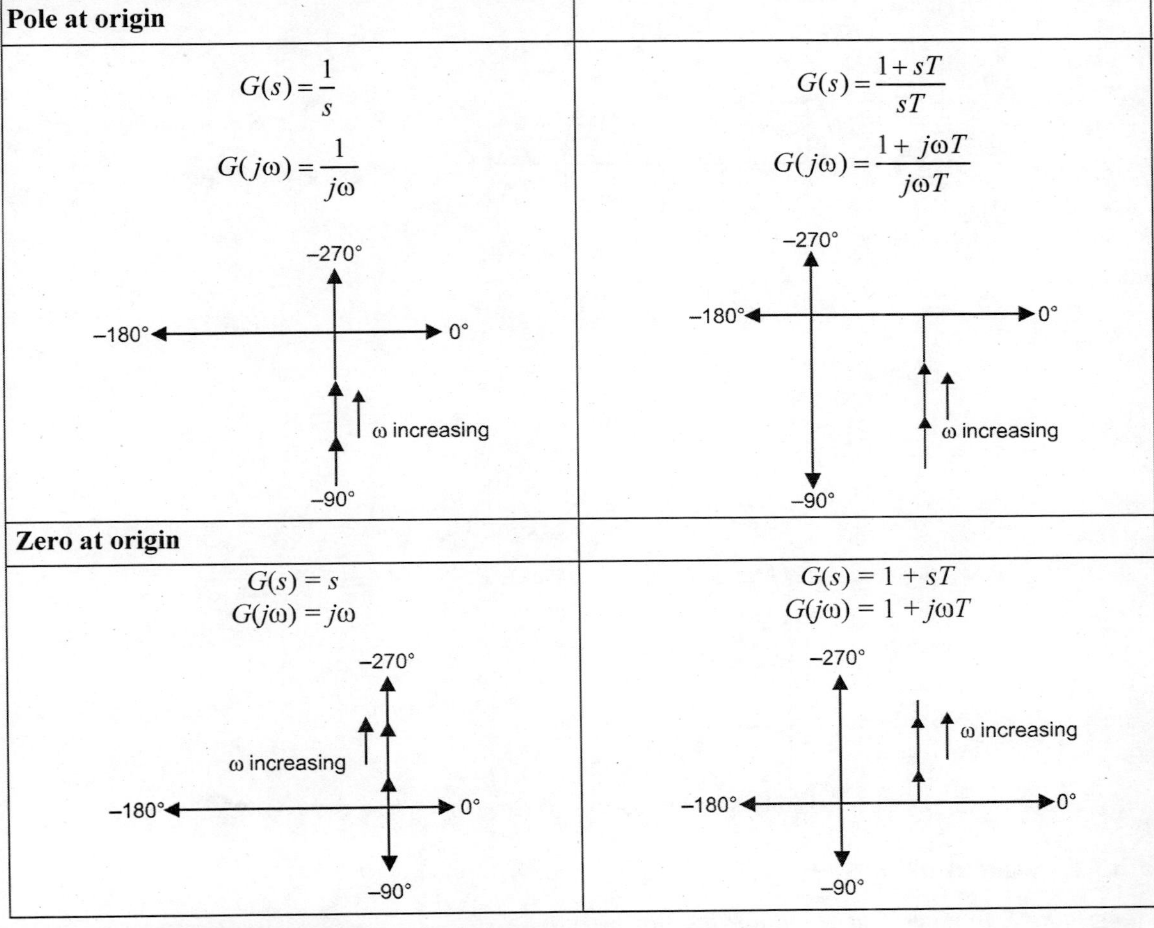

Determination of gain margin and phase margin from polar plot

The gain margin is defined as the inverse of the magnitude of $G(j\omega)$ at phase cross over frequency. The phase cross over frequency is the frequency at which the phase of $G(j\omega)$ is 180°.

Let the polar plot cut the 180° axis at point B and the magnitude circle passing through the point B be G_B. Now the gain margin $K_g = \frac{1}{G_B}$. If the point B lies within unity circle then the gain margin is positive otherwise negative. (If the polar plot is drawn in ordinary graph sheet using rectangular coordinates then the point B is the cutting point of $G(j\omega)$ locus with negative real axis and $K_g = \frac{1}{|G_B|}$ where G_B is the magnitude corresponding to point B).

The phase margin is defined as, phase margin $\gamma = 180 + \phi_{g_c}$ where ϕ_{g_c} is the phase angle of $G(j\omega)$ at gain cross over frequency. The gain cross over frequency is the frequency at which the magnitude of $G(j\omega)$ is unity.

Let the polar plot cut the unity circle at point A as shown in Fig. (4.12.3) and (4.12.4). Now the phase margin, γ is given by $\angle AOP$. i.e. if $\angle AOP$ is below –180° axis then the phase margin is positive and if it is above – 180° axis then the phase margin is negative.

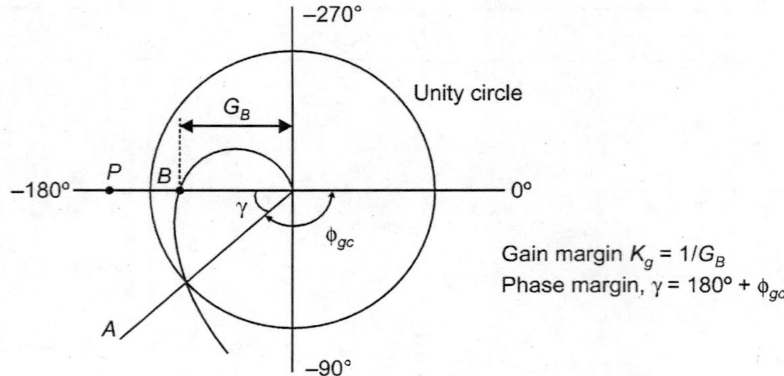

Fig. 4.39: Polar plot showing positive gain margin and phase margin.

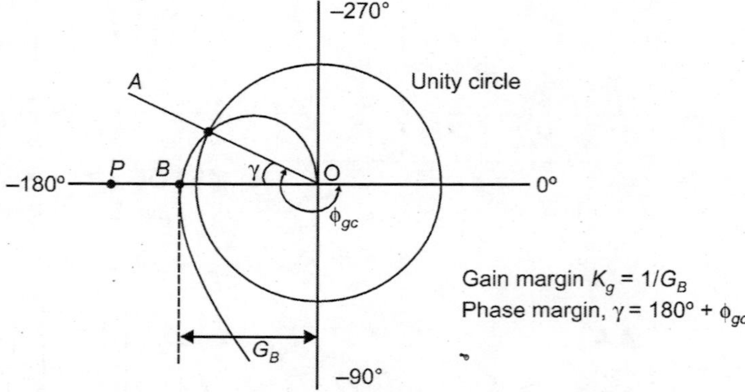

Fig. 4.40: Polar plot showing negative gain margin and phase margin.

Gain Adjustment using Polar Plot

To determine K for specified gain margin

Draw $G(j\omega)$ locus with $K = 1$. Let it cut the –180° axis at point B corresponding to again to G_B. Let the specified gain margin be x db. For this gain margin, the $G(j\omega)$ locus will cut –180° at point A whose magnitude is G_A.

Now, $$20 \log \frac{1}{G_A} = x$$

$$\log \frac{1}{G_A} = \frac{x}{20}$$

$$\frac{1}{G_A} = 10^{\frac{x}{20}}$$

∴ $$G_A = \frac{1}{10^{\frac{x}{20}}}$$

Now the value of K is given by,

$$K = \frac{G_A}{G_B}$$

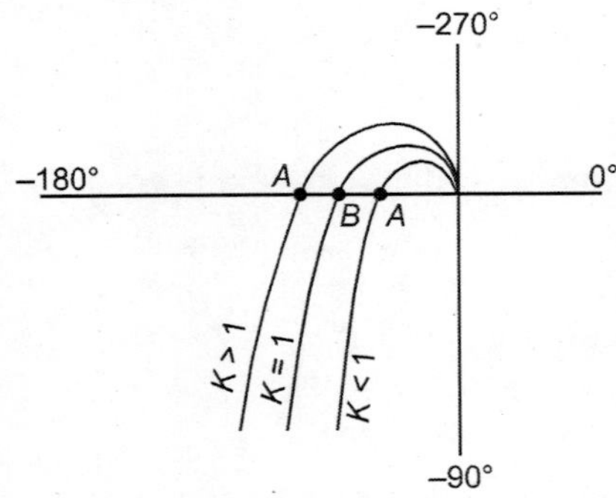

Fig. 4.41: Polar plot for different value of K.

If $K > 1$, then the system gain should be increased.

If $K < 1$, then the system gain should be decreased.

To determine K for specified phase margin

Draw $G(j\omega)$ locus with $K = 1$. Let it cut the unity circle at point B. (The gain at point B is G_B and equal to unity).

Let the specified phase margin be $x°$. For a phase margin of $x°$. Let ϕ_{gcx} be the phase angle of $G(j\omega)$ at gain cross over frequency.

$\therefore \quad x° = 180° + \phi_{gcx}$ and $\phi_{gcx} = x° - 180°$

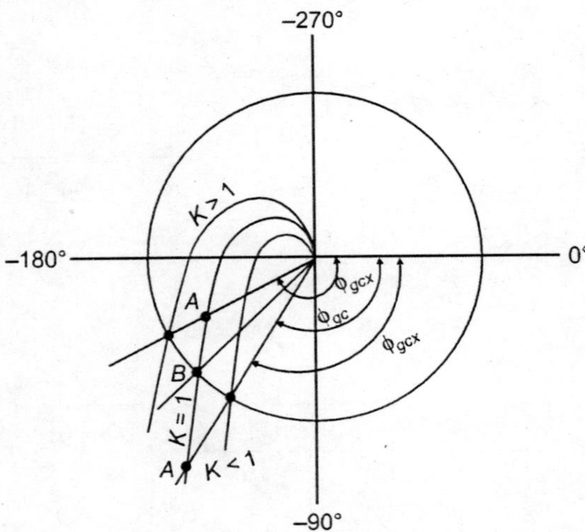

Fig. 4.42: Gain adjustment for required phase margin.

In the polar plot the radial line corresponding to ϕ_{gcx} will cut the locus of $G(j\omega)$ with $K=1$ at point A and the magnitude corresponding to the point be G_A.

Now, $$K = \frac{G_B}{G_A} = \frac{1}{G_A} \qquad [\because G_B = 1]$$

EXAMPLE 4.15

The open loop transfer function of a unity feedback system is given by $G(s) = \dfrac{1}{s(1+s)(1+2s)}$. Sketch the polar plot and determine the gain margin and phase margin.

Solution: Given that,

$$G(s) = \frac{1}{s(1+s)(1+2s)}$$

Put $$s = j\omega$$

The corner frequencies are $\omega_{c1} = \dfrac{1}{2} = .5$ rad/sec and $\omega_{c2} = 1$ rad/sec. The magnitude and phase angle of $G(j\omega)$ are calculated for the corner frequencies and for frequencies around corner frequencies and tabulated in Table (Ex. 4.15.1). Using polar to rectangular conversion, the polar coordinates listed in Table (Ex. 4.15.1) are converted to rectangular coordinates and tabulated in Table (Ex. 4.15.2). The polar plot using polar coordinates is sketched on a polar graphs sheet as shown in Fig. (Ex. 4.15.1). The polar plot using rectangular coordinates is sketched on an ordinary graph sheet shown in Fig. (Ex. 4.15.2).

$$G(j\omega) = \frac{1}{j\omega(1+j\omega)(1+j2\omega)}$$

$$= \frac{1}{\omega}\angle 90° \sqrt{1+\omega^2}\ \tan^{-1}\omega \sqrt{1+4\omega^2}\ \angle \tan^{-1} 2\omega$$

$$= \frac{1}{\omega\sqrt{1+\omega^2}\ \sqrt{1+4\omega^2}} \angle -90° - \tan^{-1}\omega - \tan^{-1} 2\omega$$

$$\therefore \quad |G(j\omega)| = \frac{1}{\omega\sqrt{(1+\omega^2)(1+4\omega^2)}} = \frac{1}{\omega\sqrt{1+4\omega^2+\omega^2+4\omega^4}}$$

$$= \frac{1}{\omega\sqrt{1+5\omega^2+4\omega^4}}$$

$$\angle G(j\omega) = -90° - \tan^{-1}\omega - \tan^{-1} 2\omega$$

Polar plot

Table Ex. 4.15.1: Magnitude and phase of $G(j\omega)$ at various frequencies

ω rad/sec	0.35	0.4	0.45	0.5	0.6	0.7	1.0
$\lvert G(j\omega)\rvert$	2.2	1.8	1.5	1.2	0.9	0.7	0.3
$\angle G(j\omega)$, deg	– 146	– 150	– 156	– 162	– 171	– 180	– 198

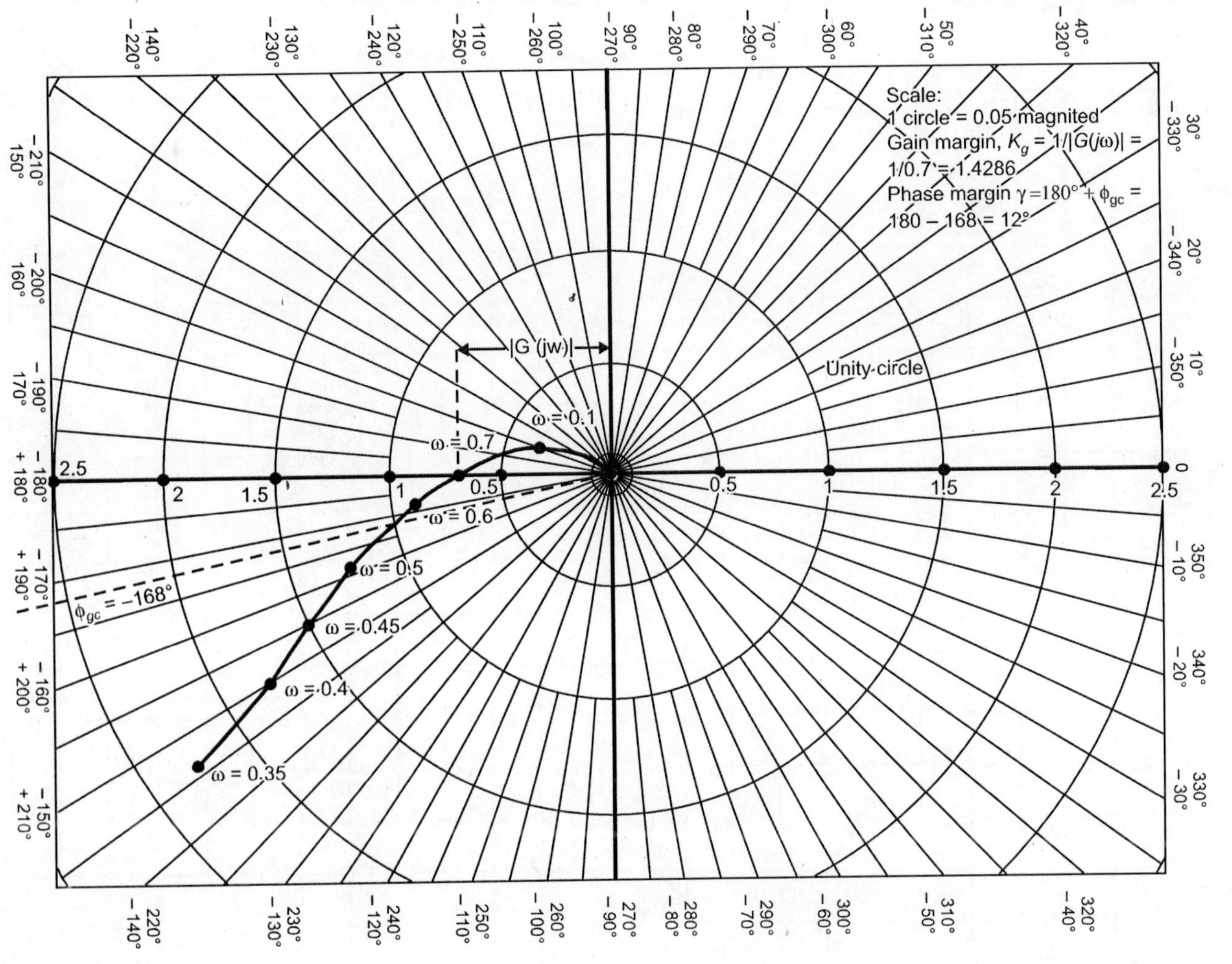

Fig. Ex. 4.1.5.1: Plar plot of $G(s) = \dfrac{1}{s(1+s)(1+2s)}$.

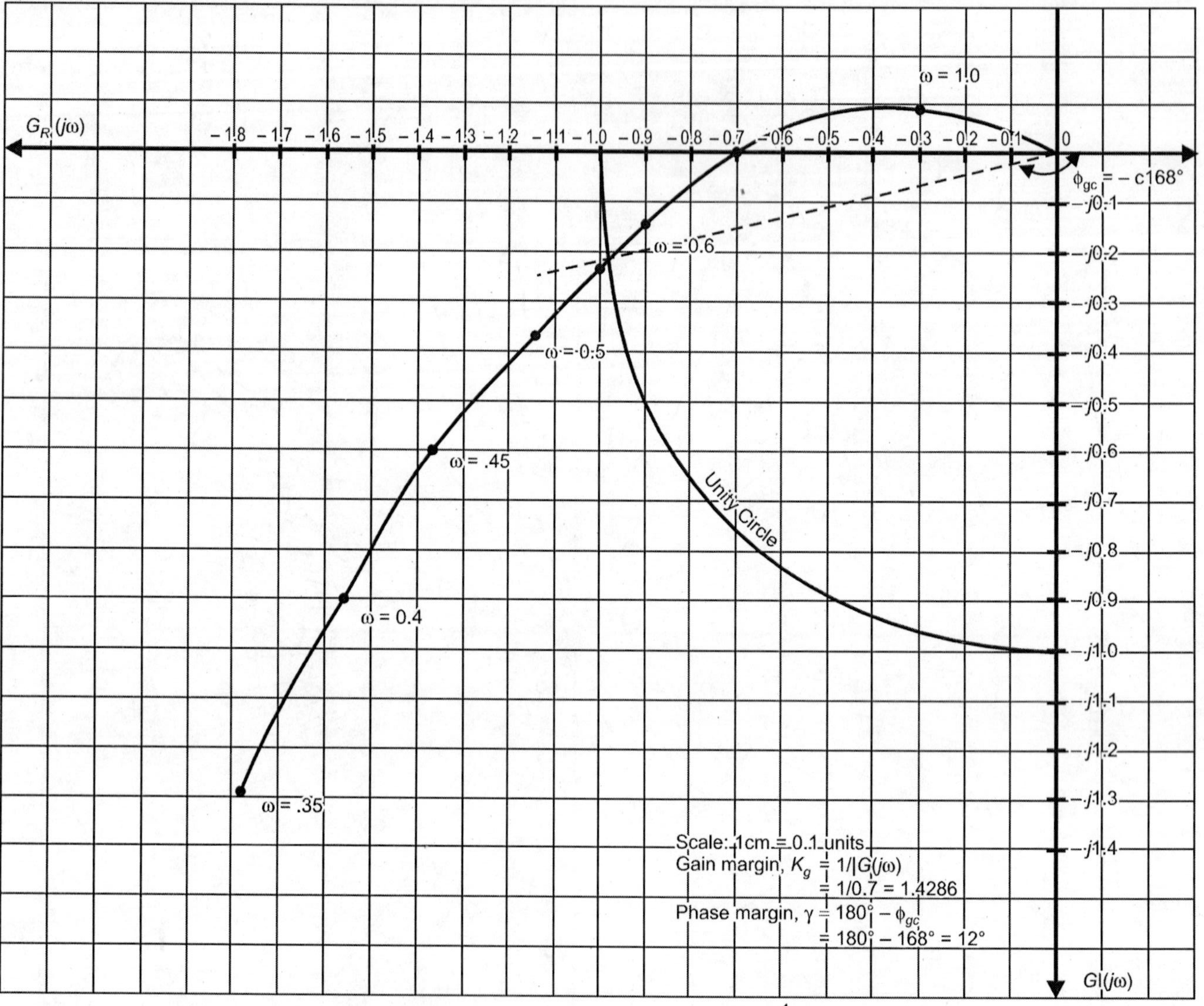

Fig. Ex. 4.15.2: Inverse polar plot of $G(s) = \dfrac{1}{s(1+s)(1+2s)}$.

Inverse polar plot

Table Ex. 4.15.2: Real and imaginary part of $G(j\omega)$ at various frequencies

ω rad/sec	0.35	0.4	0.45	0.5	0.6	0.7	1.0
Real	– 1.76	– 1.56	– 1.37	– 1.14	– 0.89	– 0.7	– 0.29
Images	– 1.29	– 0.9	– 0.61	– 0.37	– 0.14	– 0	0.09

Result: Gain margin, $K_g = 1.4286$
Phase margin, $\gamma = +12°$.

EXAMPLE 4.16

The open loop transfer function of a unity feedback system is given by $G(s) = \dfrac{1}{s^2(1+s)(1+2s)}$. Sketch the polar plot and determine the gain margin and phase margin.

Solution: Given that,

$$G(s) = \frac{1}{s^2(1+s)(1+2s)}$$

Put

$$s = j\omega$$

$$G(j\omega) = \frac{1}{(j\omega)^2\,(1+j\omega)\,(1+2j\omega)}$$

The corner frequencies are $\omega_{c1} = 0.5$ rad/sec and $\omega_{c2} = 1$ rad/sec. The magnitude and phase angle of $G(j\omega)$ are calculated for the corner frequencies and frequencies around corner frequencies are tabulated in Table (Ex. 4.16.1). The polar coordinates in Table (Ex. 4.16.1) are converted to rectangular coordinates and tabulated in Table (Ex. 4.16.2). The polar plot using polar coordinates is sketched on a polar graph sheet shown in Fig. (Ex. 4.16.1) and rectangular coordinates in sketched on an ordinary graph sheet as shown in Fig. (Ex. 4.16.2).

$$G(j\omega) = \frac{1}{(j\omega)^2\,(1+j\omega)\,(1+j2\omega)}$$

$$= \frac{1}{\omega^2\angle 180°\sqrt{1+\omega^2}\angle\tan^{-1}\omega\sqrt{1+4\omega^2}\angle\tan^{-1}2\omega}$$

$$G(j\omega) = \frac{1}{\omega^2\sqrt{1+\omega^2}\,\sqrt{1+4\omega^2}}\angle(-180\epsilon - \tan^{-1}\omega - \tan^{-1}2\omega)$$

$$|G(j\omega)| = \frac{1}{\omega^2\sqrt{1+\omega^2}\,\sqrt{1+4\omega^2}} = \frac{1}{\omega^2\sqrt{(1+\omega^2)(1+4\omega^2)}}$$

$$= \frac{1}{\omega^2\sqrt{1+5\omega^2+4\omega^4}}$$

$$<G(j\omega) = -180\epsilon - \tan^{-1}\omega - \tan^{-1}2\omega$$

CHAPTER 4

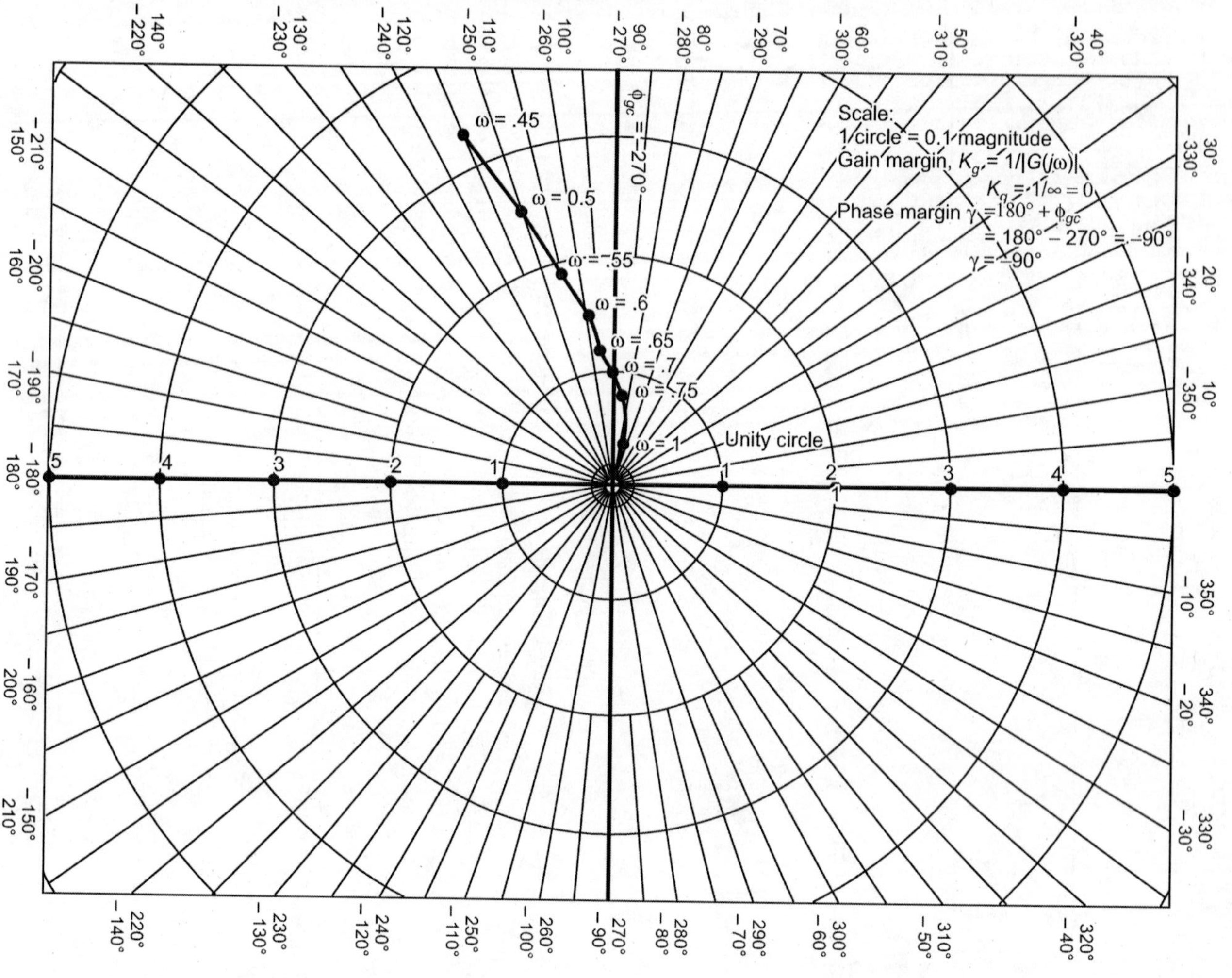

Fig. Ex. 4.16.1: Polar plot of $G(s) = 1/s^2\,(1 + s)\,(1 + 2s)$ (using polar coordinates).

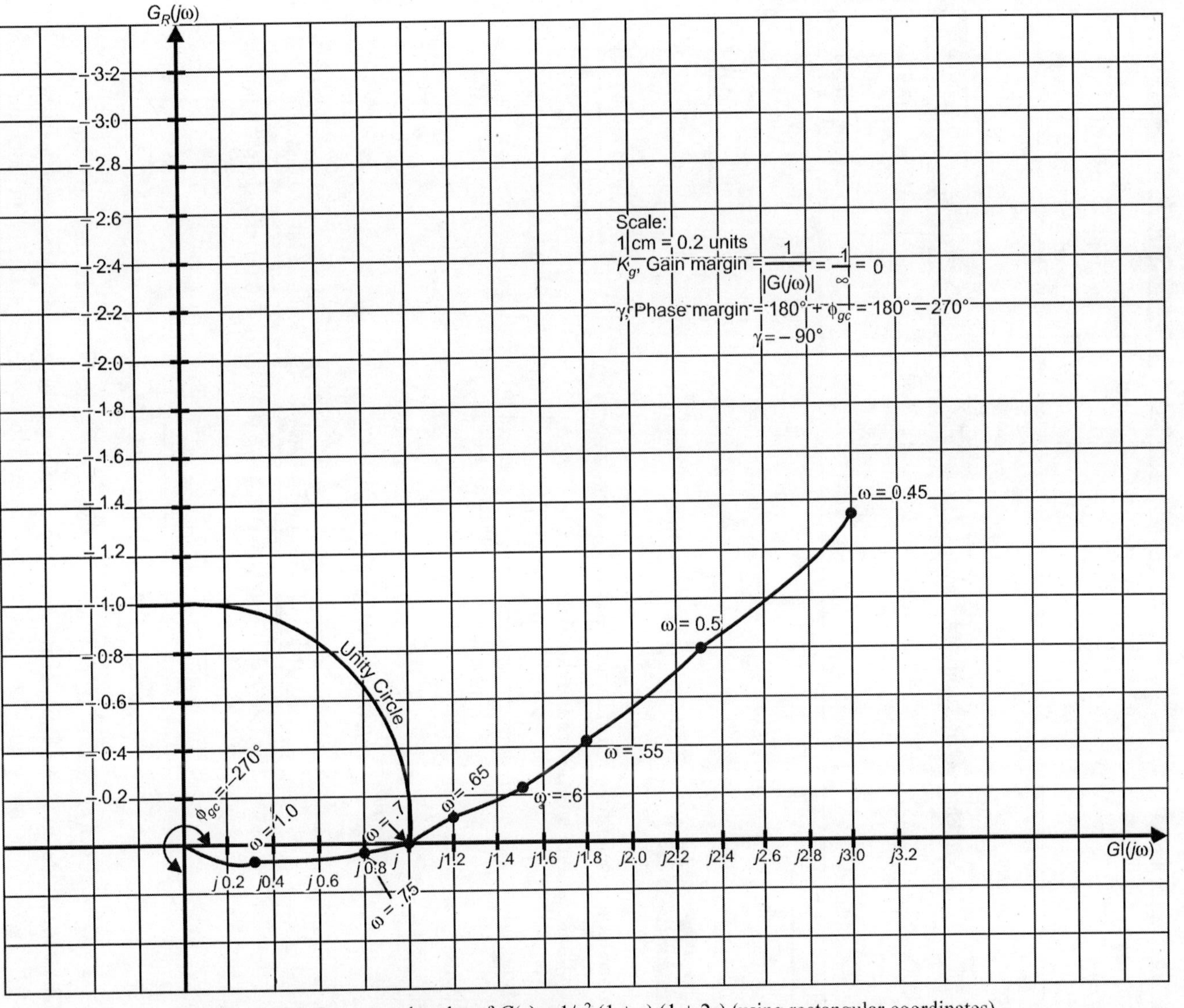

Fig. Ex. 4.16.2: Inverse polar plot of $G(s) = 1/s^2\,(1 + s)\,(1 + 2s)$ (using rectangular coordinates).

Polar plot

Table Ex. 4.16.1: Magnitude and phase of $G(j\omega)$ at varies frequencies								
ω, rad/sec	0.45	0.5	0.55	0.6	0.65	0.7	0.75	1
$\|G(j\omega)\|$	3.3	2.5	1.9	1.5	1.2	0.97 1 1	0.8	0.3
$\angle G(j\omega)$, deg	–246	–251	–256	–261	–265	–269	–273	–288

Inverse polar plot

Table Ex. 4.16.2: Real and Imaginary Part of $G(j\omega)$ at various frequencies								
ω, rad/sec	0.45	0.5	0.55	0.6	0.65	0.7	0.75	1.0
$\|G(j\omega)\|$	–1.34	–0.81	–0.46	–0.23	–0.1	–0.02	–0.04	–0.09
$\angle G(j\omega)$, deg	3.01	2.36	1.84	1.48	1.2	1.0	0.8	0.29

Result: Gain making, $K_g = 0$
Phase margin, $\gamma = -90°$.

EXAMPLE 4.17

The open loop transfer function of a unity feedback system is given by $G(s) = \dfrac{(1+0.2s)(1+0.025s)}{s^3(1+0.005s)(1+0.001s)}$. Sketch the polar plot and determine the phase margin.

Solution: Given that $G(s) = \dfrac{(1+0.2s)(1+0.025s)}{s^3(1+0.005s)(1+0.001s)}$

Put $s = j\omega$,

$$\therefore \quad G(j\omega) = \frac{(1+j0.2\omega)(1+j0.025\omega)}{(j\omega)^3(1+j0.005\omega)(1+j0.001\omega)}$$

$$= \frac{\sqrt{1+(0.2\omega)^2}\angle \tan^{-1}0.2\omega\sqrt{1+(0.025\omega)^2}\angle\tan^{-1}0.025\omega}{\omega^3\angle 270°\sqrt{1+(0.005\omega)^2}\angle\tan^{-1}0.005\omega\sqrt{1+(0.001\omega)^2}\angle\tan^{-1}0.001\omega}$$

$$|G(j\omega)| = \frac{\sqrt{1+(0.2\omega)^2}\ \sqrt{1+(0.025\omega)^2}}{\omega^3\sqrt{1+(0.005\omega)^2}\ \sqrt{1+(0.001\omega)^2}}$$

$$\angle G(j\omega) = \tan^{-1}0.2\omega + \tan^{-1}0.02\omega - 270° - \tan^{-1}0.005\,\omega - \tan^{-1}0.001\omega$$

The magnitude and phase angle of $G(j\omega)$ are calculated for various frequencies and listed in Table (Ex. 4.17.1) and rectangular coordinates are tabulated in Table (Ex. 4.17.2). The polar plot using polar coordinates is sketched as a polar graph sheet as shown in Fig. (Ex. 4.17.1) and polar plot using rectangular coordinates is sketched on a ordinary graph sheet as shown in Fig. (Ex. 4.17.2).

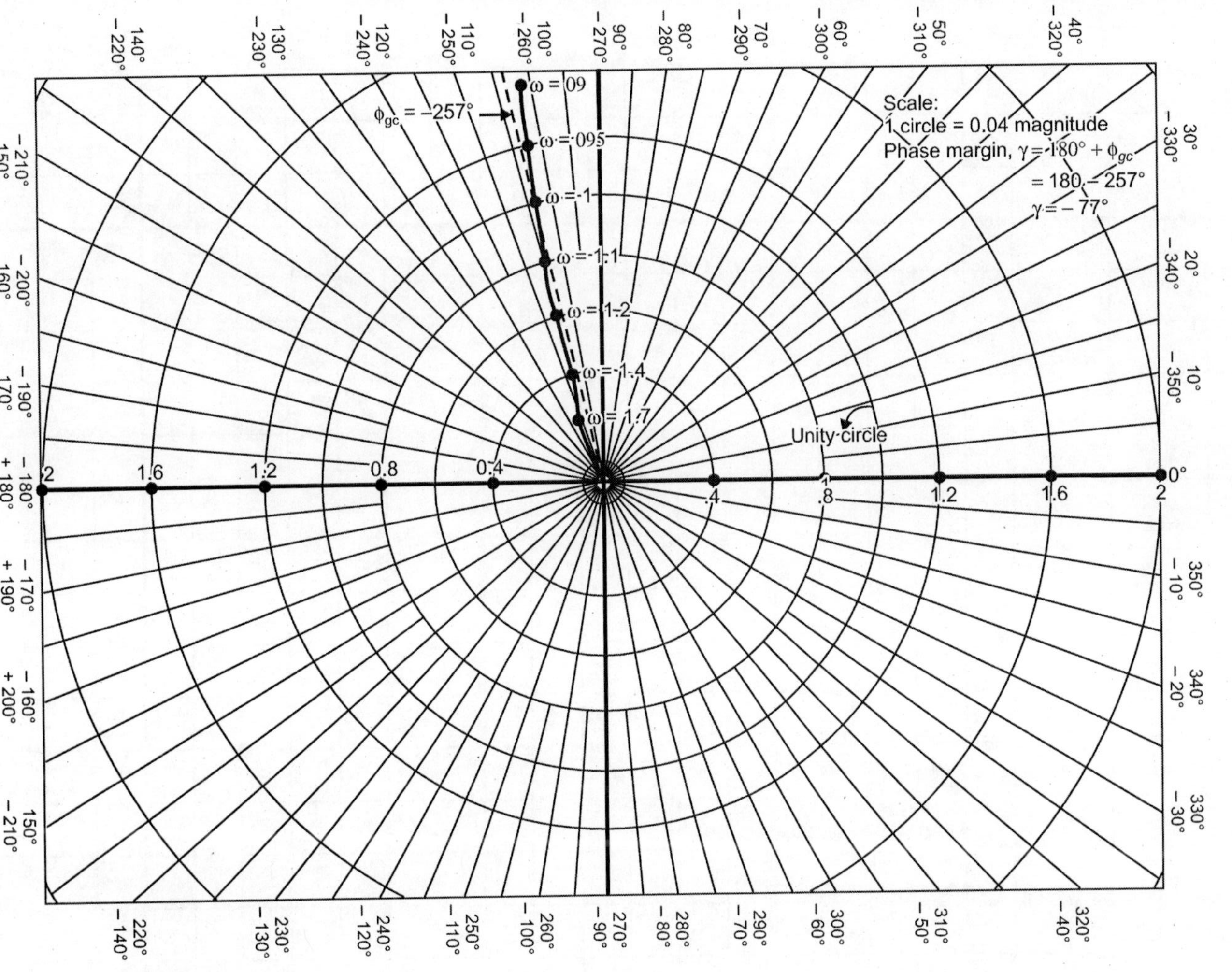

Fig. Ex. 4.17.1: Polar plot of $G(s) = (1 + .2s)\,(1 + .025s)/s3(1 + .005s)\,(1 + .001s)$ (using polar coordinates).

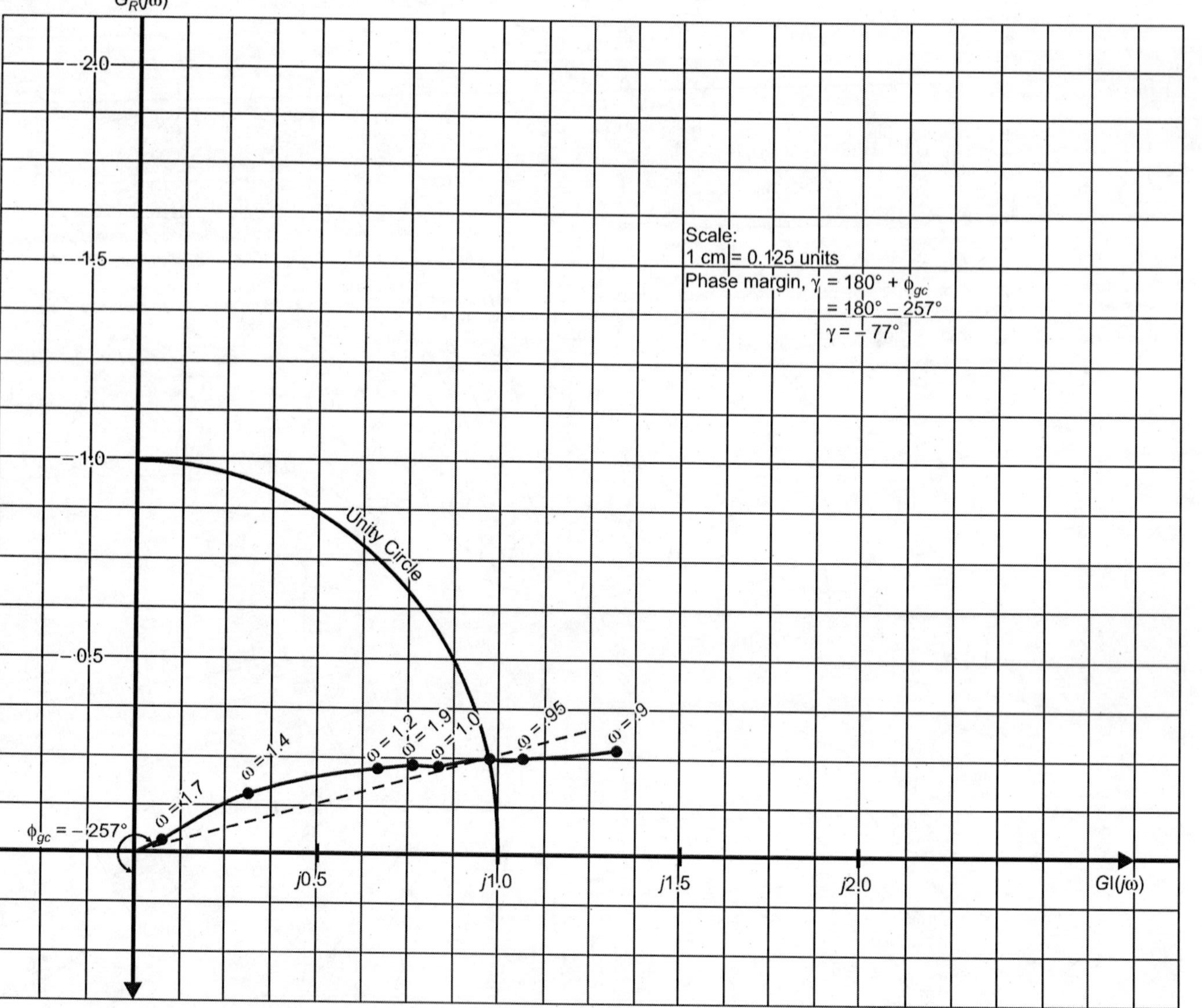

Fig. Ex. 4.17.2: Inverse polar plot of $G(s) = (1 + .2s)\,(1 + .25s)/s^3(1 + .005s)\,(1 + .001s)$ (using rectangular co-ordinates).

Polar plot

Table Ex. 4.17.1: Magnitude and phase of $G(j\omega)$

ω, rad/sec	0.9	0.95	1	1.1	1.2	1.4	1.7
$\lvert G(j\omega)\rvert$	1.4	1.2	1	0.8	0.6	0.4	0.2
$\angle G(j\omega)$, deg	–259	–258	–257	–256	–255	–253	–249

Inverse polar plot

Table Ex. 4.17.2: Real and imaginary part of $G(j\omega)$

ω, rad/sec	0.9	0.95	1	1.1	1.2	1.4	1.7
$\lvert G(j\omega)\rvert$	–0.27	–0.25	–0.22	–0.19	–0.16	–0.12	–0.07
$\angle G(j\omega)$, deg	1.37	1.17	0.97	0.78	0.58	0.38	0.19

Result: Phase margin $\gamma = -77°$

EXAMPLE 4.18

Consider a unity feedback system having an open loop transfer function $G(s) = \dfrac{K}{s(1+0.2s)(1+0.05s)}$. Sketch the polar plot and determine the value of K, so that

(i) Gain margin is 18 *db*

(ii) Phase margin is 60°.

Solution: Given that, $$G(s) = \frac{K}{s(1+0.2s)(1+0.05s)}$$

The polar plot is sketched by taking $K = 1$.

$\therefore$ put $K = 1$ and $s = j\omega$ is $G(s)$

$$\therefore \quad G(j\omega) = \frac{K}{j\omega(1+j0.2\omega)(1+j0.05\omega)}$$

The corner frequencies are $\omega_{c1} = \dfrac{1}{0.2} = 5$ rad/sec and $\omega_{c2} = \dfrac{1}{0.05} = 20$ rad/sec. The magnitude and phase angle of $G(j\omega)$ are calculated for various frequencies and tabulated in Table (Ex. 4.18.1). The rectangular coordinates and tabulated is Table (Ex. 4.18.2). The polar plot using polar coordinates is sketched on a polar graph sheet as shown in Fig. (Ex. 4.18.1). The polar plot using rectangular coordinates is sketched on an ordinary graph sheet as shown in Fig. (Ex. 4.18.2).

$$G(j\omega) = \frac{1}{j\omega(1+j0.2\omega)(1+j0.05\omega)}$$

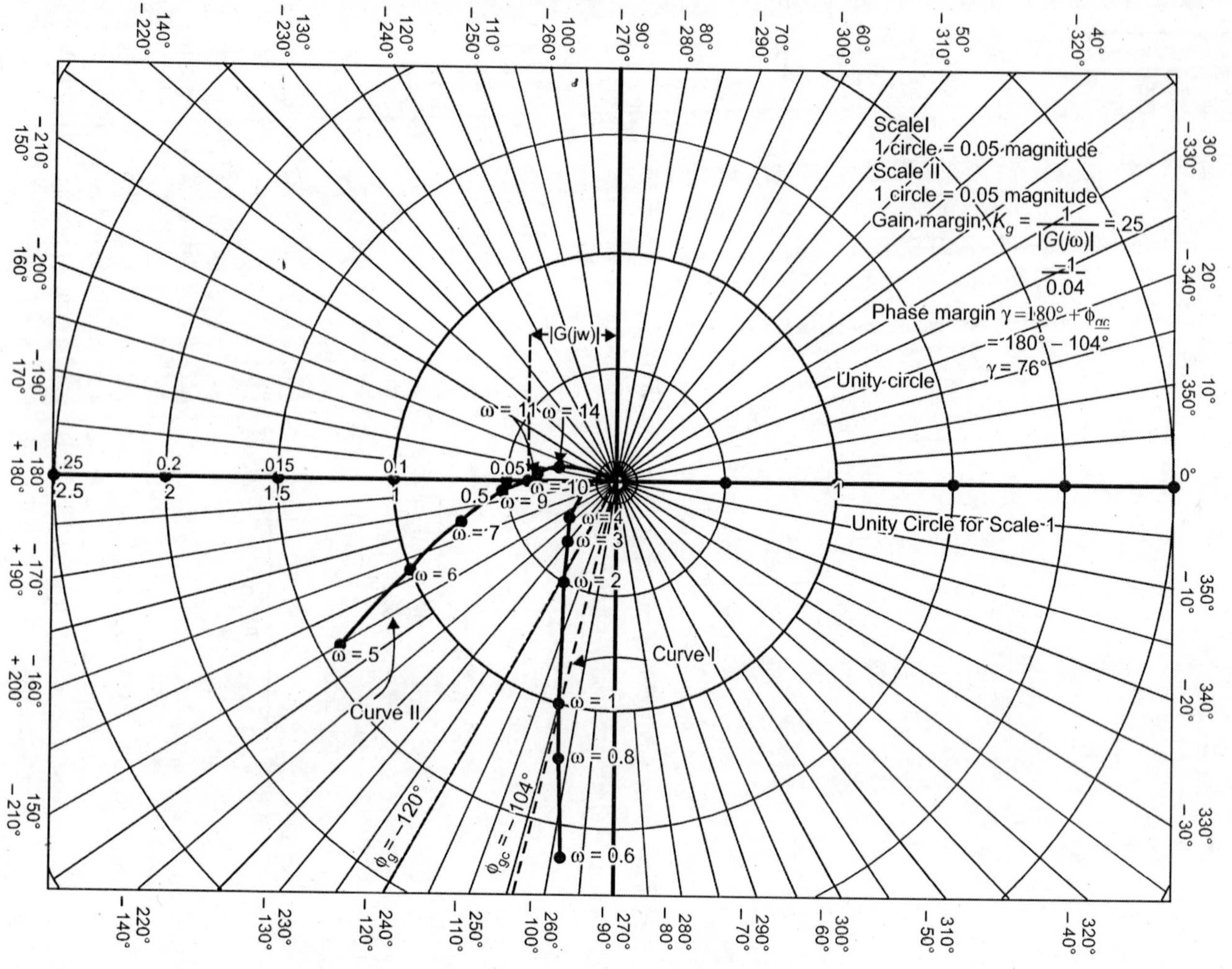

Fig. Ex. 4.18.1: Polar plot of $G(s) = K/s(1 + .2s)(1 + 0.5s)$.

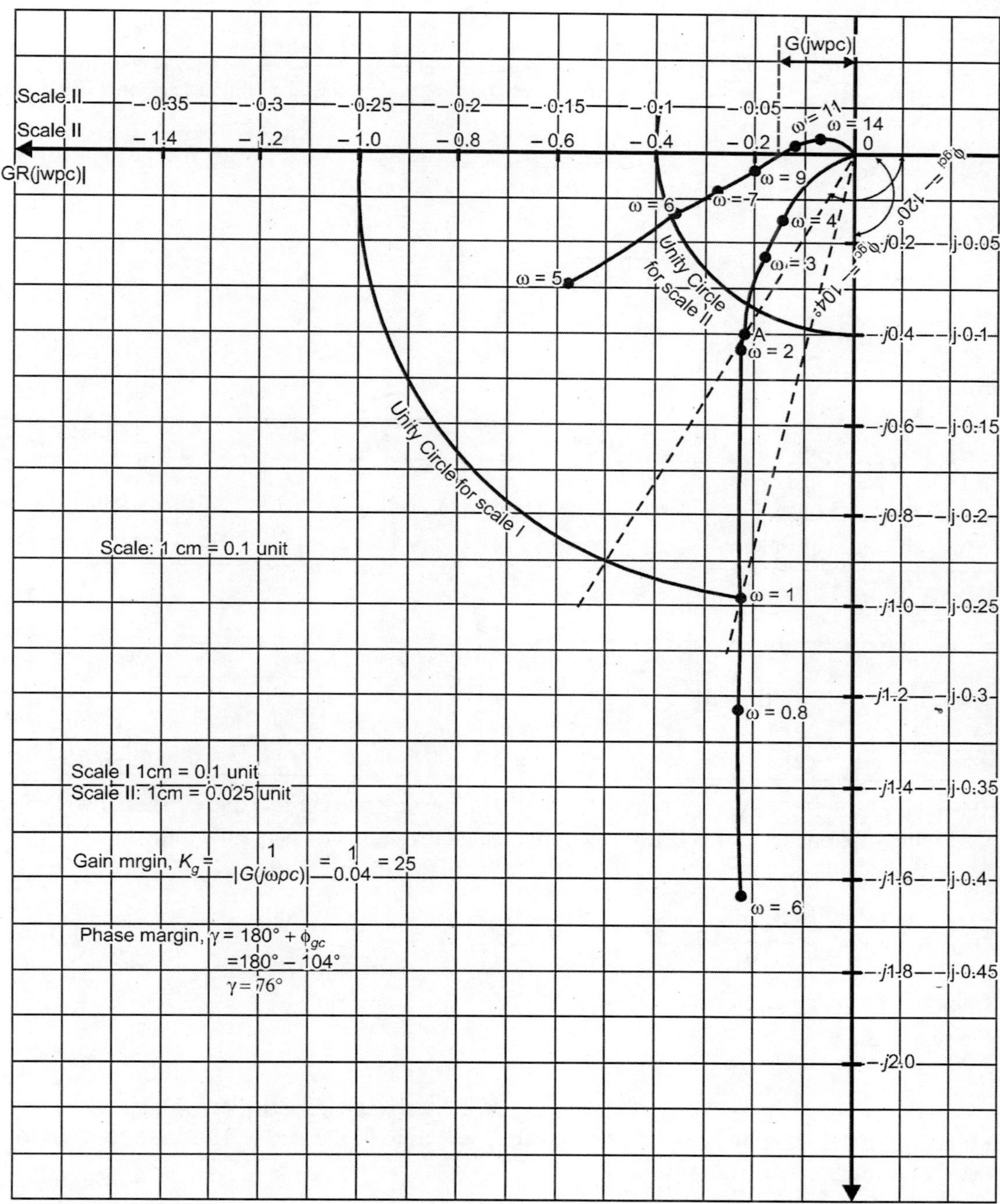

Fig. Ex. 4.18.2: Inverse polar plot of $G(s) = K/s(1 + .2s)(1 + 0.5s)$.

$$= \frac{1}{\omega \angle 90° \sqrt{1+(0.2\omega)^2} \angle \tan^{-1} 0.2\omega \sqrt{1+(0.05\omega)^2} \angle \tan^{-1} 0.05\omega}$$

$$= \frac{1}{\omega \sqrt{1+(0.2\omega)^2} \sqrt{1+(0.05\omega)^2}} < -90° - \tan^{-1} 0.02\omega - \tan^{-1} 0.05\omega$$

$$\therefore \qquad |G(j\omega)| = \frac{1}{\omega \sqrt{1+(0.2\omega)^2} \sqrt{1+(0.05\omega)^2}}$$

$$\angle G(j\omega) = -90° - \tan^{-1} .2\omega - \tan^{-1} 0.05\omega$$

Polar plot

Table Ex. 4.18.1: Polar plot

ω, rad/sec	0.6	0.8	1	2	3	4	5	6	7	9	10	11	14
$\lvert G(j\omega)\rvert$	1.65	1.23	1.0	0.5	0.3	0.2	0.14	0.1	0.07	0.05	0.04	0.03	0.02
$\angle G(j\omega)$deg	–98	–101	–104	–117.5	–129.4	–140	–149	–157	–164	–175	–180	–184	–195

Inverse polar plot

Table Ex. 4.18.2: Inverse polar plot

ω, rad/sec	0.6	0.8	1	2	3	4	5	6	7	9	10	11	14
$\lvert G(j\omega)\rvert$	–0.23	–.23	–0.24	–0.23	–0.19	–0.15	–0.12	–0.92	–0.67	–0.50	–0.04	–0.030	–0.019
$\angle G(j\omega)$, deg	–1.63	–1.21	–0.97	–0.44	–0.23	–0.13	–0.072	–0.039	–0.019	–0.044	–0	0.002	0.005

In the polar plot in Fig. (Ex. 4.18.1) and (Ex. 4.18.2). Their are two curves I and II. These two loci are sketched with different scales to clearly determine the gain margin and phase margin.

For the polar plot, with $K = 1$

$$\text{Gain margin, } K_g = \frac{1}{0.004} = 25$$

Gain margin in *db* = 20 log 25 = 28 *db*

Phase margin, $\gamma = 76°$

Case I

With $K = 1$, let $G(j\omega)$ cut the – 180° axis at point B and gain corresponding to that point be G_B. From the polar plot $G_B = 0.04$. The gain margin to 28 *db* with $K = 1$ has to be reduced to 18 *db* and so K has to be increased to a value greater than one.

Let G_A be the gain at –180° for a gain margin of 18 *db*.

$$\text{Now,} \qquad 20 \log \frac{1}{G_A} = 18$$

$$\log \frac{1}{G_A} = \frac{18}{20}$$

$$\frac{1}{G_A} = 10$$

∴ $$G_A = \frac{1}{10^{\frac{18}{20}}} = 0.125$$

The value of k is given by $K = \frac{G_A}{G_B} = \frac{.125}{0.04} = 3.125$.

Case II

With $K = 1$, the phase margin is 76°. This has to be reduced be 60°. Hence gain has to be increased.

Let ϕ_{gc_2} be the phase of $G(j\omega)$ for a phase margin of 60°.

∴ $$60° = 180° + \phi_{gc_2}$$

$$\phi_{gc_2} = 60° - 180° = -120°$$

In the polar plot the 120° line cut the locus of $G(j\omega)$ at point A and cut the unity circle at point B.

Let G_A = Magnitude of $G(j\omega)$ at point A.

G_B = Magnitude of $G(j\omega)$ at point B.

From the polar plot, $G_A = 0.425$ and $G_B = 1$.

Now, $$K = \frac{G_B}{G_A} = \frac{1}{0.425} = 2.353$$

Result:

1. When $K = 1$, gain margin, $K_g = 25$

   Gain margin is $db = 28\ db$
2. When $K = 1$, phase margin, $\delta = 76°$
3. For a gain margin of 18 db, $K = 3.125$
4. For a phase margin of 60°, $K = 2.353$

EXAMPLE 4.19

Consider a unity feedback system having an open loop transfer function $G(s) = \dfrac{K}{s(1+0.5s)(1+4s)}$.

Sketch the polar plot and determine the value of K so that

(i) Gain margin is 20 db and

(ii) Phase margin is 30°.

Solution: Given that, $$G(s) = \frac{K}{s(1+0.5s)(1+4s)}$$

The polar plot is sketched by taking $K = 1$

Put $K = 1$ and $s = j\omega$ in $G(s)$

$$\therefore \quad G(j\omega) = \frac{K}{j\omega(1+j0.5\omega)(1+j4\omega)}$$

The corner frequencies are $\omega_{c1} = \frac{1}{4} = 0.25\,\text{rad/sec}$ and $\omega_{c2} = \frac{1}{0.5} = 2\,\text{rad/sec}$. The magnitude and phase angle of $G(j\omega)$ are calculated for various frequencies and tabulated in Table (Ex. 4.19.1) and rectangular coordinates are tabulated in Table (Ex. 4.19.2). The polar plot using polar coordinates is sketched on a polar graph sheet as shown in Fig. (Ex. 4.19.1) and the polar plot using rectangular coordinates is sketched on an ordinary graph sheet as shown in Fig. (Ex. 4.19.2).

$$G(j\omega) = \frac{1}{j\omega(1+j0.5\omega)(1+j4\omega)}$$

$$= \frac{1}{\omega\angle 90^\circ\sqrt{1+(0.5\omega)^2}\angle\tan^{-1}0.5\omega\sqrt{1+(4\omega)^2}\angle\tan^{-1}4\omega}$$

$$= \frac{1}{\omega\sqrt{1+0.25\omega^2}\sqrt{1+16\omega^2}}\angle(90^\circ - \tan^{-1}0.5\,\omega - \tan^{-1}4\omega)$$

$$\therefore \quad \text{Mag } |G(j\omega)| = \frac{1}{\omega\sqrt{1+0.25\omega^2}\sqrt{1+16\omega^2}} \qquad \boxed{\text{Mag} \Rightarrow 0 - 2.5}$$

$$\text{Angle } \angle G(j\omega) = -90^\circ - \tan^{-1}0.5\,\omega - \tan^{-1}4\omega$$

Polar plot

Table Ex. 4.19.1: Magnitude and Phase of $G(j\omega)$

ω rad/sec	0.3	0.4	0.5	0.6	0.8	1.0	1.2
$\|G(j\omega)\|$ /Mag	2.11 M	1.3	0.87	0.61	0.35	0.22	0.15
Angle $\angle G(j\omega)$, deg	–149 A	–159	–167	–174	–184	–193	–199

Inverse polar plot: *conversion of Polar form to Rectangular form*

Table Ex. 4.19.2: Real and Imaginary part of $G(j\omega)$

ω, rad/sec	0.3	0.4	0.5	0.6	0.8	1.0	1.2
Real $G_R(j\omega)$	–1.8	–1.21	–0.85	–0.61	–0.35	–0.21	–0.14
Image $G_I(j\omega)$	–1.09	–0.47	–0.2	–0.06	0.02	0.05	0.05

From the polar plot, with $K = 1$

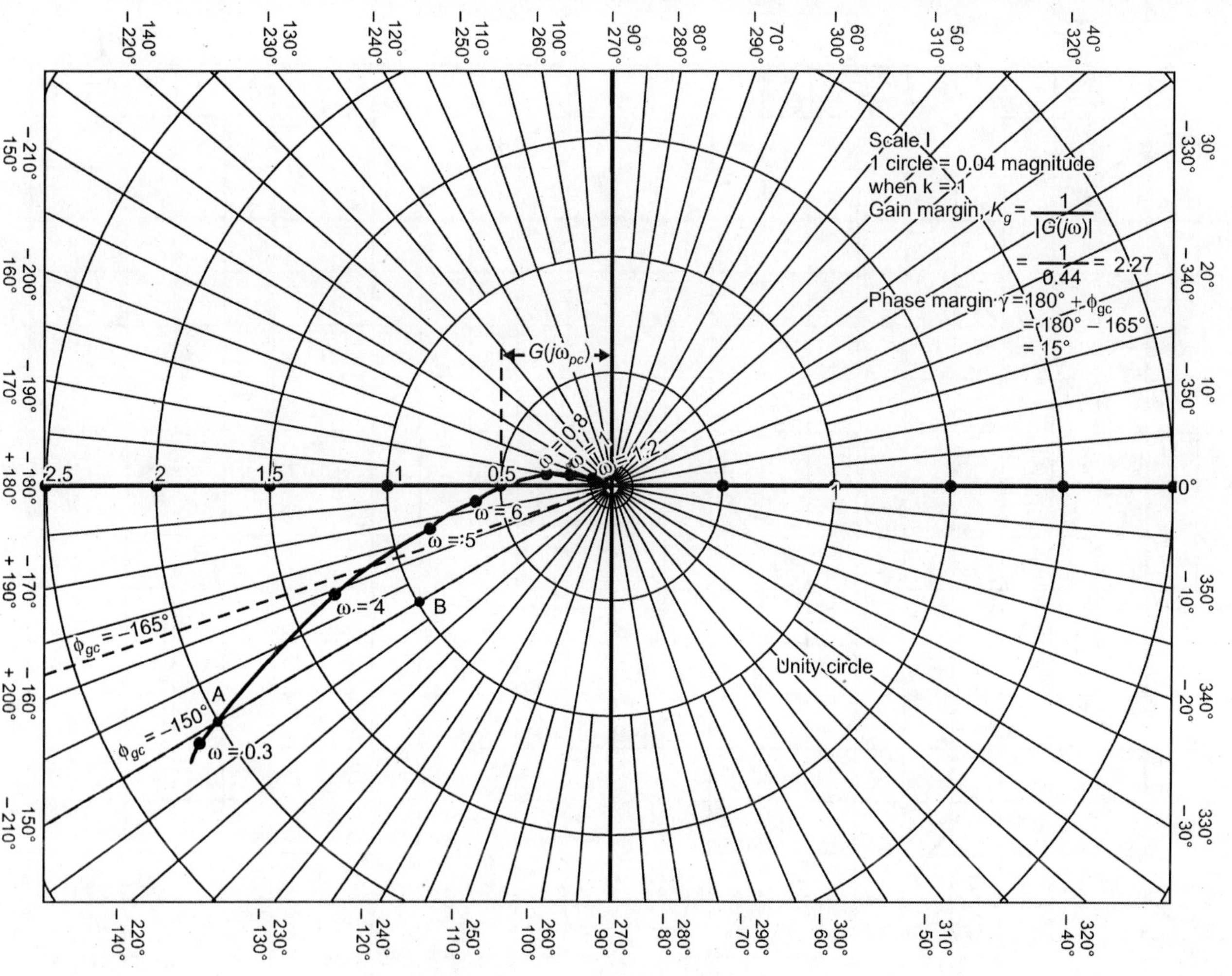

Fig. Ex. 4.19.1: Polar plot of $G(s) = 1/s(1 + .5s)(1 + 4s)$.

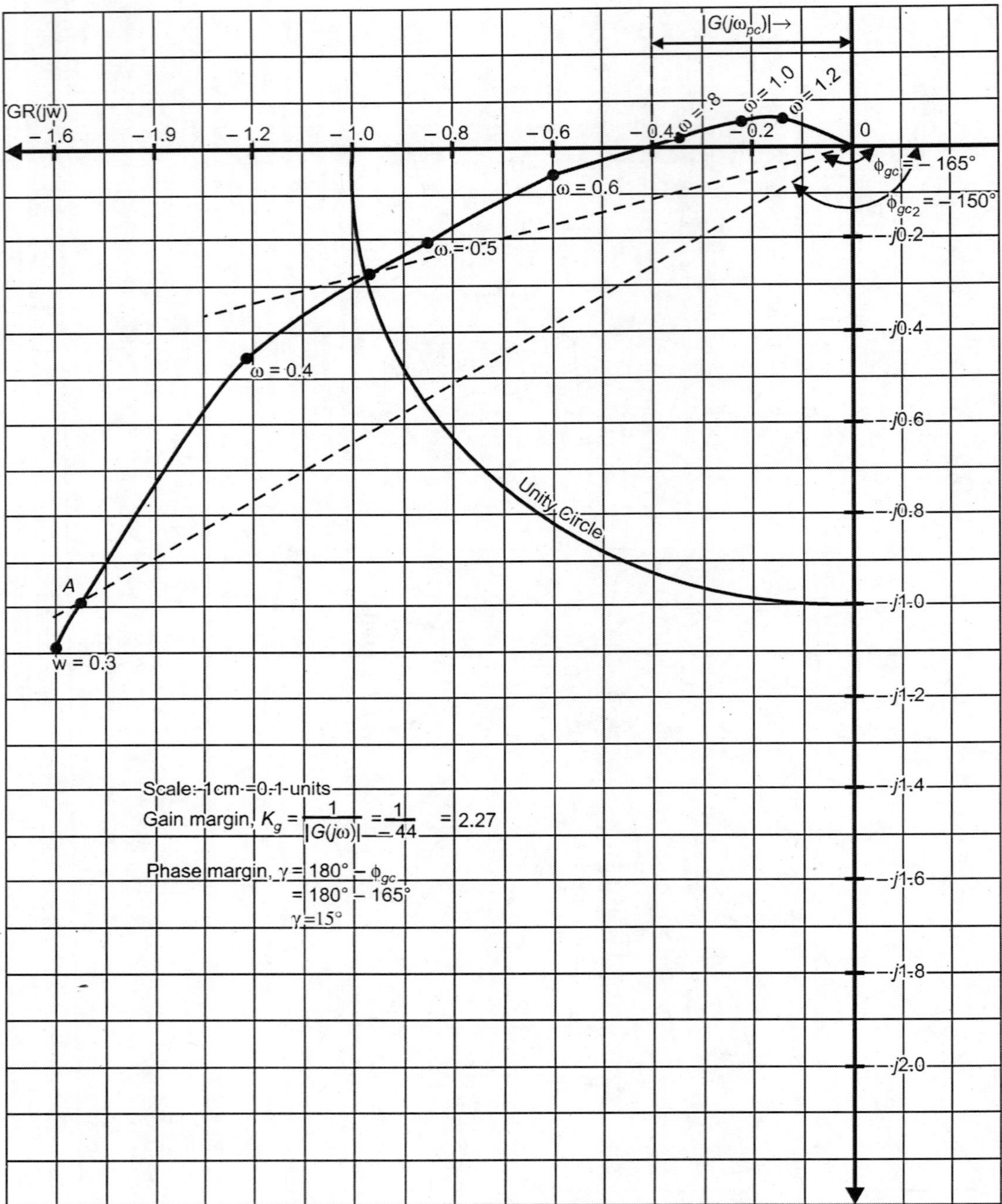

Fig. Ex. 4.19.2: Inverse polar plot of $G(s) = 1/2(1 + .5s)\ (1 + 4s)$.

Gain margin, $K_g = \dfrac{1}{0.44} = 2.27$

Gain margin in db = 20 log 2.27 = 7.12 db

Phase margin, $\gamma = 180° + \phi gc = 180° - 165° = 15°$

Case I

With $K = 1$, let $G(j\omega)$ cut the – 180° axis at point B and gain corresponding to that point be G_B. From the polar plot $G_B = 0.04$. The gain margin to 7.12 db with $K = 1$ has to be increased to 20 db and so K has to be decreased to a value less than one.

Let G_A be the gain at –180° for a gain margin of 20 db.

Now,
$$20 \log \frac{1}{G_A} = 20$$

$$\log \frac{1}{G_A} = \frac{20}{20} = 1$$

$$\frac{1}{G_A} = 10^{-1} = 10$$

$$\therefore \quad G_A = \frac{1}{10} = 0.1$$

The value of K is given by $K = \dfrac{G_A}{G_B} = \dfrac{0.1}{0.44} = 0.227$.

Case II

With $K = 1$, the phase margin is 15°. This has to be increased be 30°. Hence gain has to be decreased.

Let ϕ_{gc_2} be the phase of $G(j\omega)$ for a phase margin of 30°.

$$\therefore \quad 30° = 180° + \phi_{gc2}$$

$$\phi_{gc2} = 30° - 180° = -150°$$

In the polar plot the 150° line cut the locus of $G(j\omega)$ at point A and cut the unity circle at point B.

Let G_A = Magnitude of $G(j\omega)$ at point A.

G_B = Magnitude of $G(j\omega)$ at point B.

From the polar plot, $G_A = 2.04$ and $G_B = 1$.

Now,
$$K = \frac{G_B}{G_A} = \frac{1}{2.04} = 0.49.$$

Result:

1. When $K = 1$, gain margin, $K_g = 2.27$
2. Gain margin in db = 7.12 db
3. When $K = 1$, Phase margin, $\gamma = 15°$
4. For a gain margin of 20 db, $K = .227$
5. For a phase margin of 30°, $K = 0.49$.

SOME WORKED EXAMPLES

Note: To make the students to understand some examples are given in very detail.

EXAMPLE 1

Sketch the bode plot of the following open loop transfer function and from the bode plot determine phase margin and gain margin $G(s) = \dfrac{100(1+.1s)}{s(1+.2s)\,(1+.5s)}$

Solution:

Step 1: Put, $s = j\omega$ in $G(s)$

$$\therefore \qquad G(j\omega) = \frac{100(1+.1j\omega)}{j\omega(1+.2j\omega)\,(1+.5j\omega)}$$

Step 2: Draw magnitude plot.

Table 1

Term	Corner frequency rad/sec	Slope of each term *db*/dec	Charge in slope from one term to another term *db*/dec
$\dfrac{100}{j\omega}$	–	–20	
$\dfrac{1}{1+.5j\omega}$	$\omega_{c1} = \dfrac{1}{.5} = 2$	–20	–20 –20 = 40
$\dfrac{1}{1+.2j\omega}$	$\omega_{c2} = \dfrac{1}{.2} = 5$	–20	–40 –20 = –60
$(1+.1j\omega)$	$\omega_{c3} = \dfrac{1}{-1} = 10$	+20	–60 + 20 = –40

Let, choose low and high frequencies

The condition for choosing low frequency is

$$\omega_l < \omega_{c1}$$

The condition for choosing high frequency is

$$\omega_h > \omega_{c3}$$

$\therefore$ Let, $\omega_l = 0.1$ rad/sec

$\omega_h = 100$ rad/sec

Let, $A = |G(j\omega)|$

at, $\omega = \omega l$

$A_1 = 20 \log |\text{First term}|$

$$A_1 = 20 \log\left|\frac{100}{j\omega}\right| = 20 \log\left|\frac{100}{1}\right|$$

$$\boxed{A_1 = 60\ db}$$

at, $\omega = \omega_{c1}$

$$A_2 = 20 \log |\text{First term}|$$

$$A_2 = 20 \log\left|\frac{100}{2}\right|$$

$$\boxed{A_2 = 34\ db}$$

at, $\omega = \omega_{c2}$

A_3 = change in slope from one term to another term Ч $\log\left|\frac{\text{Present frequency}}{\text{Previous frequency}}\right|$ + Previous gain

i.e., A_3 = Change in slope from ω_{c1} to ω_{c2} Ч $\log\left|\frac{\omega c_2}{\omega c_1}\right| + A_2$

$$A_3 = -40 \log\left|\frac{5}{2}\right| + 34$$

$$\boxed{A_3 = +18\ db}$$

at, $\omega = \omega_{c3}$

A_4 = change in slope from one term to another term Ч $\log\left|\frac{\text{Present frequency}}{\text{Previous frequency}}\right|$ + Previous gain

i.e., A_4 = Change in slope from ω_{c2} to ω_{c3} Ч $\log\left|\frac{\omega_{c3}}{\omega_{c2}}\right| + A_3$

$$A_4 = -60 \log\left|\frac{10}{5}\right| + 18$$

$$\boxed{A_4 = 0\ db}$$

at $\omega = \omega_h$

A_5 = Change in slope from ω_{c3} to ω_h Ч $\log\left|\frac{\omega_h}{\omega_{c3}}\right| + A_4$

$$A_5 = -40 \log\left|\frac{100}{10}\right| + 0$$

$$\boxed{A_5 = -40\ db}$$

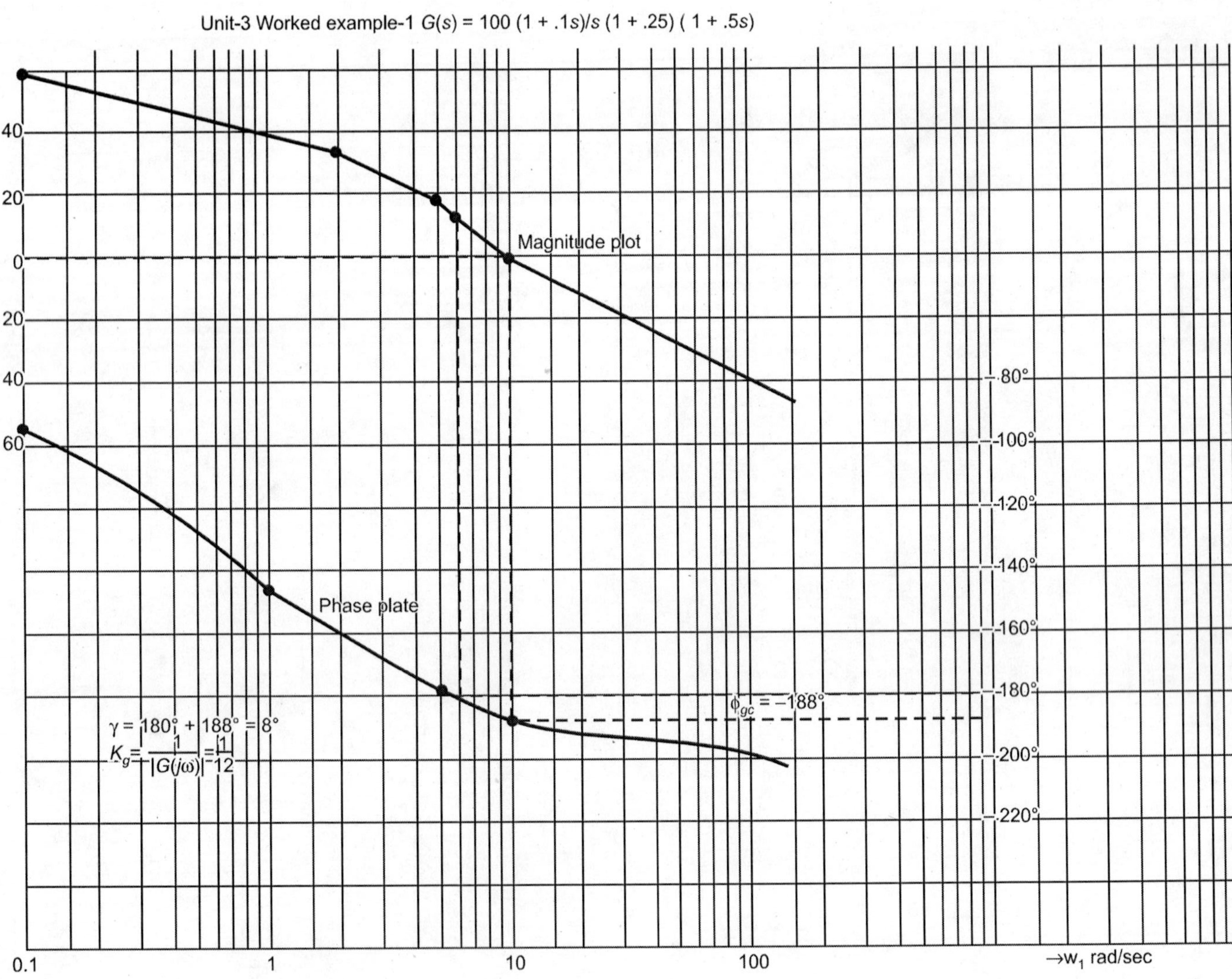
Unit-3 Worked example-1 G(s) = 100 (1 + .1s)/s (1 + .25) (1 + .5s)
+ 40
+ 20
0
– 20
– 40
– 60
Magnitude plot
Phase plate
– 80°
– 100°
– 120°
– 140°
– 160°
– 180°
– 200°
– 220°
φgc = – 188°
γ = 180° + 188° = 8°
Kg = 1/|G(jω)| = 1/12
0.1
1
10
100
→w1 rad/sec

Table 2					
ω, rad/sec	0.1	2	5	10	100
A in *db*	60	34	18	0	–40

Step 3: Draw phase plot

$$\phi = \angle G(j\omega)$$
$$\phi = -90\epsilon + \tan^{-1} .1\omega - \tan^{-1} .2\omega - \tan^{-1} .5\omega$$

Table 3					
ω, rad/sec	0.1	2	5	10	100
ϕ, in deg	–93.5є	–145.5є	–176.5є	–187є	–192є

From bode plot,

$$\text{Phase margin, } \gamma = 180\epsilon + \phi_{gc} = 180\epsilon - 188\epsilon = 8\epsilon$$

$$\text{Gain margin, } K_g = \frac{1}{|G(j\omega)|} = \frac{1}{12}$$

Result:

$$\gamma = 180\epsilon$$
$$K_g = \frac{1}{12}$$

EXAMPLE 2

The open loop transfer function of a unity feedback system is given by $G(s) = \dfrac{(1+.2s)(1+.25s)}{s^2(1+.05s)(1+.01s)}$. Sketch the polar plot and determine the phase margin.

Solution: Given that $$G(s) = \frac{(1+.2s)(1+.25s)}{s^2(1+.05s)(1+.01s)}$$

Put, $s = j\omega$ in $G(s)$

$$\therefore \quad G(j\omega) = \frac{(1+.2j\omega)(1+.25j\omega)}{(j\omega)^2(1+.05j\omega)(1+.01j\omega)} \quad (1)$$

From equation (1)

$$|G(j\omega)| = \frac{\sqrt{1+(.2\omega)^2}\,\sqrt{1+(.25\omega)^2}}{\omega^2\sqrt{1+(.05\omega)^2}\,\sqrt{(1+(.01\omega)^2}}$$

$$\angle G(j\omega) = \tan^{-1} .2\omega + \tan^{-1} .25\omega - 180\epsilon - \tan^{-1} .05\omega - \tan^{-1} .01\omega$$

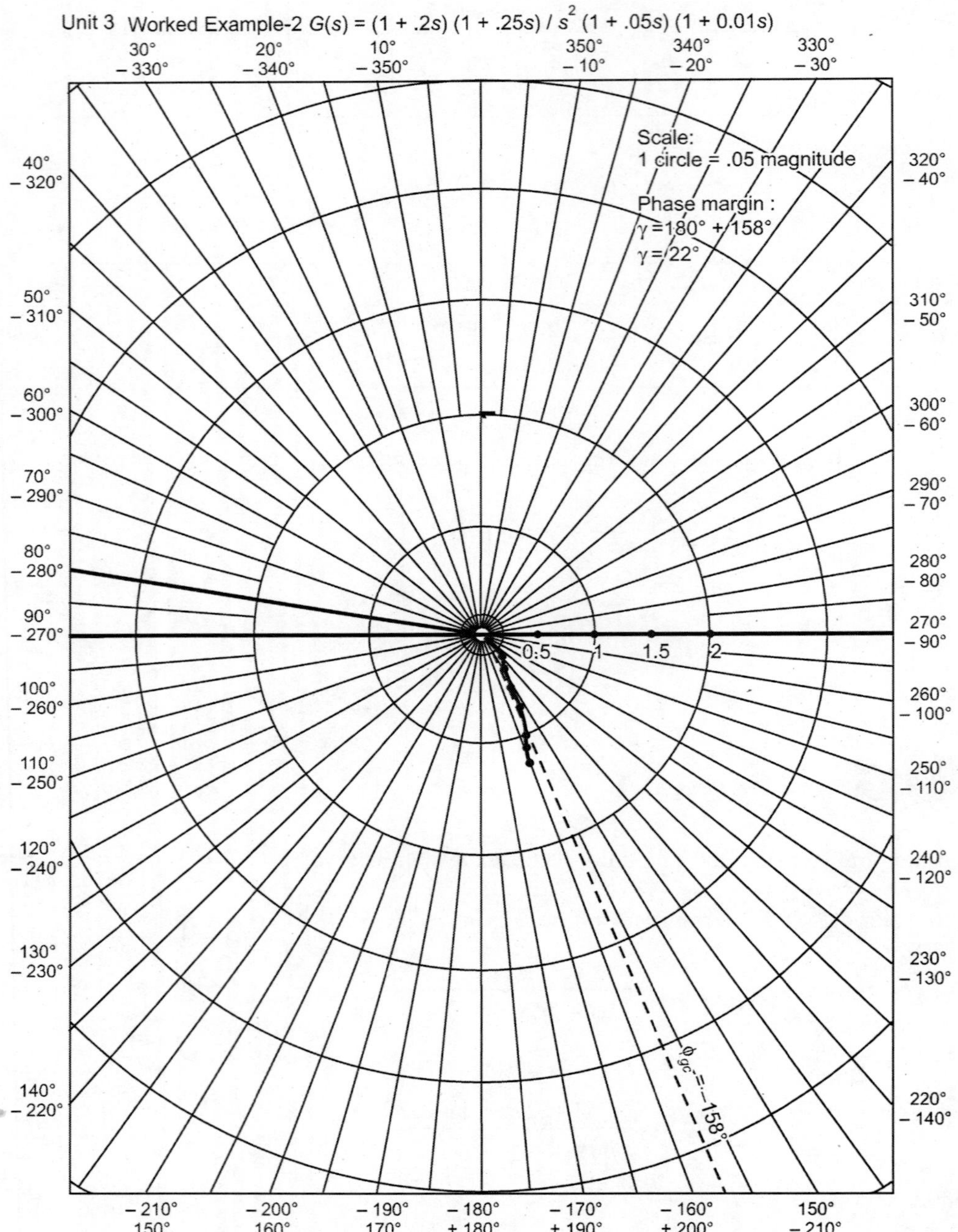

Worked example: Polar plot of $G(j\omega) = 1/j\omega\,(1 + j2\omega)$ (using polar coordinates).

The magnitude and phase angle of $G(j\omega)$ are calculated for various frequencies and listed in Table 1. The polar coordinates listed in Table 1 are converted to rectangular coordinates and tabulated in Table 2. Polar plot using polar coordinates is sketched on a polar sheet and polar plot using rectangular coordinates is sketched on a ordinary graph sheet.

Table 1: Magnitude and phase of $G(j\omega)$

ω, rad/sec	0.9	0.95	1	1.2	1.4	1.7	2
$\lvert G(j\omega)\rvert$	1.27	1.14	1	.74	.55	.39	.29
$\angle G(j\omega)$	−160ε	−159ε	−158ε	−154ε	−150ε	−144ε	−138.4ε

Convert the above values given in table 1 to real and imaginary values i.e. convert the magnitude and phase to real and imaginary values and draw it on the ordinary graph sheet (refer example problems).

Result: Phase margin, γ = 22ε.

EXAMPLE 3

Determine the gain margin of the system given below.

$$G(s) = 20/s(s + 1)\,(s^2 + 2s + 2)$$

Solution: Given that,

$$G(s) = \frac{20}{s(s+1)\,(s^2+2s+2)}$$

Let, $s = j\omega$

$$\therefore \quad G(j\omega) = \frac{20}{j\omega\,(j\omega+1)\,[(j\omega)^2 + 2(j\omega) + 2]}$$

$$= \frac{20}{j^4\omega^4 + 2j^3\omega^3 + 2j^2\omega^2 + j^3\omega^3 + 2j^2\omega^2 + 2j\omega}$$

$$= \frac{20}{\omega^4 - j2\omega^3 - 2\omega^2 - j\omega^3 - 2\omega^2 + 2j\omega}$$

$$= \frac{20}{\omega^4 - 4\omega^2 - 3j\omega^3 + 2j\omega}$$

$$= \frac{20}{\omega^2(\omega^2 - 4) + j\omega(2 - 3\omega^2)}$$

Let equate imaginary part to zero.

$$\omega_{pc}\,(2 - 3\,\omega_{pc}^2) = 0$$

Since $\omega_{pc} \neq 0$

$$2 - 3\,\omega_{pc}^2 = 0$$

$$-3\,\omega_{pc}^2 = -2$$

$$3\,\omega_{pc}^2 = 2$$

$$\omega_{pc}^2 = 2/3$$

$$\omega_{pc} = \sqrt{2/3}$$

$$\boxed{\omega_{pc} = .9}\ \text{rad/sec}$$

The gain margin, K_g is defined as the reciprocal of the magnitude of $G(j\omega)$ at phase cross over frequency.

Gain margin, $$K_g = \frac{1}{|G(j\omega)|_{\omega = \omega_{pc}}}$$

$$K_g = \frac{1}{1/|\omega^2(\omega^2 - 4)|_{\omega = \omega_{pc}}}$$

$$K_g = \omega^2(\omega^2 - 4) = 2.58$$

Gain margin in *db* $$= 20 \log Kg = 20 \log 2.5$$

$$K_g = 8\ db$$

Result: $K_g = 8$ *db*.

EXAMPLE 4

The open loop transfer function of a system in $G(s) = K/s(1 + .5s)$ (i) Determine the value of K os that gain margin is 6 *db*.

Solution: Given, that $$G(s) = K/s(1 + .1s)\,(1 + s)$$

Let, $s = j\omega$

$\therefore$ $$G(j\omega) = K/j\omega(1 + .5j\omega)\,(1 + j\omega)$$

$$= \frac{K}{j\omega\,(1 + j\omega + .5\,j\omega + .5\,j^2\omega^2)}$$

$$= \frac{K}{j\omega(1 + 1.5\,j\omega - .5\omega^2)}$$

$$= \frac{K}{j\omega(1 - .5\omega^2 + j\omega\ 1.5)} = \frac{K}{j\omega - .5\,j\omega^3 - 1.5\omega^2}$$

$$= \frac{K}{-1.5\omega^2 + j\omega - .5\,j\omega^3}$$

$$= \frac{K}{-1.5\omega^2 + j\omega\,(1 - .5\omega^2)}$$

At phase crossover frequency ω_{pc}, the $G(j\omega)$ is real and so equate the imaginary part to zero to solve for ω_{pc}.

At $\omega = \omega_{pc}$, $\omega_{pc}(1 - 5\,\omega_{pc}^2) = 0$

$$\therefore \quad 1 - .5\,\omega_{pc}^2 = 0$$

$$-.5\,\omega_{pc}^2 = -1$$

$$\omega_{pc}^2 = 1/.5$$

$$\omega_{pc}^2 = 2 \Rightarrow \omega_{pc} = \sqrt{2}$$

$$\boxed{\omega_{pc} = 1.414} \text{ rad/sec}$$

$$|G(j\omega)|_{\omega = \omega_{pc}} = \left|\frac{K}{-1.5\omega^2}\right|_{\omega = \omega_{pc}} = \frac{K}{+1.5 \times (1.414)^2}$$

$$|G(j\omega)| = \frac{K}{3}$$

$$|G(j\omega)| = K\,(.33)$$

Given that $K_g = 6\ db$

$$\therefore \quad 20 \log K_g = 6$$

$$\log K_g = 6/20$$

$$K_g = 10^{6/20} = 1.9953$$

$$K_g = \frac{1}{|G(j\omega)|}$$

$$\therefore \quad 1.9953 = \frac{1}{.33K}$$

$$\therefore \quad K = \frac{1}{(.33)(1.9953)} = \frac{1}{.658}$$

$$\boxed{K = 1.51}$$

Result: The value of K when gain margin is 6 *db* is 1.51.

QUESTIONS AND ANSWERS

Q4.1 What is frequency response?

Ans. It is a steady state output of the system, when the input is a sinusoidal signal.

CHAPTER 4

Q4.2 What are the advantages of frequency response analysis?

Ans.
1. Estimation of absolute and relative stability fo the closed loop system can be done.
2. The transfer function of complicated functions can be determined.
3. Corrective measure for parameter variation can be easily carried.

Q4.3 What are frequency domain specifications?

Ans.
1. Resonant peak
2. Resonant frequency
3. Bandwidth
4. Cut-off rate
5. Gain margin
6. Phase margin

Q4.4 What is bode plot?

Ans. The bode plot is a frequency response plot of the transfer function of a system. It consist of magnitude and phase plots. The magnitude plot is a graph between magnitude and frequency and phase plot is a graph between phase angle and frequency.

Q4.5 Define corner frequency?

Ans. The magnitude plot can be approximated by asymptotic straight lines. The frequencies corresponding to the meeting point of asymptotes are called corner frequency.

Q4.6 What are the advantages of bode plot?

Ans.
1. Frequency domain specifications can be easily determined.
2. It is used to analyse both open loop and closed loop system.

Q4.7 What is polar plot?

Ans. The polar plot is a sinusoidal transformation $G(jw)$ is a plot of the magnitude of $G(jw)$ on polar or rectangular coordinates as w is varied from zero to infinity.

Q4.8 What is minimum phase system.

Ans. The minimum phase systems are systems with minimum phase transfer functions. in minimum phase transfer functions, all poles and zeros will be on the left half of s-plane.

Q4.9 What is all-pass systems?

Ans. The all pass systems are systems with all pass transfer functions. In all pass transfer functions, the magnitude is unity at all frequencies and the transfer function will have anti-symmetric pole zero pattern.

Q4.10 What is non-minimum phase transfer function?

Ans. A transfer function which has one (or) more zeros is the right half of s-plane is known as non minimum phase transfer function.

Q4.11 In minimum phase system, how the start and end of polar plot are identified?

Ans. For minimum phase transfer functions, with only poles, the type number of the system determines the quadrant in which the polar plot starts, and the order of a system determines the quadrant in which the polar plot ends.

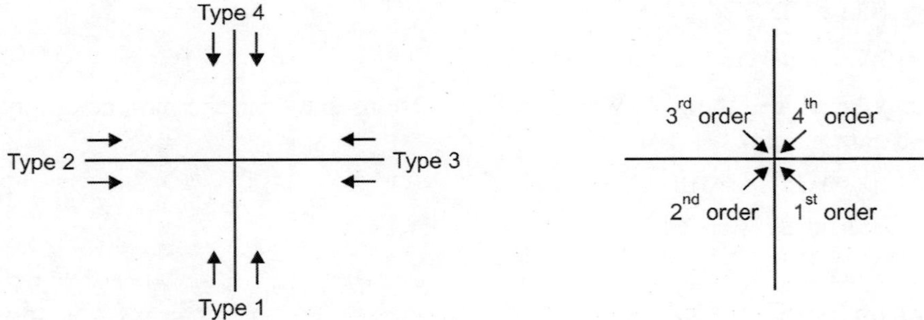

Q4.12 What are *M* and *N* circles?

Ans: The magnitude, *M* of closed loop transfer function with unity feedback will be in the form of circle in complex form for each constant value of *M*. The family of these circles are called *M* circles.

Let $N = \tan\alpha$ where a is the phase of closed loop transfer function with unity feedback. For each constant value of *N* a circle can be drawn in the complex plane. The family of these circles are called *N* circles.

Q4.13 Define BIBO stability?

Ans. A linear relaxed system is said to have BIBO stability if every bounded input results in a bounded output.

Q4.14 What is the requirement of BIBO stability?

Ans. $\int_0^\infty m(\tau)d\tau < \infty$, where $m(\tau)$ is the impulse response of the system.

Q4.15 What is impulse response?

Ans. It is the inverse laplace transform of the system transfer function.

Q4.16 What is characteristics equation?

Ans. The denominator polynomial of $C(s)/R(s)$ is the characteristic equation of the system.

Q4.17 What is Nyquist stability criterion?

Ans. If $G(s)\,H(s)$ contour in the $G(s)\,H(s)$ plane corresponding to Nyquist contour in *s*-plane encircles the point $-1 + j_0$ in the anti-clockwise direction as many times as the number of right half *s*-plane poles of $G(s)\,H(s)$. Then the system is stable.

EXERCISES

Q.1 Sketch the bode plot of the following open loop transfer functions and from the plot determine the phase margin and gain margin.

(a) $G(s) = s^2(s + 10)/(s + 5)^2\ (s + 0.1)$

(b) $G(s) = 10(1 + s)e^{-0.1s}/s\ (1 + 0.2s)$

(c) $G(s) = 30/(1 + 0.1s)/s\ (1 + .01s)\ (1 + s)$

Q.2 Sketch the polar plot of the following transfer functions and from the plot, determine the phase margin and gain margin.

(a) $G(s) = 10(s + 1)/(s + 10)^2$

(b) $G(s) = 200(s + 2)/s(s^2 + 10s + 100)$

(c) $G(s) = 1/s\ (s + 4)\ (s + 8)$

Q.3 The open loop transfer function of a system is given by $G(s) = K/s(s^2 + s + 4)$ using polar plot, determine the value of K, so that phase margin is 50є. What is the corresponding value of gain margin?

Q.4 Open loop transfer functions of certain unity feedback systems are given below. In each case determine the range of values of K for which the system is stable.

(a) $G(s) = K(s + 13)/s(s + 3)\ (s + 7)$

(b) $G(s) = K(s - 1)/s(s + 2)$

(c) $G(s) = K\ (s + 1)/s(s - 1)\ (s + 6)$

5

Compensators Design

5.1 INTRODUCTION TO DESIGN OF COMPENSATORS

The control systems are designed to satisfy the required task by applying different techniques. The basic requirements used for designing control system are generally specified as performance specifications and these are directly related to accuracy, stability, speed and less power consumption etc., for example,. In time domain, the performance specifications are rise time, maximum overshoot, settling time and damping ratio. In frequency domain, the performance specifications are phase margin, gain margin, resonant peak, cut of rate and bandwidth. The steady state requirement are given in terms of error constants.

To be more precise, when the time domain specifications are given, root locus technique is employed in designing the system. It frequency domain specifications are given, frequency response plots like bode plots are used for designing the system.

The adjustment of gain to meet the desired specification is the first step and more over the adjustment of gain alone will not be sufficient to meet the given specifications because practically, increasing the gain may lead in poor stability. In such cases, it is necessary to introduce additional devices in the system to meet the desired specifications. Such type of additional device is called compensator. The main function of compensators is to introduce pole and/or zero in open loop transfer function to modify the performance of the system.

The compensator may be electrical, mechanical, hydraulic or other type of device (or) network. In many control system, usually electric network (or) electronic device is serving as compensator. The different types of compensators are [Electrical (or) electronic].

* Lag compensator * Lead compensator * Lag-lead compensator.

The compensation schemes used for feedback control system is either series compensation (or) parallel compensation (feedback). In series compensation, the compensator with transfer function, $G_c(s)$ is placed in series with plant. In feedback compensation, the signal from some element is feedback to the input and a compensator with transfer function $G_c(s)$ is placed in the resulting inner feedback path. The series and parallel compensation schemes are shown in Fig. (5.1) and (5.2).

In control system, compensation is required in the following situations.

- When the system is absolutely unstable, then compensation is required to stabilize the system and also to meet the desired performance and specification.

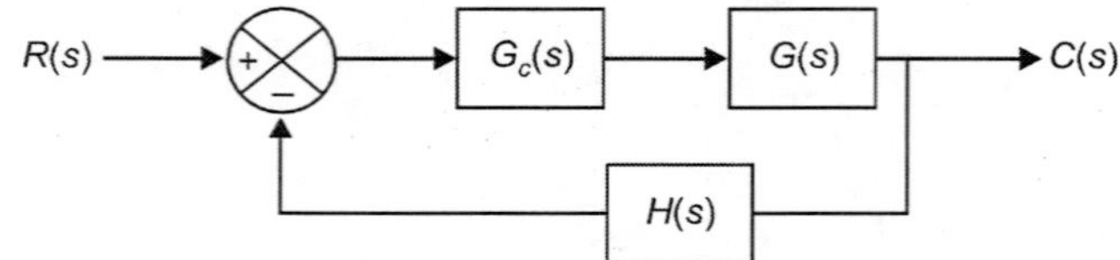

Fig. 5.1: Series compensation.

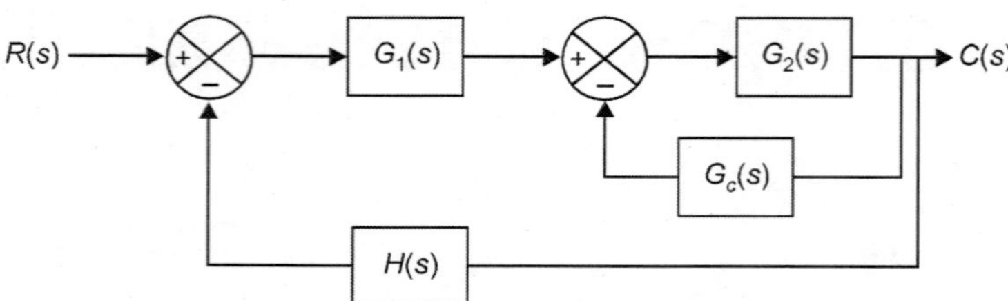

Fig. 5.2: Feedback/parallel compensation.

5.2 TIME RESPONSE APPROACH TO CONTROL SYSTEM DESIGN (or) (Root Locus Approach)

In design using root locus, the desired behaviour is specified in terms of transient response specifications and steady state error requirement. The steady state error is usually specified in terms of error constants for standard inputs, while the transient response requirement is specified in terms of peakovershoot, settling time, rise time, etc., for a step input. The transient response specifications can be translated into desired locations for a pair of dominant closed loop poles.

In order to meet the desired specifications the root loci are reshaped so that they pass through the pints where the dominant closed loop poles are located. The root loci are reshaped by introducing a compensator. The compensator will add a pole and/or a zero in the open loop transfer function of the system. The addition of a pole to the open loop transfer function has the effect of pulling the root locus to the right, which reduce the relative stability of the system and increase the settling time. The addition of a zero to the open-loop transfer function has the effect to pulling the root locus to the left which make the system more stable and reduce the settling time.

When a system is either instable (or) stable but has undesired transient response characteristics a lead compensator can be employed to modify the root locus. The transfer function of lead compensator will have a zero (compensating zero) and a pole (compensating pole).

The compensator zero can be placed on the real axis by trail-and-error to satisfy transient response specifications. The introduction of zero will amplify high frequency noise which is eliminated by the compensating pole. The compensating pole is located on real axis such that it makes negligible effect on the root locus in the region when the two dominant closed loop poles are located.

If the pole is located for away from zero then it will not be effective in reducing the noise. If the pole is too close to zero then it will not allow the zero to do its job. In order to avoid this conflict, the pole is located at 3 to 10 times the value of zero location.

The lag compensator is employed when table system has satisfactory transient response characteristics but unsatisfactory steady state characteristics, i.e., error requirement. The transfer function of lag compensator will have a zero (compensating zero) and a pole (compensating pole).

In order to preserve the transient response characteristics the compensating pole and zero should have negligible effect on shape of root locus. This is achieved by placing the compensating pole and zero very closed to each other. If the pole and zero are located close to the origin then the error constant will increase which will reduce the steady state error. The lag-lead compensator is employed when both the transient and steady-state characteristics are not satisfactory. The lead compensation will improve the transient response and lag compensation will reduce the steady state error.

The advantage in design using root locus technique is the at the information about closed loop transient response and frequency response are directly obtained from pole-zero of the system in the s-plane.

5.3 FREQUENCY RESPONSE APPROACH TO CONTROL SYSTEM DESIGN (BODE PLOT)

The main purpose of frequency domain design is to reshape (or) redesign the frequency response so that the desired specifications are obtained. The frequency domain design can be carried out using Nyquist plot, Bode plot (or) Nichols chart. But Bode plot are popularly used for design because they are easier to draw and modify.

The desired performance specifications are given in terms of frequency domain specifications and steady state are requirements.

The stability requirement specified in terms of phase margin and resonant peak. The transient response requirements are specified interms of gain cross over frequency, bandwidth and resonant frequency. The error requirement is specified in terms of static error constants (K_p, K_v and K_a).

Note: In case the transient response specifications are given in time domain, we can translate them into frequency domain specifications using the following formula.

Phase margin, $$\gamma = \tan^{-1}\frac{2\zeta}{\left[\sqrt{4\zeta^4+1}-2\,\zeta^2\right]^{1/2}} \approx 100\,\zeta$$

Gain cross over frequency, $$\omega_{gc} = \omega_n\left[\sqrt{4\zeta^4+1}-2\zeta^2\right]^{1/2}$$

Resonant peak, $$M_r = 1/\left[2\zeta\sqrt{1-\zeta^2}\right]$$

Resonant frequency, $$\omega_r = \omega_n\sqrt{1-2\zeta^2}$$

Bandwidth, $$\omega_b = \omega_n\left[(1-2\zeta^2)+\sqrt{4\zeta^4-4\zeta^2+2}\right]^{1/2}$$

The low frequency region of bode plot provides information regarding the steady state performance and high frequency region provides information regarding the transient state performance. The medium frequency (or mid-frequency) range provides information regarding relative stability. Therefore, the low frequency region of the bode plot is reshaped by lag compensation to improve steady state performance. The high frequency region of the bode plot is reshaped by lead compensation to improve transient-state performance.

When the system requires improvement in both steady-state and transient state, lag-lead compensation can be employed to alter both the low and high frequency regions of bode plot. The primary function of lag compensator is to provide attenuation in the high frequency region to achieve sufficient phase margin.

The advantage in frequency-domain design is that the effects of disturbances, sensor noise and plant uncertainties are relatively easy to visualize and assess in frequency domain. Another advantage is the ease with which experimental information can be used for design purposes.

A disadvantage of frequency-response design is that it gives us information on closed-loop system's transient response indirectly, while the root locus design gives this information directly.

5.4 LAG COMPENSATOR

A compensator having the characteristics of a lag network is called a lag compensator. When a sinusoidal signal is applied to a lag network, then it will have steady state output which will have a phase lag with respect to input.

Lag compensation results in a large improvement in steady state performance but results slow in response due to reduced bandwidth. The attenuation due to the lag compensator will shift the gain crossover frequency to lower frequency point where the phase margin is acceptable. Thus, the lag compensator will reduce the bandwidth of the system and will result in slower transient response.

Lag compensator is essentially a low pass filter and so high frequency noise signals are attenuated. If the pole introduced by the compensator is not cancelled by a zero in the system, then lag compensator increases the order of the system by one.

s-Plane representation of Lag Compensator

The lag compensator has a pole at $s = -1/\beta T$ and a zero at $s = 1/T$. The pole-zero of lag compensator is shown in Fig. (5.3).

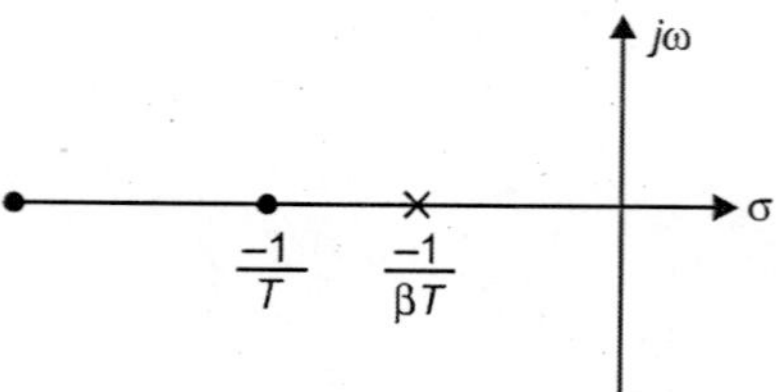

Fig. 5.3: Pole-zero plot of lag compensator.

Here, $\beta > 1$, so the zero is located to the left of the pole on the negative real axis. The general form of lag compensator transfer function is given by Eq. (5.1).

Transfer function of lag compensator,

$$G_c(s) = \frac{s + z_c}{s + p_c} = \frac{s + \dfrac{1}{T}}{s + \dfrac{1}{\beta T}} \qquad (5.1)$$

where, $T > 0$ and $\beta > 1$

The zero of lag compensator,

$$z_c = 1/T \tag{5.2}$$

The pole of lag compensator,

$$p_c = \frac{1}{\beta T} = \frac{1}{\beta} z_c \tag{5.3}$$

From Eq. (5.4.2) we get,

$$T = \frac{1}{z_c} \tag{5.4}$$

From Eq. (5.4.3) we get, $$\beta = \frac{z_c}{p_c} \tag{5.5}$$

Lag Compensator using Electrical Network

The lag compensator can be realised by the *R-C* network shown in Fig. (5.4).

Let, $E_i(s)$ = Input voltage

$E_o(s)$ = Output voltage

In the network shown in Fig. (5.4), the input voltage is applied to the series combination of R_1, R_2 and C. The output voltage is obtained across series combination of R_2 and C.

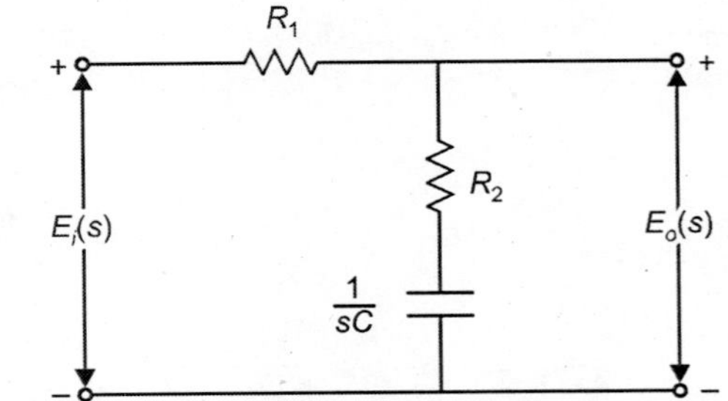

Fig. 5.4: Electrical lag compensator.

By voltage division rule,

$$E_o(s) = E_i(s)\frac{(R_2 + 1/sC)}{(R_1 + R_2 + 1/sC)}$$

$$= E_i(s)\frac{(sCR_2 + 1)/sC}{[sC(R_1 + R_2) + 1]/sC}$$

$$= E_i(s)\frac{[(sCR_2 + 1)]}{[sC(R_1 + R_2) + 1]}$$

The transfer function of the electrical network is the ratio of output voltage to input voltage.

CHAPTER 5

Transfer function of electrical network

$$\frac{E_o(s)}{E_i(s)} = \frac{CR_2(s+1/CR_2)}{C(R_1+R_2)\left[s+\dfrac{1}{C(R_1+R_2)}\right]}$$

$$= \frac{\left(s+\dfrac{1}{R_2C}\right)}{\left(\dfrac{R_1+R_2}{R_2}\right)\left[s+\dfrac{1}{\left(\dfrac{R_1+R_2}{R_2}\right)R_2C}\right]} \qquad (5.6)$$

But the transfer function of lag compensator is given by

$$G_c(s) = \frac{(s+1/T)}{(s+1/\beta T)} \qquad (5.7)$$

On comparing Eq. (5.6) and (5.7) we get

$$\frac{E_o(s)}{E_i(s)} = \frac{1}{\beta}\frac{(s+1/T)}{(s+1/\beta T)};$$

where, $T = R_2C$ and $\beta = (R_1 + R_2)/R_2$ (5.8)

The transfer function of *RC* network as given by Eq. (5.8) is similar to the general form with an attenuation of $(1/\beta)$ is not required then an amplified with gain β can be connected in cascade with *RC* network to nullify the attenuation.

Frequency Response of Lag Compensator

Consider the general form of lag compensator,

$$G_c(s) = \frac{(s+1/T)}{(s+1/\beta T)} = \frac{(sT+1)/T}{(s\beta T+1)/\beta T} = \beta\frac{(1+sT)}{(1+s\beta T)} \qquad (5.9)$$

The sinusoidal transfer function of lag compensator is obtained by letting $s = j\omega$ in Eq. (5.9).

$$\therefore \quad G_c(j\omega) = \frac{\beta(1+j\omega T)}{(1+j\omega\beta T)} \qquad (5.10)$$

when $\omega = 0,\ G_c(j\omega) = \beta$ (5.11)

From Eq. (5.11) we can say that the lag compensator provides a dc gain of β ($\beta > 1$). If the dc gain of the compensator is not desirable then it can be eliminated by a suitable attenuation. Let us assume that the gain β is eliminated by a suitable attenuation network. Now, $G_c(j\omega)$ is given by

$$G_c(j\omega) = \frac{1+j\omega T}{1+j\omega\beta T} = \frac{\sqrt{1+(\omega T)^2}\ \underline{|\tan^{-1}\omega T}}{\sqrt{1+(\omega\beta T)^2}\ \underline{|\tan^{-1}\omega\beta T}} \qquad (5.12)$$

The sinusoidal transfer function shown in Eq. (5.12) has two corner frequencies and they are denoted as ω_{c1} and ω_{c2}.

Here, $\omega_{c1} = 1/\beta T$ and $\omega_{c2} = 1/T$

Since, $\beta T > T,\ \omega_{c1} \angle \omega_{c2}$

Let $$A = |G_c(j\omega)| \text{ in } db = 20\log\frac{\sqrt{1+(\omega T)^2}}{\sqrt{1+(\omega\beta T)^2}} \quad (513)$$

At very low frequencies ie upto ω_{c1}, $\omega T << 1$ and $\omega\beta T << 1$

∴ $$A \approx 20\log 1 = 0$$

In the frequency range from ω_{c1} to ω_{c2}, $\omega T << 1$ and $\omega\beta T >> 1$

∴ $$A \approx 20\log\frac{1}{\sqrt{(\omega\beta T)^2}} = 20\log\frac{1}{\omega\beta T}$$

At very high frequencies ie after ω_{c2}, $\omega T >> 1$ and $\omega\beta T >> 1$

∴ $$A \approx 20\log\frac{\sqrt{(\omega T)^2}}{\sqrt{(\omega\beta T)^2}} = 20\log 1/\beta$$

The approximate magnitude plot of lag compensator is shown in Fig. (5.5).

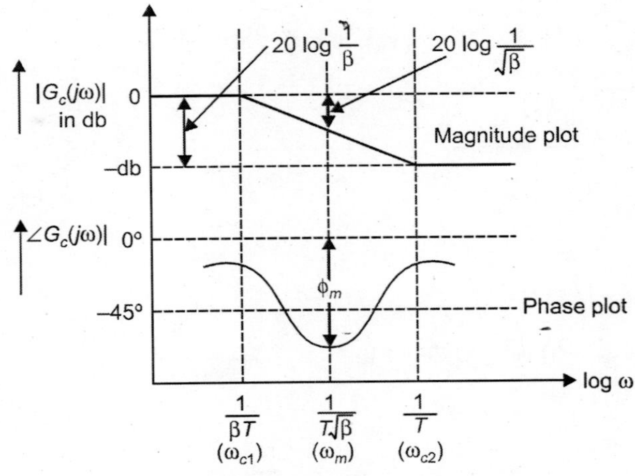

Fig. 5.5: Bode plot of lag compensator.

The magnitude plot of bode plot of $G_c(j\omega)$ is a straight line through $0db$ upto ω_{c1}, then it has a slope of $-20db$/dec upto ω_{c2} and after ω_{c2} it is a straight line with a constant gain of 20 log $(1/\beta)$.

Let, $$\phi = \angle G_c(j\omega)$$

∴ $$\phi = \tan^{-1}\omega T - \tan^{-1}\omega\beta T$$

CHAPTER 5

As $\omega \to 0$, $\phi \to 0$

As $\omega \to \infty$, $\phi \to 0$

As ω is varied from 0 to ∞, the phase angle decreases from 0 to a negative maximum value of ϕ_m at $\omega = \omega_n$, then increases from this maximum value to 0. The phase plot of lag compensator is shown in Fig. (5.5). It can be shown that the frequency at which maximum phase lag occurs in the geometric mean of the two corner frequencies.

Frequency of maximum phase lag,

$$\omega_m = \sqrt{\omega_{c1}, \omega_{c2}} = \sqrt{\frac{1}{\beta T} \cdot \frac{1}{T}}$$

$$= \frac{1}{T\sqrt{\beta}}$$

From the bode plot of lag compensator, we observe that lag compensator has a dc gain of unity while it offers a high frequency gain of $(1/\beta)$ [In decibels, it is 20 log $(1/\beta)$]. It means that the high frequency noise is attenuated in passing through the network and so the signal to noise ratio is improved. A typical choise of $\beta = 10$.

Determination of ω_m and ϕ_m

The frequency ω_m can be determined by differentiating ϕ with respect to ω and equating $d\phi/d\omega$ to zero as shown below.

From Eq. (5.12) we get,

Phase of $G_c(j\omega)$, $$\phi = \angle G_c(j\omega) = \tan^{-1}\omega T - \tan^{-1}\omega\beta T \tag{5.14}$$

On differentiating the Eq. (5.14) we get,

$$\frac{d\phi}{d\omega} = \frac{1}{1+(\omega T)^2}T - \frac{1}{1+(\omega\beta T)^2}\beta T \tag{5.15}$$

when $\omega = \omega_m$, $d\phi/d\omega = 0$

$$\left[\textbf{Note : } \frac{d}{d\theta}(\tan^{-1}\theta) = \frac{1}{1+\theta^2}\right]$$

Hence, replace ω by ω_n in Eq. (5.15) and equate to zero

$$\therefore \quad \frac{1}{1+(\omega_m T)^2}T - \frac{1}{1+(\omega_m\beta T)^2}\beta T = 0$$

$$\frac{T}{1+(\omega_m T)^2} = \frac{\beta T}{1+(\omega_m\beta T)^2}$$

On cross multiplication we get,

$$(1 + \omega_m\beta T)^2 = \beta[1 + (\omega_m T)^2]$$

$$(\omega_m\beta T)^2 - \beta(\omega_m T)^2 = \beta - 1$$

$$\beta(\omega_m T)^2 (\beta - 1) = (\beta - 1)$$

$$\omega_m^2 = \frac{1}{T^2\beta}$$

$$\therefore \quad \omega_m = \frac{1}{T\sqrt{\beta}}$$

Frequency corresponding to maximum phase lag,

$$\omega_m = \frac{1}{T\sqrt{\beta}} \tag{5.16}$$

The maximum phase angle of ϕ_m can be calculated from the knowledge of β and viceversa. The relations between ϕ_m and β are derived below.

From Eq. (5.14) we get, $G_c(j\omega) = \phi = \tan^{-1}\omega T - \tan^{-1}\omega\beta T$ on taking tan on either side we get,

$$\tan\phi = \tan[\tan^{-1}\omega T - \tan^{-1}\omega\beta T] \quad \left[\textbf{Note}: \tan(A - B) = \frac{\tan A - \tan B}{1 + \tan A \tan B}\right]$$

$$= \frac{\tan(\tan^{-1}\omega T) - \tan(\tan^{-1}\omega\beta T)}{1 + \tan(\tan^{-1}\omega T)\tan(\tan^{-1}\omega\beta T)}$$

$$= \frac{\omega T - \omega\beta T}{1 + \omega^2 T^2 \cdot \beta} = \frac{\omega T(1-\beta)}{1 + \beta(\omega T)^2}$$

At $\omega = \omega_m$, $\phi \to \phi_m$,

$$\therefore \quad \tan\phi_m = \frac{\omega_m T(1-\beta)}{1 + \beta(\omega_m T)^2}$$

From Eq. (5.16) we get, $\omega_m = \dfrac{1}{T\sqrt{\beta}}$

$$\therefore \quad \tan\phi_m = \frac{\frac{1}{T\sqrt{\beta}}T(1-\beta)}{1 + \beta\frac{1}{T^2\beta}T^2} = \frac{(1-\beta)/\sqrt{\beta}}{1+1} = \frac{(1-\beta)}{2\sqrt{\beta}} \tag{5.17}$$

CHAPTER 5

To find the value of β from ϕ_m

From the Eq. (5.17), it is evident that $(1 - \beta)$ and $2\sqrt{\beta}$ are the two sides of right angled triangle. Hence construct a right angle triangle as shown in Fig. (5.6).

With reference to Fig. (5.6), $\sin\phi_m = (1 - \beta)/(1 + \beta)$

$$\sin\phi_m(1 + \beta) = (1 - \beta)$$

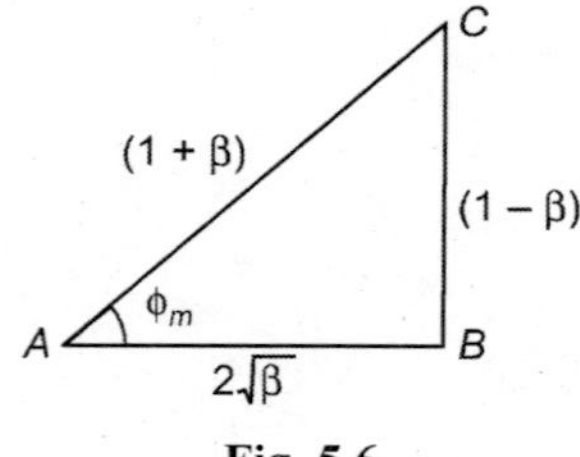

Fig. 5.6

$$\sin \phi_m + \beta \sin \phi_m = 1 - \beta$$
$$\beta \sin \phi_m + \beta = 1 - \sin \phi_m$$
$$\beta (\sin \phi_m + 1) = 1 - \sin \phi_m$$

$$\therefore \qquad \beta = \frac{1 - \sin \phi_m}{1 + \sin \phi_m} \qquad (5.18)$$

Note: From Fig. (5.6),

$$AC = \sqrt{BC^2 + AB^2} = \sqrt{(1-\beta)^2 + (2\sqrt{\beta})^2}$$
$$= \sqrt{1 + \beta^2 - 2\beta + 4\beta} = \sqrt{1 + \beta^2 + 2\beta}$$
$$= \sqrt{(1+\beta)^2} = (1 + \beta)$$

Procedure for the Design of Lag Compensator using Bode Plot

The following steps may be followed to design a lag compensator using bode plot and to be connected in series with transfer function of uncompensated system, $G(s)$.

Step 1: Choose the value of K in uncompensated system to meet the steady state error requirement.

Step 2: Sketch the bode plot of uncompensated system.

Step 3: Determine the phase margin of the uncompensated system from the bode plot. If the phase margin does not satisfy the requirement then lag compensation is required.

Step 4: Choose a suitable value for the phase margin of the compensated system.

Let γ_d = Desired phase margin of compensated system

Now, $\gamma_n = \gamma_d + \in$

where $\in$ = Additional phase lag to compensated for shift in gain crossover frequency chose an initial value of $\gamma = +5\epsilon$

Note: This $\in$ value can be assumed in the last step also.

Step 5: Determine the new gain crossover frequency, ω_{gc} the new ω_{gcn} is the frequency corresponding to a phase margin of γ_n on the bode plot of uncompensated system.

Let, ϕ_{gcn} = phase of $G(j\omega)$ at new gain crossover frequency, ω_{gcn}

Now, $\gamma_n = 180 + \phi_{gcn}$ (or) $\gamma_{gcn} = Y_n - 180\epsilon$

The new gain crossover frequency, ω_{gcn} is given by the frequency at which the phase of $G(j\omega)$ is ϕ_{gcn}.

Step 6: Determine the parameter, β of the compensator. The value of β is given by the magnitude of $G(j\omega)$ at new gain crossover frequency, ω_{gcn}. Find the *db* gain (Ag_{cn}) at new gain crossover frequency, ω_{gcn}

Now,
$$A_{gcn} = 20 \log \text{(or)} \frac{A_{gcn}}{20} = \log\beta$$

∴
$$\beta = 10\ A_{gcn}/20$$

Step 7: Determine the transfer function of log compensator. Place the zero of the compensator arbitrarily at $\frac{1}{10}$ the of new gain crossover frequency, ω_{gen}.

Zero of the log compensator,

$$z_c = \frac{1}{T} = \frac{\omega_{gcn}}{10}$$

Now
$$T = \frac{10}{\omega_{gcn}}$$

Pole of the log compensator,

$$P_c = \frac{1}{\beta T}$$

Transfer function of a long compensator

$$G_c(s) = \frac{s + \frac{1}{T}}{s + \frac{1}{\beta T}} = \beta\left(\frac{1 + sT}{1 + s\beta T}\right)$$

Step 8: Determine the open loop transfer function of compensated system. The log compensator is connected in series with the plant as shown in Fig. (5.7)

When the log compensator is inserted in series with the plant, the open loop gain of the system is amplified by the factor β (∴β>1). If the gain produced is not required then attenvator with gain $\frac{1}{\beta}$ can be introduced in series with the log compensator to nullify the gain produced by log compensator.

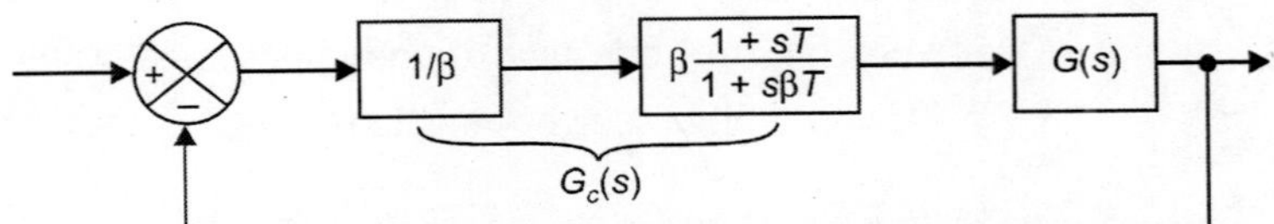

Fig. 5.7: Block diagram of lag compensated system.

The open loop transfer function of the compensated system

$$G_o(s) = \frac{1}{\beta}\cdot G_c(s)G(s)$$

$$= \frac{1}{\beta}\cdot\beta\frac{1+sT}{1+s\beta T}G(s)$$

$$= \frac{1+sT}{1+s\beta T}.G(s)$$

Step 9: Determine the actual phase margin of compensator system. Calculate the actual phase angle of the compensated system using the compensated transfer function at new gain crossover frequency, ω_{gcn}

Let, $\phi_{gco} = \phi_{gco}$ phase of $G_o(j\omega)$ at $\omega = \omega_{gcn}$

Actual phase margin of the compensated system,

$$\gamma_o = 180 + \phi_{gco}$$

If the actual phase margin satisfies the given specification then the design is accepted. Otherwise repeat the procedure from step 4 to 10 by taking $\in$ as 5ϵ more than previous design.

Note: The value of $\in$ can be of $\pm 5\epsilon$ and this can be considered either in step 4(or) in last step.

Procedure for Design of Lag Compensator using Root Locus

The following steps may be followed to design a lag compensator using root locus and be connected in series with the transfer function of uncompensated system.

Step 1: Draw the root locus of uncompensated system

Step 2: Determine the dominant pole, s_d. Draw a straight line through the origin with an angle $\cos^{-1}\zeta$ with respect to negative real axis. The intersection point of the straight line with root locus gives the dominant pole, s_d.

Step 3: Determine the open loop gain of the uncompensated system at $s = s_d$. Let this gain be K. The open loop gain K at $s = s_d$ on root locus is given by

$$K = \frac{\text{Product of vector lengths from } s_d \text{ to open loop poles}}{\text{Product of vector lengths from } s_d \text{ to open loop zeros}}$$

(Vector length should be measured with scale in cm).

Step 4: Calculate the parameter, β of the compensator

Let K_{vu} = Velocity error constant of uncompensated system.

K_{vd} = Desired velocity error constant

$$K_{vu} = \underset{s\to 0}{Lt}\ sG(s)$$

Let A be the factor by which the velocity error constant of the system has to be increased.

Where, $$A = \frac{K_{vd}}{K_{vu}}$$

Choose β such that it is 10 to 20% greater than A.

$\therefore$ $$\beta = (1.1 \text{ to } 1.2) \text{Ч} A$$

Step 5: Determine the transfer function of lag compensator. The zero of the lag compensator $\left(\frac{1}{T}\right)$ is chosen to be 10% of the second pole of uncompensated system.

$\therefore$ Zero of the compensator,

$$z_c = \left(\frac{-1}{T}\right) = 0.1 \times [\text{second pole of } G(s)]$$

Now, $$T = \frac{1}{[-0.1 \times (\text{second pole of } G(s)]}$$

Pole of the lag compensator

$$P_c = \frac{-1}{\beta T}$$

Transfer function of lag compensator,

$$G_c(s) = \frac{s + \frac{1}{T}}{s + \frac{1}{\beta T}} = \beta \frac{(1 + sT)}{(1 + s\beta T)}$$

Step 6: Determine the open loop transfer function of the compensated system. The lag compensator is connected in series with the plant as shown in Fig. (5.8)

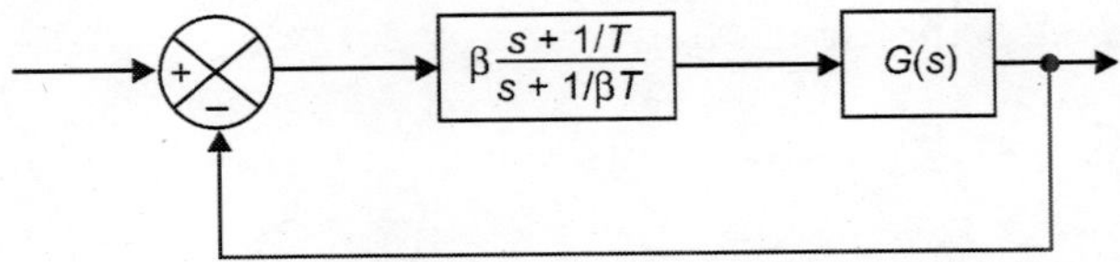

Fig. 5.8: Block diagram of lag compensated system.

Open loop transfer function of compensated system

$$G_o(s) = G_c(s)\, G(s) \frac{\left(s + \frac{1}{T}\right)}{\left(s + \frac{1}{\beta T}\right)} \cdot G(s)$$

Step 7: Verify whether the compensated system satisfies the steady state error requirement. If it is satisfied, then the design is accepted. If it is no satisfied repeat the design by modifying the location of poles and zeros of the compensator.

EXAMPLE 5.1

A unity feedback system has an open loop transfer function, $G(s) = \dfrac{K}{s(1+2s)}$. Design a suitable lag compensator so that phase margin is 40є and the steady state error for ramp input is less than or equal to 0.2 ($e_{ss} \le .2$).

Solution:

Step 1: Calculate gain K

Given that, $e_{ss} \le 0.2$ for ramp input

Let $$e_{ss} = 0.2$$

We know that $$e_{ss} = \frac{1}{K_v} \text{ for ramp input.}$$

∴ Velocity error constant

$$K_v = \frac{1}{e_{ss}} = \frac{1}{0.2} = 5$$

By definition of velocity error constant,

$$K_v = \underset{s\to 0}{Lt}\ s.G(s)H(s)$$

Since the system is unity feedback system,

$$H(s) = 1$$

∴ $$K_v = \underset{s\to 0}{Lt}\ s.G(s) = \underset{s\to 0}{Lt}\ s.\frac{K}{s(1+2s)} = K$$

∴ $$K = 5$$

Step 2: Draw Bode plot of uncompensated system.

Given that, $$G(s) = \frac{5}{s(1+2s)}$$

Let $s = j\omega$ ∴ $$G(j\omega) = \frac{5}{j\omega(1+j2\omega)}$$

(i) Magnitude plot

The corner frequency is,

$$\omega_c = \frac{1}{2} = 0.5 \text{ rad/sec.}$$

The various terms of $G(j\omega)$ are listed in Table Ex. 5.1. The table shows the slope contributed by each term and the change is slope at the corner frequency.

Note:

(i) Always first term will not have corner frequency.

(ii) For each s, the value in *db* is 20 *db*. When s is in the numerator then +20 *db* and when s is to the denominator then –20 *db*.

Table Ex. 5.1

Term	**Corner frequency rad/sec**	**Slope of each term *db*/dec**	**Change in slope from one term to another term *db*/dec**
$\frac{5}{j\omega}$	–	– 20	–
$\frac{1}{1+j2\omega}$	$\omega_c = \frac{1}{2} = 0.5$	–20	–20–20 = – 40

Choose a low frequency ω_l such that $\omega_c < \omega_l$ and choose high frequency ω_h such that $\omega_h > \omega_c$

Let $\omega_l = 0.1$ rad/sec and $\omega_h = 10$ rad /sec

Let $$A = |G(j\omega)| \text{ in } db$$

At $\omega = \omega_l$, $$A_1 = 20\log\left|\frac{5}{j\omega}\right| = 20\log\frac{5}{0.1} = 34\text{db} \quad [A_1 = 20 \log |\text{first term}|]$$

At $\omega = \omega_c$ $$A_2 = \log\left|\frac{5}{j\omega}\right| = 20\log\frac{5}{0.5} = 20\text{db} \quad [A_2 = 20 \log |\text{first term}|]$$

At $\omega = \omega_h$, $$A_3 = [\text{slope from one term to another term}] \text{ Ч } \log\left|\frac{\text{Present frequency}}{\text{Previous frequency}}\right| + \text{Previous gain } (A_2)$$

At $\omega = \omega_h$, $$A_3 = \left[\text{Slope from } \omega_c \text{ to } \omega_h \times \log\frac{\omega_h}{\omega_c}\right] + A_2(\text{at } \omega = \omega_c)$$

$$A_3 = 40 \times \log\frac{10}{0.5} + 20 = -32db$$

Let the points a, b and c be the points corresponding to frequencies ω_l, ω_c and ω_h respectively on the magnitude plot. In a semilog graph sheet choose appropriate scales and fix the points a, b and c. Join the points by straight lines and mark the slope on respective region. The magnitude plot is shown in Fig. (Ex. 5.1.1)

Phase plot

The phase angle of $G(j\omega)$ as a function of ω is given by

$$\phi = \angle G(j\omega) = -90\text{є} - \tan^{-1}2\omega$$

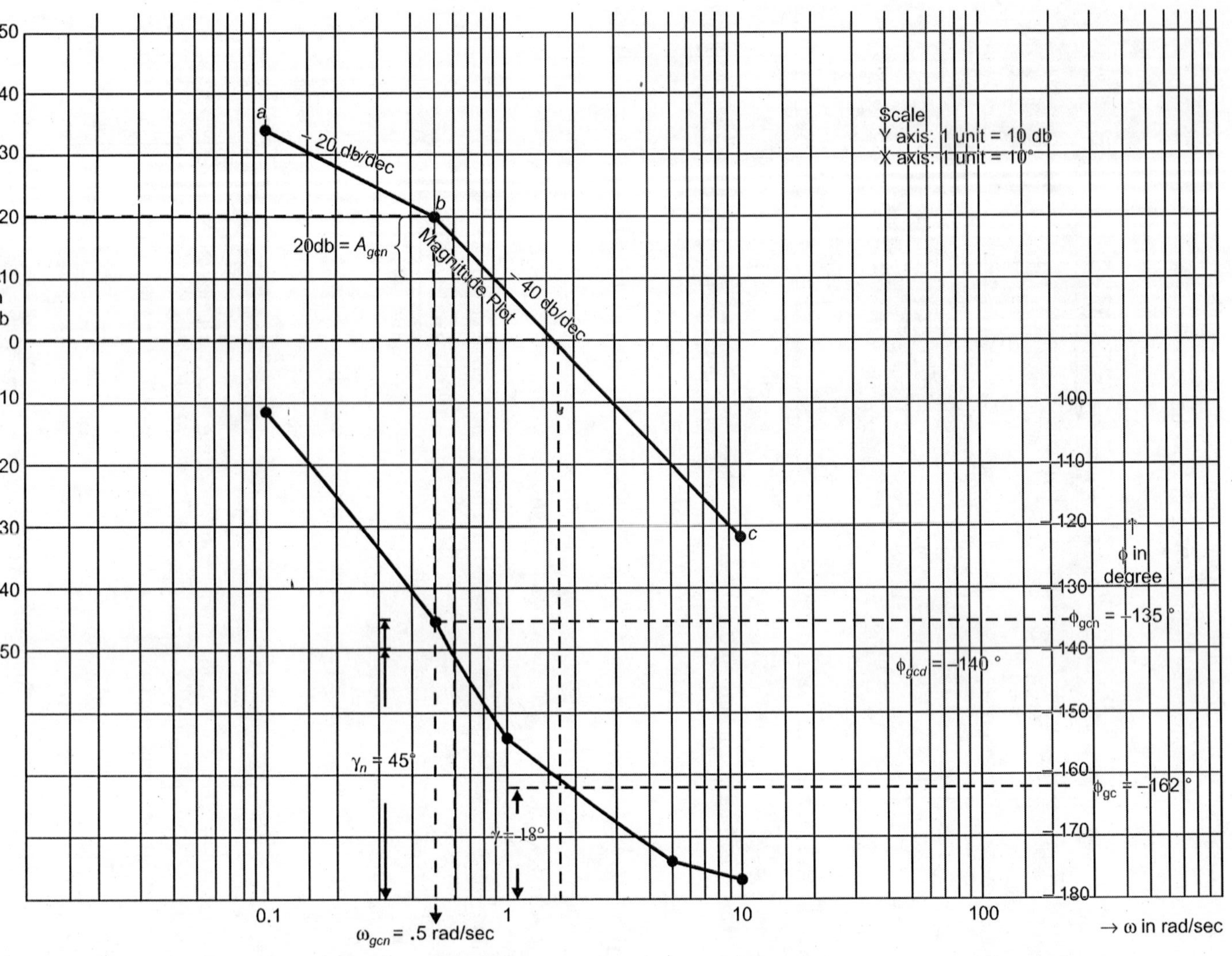

Fig. Ex. 5.1.1: Bode plot of $G(s) = K/s(1 + 2s)$ (lag compensator).

The phase angle of $G(j\omega)$ are calculated for various values of ω and listed in Table (Ex. 5.1.2).

Table Ex. 5.1.2					
rad ω /sec	0.1	0.5	1.0	5	10
$d\phi g$	–101	–135	–153	–174	–177

On the same semilog sheet take another *Y*-axis, choose appropriate scale and draw phase plot a shown in Fig. (Ex. 5.1.1).

Step 3: Determine phase margin of uncompensated system.

Let, ϕ_{gc} = phase of $G(j\omega)$ at gain crossover frequency (ω_{gc}).

γ = phase margin of uncompensated system.

From the bode plot of uncompensated system we get

$$\phi_{gc} = -162\epsilon$$

$$\gamma = 180\epsilon + \phi_{gc} = 180\epsilon - 162\epsilon = 18\epsilon$$

The system requires a phase margin of 40є, but the available phase margin is 18є so lag compensation should be employed to improve the phase margin.

Step 4: Choose a suitable value for the phase margin of compensated system

The desired phase margin, $\gamma_d = 40\epsilon$

∴ Phase margin of compensated system $\gamma_n = \gamma_d + \varepsilon$

Let initial choice of ∈ = 5є

∴ $$\gamma_n = \gamma_d + \in = 40\epsilon + 5\epsilon = 45\epsilon$$

Step 5: Determine new gain crossover frequency.

Let, ω_{gcn} = New gain crossover frequency

ϕ_{gcn} = Phase of $G(j\omega)$ at ω_{gcn}

Now, $\gamma_n = 180\epsilon + \phi_{gcn}$

∴ $\phi_{gcn} = \gamma_n - 180\epsilon = 45\epsilon - 180\epsilon = -135\epsilon$

From the bode plot we found that, the frequency corresponding to a phase of –135є is 0.5 rad/sec.

∴ New gain crossover frequency, ω_{gcn} = 0.5 rad/sec.

Step 6: Determine the parameter, β

From the bode plot we found that, the *db* magnitude at ω_{gcn} is 20 *db*

∴ $|G(j\omega)|$ in *db* at $(\omega = \omega_{gcn}) = A_{gcn} = 20db$

Also $A_{gcn} = 20 \log \beta;\ \beta = 10^{A_{gcn}/20} = 10^{20/2B} = 10$ i.e., β = 10

Step 7: Determine the transfer function of lag compensator

The zero of the compensator is placed at a frequency one tenth of ω_{gcn}

∴ zero of the lag compensation,

$$z_c = \frac{1}{T} = \omega_{gcn}/10$$

Now, $$T = \frac{10}{\omega_{gcn}} = \frac{10}{0.5} = 20$$

Pole of the lag compensation,

$$p_c = \frac{1}{\beta T} = \frac{1}{10 \times 20}$$

$$= \frac{1}{200} = 0.005$$

Transfer function of lag compensator,

$$G_c(s) = \frac{s + \frac{1}{T}}{s + \frac{1}{\beta T}} = \beta \frac{1 + sT}{1 + s\beta T} = 10 \frac{(1 + 20s)}{(1 + 200s)}$$

Step 8: Determine the open loop transfer function of compensated system

The block diagram of the compensated system is shown in Fig. (Ex. 5.1.2)

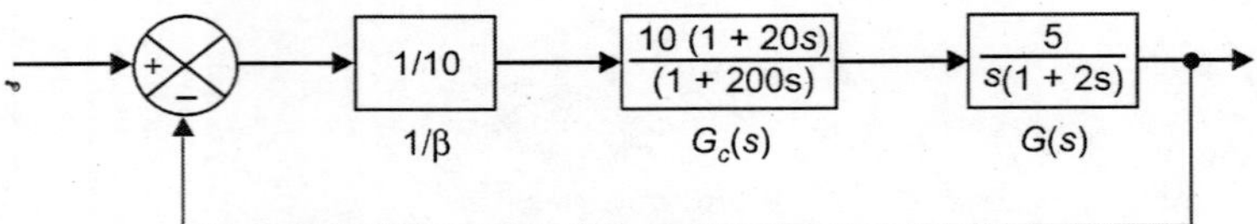

Fig. Ex. 5.1.2: Block diagram of lag compensated system.

The gain of the compensator is nullified by introducing an attenuator in series with the compensator.

Open loop transfer function of compensated system

$$G_o(s) = \frac{1}{10} \times \frac{10(1 + 20s)}{(1 + 200s)} \times \frac{5}{s(1 + 2s)}$$

$$= \frac{5(1 + 20s)}{s(1 + 200s)(1 + 2s)}$$

Step 9: Determine the actual phase margin of compensated system.

On substituting $s = j\omega$ in $G_o(s)$ we get

$$G_o(j\omega) = \frac{5(1 + j20\omega)}{j\omega(1 + j200\omega)(1 + j2\omega)}$$

Let, ϕ_0 = phase of $G_o(j\omega)$

$$\phi_{gco} = \text{phase of } G_o(j\omega) \text{ at } \omega = \omega_{gcn}$$

$$\phi_0 = \tan^{-1} 20\omega - 90\epsilon - \tan^{-1} 200\omega - \tan^{-1} 2\omega$$

At $\quad \omega = \omega_{gcn} = \phi_0 = \phi_{gco} = \tan^{-1} 20\, \omega_{gcn} - 90\epsilon - \tan^{-1} 200\omega_{gcn} - \tan^{-1} 2\, \omega_{gcn}.$

$$\phi_{gcn} = \tan^{-1} (20 \text{ Ч} 0.5) - 90\epsilon - \tan^{-1} (200 \text{ Ч} 0.5) - \tan^{-1} (2 \text{ Ч} 0.5) = -140\epsilon$$

Actual phase margin of compensated system,

$$= 180\epsilon + \phi_{gco}$$

$$= 180\epsilon - 140\epsilon = 40\epsilon$$

Conclusion: The actual phase margin of the compensated system satisfies the requirement . Hence the design is acceptable.

Result: Transfer function of lag compensator

$$G_c(s) = \frac{10(1+20s)}{(1+200s)} = \frac{(s+0.05)}{(s+0.005)}$$

Open loop transfer function of compensated system,

$$G_o(s) = \frac{5(1+20s)}{s(1+200s)(1+2s)}$$

EXAMPLE 5.2

The open loop transfer function system is given by $G(s) = K/s\,(s+4)\,(s+80)$. It is desired to have the phase margin to $K_v = 30 \text{ sec}^{-1}$. Design a phase lag series compensator.

Solution:

Step 1: Calculate gain, *K*

Given that, $\quad K_v = 30 \text{ sec}^{-1}$

By definition of velocity error constant,

$$K_v = \underset{s \to 0}{Lt}\; s \cdot G(s)\, H(s)$$

Since the system is unity feedback system, $H(s) = 1$

$$\therefore \quad K_v = \underset{s \to 0}{Lt}\; s \cdot G(s)$$

$$= \underset{s \to 0}{Lt}\; s \cdot \frac{K}{s(s+4)(s+80)} = \frac{K}{4 \times 80}$$

$$\frac{K}{4 \times 80} = 30;\; K = 30 \times 80 \times 4 = 9600$$

i.e., $\quad K = 9600$

Step 2: Draw bode plot of uncompensated system

$$G(s) = \frac{9600}{s(s+4)(s+80)}$$

$$= \frac{9600/4 \times 80}{s\left(1+\frac{s}{4}\right)\left(1+\frac{s}{80}\right)}$$

$$= 30/s(1 + 0.25s)(1 + 0.0125s)$$

Let, $s = j\omega$
$$G(j\omega) = \frac{30}{j\omega(1+j\,0.25\omega)(1+j\,0.0125\omega)}$$

(i) Magnitude plot

The corner frequencies are, $\omega_{c1} = 1/0.25 = 4$ rad/sec and $\omega_{c2} = 1/0.0125 = 80$ rad/sec. The various terms of $G(j\omega)$ are lisit in Table Ex. 5.2.1. Also the table shows the slope contributed by each term and the change in slope at the corner frequency.

Table Ex. 5.2.1

Term	Corner frequency rad/sec	Slope of each term *db*/dec	Change in slope from one term to another term *db*/dec
$\frac{30}{j\omega}$	–	–20	–
$\frac{1}{1+j\,025\omega}$	$\omega_{c1} = 1/0.25 = 4$	–20	–20 –20 = –40
$\frac{1}{1+j\,0.0125\omega}$	$\omega_{c2} = \frac{1}{0.0125} = 80$	–20	–40 –20 = –60

Choose a low frequency ω_l such that $\omega_l < \omega_{c1}$ and choose a high frequency ω_h such that $\omega_h > \omega_{c2}$. Let $A = |G(j\omega)|$ in *db*.

At $\omega = \omega_l$,
$$A = 20\log\left|\frac{30}{j\omega}\right| = 20\log\frac{30}{1} \approx 30 \text{ db} \quad [A_1 = 20\log|\text{first term}|]$$

At $\omega = \omega_{c1}$,
$$A = 20\log\left|\frac{30}{j\omega}\right| = 20\log\frac{30}{4} \approx 18 \text{ db} \quad [A_1 = 20\log|\text{first term}|]$$

At $\omega = \omega_{c2}$,
$$A_3 = [\text{slope from one term to another term}] \times \log\left|\frac{\text{Present frequency}}{\text{Previous frequency}}\right| + \text{Previous gain } (A_2)$$

$$A_3 = \left[\text{slope from } \omega_{c1} \text{ to } \omega_{c2} \times \log\frac{\omega_{c2}}{\omega_{c1}}\right] + A_2(\text{at } \omega = \omega_{c1})$$

$$= -40 \log\frac{80}{4} + 18 = -34 \text{ db}$$

At $\omega = \omega_h$, A_4 = [slope from one term to another term] Ч log

$$\left|\frac{\text{Present frequency}}{\text{Previous frequency}}\right| + \text{Previous gain } (A_3)$$

At $\omega = \omega_h$,

$$A_4 = \left[\text{slope from } \omega_{c2} \text{ to } \omega_n \times \log\frac{\omega_h}{\omega_{c2}}\right] + A_3(\text{at } \omega = \omega_{c2})$$

$$A_4 = 60 \times \log\frac{100}{80} + (-34) = -40 \text{ db}$$

Let the points *a*, *b* and *c* and *d* be the points corresponding to frequencies ω_l, ω_{c1}, ω_{c2} and ω_h respectively on the magnitude plot. in a semilog graph sheet choose appropriate scales and fix the points *a*, *b*, *c* and *d*. join the points by straight lines and mark the slope on the respective region. The magnitude plot is shown in Fig. (Ex. 5.2.1).

Phase plot: The phase angle of $G(j\omega)$ as a function of ω is given by

$$\phi = \angle G(j\omega) = -90є - \tan^{-1} 0.25\omega - \tan^{-1} 0.0125\omega$$

The phase angle of $G(j\omega)$ are calculated for various values of ω and listed in Table Ex. 5.2.2.

Table Ex. 5.2.2

ω, rad/sec	1	4	10	50	80	100
ϕ_{deg}	–104	–138	$-165 \approx -164$	$-207 \approx -208$	–222	$-229 \approx -230$

On the same samilog sheet take another *Y*-axis, choose appropriate scale and draw phase plot as shown in Fig. (Ex. 5.2.1).

Step 3: Determine of phase margin of uncompensated system.

Let, ϕ_{gc} = phase of $G(j\omega)$ at gain crossover frequency (ω_{gc})

γ = phase margin of uncompensated system.

From the bode plot of uncompensated system we found that,

$$\phi_{gc} = -168є$$

Now, $$\gamma = 180є + \phi_{gc} = 180є - 168є = 12є$$

The system requires a phase margin of at least 33є, but the available phase margin is 12є and so lag compensation should be employed to improve the phase margin.

Step 4: Choose a suitable value for the phase margin of compensated system.

The desired phase margin, γ_d = 33є

CHAPTER 5

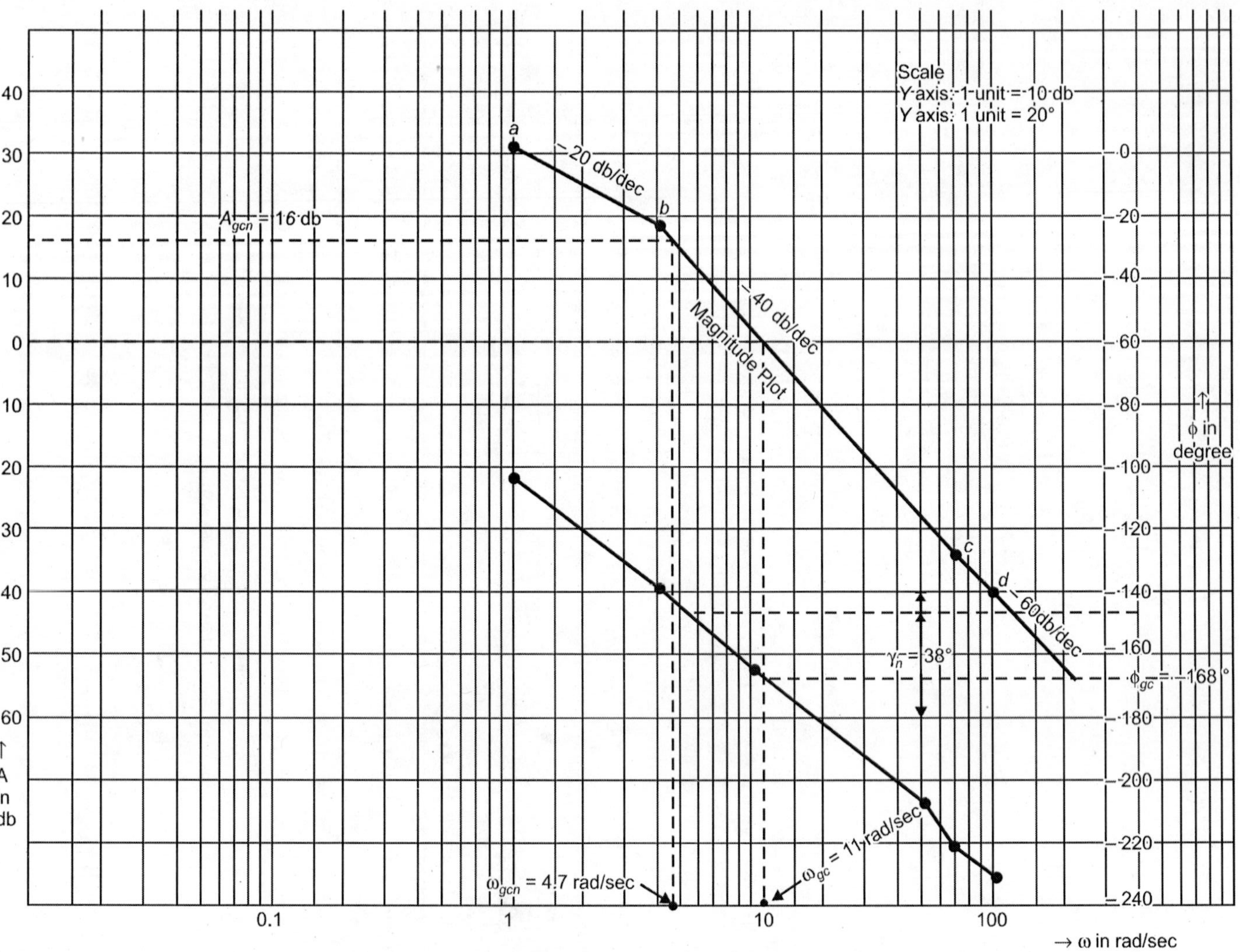

Fig. Ex. 5.2.1: Bode plot of $G(s) = K/s(s + 4)(s + 80)$ (lag compensator).

∴ Phase margin of compensated system, we found that,

$$\gamma_n = \gamma_d + \zeta.$$

Let initial choice of $\zeta = 5\epsilon$;

$$\gamma_n = \gamma_d + \zeta = 33\epsilon + 5\epsilon = 38\epsilon$$

Step 5: Determine new gain crossover frequency let ω_{gcn} = new gain crossover frequency

$$\phi_{gcn} = \text{phase of } G(j\omega) \text{ at } \omega_{gcn}$$

Now, $$\gamma_n = 180\epsilon + \phi_{gcn}$$

∴ $$\phi_{gcn} = \gamma_n - 180\epsilon = 38\epsilon - 180\epsilon = -142\epsilon$$

From the bode plot we found that, the frequency corresponding to a phase of -142ϵ is 4.7 rad sec.

∴ New gain crossover frequency,

$$\omega_{gcn} = 4.7 \text{ rad/sec}$$

Step 6: Determine the parameter, β

From the bode plot we found that, the *db* magnitude at ω_{gcn} as 16 *db*.

∴ $|G(j\omega)|$ in *db* at $(\omega = \omega_{gcn}) = A_{gcn} = 16$ *db*

Also, $$A_{gcn} = 20 \log \beta;$$

∴ $$\beta = 10^{Agcn/20} = 10(16/20) = 6.3$$

Step 7: Determine the transfer function of lag compensator.

The zero of the compensator is placed at a frequency one-tenth of ω_{gcn}

∴ Zero of the lag compensator,

$$z_c = 1/T = \omega_{gcn}/10$$

Now, $$T = 10/\omega_{gcn} = 10/4.7 = 2.13$$

Pole of the lag compensator

$$p_c = \frac{1}{\beta T} = \frac{1}{6.3 \times 2.13} = \frac{1}{13.419}$$

Transfer function of lag compensator

$$G_c(s) = \frac{s + 1/T}{s + 1/\beta T} = \beta \frac{1 + sT}{1 + s\beta T} = 6.3 \frac{(1 + 2.13s)}{(1 + 13.419s)}$$

Step 8: Determine the open loop transfer function of the compensated system.

The block diagram of the compensated system shown below.

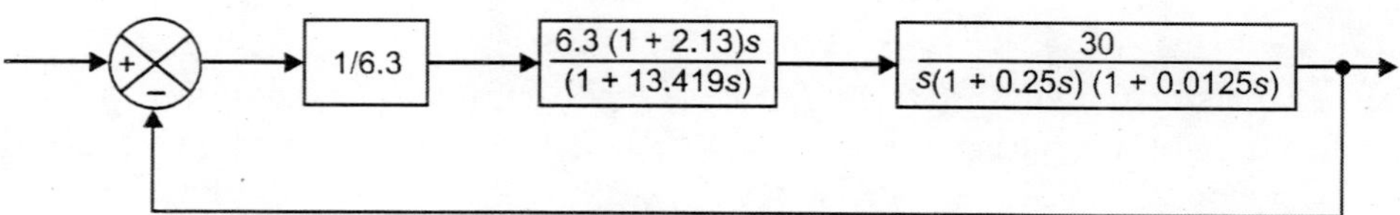

Fig. Ex. 5.2.2: Block diagram of compensated system.

The gain of the compensator is nullified by introducing an attenuator in series with the compensator, Open loop transfer function of compensated system,

$$G_o(s) = \frac{1}{6.3} \times \frac{6.3\,(1+2.13s)}{(1+13.419s)} \times \frac{30}{s(1+0.25s)\,(1+0.0125s)}$$

$$= \frac{30(1+2.13s)}{s(1+13.419s)\,(1+0.25s)\,(1+0.0125s)}$$

Step 9: Determine the actual phase margin of compensated system.

On substituting $s = j\omega$ in $G_o(s)$ we get,

$$G_o(j\omega) = \frac{30(1+j2.13\omega)}{j\omega(1+j13.419\omega)\,(1+j0.25\omega)\,(1+j0.0125\omega)}$$

Let, ϕ_0 = Phase of $G_o(j\omega)$ at $\omega = \omega_{gcn}$

$$\phi_0 = \tan^{-1} 2.13\omega - 90 - \tan^{-1} 14.419\omega - \tan^{-1} 0.25\omega - \tan^{-1} 0.0125\omega$$

At $\omega = \omega_{gcn}$, $\phi_0 = \phi_{gco} = \tan^{-1} 2.13\,\omega_{gcn} - 90є - \tan^{-1} 13.419\,\omega_{gcn} - \tan^{-1} 0.25\,\omega_{gcn} - \tan^{-1} 0.0125\,\omega_{gcn}$

∴ $\phi_{gcn} = \tan^{-1}(2.13 \text{ Ч } 4.7) - 90є - \tan^{-1}(13.419 \text{ Ч } 4.7) - \tan^{-1}(0.25 \text{ Ч } 4.7) - \tan^{-1}(0.0125 \text{ Ч } 4.7) = -147є$

Actual phase margin of compensated system.

$$\gamma_o = 180є + \phi_{gco} = 180є - 147є = 33є$$

Conclusion: The actual phase margin of the compensated system satisfies the requirement. Hence the design is acceptable.

Result: Transfer function of lag compensator

$$G_c(s) = \frac{6.3\,(1+2.13s)}{(1+13.419s)} = \frac{(s+0.469)}{(s+0.074)}$$

Transfer function of lag compensated system.

$$G_o(s) = \frac{30(1+2.13s)}{s(1+13.419s)\,(1+0.25s)\,(1+0.0125s)}$$

EXAMPLE 5.3

The forward pat transfer function of a certain unity feedback control system is given by $G(s) = K/s\,(s+2)\,(s+8)$. Design a suitable lag compensator so that the system meets the following specifications. (i) percentage overshoot ≤ 16% for unit step input, (ii) steady state error ≤ 0.125 for unity ramp input.

Solution:

Step 1: Sketch the root locus of uncompensated system.

To find poles of open loop system

Given that $G(s) = K/s\ (s + 2)\ (s + 8)$

The poles of open loop transfer function are the roots of the equation.

$$s(s + 2)\ (s + 8) = 0$$

The poles are 0, –2, –8.

To find root locus on real axis

The segments of real axis between = 0 and $s = –2$ and the segment of real axis between $s = –8$ and $s = –\infty$ will be part of root locus. Because if we choose a test point in this segment then to the right of this point we have odd number of real poles and zeros.

To find angles of asymptotes and centroid

Since there are three poles, the number of root locus branches are three. There is no finite zero and so all the three root locus branches will meet the zeros at infinity. Hence the number of asymptotes requires is three.

$$\text{Angles of asymptotes} = \frac{+180\epsilon(2q+1)}{n-m}$$

Here $n = 3$ and $m = 0$.

If $q = 0$, $\quad \text{Angles} = \dfrac{+180\epsilon}{3} = \pm\, 60\epsilon$

If $q = 1$, $\quad \text{Angles} = \dfrac{+180\epsilon(2+1)}{3} = \pm 180\epsilon$

∴ Angles of asymptotes are $\pm 60\epsilon$ and $\pm 180\epsilon$

$$\text{Centroid} = \frac{\text{Sum of poles} - \text{Sum of zeros}}{n-m} = \frac{0-2-8}{3} = -3.3$$

To find breakaway point

Closed loop transfer function,

$$\frac{C(s)}{R(s)} = \frac{G(s)}{1+G(s)} = \frac{\dfrac{K}{s(s+2)\,(s+8)}}{1+\dfrac{K}{s(s+2)\,(s+8)}} = K/s(s + 2)\ (s + 8) + K$$

The characteristics equal is, $s\ (s + 2)\ (s + 8) + K = 0$

$$K = –s\ (s + 2)\ (s + 8) = –(s^3 + 10s^2 + 16s)$$

On differentiating K with respect to s we get,

$$dK/ds = –(3s^2 + 20s + 16)$$

Put $dK/ds = 0$,

$$\therefore \quad 3s^2 + 20s + 16 = 0$$

$$s = \frac{-20 \pm \sqrt{20^2 - 4(3)(16)}}{2 \times 3} = -0.9 \text{ (or)} -5.7$$

When $s = -0.9$,
$$K = -(s^3 + 10s^2 + 16s)$$
$$= [(0.9)^3 + 10(-0.9)^2 + 16(-0.9)] = 7$$

When $s = -5.7$,
$$K = -(s^3 + 10s^2 + 16s)$$
$$= -[(5.7)3 + 10(-5.7)2 + 16(-5.7)] = -48$$

For $s = -0.9$, the value of K is positive and real so it is actual breakaway point.

To find crossing point on imaginary axis

The characteristics Eq. is,

$$s(s + 2)(s + 8) + K = 0$$
$$s^3 + 10s^2 + 16s + K = 0$$

Put $s = j\omega$,
$$(j\omega)^3 + 10(j\omega)^2 + 16(j\omega) + K = 0$$
$$(-j\omega)^2 - 10\omega^2 + j16\,\omega + K = 0$$

On equating imaginary part to zero we get,

$$-j\omega^3 + j\,16\omega = 0$$
$$-j\omega^3 = -j16\omega$$
$$\omega^2 = 16$$
$$\omega = \pm\sqrt{16} = \pm 4$$

Hence the root locus crosses the imaginary axis at $+j4$ and $-j4$. The complete root locus sketch is shown Fig. (Ex. 5.3.1).

Step 2: Determine the dominate pole s_d

Given that,
$$\%M_P = 16\%$$

We know that,
$$\%M_P = e^{-\zeta\pi\sqrt{1-\zeta^2}}$$

$$e^{-\zeta\pi\sqrt{1-\zeta^2}} = 0.16$$

On taking natural log we get,

$$\frac{-\zeta\pi}{\sqrt{1-\zeta^2}} = l_n\, 0.16 = -1.83$$

On squaring we get,
$$\frac{\zeta^2\pi^2}{1-\zeta^2} = 3.3489$$

On cross multiplication we get,

$$\zeta^2\pi^2 = 3.3489\,(1 - \zeta^2)$$

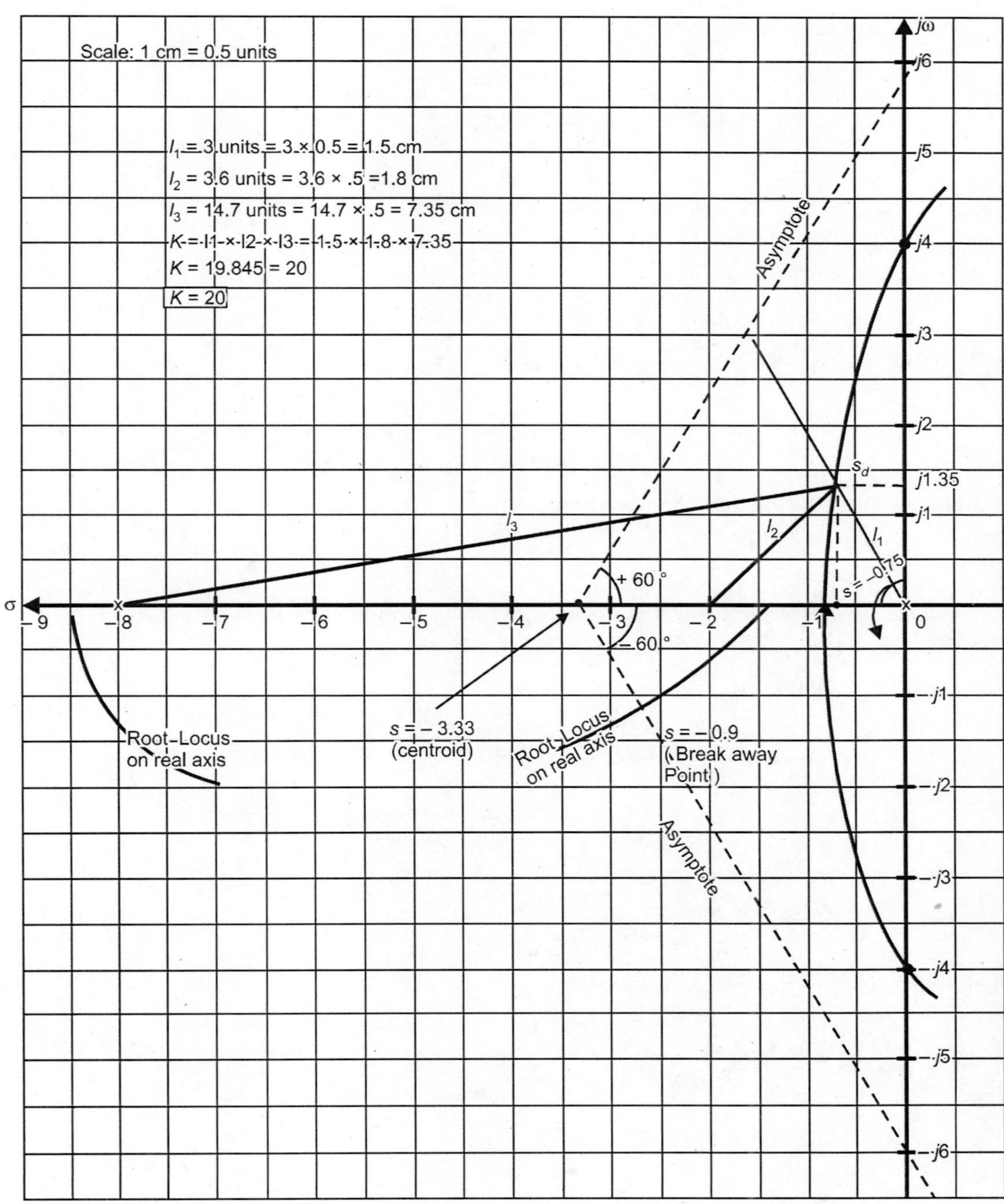

Fig. Ex. 5.3.1: Root locus of $G(s) = K/s\ (s + 2)\ (s + 8)$ (lag compensator).

$$\zeta^2\pi^2 + 3.3489\zeta^2 = 3.3489$$

$$\zeta = \sqrt{\frac{3.3489}{\pi^2 + 3.3489}} = 0.5$$

$$\cos^{-1}\zeta = \cos^{-1} 0.5 = 60\text{є}$$

Draw a straight line at an angle of 60є with respect to real axis as shown in Fig. (Ex. 5.3.1).

The meeting point of this line with root locus is the dominant pole, s_d.

From Fig. (Ex. 5.3.1) we get, $s_d = -0.75 \pm j1.35$

(Dominant poles occur as conjugate poles).

Step 3: To find gain K, at

$$s = s_d$$

$$K = \frac{\text{Product of vector lengths from open loop poles to } s_d}{\text{Product of vector lengths from open loop zeros to } s_d}$$

Product of vector lengths from poles = l_1 Ч l_2 Ч l_3

From root locus plot of Fig. (Ex. 5.3.1) we get,

$l_1 = 1.5$; $l_2 = 1.8$ and $l_3 = 7.35$ (in cm)

Since there is no finite zero, the product of vector lengths from zeros is unity.

∴ $K = l_1$ Ч l_2 Ч l_3 = 1.5 Ч 1.8 Ч 7.35 = 19.845 ≈ 20.

Step 4: To find parameter, β

Given that $G(s) = K/s(s+2)(s+8) = 20/s(s+2)(s+8)$

Velocity error constant of uncompensated system,

$$K_{vu} = \underset{s \to 0}{Lt}\ s \,.\, G(s) = \underset{s \to 0}{Lt}\ s\ \frac{20}{s(s+2)(s+8)} = \frac{20}{2 \times 8} = 1.25$$

It is given that steady state error, $e_{ss} \le 0.125$ for ramp input.

∴ The desired velocity error constant

$$K_{vd} = 1/e_{ss} = 1/0.125 = 8$$

Let A be the factor by which K_v is increased.

∴ $A = K_{vd}/K_{vu} = 8/1.25 = 6.4$

Let, β = 1.2 Ч A = 1.2 Ч 6.4 = 7.68

Step 5: Determine the transfer function of lag compensator.

Zero of the compensator, $z_c = -1/T$

= 0.1 Ч second pole of $G(s)$

= 0.1 Ч (–2) = –0.2

Now, $T = 1/0.2 = 5$

Pole of the compensator, $$p_c = \frac{-1}{\beta T} = \frac{-1}{7.68 \times 5} = \frac{-1}{38.4} = -0.026$$

$\therefore$ $1/\beta T = 0.026$ and $\beta T = 0.026$ and $\beta T = 38.4$

Transfer function of lag compensator,

$$G_c(s) = \frac{s + 1/T}{s + 1/\beta T} = \frac{(s + 0.2)}{(s + 0.026)}$$

Step 6: Transfer function of compensated system.

The lag compensator is connected in series with $G(s)$ as shown below. Fig. (Ex. 5.3.2).

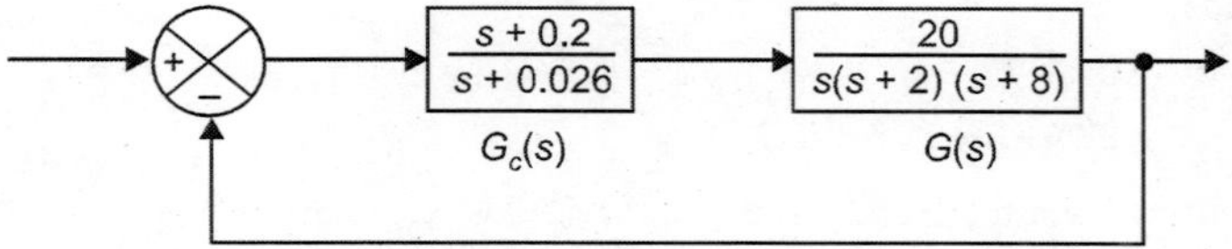

Fig. Ex. 5.3.2: Block diagram of compensated system.

Open loop transfer function of compensated system.

$$G_o(s) = \frac{(s + .2)}{(s + 0.026)} \times \frac{20}{s(s + 2)(s + 8)}$$

$$= \frac{20(s + 0.2)}{s(s + 2)(s + 8)(s + 0.06)}$$

Step 7: Check steady state e rror of compensated system velocity error constant of compensated system,

$$K_{vc} = \underset{s \to 0}{Lt}\ s \cdot G_o(s)$$

$$= \underset{s \to 0}{Lt}\ s \frac{20(s + 0.2)}{s(s + 2)(s + 8)(s + 0.06)}$$

$$= \frac{20 \times 0.2}{2 \times 8 \times 0.026} = 96.5$$

Steady state error of compensated system for unit ramp input,

$$e_{ss} = 1/K_{vc} = 1/96.5 = 0.104$$

Conclusion: Since the steady state error of the compensated system is less than 0.125, the design is acceptable.

Result: Transfer function of lag compensator,

$$G_c(s) = (s + 0.2)/(s + 0.026)$$

Open loop transfer function of lag compensated system,

$$G_o(s) = 20(s + 0.2)/\ s(s + 2)(s + 8)(s + 0.026).$$

EXAMPLE 5.4

A plant of a unity feedback system is $G(s) = K/s(s + 10)^2$. It is specified that velocity error constant of the system be equal to 20, while the damping ratio of the dominant roots be 0.707. Design a suitable cascade compensation scheme to meet the above specifications.

Solution:

Step 1: Sketch the root locus of uncompensated system.

To find poles of open loop system

Given that, $G(s) = K/s(s + 10)^2$

The poles of open loop transfer function are the roots of the equation $s(s + 10)^2 = 0$

∴ The poles are 0, –10, –10.

To find root locus on real axis

The entire negative real axis will be a part of root locus. because if we choose a test point on negative real axis then to the right of this point we have odd number of real poles and zeros.

To find angles of asymptotes and centroid

Since there are three poles, the number of root locus branches are three. There is no finite zero and so all the three root locus branches will meet the zeros at infinity. Hence the number of asymptotes required is three.

$$\text{Angles of asymptotes} = \frac{\pm 180є(2q+1)}{n-m}$$

Here $n = 3$, $m = 0$.

∴ If $q = 0$, Angles $= \pm 180є/3 = \pm 60є$

If $q = 1$, Angles $= \pm 180є(2 + 1) / 3 = \pm 180є$

∴ Angles of asymptotes are $\pm 60є$ and $\pm 180є$

controid = sum of poles –sum of zeros /*n*-*m*.

= 0 – 10 – 10/3 = –20/3 = –6.6.

To find breakaway point

Closed loop transfer function,

$$\frac{C(s)}{R(s)} = \frac{G(s)}{1+G(s)} = \frac{K/s(s+10)^2}{1+\dfrac{K}{s(s+10)^2}} = K/s(s+10)^2 + K \quad [\text{Œ}H(s) = 1 \text{ for unity feedback system}]$$

The characteristics equation is, $s(s + 10)^2 + K = 0$

∴ $$K = -s(s + 10)^2 = -(s^3 + 20^2 + 100s)$$

On differentiating K with respect to s we get.

$$\frac{dK}{ds} = -(3s^2 + 40s + 100)$$

Put $dK/ds = 0$,

$$\therefore \quad 3s^2 + 40s + 100 = 0$$

$$s = \frac{-40 \pm \sqrt{40^2 - 4(3)(100}}{2 \times 3} = \frac{-40 \pm 20}{6} = -3.33, -10$$

When $\quad s = -3.33, K = -\left[(-3.33)^3 + 20(-3.33)^2 + 100(-3.33)\right]$

$$= 148.14$$

When $\quad s = -10, K = -\left[(-10)^3 + 20(-10)^2 + 100(-10)\right] = 0$

For $s = -3.33$, the value of K is positive and real and so it is actual breakaway point.

To find crossing point of imaginary axis:

The characteristics equation is,

$$s(s + 10)^2 + K = 0$$

$$s^3 + 20s^2 + 100s + K = 0$$

Put $s = j\omega$

$$(j\omega)^3 + 20(j\omega)^2 + 100(j\omega) + K = 0$$

$$-j\omega^3 - 20\omega^2 + j100\omega + K = 0$$

On equating imaginary part to zero we get,

$$-j\omega^3 + j100\omega = 0$$

$$-j\omega^3 = j100\omega$$

$$\omega^2 = 100$$

$$\omega = \pm\sqrt{100} = \pm 10$$

Hence the root locus crosses the imaginary axis at $+j10$ and $-j10$. The complete root locus sketch is shown in Fig. (Ex. 5.4.1).

Step 2: Determine the dominant pole s_d.

Given that, $\quad \zeta = 0.707$

$$\therefore \quad \cos^{-1}\zeta = \cos - 0,707$$

$$= 45.008 \approx 45\epsilon$$

Draw a straight line at an angle of 45є with respect to real axis as shown in Fig. (Ex. 5.4.1). The meeting point of this line with root locus is the dominant pole, s_d.

From Fig. (Ex. 5.4.1), we get, $s_d = -2.9 \pm 2.9$ (Dominant poles occur as conjugate poles).

Step 3: To find gain K, at $s = s_d$

$$K = \frac{\text{Product of vector lengths from open loop poles to } s_d}{\text{Product of vector lengths from open loop zeros to } s_d}$$

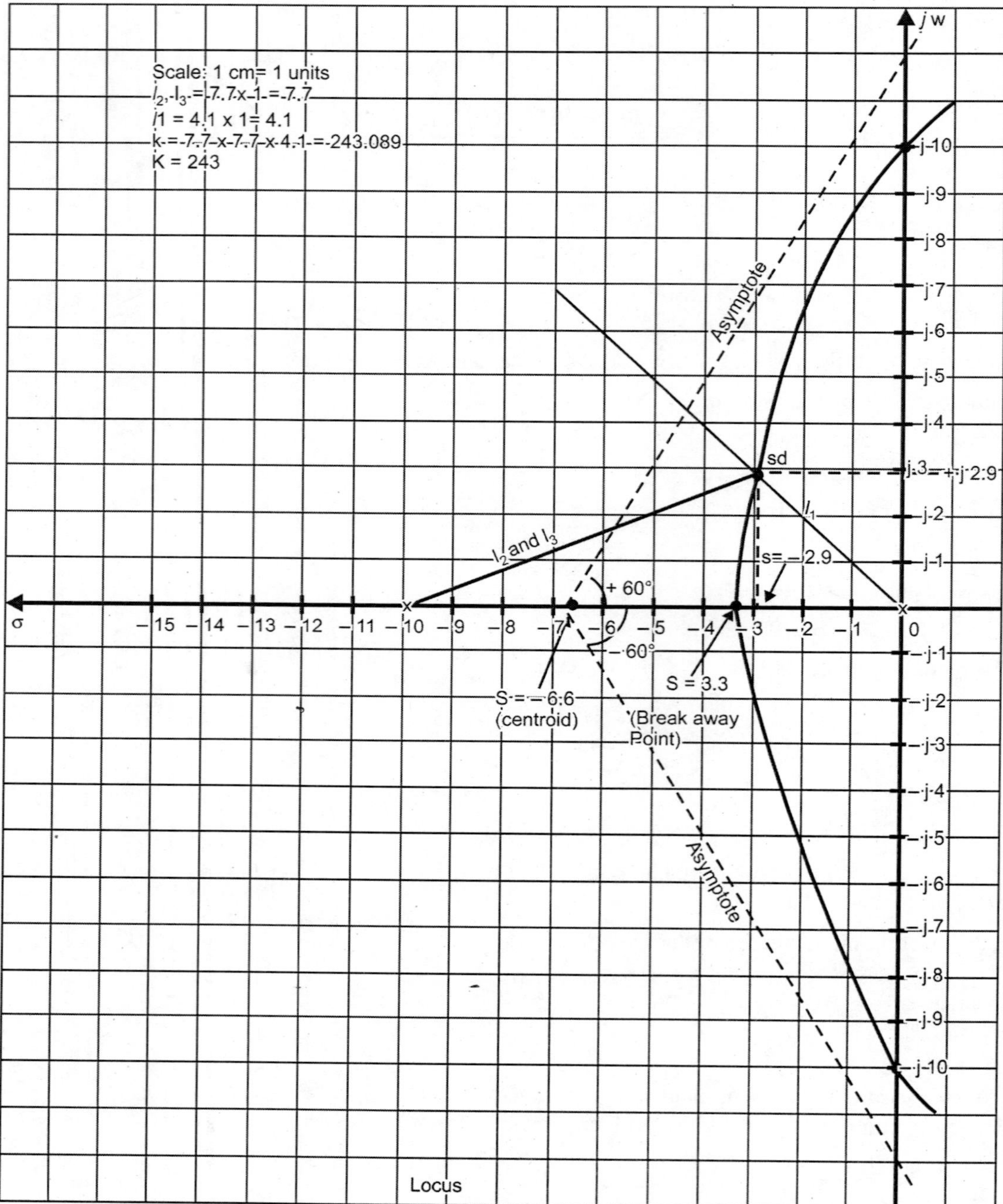

Fig. Ex. 5.4.1: Root locus of $G(s) = K/s(s + 10)^2$ (lag compensator).

Product of vector lengths from poles = l_1 Ч l_2 Ч l_3

From root locus plot of Fig. (Ex. 5.5.1) we get

$$l_1 = 4.1\ l_2 = l_3 = 7.7$$

Since there is no finite zero, the product of vector lengths from zeros is unity

$\therefore$ $K = l_1$ Ч l_2 Ч l_3 = 4.1 Ч 7.7 Ч 7.7 = 243.089 ≈ 240

Step 4: To find parameter, β

Given that, $G(s) = K/s\,(s+10)^2 = 240/s\,(s+10)^2$

Velocity error constant of uncompensated system,

$$K_{vu} = \underset{s\to 0}{Lt}\ s.G(s) = \underset{s\to 0}{Lt}\ s\cdot\frac{240}{s(s+10)^2} = \frac{240}{10^2} = 2.4$$

It is given that desired velocity error constant, K_{vu} should be 20.

Let A be the factor by which K_v is increased.

$$\therefore \qquad A = K_{vd}/K_{vu} = \frac{20}{2.4} = 8.33$$

Let, β = 1.2 Ч A = 1.2 Ч 8.33 = 9.996 ≈ 10

Step 5: Determine the transfer function of lag compensator zero of the compensator

z_c = – 1/T = 0.1 Ч second pole of $G(s)$

0.1 Ч (– 10) = – 1.0 1/T = 1 and T = 1.0

Pole of the compensator

p_c = – 1/βT = – 1/10 Ч1 = – 0.1 1/βT = 0.1 and βT = 10

Transfer function of log compensator,

$$G_c(s) = \frac{s+\frac{1}{T}}{s+\frac{1}{\beta T}} = \frac{(s+1)}{(s+0.1)}$$

Step 6: Determine the open loop transfer function of compensated system.

The block diagram of lag compensated system is shown below:

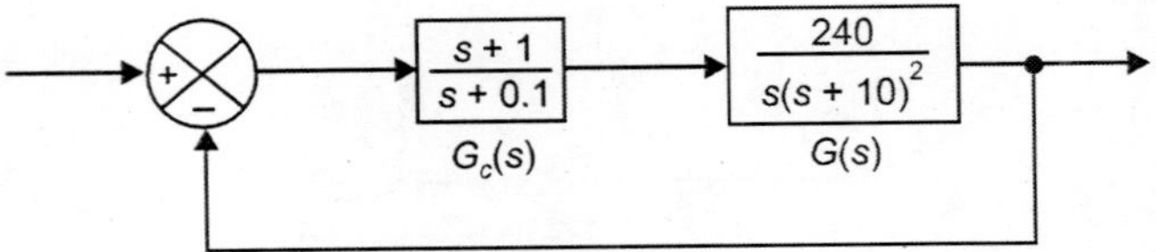

Fig. Ex. 5.4.2

Open loop transfer function of compensated system,

$$G_o(s) = \frac{(s+1)}{(s+0.1))} \times \frac{240}{s(s+10)^2} = \frac{240(s+1)}{s(s+10)^2(s+0.1)}$$

Step 7: Verify K_v of compensated system velocity error constant of compensated system,

$$K_{vc} = \underset{s\to 0}{Lt}\ s \cdot G_o(s) = \underset{s\to 0}{Lt}\ s \cdot \frac{240(s+1)}{s(s+10)^2(s+0.1)}$$

$$= \frac{240}{10^2 \times 0.1} = 24$$

Conclusion: Since the velocity error constant of the compensated system is greater than the desired value, the design is accepted.

Result: Transfer function of lag compensator,

$$G_c(s) = \frac{240(s+1)}{s(s+10)^2(s+0.1)}$$

5.5 LEAD COMPENSATOR

A compensator having the characteristics of a lead network is called a lead compensator. If a sinusoidal signal is applied to the lead network, then in steady state the output will have a phase lead with respect to the input.

The lead compensation increases the band width, which improves the speed of response and also reduces the amount of overshoot. Lead compensation appreciably improves the transient response, whereas there is small change in steady state accuracy. Generally, lead compensation is provided to make unstable system as a stable system.

A lead compensator is basically a high pass filter and so it amplifies high frequency noise signals. If the pole introduced by the compensator is not cancelled by a zero in the system, then lead compensation increases the order of the system by one.

s-plane Representation of Lead Compensator

The lead compensator has a zero at $s = -1/T$ and a pole at $s = -1/\alpha T$. The pole – zero plot of lead compensator is shown below, Fig. (5.9)

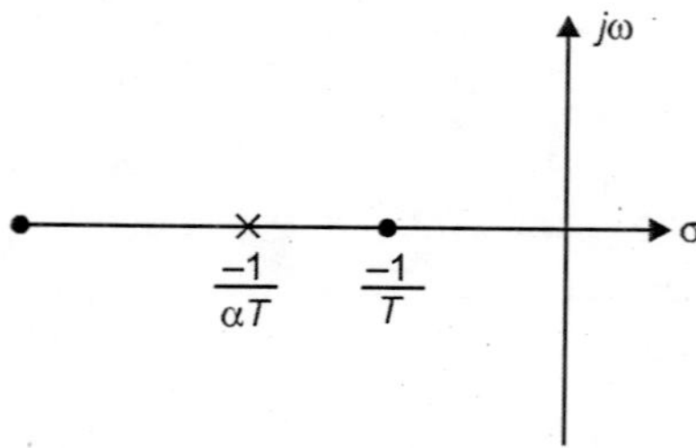

Fig. 5.9: Pole zero of lead compensator.

Here, ∝ < 1, so the zero is closer to the origin than the pole. The general form of lead compensator transfer function is given by Eq. (5.19)

$$G_c(s) = \frac{s + z_c}{s + P_c} = \frac{\left(s + \frac{1}{T}\right)}{s + \frac{1}{\alpha T}} \tag{5.19}$$

where $T > 0$ and $\alpha < 1$

The zero of lead compensator,

$$z_c = 1/T \tag{5.20}$$

The pole of lead compensator,

$$P_c = 1/\beta T \tag{5.21}$$

From Eq. (5.20) we get,

$$T = 1/z_c \tag{5.22}$$

From Eq. (5.21) we get,

$$\alpha = z_c / P_c \tag{5.23}$$

Lead Compensator using Electrical Network

The lead compensator can be realised by the *RC* network shown below Fig. (5.10)

Let, $E_i(s)$ = Input Voltage

$E_o(s)$ = Output Voltage

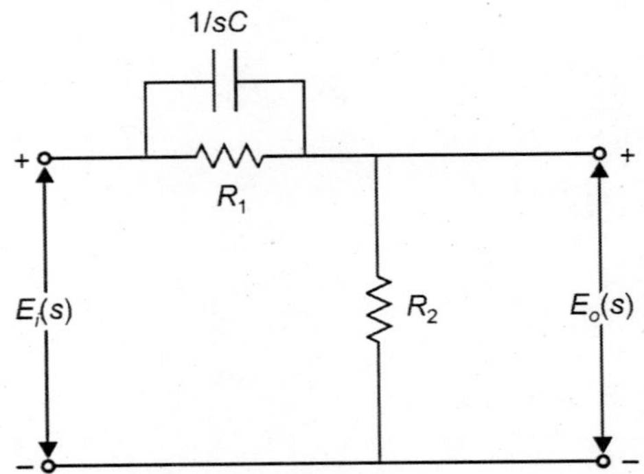

Fig. 5.10: Electrical lead compensator.

In the network shown in Fig. (5.10), the input voltage is applied to the series combination of $(R_1 || C)$ and R_2. The output voltage is obtained across R_2.

By voltage division rule,

Output voltage, $$E_o(s) = E_i(s)\frac{R_2}{R_2 + \frac{\left(R_1 \times \frac{1}{sC}\right)}{\left(R_1 + \frac{1}{sC}\right)}}$$

$$E_o(s) = E_i(s)\frac{R_2}{R_2 + \dfrac{R_1}{(R_1Cs+1)}}$$

$$= E_i(s)\frac{R_2}{\dfrac{R_2(R_1Cs+1)+R_1}{R_1Cs+1}}$$

The transfer function of the electrical network is the ratio of output voltage to input voltage.

Transfer function of electrical network

$$\frac{E_o(s)}{E_i(s)} = \frac{R_2(R_1Cs+1)}{[R_1R_2Cs + R_2 + R_1]} = \frac{R_1CR_2\left[s+\dfrac{1}{R_1C}\right]}{R_1CR_2\left[s+\dfrac{(R_1+R_2)}{R_1CR_2}\right]}$$

$$= \frac{\left[s+\dfrac{1}{R_1C}\right]}{\left[s+\left(\dfrac{1}{\dfrac{R_2}{(R_1+R_2)}}\right)\dfrac{1}{R_1C}\right]} \tag{5.24}$$

The general form of lead compensator transfer function is

$$G_o(s) = \frac{\left(s+\dfrac{1}{T}\right)}{s+\dfrac{1}{\alpha T}} \tag{5.25}$$

On comparing Eqs. (5.24) and (5.25) we get,

$$\frac{E_o(s)}{E_i(s)} = \frac{s+\dfrac{1}{T}}{\left(s+\dfrac{1}{\alpha T}\right)} \tag{5.26}$$

where $T = R_1C$ and $\alpha = R_2 / R_1 + R_2$

The transfer function of the *RC* network is similar to the general form of transfer function of lead compensator.

Frequency Response of Lead Compensator

Consider the general form of lead compensator.

$$G_o(s) = \frac{s+\dfrac{1}{T}}{\left(s+\dfrac{1}{\alpha T}\right)} = \frac{\dfrac{(1+sT)}{T}}{\dfrac{(s\alpha T+1)}{\alpha T}} = \alpha\frac{(1+sT)}{(1+\alpha sT)} \tag{5.27}$$

The sinusoidal transfer function of lead compensator is obtained by letting $s = j\omega$ in Eq. (5.27)

$$\therefore \quad G_c(j\omega) = \alpha \frac{(1 + j\omega T)}{(1 + j\omega\alpha T)} \tag{5.28}$$

When $\omega = 0$, $\quad G_c(j\omega) = \alpha$ (5.29)

From equation (5.30) we can say that the lead compensator provides an attenuation of α ($\alpha < 1$). If the attenuation of the compensator is not desirable then it can be eliminated by a suitable amplifier.

Let us assume that the attenuation α is eliminated by a suitable amplifier network. Now, $G_c(j\omega)$ is given by,

$$G_c(j\omega) = \frac{(1 + j\omega T)}{(1 + j\omega\alpha T)} = \frac{\sqrt{1+(\omega T)^2}\angle\tan^{-1}\omega T}{\sqrt{1+(\omega\alpha T)^2}\angle\tan^{-1}\omega T} \tag{5.31}$$

The sinusoidal transfer function shown in equation (4.5.12) has two corner frequencies ω_{ci} and ω_{c2}

Here, $\omega_{c1} = 1/T$ and $\omega_{c2} = 1/\alpha\, T$

Since, $T > \alpha T$, $\omega c_1, < \omega_{c2}$

Let $\quad A = |G_c(j\omega)|$ in db = $20 \log \dfrac{\sqrt{1+(\omega t)^2}}{\sqrt{1+(\omega\alpha T)^2}}$ (5.32)

At very low frequencies i.e upto ω_{c1}, $\omega_T << 1$ and $\omega\, \alpha\, T << 1$

$$\therefore \quad A \approx 20 \log 1 = 0$$

In the frequency range from ω_{c1} to ω_{c2}, $\omega_T >> 1$ and ω □ $\alpha\, T << 1$

$$\therefore \quad A \approx 20 \log\sqrt{(\omega T)^2} = 20 \log (\omega_T)$$

At very high frequencies i.e after ω_{c2}, $\omega T >> 1$ and $\omega\, \alpha\, T >> 1$

$$\therefore \quad A \approx 20 \log \frac{\sqrt{(\omega T)^2}}{(\omega\alpha T)^2} = 20 \log 1/\alpha$$

The approximate magnitude plot of lead compensator is shown in Fig. (5.11).

The magnitude plot of Bode plot of $G_c(j\omega)$ is a straight line through $0db$ upto ω_{c1}, then it has a slope of +20 db/decode upto ω_{c2} and after ω_{c2} it is a straight line with a constant gain of 20 log $(1/\alpha)$

Let $\phi = \angle\, G_c\,(j\omega)$

$$\therefore \quad \phi = \tan^{-1}\omega T - \tan^{-1}\omega\alpha T$$

As $\quad \omega \to 0, \quad \phi = \to 0$

As $\quad \omega \to \infty, \quad \phi = \to 0$

As ω is varied from 0 to ∞, the phase angle increases from 0 to a maximum value of ϕ_m at $\omega = \omega_m$, then decreases from maximum value to 0.

It can be shown that the frequency at which maximum phase lead occurs in the geometric mean of the two corner frequencies.

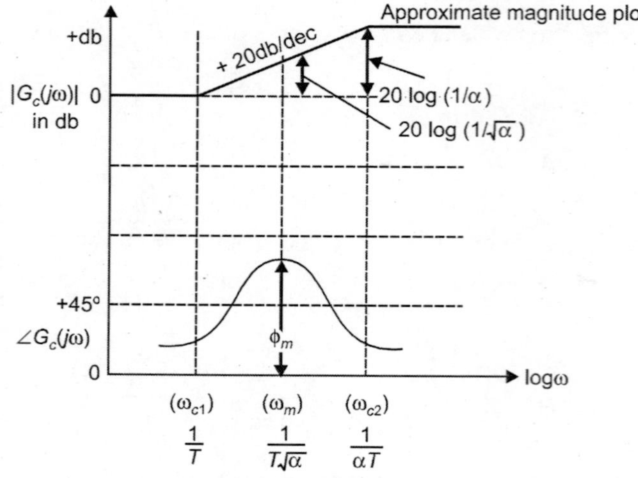

Fig. 5.11: Bode plot of lead compensator.

Frequency of maximum phase lead,

$$\omega_m = \sqrt{\omega_{c1} \cdot \omega_{c2}} = \sqrt{\frac{1}{T} \cdot \frac{1}{\alpha T}} = \frac{1}{T\sqrt{\alpha}}$$

The choice of α is governed by the inherent noise in control systems. From the Bode plot of the lead network, we observe that the high frequency noise signals are amplified by a factor $\frac{1}{\alpha}$, while the low frequency control signals undergo unit amplification. Thus the signal to noise ratio at the output of the lead compensator is poorer than at its input. To prevent the signal to noise ratio at the output from deteriorating excessively, it is recommended that the value of α should not be elss than 0.07. A typical choice of $\alpha = 0.1$. Also it is advisable to provide two cacaded lead networks when ϕ_m required (i.e., phase lead required) is more than 60є

Determination of ω_m, ϕ_m and α

The frequency ω_m can be determined by differentiating ϕ with respect to ω and equating $\frac{d\phi}{d\omega}$ to zero.

$$\phi = \angle G_c(j\omega) = \tan^{-1} \omega T - \tan^{-1} \alpha\, \omega T \tag{5.32}$$

On differentiating ϕ W.r. to ω and equally to zero, we get,

$$\omega_m = 1/T\sqrt{\alpha}$$

$\therefore$ frequency corresponding to maximum phase lead,

$$\omega_m = \frac{1}{T\sqrt{\alpha}} \tag{5.33}$$

ϕ_m in terms of α $\qquad \phi_m = \tan^{-1}\left[\dfrac{1-\alpha}{2\sqrt{\alpha}}\right]$

α in terms of ϕ_m $\qquad \alpha = \dfrac{1-\sin\phi_m}{1+\sin\phi_m}$

Procedure for Designing Lead Compensator using Bode Plot

Step 1: The open loop gain K of the given system is determined to satisfy the requirement of the error constant.

Step 2: After determining the value of K, draw bode plot of uncompensated system.

Step 3: The phase margin of the uncompensated system determined from the bode plot.

Step 4: Determine the amount of phase angle to be contributed by the lead network by using formula given below:

$$\phi_m = \gamma_d - \gamma + \zeta$$

where, $\phi_m \rightarrow$ Maximum phase lead angle

$r_d \rightarrow$ Desired phase margin

$\gamma \rightarrow$ Phase margin of uncompensated system

$\zeta \rightarrow$ Additional phase lead to compensate for shift in gain crossover frequency.

Choose an initial choice of ζ as$\pm 5\epsilon$

Note: If ϕ_m is more than 60єthen realize the compensator as cascade of two lead compensator with each compensator contributing half of the required angle.

Step 5: Determine the transfer function of lead compensator. Calculate α using the equation,

$$\alpha = \frac{1-\sin\phi_m}{1+\sin\phi_m}$$

From the bode plot, determine the frequency at which the magnitude of $G(j\omega)$ is $-20\log\dfrac{1}{\sqrt{\alpha}}$ *db*.

This frequency is ω_m

Calculate T,

$$\omega_m = \frac{1}{T\sqrt{\alpha}}$$

$$\therefore \qquad T = \frac{1}{\omega_m\sqrt{\alpha}}$$

Transfer function of lead compensatio n,

$$G_c(s) = \frac{s+\dfrac{1}{T}}{s+\dfrac{1}{\alpha T}} = \frac{\alpha(1+sT)}{(1+\alpha sT)}$$

Step 6: Determine the open loop transfer function of compensated system:

The lag compensator is connected in series with $G(s)$ as shown in Fig. (5.12). When the lead network is inserted in series with the plant, the open loop gain of the system is attenuated by the factor α ($\because \alpha < 1$), so an amplifier with the gain of $1/\alpha$ has to be introduced in series with the compensator to nullify the attenuation caused by the lead compensator.

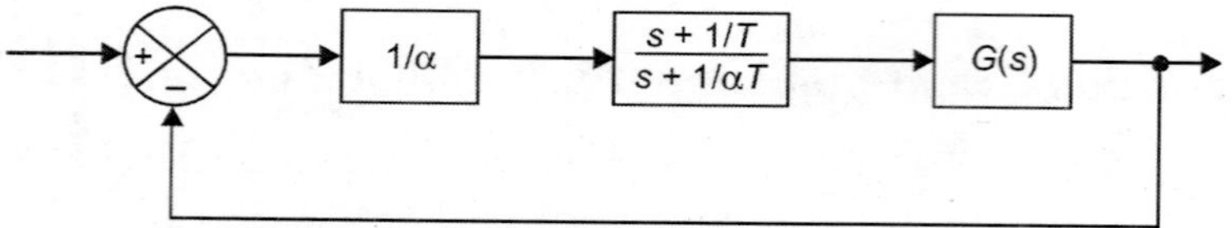

Fig. 5.12: Block diagram of lead compensated system.

Open loop transfer function of the overall system

$$G_o(s) = \frac{1}{\alpha} \frac{s + \frac{1}{T}}{s + \frac{1}{\alpha T}} \cdot G(s)$$

$$= \frac{1}{\alpha} \cdot \frac{\alpha(1 + sT)}{(1 + s\alpha T)} G(s) = \frac{(1 + sT)}{(1 + s\alpha T)} \cdot G(s)$$

Step 7: Verify whether it satisfies the given specifications. If the phase margin of the compensated system is less than the required phase margin then repeat step 4 to 10 by taking $\in$ as 5ϵ more than previous design.

Procedure for Designing Lead Compensator using Root Locus

The following steps may be followed to design a lead compensator using root locus and to be connected in series with transfer function of uncompensated system, $G(s)$.

Step 1: Determine the dominant pole,

$$s_d = -\zeta\omega_n \pm j\omega_n \sqrt{1 - \zeta^2}$$

[If ζ along is specified and ω_n is not available, then draw the root locus and from the root locus find the dominant pole]

Step 2: Mark the poles and zeros of open loops transfer function and the dominant pole on the s-plane. Let the dominant pole be point P.

Step 3: Determine the angle to be contributed by lead network.

Let, ϕ = Angle to be contributed by lead network to make the point P on root locus.

Draw vectors from all open loop poles and zeros to point P. Measure the angle contributed by the vectors.

$$\text{Now,} \qquad \phi = \begin{bmatrix} \text{Sum of angles contributed} \\ \text{by poles of uncompensated} \\ \text{system} \end{bmatrix} - \begin{bmatrix} \text{Sum of angles contributed} \\ \text{by zeros of uncompensated} \\ \text{system} \end{bmatrix} \pm n\,180є$$

Where n is an odd integer so that n 180єis nearest to the difference between angles contributed by poles and zeros.

[If the angle to be contributed is more than 60є then the compensator as cascade of two lead compensators with each compensator contributing half of the required angle].

Step 4: Determine the pole and zero of the lead compensator. Let point O be the origin of s-plane and point P be the dominant pole. Draw straight lines OP and AP such that AP is parallel to x-axis as shown in Fig. (5.13).

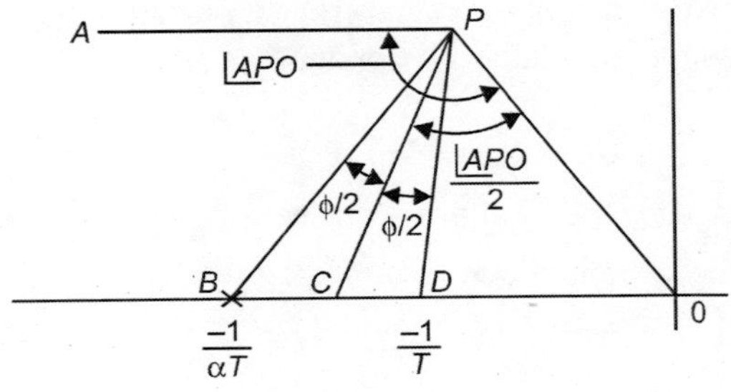

Fig. 5.13

Draw a line PC so as to bisect the angle APO [∠APO] where the point C is on the real axis. With line PC as reference, draw angles BPC and CPD such that each equal to $\phi/2$. Here the points B and D are located on the real axis. Now the point B is the location of the pole of the compensator ($-1/\alpha T$) and the point D is the location of the zero of the compensator ($-1/T$). From the values of point D and B compute T and α.

Step 5: Determine the transfer function of lead compensator.

Transfer function of lead compensator

$$G_c(s) = \frac{\left(s + \dfrac{1}{T}\right)}{\left(s + \dfrac{1}{\alpha T}\right)} = \alpha\,\frac{(1 + sT)}{(1 + s\alpha T)}$$

Step 6: Determine open loop transfer function of lead compensated system.

The lead compensator is connected in series with the plant as shown in Fig. (5.14).

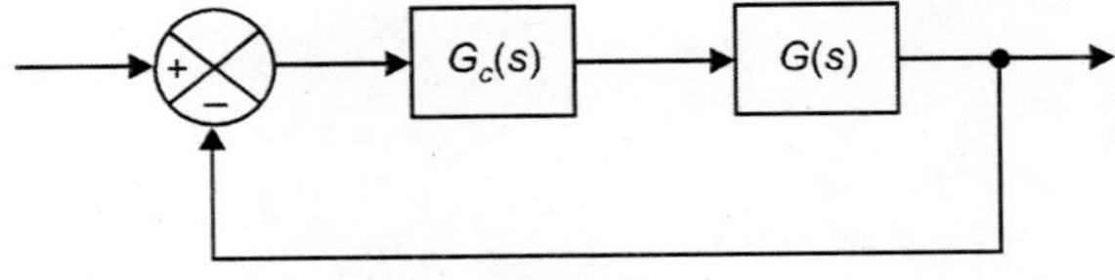

Fig. 5.14

Open loop transfer function of compensated system

$$G_o(s) = G_c(s) \cdot G(s)$$

The open loop gain K is given by the value of gain of $s = s_d$. The value of gain, K is detrmined from pole zero plot of lead compensated system and by using the magnitude condition given below.

$$K = \frac{\text{Product of vector lengths from all poles to } s = s_d}{\text{Product of vector lengths from all zeros to } s = s_d}$$

Step 7: Verify whether the compensated system satisfies the error requirement. If the error requirement is satisfied then the design is accepted. Otherwise repeat the design by altering the location of poles and zeros by trial and error, without changing the value of ϕ.

[If the open loop gain K is specified in the problem, then take the gain of $s = s_d$ as K_1. Find a parameter, A where $A = K_1 / K$. Now introduce an amplifier with gain, A is cascade with compensator to account for reduction in gain due to attenuation by parameter, α. Now, $G_o(s) = A \,.\, G_c(s).\, G(s)$.

EXAMPLE 5.5

Design a phase lead compensator for the system shown below to satisfy the following specifications

(i) The phase margin of the system $\geq$ 45є

(ii) Steady state error for a unit ramp input $\leq 1/15$

(iii) The gain crossover frequently of the system must be less than 7.5 rad/sec

Solution:

Step 1: Determine K

Given that. steady state error,

$$e_{ss} \leq 1/15 \text{ for unit ramp input.}$$

When the input is unit ramp,

$$e_{ss} = 1/K_v$$

$$e_{ss} = 1/15$$

$$\therefore \quad K_v = 15$$

By definition of velocity error constant, K_v we get,

$$K_v = \underset{s \to 0}{Lt}\, s.G(s).H(s)$$

Here, $G(s) = K/s(s + 1)$ [from Fig. (Ex. 5.5.1)]

and $H(s) = 1$ [for unity feedback system]

$$\therefore \quad K_v = \underset{s \to 0}{Lt}\, s \cdot \frac{K}{s(s+1)} = K$$

Step 2: Draw bode plot

Given that, $G(s) = K/s(s + 1) = 15/s(s + 1)$

Let $s = j\omega$

$\therefore$ $G(j\omega) = 15/j\omega\,(1 + j\omega)$

(i) Magnitude plot

The corner frequency is. $\omega_{c1} = 1$ rad/sec. The various terms of $G(j\omega)$ are listed in Table Ex. 5.5.1. Also the table shows the slope contributed by each term and the change in slope at the corner frequency.

Table Ex. 5.5.1

Term	Corner frequency rad/sec	Slope of each term *db*/dec	Change in slope from one term to another term *db*/dec
$15/j\omega$	–	– 20	–
$1/1 + j\omega$	$\omega_{c1} = 1$	– 20	– 20 + – 20 = – 40

Choose a low frequency ω_l such that $\omega_l \angle \omega c_1$ and choose a high frequency ω_h such, that $\omega_h > \omega_{c1}$

Let $\omega_l = 0.1$ rad/sec and $\omega_h = 10$ rad/sec

Let $A = |G(j\omega)|$ in *db*

Let us calculate A at ω_l, $\omega_{c1}$and ω_n

At $\omega = \omega_l$, $A_1 = 20 \log |\text{first term}|$

i.e., $$A_1 = 20 \log \left|\frac{15}{j\omega}\right| = 20 \log \frac{15}{0.1}$$

$$A_1 = 43.5 \approx 44\ db$$

At $\omega = \omega_{c1} = 1$ rad/sec, $A_2 = 20 \log |\text{first term}|$

i.e., $$A_2 = 20 \log \left|\frac{15}{j\omega}\right| = 20 \log \frac{15}{1}$$

$$= 23.5 \approx 24\ db.$$

At $\omega = \omega_h = \omega$ rad/sec, $A_3 = [\text{slope from one term to another term}]$ Ч$\log \left|\frac{\text{Present frequency}}{\text{Previous frequency}}\right| +$ Previous gain (A_2)

$$A_3 = \left[\text{slope from } \omega_{c1} \text{ to } \omega_h \times \log \frac{\omega_h}{\omega_{c1}}\right] + A_2\ (\text{at } \omega = \omega_{c1})$$

$$A_3 = -40 \text{ Ч lag } 10/1 + 24 = -16\ db.$$

CHAPTER 5

Let the points a, b and c be the points corresponding to frequencies ω_l, ω_{c1} and ω_h respectively on the magnitude plot. In a semi log graph sheet choose appropriate scales and fix the points a, b and c. Join the points by straight lines and mark the slope on the respective region. The magnitude plot show in Fig. (Ex. 5.5.1).

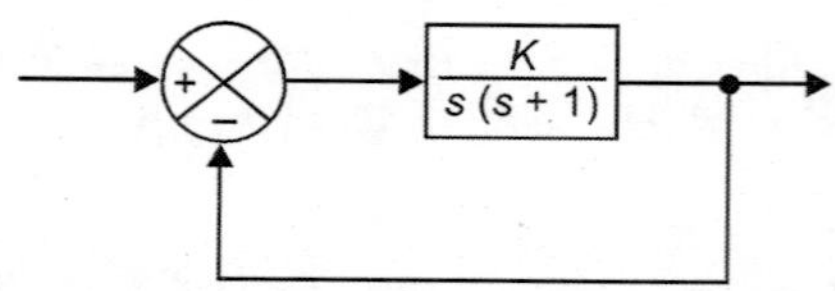

Fig. Ex. 5.5.1

Phase plot

The phase angle of $G(j\omega)$ as a function of ω is given by

$$\phi = \angle G(j\omega) = -90\epsilon - \tan^{-1}\omega$$

The phase angle of $G(j\omega)$ are calculated for various values of ω and listed in Table Ex. 5.5.2.

Table Ex. 5.5.2

ω, rad/sec	0.1	0.5	1	2	5	10
ϕ, deg	– 96	– 117	–135	– 153	– 169	– 174

On the same semi-log sheet-take another y-axis, choose appropriate scale and draw phase plot as shown in Fig. Ex. 5.5.1.

Step 3: Determine the phase margin of uncompensated system

Let, ϕ_{gc} = phase of $G(j\omega)$ at gain crossover frequency (ω_{gc})

γ = phase margin of compensated system.

From bode plot of uncompensated system we get,

$$\phi_{gc} = -167\epsilon$$

Now, $\gamma = 180\epsilon + \phi_{gc} = 180\epsilon - 167\epsilon = 13\epsilon$

The system requires a phase margin of 45є but the available phase margin is 13є and so lead compensation should be employed to improve the phase margin.

Step 4: Find ϕ_m

The desired phase margin, $\gamma_d \geq 45\epsilon$

Let additional phase lead required, $\in = 5\epsilon$

Maximum lead angle, $\phi_m = +\gamma_d - \gamma + \zeta = 45\epsilon - 13\epsilon + 5\epsilon = 37\epsilon$

Step 5: Determine the transfer function of lead compensator.

$$\alpha = \frac{1 - \sin\phi_m}{1 + \sin\phi_m} = \frac{1 - \sin 37\epsilon}{1 + \sin 37\epsilon} = .2486 \approx .25$$

The db magnitude corresponding to ω_m

$$= -2 \log \frac{1}{\sqrt{\alpha}}$$

$$= -20 \log \frac{1}{\sqrt{0.25}} = -6db$$

From the bode plot of uncompensated system the frequency ω_m corresponding to a db gain of –6 db is found to be 5.6 rad / sec.

$\therefore$ $$\omega_m = 5.6 \text{ rad/sec.}$$

Now, $$T = \frac{1}{\omega_m \sqrt{\alpha}} = \frac{1}{5.6\sqrt{0.25}} = 0.357 \approx 0.36$$

Transfer function of the lead compensator

$$G_c(s) = \frac{s + \frac{1}{T}}{s + \frac{1}{\alpha T}}$$

$$= \frac{\alpha(1 + sT)}{(1 + s\alpha T)} = 0.25\frac{(1 + 0.36s)}{(1 + 0.09s)}$$

Step 6: Open loop transfer function of compensated systems. Block diagram of lead compensated given below:

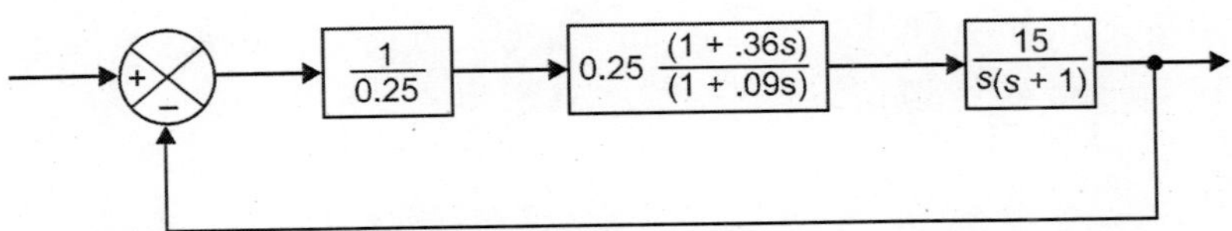

Fig. Ex. 5.5.2

The compensator will provide an attenuation of α. To compensate for that, an amplifier of gain $1/\alpha$ is introduced in series with compensator.

Open loop transfer function of compensated system

$$G_o(s) = \frac{1}{0.25} \times \frac{0.25(1 + 0.36s)}{(1 + 0.09s)} \times \frac{15}{s(s + 1)}$$

$$= \frac{15(1 + 0.36s)}{s(1 + 0.09s)(1 + s)}$$

Step 7: Draw the bode plot of compensated system to verify the design.

Put $$s = j\omega \text{ in } G_o(s)$$

$$\therefore \qquad G_o(j\omega) = \frac{15(1+j36\omega)}{j\omega(1+0.09\omega)(1+j\omega)}$$

Magnitude plot

The corner frequencies are ω_{c1}, ω_{c2} and ω_{c3}

$$\omega_{c1} = 1/1 = 1 \text{ rad/sec.}$$

$$\omega_{c2} = 1/0.36 = 2.8 \text{ rad/sec.}$$

$$\omega_{c3} = 1/0.09 = 11.1 \text{ rad/sec}$$

The various terms of $G_o(j\omega)$ are listed in Table Ex. 5.5.3. Also the table shows the slope contributed by each ter and the change in slope at the corner frequency.

Table Ex. 5.5.3

Term	Corner frequency rad/sec	Slope of each term *db*/dec	Change in slope from one term to another term *db*/dec
$15/j\omega$	–	–20	–
$1/1 + j\omega$	$\omega_{c1} = 1$	–20	– 20 – 20 = – 40
$1 + j\,0.6\omega$	$\omega c_2 = 1/0.36 = 2.8$	+20	– 40 +20 = – 20
$\frac{1}{1+j0.09\omega}$	$\omega_{c3} = 1/0.09 = 1.11$	– 20	– 20 – 20 = – 40

Choose a low frequency ω_l such that $\omega_l \angle \omega_{c1}$ and choose a high frequency ω_h such that $\omega_h > \omega_{c3}$. Let $\omega_l = 0.1$ rad/sec and $\omega_h = 50$ rad/sec

Let, $A_o = |G_o(j\omega)|$ in *db*

At $\omega = \omega_l = 0.1$ rad/sec, $A_{o1} = 20\log\left|\frac{15}{j\omega}\right| = 20\log\frac{15}{0.1} = 44db$ $[A_{o1} = 20 \log |\text{first term}|]$

At $\omega = \omega_{c1} = 1$ rad/sec, $A_{o2} = 20\log\left|\frac{15}{j\omega}\right| = 20\log\frac{15}{1} = 24db$ $[A_{o2} = 20 \log |\text{first term}|]$

At $\omega = \omega_{c2}$, 2.8 rad/sec,

$$A_{o3} = [\text{slope from one term to another term}] \text{Ч} \log\left|\frac{\text{Present frequency}}{\text{Previous frequency}}\right| + \text{previous gain}$$

$$A_{o3} = \left[\text{slope from } \omega_{c1} \text{ to } \omega_{c2} \times \log\frac{\omega_{c2}}{\omega_{c1}}\right] + [\text{gain at } \omega = \omega_{cl}\ (A_{o2})]$$

$$A_{o3} = -40 \text{ Ч} \log\frac{2.8}{1} + 24 = 6\ db$$

At $\omega = \omega_{c3} = 11.1$ rad/sec,

$$A_{o4} = [\text{slope from one term to another term}]\ \text{Ч}\ \log\left|\frac{\text{Present frequency}}{\text{Previous frequency}}\right| + \text{previous gain}$$

i.e., $$A_{o4} = \left[\text{slope from } \omega_{c2} \text{ to } \omega_{c3} \times \log\frac{\omega_{c3}}{\omega_{c2}}\right] + [\text{gain at } \omega = \omega_{c2}\ (A_{o3})]$$

$$A_{o4} = -20\ \text{Ч}\log\frac{11.1}{2.8} + 6 = -\ 6db$$

At $\omega = \omega_h = 50$ rad/sec,

$$A_{o5} = \left[\text{slope from } \omega_{c3} \text{to } \omega_h \times \log\frac{\omega_h}{\omega_{c3}}\right] + [\text{gain at } \omega = \omega_{c3}(A_{o4})]$$

$$A_{o5} = -\ 40 \times \log\frac{50}{11.1} + (-\ 6) = -\ 32db$$

Let the points *a, b, c, d, e* and *f* be the points corresponding to frequencies ω_l, ω_{c1}, ω_{c2}, ω_{c3} and wn respectively on the magnitude plot of compensated system.

The magnitude plot of compensated system is drawn on the same semilog sheet by using the same scales as in Fig. Ex. 5.5.1

Phase plot

The phase angle of $G_o(j\omega)$ as a function of ω is given by

$$\phi_o = [G_o(j\omega) = \tan^{-1} 0.36\omega - 90є - \tan^{-1} 0.09\omega - \tan^{-1}\omega.$$

The phase angle of $G_o(j\omega)$ are calculated for various values of ω and listed in Table Ex. 5.5.4.

Table Ex. 5.5.4

ω, rad/sec	0.1	0.5	1	2	5	10
ϕ, deg	– 94	– 109	–120	– 128	– 132	– 142

In the same semilog sheet and by using scales, the phase plot of compensated system is sketched.

Let, ϕ_{gc} = phase of $G_o(j\omega)$ at new gain crossover frequency

γ_o = phase margin of compensated system.

From the bodo plot of compensated system we get,

$$\phi_{gco} = -134є$$

$$\gamma_o = 180є + \phi_{gco}$$

$$= 180є - 134є = 46є$$

Conclusion: The phase margin of the compensated system is satisfactory. Hence the design is acceptable.

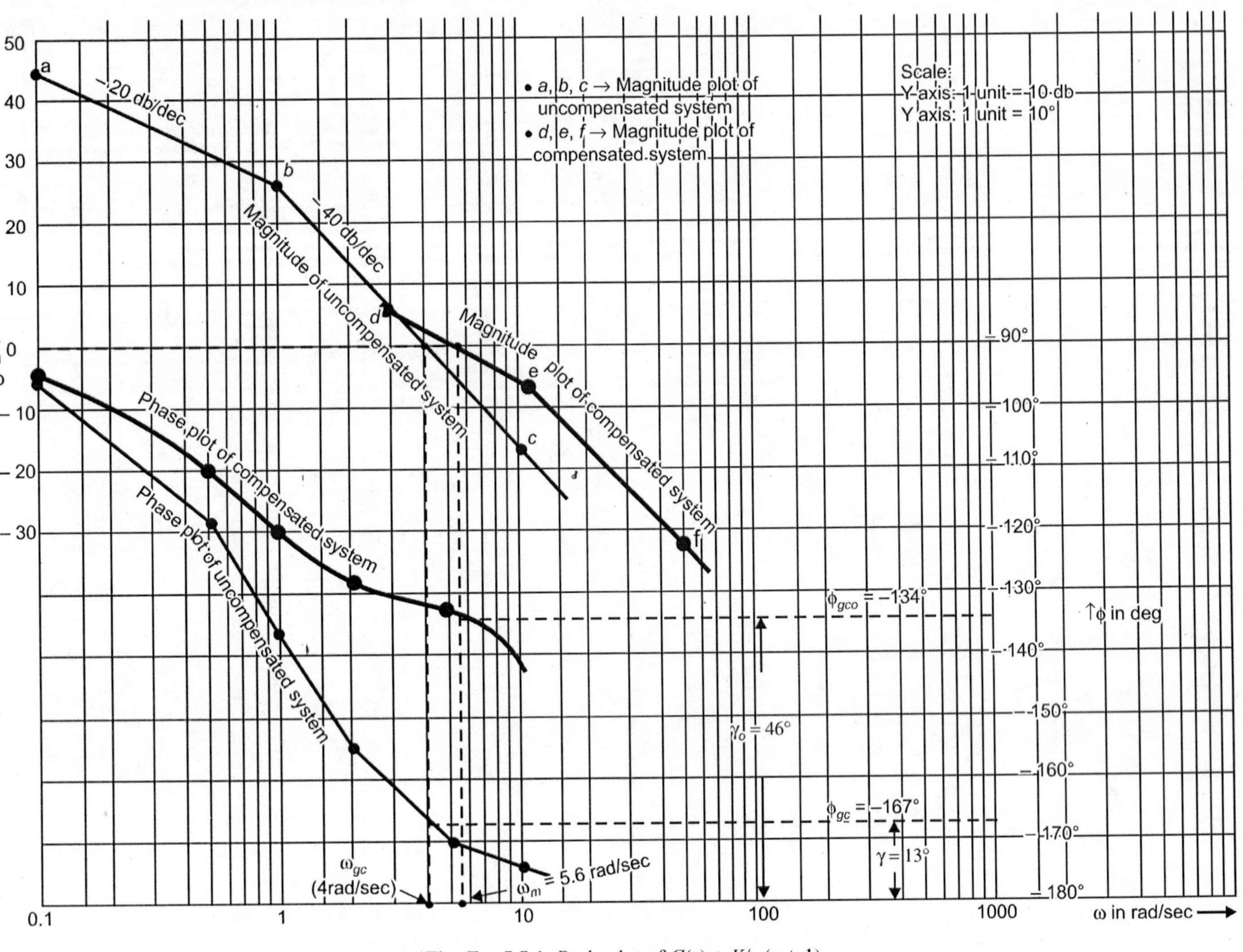

Fig. Ex. 5.5.1: Bode plot of $G(s) + K/s\,(s + 1)$.

Result: The transfer function of lead compensator

$$G_c(s) = \frac{0.25(1+0.36s)}{(1+0.09s)} = \left(\frac{s+2.78}{s+11.11}\right)$$

Open loop transfer function of lead compensated system

$$G_o(s) = \frac{15(1+0.36s)}{s(1+0.09s)\,(1+s)}$$

EXAMPLE 5.6

Design a lead compensator for a unity feedback system with open loop transfer function $G(s) = K/s$ $(s + 1)\,(s + 5)$ to satisfy the following specifications:

(i) Velocity error constant, $K_v \geq 50$; (ii) Phase margin is $\geq$ 20.

Solution:

Step 1: Determine K

Given that, $K_v \geq 50$,

Let $K_v = 50$

By definition of velocity error constant K_v we get,

$$\underset{s\to 0}{Lt}\; s\cdot G(s) = \underset{s\to 0}{Lt}\; s\cdot\frac{K}{s(s+1)(s+5)} = \frac{K}{5} \quad [\text{Œ}H(s) = 1 \text{ for unity feedback system}]$$

$\therefore$ $K = 5\ \text{Ч}\ K_v = 5\ \text{Ч}\ 50 = 250$

Step 2: Draw bode plot

Given that,

$$G(s) = \frac{K}{s(s+1)\,(s+5)} = \frac{250}{s(s+1)(s+5)}$$

$$= \frac{250}{s(1+s)\times 5\times\left(1+\frac{s}{5}\right)} = \frac{50}{s(1+s)(1+0.2s)}$$

Let $s = j\omega$

$$\therefore \quad G(j\omega) = \frac{50}{j\omega(1+j\omega)\,(1+j.0.2\omega)}$$

(i) Magnitude plot

The corner frequency are,

$\omega_{c1} = 1$ rad/sec and

$\omega_{c2} = 1/0.2 = 5$ rad/sec.

The various terms of $G(j\omega)$ are listed in below Table Ex. 5.6.1. Also the table shows the slope contributed by each term and the change in slope at the corner frequencies.

Table Ex. 5.6.1

Term	Corner frequency rad/sec	Slope of each term *db*/dec	Change in slope from one term to another term *db*/dec
$\frac{50}{j\omega}$	–	– 20	
$\frac{1}{1+j\omega}$	$\omega_{c1} = 1$	– 20	$-2 + (-20) = 40$
$\frac{1}{1+j.2\omega}$	$\omega_{c2} = 1/.2 = 5$	– 20	$-40 - 20 = -60$

Choose a low frequency ω_l such that $\omega_l < \omega_{c1}$ and choose a high frequency ω_h such that $\omega_h > \omega_{c2}$

Let $\omega_l = 0.5$ rad/sec and $\omega_h = 10$ rad/sec.

At $\omega = \omega_l = .5$ rad/sec, $\quad A_1 = 20 \log |\text{first term}|$

i.e., $$A_1 = 20 \log \left|\frac{50}{j\omega}\right| = 20 \log\left[\frac{50}{.5}\right] = 40db$$

At $\omega = \omega_{c1} = 1$ rad/sec, $\quad A_2 = 20 \log |\text{first term}|$

i.e., $$A_2 = 20 \log \left|\frac{50}{j\omega}\right| = 20 \log\left[\frac{50}{1}\right] = 34db$$

At $\omega = \omega_{c2} = 5$ rad/sec,

$$A_3 = [\text{slope from one term to another term}] \text{ Ч} \log \left|\frac{\text{Present frequency}}{\text{Previous frequency}}\right| + \text{Previous gain}$$

i.e., $$A_3 = \left[\text{slope from } \omega_{c1} \text{ to } \omega_{c2} \times \log\left(\frac{\omega_{c2}}{\omega_{c1}}\right)\right] + A_2$$

$$A_3 = 40 \text{ Ч} \log 5/1 + 34 = 6\ db$$

At $\omega = \omega_h = 10$ rad/sec,

$$A_4 = [\text{slope from one term to another term}] \text{ Ч} \log \left|\frac{\text{Present frequency}}{\text{Previous frequency}}\right| + \text{Previous gain}$$

i.e., $$A_4 = \left[\text{slope from } \omega_{c2} \text{ to } \omega_n \times \log\frac{\omega_h}{\omega_{c2}}\right] + A_3$$

$$A_4 = 60 \text{ Ч} \log\left(\frac{10}{5}\right) + 6 = -12db$$

Let the points *a*, *b*, *c* and *d* be the points corresponding to frequency ω_l, ω_{c1}, ω_{c2} and ω_n respectively on the magnitude plot. In a semilog graph sheet choose appropriate scales and fix the points *a*, *b*, *c* and *d*. Join the points by straight lines and mark the slope on the respective region. The magnitude plot is shown in Fig. (Ex. 5.6.1)

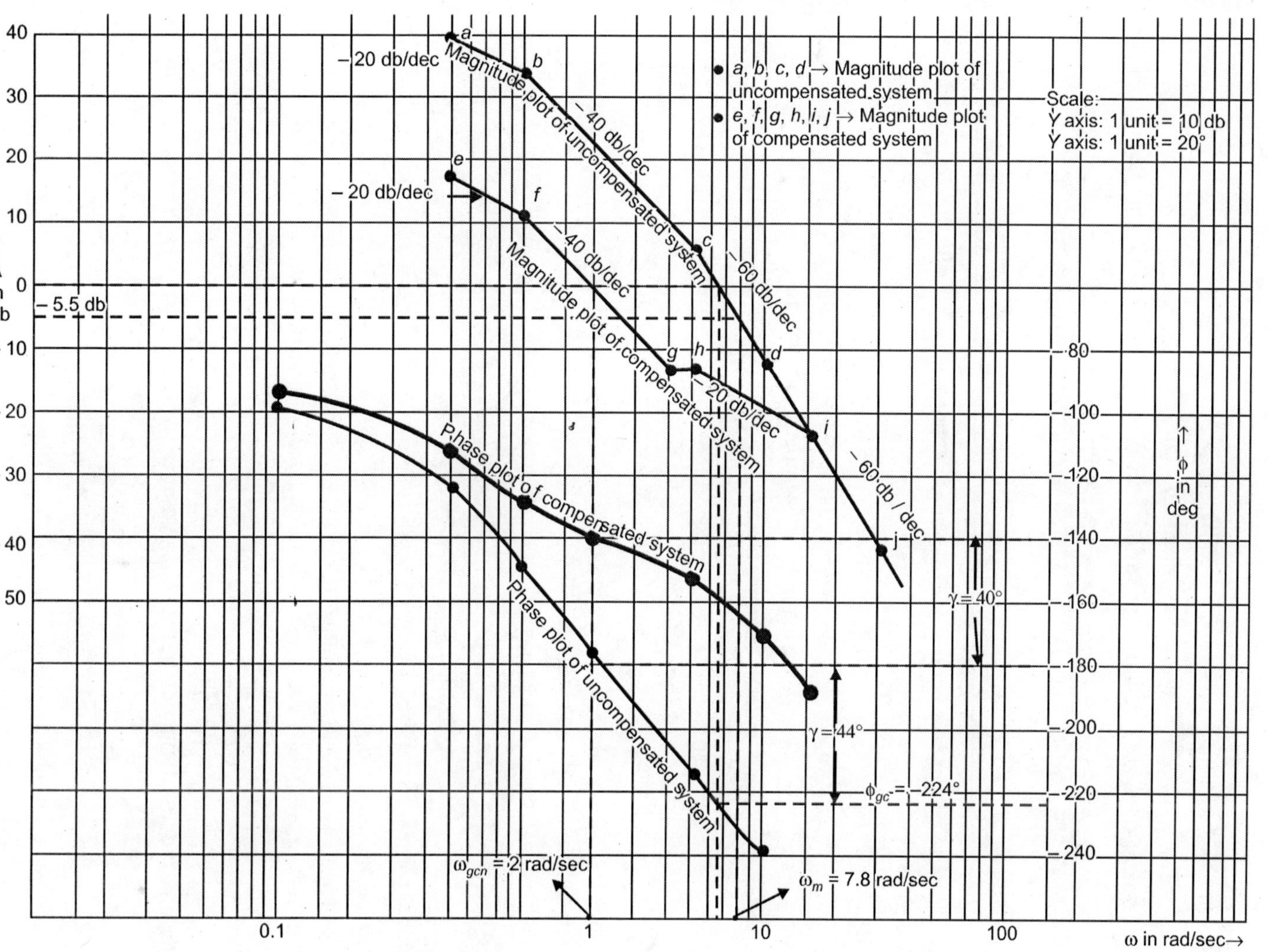

Fig. Ex. 5.6.1: Bode plot of $G(s) = K/s(s + 1)\ (s + 5)$.

Phase plot

The phase angle of $G(j\omega)$ as a function of ω is given by

$$\phi = G(j\omega) = -\,90є\tan^{-1}\,\omega - \tan^{-1}.\,2\,\omega.$$

The phase angle of $G(\omega)$ are calculated for various values of ω and listed in Table (Ex. 5.6.2).

Table Ex. 5.6.2

ω, rad/sec	0.1	0.5	1.0	5	10
ϕ, deg	– 96	– 122	–146	– 214	– 238

On the same semilog sheet take another y-axis, choose appropriate scale and draw phase plot as shown in Fig. (Ex. 5.6.1)

Step 3: Determine Phase margin:

Let, ϕ_{gc} = phase of $G(j\omega)$ at gain cross over frequency

and γ = phase margin of uncompensated system.

From bode plot of uncompensated system we get,

$$\phi_{gc} = 224є$$

Now,

$$g = 180є + \phi_{gc}$$

$$= 180є - 224є = -\,44є$$

Since the phase margin is negative the system is unstable, so we used lead compensation to make the system stable and to have phase margin of 20є.

Step 4: Find ϕ_m

The desired phase margin,

$$\gamma_d \geq 20є$$

Let additional phase lead required ! $\in$ = 5є

Maximum lead angle, $\phi_m = \gamma_d - \gamma + \in = 20 + 44є + 5є = 69є$

Since the lead angle required is greater than 60є, we have to realise the lead compensator as cascade of two lead compensator with each compensator providing half of the required phase lead angle.

$$\therefore \qquad \phi_m = \frac{69є}{2} = 34.5є$$

Step 5: Determine the transformation of lead compensator.

$$\alpha = \frac{1 - \sin\phi_m}{1 + \sin\phi_m} = \frac{1 - \sin 39.5є}{1 + \sin 34.5є} = \cdot 28$$

The db magnitude corresponding to ω_m

$$= -\,20\log\frac{1}{\sqrt{\alpha}}$$

$$= -20\log\frac{1}{\sqrt{0.28}} = -5.5\ db$$

From the bode plot of uncompensated system the frequency $\omega_m$corresponding to a *db*gain at –5.5 *db* is found to be 7.8 rad/sec.

$\therefore$ $$\omega_m = 7.8 \text{ rad/sec}$$

Now, $$T = \frac{1}{\omega_m\sqrt{\alpha}} = \frac{1}{7.8\sqrt{.28}} = .24$$

Transfer function of the lead compensation

$$G_c(s) = \frac{\left(s+\frac{1}{T}\right)^2}{\left(s+\frac{1}{\alpha T}\right)^2} = \alpha^2\frac{(1+sT)^2}{(1+s\alpha T)^2}$$

$$= (.28)^2\frac{(1+.24s)^2}{(1+.28\times .24s)^2} = \cdot 0784\frac{(1+\cdot 24s)^2}{(1+\cdot 067s)^2}$$

$$G_c(s) = \cdot 0784\frac{(1+\cdot 24s)^2}{(1+\cdot 067s)^2}$$

Step 6: Open loop transfer function of compensated system

The block diagram of the compensated system is shown in Fig. (Ex. 5.6.2)

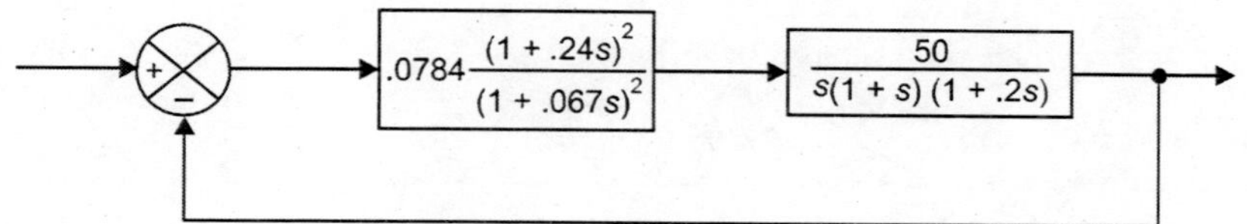

Fig. Ex. 5.6.2: Block diagram of compensated system.

The attenuation provided by the compensator can be retained to reduce the large value of open loop gain, so that the unstable system can be easily brought to stable region.

Open loop transfer function of compensated system

$$G_o(s) = \frac{(.0784)(1+\cdot 24s)^2}{(1+\cdot 067s)^2}\times\frac{50}{s(1+s)(1+\cdot 25)}$$

$$\boxed{G_o(s) = \frac{4(1+\cdot 24s)^2}{s(1+s)(1+\cdot 2s)(1+\cdot 067s)^2}}$$

Step 7: Draw the bode plot of compensated system to verify the design

Put $$s = j\omega \text{ in } G_o(s)$$

$$G_o(j\omega) = \frac{4(1 + j \cdot 24\omega)^2}{j\omega(1 + j\omega)(1 + j0.2\omega)(1 + j0.067\omega)^2}$$

(i) Magnitude plot

The corner frequencies are ω_{c1}, ω_{c2}, ω_{c3} and ω_{c4}

$$\omega_{c1} = 1$$

$$\omega_{c2} = \frac{1}{.24} = 4.2$$

$$\omega_{c3} = \frac{1}{.2} = 5$$

$$\omega_{c4} = \frac{1}{0.067} = 15$$

The various terms of $G_o(i\omega)$ are listed in Table (Ex. 5.6.3)

Choose a low frequency ω_l such that $\omega_l < \omega_{c1}$ and choose a high frequency ω_h such that $\omega_h > \omega_{c4}$

Let, $\omega_l = 0.5$ rad/sec and $\omega_h = 30$ rad/sec.

Let $A_o = |G_o(j\omega)|$ in db

Table Ex. 5.6.3

Term	Corner frequency rad/sec	Slope of each term *db*/dec	Change in slope from one term to another term *db*/dec
$\frac{4}{j\omega}$	–	– 20	
$\frac{1}{1 + j\omega}$	$\omega_{c1} = 1$	– 20	– 20 – 20 = – 40
$(1 + .24\,\omega)^2$	$\omega_{c2} = 4.2$	+ 40	– 40 + 40 = 0
$\frac{1}{(1 + j0.2\omega)}$	$\omega_{c3} = 5$	– 20	0 – 20 = – 20
$\frac{1}{(1 + j0.067\omega)^2}$	$\omega_{c4} = 15$	– 40	– 20 – 40 = – 60

At $\omega = \omega l$, $\quad A_{o1} = 20 \log |\text{first term}|$

i.e., $\quad A_{o1} = 20 \log \left|\dfrac{4}{j\omega}\right| = 20 \log \dfrac{4}{-5} = 18db$

At $\omega = \omega_{c1}$ $\quad A_{o2} = 20 \log |\text{first term}|$

i.e., $\quad A_{o2} = 20 \log \left|\dfrac{4}{j\omega}\right| = 20 \log \dfrac{4}{1} = 12db$

At $\omega = \omega_{c2}$,

$$A_{o3} = [\text{slope from one term to another term}]\ \text{Ч}\ \log \left|\frac{\text{Present frequency}}{\text{Previous frequency}}\right| + \text{Previous gain}$$

i.e.,

$$A_{o3} = \left[\text{slope from } \omega_{c1} \text{ to } \omega_{c2} \times \log \frac{\omega_{c2}}{\omega_{c1}}\right] + A_{o2} \text{ at } (\omega = \omega_{c1})$$

$$= -40 \times \log \left(\frac{4.2}{1}\right) + 12 = -13db$$

At $\omega = \omega_{c3}$,

$$A_{o4} = [\text{slope from one term to another term}]\ \text{Ч}\ \log \left|\frac{\text{Present frequency}}{\text{Previous frequency}}\right| + \text{Previous gain}$$

i.e.,

$$A_{o4} = \left[\text{slope from } \omega_{c3} \text{to } \omega_{c4} \times \log \frac{\omega_{c3}}{\omega_{c2}}\right] + A_o\,(\text{at } \omega = \omega_{c3})$$

$$= 0 \times \text{dog} \frac{5}{4.2} + (-13) = -13db$$

At $\omega = \omega_{c4}$,

$$A_{o5} = [\text{slope from one term to another term}]\ \text{Ч}\ \log \left|\frac{\text{Present frequency}}{\text{Previous frequency}}\right| + \text{Previous gain}$$

i.e.,

$$A_{o5} = \left[\text{slope from } \omega_{c3} \text{to } \omega_{c4} \times \log \frac{\omega_{c4}}{\omega_{c3}}\right] + 40\ (\text{at } \omega = \omega_{c4})$$

$$= -20 \log \left(\frac{15}{5}\right) + (-13) = -22.5 \approx -23db$$

At $\omega = \omega_h$

$$A_{o6} = [\text{slope from one term to another term}]\ \text{Ч}\ \log \left|\frac{\text{Present frequency}}{\text{Previous frequency}}\right| + \text{Previous gain}$$

$$A_{o6} = \left[\text{slope from } \omega_4 \text{ to } \omega_n \text{ Ч } \log \frac{\omega_h}{\omega_{c4}}\right] + A_{o5} (\text{ at } \omega = \omega_h)$$

$$= -20 \log\left(\frac{30}{15}\right) + (-23) = -41\ db$$

Let the points *e, f, g, h, i* and *j* be the points corresponding to frequency ω_l, ω_{c1}, ω_{c2}, ω_{c3}, ω_{c4} and ω_h respectively on the magnitude plot of compensated system. The magnitude plot of compensated system. is drawn on the same semilog sheet using same scales as shown in Fig. (Ex. 5.6.1)

Phase plot

The phase angle of $G_o(j\omega)$ as a function ω is given by

$$\phi_o = \angle G_o(j\omega) = 2\tan^{-1} \cdot 24\omega - 90є - \tan^{-1}\omega - \tan^{-1} \cdot 2\omega - 2\tan^{-1} \cdot 067\omega$$

The phase angle of $G_o(j\omega)$ are calculated for carious values of ω and listed in Table Ex. (5.6.4)

Table Ex. 5.6.4

ω, rad/sec	0.1	0.5	1.0	2.0	5	10	15
φ, deg	– 94	– 112	– 126	– 140	– 150	– 170	– 188

In the same semilog sheet and by using same scales, the phase plot of compensated system is sketched as shown in Fig (Ex. 5.6.1).

Let, ϕ_{gco} = phase of $G_o(j\omega)$ at new gain cross over frequency (ω_{gcn}).

and ϕ_o = Phase margin of compensated system.

From the bode plot of compensated system we get,

$$\phi_{gco} = -140є$$

Now, $$\gamma_o = 180є - 140є = 40є$$

Conclusion: The phase margin of the compensated system is satisfactory. Hence the design is acceptable.

Result: The transfer function of lead compensation

$$G_c(s) = \frac{\cdot 0784(1 + \cdot 24s)^2}{(1 + \cdot 067s)^2}$$

$$= \frac{(s + 4 \cdot 17)^2}{(s + 14 \cdot 92)^2}$$

The open loop transfer function of lead compensated system

$$\boxed{G_o(s) = \frac{4(1 + \cdot 24s)^2}{s(1+s)(1+0 \cdot 2s)(1 + \cdot 067s)^2}}$$

EXAMPLE 5.7

Consider a unity feedback system with open loop transfer function. $G(s) = K/s(s + 8)$. Design a lead compensator to meet the following specifications.

(i) Percentage peak overshoot = 9.5%

(ii) Natural frequency of oscillation, $\omega_n = 12$ rad/sec

(iii) Velocity error constant, $K_v \geq 10$.

Solution:

Step 1: Determine the dominant pole, s_d,

Dominant pole, $s_d = \zeta\omega_n \pm j\omega_n\sqrt{1-\zeta^2}$

Given that, $\omega_n = 12$ rad/sec and

$$\%M_p = 9.5\%$$

We know that, $$\%M_p = e^{-\zeta\pi/\sqrt{1-\zeta^2}} \times 100$$

$$\therefore \quad e^{-\zeta\pi/\sqrt{1-\zeta^2}} = .095$$

On taking natural log we get,

$$-\zeta\pi/\sqrt{1-\zeta^2} = l_n\,(.095)$$

On squaring we get,

$$\frac{\zeta^2\pi^2}{1-\zeta^2} = (ln.095)^2 = 5.54$$

$$\therefore \quad \zeta^2\pi^2 = 5.54 - 5.54\,\zeta^2$$

$$\zeta^2\pi^2 + 5.54\,\zeta^2 = 5.54$$

$$\zeta^2(\pi^2 + 5.54) = 5.54$$

$$\therefore \quad \zeta = \sqrt{\frac{5.54}{\pi^2 + 5.54}} = 0.6$$

$$\zeta = .6$$

$$\therefore \quad s_d = -(0.6 \times 12) \pm j12 \times \sqrt{1-0.6^2} = -7.2 \pm j\,9.6$$

$$\boxed{s_d = -7.2 \pm j9.6}$$

Step 2: Draw the pole zero plot.

The pole-zero plot of open loop transfer function is shown in Fig (Ex. 5.7.1). Poles are represented by the symbol "X". The pole at point P in the dominant pole, sd.

Step 3: To find the angle to be contributed by lead network.

Let ϕ = Angle to be contributed by lead network to make point, P as a point in root locus.

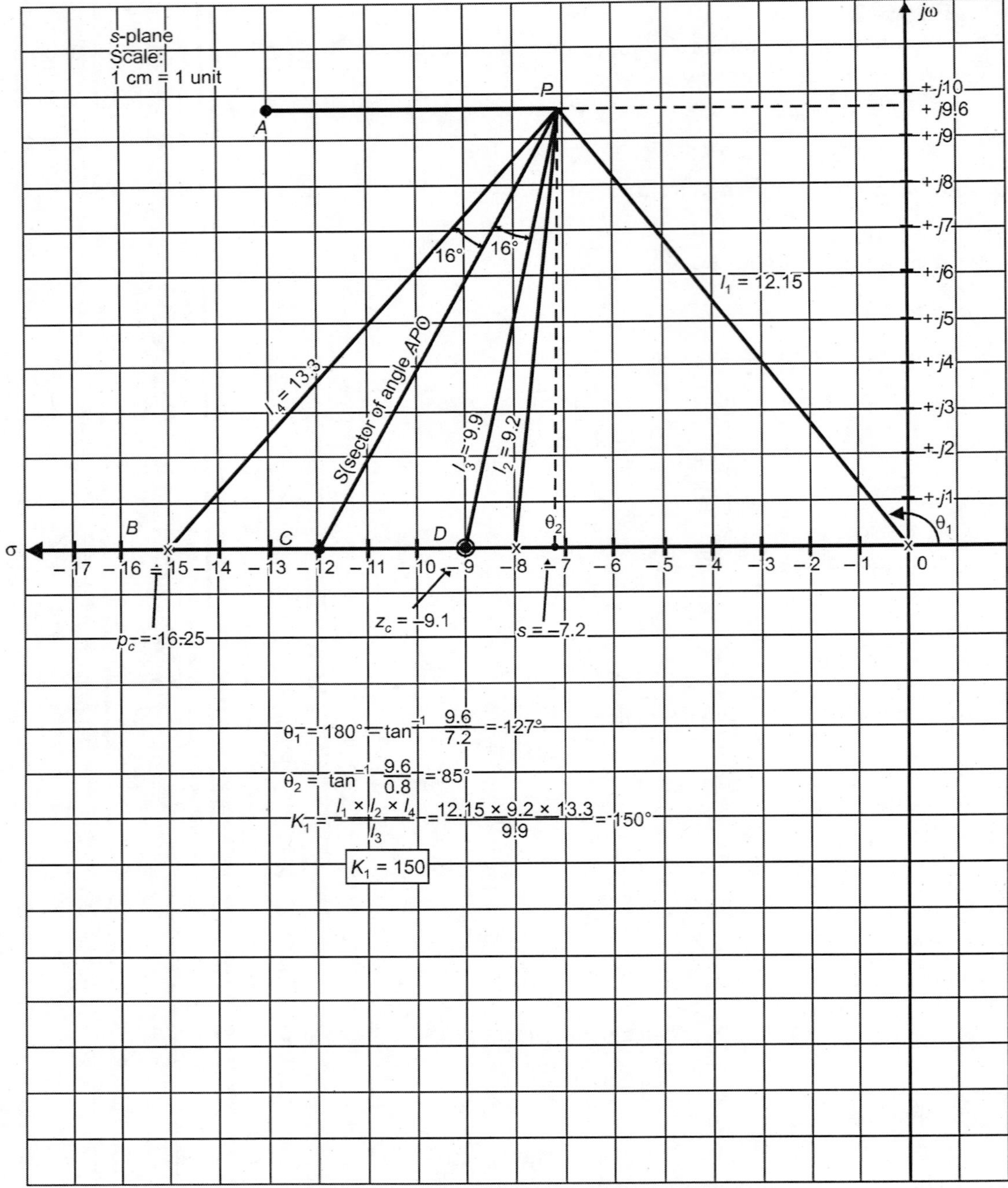

Fig. Ex. 5.7.1: Pole zero plot of $G(s) = K/s\,(s + 8)$ lead compensator.

Now, $\phi = \begin{bmatrix} \text{Sum of angles contributed} \\ \text{by poles of uncompensated} \\ \text{system} \end{bmatrix} - \begin{bmatrix} \text{aim of angles} \\ \text{contributed by zeros} \\ \text{of uncompensated system} \end{bmatrix} \pm n180^o$

From Fig. (Ex. 5.7.1), we get

Sum of angles,contributed by poles of uncompensated system

$$\theta_1 + \theta_2 = 127\epsilon + 85\epsilon = 212\epsilon$$

Since there is no finite zero is uncompensated system, there is no angle contribution by zeros

$$\phi = 212\epsilon \pm n180\epsilon$$

Let, $n = 1$

$\therefore$ $\phi = 212\epsilon - 180\epsilon = 32\epsilon$

Step 4: To find the pole and zero of the compensator.

Draw a line *AP* parallel to X-axis shown in Fig (Ex. 5.7.1). the bisector *PC* is drawn to bisect the angle *APO.* The angles *CPD* and *BPC* are constructed as shown in Fig (Ex. 5.7.1). Here

$$\angle CPD = \angle BPC = \phi/2 = 32\epsilon 12 = 16\epsilon$$

From Fig (Ex. 5.7.1)

Pole of the compensator,

$$p_c = -16.25$$

Zero of the compensator,

$$z_c = -9.1$$

We know that,

$$z_c = -1/T$$

$\therefore$ $T = 1/9.1 = 0.11$

We know that,

$$p_c = -1/\alpha T$$

$\therefore$ $\alpha \mathrm{T} = 1/16.25$ (or)

$$\alpha = 1/(T Ч 16.25) = .56.$$

Step 5: Determine the transfer function of lead compensator

Transfer function of lead compensator,

$$G_c(s) = \frac{\left(s + \dfrac{1}{T}\right)}{\left(s + \dfrac{1}{\alpha T}\right)} = \frac{(s + 9.1)}{(s + 16.25)}$$

Step 6: Determine the open loop transfer function of lead compensated system

The block diagram of lead compensated system is shown in Fig. Ex. 5.7.2.

Open loop transfer function of lead compensated system

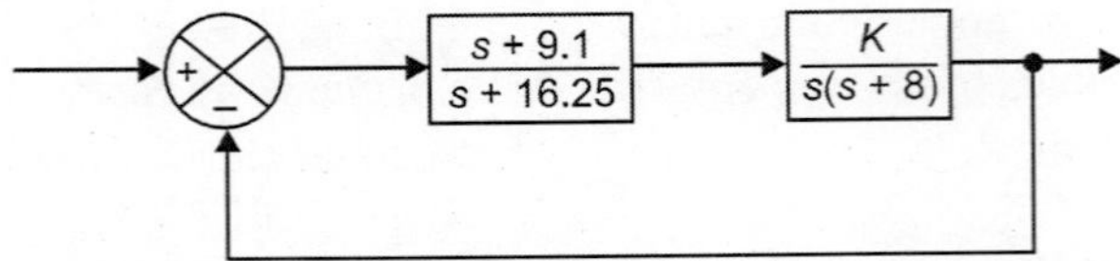

Fig. Ex. 5.7.2: Block diagram of lead compensated system.

$$G_o(s) = \frac{(s+9.1)}{(s+16.25)} \times \frac{K}{s(s+8)}$$

$$= \frac{K(s+9.1)}{s(s+8)\,(s+16.2s)}$$

Here the value of K is given by, the value of gain at the dominant pole, s_d on the root locus. From magnitude condition K is given by,

$$K = \frac{\text{Product of vector lengths from all poles to } s = s_d}{\text{Product of vector lengths from all zeros to } s = s_d}$$

From Fig. (4.7.1), we get,

$$K = \frac{l_1 \times l_2 \times l_4}{l_3} = \frac{12.5 \times 9.2 \times 13.3}{9.9} = 150$$

$$\therefore \qquad G_o(s) = \frac{150(s+9.1)}{s(s+8)(s+16.25)}$$

Step 7: Verify for error requirement for the compensated system, the velocity error constant is given by,

$$K_v = \underset{s \to 0}{Lt}\; sG_o(s)$$

$$= \underset{s \to 0}{Lt}\; s\frac{150(s+9.1)}{s(s+8)(s+16.25)}$$

$$= \frac{150 \times 9.1}{8 \times 16.25} = 10.5$$

Conclusion: Since the velocity error constant of the compensated system, satisfies the requirement, the design is acceptable.

Result: Transfer function of lead compensator

$$G_c(s) = \frac{(s+9.1)}{(s+16.25)} = .56\frac{(1+.11s)}{(1+\cdot 06s)}$$

Transfer function of lead compensated system

$$G_o(s) = \frac{150(s+9.1)}{s(s+8)(s+16.25)}$$

$$\boxed{G_o(s) = \frac{10.5(1+0.11s)}{S(1+\cdot 125s)(1+0\cdot 06s)}}$$

EXAMPLE 5.8

Design a lead compensator for a unity feedback system with open loop transfer function $G(s) = K/s$ $(s + 4)$ $(s + 7)$ to meet the following specifications:

(i) % Peak overshoot = 12.63%

(ii) Natural frequency of ocillation, $\omega_n = 8$ rad/sec.

(iii) Velocity error constant, $K_v \geq 2.5$.

Solution:

Step 1: Determine the dominant Pole, s_d

Dominant Pole,

$$s_d = -\zeta\omega_n \pm j\omega_n\sqrt{1-\zeta^2}$$

Given that,

$$\omega_n = 8 \text{ rad/sec and } \%M_p = 12.63\%$$

We know that,

$$\%M_p = e^{-\zeta\pi/\sqrt{1-\zeta^2}} \times 100$$

$$\therefore \quad e^{-\zeta\pi/\sqrt{1-\zeta^2}} = .1263$$

On taking natural log we get,

$$-\zeta\pi/\sqrt{1-\zeta^2} = l_n(.1263)$$

On squaring we get,

$$\frac{\zeta^2\pi^2}{1-\zeta^2} = (l_n\, 0.1263)^2 = 4.28$$

$$\therefore \quad \zeta^2\pi^2 = 4.28 - 4.28\,\zeta^2$$

$$\zeta^2\pi^2 + 4.28\,\zeta^2 = 4.28$$

$$\zeta^2(\pi^2 + 4.28) = 4.28$$

$$\therefore \quad \zeta = \sqrt{\frac{4.28}{\pi^2 + 4.28}} = 0.55$$

∴ $$s_d = -(0.55 \text{ Ч } 8) \pm j\, 8 \text{ Ч } \sqrt{1-0.55^2}$$

$$= -4.4 \pm j\, 6.68 = -4.4 \pm j\, 6.7$$

Step 2: Draw the pole zero plot

The pole zero plot of open -loop transfer function is shown in Fig (Ex. 5.8.1). Poles are represented by the symbol "*X*". The pole at point *P* is the dominant pole, s_d.

Step 3 To find the angle to be contributed by lead network to make point, *P* as a point on root locus.

Now, $$\phi = \begin{bmatrix} \text{Sum of angles contributed by} \\ \text{poles of uncompensated system} \end{bmatrix} - \begin{bmatrix} \text{Sum of angles contributed by} \\ \text{zeros of uncompensated system} \end{bmatrix} \pm n\, 180\text{є}$$

From Fig. 4.8.1, we get,

Sum of angles contributed by poles of uncompensated system

$$\theta_1 + \theta_2 + \theta_3 = 123\text{є} + 93\text{є} + 69\text{є} = 285\text{є}$$

Since there is no finite zero in uncompensated system, there is no angle contribution by zeros.

∴ $$\phi = 285\text{є} \pm n 180\text{є}$$

Let $n = 1$, $$\phi = 285\text{є} - 180\text{є} = 105\text{є}$$

Since the angle contribution is more than 60є, the lead compensator is realised as cascade of two compensators with each compensator, contributing half of the required angle.

∴ $$\phi = \frac{105}{2} = 52.5 \approx 52\text{є}$$

Step 4: To find the poles and zeros of the compensator

Draw a line *AP* parallelto *X*-axis as shown in Fig. (Ex. 5.8.2). The bisector *PC* is drawn to bisect the ∠*APO*. The angles ∠*CPD* and ∠*BPC* are constructed as shown in Fig. (Ex. 5.8.2).

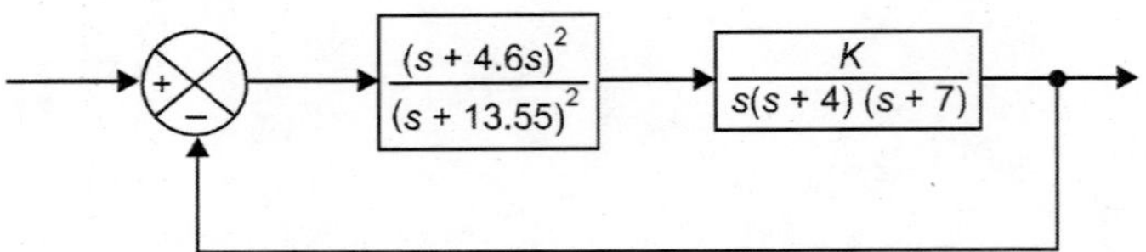

Fig. Ex. 5.8.2: Block diagram of lead compensator system.

Here ∠CPD = ∠BPC = ϕ/2 = 52є/2 = 26є

From Fig (Ex. 5.8.2).

Pole of the compensator,

$$p_c = -13.55$$

Zero of the compensator,

$$z_c = -4.65$$

We know that, $$z_c = \frac{-1}{T}$$

∴ $$T = \frac{1}{4.65} = 0.215$$

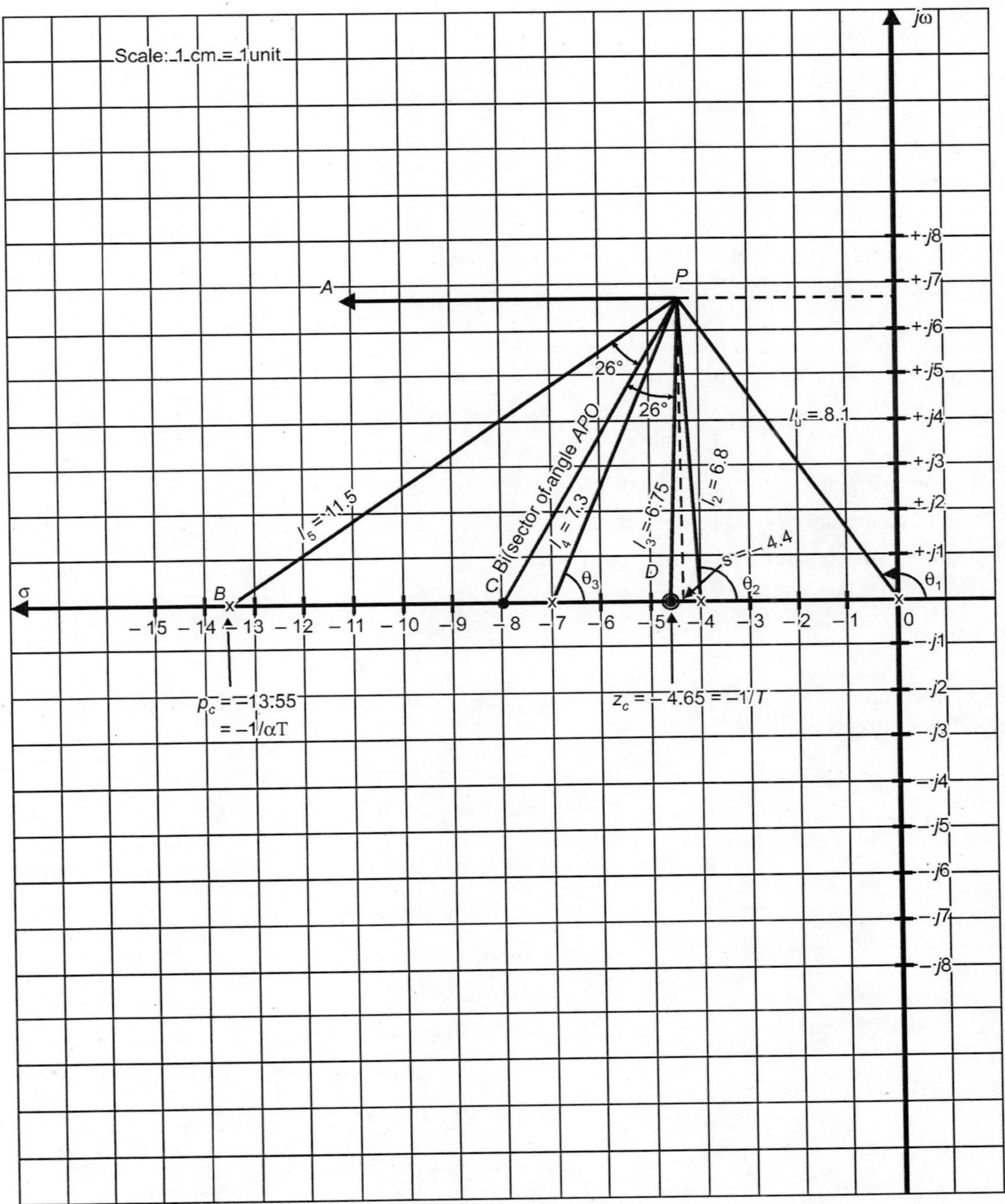

Fig. Ex. 5.8.1: Pole zero plot of $G(s) = K/s\,(s + 4)\,(s + 7)$ (lead compensator).

We know that, $$p_c = \frac{-1}{\alpha T}$$

$$\therefore \quad \alpha T = \frac{1}{13.55} \text{ (or)}$$

$$\alpha = \frac{1}{T} \times 13.55$$

$$\alpha T = \cdot 343$$

Step 5: Determine the transfer function of lead compensator

Transfer function of lead compensator

$$G_c(s) = \frac{\left(s + \frac{1}{T}\right)^2}{\left(s + \frac{1}{\alpha T}\right)^2} = \frac{(s+4.65)^2}{(s+13.55)^2}$$

Step 6: Determine the open loop transfer function of lead compensated system.

The block diagram of lead compensated system is shown in Fig. (Ex. 5.8.2)

Open loop transfer function of lead compensated system

$$G_o(s) \doteq \frac{(s+4.65)^2}{(s+13.55)^2} \times \frac{K}{s(s+4)(s+7)}$$

Here the value of K is giveñ by the value of gain out the dominant pole, s_d on the root locus. From the magnitude condition K is given by,

$$K = \frac{\text{Product of vector lengths from all poles to } s = sd}{\text{Product of vector lengths from all zeros to } s = sd}$$

From Fig.(Ex. 5.8.1) we get,

$$K = l_1 \text{ Ч } l_2 \text{ Ч } l_4 \text{ Ч } 2l_5/2l_3 = 8.1 \text{ Ч } 6.8 \text{ Ч } 7.3 \text{ Ч } 2 \text{ Ч } 11.5 / 2 \text{ Ч } 6.75$$

$$K = 685$$

$$\therefore \quad G_o(s) = \frac{685(s+4.65)^2}{s(s+4)(s+7)(s+13.55)^2}$$

Step 7: Check for error requirement for the compensated system, the velocity error constant is given by

$$K_v = \underset{s \to 0}{Lt}\ sG_o(s) = \underset{s \to 0}{Lt}\ s \frac{685(s+4.65)^2}{s(s+4)(s+7)(s+13.55)^2}$$

$$= \frac{685 \times 4.65^2}{4 \times 7 \times 13.55^2} = 2.88$$

Conclusion: Since the velocity error constant of the compensated system, satisfies the requirement, the design is acceptable.

Result: Tranfer function of lead compensator

$$G_c(s) = \frac{(s+4.65)^2}{(s+13.55)^2} = 0.1178\,\frac{(1+.2515s)^2}{(1+.0738s)^2}$$

Transfer function of lead compensated system

$$G_o(s) = \frac{685(s+4.65)^2}{s(s+4)(s+7)(s+13.55)^2}$$

$$\boxed{G_o(s) = \frac{2.88(1+\cdot 215s)^2}{s(1+\cdot 25s)(1+\cdot 143s)(1+\cdot 07385)^2}}$$

5.6 LAG-LEAD COMPENSATOR

A compensator having the characteristics of lag-lead network is called Lag lead compensator. In this, when sinusoidal signal is applied, both phase lag and phase lead occurs in the output, but in the different frequency regions. Phase lag occurs in the low frequency region and phase lead occurs in the high frequency region i.e., the phase angle varies from lag to lead as the frequency is increased from zero to infinity.

- It increases the band width (Lead)
- Fast response (Lead)
- Decreases the maximum overshoot in the step response (Lead)
- Increases the low frequency gain (Lag)
- Improves steady state accuracy (Lag)
- Reduced band width (Lag)
- Reduces the speed (Lag)

A Lag-lead compensator passes two poles and two zeros and so such a compensation increases the order of the system by two, unless cancellation of poles and zeros occurs in the compensated system.

s-plane representation of lag-Lead compensator

The s-Plane representation of Lag-Lead compensator is shown in Fig. 5.15, the lag section has one real pole and real zero with pole to the right of zero. The lead section also has one real pole and one real zero but the zero is to the right of the pole.

Transfer function of Lag-Lead compensator

$$G_c(s) = \underbrace{\frac{\left(s+\frac{1}{T_1}\right)}{\left(s+\frac{1}{\beta T_1}\right)}}_{\text{Lag}}\;\underbrace{\frac{\left(s+\frac{1}{T_2}\right)}{\left(s+\frac{1}{\alpha T_2}\right)}}_{\text{Lead}}$$

where $\beta > 1$ and $0 < \alpha < 1$.

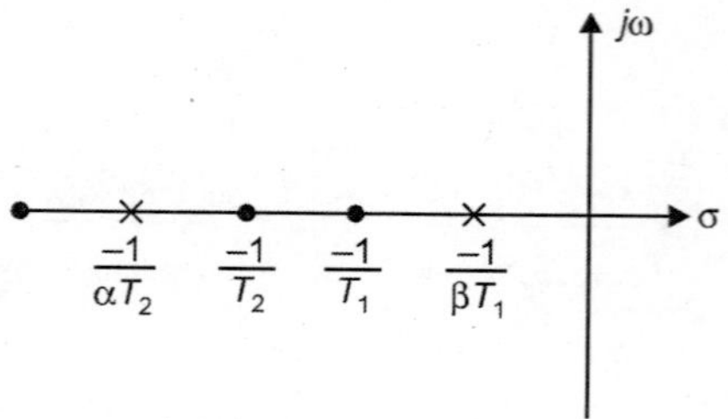

Fig. 5.15: Pole zero plot of Lag-lead compensator.

Lag-lead compensator using electrical network

The Lag-lead compensator can be realised by the *R-C* network shown in Fig. 5.16

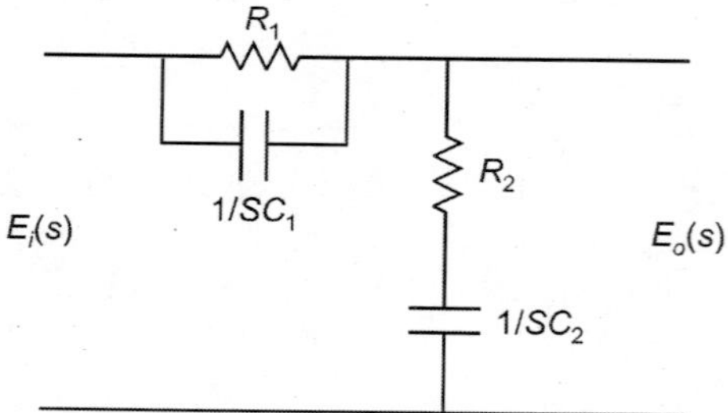

Fig. 5.16: Electrical lag lead compensator.

Let $E_i(s)$ = Input voltage

$E_o(s)$ = Output voltage

In the network shown in Fig. 5.16 , the input voltage is applied to the series combination of $R_1 \| C_1$, R_2 and C_2. The output voltage, is obtained across series combination of R_2 and C_2.

By voltage division rule,

$$E_o(s) = E_i(s) \frac{R_2 + \dfrac{1}{sC_2}}{\left(R_1 \| \dfrac{1}{sC_1}\right) + R_2 + \dfrac{1}{sC_2}}$$

$$\therefore \quad \frac{E_o(s)}{E_i(s)} = \frac{\dfrac{sR_2C_2+1}{sC_2}}{\dfrac{R_1/sC_1}{R_1+\dfrac{1}{sC_1}} + \dfrac{sR_2C_2+1}{sC_2}} = \frac{\dfrac{sR_2C_2+1}{sC_2}}{\dfrac{R_1}{sR_1C_1+1} + \dfrac{sR_2C_2+1}{sC_2}}$$

$$\frac{E_o(s)}{E_i(s)} = \frac{R_1C_1R_2C_2\left(s+\dfrac{1}{R_1C_1}\right)\left(s+\dfrac{1}{R_2C_2}\right)}{sR_1C_2 + R_1C_1R_2C_2\left(s+\dfrac{1}{R_1C_1}\right)\left(s+\dfrac{1}{R_2C_2}\right)}$$

On dividing the numerator and denominator by $R_1C_1\ R_2C_2$ we get,

$$\frac{E_o(s)}{E_i(s)} = \frac{\left(s+\dfrac{1}{R_1C_1}\right)\left(s+\dfrac{1}{R_2C_2}\right)}{\dfrac{s}{R_2C_1}+\left(s+\dfrac{1}{R_1C_1}\right)\left(s+\dfrac{1}{R_2C_2}\right)} \tag{5.34}$$

$$= \frac{\left(s+\dfrac{1}{R_1C_1}\right)\left(s+\dfrac{1}{R_2C_2}\right)}{s^2+s\left(\dfrac{1}{R_1C_1}+\dfrac{1}{R_2C_2}+\dfrac{1}{R_2C_1}\right)+\dfrac{1}{R_1R_2C_1C_2}}$$

The transfer function of lag-lead compensator is given by

$$G_c(s) = \frac{\left(s+\dfrac{1}{T_1}\right)\left(s+\dfrac{1}{T_2}\right)}{\left(s+\dfrac{1}{\beta T_1}\right)\left(s+\dfrac{1}{\alpha T_2}\right)}$$

$$= \frac{\left(s+\dfrac{1}{T_1}\right)\left(s+\dfrac{1}{T_2}\right)}{s^2+s(1/\beta T_1)+\left(\dfrac{1}{\alpha T_2}\right)+\dfrac{1}{\alpha\beta T_1T_2}} \tag{5.35}$$

On comparing Eqs. (5.34) and (5.35) we get,

$$\left.\begin{aligned} T_1 &= R_1\ C_1 \\ T_2 &= R_2\ C_2 \end{aligned}\right\} \tag{5.36}$$

$$R_1\ R_2\ C_1\ C_2 = \alpha\ \beta\ T_1\ T_2 \tag{5.37}$$

$$\frac{1}{R_1C_1}+\frac{1}{R_2C_2}+\frac{1}{R_2C_1} = \frac{1}{\beta T_1}+\frac{1}{\alpha T_2} \tag{5.38}$$

From Eq. (5.37) we get

$$\alpha\beta = \frac{T_1T_2}{R_1R_2C_1C_2} \tag{5.39}$$

From Eq. (5.36) and (5.39) we can say that

$$\alpha\beta = 1 \tag{5.40}$$

The Eq. (5.40) implies that a single lag -lead network does not allow an independent choice of α and β. Hence in the transfer function of electrical lag lead compensator replace α by $1/\beta$ as shown below:

$$\frac{E_o(s)}{E_i(s)} = G_c(s) = \frac{\left(s+\frac{1}{T_1}\right)\left(s+\frac{1}{T_2}\right)}{\left(s+\frac{1}{\beta T_1}\right)(s+\beta/T_2)} \tag{5.41}$$

where $\beta > 1,\ T_1 = R_1 C_1$

$T_2 = R_2 C_2$

and $\frac{1}{R_1C_1} + \frac{1}{R_2C_2} + \frac{1}{R_2C_1} = \frac{1}{\beta T_1} + \frac{\beta}{T_2}$

Frequency response of lag-lead network

Consider the transfer function of lag-lead compenation

$$G_c(s) = \frac{\left(s+\frac{1}{T_1}\right)\left(s+\frac{1}{T_2}\right)}{\left(s+\frac{1}{\beta T_1}\right)\left(s+\frac{1}{\alpha T_2}\right)}$$

$$= \alpha\beta\frac{(1+sT_1)(1+sT_2)}{(1+s\beta T_1)(1+s\alpha T_2)} \tag{5.42}$$

The sinusoidal transfer function of Lag-lead compensator is obtained by letting $s = j\omega$ in Eq. (5.42).

$\therefore$
$$G_c(j\omega) = \alpha\beta\frac{(1+j\omega T_1)(1+j\omega T_2)}{(1+j\omega\beta T_1)(1+j\omega\alpha T_2)} \tag{5.43}$$

For a single lag-lead compensator, $\alpha\beta = 1$. Hence from Eq. (5.43) we can say tht the lag-lead conpensator provides a dc gain of unity,

$\therefore$
$$G_c(j\omega) = \frac{(1+j\omega T_1)(1+j\omega T_2)}{C_1 + j\omega PT_1\,(1+j\omega\alpha T_2)} \tag{5.44}$$

The sinusoidal transfer function shown in Eq. (5.44) has four corner frequencies and they are ω_{c1} ω_{c2}, ω_{c3} and ω_{c4} where $\omega_{c1} < \omega_{c2} < \omega_{c3} <$ and ω_{c4} .

Here, $\omega_{c1} = 1/\beta T$; $\omega_{c2} = 1/T$; $\simeq \omega_{c3}\ \frac{1}{T_2}$ and $\omega_{c4} = \frac{1}{\alpha T_2}$

By an analysis similar to that of lag compensator the bode plot of lag lead compensator is sketched in Fig. (Ex. 5.17)

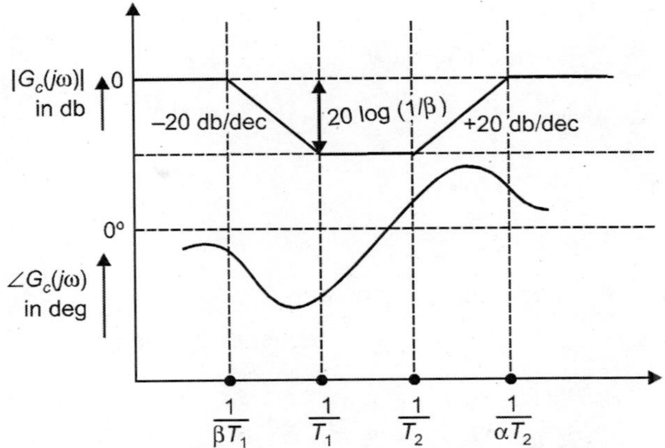

Fig. 5.17: Bode plot of lag-lead compensator.

Procedure for designing Lag-lead compensator using bode plot

Step 1: Determine the openloop gain K of the uncompensated system to satisfy specified error requirement.

Step 2: Draw the bode plot of uncompensated system

Step 3: From the bode plot determine the gain margin of the uncompensated system.

Let, ϕ_{gc} = Phase of $G(j\omega)$ at gain cross over frequency.

$\gamma \Rightarrow$ Phase margin of uncompensated system.

Now, $\gamma = 180\epsilon + \phi_{gc}$

If the gain margin is not satisfactory then compensation is required.

Step 4: Choose a new phase margin

Let, γ_d = Desired phase margin

Now, new phase margin,

$$\gamma_n = \gamma_d + \in$$

Choose an initial value of $\in = \pm 5\epsilon$

Step 5: From bode plot, determine the new gain cross over frequency, which is the frequency corresponding to a phase margin of γ_n.

Let,

ω_{gcn} = New gain cross over frequency

and ϕ_{gcn} = Phase of $G(j\omega)$ at ω_{gcn}

$$\gamma_n = 180\epsilon + \phi_{gcn}$$

or $\phi_{gcn} = \gamma_n - 180\epsilon$

In the phase plot of uncompensated system, the frequency corresponding to a phase of ϕ_{gcn} is the new gain crossover frequency ω_{gcn}. Choose the gain crossover frequency of the lag compensator, ω_{gcl}, some what greater than ω_{gcn} (i.e. choose ω_{gcl} such that $\omega_{gcl} > \omega_{gcn}$) .

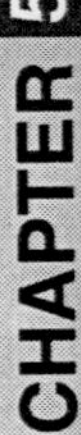

Step 6: Calculate β of Lag compensator.

Let, $A_{gcl} = |G(j\omega)|$ in db at $\omega = \omega_{gcl}$

From the bode plot find A_{gcl}

Now,
$$[A_{gcl} / 20]$$
$$A_{gcl} = 20 \log \beta$$

or
$$\beta = 10^{(A_{gcl}/20)}$$

Step 7: Determine the transfer function of Lag section.

The zero of the lag compensation is placed at a frequency one-tengthof ω_{gcl}

∴ zero of lag compensator,

$$z_{c1} = 1/T_1 = \omega_{gcl10}$$

Now,
$$T_1 = 10/\omega_{gcl}$$

Pole of lag compensator,

$$p_{c1} = \frac{1}{\beta T_1}$$

Transfer function of lag section

$$G_1(s) = \frac{(s + 1/T_1)}{(s + 1/\beta T_1)} = \beta \frac{(1 + sT_1)}{(1 + s\beta T_1)}$$

Step 8: Determine the transfer function of lead section.

Take, $\alpha = 1/\beta$

From the bode plot find ω_m which is the frequency at which the db gain is $-20 \log (1/\sqrt{\alpha})$.

Now
$$T_2 = 1/\omega_m\sqrt{\alpha}$$

Transfer function of lead section

$$G_2(s) = \frac{(s + 1/T_2)}{(S + 1/\alpha T_2)} = \alpha \frac{(1 + sT_2)}{(1 + s\alpha T_2)}$$

Step 9: Determine the tansfer function of lag lead compensator.

Transfer function of lag-lead compensator,

$$G_c(s) = G_1(s) \text{ Ч } G_2(s)$$

$$= \beta \frac{(1 + sT_1)}{(1 + s\alpha T_1)} \times \alpha \frac{(1 + sT_2)}{(1 + s\alpha T_2)}$$

Since
$$\alpha = 1/\beta$$

$$G(s) = \frac{(1 + sT_1)}{(1 + s\beta T_1)} \cdot \frac{(1 + sT_2)}{(1 + s\alpha T_2)}$$

Step 10: Determine the open loop transfer function of compensated system.

The lag lead compensator is connected in series with $G(s)$ as shown in Fig. (5.18).

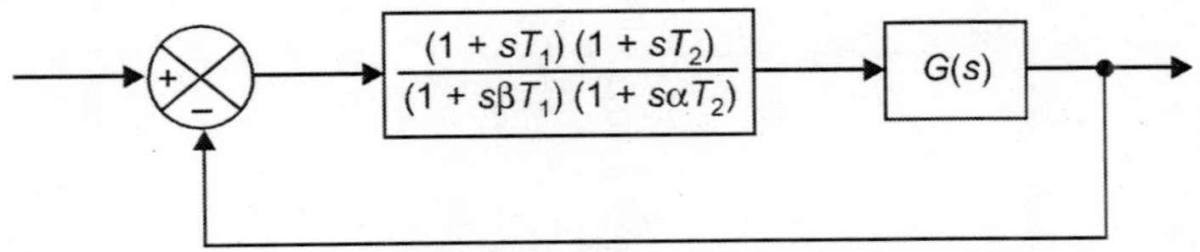

Fig. 5.18: Block diagram of lag-lead compensated system.

Open loop transfer function of compensated system

$$G_o(s) = \frac{(1+sT_1)(1+sT_2)}{(1+s\beta T_1)(1+s\alpha T_2)} \times G(s)$$

Step 11: Draw the plot of compensated system and verify whether the specifications are satisfied (or) not. If the specifications are not satisfied then choose another choice of $\alpha \angle 1/\beta$ and repeat the step 8 to 11 till design get satisfied.

Procedure for design of lag-lead compensator using root locus

The lag-lead compensation is employed to improve both the transient and steady state response of a system. First design a lead section to realize the required ζ and ω_n for the dominant closed loop poles. Then determine the error constant of lead compensated system. If it is satisfactory then only lead compensation will meet the requirement. If the error constant has to be increased then design a lag section the step-by-step procedure for the design of lag-lead compensator is given below.

Step 1: Determine the dominant pole, s_d

$$s_d = -\zeta\pi \pm \omega_n \sqrt{\zeta^2 - 1}$$

where, ζ = Damping ratio

ω_n = Natural frequency of oscillation, rad/sec

Step 2: Mark the poles and zeros of open loop transfer function and the dominant pole on the s-plane let the dominant pole be point P.

Step 3: Find the angle to be contributed by lead network to make the point P as a point on root lcus.

Let, ϕ = Angle to be contributed by lead network to make point, P as a point on root locus. Draw vectors from all open loop poles and zeros to point P. Measure the angle contributed by the vectors.

Now, $$\phi = \begin{bmatrix}\text{Sum of angles contributed by} \\ \text{poles of uncompensated system}\end{bmatrix} - \begin{bmatrix}\text{Sum of angles contributed by} \\ \text{zeros of uncompensated system}\end{bmatrix} \pm n\,180\epsilon$$

where n is an odd integer, so that n 180ϵ is nearest to the difference between angles contributed by poles and zeros.

Step 4: Determine the pole and zeros of the lead section.

Let point O be the origin of s-plane and point P be the dominant pole. Draw straight lines OP and AP such that AP is parallel to X-axis as shown in Fig. (5.19). Draw a line PC so as to bisect the angle APO [$\angle APO$] where the point is on the real axis with line PC so as to bisect the angle APO [$\angle APO$]

CHAPTER 5

where the point C on the real axis with the PC as reference, draw angles BPC and CPD such that each equal to $\frac{\phi}{2}$. Here the points B and D are located on the real axis.

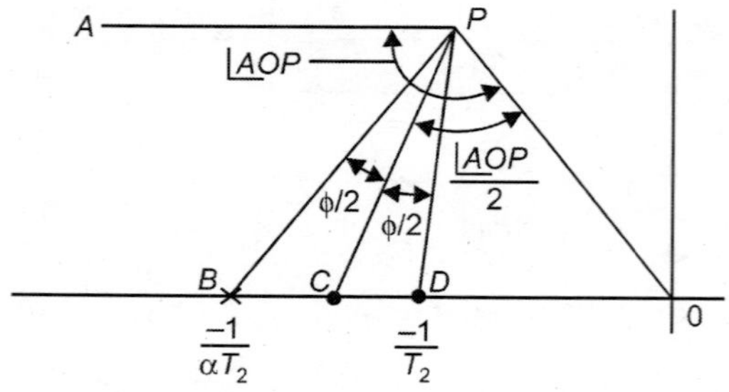

Fig. 5.19: Determination of poles and zeros of lead section.

Now the point B is the location of the pole of the compensator $(-1/\alpha T_2)$ and the point D is the location of the zero of the compensator $(-1/T_2)$. Compute T_2 and a from the values of point D and B.

Step 5: Determine the transfer function of lead section:

Transfer function of lead section

$$G_2(s) = \frac{(s + 1/T_2)}{(s + 1/\alpha T_2)}$$

Step 6: Determine the open loop gain, K

The open loop gain K is the value of gain at $s = s_d$ the value of gain, K is determined from pole-zero plot of lead compensated system and by using the magnitude condition given below.

$$K = \frac{\text{Product of vector lengths from all poles to } s = s_d}{\text{Product of vector lengths from zeros to } s = s_d}$$

Note: The length vectors should be measured to scale.

Open loop transfer function of lead compensated system

$$G_o(s) = G_2(s) \text{ ч } G(s)$$

Step 7: Determine the velocity error constant of lead compensated system.

Velocity error constant of lead compensated system

$$K_{v2} = \underset{s \to 0}{Lt}\, sG_{o2}(s)$$

If K_{v2} satisfies the requirement then only lead compensation of sufficient but if K_{v2} is less than the desired value then provide lag compensation.

Step 8: Determine the parameter, β of lag section.

Lag $\quad K_{vd}$ = Desired velocity error constant

A = The factor by which K_v is increased

Now $\quad A = K_{vd}/K_{v2}$

Select β, such that $\beta > A$ [i.e., $b = (1.1 \text{ to } 1.2) \text{ ч } A$)

Step 9: Determine the transfer function of lag section. Choose the zero of lag section at 10% of the second pole of uncompensated system.

∴ Zero of lag section,

$$z_{c1} = 0.1 \text{ Ч second pole of } G(s)$$

Also, $$z_{c1} = -1/T_2;$$

$$T_1 = -1/z_{c1}$$

Pole of lag section, $$p_{c1} = -1/\beta T_1$$

Transfer function of lag section,

$$G_1(s) = \frac{(s + 1/T_1)}{(s + 1/\beta T_1)}$$

Step 10: Determine the transfer function of lag-lead compensator and compensate system.

Transfer function of lag-lead compensator

$$G(s) = G_1(s) \text{ Ч } G_2(s)$$

$$= \frac{(s + 1/T_1)(s + 1/T_2)}{(s + 1/\beta T_1)(s + 1/\alpha T_2)}$$

Fig. 5.20: Block diagram of lag-lead system.

Open loop transfer function of lag-lead compensated system

$$G_o(s) = \frac{(s + 1/T_1)(s + 1/T_2)}{(s + 1/\beta T_1)(s + 1/\alpha T_2)} \times G(s)$$

Step 11: Check the velocity error constant of compensated system, if it is satisfactory then the design is accepted, otherwise repeat the design by modifying the location of poles and zeros of the compensator.

EXAMPLE 5.9

Consider the unity feedback system whose open loop transfer function is $G(s) = K/s(s + 3)(s + 6)$. Design a lag-lead compensator to meet the following specifications (i) velocity error constant, $K_v = 80$ (ii) phases margin, $\gamma \leq 35$є.

Solution:

Step 1: Determine K

For unity feedback system, velocity error constant,

$$K_v = \underset{s \to 0}{Lt}\ sG(s) \quad [\text{Œ}H(s) = 1 \text{ for unity feedback system}]$$

Given that, $K_v = 80$

$\therefore$ $$\underset{s \to 0}{Lt}\, sG(s) = \underset{s \to 0}{Lt}\, s \cdot \frac{K}{s(s+3)(s+6)} = 80$$

$$\frac{K}{3 \times 6} = 80 \text{ (or) } K = 80 \text{ Ч } 3 \text{ Ч } 6 = 1440$$

$\therefore$ $$G(s) = 1440/s(s+3)(s+6)$$

$$= 1440/s \text{ Ч } 3\,(1 + s/3) \text{ Ч } 6(1 + s/6)$$

$$= 80/s\,(1 + 0.33s)(1 + .167s)$$

Step 2: Draw bode plot of uncompensated system.

In $G(s)$, put $s = j\omega$

$\therefore$ $$G(j\omega) = 80/j\omega\,(1 + j\,0.33\omega)(1 + j0.167\omega)$$

(i) Magnitude plot

The corner frequencies are ω_{c1} and ω_{c2}.

here $\omega_{c1} = 1.33 = 3$ rad/sec

and $\omega_{c2} = 1.67 = 6$ rad/sec

The various terms of $G(j\omega)$ are listed in Table Ex. 5.9.1 also table shows the slope contributed by each term and the change is slope at the corner frequency.

Table Ex. 5.9.1

Term	**Corner frequency rad/sec**	**Slope of each term *db*/dec**	**Change in slope from one term to another term *db*/dec**
$80/j\omega$	–	– 20	
$\frac{1}{1 + j0.33\omega}$	$\omega_{c1} = \frac{1}{.33} = 3$	– 20	$-20 - 20 = -40$
$\frac{1}{1 + j.167\omega}$	$\omega_{c2} = \frac{1}{.167} = 6$	–20	$-40 - 20 = -60$

Choose a low frequency ω_l such that $\omega_l < \omega_{c1}$ and choose a high frequency ω_h such that $\omega_h > \omega_{c2}$

Let $\omega_l = .5$ rad/sec and $\omega_h = 20$ rad/sec

Let $A = |G(j\omega)|$ in *db*

At $\omega = \omega_l$, $A_1 = 20 \log |\text{first term}|$

i.e., $$A = 20 \log \frac{80}{\omega} = 20 \log \frac{80}{.5} = 44 \text{ db}$$

At $\omega = \omega_{c1}$ $A_2 = 20 \log |\text{first term}|$

i.e., $$A_2 = 20 \log \frac{80}{\omega} = 20 \log \frac{80}{3} = 28.5 = 28\ db$$

At $\omega = \omega_{c2}$,

$$A_3 = [\text{slope from one term to another term}]\ Ч \log \left|\frac{\text{Present frequency}}{\text{Previous frequency}}\right| + \text{Previous gain}$$

i.e., $$A_3 = \left[\text{slope from } \omega_{c1} \text{ to } \omega_{c2} \times \log\frac{\omega_{c2}}{\omega_{c1}}\right] + A_2 \text{ at } (\omega = \omega_{c1})$$

$$A_3 = -40\ Ч \log(6/3) + 28 = 16\ db$$

At $\omega = \omega_h$,

$$A_4 = [\text{slope from one term to another term}]\ Ч \log \left|\frac{\text{Present frequency}}{\text{Previous frequency}}\right| + \text{Previous gain}$$

i.e., $$A_4 = \left[\text{slope from } \omega_{c2} \text{ to } \omega_n \times \log\frac{\omega_n}{\omega_{c1}}\right] + A_3 \text{ (at } \omega = \omega_n)$$

$$A_4 = -60\ Ч \log(20/6) + 16 = -15\ db$$

Let the points *a*, *b*, *c* and *d* be the points corresponding to frequencies ω_l, ω_{c1}, ω_{c2} and ω_h respectively on the magnitude plot. The magnitude plot is shown in Fig. (Ex. 5.9.1).

Phase plot

The phase angle of $G(j\omega)$ as function of ω is given by

$$\phi = \angle G(j\omega) = -90 - \tan^{-1} \cdot 33j\omega - \tan^{-1} \cdot .167\ \omega$$

The phase angle of $G(j\omega)$ are calculated for various values of ω and listed in Table Ex. 5.9.2.

Table Ex. 5.9.2

ω, rad/sec	0.5	1	3	6	10	20
ϕ, deg	–104	–118	–160	–198	–22	–244

On the same semilog sheet take another *y* axis, choose appropriate scale and draw phase plot as shown in Fig. (Ex. 5.9.1).

Step 3: Find phase margin of uncompensated system.

Let, $\phi_{gc} \rightarrow$ phase of $G(j\omega)$ at gain cross over frequency.

and $\gamma \rightarrow$ phase marging at uncompensated system.

From the bode plot of uncompensated system, we get

$$\phi_{gc} = -220\epsilon$$

Now, $$\gamma = 180\epsilon + \phi_{gc}$$
$$= 180\epsilon - 226\epsilon = -46\epsilon$$

CHAPTER 5

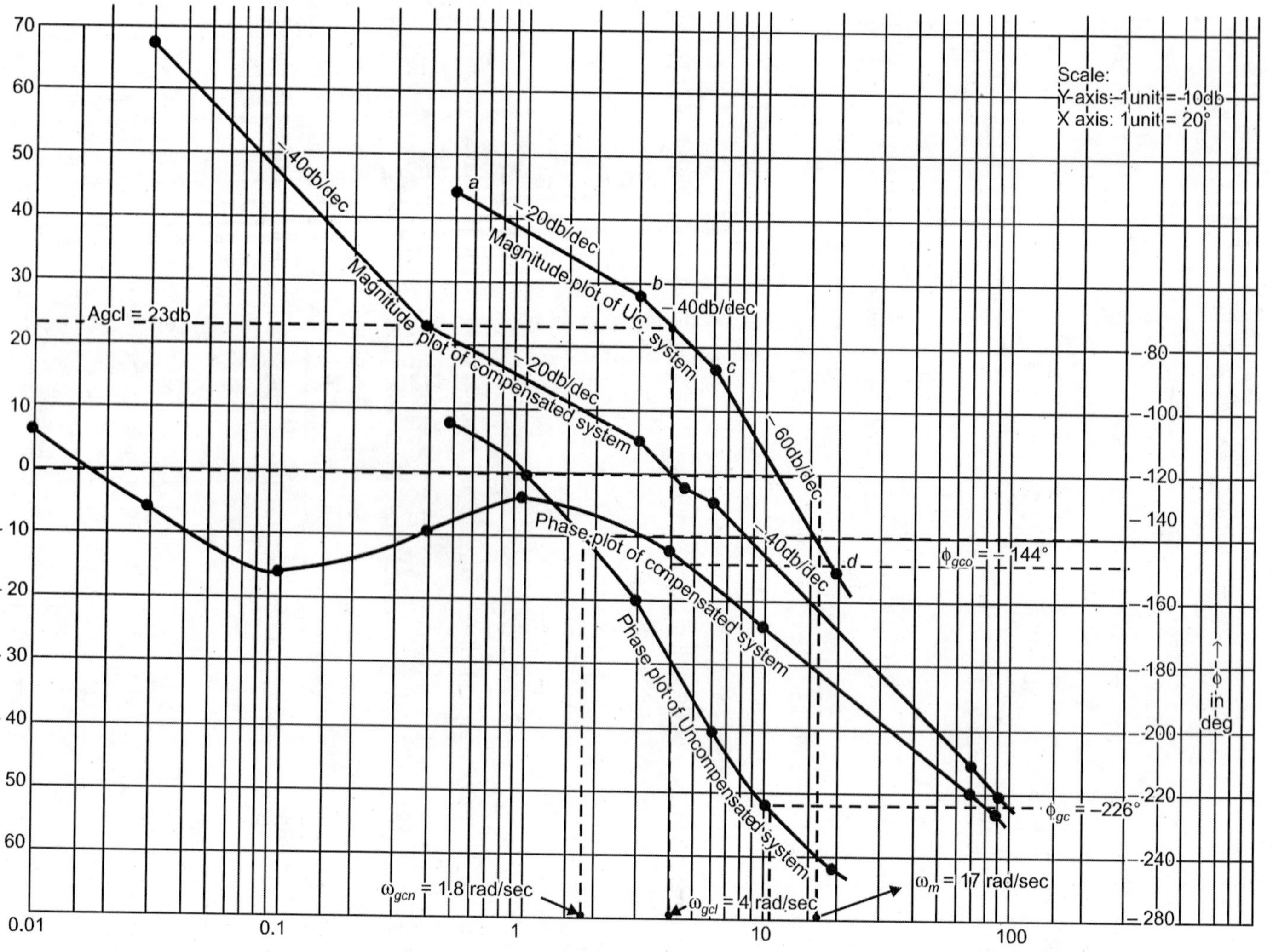

Fig. Ex. 5.9.1: Bode plot of $G(s) = K/s\,(s + 3)\,(s + 6)$ (lag lead compensator).

Step 4: Choose a new phase margin

The desired phase margin of compensated system,

$$\gamma_n = \gamma_d + \in$$

Let initial choice of $\in = 5\epsilon$

$$\therefore \qquad \gamma_n = \gamma_d + \in = 35\epsilon + 5\epsilon = 40\epsilon$$

Step 5: Determine new gain crossover frequency

Let ω_{gcn} = new gain cross over frequency

ϕ_{gcn} = phase of $G(j\omega)$ at ω_{gcn}

Now, $\gamma_n = 180\epsilon + \phi_{gcn}$, $\phi_{gcn} = \gamma_n - 180\epsilon = 40\epsilon - 180\epsilon = -140\epsilon$

From the bode plot we found that the frequency corresponding to a phase of -140ϵ is 1.8 rad/sec

Let ω_{gcl} = gain cross over frequency of lag compensator.

Choose ω_{gcl} such that, $\omega_{gcl} > \omega_{gcn}$

Let ω_{gcl} = 4 rad/sec.

Step 6: Calculate β of lag compensator.

From the bode plot we found that the *db* magnitude at ω_{gcl} is 23 *db*.

$\therefore$ $|G(j\omega)|$ in *db* at $(\omega = \omega_{gcl}) = A_{gcl} = 23$ *db*

Also, $A_{gcl} = 20 \log \beta$;

$$\therefore \qquad \beta = 10^{A_{gcl}/20} = 10^{23/20} = 14$$

Step 7: Determine the transfer function of lag section.

$\therefore$ The zero of lag compensator is placed at a frequency one-length of ω_{gcl}.

$\therefore$ Zero of lag compensator,

$$z_{c1} = \frac{-1}{T_1} = \frac{\omega_{gcl}}{10}$$

Now,

$$T_1 = \frac{10}{\omega_{gcl}} = \frac{10}{4} = 2.5$$

Pole of lag compensator

$$p_{c1} = \frac{1}{\beta T_1} = \frac{1}{14 \times 2.5} = \frac{1}{35}$$

Transfer function of lag section

$$G_1(s) = \frac{\beta(1 + sT_1)}{(1 + s\beta T_1)} = 14\frac{(1 + 2.5s)}{(1 + 35s)}$$

Step 8: Determine the transfer function of lead section.

Let $\alpha = 1/(\beta)$; $\alpha = 1/14 = .07$

The *db* gain corresponding to $\omega_n = -20 \log \dfrac{1}{\sqrt{\alpha}}$

$$= -20 \log \frac{1}{\sqrt{.07}} = -11.5 \approx -12\text{db}$$

The *db* gain (magnitude) corresponding to ω_m

$$= -20 \log \frac{1}{\sqrt{\alpha}} = -20 \log \frac{1}{\sqrt{.07}} = -11.5$$

1 –12 *db*.

From the bode plot of uncompensated system the frequency ω_m corresponding to *db* pair of –12 *db* in found to be 17 rad/sec.

$$\therefore \qquad \omega_m = 17 \text{ rad/sec}$$

$$\therefore \qquad T_2 = \frac{1}{\omega_m \sqrt{\alpha}} = \frac{1}{17\sqrt{.07}} = .22$$

Transfer function of lead section $(G_2 s) = \dfrac{\alpha(1+sT_2)}{(1+s\alpha T_2)} = .07 \dfrac{(1+.22s)}{(1+.0514s)}$

Step 10: Determine the transfer function of lag-lead compensator.

Transfer function of lag-lead compensator

$$G(s) = G_1(s) \text{ Ч } G_2(s)$$

$$= 14 \frac{(1+2.65s)}{(1+35s)} \times .07 \frac{(1+.22s)}{(1+.0154s)}$$

$$G_c(s) = \frac{(1+2.5s)(1+\cdot 22s)}{(1+35s)(1+\cdot 0154s)}$$

Step 11: Determine open loop transfer function of compensated system.

The Lag Lead compensator is connected in series with $G(s)$ shown in Fig. (Ex. 5.9.2)

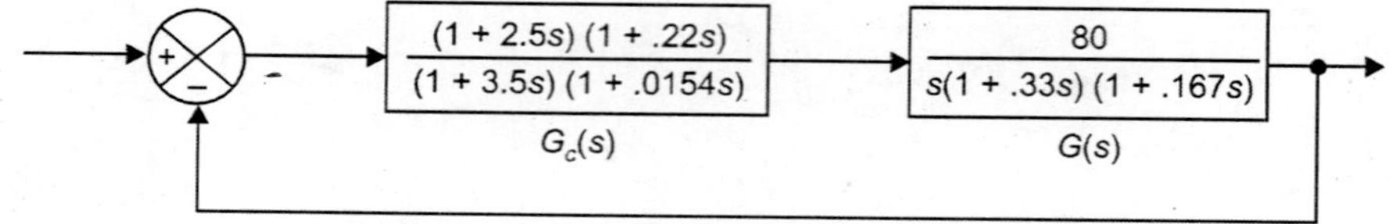

Fig. Ex. 5.9.2

The open loop transfer function of compensated system

$$G_o(s) = \frac{80(1+2.5s)(1+\cdot 22s)}{s(1+35s)(1+\cdot 0154s)(1+\cdot 33s)(1+\cdot 167s)}$$

Step 2: Bode plot of compensated system

Put $\qquad s = j\omega$ in $G_o(s)$

$$\therefore \qquad G_o(j\omega) = \frac{80(1+j2.5\omega)(1+j.22\omega)}{j\omega(1+j35\omega)(1+j0.0154\omega)(1+j.33\omega)(1+j.167\omega)}$$

Magnitude plot

There are six corner frequencies, which are given below.

$$\omega_{c1} = 1/35 = .03 \text{ rad/ec ;}$$

$$\omega_{c2} = \frac{1}{2.5} = .4 \text{ rad/sec}$$

$$\omega_{c3} = \frac{1}{.33} = 3 \text{ rad/sec.}$$

$$\omega_{c4} = \frac{1}{.22} = 4.5 \text{ rad/sec.}$$

$$\omega_{c5} = \frac{1}{.167} = 6 \text{ rad/sec,}$$

$$\omega_{c6} = \frac{1}{.0154} = 65 \text{ rad/sec.}$$

The various terms of $G_o(j\omega)$ are listed in Table Ex. 5.9.3. Also table shows the slope contributed by each term and the change in slope at the corner frequency.

Table Ex. 5.9.3

Term	Corner frequency rad/sec	Slope of each term *db*/dec	Change in slope from one term to another term *db*/dec
$\frac{80}{j\omega}$	–	–20	
$\frac{1}{1+j35\omega}$	$\omega_{c1} = 1/35 = .03$	–20	– 20 – 20 = – 40
$1+j\,2.5\omega$	$\omega_{c2} = 1/2.5 = .4$	+ 20	– 40 + 20 = – 20
$1/(1+j.33\omega)$	$\omega_{c3} = 1/.33 = 3$	– 20	– 20 –20 = – 40
$1+j.22\omega$	$\omega_{c4} = 1/.22 = 4.5$	+ 20	– 40 + 20 = – 20
$\frac{1}{1+j0.167\omega}$	$\omega_{c5} = 1/.167 = 6$	– 20	– 20 – 20 – 40
$1/(1+j\,0.0154\omega)$	$\omega_{c6} = 1/.0154 = 65$	– 20	– 40 –20 = –60

CHAPTER 5

Choose a low frequency ω_l such that $\omega_l < \omega_{c1}$ and choose a high frequency ω_h such that $\omega_h > \omega_{c6}$.

Let, $\omega_l = 0.01$ rad/sec and $\omega_h = 80$ rad/sec

Let, $A_o = |G_o(j\omega)|$ in *db*

At $\omega = \omega_l$, $A_{o1} = 20 \log |\text{first term}|$

i.e., $$A_o = 20 \log \frac{80}{\cdot 01} = 78 \, db$$

At $\omega = \omega_{c1}$,

$$A_{o2} = 20 \log |\text{first term}|$$

i.e., $$A_{o2} = 20 \log \frac{80}{\cdot 03} = 68.5 \approx 68 \text{ db}$$

At $\omega = \omega_{c2}$,

$$A_{o3} = [\text{slope from one term to another term}] \text{ Ч } \log \left|\frac{\text{Present frequency}}{\text{Previous frequency}}\right| + \text{Previous gain}$$

i.e., $$A_{o3} = -40 \times \log \frac{0.4}{\cdot 03} + 68 = 23 \, db$$

Similarly,

At $\omega = \omega_{c3}$, $$A_{o4} = -20 \times \log \frac{3}{\cdot 4} + 23 = 50 \, db$$

At $\omega = \omega_{c4}$, $$A_{o5} = -40 \log \frac{4.5}{3} + 5 = -2 \, db$$

At $\omega = \omega_{c5}$, $$A_{o6} = -20 \log \frac{6}{4 \cdot 5} + (-2) = -4 \, db$$

At $\omega = \omega_{c6}$, $$A_{o7} = -40 \log \frac{65}{6} + (-4) = -45 \, db$$

At $\omega = \omega_n$, $$A_{o8} = -60 \times \log \left(\frac{80}{6.5}\right) + (-45) = -50 \, db$$

Using the values of A_o at various frequencies the magnitude plot of compensataed system is drawn in Fig. (Ex. 5.9.1).

Phase plot

The phase angle of $G_o(j\omega)$ as a function of ω is given by

$$\phi = \angle G_o(j\omega)$$

$$= \tan^{-1} 2.5\omega + \tan^{-1} \cdot 22\omega - 90\epsilon - \tan^{-1} 35\,\omega - \tan^{-1} \cdot 0154\omega - \tan^{-1} \cdot 33\omega - \tan^{-1} \cdot 167\omega$$

The phase angle of $G_o(\omega)$ are calculated for various values of ω and listed in Table Ex. 5.9.4.

Table Ex. 5.9.4									
ω, rad/sec	.01	.03	.1	.4	1	4	10	65	80
φ, deg	– 108	–132	–152	–138	–126	–144	–168	–220	–228

Using the values of of ϕ_0 listed in Table Ex. 5.9.4 the phase plot of compensated system is sketched as shown in Fig. (4.9.1)]

Let, ϕ_{go} = phase of $G_o(j\omega)$ at the gain zeros over frequency of compensator system

and γ_o = phase margin of compensated system.

From the bode plot of compensated system, we get,

$$\phi_{gco} = -144\epsilon$$

Now, $$\gamma_o = (80\epsilon + \phi_{go}) = 180\epsilon - 144\epsilon = 36\epsilon$$

Conclusion: The phase margin of the compensated system is satisfactory. Hence the design is acceptable.

Result: Transfer function of lag-lead compensator,

$$G_c(s) = \frac{(1+2.5s)(1+\cdot 22s)}{(1+35s)(1+\cdot 0154s)}$$

Open loop transfer function of compensated system

$$G_o(s) = \frac{80(1+2.5s)(1+\cdot 22s)}{s(1+35s)(1+\cdot 0154s)(1+\cdot 33s)(1+\cdot 167s)}$$

EXAMPLE 5.10

Design a lag-lead compensator for a system with open loop transfer function $G(s) = \frac{K}{s(s+.5)}$ to satisfy the following specifications:

(i) Damping ratio of dominant closed loop poles, $\zeta = \cdot 5$

(ii) Undamped natural frequency of dominant closed loop poles, $\omega_n = 5$ rad/sec.

(iii) Velocity error constant, $K_v = 80$ sec^{-1}

Solution:

Step 1: Determine dominant pole s_d

Dominant pole, $$s_d = \zeta\omega_n \pm j\omega n\sqrt{1-\zeta^2}$$

Given that, $$\zeta = \cdot 5$$

and $$\omega_n = 5 \text{ rad/sec.}$$

∴ $$s_d = -\cdot 5 \times 5 \pm j5 \times \sqrt{1-\cdot 5^2}$$
$$= -2.5 \pm j4.3$$

CHAPTER 5

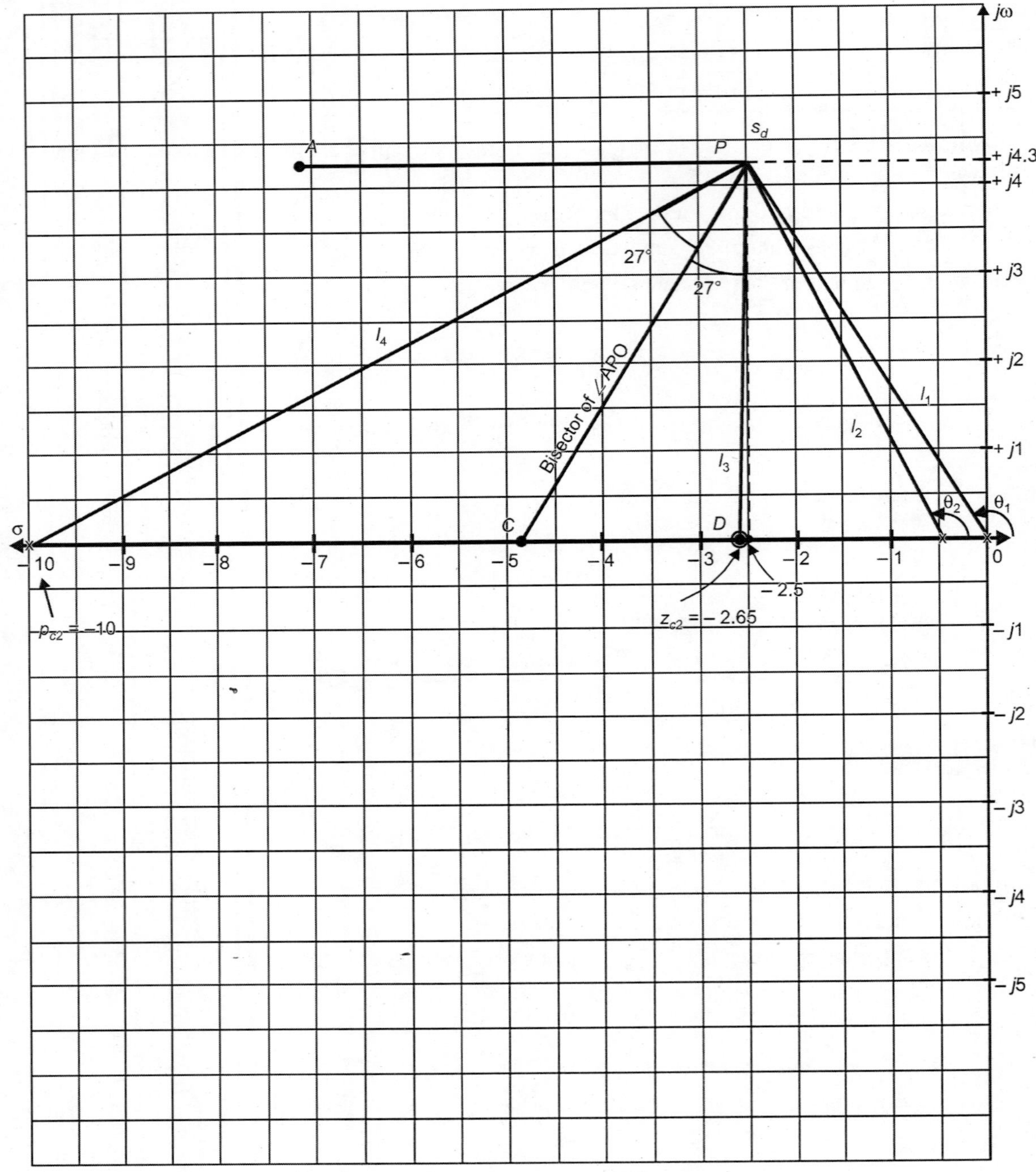

Fig. Ex. 5.10.1: Pole zero plot of $G(s) = K/s\ (s + 0.5)$.

Step 2: Draw the pole zero plot

The pole zero plot of open loop transfer function is shown in Fig. (Ex. 5.10.1). The pole at point P is the dominant pole, s_d

Step 3: To find angle to be contributed by lead network

Let

ϕ = Angle to be contributed by lead network to make point, P as a point on root locus.

Now, $$\phi = \begin{bmatrix} \text{Sum of angle contributed by} \\ \text{poles of uncompensated system} \end{bmatrix} - \begin{bmatrix} \text{Sum of angles contributed by} \\ \text{zeros of uncompensated system} \end{bmatrix} \pm n\,180\epsilon$$

from Fig. (4.10.1), we get

Sum of angles contributed by poles of uncompensated system

$$\theta_1 + \theta_2 = 120\epsilon + 115\epsilon = 235\epsilon$$

Since there is no finite zeor in uncompensated system, there is no angle contribution by zeros.

$\therefore$ $$\phi = 235\epsilon \pm n\ 180\epsilon$$

Let $n = 1$, $$\phi = 235\epsilon - 180\epsilon = 55\epsilon$$

Step 4: To find the pole and zero of the lead section

Draw a line AP parallel to X-axis as shown in Fig. (Ex. 5.10.1). The bisector PC is drawn to bisect the angle APO. The angles CPD and BPC are constructed as shown in Fig. (Ex. 5.10.1),

Here $$\angle CPD = \angle BPC = \phi/2 = 55\epsilon/2 = 27.5\epsilon$$

$$1\ 27\epsilon$$

Lead section, $p_{c2} = -10$

Zero of the lead section, $z_{c2} = -2.65$

From Fig. (Ex. 5.10.1), pole of the lead section, $p_{c2} = -10$

Zero of the lead section, $z_{c2} = -2.65$

We know that, $$z_{c2} = 1/T_2$$

$\therefore$ $$T_2 = 1/2.65 = .377$$

We know that, $$p_{c2} = -1/\alpha T_2;$$

$\therefore$ $$\alpha T_2 = 1/10$$

or $$\alpha = 1/T_2 \text{ Ч } 10 = 0.265$$

Step 5: Transfer function of lead compensator transfer function at lead section,

$$G_2(s) = \frac{(s + 1/T_2)}{(s + 1/\alpha T_2)} = \frac{(s + 2.65)}{(s + 10)}$$

Step 6: To find gain, K

Open loop transfer function of lead compensated system

$$G_{o2}(s) = G_2(s) \text{ Ч } G(s)$$

$$= \frac{(s + 2.65)}{(s + 10)} \times \frac{K}{s(s + .5)}$$

$$G_{o2}(s) = \frac{K(s + 2.65)}{s(s + .5)\,(s + 10)}$$

Here the value of K is gvien by the value of gain at the dominant pole s_d on the root locus from magnitude condition K is gvien by

$$K = \frac{\text{Product of vector lengths from all poles to } s = s_d}{\text{Product of vector lengths from all zeros to } s = s_d}$$

From Fig. (Ex. 5.10.1), we get,

$$K = l_1 \text{ Ч } l_2 \text{ Ч } l_4/l_3 = 5 \text{ Ч } 4.75 \text{ Ч } 7.75/4.3 = 42.8$$

$$G_{o2}(s) = \frac{42.8\,(s + 2.65)}{s(s + .5)\,(s + 10)}$$

Step 7: To find velocity error constant of lead compensated system.

Let $\quad K_{v2}$ = velocity error constant of lead compensated system

$$\therefore \quad K_{v2} = \underset{s\,0}{Lt}\; sG_{o2}(s) = \underset{s \to 0}{Lt}\; s \,.\, \frac{42.8(s + 2.65)}{s(s + .5)\,(s + 10)} = \frac{42.8 \times 2.65)}{.5 \times 10}$$

$$K_{v2} = 22.684$$

Step 8: To find the parameter, β

Let, $\quad K_{vd}$ = desired velocity error constant

A = the factor by which K_v is increased

Now, $\quad A = K_{vd}/K_{v2} = 80/22.684 = 3.5267$

Selec β, such that $\beta > A$

Let $\beta = 4$

Step 9: To find the transfer function of lag section

Let zero at lag section,

$$z_{c1} = .1 \text{ Ч second pole of } G(s) = 0.1 \text{ Ч } (-0.5) = -.05$$

Also, $\quad z_{c1} = -1/T_1;$

$$\therefore \quad T = 1/.05 = 20$$

Pole of lag section, $\quad p_{c1} = \dfrac{-1}{\beta T_1} = \dfrac{-1}{4 \times 20} = \dfrac{-1}{80} = -0.0125$

Transfer function of lag section,

$$G_1(s) = \frac{(s + 1/T_1)}{(s + 1/\beta T_1)} = \frac{(s + .05)}{(s + .0125)}$$

Step 10: Transfer function of compensated system

Transfer function of lag-lead compensator

$$G(s) = G_1(s) \times G_2(s) = \frac{(s+.05)(s+2.65)}{(s+.0125)(s+10)}$$

The lag-lead compensator is conneted in series with $G(s)$ as shown in Fig. (Ex. 5.10.2)

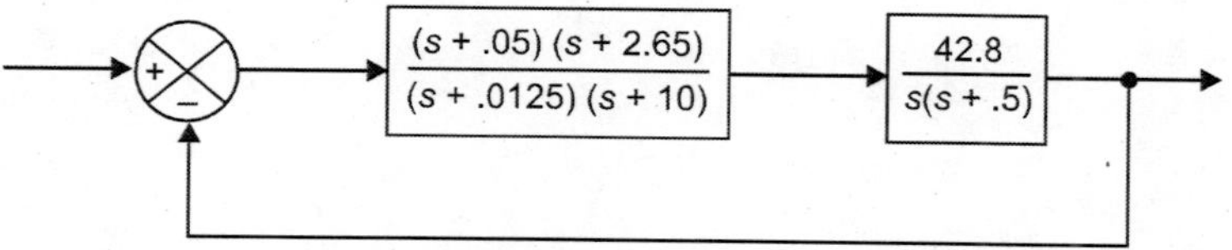

Fig. Ex. 5.10.2: Block diagram of lag-lead compensated system.

Open loop transfer function of compensated system

$$G_o(s) = \frac{42.8(s+.05)(s+2.65)}{s(s+.0125)(s+.5)(s+10)}$$

Step 11: Check velocity error

$$K_{vc} = \underset{s \to 0}{Lt}\ sG_o(s)$$

$$K_{vc} = 90.7$$

Conclusion: K_v of compensated system satisfies the requirement. Hence the deisgn is accepted.

Result: Transfer function of lag-lead compensator

$$G_c(s) = \frac{(s+.05)(s+2.65)}{(s+0.0125)(s+10)}$$

Open loop transfer function of lag-lead compensated system

$$G_o(s) = \frac{42.8(s+0.05)(s+2.65)}{s(s+0.0125)(s+.5)(s+10)}$$

SOME WORKED EXAMPLES

Note: To make the students to understand about the design of compensators some more problems are solved in very detail.

EXAMPLE 1

The open loop transfer function of a certain unity feedback control system is given by $G(s = K/s(s+1)$.

It is desired to have the velocity error. Constant, $K_V = 10$ and the phase margin to be atleast 60є. Design a phase lag compensator.

Solution:

Step 1: Calculator of gain, K

Given that, $K_v = 10$

We know, $K_v = \underset{3 \to 0}{Lt}\ s \cdot G(s)\, H(s) = \underset{s \to 0}{Lt}\ s \cdot G(s)$ $[\because H(s) = 1]$

$$\therefore \quad K_v = \underset{s \to 0}{Lt}\ s \cdot \frac{K}{s(s+1)} = K \Rightarrow K_v = K$$

$$\therefore \quad \boxed{K = 10}$$

$$\therefore \quad G(s) = 10/s\ (s + 1)$$

Step 2: Find data's to draw bode plot [without compensator]

$$G(s) = 10/s(s + 1)$$

Put, $s = j\omega \Rightarrow G(j\omega) = 10/j\omega\ (1 + j\omega)$

Magnitude plot

From given $G(s)$ equation $\omega_{c1} = 1$ rad/sec.

Assume low and high frequencies.

The condition for assuming low frequency is

$$\omega_l < \omega_{c1}$$

The condition for assuming high frequency is

$$\omega_h > \omega_{c1}$$

$\therefore$ Let, $\omega_l = .1$ rad/sec

$\omega_n = 10$ rad/sec

Table 1

Term	Corner frequency rad/sec	Slope of each term *db*/dec	Change in slope from one term to another term *db*/dec
$\frac{10}{j\omega}$	–	–20	
$\frac{1}{(1+j\omega)}$	$\omega_{c1} = 1/1 = 1$ rad/sec	–20	–20–20 = –40

At $\omega = \omega_l$,

$$A_1 = 20 \log |\text{first term}|$$

$$A_1 = 20 \log \left|\frac{10}{j\omega}\right| \Rightarrow A_1 = 20 \log \left|\frac{10}{\omega_l}\right|$$

$$A_1 = 20 \log \left|\frac{10}{.1}\right| \Rightarrow +40\text{db}$$

At $\omega = \omega_{c1}$

$$A_2 = 20 \log |\text{First term}|$$

$$= 20 \log \left|\frac{10}{j\omega}\right| \Rightarrow A_2 = 20 \log \left|\frac{10}{\omega_{c1}}\right|$$

$$A_2 = 20 \log \left|\frac{10}{1}\right| \Rightarrow +20\text{ db}$$

At $\omega = \omega_h$,

$$A_3 = \text{Change in slope from previous point to preset point Ч}$$

$$\log \left|\frac{\text{Present frequency}}{\text{Previous frequency}}\right| + \text{Previous gain}$$

ie.

$$A_3 = \text{Change in slope from } \omega_{c2} \text{ Ч} \log \left|\frac{\omega_h}{\omega_{c1}}\right| + A_2$$

$$A_3 = -40 \times \log \left|\frac{10}{1}\right| + A_2$$

$$A_3 = -40 \log 10 + 20 \Rightarrow -20\text{ db}$$

Phase plot

The phase angle of $G(j\omega)$ as a function of ω is given by,

$$\phi = \angle G(j\omega) = -90∈ - \tan^{-1}\omega$$

The value of ϕ for various ω value are

Step 3: Using Table 2 and table 3 draw bode plot.

Table 2

ω	A in db
$\omega = \omega_l = \cdot 1$	+ 40 db
$\omega = \omega_{c1} = 1$	+ 20 db
$\omega = \omega_n = 10$	–20 db

Table 3

ω	ϕ
0.1	–96∈
1	–135∈
5	–169∈
10	–174∈

Step 4: From bode plot determine phase margin of uncompensated system.

Let, ϕ_{gc} = phase of $G(j\omega)$ at gain cross over frequency (ω_{gc})

and γ = phase marging of uncompensated system

From the bode plot

$$\phi_{gc} = -162\epsilon$$

Now, $$\gamma = 180\epsilon + \phi_{gc} = 180\epsilon - 162\epsilon = 18\epsilon$$

$$\boxed{\gamma = 18\epsilon}$$

But the desired phase margin is 60ε

i.e., $$\gamma_d = 60\epsilon$$

Step 5: Determine desired gain cross over frequency (ω_{gcd})

$$\gamma_d = 180\epsilon + \gamma_{gcd} \Rightarrow 60\epsilon = 180\epsilon + \phi_{gcd}$$

$$\phi_{gcd} = 60\epsilon - 180\epsilon = -120\epsilon$$

From bode plot, the frequency at -120ϵ is .5 rad/sec

∴ $$\omega_{gcd} = .5 \text{ rad/sec}$$

Desired gain crossover frequency is $\omega_{gcd} = .5$ rad/sec

Step 6: Determine β

From bode plot, gain at .5 rad/sec is $A_{gcd} = 2\ db$

∴ $$A_{gcd} = 20 \log \beta \Rightarrow \beta = 10^{Agcd/20} = 10^{26/20}$$

Step 7: Determine the transfer function of lag compensator.

The zero of the compensator is placed at a frequency one-tenth of ω_{gcd}.

∴ Zero of the lag compensator,

$$Z_c = 1/T = \frac{\omega_{gcd}}{10}$$

Now, $$T = \frac{10}{\omega_{gcd}} = \frac{10}{.5} = 20$$

$$\boxed{T = 20}$$

Pole of the lag compensator,

$$P_c = \frac{1}{\beta T} = \frac{1}{20 \times 20} = \frac{1}{400}$$

Transfer function of lag compensator

$$G_c(s) = \frac{s + 1/T}{s + 1/\beta T}$$

∴ $$G(s) = \beta \frac{(1 + sT)}{(1 + s\beta T)} = 20 \frac{(1 + 20s)}{(1 + 40s)}$$

Step 8: Determine the open loop transfer function of the compensated system.

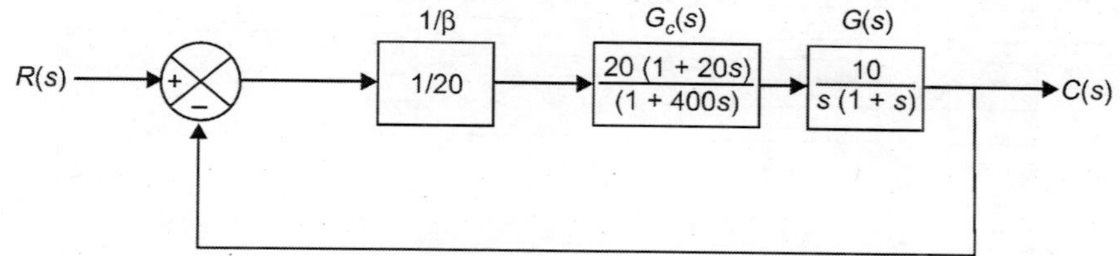

Open loop transfer function of compensated system.

$$G_c(s) = \frac{1}{\beta} \times G_c(s) \times G(s)$$

$$G_o(s) = \frac{20\phi\,(1 + 20s)}{2\phi s(1 + 200s)\,(1 + s)}$$

$$G_o(s) = \frac{10(1 + 20s)}{s(1 + s)\,(1 + 400s)}$$

Step 9: Determine the phase margin of compensated system and verify the design.

Put $s = j\omega$,

$$G_o(j\omega) = \frac{10(1 + 20\,j\omega)}{j\omega\,(1 + j\omega)\,(1 + 400\,j\omega)}$$

Let,

$$\phi_0 = -90\epsilon + \tan^{-1} 20\omega - \tan^{-1}\omega - \tan^{-1} 400\,\omega$$

where,

$$\omega = \omega_{gcd} = .5 \text{ rad/sec}$$

$$\phi_0 = -90\epsilon + \tan^{-1} 20(.5) - \tan^{-1} .5 - \tan^{-1} 400\,(.5)$$

$$\phi_0 = -90\epsilon + 84.28 - 26.56 - 89\epsilon$$

$$\phi_0 = 121.56 \approx -122\epsilon$$

∴

$$\phi_0 = 180\epsilon - 122\epsilon = 58\epsilon \Rightarrow \boxed{\gamma_0 = 58\epsilon}$$

$$\gamma_n = \gamma_o + \in \Rightarrow \gamma_o + 2\epsilon$$

$$\boxed{\gamma_n = 60\epsilon}$$

Note: $\in$ value can be assumed from 1ϵ to 5ϵ according to the desired value. [This $\in$ value can be assumed in the begining (or) in the last step].

Conclusion: The γ_n satisfies the requirement. Hence the design is acceptable.

Result: Transfer function of lag compensator

$$G_c(s) = \frac{2.(1 + 20s)}{(1 + 400s)}$$

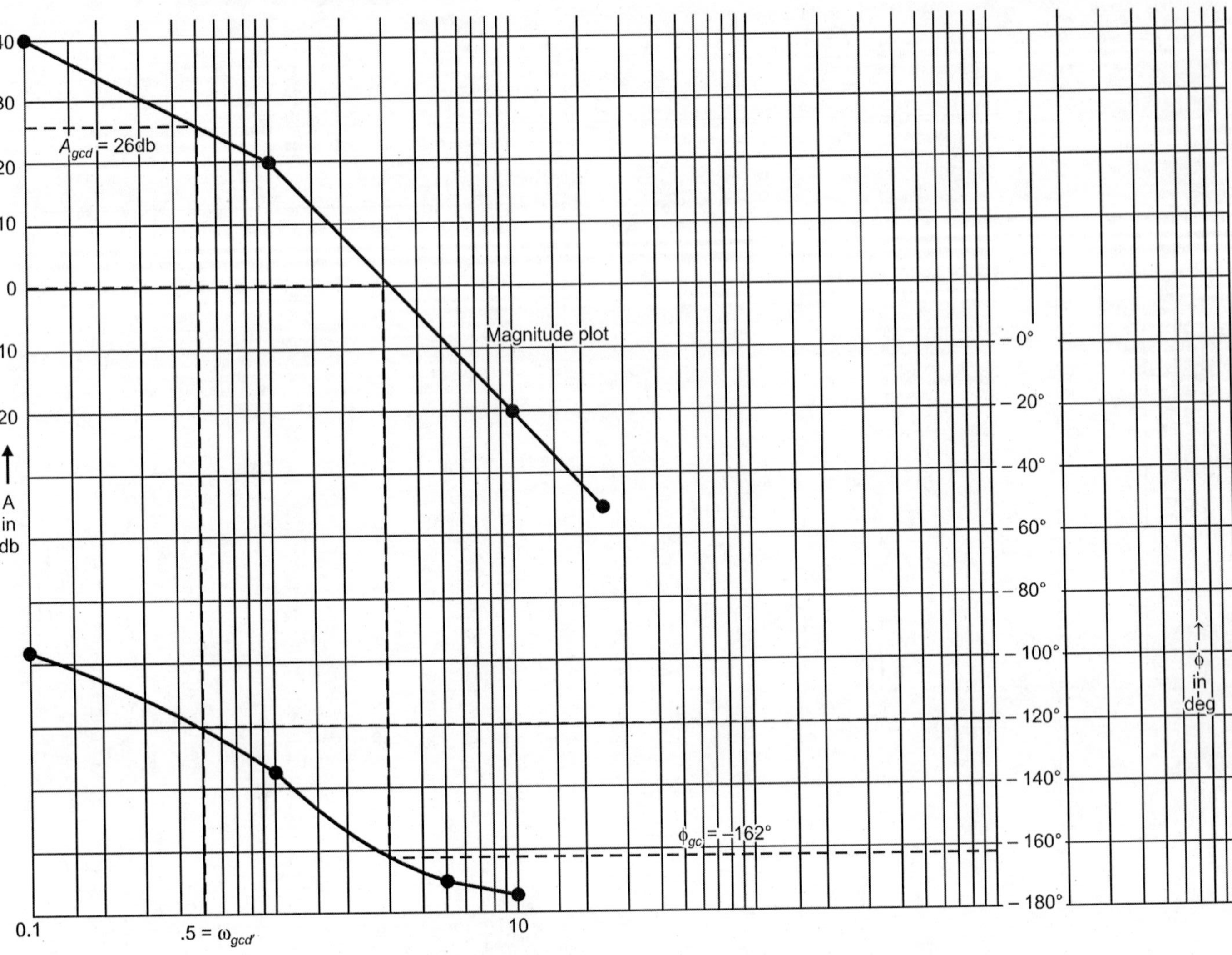

Fig. Worked Example 1: Bode plot of $G(s) = K/s\,(s+1)$ (lag compensator).

Transfer function of lag compensated system

$$G_o(s) = \frac{10(1+20s)}{s(1+s)(1+400s)}$$

EXAMPLE 2

The controlled plant of a unity feedback system is $G(s) = K/s\ (s + 5)$. It is specified that velocity error constant of the system be equal to 15. Design a lag compensator. So that phase marging will be at least 50є.

Solution:

Step 1: Calculation of gain K

$$K_v = \underset{s\to 0}{Lt}\ s \cdot G(s)\,H(s)$$

$$\pm \underset{s\to 0}{Lt}\ s \cdot G(s) \qquad [\because H(s) = 1]$$

$$= \underset{s\to 0}{\text{Lt}}\ s \cdot \frac{K}{s(s+5)} = \frac{K}{5} = K_v$$

$$K = 5 \text{ Ч } 15$$

$$\boxed{K = 75}$$

$\therefore$ $$G(s) = 75/s\ (s+5) = \frac{75}{5s(1+5/s)} = \frac{75}{5s\ (1+.2s)}$$

$\therefore$ $$G(s) = 15/3(1 + .2s)$$

Step 2: Find the data's to draw bode plot.

Magnitude plot

Put, $s = j\omega$ is $G(s)$

$\therefore$ $$G(j\omega) = 15/j\omega\ (1 + .2j\omega)$$

Table 1

Term	**Corner frequency rad/sec**	**Slope of each term *db*/dec**	**Change in slope from one term to another term *db*/dec**
$\frac{15}{j\omega}$	–	–20	
$\frac{1}{(1+.2j\omega)}$	$\omega_{c1} = 1/.2 = 5$ rad/sec	–20	–20–20 = –40

Assume low and high frequencies.

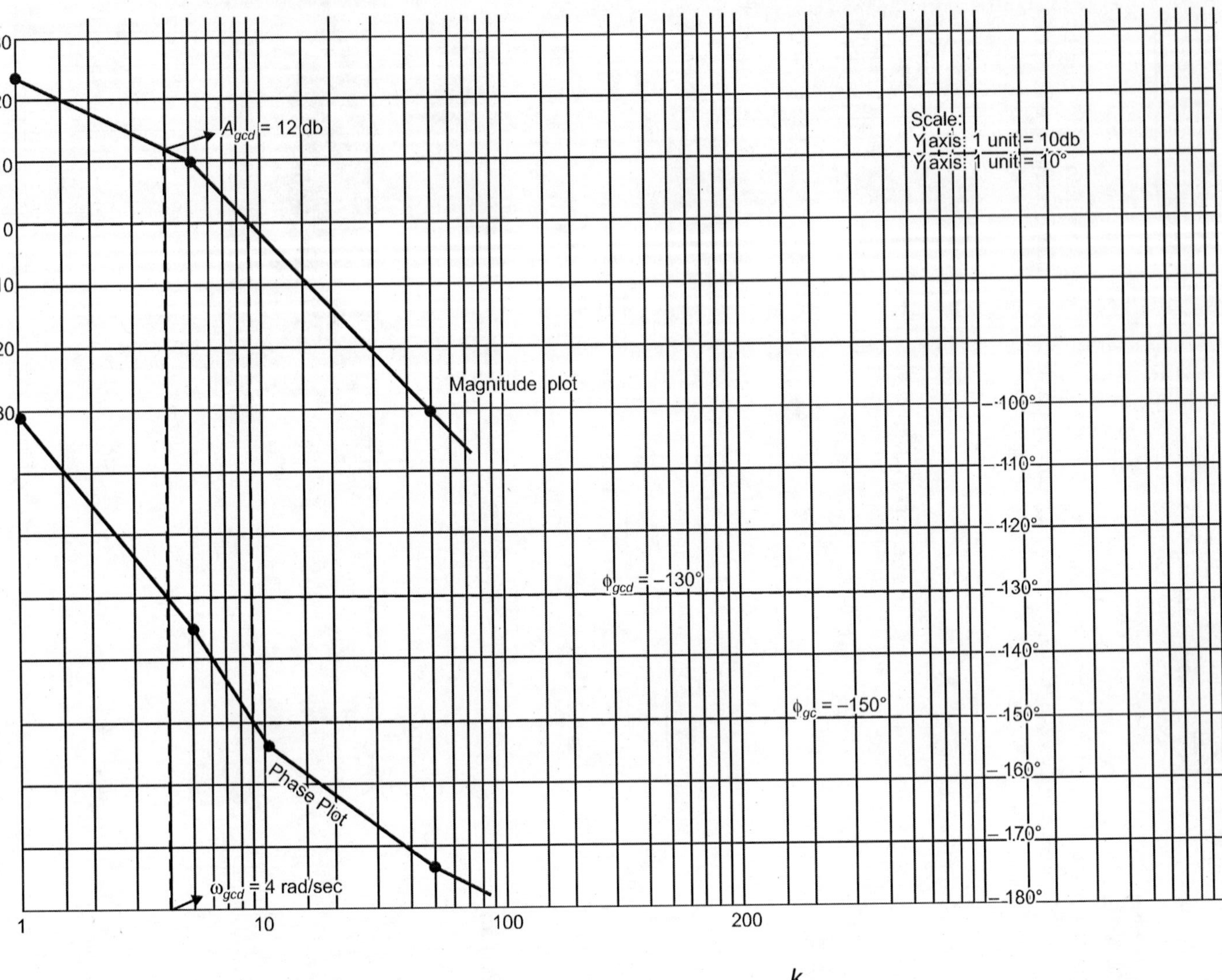

Fig. Worked Example 2: Bode plot of $G(s)=\dfrac{k}{s(s+5)}$.

The condition for low frequency is $\omega_l < \omega_{c1}$

The condition for high frequency is $\omega_h > \omega_{c1}$

∴ Let, $\omega_l = 1$ rad/sec and $\omega_h = 50$ rad/sec

At, $\omega = \omega_l$

$$A_1 = 20 \log |\text{first term}|$$

$$= 20 \log \left|\frac{15}{j\omega}\right| = 20 \log \left|\frac{15}{1}\right|$$

$$A_1 = +24\ db$$

At $\omega = \omega_{c1}$

$$A_2 = 20 \log |\text{first term}|$$

$$= 20 \log \left|\frac{15}{j\omega}\right| = 20 \log \left|\frac{15}{5}\right|$$

$$= +10\ db$$

At, $\omega = \omega_n$,

$$A_3 = [\text{change in slope from previous to present term}] \text{ Ч } \log\left(\frac{\text{Present frequency}}{\text{Previous frequency}}\right) + \text{Previous gain}$$

ie $$A_3 = \text{change in slope from } \omega_{c1} \text{ to } \omega_n \text{ Ч } \log\left(\frac{\omega_n}{\omega_{c1}}\right) + A_2$$

$$A_3 = -40 \times \log \left|\frac{50}{5}\right| + A_2$$

$$A_3 = -40 \log |10| + 10\ db$$

$$A_3 = -40 + 10 = -30\ db$$

Phase plot

The phase angle of $G(j\omega)$ as a function of ∞ is given by

$$\phi = \angle G(j\infty) = -90\epsilon - \tan^{-1} .2\omega$$

Step 3: Using table 2 and table 3, draw the bode plot.

Table 2	
ω	***A* in *db***
1	+ 24 *db*
5	+ 10 *db*
50	–30 *db*

Table 3				
ω, rad/sec	1	5	10	50
ϕ, deg	–101	–135є	–153є	–174є

Step 4: From bode plot, determine phase marging at uncompensated system.

Let, ϕ_{gc} = phase of $G(j\omega)$ at gain crossover frequency ω_{gc}

and γ = phase marging of uncompensated system

From bode plot,

$$\phi_{gc} = -150\epsilon$$

$$\gamma = 180\epsilon + \phi_{gc} = 180 - 158\epsilon = 30\epsilon$$

$$\boxed{\gamma = 30\epsilon}$$

But the desired phase marging is 50є

Step 5: Determine gain cross over frequency (ω_{gcd})

$$\gamma_d = 180\epsilon + \gamma_{gcd}$$

$$50\epsilon = 180\epsilon + \phi_{gcd} \Rightarrow \phi_{gcd} = 50\epsilon - 180\epsilon$$

$$\boxed{\phi_{gcd} = -130\epsilon}$$

From bode plot, the frequency at –130є is 4.3 rad/sec

∴ ω_{gcd} = 4 rad/sec

Desired gain crossover frequency is 4 rad/sec

Step 6: Determine β

From bode plot, gain at 4.3 rad/sec is 12 *db*

ie $A_{gcd} = 12\ db$

∴ $A_{gcd} = 20 \log \beta \Rightarrow \Rightarrow \beta = 10^{A_{gcd}/20} = 10^{12/20} \Rightarrow \beta = 4$

Step 7: Determine the transfer function of lag compensator.

The zero of the compensator is placed at a frequency one tenth of ω_{gcd}

∴ Zero of the lag compensator,

$$Z_c = \frac{1}{T} = \frac{\omega_{gcd}}{10}$$

Now,

$$T = \frac{10}{\omega_{gcd}} = \frac{10}{4} = 2.5$$

$$\boxed{T = 2.5}$$

Pole of the lag compensator,

$$P_c = \frac{1}{\beta T} = \frac{1}{4 \times 2.5} = \frac{1}{10} = .1$$

Transfer function of lag compensator

$$G_c(s) = \frac{s + 1/T}{s + 1/\beta T}$$

$$\therefore \qquad G_c(s) = \beta\frac{(1 + sT)}{(1 + s\beta T)} = 4\frac{(1 + 2.5)}{(1 + 10s)}$$

Step 8: Determine the open loop transfer function of the compensated system.

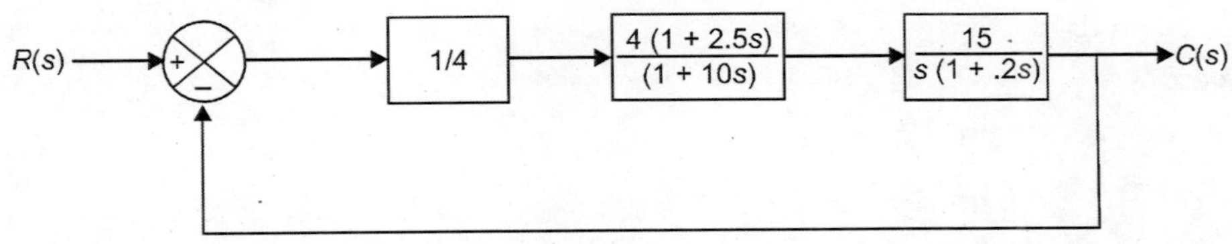

$$G_o(s) = \frac{4 \times 15(1 + 2.5\ 2s)}{4s(1 + 10s)\,(1 + .2s)}$$

$$G_o(s) = \frac{15(1 + 2.5\ 2s)}{s(1 + 10s)\,(1 + .2s)}$$

$$\phi_0 = \angle G(j\omega) = \tan^{-1} 2.5\omega - 90\epsilon - \tan^{-1} 10\omega - \tan^{-1} .2\omega$$

where, $\omega = 4$ rad/sec

$$\phi_0 = \tan^{-1} 2.5\,(4) - 90 - \tan^{-1} 10(4) - \tan{-1}\ .2(4)$$

$$\phi_0 = 85\epsilon - 90\epsilon - 88\epsilon - 40\epsilon$$

$$\phi_0 = -134\epsilon$$

$$\gamma_o = 180\epsilon + \phi_0 \rightarrow 180\epsilon - 134\epsilon = 46\epsilon$$

$$\gamma_n = \phi + \zeta \Rightarrow 46\epsilon + 4\epsilon = 50\epsilon$$

Note: The ζ value can be assumed from 1ϵ to 5ϵ according to the desired phase margin.

Conclusion: The γ_o satisfies the requirement. hence the design is acceptable.

Result: Transfer function of lag compensator.

$$G_c(s) = \frac{4(1 + 2.32s)}{(1 + 9.3s)}$$

Transfer function of lag compensated system.

$$G_o(s) = \frac{15(1 + 2.32s)}{s(1 + 9.3s)\,(1 + .2s)}$$

EXAMPLE 3

The forward path transfer function of unity feedback control system is $G(s) = K/s\,(s+2)(s+4)$. Design a suitable lag compensator so that the system meets the following specifications. (i) Percentage overshoot ≤ 180% for unit step input (ii) Steady state error ≤ .125 for unit ramp input.

Solution:

Step 1: Find poles of open loop system

Given,
$$G(s) = K/s\,(s+2)\,(s+4)$$

∴
$$s\,(s + 20\,(s+4) = 0$$

∴ The pole are, $s = 0$, $s = -2$ and $s = -4$.

Step 2: Find root locus on real axis

There is two root locus paths.

One in from $s = -0$ to $s = -2$ and second one is from $s = -4$ to $s = -\infty$.

Note: If we choose a test point in this segment then to the right of this point we have odd number of real poles and zeros.

Step 3: Find angles of asymptotes

There are three poles, the number of root locus branches are three. There is no finite zero and so all the three root locus branches will meet the zeros at infinity.

Hence we require three asymptotes.

Hence, $n = 3$ and $m = 0$

∴ If $q = 0$,
$$\text{Angles} = \pm\frac{180\epsilon(1)}{3} = \pm 60\epsilon$$

If $q = 1$,
$$\text{Angles} = \pm\frac{180\epsilon(2+1)}{3} = \pm 180\epsilon$$

∴ If $q = 1$,
$$\text{Angles} = \pm\frac{180\epsilon(3+1)}{3} = \pm 240\epsilon[240\epsilon - 180\epsilon]$$

$$\text{Angles} = \pm 60\epsilon$$

Since ±60є repeated angle, we can choose ±60є as angle of asymptotes.

Step 4: Find centroid

$$\text{Centroid} = \frac{\text{Sum of poles} - \text{Sum of zeros}}{n-m} = \frac{0-2-4}{3} = -\frac{6}{3} = -2$$

$$\boxed{\text{Centroid} = -2}$$

Step 5: Find breakaway point

Closed loop transfer function,

$$\frac{C(s)}{R(s)} = \frac{G(s)}{1+G(s)}$$

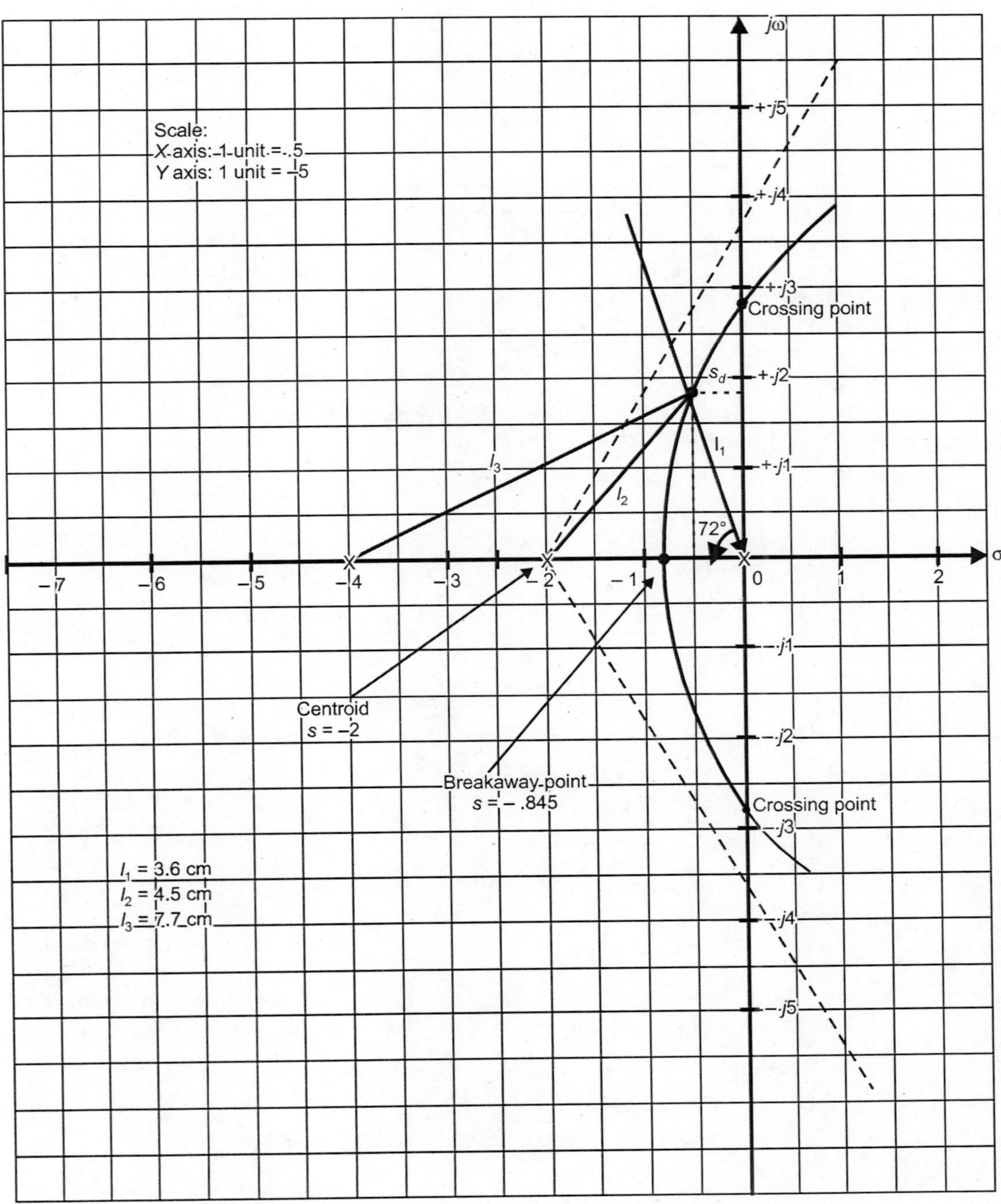

Fig. Worked Example 3: $G(s) = K/s\,(s + 2)\,(s + 4)$ lag compensator.

$$= \frac{\dfrac{K}{s(s+2)(s+4)}}{1+\dfrac{K}{s(s+2)(s+4)}}$$

$$= \frac{\dfrac{K}{s(s+2)(s+4)}}{\dfrac{s(s+2)(s+4)+K}{s(s+2)(s+4)}} = \frac{K}{s(s+2)(s+4)+K}$$

The characteristic equation is $s(s+2)(s+4)+K=0$

$$\therefore \quad K = -s(s+2)(s+4)$$

$$= (-s^2 - 2s)(s+4) = -s^3 - 4s^2 - 2s^2 - 8s$$

$$K = -s^3 - 6s^2 - 8s$$

on differentiating with respect to s we get,

$$dK/ds = -(3s^2 + 12s + 8)$$

Put $dK/ds = 0$

$$\therefore \quad 3s^2 + 12s + 8 = 0$$

$$s = \frac{-12 \pm \sqrt{(-12)^2 - 4 \times 3 \times 8}}{2 \times 3} = \frac{-12 \pm \sqrt{144 - 96}}{6}$$

$$s = -0.845 \quad \text{or} \quad -3.15$$

when $s = -.845$, $K = -(s^3 + 6s^2 + 8s) = -(-.345)^3 + 6(-.845)^2 + 8(-.845)]$

$$= -[-.603 + 4.28 - 6.76]$$

$$K = -[-3.083] = 3.083$$

when $s = -3.15$, $K = -(s^3 + 6s^2 + 8s)$

$$= -[(-3.15)^3 + 6(-3.15)^2 + 8(-3.15)]$$

$$= -[-31.25 + 59.535 - 25.2]$$

$$K = -[+3.0850 = -3.085$$

For $s = -.845$ the value of K is positive and real so $s = -.845$ is the actual break away point.

Step 6: Find crossing point on imaginary axis

The characteristic equation is,

$$s(s+2)(s+4) + K = 0$$

$$s^3 + 6s^2 + 8s + K = 0$$

Put, $s = j\omega$

$$(j\omega)^3 + 6(j\omega)^2 + 8(j\omega) + K = 0$$

$$-j\omega^3 - 6\omega^2 + j8\omega + K = 0$$

on equating imaginary part to zero we get,

$$-j\omega^3 + j8\omega = 0$$
$$\omega^2 = 8$$
$$\omega = \sqrt{8}$$
$$\omega = \pm 2.82$$

Root locus crosses the imaginary axis at $+j\,2.82$ and $-j\,2.82$.

Step 7: Determine the dominant pole, s_d:

Given that, $\%M_p = 18\%$

We know that, $\%M_p = e^{\zeta\pi/\sqrt{1-\zeta^2}}$

$\therefore$ $e^{-\zeta\pi/\sqrt{1-\zeta^2}} = .18$

on taking natural log we get,

$$\frac{-\zeta\pi}{\sqrt{1-\zeta^2}} = \ln 0.18 = -1.71$$

on squaring we get,

$$\frac{\zeta^2\pi^2}{1-\zeta^2} = 2.92$$

on cross multiplication we get,

$$\zeta^2\pi^2 = 2.92\,(1-\zeta^2)$$
$$\zeta^2\pi^2 + 2.92\zeta^2 = 2.92$$
$$\zeta^2(\pi^2 + 2.92) = 2.92$$

$\therefore$
$$\zeta = \sqrt{\frac{2.92}{\pi^2 + 29}} = \sqrt{\frac{2.92}{28.59}} = .31$$

$$\boxed{\zeta = .31}$$

$$\cos^{-1}\zeta = \cos{-1}(.31) = 71.9\epsilon \approx 72\epsilon$$

$$\boxed{\cos^{1}\zeta = -72\epsilon}$$

Draw a straight line at an angle of 72ϵ with respect to real axis. The meeting point of this line with root locus is the dominant pole, s_d.

From root locus Fig. we get,

$s_d =$ (Dominant poles occur as conjugate poles)

Step 8: Find gain K, at $s = s_d$

$$K = \frac{\text{Product of vector lengths from open loop poles to } s_d}{\text{Product of vector lengths from open loop zeros to } s_d}$$

Product of vector lengths from open loop poles = l_1 Ч l_2 Ч l_3

From root locus plot we get,

$l_1 = 3.6$ cm $l_2 = 4.5$ cm $l_3 = 7.7$cm

since there is no finite zero, the product of vector lengths from zero is unity.

∴ $K = l_1\, l_2\, l_3$ = 3.6 Ч 4.5 Ч 7.7 ≈ 125

Step 9: Find β

Given, $G(s) = K/s(s + 2)(s + 4)$

velocity error constant of uncompensated system

$$K_{vu} = \underset{s\to 0}{Lt}\ s \cdot G(s)$$

$$K_{vu} = \underset{s\to 0}{Lt}\ s \cdot \frac{125}{s(s+2)(s+4)} = \frac{125}{2\times 4} = 15.5 \approx 16$$

$e_{ss} \le .5$ for ramp inpuit (given)

The desired velocity error constant is

$$K_{vd} = 1/e_{ss} = \frac{1}{.125} = 8$$

Let A be the factor by which K_v is increased

$$\therefore \quad A = \frac{K_{vd}}{K_{vu}} = \frac{8}{16} = .5$$

β = 1.2 Ч .5 = 0.6

$$\boxed{\beta = 6}$$

Note: Always $\beta = 1.2A$

Step 10: Determine the transformation of lag compensator zero of the compensator

$z_c = 1/T$ = 0.1 Ч second pole of $G(s)$

= 0.1 Ч (–2) = –0.2

Now, $T = 1/0.2 = 5$

Pole of the compensator, $p_c = -\dfrac{1}{\beta T} = \dfrac{-1}{.6\times 5} = -\dfrac{1}{3}$

$\boxed{p_c = -.33}$ ∴ $\boxed{\beta T = 3}$

Transfer function of lag compensator

$$G_c(s) = \frac{s + 1/T}{s + 1/\beta T} = \frac{(s+0.2)}{(s+.33)}$$

Step 11: Transfer function of compensated system.

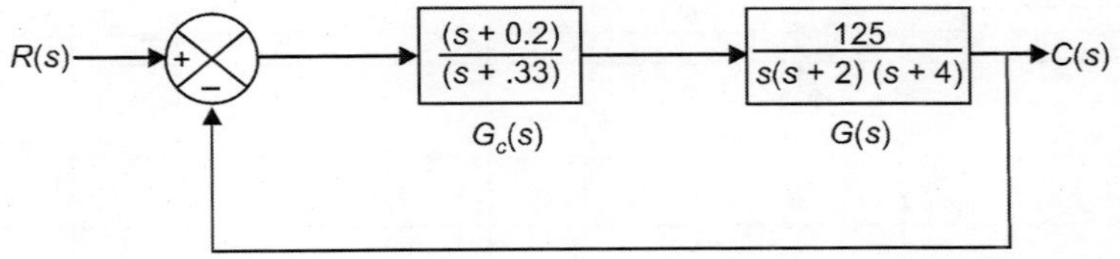

Open loop transfer function of compensated system

$$G_o(s) = \frac{(s+0.2)}{(s+.33)} \times \frac{125}{s(s+2)(s+4)}$$

$$G_o(s) = \frac{125(s+.2)}{s(s+2)(s+4)(s+.33)}$$

Step 12: Check steady state error of compensated system:

$$K_{vc} = \underset{s \to 0}{Lt}\ s \cdot G_o(s) = \underset{s \to 0}{Lt}\ s \cdot \frac{125(s+0.2)}{s(s+2)(s+4)(s+.33)}$$

$$K_{vc} = \frac{25}{2.64}$$

$$K_{vc} = 9.46$$

$$\therefore \quad e_{ss} = \frac{1}{K_{vc}} = \frac{1}{9.46} = .105$$

$$\boxed{e_{ss} = .105}$$

Conclusion: Since e_{ss} value of compensated system is less than .125 (given), the design is acceptable.

Result: Transfer function of lag compensator,

$$G_c(s) = (s + 0.2)/(s + 0.33)$$

Transfer function of compensated system.

$$G_o(s) = \frac{125(s+0.2)}{s(s+2)(s+4)(s+0.33)}$$

CHAPTER 5

EXAMPLE 4

The open loop transfer function of unity feedback control system is given by $G(s) = 10/s\ (1 + 0.4s)(1 + 0.1s)$. It is desired to have the phase margin to be atleast 20є. Design a suitable lead compensator.

Solution:

Step 1: Determine the data's which is required to draw bode plot.

$$G(s) = 10/s\ (1 + 0.4s)\ (1 + 0.8s)\ (1 + 0.1s)$$

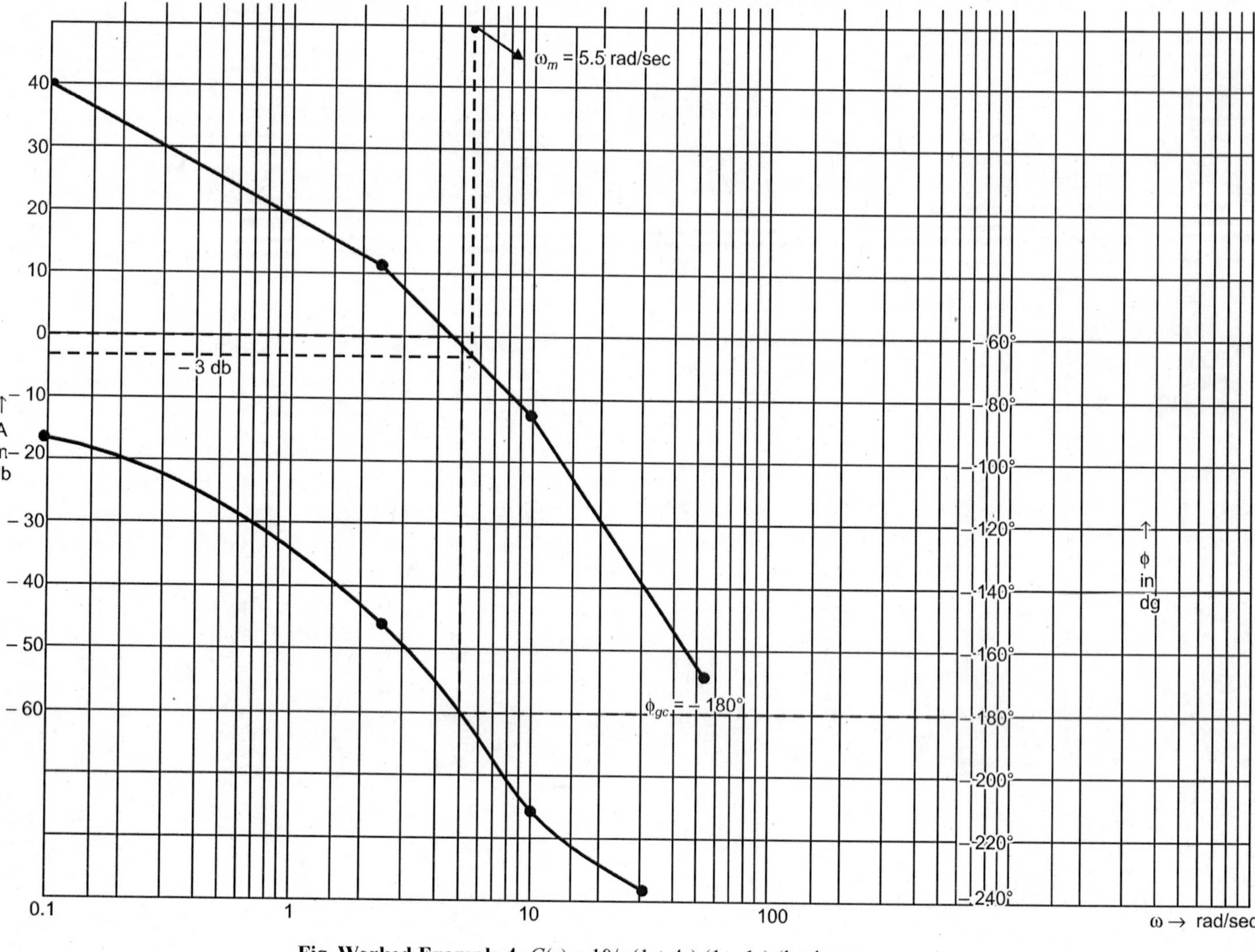

Fig. Worked Example 4: $G(s) = 10/s\ (1 + .4s)\ (1 + .1s)$ (lead compensator).

Put, $s = j\omega$

$\therefore \quad G(j\omega) = 10/j\omega\,(1 + 0.4j\omega)\,(1 + 0.1\,j\omega)$

Magnitude plot

The corner frequencies are,

$$\omega_{c1} = 1/0.4 = 2.5 \text{ rad/sec}, \ \omega_{c2} = 1/0.1 = 10 \text{ rad/sec}$$

Table 1

Term	Corner frequency rad/sec	Slope of each term *db*/dec	Change in slope from one term to another term *db*/dec
$\frac{10}{j\omega}$	–	–20	
$\frac{1}{1+j0.4\omega}$	$\omega_{c1} = 1/0.4 = 2.5$	–20	–20 –20 = –40
$\frac{1}{1+0.1\,j\omega}$	$\omega_{c2} = 1/0.1 = 10$	–20	–40 –20 = –60

Choose low and high frequencies.

The condition for choosing low frequency is $\omega_l < \omega_{c1}$.

The condition for choosing high frequency is $\omega_h > \omega_{c2}$

Let, $\omega_l = 0.1$ rad/sec

and $\omega_h = 50$ rad/sec

Let, $A = |G(j\omega)|$ in *db*

at, $\omega = \omega_l$

$A_1 = 20 \log |\text{first term}|$

$$A_1 = 20 \log \left|\frac{10}{\omega}\right| = 20 \log \left|\frac{10}{0.1}\right|$$

$$\boxed{A_1 = 40 \text{ db}}$$

At $\omega = \omega_{c1}$

$A_2 = 20 \log |\text{first term}|$

$$A_2 = 20 \log \left|\frac{10}{\omega}\right| = 20 \log \left|\frac{10}{2.5}\right| = 22 \text{ db}$$

$$\boxed{A_2 = 12 \text{ db}}$$

at $\omega = \omega_{c2}$

A_3 = change in slope one term to another term Ч log $\left|\frac{\text{Present frequency}}{\text{Previous frequency}}\right|$ + Previous gain (A_2)

i.e. $A_3 = -40 \log\left(\frac{10}{2.5}\right) + A_2 \Rightarrow -40 \log 4 + 12$

$\boxed{A_3 + -12 \text{ db}}$

at $\omega = \omega_h$

A_4 = change in slope form or term to another term Ч log $\left|\frac{\text{Present frequency}}{\text{Previous frequency}}\right|$ Previous gain (A_3)

$A_4 = -60 \log\left|\frac{50}{10}\right| + A_3$

$A_4 = -60 \log 5 - 12$

$\boxed{A_4 = -54 \text{ db}}$

Table 2

ω, rad/sec	0.1	2.5	0	50
A, db	40 *db*	12 *db*	–12 *db*	–54 *db*

Phase plot

The phase angle of $G(j\omega)$ as a function of ω is given by

$$\phi = -90\epsilon - \tan^{-1} - 4\omega - \tan^{-1} 0.1\omega$$

Draw the bode plot using table 2 and 3.

Table 3

ω, rad/sec	0.1	2.5	10	20
φ, deg	–92є	–150є	–210є	236

Step 2: Determine phase marging of uncompensated system

Let, $\phi_{gc} \rightarrow$ Phase of $G(j\omega)$ as gain crossover frequency (ω_{gc})

$\gamma \rightarrow$ phase marging of uncompensated system

From bode plot,

$$\gamma = 180\epsilon + \phi_{gc} = 180\epsilon - 180\epsilon = 0\epsilon$$

But the system required phase marging of 20є, so lead compensator should be employed to improve the phase margin.

Step 3: Find ϕ_m(maximum lead angle)

The desired phase marging is $\gamma = 20\epsilon$

$$\therefore \quad \phi_m = \gamma_d - \gamma = 20 - 0 = 20\epsilon$$

Step 4: Determine the transfer function of lead compensator.

$$\gamma = \frac{1-\sin\phi_m}{1+\sin\phi_m} = \frac{1-\sin 20\epsilon}{1+\sin 20\epsilon} = 1-.342$$

$$\boxed{\alpha = .5}$$

The *db* magnitude corresponding to ϕ_m (or) ω_m

$$= -20\log\frac{1}{\sqrt{\alpha}} = -20\log\frac{1}{\sqrt{.5}}$$

$$= -3\ db$$

From bode plot of uncompensated system the frequency ω_m corresponding to $-3\ db$ is found to be 5.5 rad/sec

$$\therefore \qquad \omega_m = 4.2 \text{ rad/sec} \qquad \omega_m = 5.5 \text{ rad/sec}$$

Now,

$$T = \frac{1}{\omega_m\sqrt{\alpha}} = \frac{1}{6.5\sqrt{.5}} = .25$$

$$\boxed{T = .25}$$

Transfer function of the lead compensator

$$G_c(s) = \frac{s+1/T}{s+1/\alpha T} = \alpha\frac{(1+sT)}{(1+s\alpha T)}$$

$$G_c(s) = .5\frac{(1+.25s)}{(1+.125s)}$$

Step 5: The open loop transfer function of compensated system.

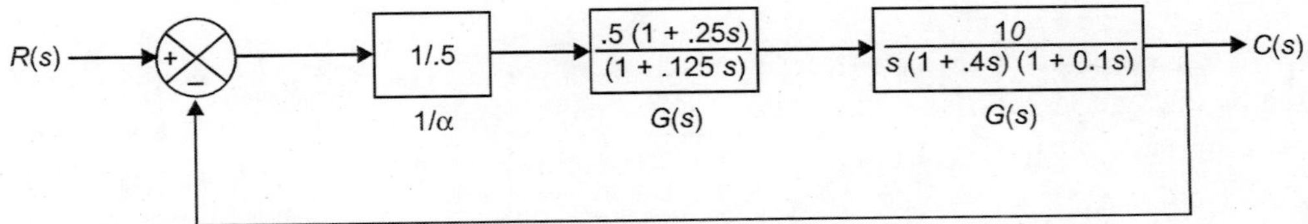

Open loop transfer function of compensated system

$$G_o(s) = \frac{1}{.5}\times\frac{.5(1+.25s)}{1+.125s)}\times\frac{10}{s(1+.4s)(1+.1s)}$$

$$G_o(s) = 10(1+.25s)/sc(1+.4s)(1-0.1s)(1+.125s)$$

Put, $s = j\omega$

$$G_o(j\omega) = \frac{10(1+.35j\omega)}{j\omega(1+.4j\omega)(1+.1\omega)(1+.175j\omega)}$$

The phase angle of $G_o(j\omega)$ as a function of ω is given by,

$$\phi_0 = -90\epsilon + \tan^{-1}.25\omega - \tan^{-1}.4\omega - \tan^{-1}.1\omega - \tan^{-1}.125\omega$$

where, $\omega = \omega_m = 5.5$ rad/sec

$$\phi_0 = -164є$$

The phase margin of compensated system is

$$\gamma_0 = 190є - \phi_0$$

$$\gamma_0 = 180є - 164 = 16є = 16є$$

$$\gamma_d = \gamma_0 \pm \zeta \therefore \quad \gamma_d = 20є \text{ (where } \zeta = 4є)$$

Note: ζ value can be assumed $\pm 5є$ according to desired value of phase margin.

The verification of design can be done in another way also (refer example sum 5 and 6)

EXAMPLE 5

A unity feedback system with open loop transfer function $G(s) = K/s(s + 1.5)$ is to be lead compensated to satisfy the following specifications.

(i) Damping ratio (ζ) = .45

(ii) Undamped natural frequency, $\omega_n = 2.2$ rad/sec

(iii) Velocity error constants, $K_v \geq 6$

Solution:

Step 1: Determine the dominant pole, s_d

Dominant pole, $s_d = -\zeta\omega_n \pm j\omega_n\sqrt{1-\zeta^2}$

Given than, $\omega_n = 2.2$ rad/sec and $\zeta = .45$

$$\therefore \quad s_d = (-.45)(2.2) \pm j\,2.2\sqrt{1-(.45)^2}\ .99 \pm j\,1.96$$

$$s_d = -1 \pm j2$$

Step 2: Draw the pole zero plot

The pole zero plot of open loop transfer function shown in Fig. worked example 5 poles are represented by the symbol "*X*". The pole at point *P* is the dominant pole, s_d.

Step 3: To find the angle to be contributed by lead network.

Let ϕ = Angle to be contributed by lead network to make point, *P* as a point on root locus.

$$\text{Now,} \quad \phi = \begin{bmatrix}\text{Sum of angles contributed by} \\ \text{poles of uncompensated system}\end{bmatrix} - \begin{bmatrix}\text{Sum of angles contributed by} \\ \text{zeros of uncompensated system}\end{bmatrix} \pm n\,180є$$

From Fig. worked example 5, we get

Sum of angles contributed by poles of uncompensated system

$$\theta_1 + \theta_2 = 193є$$

There is no angle contribution by zeros because there is no finite zero is uncompensated system.

$$\phi = 193є \pm n\,180є$$

Let, $n = 1$

$$\therefore \quad \phi = 193є \pm 180є = 13є$$

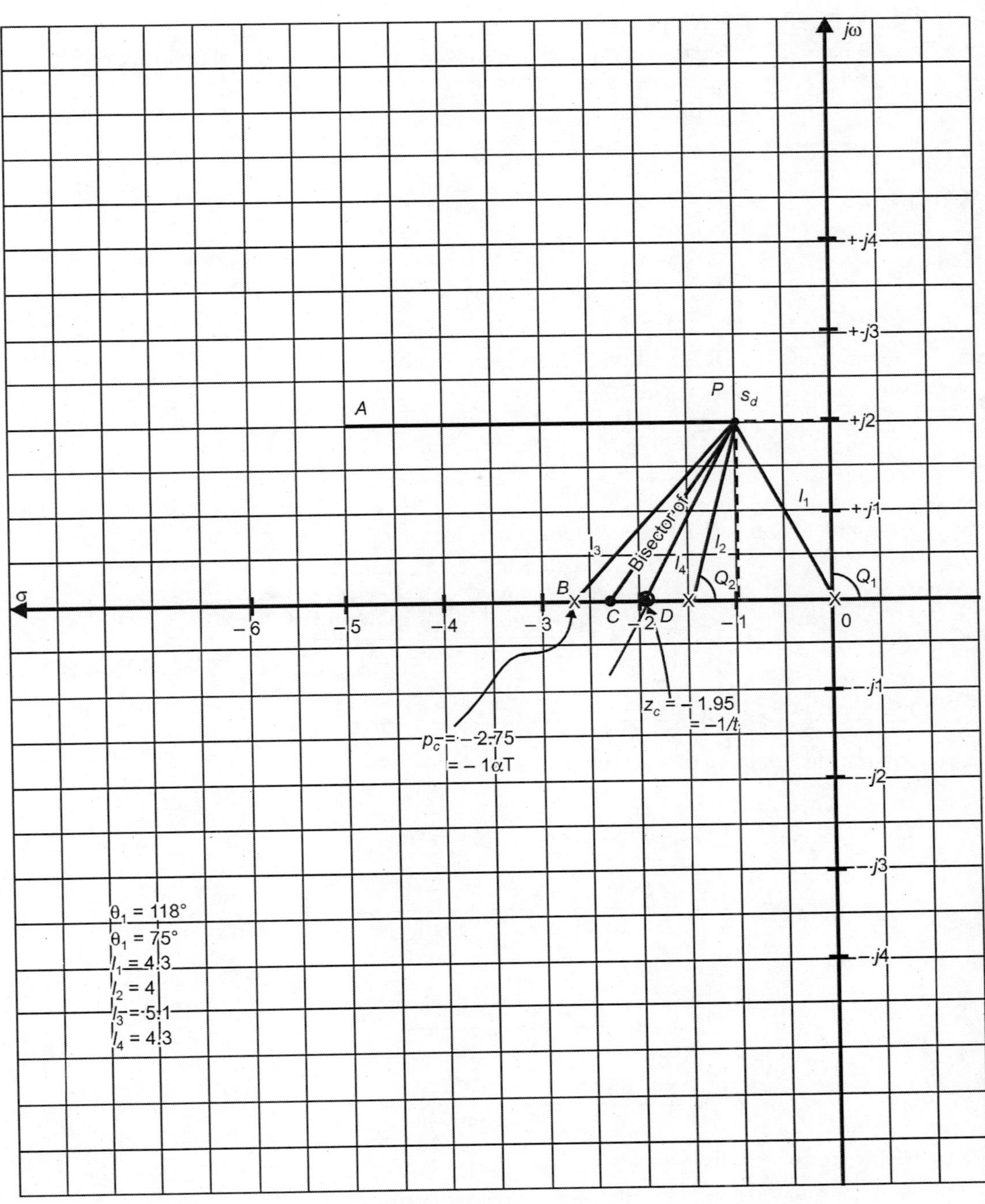

Fig. Worked Example 5

Step 4: Find poles and zeros of the compensator.

Draw a line AP parallel to X-axis as shown in Fig. worked example 5. The bisector PC is drawn to bisect the angle APO. The angles CPD and BPC are constructed with angle of 7є (∵ ∠CPD = ∠BPC = $\phi/2$ = 6.5є ≈ 7є) From Fig. worked example 5.

$$\text{Poles of compensator } p_C = -2.75$$

$$\text{Zero of compensator } z_C = -1.95$$

We know,

$$z_C = -1/T$$

$$p_C = -1/\alpha T$$

$$\therefore \quad T = 1/1.95 = .51$$

$$\therefore \quad \alpha_T = 1/2.75 = .36 \quad \text{(or)} \quad a = 1/T \text{ Ч} 2.75$$

$$\alpha = .714$$

Step 5: Determine the transfer function of lead compensator.

Transfer function of lead compensator

$$G_c(s) = \frac{(s+1/T)^2}{(s+1/\alpha T)^2} = \frac{(s+1.95)^2}{(s+2.75)^2}$$

Step 6: Determine the open loop transfer function of lead compensated system.

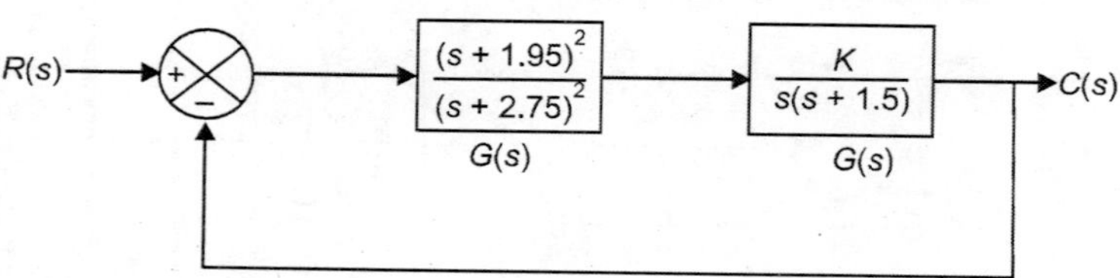

Open loop transfer function of lead compensated system

$$G_o(s) = \frac{(s+1.95)^2}{(s+2.75)^2} \times \frac{K}{s(s+1.5)}$$

where,

$$K = \frac{\text{Product of vector lengths from all poles to } s = s_d}{\text{Product of vector lengths from all zeros to } s = s_d}$$

From Fig. worked example 2 we get,

$$K = \frac{l_1 \times l_2 \times l_3}{l_4} = \frac{4.3 \times 4 \times 5.1}{4.3} = 20.4 \approx 20$$

$$\therefore \quad G_o(s) = \frac{20(s+1.9s)^2}{s(s+1.5)\,(s+2.75)^2}$$

Step 7: Check for error requirement

For the compensated system, the velocity error constant is given by, (Given $K_v \geq 6$)

$$K_v = \underset{s\to 0}{Lt}\; s \cdot G_o(s) = \underset{s\to 0}{Lt}\; s \cdot \frac{20(s+1.9s)^2}{s(s+1.5)\,(s+2.75)^2}$$

$$= \frac{20 \times 195^2}{(1.5)\,(2.75)^2} = \frac{76}{11.34}$$

$$\boxed{K_v = 6.7}$$

Conclusion: Since the velocity error constant of the compensated system, satisfies the requirement, the design is acceptable. (Refer examples 7 and 8).

Result: Transfer function of compensator system

$$G_c(s) = \frac{(s + 1.95)^2}{(s + 2.75)^2}$$

Transfer function of lead compensated system

$$G_o(s) = \frac{20(s + 1.95)^2}{s(s + 1.5)\,(s + 2.75)^2}$$

EXAMPLE 6

Consider a unity feedback control system whose forward transfer function is $G(s) = K/s(s + 2)\,(s + 8)$. Design a lag-lead compensator so that $K_v = 80s^{-1}$ and dominant closed loop poles are located at $-2 + j\,2\sqrt{3}$.

Solution:

Step 1: Draw the pole zero plot

The pole zero plot of open loop transfer function shown in Fig. worked example 6. Poles are represented by the symbol "*X*". The pole at point *P* is the dominant pole *sd*.

$$s_d = -2 + j2\sqrt{3} \rightarrow -2 \pm j\,3.5$$

Step 2: Find the angle to be contributed by lead network.

Let, ϕ = angle to be contributed by lead network to make point, *P* as a point on root locus.

Now, $$\phi = \begin{bmatrix}\text{Sum of angles contributed by} \\ \text{poles of uncompensated system}\end{bmatrix} - \begin{bmatrix}\text{Sum of angles contributed by} \\ \text{zeros of uncompensated system}\end{bmatrix}$$

From Fig. worked example 3, we get

Sum of angles contributed by poles of uncompensated system

$$\theta_1 + \theta_2 + \theta_3 = 120є + 90є + 30є$$

$$= 240є$$

Since there is no finite zero in uncompensated system, there is no angle contribution by zeros.

$$\therefore \quad \phi = 240є \pm n\,180є$$

Let, $n = 1$, $$\boxed{\therefore \phi = 60є}$$

CHAPTER 5

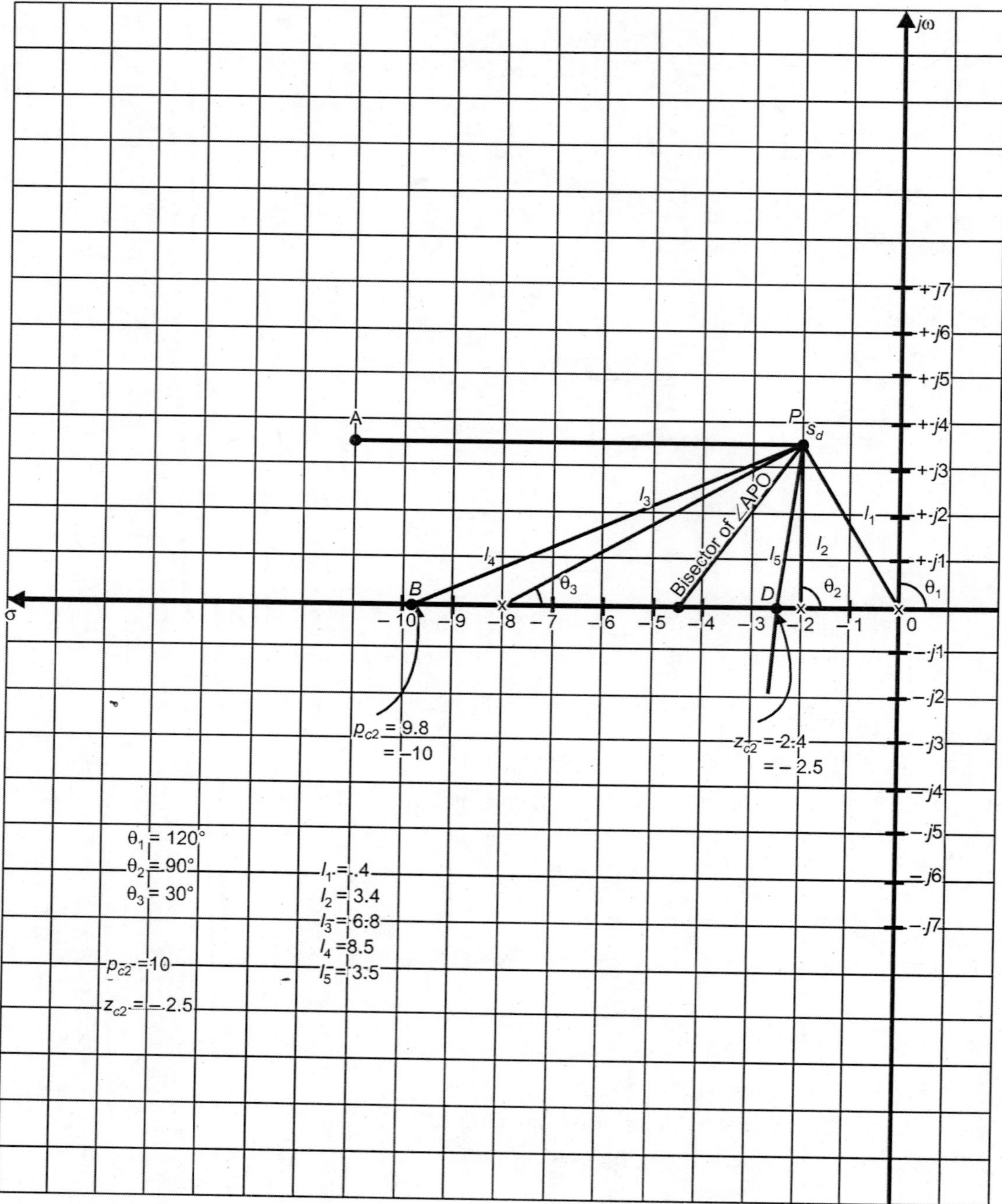

Fig. Worked Example 6: $G(s) = K/s\,(s + 2)\,(s + 8)$ lag lead compensator.

Step 3: Find pole and zero of load section.

Draw a line AP parallel to X-axis as shown in Fig. worked example 6 the bisector PC is drawn to bisect the angle APO. The angles CPD and BPC are constructed as shown in Fig. worked example 6. Here $\angle CPD = \angle BPC = \phi/2 = 60°/2 = 30$є.

Pole of the lead reaction

$$p_{C2} = -10$$

Zero of the lead section $z_{C2} = -2.5$

We know that, $z_{C2} = -1/T_2 \quad \therefore\ T_2 = 1/2.5 = .4$

We know that, $p_{C2} = -1/\alpha T_2 \ \therefore\ \alpha T_2 = 1/10$

or $\alpha = .25$

Step 4: Transfer function of lead compensator:

Transfer function of lead section

$$G_2(s) = \frac{(s+1/T_2)}{(s+1/\alpha T_2)} = \frac{(s+2.5)}{(s+10)} \times \frac{K}{s(s+2)(s+8)}$$

$$= \frac{K(s+2.5)}{s(s+2)(s+8)(s+10)}$$

Here the value of K is given by

$$K = \frac{\text{Product of vector lengths from all poles to } s = s_d}{\text{Product of vector lengths from all zeros to } s = s_d}$$

From Fig. worked example 6 we get,

$$K = \frac{l_1\, l_2\, l_3\, l_4}{d_5} = \frac{4 \times 3.4 \times 6.8 \times 8.5}{3.5}$$

$$\boxed{K = 225}$$

$$\therefore \qquad G_2(s) = \frac{225(s+2.5)}{s(s+2)(s+8)(s+10)}$$

Step 5: To find velocity error constant of lead compensated system.

$$K_{v2} = \underset{s\to 0}{Lt}\ s\,.\,G_{o2}(s) = \underset{s\to 0}{Lt}\ s \cdot \frac{225\,(s+2.5)}{s\,(s+2)\,(s+8)\,(s+10)}$$

$$= \underset{s\to 0}{Lt}\ \frac{225(2.5)}{160} = 3.5$$

$$\boxed{K_{v2} = 3.5}$$

But $K_v = 80\ s^{-1}$ is required value, so leg compensator has to be implanted to improve the value.

Step 6: Find β

$$K_{vd} = \text{Desired error constant}$$

$$A = K_{vd}/K_{v2} = \frac{80}{3.5} = 22.85 \Rightarrow \boxed{A = 22.85}$$

CHAPTER 5

$$A = \text{Factor by which } KV \text{ is increased}$$

select β, such that $\beta > A$ let, $\beta = 25$.

Step 7: Find the transfer function of lag section.

Let zero fo lag section, $z_{C1} = 0.1$ Ч second pole of $G(s)$

$$z_{C1} = 0.1 \text{ Ч} (-2) = -0.2$$

Also, $z_{C1} = -1/T_1, \quad \therefore\ T_1 = 1/{-2} = 5 \Rightarrow T_1 = 5$

Pole of lag section, $p_{C1} = \dfrac{-1}{\beta T_1} = \dfrac{-1}{25 \times 5} = 8 \times 10^{-3}$

Transfer function of lag section

$$G_1(s) = \frac{(s + 1/T_1)}{(s + 1/\beta T_1)} = \frac{(s + .2)}{(s + 8 \times 10^{-3})}$$

Step 8: Transfer function of compensated system.

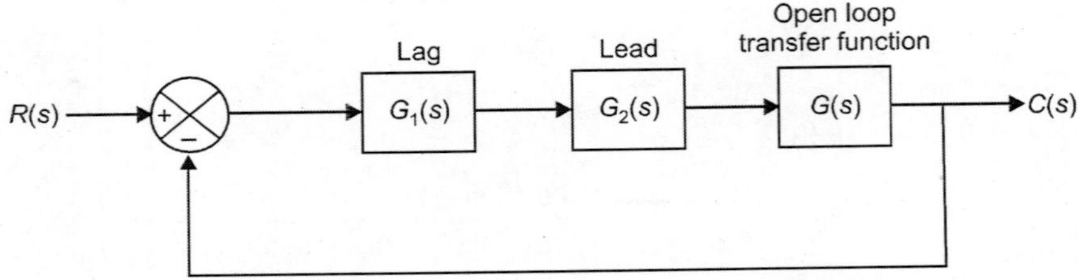

The open loop transfer function of compensated system

$$= \frac{225(s + 2.5)(s + .2)}{s(s + 2)(s + 8)(s + 10)(s + 8 \times 10^{-3})}$$

Step 9: Check the value of K_v

We know, $K_{v_0} = \underset{s \to 0}{Lt}\ s \cdot G(s)\, H(s)$

$$K_{v_0} = \underset{s \to 0}{Lt}\ s \cdot \frac{225\,(s + 2.5)(s + .2)}{s(s + 2)(s + 8)(s + 10)(s + 8 \times 10^{3})}$$

$$K_{v_0} = \underset{s \to 0}{Lt} \frac{(225)(2.5)(.2)}{(2)(8)(10)(8 \times 10^{-3})} = \frac{112.5}{1.28}$$

$$K_{v_0} = 87.89 = 88$$

$$\boxed{K_v = 88}$$

Conclusion: The velocity error constant of compensated system satisfied the requirement. Hence the design is acceptable.

Result:

1. Transfer function of lead compensator

$$G_2(s) = \frac{(s+2.5)}{(s+10)}$$

2. Transfer function of lag section

$$G_1(s) = \frac{(s+.2)}{(s+8\times10^{-3})}$$

3. Transfer function of compensated system

$$G_o(s) = \frac{225(s+2.5)(s+.2)}{s(s+2)(s+8)(s+10)(s+8\times10^{-3})}$$

5.7 STATE VARIABLE ANALYSIS

The state variable approach is powerful tool or technique for the analysis and design of control systems. The analysis and design of the following systems can be carried using state space method.

1. Linear system
2. Non-linear system
3. Time invariant system
4. Time varying system
5. Multiple input and multiple output system.

The state space analysis is modern approach and also easier for analysis using digital computers. The conventional methods of analysis employs the transfer function of the system. The drawbacks in the transfer function model and analysis are of conventional method are,

1. Transfer function is defined under zero initial conditions.
2. Transfer function is applicable to linear time invariant systems.
3. Transfer function analysis is restricted to single input and single output systems.
4. Does not provide information regarding the internal state of the system.

The state variable analysis can be applied for any type of systems. The analysis can be carried with initial conditions and can be carried on multiple input and multiple output systems. In this method of analysis, it is not necessary that the state variables represent physical quantities of the systems, but variables that do not represent physical quantities and those that are neither measurable nor observable may be chosen as state variables.

5.8 STATE SPACE FORMULATION OR STATE SPACE REPRESENTATION

The state of a dynamic system is minimal set of variables (known as state variables) such that the knowledge of these variables at $t = t_0$ together with the knowledge of the inputs for $t \geq t_0$, completely determines the behaviour of the system for $t > t_0$ (or) A set of variables which describes the system at any time instant are called STATE VARIABLES.

In the state variables formulation of a system in general, consists of m-inputs, P-outputs and n-state variables. The state space representation of the system may be utilised in shown in Fig. (5.2.1).

Let, State variables = $X_1(t), X_2(t), X_3(t) \ldots X_n(t)$

Input variables = $u_1(t), u_2(t), u_3(t) \ldots u_m(t)$

Output variables = $y_1(t), y_2(t), y_3(t) \ldots y_p(t)$

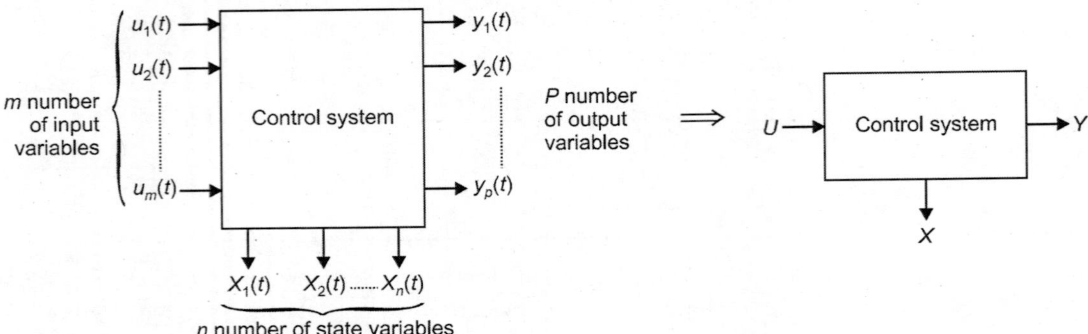

Fig. 5.2.1: State space representation of system.

The different variables may be represented by the vectors (column matrix) as shown below.

$$\text{Input vector, } U(t) = \begin{bmatrix} u_1(t) \\ u_2(t) \\ \vdots \\ u_m(t) \end{bmatrix}$$

$$\text{Output vector, } Y(t) = \begin{bmatrix} y_1(t) \\ y_2(t) \\ \vdots \\ y_p(t) \end{bmatrix}$$

$$\text{State variable vector, } X(t) = \begin{bmatrix} x_1(t) \\ x_2(t) \\ \vdots \\ x_n(t) \end{bmatrix}$$

State equations

The state variable representation can be arranged in the form of n number of first order differential equations as shown below.

$$\frac{dX_1}{dt} = \dot{X}_1 = f_1(X_1, X_2, \ldots X_n; u_1, u_2, \ldots u_m)$$

$$\frac{dX_2}{dt} = \dot{X}_2 = f_2(X_1, X_2, \ldots X_n; u_1, u_2, \ldots u_m)$$

$$\vdots$$

$$\frac{dX_n}{dt} = \dot{X}_n = f_n(X_1, X_2 \ldots X_n; u_1, u_2, u_m) \qquad (5.2.1)$$

The numbers of differential equations may be written in vector notations as

$$\dot{X}(t) = f[(X(t), U(t)] \qquad (5.2.2)$$

The set of all possible values which the input vector $U(t)$ can have (assume) at time t forms the input space of the system. Similarly, the set of all possible values which the output vector of $Y(t)$ can have at time t forms the output space of the system and the set of all possible values which the state vector $X(t)$ can have at time t forms the state space of the system.

5.9 STATE MODEL OF LINEAR SYSTEM

The state model of a system consist of the state equation and output equation. The state equation of a system is a function of state variable sand inputs as defined by equation (5.2.2). For linear time invariant systems the first derivative of state variables can be expressed as a linear combination of state variables and inputs.

$$\dot{X}_1 = a_{11}X_1 + a_{12}X_2 + \ldots. + a_{1n}X_n + b_{11}u_1 + b_{12}u_2 + +\ldots b_{1m}u_m$$

$$\dot{X}_2 = a_{21}X_1 + a_{22}X_2 + \ldots. + a_{2n}X_n + b_{21}u_1 + b_{22}u_2 + +\ldots b_{2m}u_m$$

$$\vdots$$

$$\dot{X}_n = a_{n1}X_1 + a_{n2}X_2 + \ldots. + a_{nn}X_n + b_{n1}u_1 + b_{n2}u_2 + +\ldots b_{nm}u_m \qquad (5.3.1)$$

where the coefficients a_{ij} and b_{ij} are constants.

In the matrix form the above equations can be expressed as,

$$\begin{bmatrix} \dot{X}_1 \\ \dot{X}_2 \\ \vdots \\ \dot{X}_n \end{bmatrix} = \begin{bmatrix} a_{11} & a_{12} & \cdots & a_{1n} \\ a_{21} & a_{22} & \cdots & a_{2n} \\ \vdots & \vdots & \vdots & \vdots \\ a_{n1} & a_{n2} & \cdots & a_{nn} \end{bmatrix} \begin{bmatrix} X_1 \\ X_2 \\ \vdots \\ X_n \end{bmatrix} + \begin{bmatrix} b_{11} & b_{12} & \cdots & b_{1m} \\ b_{21} & b_{22} & \cdots & b_{2m} \\ b_{31} & b_{32} & \vdots & b_{3m} \\ \vdots & \vdots & \vdots & \vdots \\ b_{n1} & b_{n2} & \cdots & b_{nm} \end{bmatrix} \begin{bmatrix} u_1 \\ u_2 \\ u_3 \\ \vdots \\ u_m \end{bmatrix} \qquad (5.3.2)$$

the matrix equation (5.3.2) can also be written as

$$\dot{X}(t) = AX(t) + BU(t) \qquad (5.2.3)$$

where, $X(t)$ = State vector of order (n Ч 1)

$U(t)$ = Input vector of order (m Ч 1)

A = system matrix of order (n Ч n)

B = Input matrix of order (n Ч m)

Note: $u_1, u_2 ...; y_1, y_2 ...$ and $X_1, X_2 ...$ one functions of time t, the equation $\dot{X}(t) = AX(t) + BU(t)$ is called the state equation of linear time invariant (LTI) system. The output any time are functions of state variables and inputs.

$\therefore$ Output vector, $Y(t) = f[(X(t), U(t)]$ (5.3.3)

Hence the output variables can be expressed as a linear combination of state variables and inputs.

$$y_1 = c_{11}X_1 + c_{12}X_2 + ... + c_{1n}X_n + d_{11}u_1 + ... d_{1m}u_m$$

$$y_2 = c_{21}X_1 + c_{22}X_2 + ... + c_{2n}X_n + d_{21}u_1 + ... d_{2m}u_m$$

$$\vdots$$

$$y_p = c_{p1}X_1 + c_{p2}X_2 + ... + c_{pn}X_n + d_{p1}u_1 + d_{p2}u_2 + .. + d_{pm}u_m \quad (5.3.4)$$

where the co-efficients cij and dij are constants. In the matrix form the above equations can be expressed as,

$$\begin{bmatrix} y_1 \\ y_2 \\ \vdots \\ y_p \end{bmatrix} = \begin{bmatrix} c_{11} & c_{12} & \cdots & c_{1n} \\ c_{21} & c_{22} & \cdots & c_{2n} \\ \vdots & \vdots & & \\ c_{p2} & c_{p2} & \cdots & c_{pn} \end{bmatrix} \begin{bmatrix} X_1 \\ X_2 \\ \vdots \\ X_n \end{bmatrix} + \begin{bmatrix} d_{11} & d_{12} & \cdots & d_{1m} \\ d_{21} & d_{22} & \cdots & d_{2m} \\ \vdots & \vdots & & \\ d_{p1} & d_{p2} & \cdots & d_{pm} \end{bmatrix} \begin{bmatrix} u_1 \\ u_2 \\ \vdots \\ u_m \end{bmatrix} \quad (5.3.5)$$

the matrix equation (5.3.5) can be written as

$$Y(t) = [CX(t) + DU(t)] \quad (5.3.6)$$

where, $X(t)$ = state vector of order (n Ч 1)

$U(t)$ = input vector of order (m Ч 1)

$Y(t)$ = output vector of order (p Ч 1)

C = output matrix of order (p Ч n)

D = transmission matrix of order (p Ч m)

the equation $Y(t) = [X(t) + DU(t)]$ is called the output equation of linear time invariant (LTI) system.

The state model of a system consists of state equation and output equation (or) the state equation and output equation together called state model of the system. Hence the state model of a linear time invariant system (LTI) system is given by the following equations.

$$\dot{X}(t) = AX + BU(t) \text{ state equation}$$

$$Y(t) = CX(t) + DU(t) \text{ output equation}$$

5.10 STATE DIAGRAM

The pictorial representation of the state model of the system is called state diagram. The state diagram of the system can be either in block diagram form or in signal flow graph form.

The state diagram describes the relationship among the state variables and provides physical interpretations of the state variables. The time domain state diagram may be obtained directly from the differential equation governing the system and this diagram can be used for simulation of the system in analog computers.

The S-domain state diagram can be obtained from the transfer function of the system. The state diagram provides direct relation between time domain and S-domain [i.e., the time domain equations can be directly obtained from the s-domain state diagram].

The state diagram (block diagram and signal flow graph) of state model is constructed using three basic elements scalar, Adder and Integrator.

Scalar: The scalar is used to multiply a signal by a constant. The input signal $X(t)$ is multiplied by the scalar a to given the output, $aX(t)$.

Adder: The adder is used to add two (or) more signals. The output of the adder is the sum of the incoming signals.

Integrator: The integrator is used to integrate the signal. they are used to integrate the derivatives of state variables to get the state variables. The initial conditions of the state variables can be added by using an adder after integrator.

The time domain and s domain elements of block diagram are shown in Table (5.4.1). The time domain and s-domain elements of signal flow graph are shown in Table (5.4.2).

The state model of linear time invariant system is given by the equations.

$$\dot{X}(t) = AX(t) + BU(t) \text{ state equation} \tag{5.4.1}$$

$$Y(t) = CX(t) + DU(t) \text{ output equation} \tag{5.4.2}$$

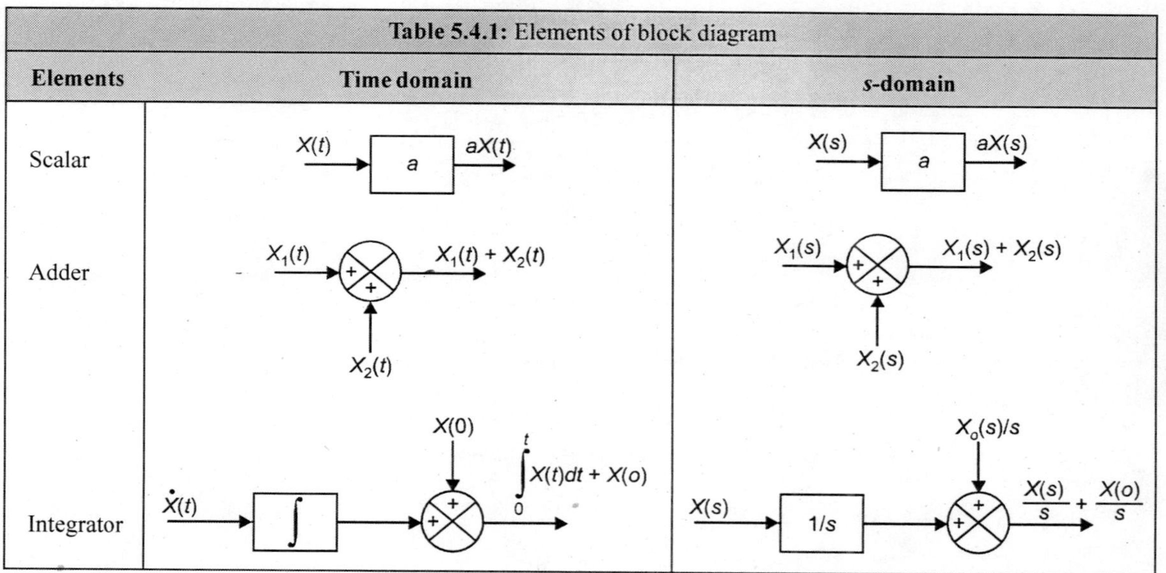

Table 5.4.2: Elements of signal flow graph

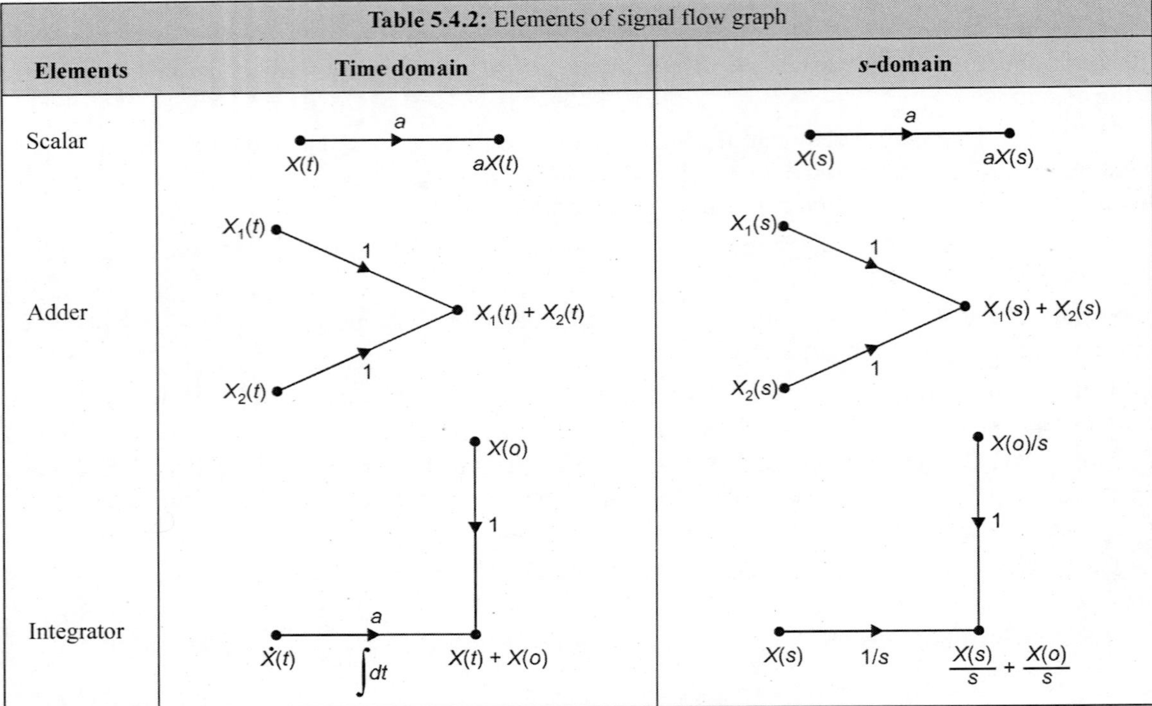

Elements	Time domain	s-domain
Scalar	$X(t)$ —a→ $aX(t)$	$X(s)$ —a→ $aX(s)$
Adder	$X_1(t)$ —1→, $X_2(t)$ —1→ $X_1(t) + X_2(t)$	$X_1(s)$ —1→, $X_2(s)$ —1→ $X_1(s) + X_2(s)$
Integrator	$\dot{X}(t)$ —a, $\int dt$→ $X(t) + X(o)$; $X(o)$ —1→	$X(s)$ —$1/s$→ $\frac{X(s)}{s} + \frac{X(o)}{s}$; $X(o)/s$ —1→

The time domain in block diagram representation of the state model of equation (5.4.1) is shown in Fig. (5.4.1) and the time domain in signal flow graph representation of the system equation (5.4.2) is shown in Fig. (5.4.2).

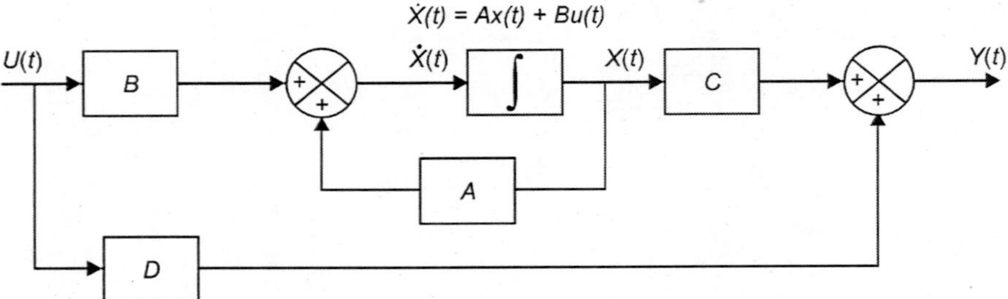

Fig. 5.4.1: Block diagram of state model of equation (5.4.1).

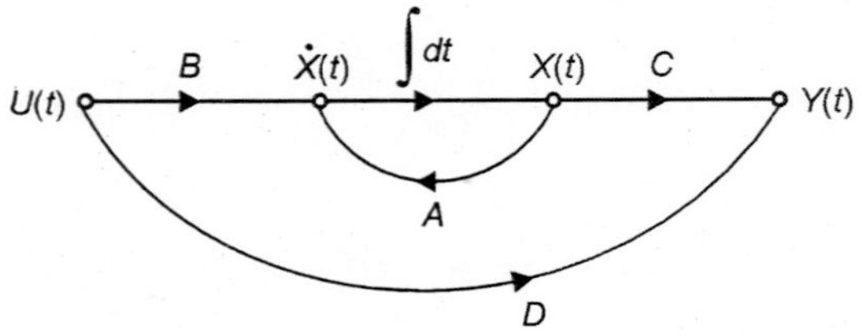

Fig. 5.4.2: Signal flow graph of state model of equation (5.4.2).

5.11 CONSTRUCTION OF TIME DOMAIN STATE DIAGRAM

In state space modelling, n-numbers of first order differential equations are formed for a nth order system. In order to integrate n-numbers of first derivatives, the state diagram requires n-number of integrators, Therefore the first step in constructing the state diagram is to draw n-numbers of integrators. Mark the input to the integrators as first derivatives of state variables and so the output of the integrators are state variables. [If initial conditions are given, then they can be added at the output of integrators using adders].

In each state equation, the first derivative of state variable is expressed as a function of state variables and inputs. Therefore from the knowledge of a state equation, the state variables and inputs and multiplied by appropriate scalars and then added to get the first derivative of a state variables. Now, the first derivative of the state variables is given as input to the corresponding integrator. Similarly the input of all other integrators are obtained by considering the state equations one by one.

Each output equation is function of state variables and inputs. Therefore from the knowledge an output equation, the state variables and inputs are multiplied by appropriate scalars and then added to get an output. Similar procedure is followed to generate all others output.

5.12 STATE-SPACE REPRESENTATION USING PHYSICAL VARIABLES

In state-space modelling of systems, the choice of state variables in arbitrary one of the possible choice of state variables is the physical variables. The physical variables of electrical systems are current (or) voltage in the R, L and C elements. The physical variables of mechanical systems are displacement, velocity and acceleration. The advantages of choosing the physical variables (or quantities) of the system as state variables are the following:

1. The state variable can be utilized for the purpose of feedback.
2. The implementation of design with state variable feedback become straight forward.
3. The solution of state equation gives time variation of variables which have direct relevance to the physical system.

The drawback in choosing the physical quantities as state variables is that the solution of state equation amy become or difficult task.

In state space modelling using physical variables, the state equations are obtained from the differential equations governing the system. The differential equations governing a system are obtained from a basic model of the system which is developed using the fundamental elements of the system.

5.12.1 State-space Representation using Electrical Systems

The basic model of an electrical system can be obtained by using the fundamental elements resistor, capacitor and Inductor. Using these elements the electrical network (or) equivalent circuit of the system is drawn. Then the differential equations governing the electrical systems can be formed by writing KCL equations by choosing various nodes in the network (or) KVL by choosing various closed path in the network. The current voltage relation of the basic elements R, L and C and came given in Table (5.6.1.1).

Table 5.6.1

Element	Voltage across the element	Current through the element
$i(t)$, R, $V(t)$ (+ −)	$V(t) = Ri(t)$	$i(t) = V(t)/R$
$i(t)$, L, $V(t)$ (+ −)	$V(t) = L\frac{di}{dt}(it)$	$i(t) = \frac{1}{L}\int v(t)\,dt$
$i(t)$, C, $V(t)$ (+ −)	$V(t) = \frac{1}{C}\int i(t)\,dt$	$i(t) = C\frac{dv(t)}{dt}$

A minimal number of state variables are chosen for obtaining the state model of the system. The best choice of state variables in electrical system are currents and voltages in energy storage elements. The energy storage elements are inductance and capacitance. The physical variables in the differential equations are replaced by state variables and the equations are rearranged as first order differential equations. These set of first order equations constitutes the state equation of the system.

The input to the system are voltage source (or) current source. The outputs in electrical system are usually volates (or) current in energy dissipating element. The resistance is a energy dissipating element in electrical network. In general the output variables can be voltage (or) current in the network.

5.12.2 State-space Representation using Mechanical Translational System

The basic model of mechanical translational system can be obtained by using three basic elements mass, spring and dash-pot. When a force is applied to a mechanical translational system, it is opposed by opposing forces due to mass, friction and elasticity of the system. The forces acting on a body are governed by newton's second law of motion.

The differential equations governing the system are obtained by writing force balance equations at various nodes in the system.

A node is a meeting point of elements. Table (5.6.2.1) shows the force balance equations of idealized elements,

List of symbols used in mechanical translational system are

x or y = displacement, n

$v = dy/dt$ = velocity, m/sec

$a = dv/dt = d^2y/dt^2$ = acceleration, m/sec^2

f = applied force, N (newton)

f_m = opposing force offered by mass of the body, N

f_k = opposing force offered by elasticity of the body (spring), N

f_b = opposing force offered by the friction of the body (dash-pot), N

M = Mass, kg

K = stiffness of spring, $N/_m$

B = viscuss friction coefficient, $N/(m/\text{sec})$

Table 5.6.2.1

Elements	Force balanced equation
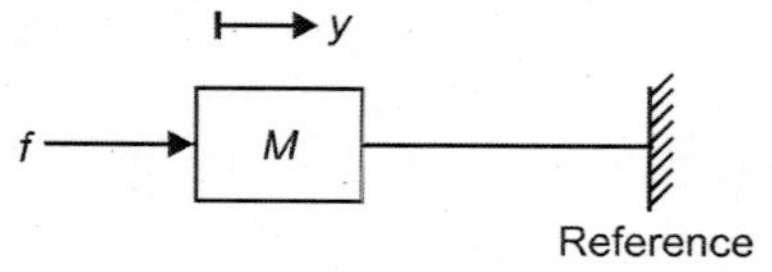 f → M (displacement y), Reference	$f = f_m = M\,d^2y/dt^2$
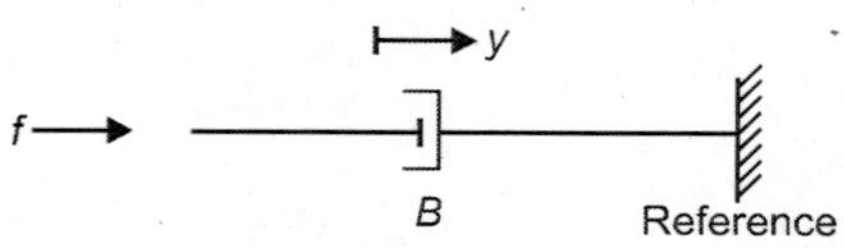 f → B (displacement y), Reference	$f = f_b = Bdy/dt$
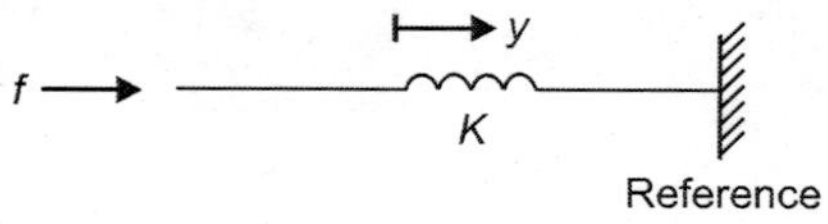 f → K (displacement y), Reference	$f = f_k = Ky$
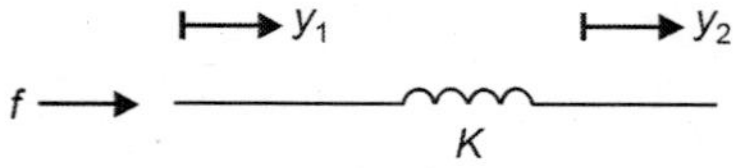 f → K (displacements y_1, y_2)	$f = f_k = K(y_1 - y_2)$
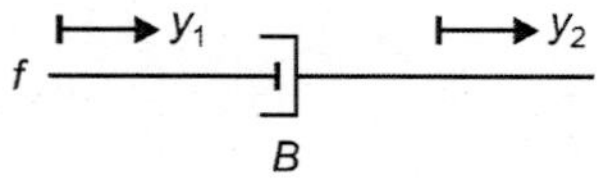 f → B (displacements y_1, y_2)	$f = f_b = B\frac{d}{dt}(y_1 - y_2)$

Guidelines to form the state model of mechanical translational systems

1. For each node in the system one differential equation can be formed by equating the sum of applied forces to the sum of opposing forces. Generally the nodes are mass elements of the system, but in some cases the nodes may be without mass element.
2. Assign a displacement to each node and draw a free body diagram for each node. The free body diagram is obtained by drawing each mass of node separately and then marking all the forces acting on it.
3. In the free body diagram, the opposing forces due to mass, spring and dash-pot are always act in a direction opposite to applied force. The displacement, velocity and acceleration will be in the direction of applied force (or) in the direction opposite to that of opposing force.
4. For each free body diagram write one differential equation by equating the sum of applied forces to the sum of opposing forces.
5. Choose a minimum number of state variables. The choice of state variables are displacement, velocity (or) acceleration.

6. The physical variables in differential equations are replaced by state variables and the equations are rearranged as first order differential equations. These set of first order equations constitute the state equation of the system.
7. The inputs are the applied forces and the outputs are the displacement, velocity (or) acceleration of the desired nodes.

5.12.3 State-space Representation using Mechanical Rotational System

the basic model of mechanical rotational system can be obtained by using three basic elements moment of inertia of mass, rotational dash-pot and rotational spring when a torque is applied to a mechanical rotational system, it is opposed by opposing torques due to moment of inertia, friction and elasticity of the system. The torque acting on a body are governed by newton's second law of motion.

The differential equations governing the system are obtained by writing torque balance equations at various nodes in the system. A node is a meeting point of elements. The table (5.6.3.1) shows the torque balance equations of the idealized elements.

Table 5.6.3.1: Torque balance equations of idealized elements

Elements	Torque balance equations
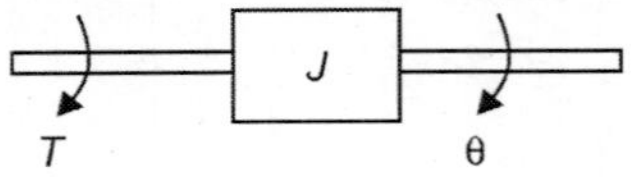 T J θ	$T = T_j = Jd^2\theta/dt^2$
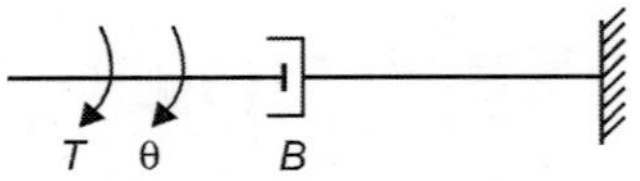 T θ B	$T = T_b = B\dfrac{d\theta}{dt}$
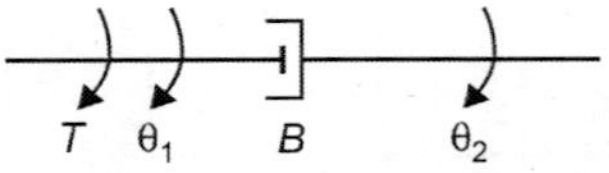 T θ_1 B θ_2	$T = T_b = Bd/dt(\theta_1 - \theta_2)$
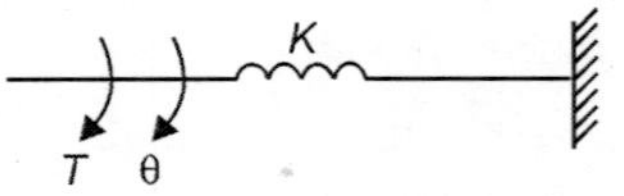 T θ K	$T = T_k = k\theta$
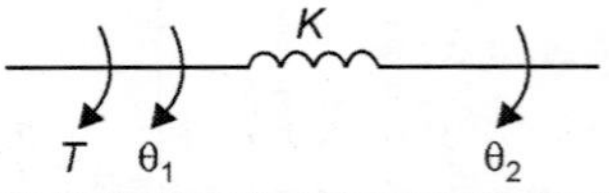 T θ_1 K θ_2	$T = T_k = k\,(\theta_1 - \theta_2)$

List of symbols used in mechanical rotational system

θ = Angular displacement, rad

$d\theta/dt$ = Angular velocity, rad/sec

$\dfrac{d^2\theta}{dt^2}$ = Angular acceleration, rad/sec^2

T = Applied torque, N-m

J = Moment of inertia, kgm^2/rad

B = Rotational frictional coefficient, N-m/rad/sec)

K = Stiffness of the spring, N/m/rad

Guidelines to form the state model of mechanical rotational systems:

1. For each node in the system one differential equation can be formed by equating the sum of applied torques to the sum of opposing torques. Generally the nodes one mass elements but in some cases the nodes may be without mass element.
2. Assign an angular displacement to each node and draw a free body diagram for each node. The free body diagram is obtained by marking each node separately and then drawing all the torques acting on it.
3. In the freebody diagram, the opposing torques due to moment of inertia, spring and dash-pot are always act in a direction opposite to applied force. The angular displacement velocity and acceleration will be in the direction of applied torque or in the direction opposite to that of opposing torque.
4. For each free body diagram write one differential equation by equating the sum of applied torques to the sum of opposing torques.
5. Choose a minimum number of state variables. The choice of state variables are angular displacement, velocity (or) acceleration.
6. The physical variables in differential equations are replaced by state variables and the equations are rearranged as first order differential equations. These set of first order equations constitute the state equation of the system.
7. The inputs are the applied torques and the outputs are the angular displacement, velocity (or) acceleration of the the desired nodes.

SOME WORKED EXAMPLES

EXAMPLE 5.11

Obtain the state model of the electrical network shown in Fig. (Ex. 5.11.1) by choosing minimal number of state variables.

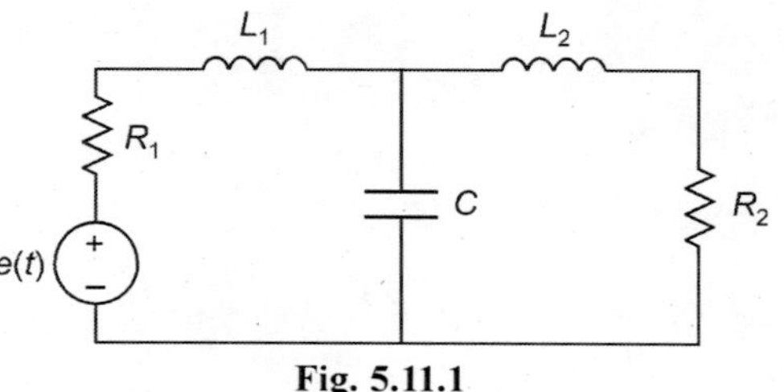

Fig. 5.11.1

Solution: Let us choose the current through the inductances i_1, i_2 and voltage across the capacitor V_c as state variables. The assumed directions of currents and polarity of the voltage are shown in Fig. (Ex. 5.11.2)

CHAPTER 5

Note: The suitable and best choice of state variables in electrical network are currents and voltages in energy storage elements.

Let the three state variables X_1, X_2 and X_3 are related to physical quantities as shown below.

$$X_1 = i_1 = \text{current through } L_1$$

$$X_2 = i_2 = \text{current through } L_2$$

$$X_3 = V_c = \text{voltage across capacitor}$$

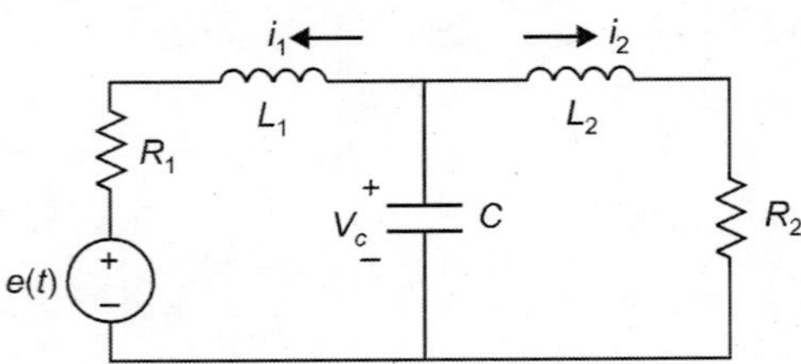

Fig. Ex. 5.11.2

At node A, by Kirchoff's curved law [refer Fig. (Ex. 5.11.3)]

$$i_1 + i_2 + C\frac{dv_c}{dt} = 0 \tag{5.1.1}$$

Fig. Ex. 5.11.3

on substituting the state variables for physical variables in equation (5.1.1) we get,

$\left(\text{i.e., } i_1 = X_1,\ i_2 = X_2 \text{ and } \frac{dv_c}{dt} = \dot{X}_3\right)$

$$X_1 + X_2 + C\,\dot{X}_3 = 0$$

$$C\dot{X}_3 = -X_1 - X_2$$

$$\dot{X}_3 = -\frac{1}{C}X_1 - \frac{1}{C}X_2 \tag{5.1.2}$$

By KVL in the closed path shown in Fig. (Ex. 5.11.4) we get,

$$e(t) + i_1R_1 + L_1\frac{di_1}{dt} = V_c \tag{5.1.3}$$

on substituting the state variables for physical variables in eq. (5.1.3) we get,

$\left(\text{i.e., } i_1 = X_1, \frac{di_1}{dt} = \dot{X}_1 \text{ and } V_c = X_3\right)$

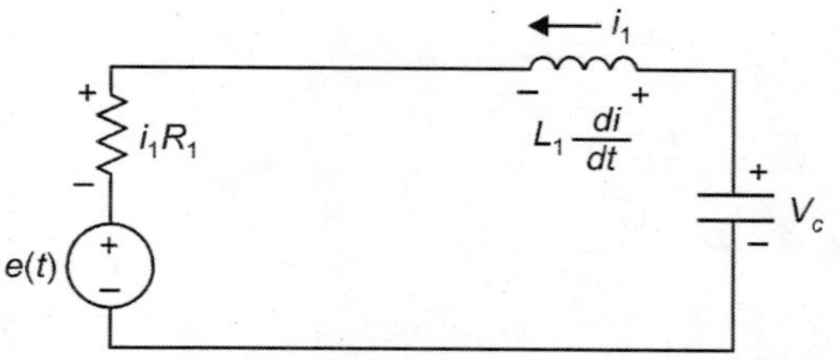

Fig. 5.11.4

$$e(t) + X_1R_1 + L_1\dot{X}_1 = X_3$$

Also, let $u(t) = e(t)$ = input to the system

$$\therefore \quad u + X_1R_1 + L_1\dot{X}_1 = X_3$$

$$L_1\dot{X}_1 = X_3 - X_1R_1 - u$$

$$\dot{X}^1 = -\frac{R_1}{L_1}X_1 + \frac{1}{L_1}X_3 - \frac{1}{L_1}u \qquad (5.1.4)$$

By KVL in the closed path shown in Fig (Ex. 5.11.5) we get

$$V_c = L_2\frac{di_2}{dt} + i_2R_2 \qquad (5.1.5)$$

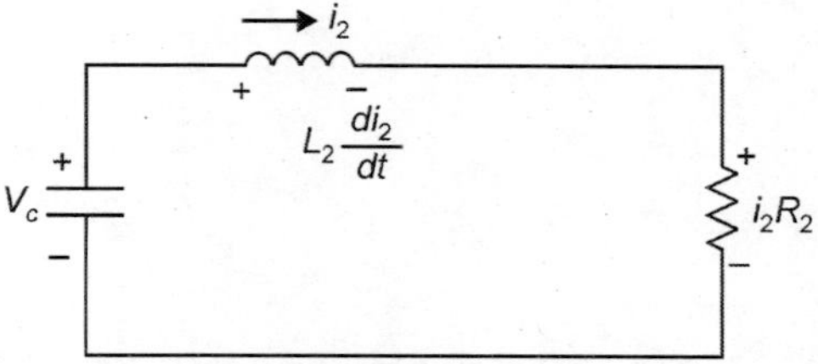

Fig. Ex. 5.11.5

on substituting the state variables for physical variables in equation (5.1.5) we get

$$\left(\text{ie } i_2 = X_2, \frac{di_2}{dt} = \dot{X}_2 \text{ and } V_c = X_3\right)$$

$$X_3 = L_2\dot{X}_2 + X_2R_2$$

$$\therefore \quad L_2\dot{X}_2 = -X_2R_2 + X_3$$

$$\dot{X}^2 = -\frac{R_2}{L_2}X_2 + \frac{1}{L_2}X_3 \qquad (5.1.6)$$

The equation (5.1.2) (5.1.4) and (5.1.6) are the state equations of the system. Hence the state equations of the system are,

$$\dot{X}_1 = \frac{-R_1}{L_1}X_1 + \frac{1}{L_1}X_3 - \frac{1}{L_1}u$$

$$\dot{X}_2 = \frac{-R_2}{L_2}X_2 - \frac{1}{L_2}X_3$$

$$\dot{X}_3 = -\frac{1}{C}X_1 - \frac{1}{C}X_2$$

on arranging the state equations in the matrix form we get,

$$\begin{bmatrix} \dot{X}_1 \\ \dot{X}_2 \\ \dot{X}_3 \end{bmatrix} = \begin{bmatrix} -\frac{R_1}{L_1} & 0 & \frac{1}{L_1} \\ 0 & -\frac{R_2}{L_2} & \frac{1}{L_2} \\ -\frac{1}{C} & -\frac{1}{C} & 0 \end{bmatrix} \begin{bmatrix} X_1 \\ X_2 \\ X_3 \end{bmatrix} + \begin{bmatrix} -\frac{1}{L_1} \\ 0 \\ 0 \end{bmatrix} \{u\} \quad \text{state equation (5.1.7)}$$

Let us choose the voltage across the resistances as output variables and the output variables are denoted by y_1 and y_2

$\therefore$ $$y_1 = i_1R_1$$

and $$y_2 = i_2R_2 \qquad (5.1.9)$$

on substituting the state variables in equation (5.1.8 and 5.1.9) we get, (ie, $i_1 = X_1$ and $i_2 = X_2$)

$$y_1 = X_1R_1,\ y_2 = X_2R_2$$

on arranging the above equations in the matrix form we get,

$$\begin{bmatrix} y_1 \\ y_2 \end{bmatrix} = \begin{pmatrix} R_1 & 0 \\ 0 & R_2 \end{pmatrix} \begin{pmatrix} X_1 \\ X_2 \end{pmatrix} \quad \text{output equation} \qquad (5.1.10)$$

Result: The state equation (5.1.7) and output equation (5.1.10) together constitute the state model of the system.

EXAMPLE 5.12

Obtain the state model of the electrical network shown in Fig. (Ex. 5.12.1) by choosing $V_1(t)$ and $V_2(t)$ as shown variables.

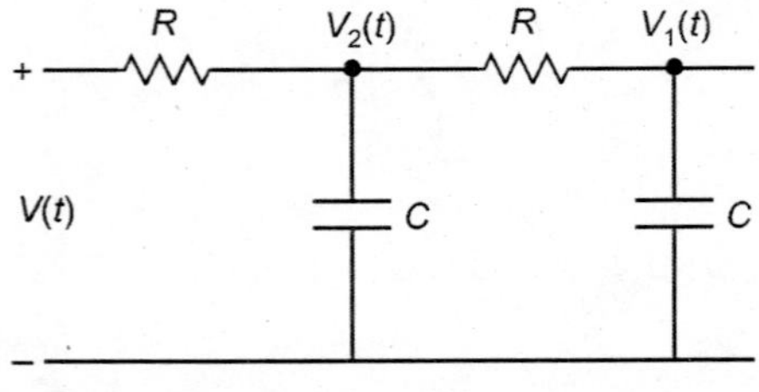

Fig. Ex. 5.12.1

Solution: Connect a voltage source at the input shown in Fig. (Ex. 5.12.2).

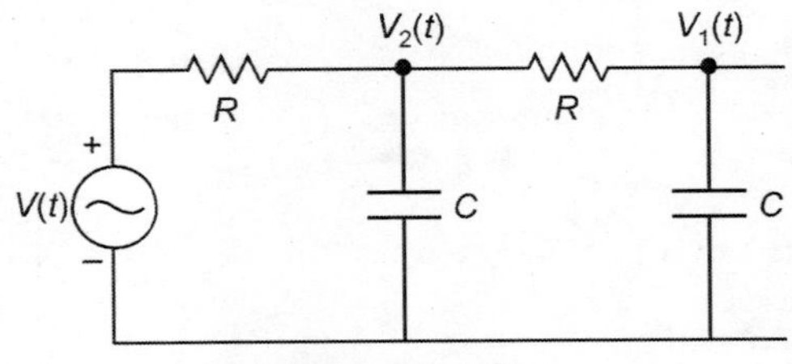

Fig. Ex. 5.12.2

Convert the voltage source to current source as shown in Fig. (Ex. 5.12.3)

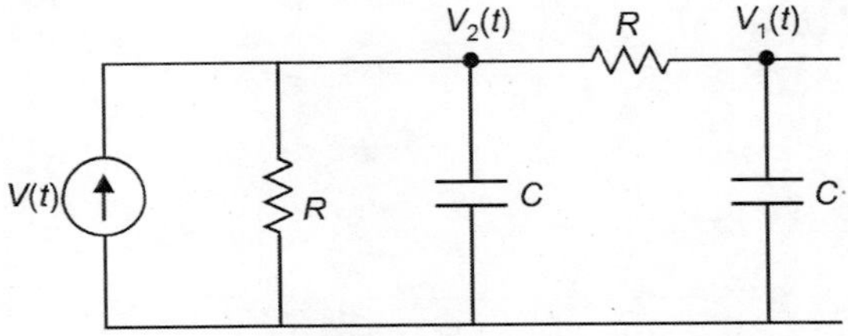

Fig. Ex. 5.12.3

At node 1, by KCL we can write (refer Fig. Ex. 5.12.4)

$$\frac{V_1 - V_2}{R} + C\frac{dV_1}{dt} = 0 \tag{5.2.1}$$

Fig. Ex. 5.12.4

At node 2, by KCL, we can write (refer Fig. Ex. 5.12.5)

$$\frac{V_2 - V_1}{R} + \frac{V_2}{R} + C\frac{dV_2}{dt} = \frac{V(t)}{R} \tag{5.2.2}$$

Fig. Ex. 5.12.5

Let the state variables be X_1 and X_2 and they are related to physical variable as shown below

$$V_1 = X_1 \text{ and } V_2 = X_2$$

Also, let
$$V(t) = u = \text{input}$$

On substituting the state variables in equations (5.2.1) and (5.2.2) we get,

$$\frac{X_1 - X_2}{R} + C\frac{dX_1}{dt} = 0 \tag{5.2.3}$$

$$\frac{X_2 - X_1}{R} + \frac{X_2}{R_2} + C\frac{dX_2}{dt} = \frac{u}{R} \tag{5.2.4}$$

From equation (5.2.3) we get,

$$\frac{X_1}{R} - \frac{X_2}{R} + C\dot{X}_1 = 0$$

$$\therefore \quad C\dot{X}_1 = -\frac{X_1}{R} + \frac{X_2}{R}$$

$$\dot{X}_1 = -\frac{1}{RC}X_1 + \frac{1}{RC}X_2 \tag{5.2.5}$$

From equation (5.2.4) we get

$$\frac{X_2}{R} - \frac{X_1}{R} + \frac{X_2}{R} + C\dot{X}_2 = \frac{u}{R}$$

$$\therefore \quad C\dot{X}_2 = \frac{X_1}{R} - \frac{X_2}{R} - \frac{X_2}{R} + \frac{u}{R}$$

$$\dot{X}_2 = \frac{1}{RC}X_1 - \frac{2}{RC}X_2 + \frac{1}{RC}u \tag{5.2.6}$$

The equation (5.2.4) and (5.2.6) are state equations of the system. Hence the state equation of the system are

$$\dot{X}_1 = -\frac{1}{RC}X_1 + \frac{1}{RC}X_2$$

$$\dot{X}_2 = \frac{1}{RC}X_1 - \frac{2}{RC}X_2 + \frac{1}{RC}u$$

On arranging the state equations in the matrix form,

$$\begin{bmatrix} \dot{X}_1 \\ \dot{X}_2 \end{bmatrix} = \begin{bmatrix} -\frac{1}{RC} & \frac{1}{RC} \\ \frac{1}{RC} & -\frac{2}{RC} \end{bmatrix} \begin{bmatrix} X_1 \\ X_2 \end{bmatrix} + \begin{bmatrix} 0 \\ \frac{1}{RC} \end{bmatrix} [u] \tag{5.2.7}$$

The output
$$y = V_1(t) = X_1$$

$\therefore$ The output equation is

$$y = [1 \quad 0]\begin{bmatrix} X_1 \\ X_2 \end{bmatrix} \tag{5.2.8}$$

Result: The state equation [Eq. (5.2.7) and output equation [eq. (5.2.8)] together constitute the state model of the system.

EXAMPLE 5.3

Construct the state model of mechanical system shown in Fig. (Ex. 5.13.1).

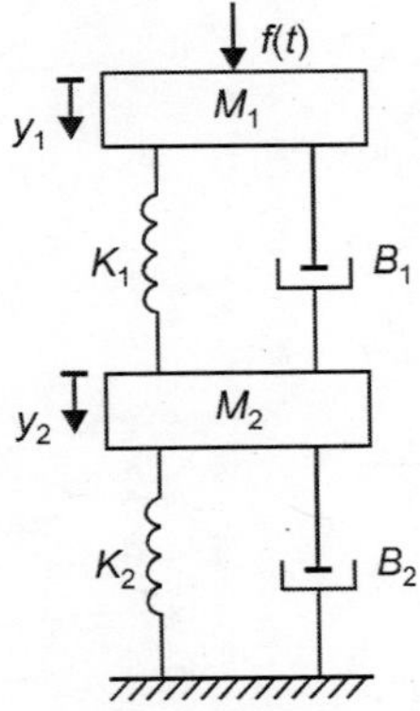

Fig. Ex. 5.13.1

Solution: Free body diagram of M_1 is a shown in Fig. (Ex. 5.13.2).

$$f_{m1} = M_1\frac{d^2 y_1}{dt^2};$$

$$f_{b1} = B_1\frac{d}{dt}(y_1 - y_2)$$

$$f_{k1} = K(y_1 - y_2)$$

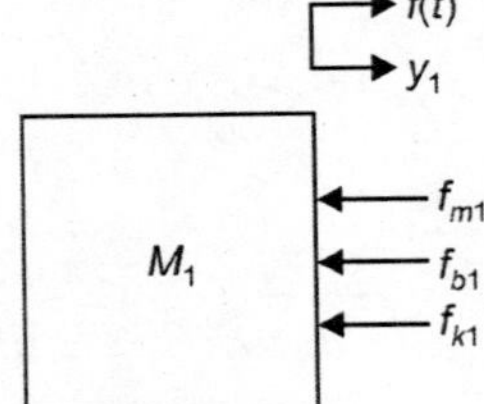

Fig. Ex. 5.13.2

By Newton's second case, the force balance equation at node M_1 is

$$f(t) = f_{m1} + f_{b1} + f_{k1}$$

$$f(t) = M_1 \frac{d^2 y_1}{dt^2} + B_1 \frac{d}{dt}(y_1 - y_2) + K_1(y_1 - y_2)$$

$$f(t) = M_1 \frac{d^2 y_1}{dt^2} + B_1 \frac{dy_1}{dt} - B_1 \frac{dy_2}{dt} + K_1 y_1 - K_1 y_2 \qquad (5.3.1)$$

Free body diagram of M_2 is shown in Fig. (Ex. 5.13.3)

$$f_{m2} = M_2 \frac{d^2 y_2}{dt^2}$$

$$f_{b2} = B_2 \frac{dy_2}{dt}$$

$$f_{b1} = B_2 \frac{d}{dt}(y_2 - y_1)$$

$$f_{k2} = K_2 y_2$$

$$f_{k1} = K_1(y_2 - y_1)$$

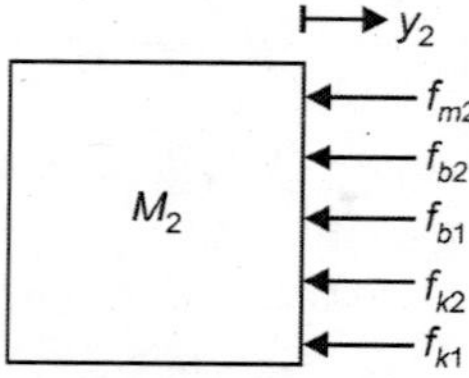

Fig. Ex. 5.13.3

By Newton's second law, the free balance equation at node M_2 is

$$f_{m2} + f_{b2} + f_{b1} + f_{k2} + f_{k1} = 0$$

$$\therefore \qquad M_2 \frac{d^2 y_2}{dt^2} + B_2 \frac{dy^2}{dt} + B_1 \frac{d}{dt}(y_2 - y_1) + K_2 y_2 + K_1(y_2 - y_1) = 0$$

$$M_2 \frac{d^2 y_2}{dt^2} + B_2 \frac{dy_2}{dt} + B_1 \frac{dy_2}{dt} - B_1 \frac{dy_1}{dt} + K_2 y_2 + K_1 y_2 - K_1 y_1 = 0 \qquad (5.3.2)$$

Let us choose for state variable X_1, X_2, X_3 and X_4. Also, let the input $f(t) = u$. The state variables as follows.

$$X_1 = y_1;\ X_2 = y_2;\ X_3 = \frac{dy_1}{dt};\ X_4 = \frac{dy_2}{dt};\ \dot{X}_3 = \frac{d^2 y_1}{dt^2};\ \dot{X}_4 = \frac{d^2 y_2}{dt^2}$$

on substituting $y_1 = X_1;\ y_2 = X_2;\ \frac{dy_1}{dt} = X_3;\ \frac{dy_2}{dt} = X_4;\ \frac{d^2 y_1}{dt} = \dot{X}_3$ and $f(t) = u$ in equation (5.3.1) we get

$$u = M_1 \dot{X}_3 + B_1 X_3 - B_1 X_4 + K_1 X_1 - K_1 X_2$$

$$M_1\dot{X}_3 = -B_1X_3 + B_1X_4 - K_1X_1 + K_1X_2 + u$$

$$\therefore \quad \dot{X}_3 = -\frac{K_1}{M_1}X_1 + \frac{K_1}{M_1}X_2 - \frac{B_1}{M_1}X_3 + \frac{B_1}{M_1}X_4 + \frac{1}{M_1}u \qquad (5.3.3)$$

on substituting $y_1 = X_1; y_2 = X_2; \frac{dy_1}{dt} = X_3; \frac{dy_2}{dt} = X_4$ and $\frac{d^2y_2}{dt} = \dot{X}_4$ is equation (5.3.2) we get,

$$M_2\dot{X}_4 + B_2X_4 + B_1X_4 - B_1X_3 + K_1X_2 + K_2X_2 - K_1X_1 = 0$$

$$M_2\dot{X}_4 = -B_2X_4 - B_1X_4 + B_1X_3 - K_2X_2 - K_1X_2 + K_1X_1$$

$$= -(B_2 + B_1)X_4 + B_1X_3 - (K_2 + K_1)X_2 + K_1X_1$$

$$\therefore \quad \dot{X}_4 = \frac{K_1}{M_2}X_1 - \frac{(K_1 + K_2)}{M_2}X_2 + \frac{B_1}{M_2}X_3 - \frac{(B_1 + B_2)}{M_2}X_4 \qquad (5.3.4)$$

the state variable $X_1 = y_1$

on differentiating $X_1 = y_1$ with respect to t we get, $\frac{dX_1}{dt} = \frac{dy_1}{dt}$

$$\text{Let, } \left[\frac{dX_1}{dt} = \dot{X}_1 \text{ and } \frac{dy_1}{dt} = X_3 \qquad \therefore \dot{X}_1 = X_3\right] \qquad (5.3.5)$$

The state variable, $X_2 = Y_2$

on differentiating $X_2 = y_2$ with respect to t we get, $\frac{dX_2}{dt} = \frac{dy_2}{dt}$

$$\frac{d}{dt}(X_2) = \frac{d}{dt}(y_2)$$

$$\text{Let, } \left[\frac{dX_2}{dt} = \dot{X}_2 \text{ and } \frac{dY_2}{dt} + X_4 \therefore \dot{X}_2 = X_4\right] \qquad (5.3.6)$$

The equations (5.3.3) to (5.3.6) are state equations of the mechanical system. Hence the state equations of the mechanical system are,

$$\dot{X}_1 = X_3$$

$$\dot{X}_2 = X_4$$

$$\dot{X}_3 = -\frac{K_1}{M_1}X_1 + \frac{K_1}{M_1}X_2 - \frac{B_1}{M_1}X_3 + \frac{B_1}{M_1}X_4 + \frac{1}{M_1}u$$

$$\dot{X}_4 = \frac{K_1}{M_1}X_1 - \frac{(K_1 + K_2)}{M_1}X_2 + \frac{B_1}{M_1}X_3 - \frac{(B_1 + B_2)}{M_2}X_4$$

On arranging the state equations in the matrix form, we get

$$\begin{bmatrix} \dot{X}_1 \\ \dot{X}_2 \\ \dot{X}_3 \\ \dot{X}_4 \end{bmatrix} = \begin{bmatrix} 0 & 0 & 1 & 0 \\ 0 & 0 & 0 & 1 \\ -\dfrac{K_1}{M_1} & \dfrac{K_1}{M_1} & -\dfrac{B_1}{M_1} & \dfrac{B_1}{M_1} \\ \dfrac{K_1}{M_2} & -\dfrac{(K_1+K_2)}{M_2} & \dfrac{B_1}{M_2} & -\dfrac{(B_1+B_2)}{M_2} \end{bmatrix} \begin{bmatrix} X_1 \\ X_2 \\ X_3 \\ X_4 \end{bmatrix} + \begin{bmatrix} 0 \\ 0 \\ \dfrac{1}{M_1} \\ 0 \end{bmatrix} [u] \qquad (5.3.7)$$

Let the discplacements y_1 and y_2 be the output of the system

$\therefore \quad y_1 = X_1$ and $y_2 = X_2$.

The output equation in matrix form is given by,

$$\begin{bmatrix} y_1 \\ y_2 \end{bmatrix} = \begin{bmatrix} 1 & 0 & 0 & 0 \\ 0 & 1 & 0 & 0 \end{bmatrix} \begin{bmatrix} X_1 \\ X_2 \\ X_3 \\ X_4 \end{bmatrix} \qquad (5.3.8)$$

Result: The state equation (5.3.7) and the output equation (5.3.8) together called state model of the system.

EXAMPLE 5.14

Obtain the state model of the mechanical system Fig. (Ex. 5.14.1) given below by choosing a minimum of three state variables.

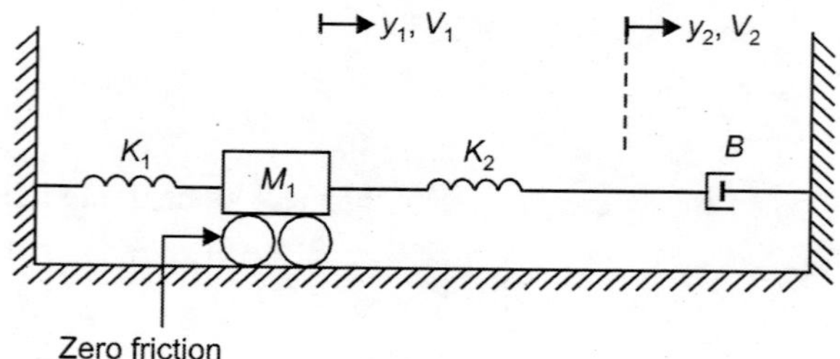

Fig. Ex. 5.14.1

Solution: Let the three state variables be X_1, X_2 and X_3 and they are related to physical variables as shown below.

$$X_1 = y_1;\ X_2 = y_2;\ X_3 = \frac{dy_1}{dt} = V_1$$

Free body diagram of mass M_1 is shown Fig. (Ex. 5.14.2).

$$f_m = M\frac{d^2y_1}{dt^2}$$

$$f_{k1} = K_1Y_1$$

$$f_{k2} = K_2(y_1 - y_2)$$

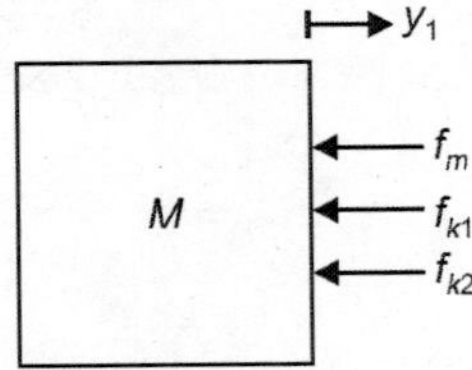

Fig. Ex. 5.14.2

By Newton's second law, the force balance equation at node M is,

$$f_m + f_{k1} + f_{k2} = 0$$

$$M\frac{d^2y_1}{dt^2} + K_1y_1 + K_2(y_1 - y_2) = 0$$

$$M\frac{d^2y_1}{dt^2} + K_1y_1 + K_2y_1 - K_2y_2 = 0 \qquad (5.4.1)$$

Put $\quad M\frac{d^2y_1}{dt^2} = \dot{X}_3;\; y_1 = X_1;\; y_2 = X_3$ in eq. (5.4.1)

$$M\dot{X}_3 + K_1X_1 + K_2X_1 - K_2X_2 = 0$$

$$M\dot{X}_3 + (K_1 + K_2)X_1 - K_2X_3 = 0$$

$$\dot{X}_3 = -\frac{K_1 + K_2}{M}X_1 + \frac{K_2}{M}X_3 \qquad (5.4.2)$$

The free body diagram of node 2 (meeting point of k_2 and B) is shown in Fig. (Ex. 5.14.3).

$$f_b = B\frac{dy_2}{dt}$$

$$f_{k2} = K_2(y_2 - y_1)$$

Fig. Ex. 5.14.3

writing force balance equation at the meeting point of K_2 and B, we get

$$f_b + f_{k2} = 0$$

$$B\frac{dy_2}{dt} + K_2(y_2 - y_1) = 0$$

$$\therefore \quad \frac{dy_2}{dt} = \frac{K_2}{B}y_1 - \frac{K_2}{B}y_2$$

Put $\frac{dy_2}{dt} = \dot{X}_2, \quad y_1 = X_1$ and $y_2 = X_2$,

$$\therefore \quad \dot{X}_2 = +\frac{K_2}{B}X_1 - \frac{K_2}{B}X_2 \qquad (5.4.3)$$

the state variable, $X_1 = y_1$

$$\frac{d}{dt}(X_1) = \frac{d}{dt}(y_1)$$

Let, $\frac{dX_1}{dt} = \dot{X}_1$ and $\frac{dy_1}{dt} = X_3$

$$\therefore \quad \dot{X}_1 = X_3 \qquad (5.4.4)$$

The state equations are given by equations (5.4.4), (5.4.3) and (5.4.2)

$$\dot{X}_1 = X_3$$

$$\dot{X}_2 = \frac{K_2}{B}X_1 - \frac{K_2}{B}X_2$$

$$\dot{X}_3 = -\frac{K_1 + K_2}{M}X_1 - \frac{K_2}{M}X_3$$

On arranging the state equations in the matrix form,

$$\begin{bmatrix} \dot{X}_1 \\ \dot{X}_2 \\ \dot{X}_3 \end{bmatrix} = \begin{bmatrix} 0 & 0 & 1 \\ \frac{K_2}{B} & -\frac{K_2}{B} & 0 \\ -\frac{K_1 + K_2}{M} & 0 & \frac{K_2}{M} \end{bmatrix} \begin{bmatrix} X_1 \\ X_2 \\ X_3 \end{bmatrix} \qquad (5.4.5)$$

If the desired output are y_1 and y_2, then $y_1 = X_1$ and $y_2 = X_2$.

The output equation in the matrix form is given by

$$\begin{bmatrix} y_1 \\ y_2 \end{bmatrix} = \begin{bmatrix} 1 & 0 \\ 0 & 1 \end{bmatrix} \begin{bmatrix} X_1 \\ X_2 \end{bmatrix} \qquad (5.4.6)$$

Result: The state equation (5.4.5) and the output equation (5.4.6) together constitute the state model of the system.

EXAMPLE 5.15

Determine the state model of armature controlled dc motor.

Solution: The speed of DC motor is directly proportional to armature voltage and inversely proportional to flux. In armature controlled DC motor the desired speed is obtained by varying the armature voltage. This speed control system is an electro-mechanical control system. The electrical system consists of armature and field circuit but for analysis purpose, only the armature circuit is considered because the field is excited by a constant voltage. The mechanical system consists of rotating part of the motor and load connected to the shaft of the motor. The armature controlled DC motor speed control system is shown in Fig. (Ex. 5.15.1).

Let R_a = Armature resistance (Ω)

L_a = Armature inductance (H)

i_a = Armature current, (A)

V_a = Armature voltage, (V)

e_b = Back emf, (V)

k_t = Torque constant, ($N-m/A$)

T = Torque developed by motor, ($N-m$)

θ = Angular displacement of shaft, (rad)

$\omega = d\theta/dt$ = Angular velocity of the shaft, (rad)

J = Moment of inertia of motor and load, (kg m^2/rad)

B = Frictional coefficient of motor and load , ($N-m$/(rad/sec)

K_b = Back emf constant (V/rad/sec)

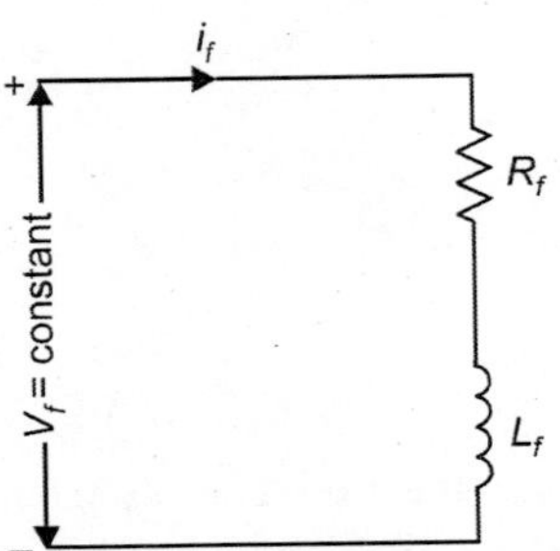

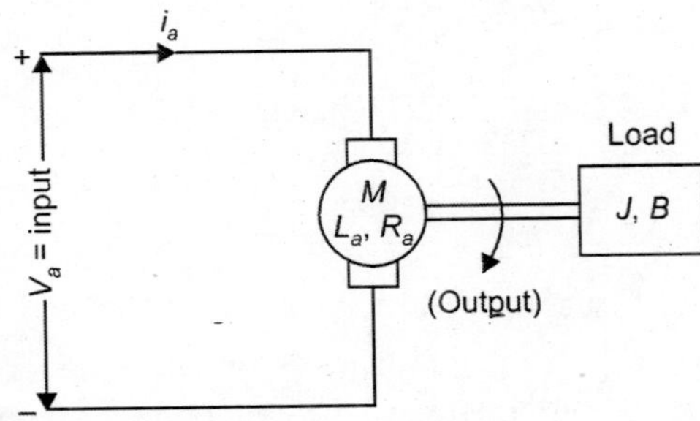

Fig. Ex. 5.15.1: Armature controlled DC motor.

The equivalent circuit of armature is shown Fig. (Ex. 5.15.2).

By applying KVL to the circuit shown in Fig. (5.5.2), we can write

$$i_a R_a + L_a \frac{di_a}{dt} + e_b = V_a \tag{5.5.1}$$

Torque of DC motor is proportional to the product of flux and current. Since flux is constant in this system, the torque is proportional to i_a alone.

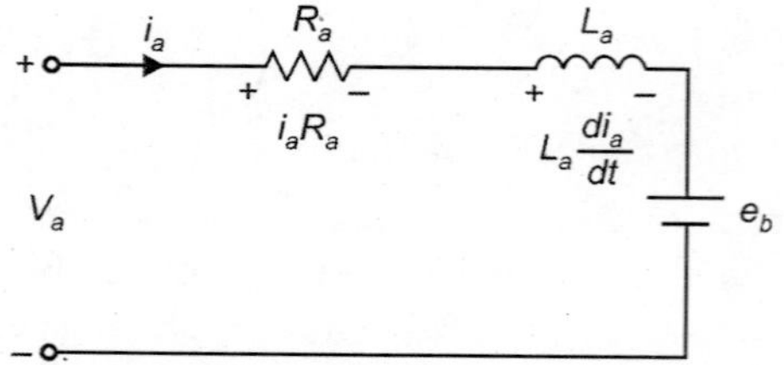

Fig. Ex. 5.15.2: Equivalent circuit of armature.

$$T \,\alpha\, i_a$$

$\therefore$ Torque, $T = K_t\, i_a$ (5.5.2)

The mechanical system of the motor is shown in Fig. (Ex. 5.15.3). The differential equation governing the mechanical system of motor is given by,

$$J\frac{d^2\theta}{dt^2} + B\frac{d\theta}{dt} = T \tag{5.5.3}$$

Fig. Ex. 5.15.3

The back emf of DC machine is proportional to speed (angular velocity) of shaft

$$\therefore \quad e_b \,\alpha\, \frac{d\theta}{dt}; \quad \text{Back emf,} \quad e_b = K_b\frac{d\theta}{dt} \tag{5.5.4}$$

From eq. (5.5.1) and (5.5.4) we get,

$$i_aR_a + L_a\frac{di_a}{dt} + K_b\frac{d\theta}{dt} = V_a \tag{5.5.5}$$

From eq. (5.5.2) and (5.5.3) we get,

$$J\frac{d^2\theta}{dt^2} + B\frac{d\theta}{dt} = K_t i_a \tag{5.5.6}$$

The eq. (5.5.5) and (5.56) are the differential equations governing the armature controlled dc motor.

Let us choose i_a, ω and θ as state variables to model the armature controlled dc motor. The physical variables i_a, ω and θ are related to the general notation of state variables X_1, X_2 and X_3 as shown below

$$X_1 = i_a; \; X_2 = \omega = \frac{d\theta}{dt} \text{ and } X_3 = \theta$$

The input to the motor is the armature voltage, V_a and let $V_a = u$, where u is the general notation for input variable.

On substituting the state variables for the physical variables in eq. (5.5.5) we get,

$$X_1R_a + L_a\frac{dX_1}{dt} + K_bX_2 = u$$

Let, $$\frac{dX_1}{dt} = \dot{X}_1;$$

$$\therefore \quad X_1 R_a + L_a\dot{X}_1 + K_b X_2 = u$$

$$\dot{X}_1 = -\frac{R_a}{L_a}X_1 - \frac{K_b}{L_a}X_2 + \frac{1}{L_a}u \qquad (5.5.7)$$

On substituting the state variables for physical variables in eq. (5.5.6) we get,

$$J\frac{d^2X_3}{dt^2} + B\frac{dX_3}{dt} = K_t X_1$$

Let $\frac{d^2X_3}{dt^2} = \dot{X}_2$ and $\frac{dX_3}{dt} = X_2;$

$$\therefore \quad J\dot{X}_2 + BX_2 = K_t X_1$$

$$\dot{X}_2 = \frac{K_t}{J}X_1 - \frac{B}{J}X_2 \qquad (5.5.8)$$

The state variables $X_3 = \theta$. On differentiating $X_3 = \theta$ with respect to "t" we get,

$$\frac{dX_3}{dt} = \frac{d\theta}{dt}$$

Put $$\frac{dX_3}{dt} = \dot{X}_3 \quad \text{and} \quad \frac{d\theta}{dt} = X_2$$

$$\therefore \quad \dot{X}_3 = X_2 \qquad (5.5.9)$$

The equations (5.5.7), (5.5.8) and (5.5.9) are the state equations of the system.

$$\dot{X}_1 = -\frac{R_a}{L_a}X_1 - \frac{K_b}{L_a}X_2 + \frac{1}{L_a}u$$

$$\dot{X}_2 = \frac{K_t}{J}X_1 - \frac{B}{J}X_2$$

$$\dot{X}_3 = X_2$$

Arranging the state equations in the matrix form,

$$\begin{bmatrix} \dot{X}_1 \\ \dot{X}_2 \\ \dot{X}_3 \end{bmatrix} = \begin{bmatrix} -\frac{R_a}{L_a} & -\frac{K_b}{L_a} & 0 \\ \frac{K_t}{J} & -\frac{B}{J} & 0 \\ 0 & 1 & 0 \end{bmatrix} \begin{bmatrix} X_1 \\ X_2 \\ X_3 \end{bmatrix} + \begin{bmatrix} \frac{1}{L_a} \\ 0 \\ 0 \end{bmatrix} [u] \qquad (5.5.10)$$

Let the desired outputs be i_a, ω and θ. Let us equate the desired output quantities to standard notation y_1, y_2 and y_3 as given below,

$$y_1 = i_a;\ y_2 = \omega = d\theta/dt;\ y_3 = \theta$$

On relating the outputs to state variable, we get,

$$y_1 = X_1;\ y_2 = X_2;\ y_3 = X_3$$

∴ The output equation in the matrix form is

$$\begin{bmatrix} y_1 \\ y_2 \\ y_3 \end{bmatrix} = \begin{bmatrix} 1 & 0 & 0 \\ 0 & 1 & 0 \\ 0 & 0 & 1 \end{bmatrix} \begin{bmatrix} X_1 \\ X_2 \\ X_3 \end{bmatrix} \quad (5.5.1)$$

Result: The state equation [5.5.10] and the output equation [5.5.11] together constitute the state model of the armature controlled dc motor.

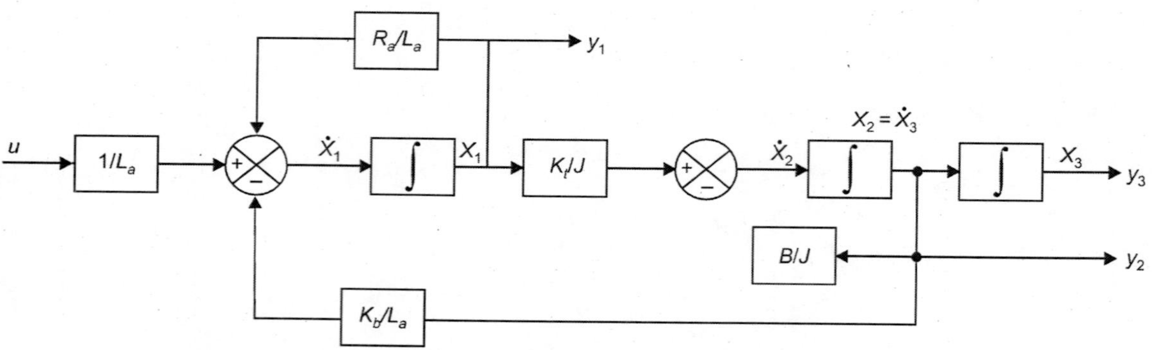

Fig. Ex. 5.15.4: Block diagram representation of the state model of armature controlled dc motor.

EXAMPLE 5.16

Determine the state model of field controlled dc motor.

Solution: The speed of a DC motor is directly proportional to armature voltage and inversely proportional to flux. In field controlled DC motor the armature voltage is kept constant and the speed is varied by varying the flux. Since flux is directly proportional to field current, the flux is varied by varying field current. The speed control system is an electromechanical control system. The electrical system consists of armature field circuit but for analysis purpose, only field circuit is considered because the armature is excited by a constant voltage. The mechanical system consists of the rotating part of the motor and the load connected to the shaft of the motor. The field controlled DC motor speed control system is shown in Fig. (Ex. 5.16.1).

Let, R_f = Field resistance (Ω)

L_f = Field inductance, (H)

i_f = Field current, (A)
V_f = Field voltage, (V)
θ = Angular displacement of the motor shaft (rad)
$\omega = d\theta/dt$ = Angular velocity of the motor shaft (rad/sec)
T = Torque developed by motor, (N-m)
k_{tf} = Torque constant (N-m/A)
J = Moment of inertia of rotor and load (kg-m^2)/rad)
B = Frictional co-efficient of rotor and load, (N-m/(rad/sec)

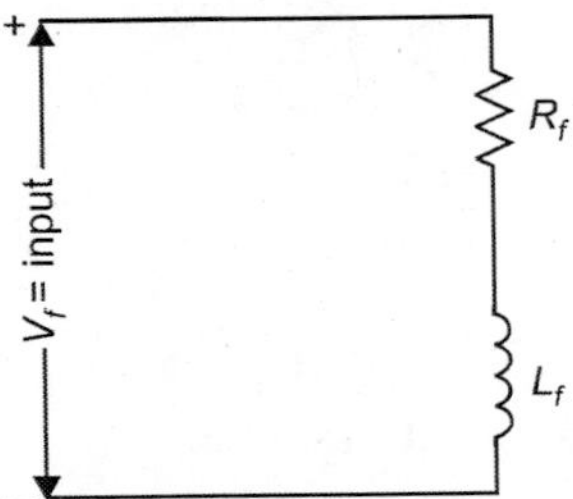

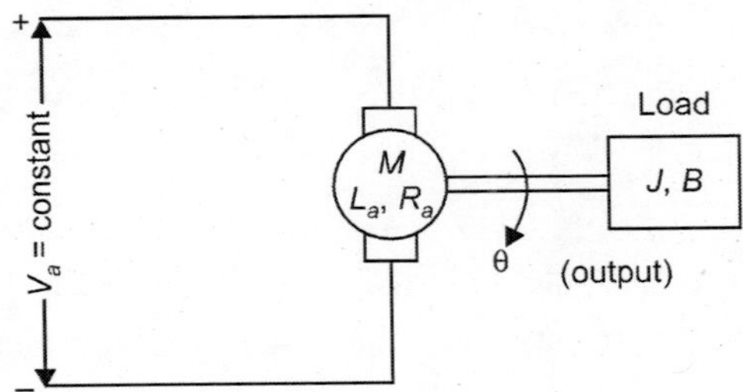

Fig. Ex. 5.16.1

The equivalent circuit of field is shown in Fig. (Ex. 5.16.2),

By applying KVL to the circuit shown in Fig. (Ex. 5.16.2), we can write

$$R_f i_f + L_f \frac{d_{if}}{dt} = V_f \qquad (5.6.1)$$

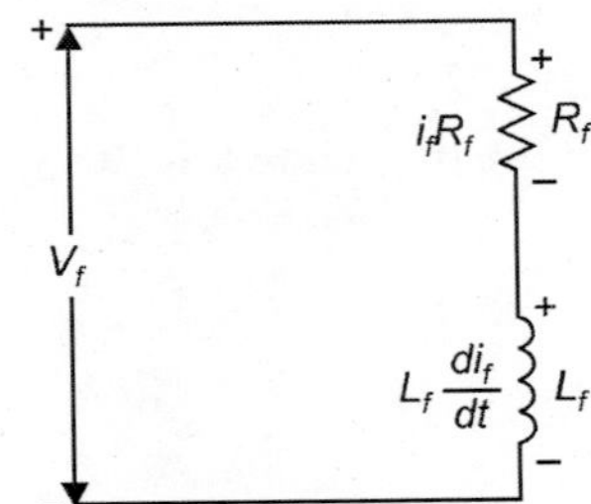

Fig. 5.6.2: Equivalent circuit of field.

The torque of DC motor is proportional to product of flux and armature current. Since armature current is constant in this system, the torque is proportional to flux alone, but flux is proportional to field current,

$\therefore \qquad T \alpha I_f$; Torque, $T = K_{tf}\, i_f \qquad (5.6.2)$

The mechanical system of the motor is shown in Fig. (Ex. 5.16.3). The differential equation governing the mechanical system of the motor is given by

$$J\frac{d^2\theta}{dt^2} + B\frac{d\theta}{dt} = T \qquad (5.6.3)$$

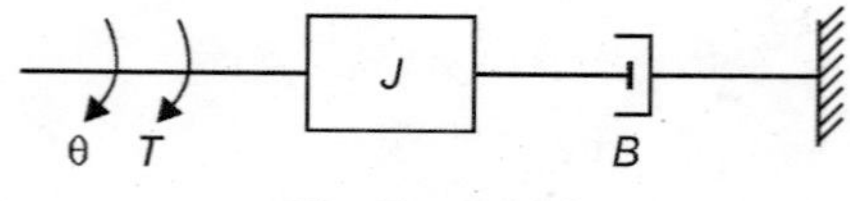

Fig. Ex. 5.16.3

From eq. (5.6.2) and (5.6.3) we get,

$$J\frac{d^2\theta}{dt^2} + B\frac{d\theta}{dt} = K_{tf}\, i_f \qquad (5.6.2)$$

The eq. (5.6.1) and (5.6.4) are the differential equations governing the field controlled dc motor.

Let us choose i_f, ω and θ as state variable to model the field controlled dc motor. The physical variables *if*, ω and θ are related to the general notation of state variables X_1, X_2 and X_3 as shown below.

$$X_1 = i_f;\ X_2 = \omega = \frac{d\theta}{dt} \text{ and } X_3 = \theta$$

The input to the system is the field voltage V_f. Let $V_f = u$, where u is the general notation for input. On substituting the state variables and input variables for the physical variables in eq. (5.6.1) we get,

$$R_f X_1 + L_f\frac{dX_1}{dt} = u$$

Let $\frac{dX_1}{dt} = \dot{X}_1$,

$$\therefore \quad R_f X_1 + Lf\dot{X}_1 = u$$

$$\dot{X}_1 = -\frac{R_f}{L_f}X_1 + \frac{1}{L_f}u \qquad (5.6.5)$$

On substituting the state variables for the physical variables in eq. (5.6.1) we get,

$$J\frac{d^2X_3}{dt^2} + B\frac{dX_3}{dt} = K_{tf}X_1$$

Let, $\frac{d^2X_3}{dt^2} = \dot{X}_2$ and $\frac{dX_3}{dt} = X_2$,

$$\therefore \quad J\dot{X}_2 + BX_2 = K_{tf}X_1$$

$$\dot{X}_2 = \frac{K_{tf}}{J}X_1 - \frac{B}{J}X_2 \qquad (5.6.6)$$

The state variables $X_2 = \theta$. On differentiating $X_3 = \theta$ with respect to t we get, $\frac{dX_3}{dt} = \frac{d\theta}{dt}$

Put $\frac{dX_3}{dt} = \dot{X}_3$ and $\frac{d\theta}{dt} = X_2$

$$\therefore \quad \dot{X}_3 = X_2 \qquad (5.6.7)$$

The eq. (5.6.5), (5.6.6) and (5.6.7) or the state equations of the system.

$$\dot{X}_1 = -\frac{R_f}{L_f}X_1 + \frac{1}{L_f}u$$

$$\dot{X}_2 = +\frac{K_{tf}}{J}X_1 - \frac{B}{J}X_2$$

$$\dot{X}_3 = X_2$$

$$\begin{bmatrix} \dot{X}_1 \\ \dot{X}_2 \\ \dot{X}_3 \end{bmatrix} = \begin{bmatrix} -\frac{R_f}{L_f} & 0 & 0 \\ \frac{K_{tf}}{J} & -\frac{B}{J} & 0 \\ 0 & 1 & 0 \end{bmatrix} \begin{bmatrix} X_1 \\ X_2 \\ X_3 \end{bmatrix} + \begin{bmatrix} \frac{1}{L_f} \\ 0 \\ 0 \end{bmatrix} [u] \tag{5.6.8}$$

Let the desired output be ω and θ. Let us equate the desired quantities to standard notation y_1 and y_2 as shown below.

$y_1 = \omega$; $y_2 = \theta$

On relating the outputs to state variable we get,

$y_1 = X_2$; $y_2 = X_3$

$$\begin{bmatrix} y_1 \\ y_2 \end{bmatrix} = \begin{bmatrix} 0 & 1 & 0 \\ 0 & 0 & 1 \end{bmatrix} \begin{bmatrix} X_1 \\ X_2 \\ X_3 \end{bmatrix} \tag{5.6.9}$$

Result: The state equation (5.6.8) and the output eq. (5.6.9) together constitute the state model of the system.

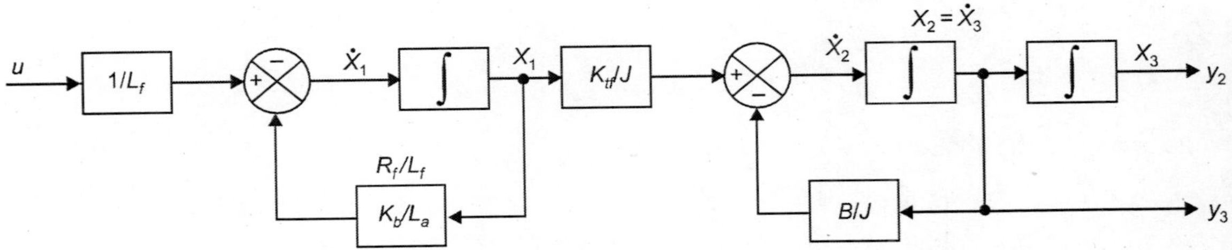

Fig. Ex. 5.16.4: Block diagram representation of the state model field controlled dc motor.

5.13 STATE SPACE REPRESENTATION USING PHASE VARIABLES

The phase variables are defined as those particular state variables which are obtained from one of the system variable and its derivatives. Usually the variable used in the system output and the remaining state variables are then derivatives of the output. The state model using phase variables can be easily determined if the system model is already known in the differential equation or transfer function form.

There are three methods of modelling a system using phase variables and they are explained in the following sections.

Method 1: Consider the following n^{th} order linear differential equation relating the output $y(t)$ to the input $u(t)$ of a system.

$$\overset{n\cdot}{Y} + a_1 \overset{(n-1)\cdot}{y} + a_2 \overset{(n-2)\cdot}{y} + \ldots + a_{(n-2)}\ddot{y} + a_{(n-1)}\dot{y} + a_n y = bu \tag{5.10}$$

By choosing the output Y and their derivatives as state variables, we get,

$$X_1 = Y$$

$$X_2 = \dot{Y}$$

$$X_3 = \ddot{Y}$$

$$\vdots$$

$$X_n = \overset{(n-1)\cdot}{Y};$$

$$\dot{X}_n = \overset{n\cdot}{y}$$

On substituting the state variables in the differential equation governing the system eq. (5.10) we get,

$$\dot{X}_n + a_1 X_n + a_2 X_{n-1} + \ldots a_{n-2} X_3 + a_{n-1} X_2 + a_n X_1 = bu$$

$$\therefore \qquad \dot{X}_n = -a_n X_1 - a_{n-1} X_2 - a_{n-2} X_3 - \ldots - a_2 X_{n-1} - a_1 X_n + bu$$

The state equations of the system are,

$$\dot{X}_1 = X_2$$

$$\dot{X}_2 = X_3$$

$$\vdots$$

$$\dot{X}_{n-1} = X_n$$

$$\dot{X}_n = -a_n X_1 - a_{n-1} X_2 - a_{n-2} X_3 - \ldots - a_2 X_{n-1} - a_1 X_n + bu$$

On arranging the above equations in the matrix form, we get,

$$\begin{bmatrix} \dot{X}_1 \\ \dot{X}_2 \\ \dot{X}_3 \\ \vdots \\ \dot{X}_3 \end{bmatrix} = \begin{bmatrix} 0 & 1 & 0 & 0 & \cdots\cdots & 0 \\ 0 & 0 & 1 & 0 & \cdots\cdots & 0 \\ 0 & 0 & 0 & 1 & \cdots\cdots & 0 \\ \vdots & & & & & \\ 0 & 0 & 0 & 0 & \cdots\cdots & 1 \\ -a_n & -a_{n-1} & -a_{n-2} & -a_{n-3} & \cdots\cdots & -a_1 \end{bmatrix} \begin{bmatrix} X_1 \\ X_2 \\ X_3 \\ \vdots \\ X_{n-1} \\ X_n \end{bmatrix} + \begin{bmatrix} 0 \\ 0 \\ 0 \\ \vdots \\ 0 \\ b \end{bmatrix} [u] \tag{5.11}$$

or $\qquad \dot{X} = AX + BU$

Here the matrix A (System matrix) has a very special form. It has all 1's in the upper off-diagonal, its last row is comprised of the negative of the coefficients of the original differential equation and all other elements are zero. This form of matrix A is known as Bush form (or) companion form.

Also note that B matrix has the speciality that all its elements except the last element are zero. The output being $y = X_1$, the output equation is given by,

$$Y = [1\ 0\ 0\ 0\ \ldots\ 0]\begin{bmatrix} X_1 \\ X_2 \\ X_3 \\ \vdots \\ X_n \end{bmatrix} \quad (5.12)$$

or

$$Y = CX$$

The advantage in using phase variables for state space modelling is that the system state model can be written directly by inspection from the differential equation governing the system.

Method 2: Consider the following n^{th} order differential equation governing the output $y(t)$ to the input $u(t)$ of the system

$$\overset{n\cdot}{y} + a_1 \overset{(n-1)\cdot}{y} + \ldots + a_{n-1}\dot{y} + a_n y = b_0 \overset{m\cdot}{u} + b_1 \overset{(m-1)\cdot}{u} + \ldots + b_{m-1}\dot{u} + b_m u \quad (5.13)$$

Let

$$n = m = 3$$

$$\therefore \quad \dddot{y} + a_1\ddot{y} + a_2\dot{y} + a_3 y = b_0\dddot{u} + b_1\ddot{u} + b_2\dot{u} + b_3 u \quad (5.14)$$

On taking Laplace transform of eq. (5.14) with zero initial conditions we get,

$$s^3Y(s) + a_1s^2Y(s) + a_2sY(s) + a_3Y(s) = b_0s^3U(s) + b_1s^2U(s) + b_2sU(s) + b_3U(s)$$

$$(s^3 + a_1s^2 + a_2s + a_3)\ Y(s) = (b_0s^3 + b_1s^2 + b_2s + b_3)\ U(s)$$

$$\therefore \quad \frac{Y(s)}{U(s)} = \frac{b_0s^3 + b_1s^2 + b_2s + b_3}{s^3 + a_1s^2 + a_2s + a_3}$$

$$= \frac{s^3\left(b_0 + \frac{b_1}{s} + \frac{b_2}{s^2} + \frac{b_3}{s^3}\right)}{s^3\left(1 + \frac{a_1}{s} + \frac{a_2}{s^2} + \frac{a_3}{s^3}\right)} = \frac{b_0 + \frac{b_1}{s} + \frac{b_2}{s^2} + \frac{b_3}{s^3}}{1 - \left(-\frac{a_1}{s} - \frac{a_2}{s^2} - \frac{a_3}{s^3}\right)} \quad (5.15)$$

From the Mason's gain formula, the transfer function of the system is given by

$$T(S) = \frac{1}{\Delta}\sum_K P_K \Delta_K \quad (5.16)$$

where, P_K - path gain of Kth forward path.

$\Delta =$ 1 – (Sum of loop gain of all individual loops) + (Sum of gain products of all possible combinations of two non-touching loops) ...

$\Delta_K =$ Δ for that part of the graph which is not touching K^{th} forward path.

CHAPTER 5

The transfer function of a system with four forward paths and with three feedback loops (touching each other) is given by.

$$T(S) = \frac{P_1 + P_2 + P_3 + P_4}{1 - (P_{11} + P_{12} + P_{13})} \tag{5.17}$$

On comparing eq. (5.15) and (5.17) we get,

$$P_1 = b_0; \quad P_2 = \frac{b_1}{s}; \quad P_3 = \frac{b_2}{s^2} \quad \text{and} \quad P_4 = \frac{b_3}{s^3}$$

$$P_{11} = -\frac{a_1}{s}; \quad P_{12} = -\frac{a_2}{s^2} \quad \text{and} \quad P_{13} = -\frac{a_3}{s^3}$$

Hence for the system represented by the transfer function as that of equation (5.15), a signal flow is constructed such that all $\Delta K = 1$ and all loops are touching loops.

Let us assign state variables at the output of each integration in the signal flow graph. Hence at the input of each integrator, the first derivative of the state variable will be available. The state equations are formed by summing all the incoming signals to the nodes, whose values corresponds to first derivative of state variables.

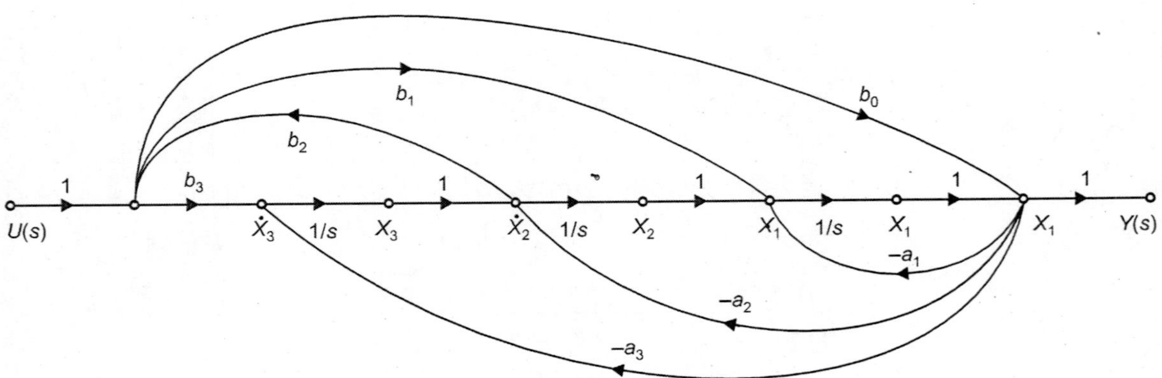

Fig. 5.4: Signal flow graph of the system represented by the eq. (5.15).

By summing UP the incoming signals to node $\dot{X}_1$ we get, refer Fig. (5.4.a)

$$\dot{X}_1 = -a_1 (X_1 + b_0 u) + X_2 + b_1 u$$

$$\dot{X}_1 = -a_1 X_1 + X_2 + (b_1 - a_1 b_0)u \tag{5.180}$$

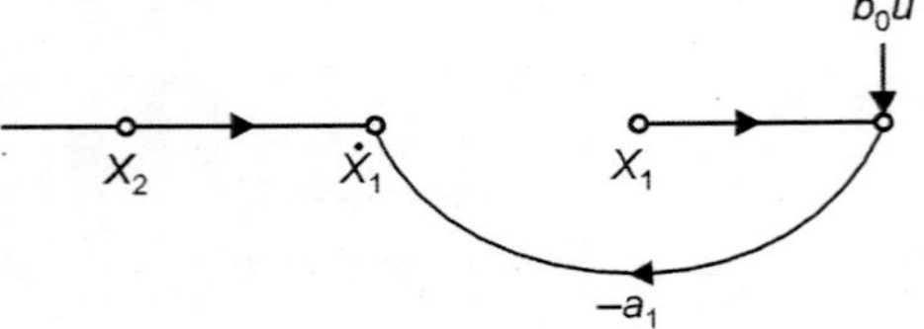

Fig. 5.4a

By summing UP the incoming signals to node $\dot{X}_2$ we get, (refer Fig. 5.4b).

$$\dot{X}_2 = -a_2(X_1 + b_0 u) + X_3 + b_2 u$$

$$\dot{X}_2 = -a_2 X_1 + X_3 + (b_2 - a_2 b_0)u \qquad (5.19)$$

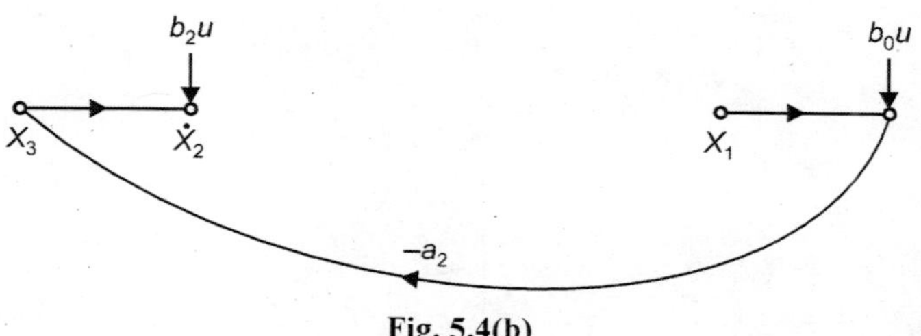

Fig. 5.4(b)

By summing UP the incoming signals to node $\dot{X}_3$ we get. (Refer Fig. 5.4.c).

$$\dot{X}_3 = -a_3(X_1 + b_0 u) + b_3 u$$

$$\dot{X}_3 = -a_3 X_1 + (b_3 - a_3 b_0)u \qquad (5.20)$$

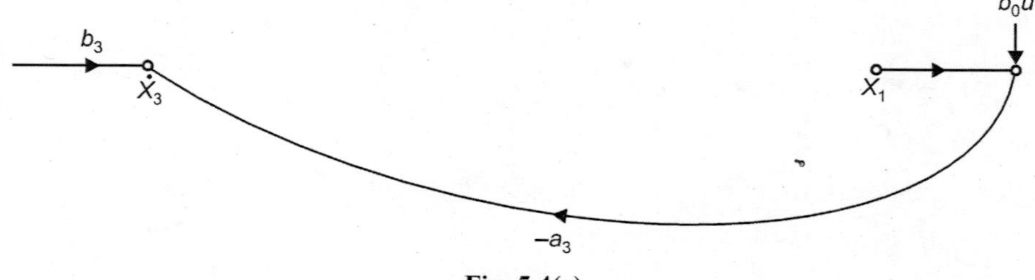

Fig. 5.4(c)

The output equation is given by the sum of incoming signals to output node.

$$\therefore \quad Y = X_1 + b_0 u \qquad (5.21)$$

On arranging the state equations and the output equations in the matrix form, we get

$$\begin{bmatrix} \dot{X}_1 \\ \dot{X}_2 \\ \dot{X}_3 \end{bmatrix} = \begin{bmatrix} -a_1 & 1 & 0 \\ -a_2 & 0 & 1 \\ -a_3 & 0 & 0 \end{bmatrix} \begin{bmatrix} X_1 \\ X_2 \\ X_3 \end{bmatrix} + \begin{bmatrix} b_1 & -a_1 b_0 \\ b_2 & -a_2 b_0 \\ b_3 & -a_3 b_0 \end{bmatrix} [u] \qquad (5.22)$$

$$y = \begin{bmatrix} 1 & 0 & 0 \end{bmatrix} \begin{bmatrix} X_1 \\ X_2 \\ X_3 \end{bmatrix} + b_0 u \qquad (5.23)$$

The above results can be generalized for an nth order differential equation, and the general state model form = n is given below.

$$\begin{bmatrix} \dot{X}_1 \\ \dot{X}_2 \\ \vdots \\ \dot{X}_{n-1} \\ \dot{X}_n \end{bmatrix} = \begin{bmatrix} -a_1 & 1 & 0 & \cdots & 0 \\ -a_2 & 0 & 1 & \cdots & 0 \\ \vdots & \vdots & \vdots & & \vdots \\ -a_{n-1} & 0 & 0 & \cdots & 1 \\ -a_n & 0 & 0 & \cdots & 0 \end{bmatrix} \begin{bmatrix} X_1 \\ X_2 \\ \vdots \\ X_{n-1} \\ X_n \end{bmatrix} + \begin{bmatrix} b_1 - a_1 b_0 \\ b_2 - a_2 b_0 \\ \\ b_{n-1} - a_{n-1} b_0 \\ b_n - a_n b_0 \end{bmatrix} [u] \quad (5.24)$$

$$y = [1 \;\; 0 \;\; 0 \ldots. 0] \begin{bmatrix} X_1 \\ X_2 \\ X_3 \\ \vdots \\ X_n \end{bmatrix} + b_0 u \quad (5.25)$$

Method 3: Consider the following nth order differential equation governing the output $Y(t)$ to the input $u(t)$ of a system.

$$\overset{n\cdot}{y} + a_1 \overset{(n-1)\cdot}{y} + \ldots + a_{n-1}\dot{y} + a_n y = b_0 \overset{m\cdot}{u} + b_1 \overset{(m-1)\cdot}{u} + \ldots + b_{m-1}\dot{u} + b_m u \quad (5.26)$$

Let $n = m = 3$,

$$\dddot{y} + a_1 \ddot{y} + a_2 \dot{y} + a_3 y = b_0 \ddot{u} + b_1 \ddot{u} + b_2 \dot{u} + b_3 u \quad (5.27)$$

On taking Laplace transform of eq. (5.27) with zero initial conditions, we get,

$$s_2 Y(s) + a_1 s_2 Y(s) + a_2 s Y(s) + a_3 Y(s) = b_0 s^3 U(s) + b_1 s^2 U(s) + b_2 s U(s) + b_3 U(s)$$

$$(s_3 + a_1 s^2 + a_2 s + a_3) Y(s) = (b_0 s^3 + b_1 s^2 + b_2 s + b_3) U(s)$$

$$\therefore \qquad \frac{Y(s)}{U(s)} = \frac{b_0 s^3 + b_1 s^2 + b_2 s + b_3}{s^3 + a_1 s^2 + a_2 s + a_3}$$

Let $$\frac{Y(s)}{U(s)} = \frac{X_1(s)}{U(s)} \cdot \frac{Y(s)}{X_1(s)}$$

where $$\frac{X_1(s)}{U(s)} = \frac{1}{s^3 + a_1 s^2 + a_2 s + a_3} \quad (5.28)$$

and $$\frac{Y(s)}{X_1(s)} = b_0 s^3 + b_1 s^2 + b_2 s + b_3 \quad (5.29)$$

on cross multiplying the eq. (5.28), we get

$$X_1(s)\,[s^3 + a_1 s^2 + a_2 s + a_3] = U(s)$$

$$s^3 X_1(s) + a_1 s^2 X_1(s) + a_2 s X_1(s) + a_3 X_1(s) = U(s) \quad (5.30)$$

On taking inverse Laplace transform of eq. (5.30), we get

$$\dddot{X} + a_1\ddot{X}_1 + a_2\dot{X}_1 + a_3X_1 = U \tag{5.31}$$

Let the state variable be, X_1, X_2 and X_3

where, $$X_2 = \dot{X}_1$$

and $$X_3 = \ddot{X}_1 = \dot{X}_2; \quad \therefore \dot{X}_3 = \dddot{X}_1$$

On substituting the state variables in eq. (5.31), we get,

$$\dot{X}_3 + a_1X_3 + a_2X_2 + a_3X_1 = u$$

$\therefore$ $$\dot{X}_3 = -a_3X_1 - a_2X_2 - a_1X_3 + u$$

The state equations are,

$$\dot{X}_1 = X_2$$

$$\dot{X}_2 = X_3$$

$$\dot{X}_3 = -a_3X_1 - a_2X_2 - a_1X_3 + u$$

On cross multiplying the eq. (5.29) we get,

$$Y(s) = b_0s^3X_1(s) + b_1s^2\, X_1(s) + b_2s\, X_1(s) + b_3\, X_1\,(s) \tag{5.32}$$

On taking inverse Laplace transform of eq. (4.32), we get,

$$y = b_0\dddot{X}_1 + b_1\ddot{X}_1 + b_2\dot{X}_1 + b_3X_1 \tag{5.33}$$

On substituting the state variables in eq. (5.33), we get

$$y = b_0\dot{X}_3 + b_1X_3 + b_2X_2 + b_3X_1 \tag{5.34}$$

Put $$\dot{X}_3 = -a_3X_1 - a_2X_2 - a_1X_3 + u \text{ in eq. (5.34)}$$

$$y = b_0(-a_3X_1 - a_2X_2 - a_1X_3 + u) + b_1X_3 + b_2X_2 + b_3X_1$$

$$y = (b_3 - a_3b_0)X_1 + (b_2 - a_2b_0)X_2 + (b_1 - a_1b_0)\, X_3 + b_0u \tag{5.36}$$

The eq. (5.35) is the output equation.

On arranging the state eq. and output eq. in the matrix form we get,

$$\begin{bmatrix} \dot{X}_1 \\ \dot{X}_2 \\ \dot{X}_3 \end{bmatrix} = \begin{bmatrix} 0 & 1 & 0 \\ 0 & 0 & 1 \\ -a_3 & -a_2 & -a_1 \end{bmatrix} \begin{bmatrix} X_1 \\ X_2 \\ X_3 \end{bmatrix} + \begin{bmatrix} 0 \\ 0 \\ 1 \end{bmatrix} [u] \tag{5.36}$$

$$y = [(b_3 - a_3b_0)\,(b_2 - a_2b_0)\,(b_1 - a_1b_0)] \begin{bmatrix} X_1 \\ X_2 \\ X_3 \end{bmatrix} + [b_0]\, u \tag{5.37}$$

The above results can be generalized for an nth order differential equation and the general state model for $m = n$ is given below.

$$\begin{bmatrix} \dot{X}_1 \\ \dot{X}_2 \\ \dot{X}_{n-1} \\ \dot{X}_n \end{bmatrix} = \begin{bmatrix} 0 & 1 & 0 & \cdots \\ 0 & 0 & 1 & \cdots \\ \vdots & \vdots & \vdots & \vdots \\ 0 & 0 & 0 & 1 \\ -a_n & -a_{n-1} & -a_{n-2} & -a_1 \end{bmatrix} \begin{bmatrix} X_1 \\ X_2 \\ \vdots \\ X_{n-1} \\ X_n \end{bmatrix} \begin{bmatrix} 0 \\ 0 \\ \vdots \\ 0 \\ 1 \end{bmatrix} [u] \qquad (5.38)$$

$$y = \begin{bmatrix} (b_n - a_n b_o)(b_{n-1} - a_{n-1} b_o) \cdots (b_2 - a_2 b_o)(b_1 - a_1 b_o) \end{bmatrix} \begin{bmatrix} X_1 \\ X_2 \\ \vdots \\ X_{n-1} \\ X_n \end{bmatrix} + b_o^{\,u} \qquad (5.3.9)$$

5.13.1 Advantages and Disadvantages of Phase Variables

(i) Advantages of phase variables: The state space model can be directly formed by inspection from the differential equations ,governing the system. The phase variables provides a link between the transfer function design approach and time-domain design approach

(ii) Disadvantage of phase variables: The phase variables are not physical variables of the system and therefore are not available for measurement and control purposes.

EXAMPLE 5.17

Construct a state model for a system characterized by the differential equation, $\frac{d^3y}{dt^3} + 6\frac{d^2y}{dt^2} + 11\frac{dy}{dt} + 6y + u = 0$. Give the block diagram representation of the state model.

Solution: Let us choose y and their derivatives as state variables. The system is governed by third order differential equation and so the number of state variables are three.

The state variables X_1, X_2 and X_3 are related to phase variables as follows:

$$X_1 = y$$

$$X_2 = \frac{dy}{dt} = \dot{X}_1$$

$$X_3 = \frac{d^2y}{dt^2} = \dot{X}_2 \Rightarrow \therefore \frac{dy}{dt} = x_2 \;\; and \;\; \frac{d^2y}{dt^2} = x_3 \;\; \& \;\; \frac{d^3y}{dt^3} = \dot{x}_3$$

Put $y = X_1$; $\frac{dy}{dt} = X_2$ and $\frac{d^2y}{dt^2} = X_3$ and $\frac{d^3y}{dt^3} = \dot{X}_3$ in the given equation,

$$\therefore \quad \dot{X}_3 + 6X_3 + 11X_2 + 6X_1 + u = 0$$

or

$$\dot{X}_3 = -6X_1 - 11X_2 - 6X_3 - u$$

The state equations are,

$$\dot{X}_1 = X_2$$

$$\dot{X}_2 = X_3$$

$$\dot{X}_3 = -6X_1 - 11X_2 - 6X_3 - u$$

On arranging the state equations in the matrix form

We get, $$\begin{bmatrix} \dot{X}_1 \\ \dot{X}_2 \\ \dot{X}_3 \end{bmatrix} = \begin{bmatrix} 0 & 1 & 0 \\ 0 & 0 & 1 \\ -6 & -11 & -6 \end{bmatrix} \begin{bmatrix} X_1 \\ X_2 \\ X_3 \end{bmatrix} + \begin{bmatrix} 0 \\ 0 \\ -1 \end{bmatrix} [u]$$

Here, y = Output
But, $y = X_1$

The Output equation is, $$y = [1\,0\,0] \begin{bmatrix} X_1 \\ X_2 \\ X_3 \end{bmatrix}$$

The state equation and output equation, constitutes the state model of the system

The block diagram of the state diagram of the system is shown in Fig. (Ex. 5.17.1).

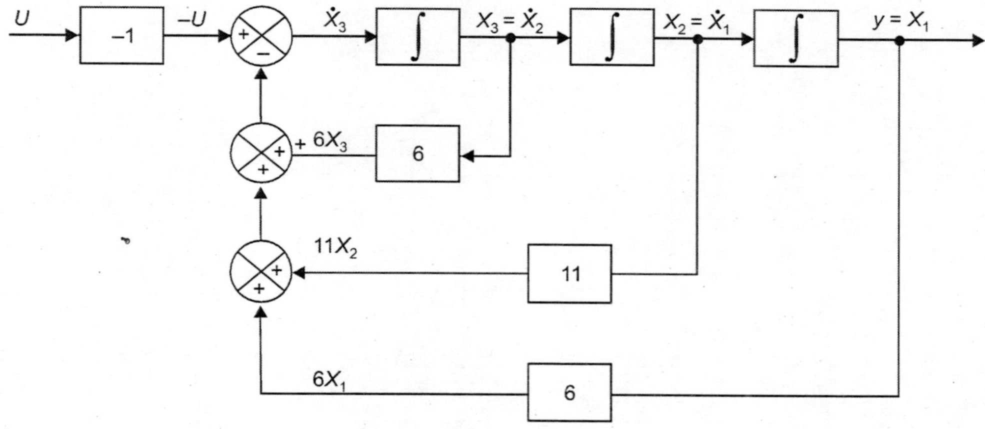

Fig. Ex. 5.17.1: Block diagram form of state diagram.

EXAMPLE 5.18

The state diagram of a system is shown in Fig. (Ex. 5.18.1). Assign state variables and obtain the state model of the system.

Solution: Since there are 4-integrators in the state diagram we can assign, 4 State variables. The state variables can be assigned at the output of the integrators as show in Fig. (Ex. 5.18.2). Hence at the input of the integrator, the firs derivative of the state variable will be available. The state equations are formed by summing all the incoming signals to the input of the integrator and equating to the corresponding first derivative of the state variable.

On adding the signals coming to the 1st Integrator

We get, (refer Fig. Ex. 5.18.3)

$$\dot{X}_1 = -4X_4 + 3u_1$$

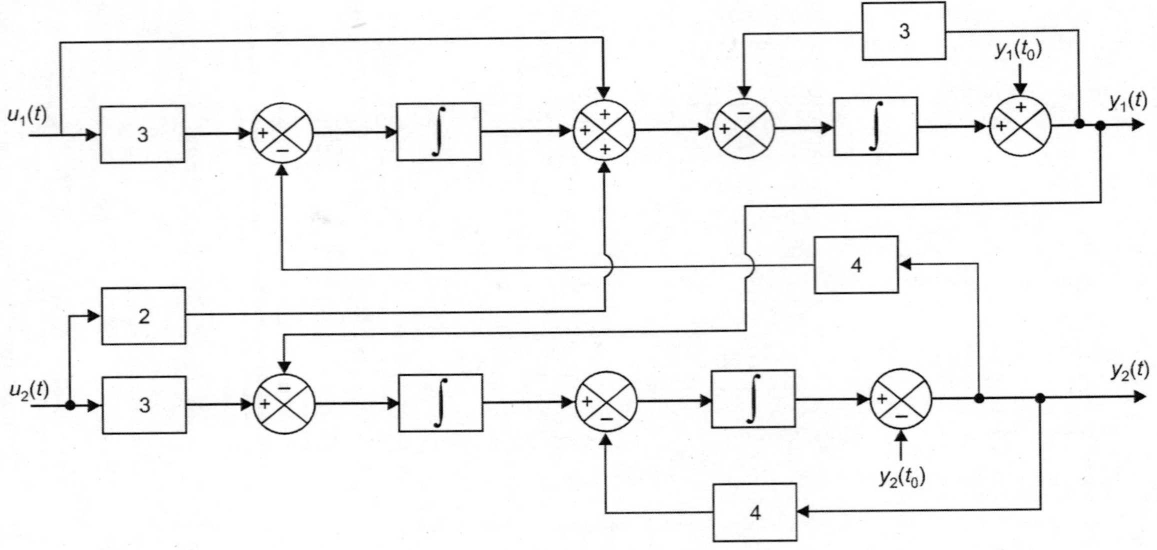

Fig. Ex. 5.18.1

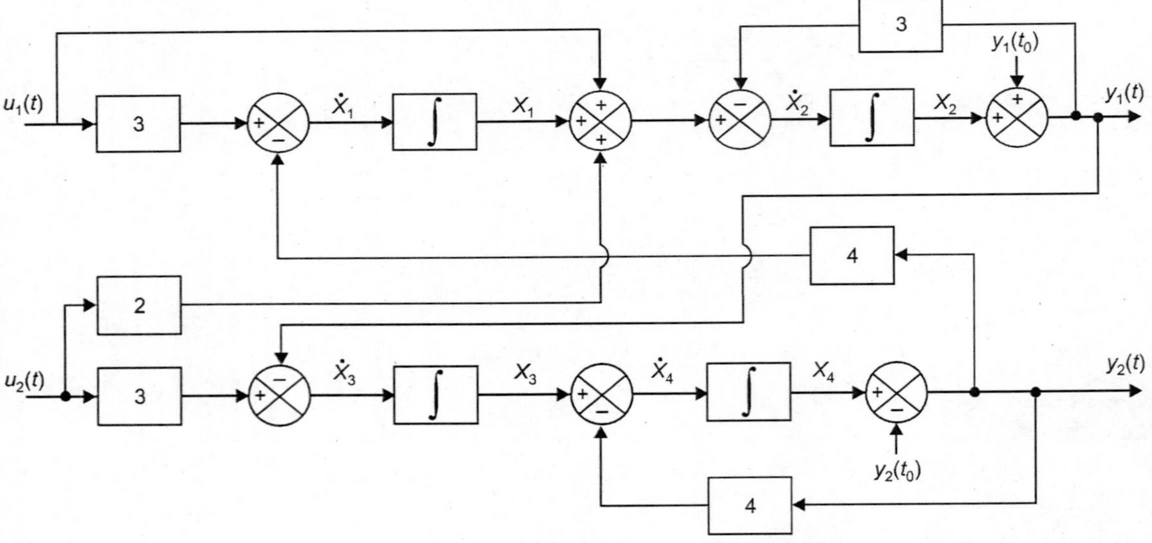

Fig. Ex. 5.18.2

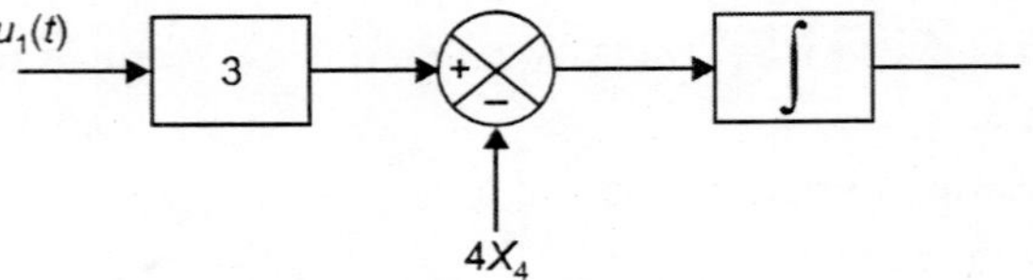

Fig. Ex. 5.18.3

On adding the signals coming to the 2nd integrator we get, (refer Fig Ex. 5.18.4)

$$\dot{X}_2 = X_1 - 3X_2 + u_1 + 2u_2$$

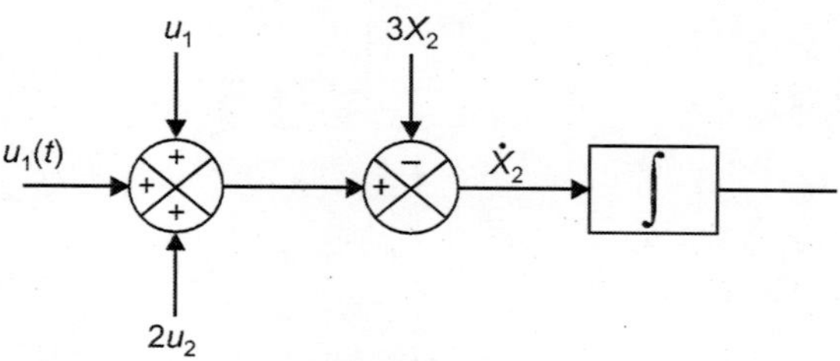

Fig. Ex. 5.18.4

On adding the signals coming to the 3rd integrator we get, (refer Fig. Ex. 5.18.5)

$$\dot{X}_3 = -X_2 + 3u_2$$

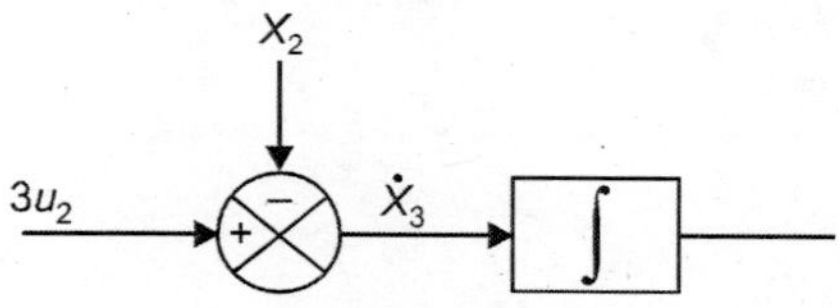

Fig. Ex. 5.18.5

On adding the signals coming to the 4th integrator we get, (refer Fig. Ex. 5.18.6)

$$\dot{X}_4 = X_3 - 4X_4$$

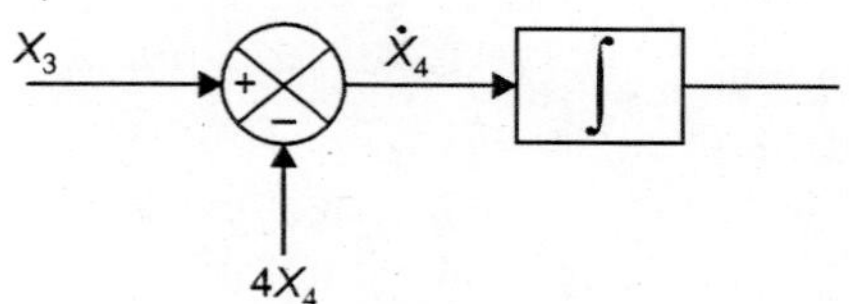

Fig. Ex. 5.18.6

The state equations are,

$$\dot{X}_1 = -4X_4 + 3u_1$$

$$\dot{X}_2 = X_1 - 3X_2 + u_1 + 2u_2$$

$$\dot{X}_3 = -X_2 + 3u_2$$

$$\dot{X}_4 = X_3 - 4X_4$$

The output equations are, $y_1 = X_2$ and $y_2 = X_4$.

Result: The state equations and output equations are arranged in the matrix form as shown below. The state equations and output equations together constitute the state model of the system.

$$\begin{bmatrix} \dot{X}_1 \\ \dot{X}_2 \\ \dot{X}_3 \\ \dot{X}_4 \end{bmatrix} = \begin{bmatrix} 0 & 0 & 0 & -4 \\ 1 & -3 & 0 & 0 \\ 0 & -1 & 0 & 0 \\ 0 & 0 & 1 & -4 \end{bmatrix} \begin{bmatrix} X_1 \\ X_2 \\ X_3 \\ X_4 \end{bmatrix} + \begin{bmatrix} 3 & 0 \\ 1 & 2 \\ 0 & 3 \\ 0 & 0 \end{bmatrix} \begin{bmatrix} u_1 \\ u_2 \end{bmatrix}$$

$$\begin{bmatrix} Y_1 \\ Y_2 \end{bmatrix} = \begin{bmatrix} 0 & 1 & 0 & 0 \\ 0 & 0 & 0 & 1 \end{bmatrix} \begin{bmatrix} X_1 \\ X_2 \\ X_3 \\ X_4 \end{bmatrix}$$

EXAMPLE 5.9

The state diagram of a linear system is given below. Assign state variables and obtain the state model.

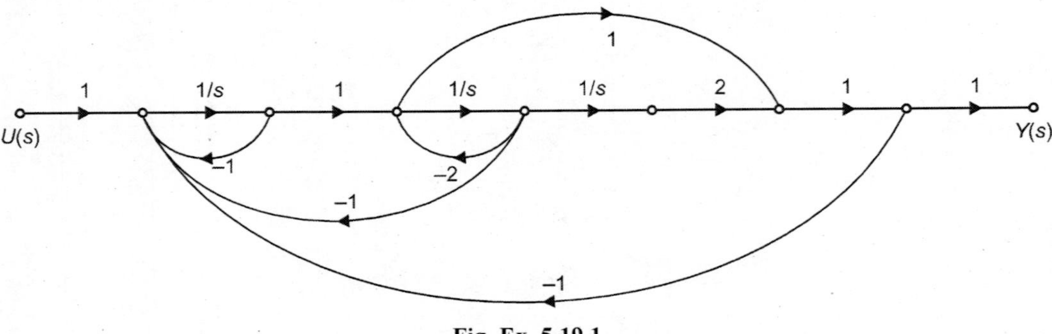

Fig. Ex. 5.19.1

Solution: Since there are three integrators ($1/s$) we can assign three state variables. The state variables are assigned at the output of the integrator as shown in (Fig. Ex. 5.19.2). At the input of the integrator we have the first derivative of the state variable. The state equations are formed by summing all the signals at the input of integrator and equating to the corresponding first derivatives of state variable.

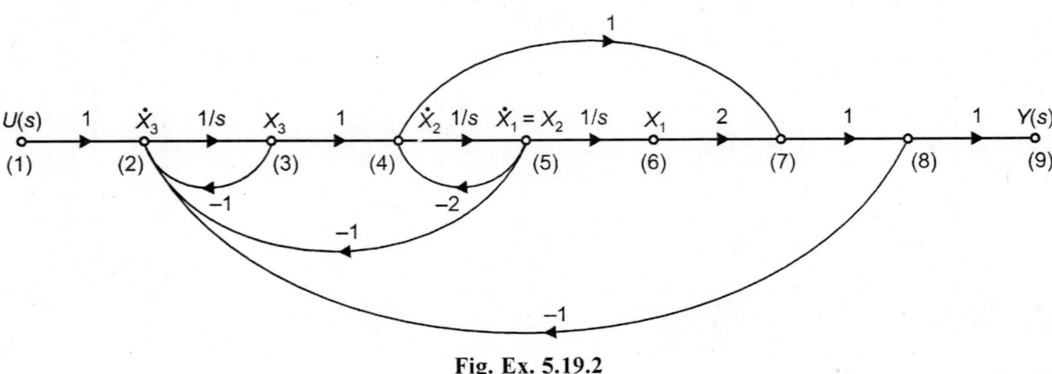

Fig. Ex. 5.19.2

On adding the signals coming to node -5, (refer Fig. 5.19.3)

$$\dot{X}_1 = X_2$$

Fig. Ex. 5.19.3

On adding the signals coming to node-4, (refer Fig. Ex. 5.19.4)

$$\dot{X}_2 = -2X_2 + X_3$$

Fig. Ex. 5.19.4

On adding the signals coming to node –2, we get (Fig. Ex. 5.19.5)

$$\dot{X}_3 = -\left(\dot{X}_2 + 2X_1\right) - X_2 - X_3 + u = -\dot{X}_2 - 2X_1 - X_2 - X_3 + u$$

Put $$\dot{X}_2 = -2X_2 + X_3$$

$$\therefore \quad \dot{X}_3 = 2X_2 - X_3 - 2X_1 - X_2 - X_3 + u = -2X_1 + X_2 - 2X_3 + u$$

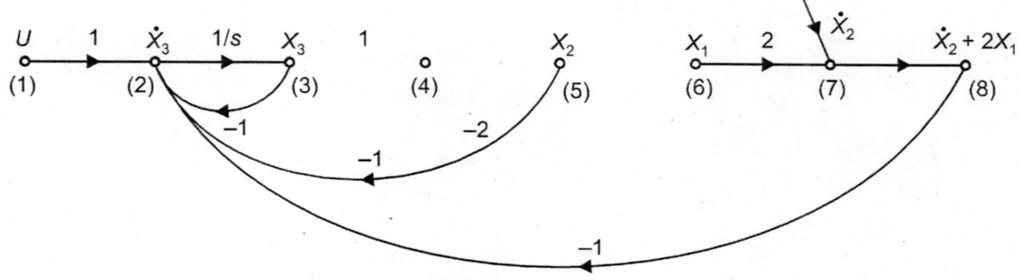

Fig. Ex. 5.19.5

The state equations are,

$$\dot{X}_1 = X_2$$

$$\dot{X}_2 = -2X_2 + X_3$$

$$\dot{X}_3 = -2X_1 + X_2 - 2X_3 + u$$

The output equation is obtained by adding the signals coming to output made (refer Fig. Ex. 5.19.6)

$$y = 2X_1 + \dot{X}_2$$

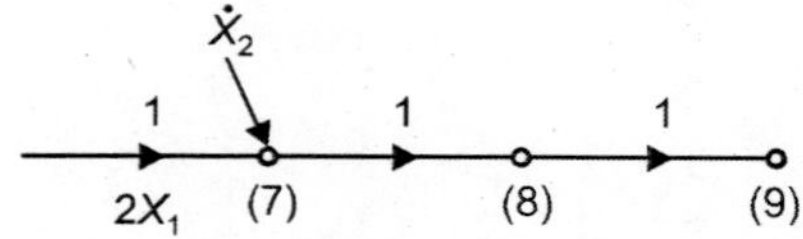

Fig. Ex. 5.19.6

CHAPTER 5

Put
$$\dot{X}_2 = -2X_2 + X_3$$
$$y = 2X_1 + (-2X_2 + X_3)$$
$$y = 2X_1 - 2X_2 + X_3$$

Result: The state equations and the output equations are arranged in the matrix form as shown below:

$$\begin{bmatrix} \dot{X}_1 \\ \dot{X}_2 \\ \dot{X}_3 \end{bmatrix} = \begin{bmatrix} 0 & 1 & 0 \\ 0 & -2 & 1 \\ -2 & 1 & -2 \end{bmatrix} \begin{bmatrix} X_1 \\ X_2 \\ X_3 \end{bmatrix} + \begin{bmatrix} 0 \\ 0 \\ 1 \end{bmatrix} [u]$$

$$y = \begin{bmatrix} 2 & -2 & 1 \end{bmatrix} \begin{bmatrix} X_1 \\ X_2 \\ X_3 \end{bmatrix}$$

EXAMPLE 5.20

Obtain the state model of the system whose transfer function is given as, $\frac{Y(s)}{U(s)} = \frac{10}{s^3 + 4s^2 + 2s + 1}$.

Solution:

Method 1: Given that,
$$\frac{Y(s)}{U(s)} = \frac{10}{s^3 + 4s^2 + 2s + 1} \qquad (5.10.1)$$

$$Y(s)\,[s^3 + 4s^2 + 2s + 1] = U(s).10$$
$$s^3Y(s) + 4ss^2\,Y(s) + 2s\,Y(s) + Y(s) = 10.U(s) \qquad (5.10.2)$$

On taking inverse laplace transform of equation (5.10.2) we get,
$$\dddot{y} + 4\ddot{y} + 2\dot{y} + y = 10u \qquad (5.10.3)$$

Let us define state variables as follows:

$$X_1 = y;\ X_2 = \dot{y};\ X_3 = \ddot{y}$$

Put $\dddot{y} = \dot{X}_3$; $\ddot{y} = X_3$; $\dot{y} = X_2$ and $y = X_1$ in equation (5.10.3)

$\therefore$
$$\dot{X}_3 + 4X_3 + 2X_2 + X_1 = 10u$$

or
$$\dot{X}_3 = -X_1 - 2X_2 - 4X_3 + 10u$$

The state equations are,

$\dot{X}_1 = X_2;\ \dot{X}_2 = X_3;\ \dot{X}_3 = -X_1 - 2X_2 - 4X_3 + 10u$

The output equation is $y = X_1$

The state model in the matrix form is,

$$\begin{bmatrix} \dot{X}_1 \\ \dot{X}_2 \\ \dot{X}_3 \end{bmatrix} = \begin{bmatrix} 0 & 1 & 0 \\ 0 & 0 & 1 \\ -1 & -2 & -4 \end{bmatrix} \begin{bmatrix} X_1 \\ X_2 \\ X_3 \end{bmatrix} + \begin{bmatrix} 0 \\ 0 \\ 10 \end{bmatrix} [u]$$

$$Y = \begin{bmatrix} 1 & 0 & 0 \end{bmatrix} \begin{bmatrix} X_1 \\ X_2 \\ X_3 \end{bmatrix}$$

Method 2:

$$\frac{Y(s)}{U(s)} = \frac{10}{s^3 + 4s^2 + 2s + 1} = \frac{10}{s^3\left(1 + \frac{4}{5} + \frac{2}{s^2} + \frac{1}{s^3}\right)} = \frac{\frac{10}{s^3}}{1 - \left(-\frac{4}{5} - \frac{2}{s^2} - \frac{1}{s^3}\right)}$$

The signal flow graph for the above transfer function can be constructed as shown in Fig. (5.10.1). This signal flow graph has a single forward path consisting of three integrators and with path gain $\frac{10}{s^3}$. The graph will have three individual loops with loop gains $\frac{-4}{s}, \frac{-2}{s^2}$ and $\frac{-1}{s^3}$

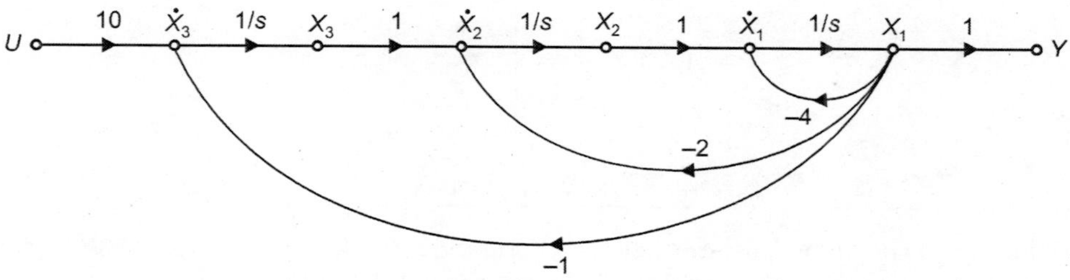

Fig. Ex. 5.20.1

Assign state variables t the output of the integrator $\left(\frac{1}{s}\right)$. The state equations are obtained by summing the incoming signals to the input of the integrators and equating them to the corresponding first derivative of the state variable refer Fig. (Ex. 5.20.2, 5.10.3 and 5.10.4)

The state equations are,

$$\dot{X}_1 = -4X_1 + X_2$$

Fig. Ex. 5.10.2

$$\dot{X}_2 = -2X_1 + X_3$$

Fig. Ex. 5.10.3

CHAPTER 5

$$\dot{X}_3 = -X_1 + 10u$$

10u
$\dot{X}_3$
$-X_1$

Fig. Ex. 5.10.4

The output equation is, $y = X_1$

The state model in the matrix form is,

$$\begin{bmatrix} \dot{X}_1 \\ \dot{X}_2 \\ \dot{X}_3 \end{bmatrix} = \begin{bmatrix} -4 & 1 & 0 \\ -2 & 0 & 1 \\ -1 & 0 & 0 \end{bmatrix} \begin{bmatrix} X_1 \\ X_2 \\ X_3 \end{bmatrix} + \begin{bmatrix} 0 \\ 0 \\ 10 \end{bmatrix} [u];$$

$$y = \begin{bmatrix} 1 & 0 & 0 \end{bmatrix} \begin{bmatrix} X_1 \\ X_2 \\ X_3 \end{bmatrix}$$

5.14 STATE SPACE REPRESENTATION USING CANONICAL VARIABLES

The Canonical form (or normal form) of state model, the system matrix A will be a diagonal matrix. The elements on the diagonal are the poles of the transfer function of the system.

By partial fraction expansion, the transfer function $Y(s) / U(s)$ of the nth order system can be expressed or in equation (5.40).

$$\frac{Y(s)}{U(s)} = b_O + \frac{C_1}{b + \lambda_1} + \frac{C_2}{s + \lambda_2} + \ldots + \frac{C_n}{s + \lambda_n} \tag{5.40}$$

where, $C_1, C_2, C_3 \ldots C_n$ - Residues

$\lambda_1, \lambda_2, \lambda_3, \ldots. \lambda_n$ - Roots of denominator polynomial [(or) poles of the system]

The equation (5.40) can be rearranged as shown below:

$$\frac{Y(s)}{U(s)} = b_o + \frac{C_1}{s\left(1 + \frac{\lambda_1}{s}\right)} + \frac{C_2}{s\left(1 + \frac{\lambda_2}{s}\right)} + \ldots + \frac{C_n}{s\left(1 - \frac{\lambda_n}{s}\right)} = b_o + \frac{\frac{C_1}{s}}{1 + \frac{\lambda_1}{s}} + \frac{\frac{C_2}{s}}{1 + \frac{\lambda_2}{s}} + \ldots + \frac{\frac{C_n}{s}}{1 + \frac{\lambda_n}{s}}$$

$$\therefore \quad Y(s) = b_o u(s) \left[\frac{(1/s)}{1 + (1/s) \times \lambda_1} \times C_1 \right] U(s) + \left[\frac{(1/s)}{1 + (1/s) \times \lambda_2} \times C_2 \right] U(s) + \ldots$$

$$+ \left[\frac{(1/s)}{1 + (1/s) \times \lambda_n} \times C_n \right] U(s) \tag{5.41}$$

The equation (5.41) can be represented by a block diagram as shown in Fig. (5.5)

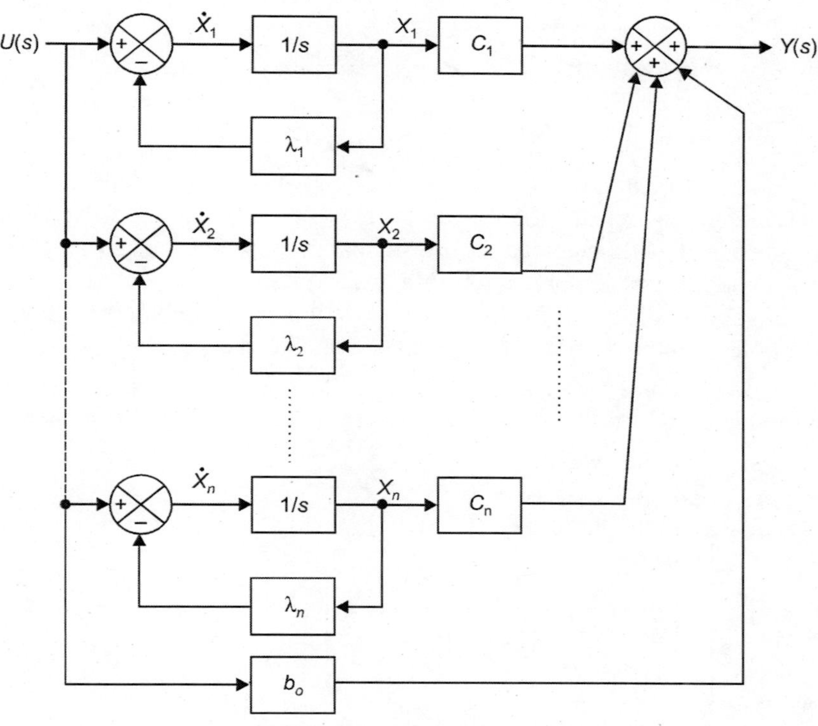

Fig. 5.5: Block diagram of canonical state model.

Assign state variables at the output of integrator. The input of the integrator will be first derivative of state variable. The state equations are formed by adding the incoming signals to the integrator and equating of first derivative of state variable. The state equations are,

$$\dot{X}_1 = -\lambda_1 X_1 + u$$

$$\dot{X}_2 = -\lambda_2 X_2 + u$$

$$\vdots$$

$$\dot{X}_n = -\lambda_n X_n + u$$

The output equation is, $y = C_1 X_1 + C_2 X_2 + \dots C_n X_n + b_o u$.

The canonical form of state model in the matrix form is given below:

$$\begin{bmatrix} \dot{X}_1 \\ \dot{X}_2 \\ \dot{X}_3 \\ \vdots \\ \dot{X}_n \end{bmatrix} = \begin{bmatrix} -\lambda_1 & 0 & 0 & \cdots & 0 \\ 0 & -\lambda_2 & 0 & \cdots & 0 \\ 0 & 0 & -\lambda_3 & \cdots & 0 \\ \vdots & \vdots & \vdots & & \vdots \\ 0 & 0 & 0 & 0 & -\lambda_n \end{bmatrix} \begin{bmatrix} X_1 \\ X_2 \\ X_3 \\ \vdots \\ X_n \end{bmatrix} \begin{bmatrix} 1 \\ 1 \\ 1 \\ \vdots \\ 1 \end{bmatrix} [u] \tag{5.42}$$

$$Y = \begin{bmatrix} C_1 & C_2 & C_3 & \cdots & C_n \end{bmatrix} \begin{bmatrix} X_1 \\ X_2 \\ X_3 \\ \vdots \\ X_n \end{bmatrix} + [b_o][u] \qquad (5.43)$$

The advantage of canonical form is that the state equations are independent of each other. The disadvantage is that the canonical variables are not physical variables and so they are not available for measurement and control. when a pole of the transfer has multiplicity, the canonical state model will be in a special form called Jordan Canonical form. In this form the system matrix A will have a Jordan block size $q \times q$, correspond to a pole of value. λ_i with multiplicity q. In the Jordan block the diagonal element will be the poles and the element just above the diagonal is one.

Consider a system with poles $\lambda_1, \lambda_2, \lambda_3, \lambda_4, \lambda_5, ...\lambda_n$ where λ_1 has multiplicity of three. The input matrix (B) and system matrix for this case will be as shown in equation (5.41)(a)

The system matrix is also denoted as J

Jordan block of size 3 × 3

$$B = \begin{bmatrix} 0 \\ 0 \\ 1 \\ 1 \\ \vdots \\ 1 \end{bmatrix} ; \; A = J = \begin{bmatrix} -\lambda_1 & 1 & 0 & 0 & \cdots & 0 \\ 0 & -\lambda_1 & 1 & 0 & \cdots & 0 \\ 0 & 0 & -\lambda_1 & 0 & \cdots & 0 \\ 0 & 0 & 0 & \lambda_4 & \cdots & 0 \\ \vdots & \vdots & \vdots & \vdots & & \\ 0 & 0 & 0 & 0 & \cdots & -\lambda_n \end{bmatrix} \qquad (5.41)\,(a)$$

The transfer function of the system for this case is given by equation (5.40a) and the block diagram is shown in Fig. 5.5(a). See Fig. [5.5(a)] in page no. 645.

$$\frac{Y(s)}{U(s)} = b_o + \frac{C_1}{(s+\lambda_1)^3} + \frac{C_2}{(s+\lambda_1)^2} + \frac{C_3}{(s-\lambda_1)} + \frac{C_n}{(s+\lambda_4)} + .. + \frac{C_n}{s+\lambda_n} \qquad (5.40a)$$

EXAMPLE 5.21

A feedback system has a closed-loop transfer function $\dfrac{Y(s)}{U(s)} = \dfrac{10(s+4)}{s(s+1)(s+3)}$.

Construct three different state models for this system and give block diagram representation for each state model.

Solution: Model 1:

$$\frac{Y(s)}{U(s)} = \frac{10(s+4)}{s(s+1)(s+3)} = \frac{10s+40}{s(s^2+4s+3)} = \frac{10s+40}{s^3+4s^2+3s} = \frac{10s+40}{s^3\left(1+\frac{4}{s}+\frac{3}{s^2}\right)}$$

$$\frac{Y(s)}{U(s)} = \frac{\frac{10}{s^2}+\frac{40}{s^3}}{1-\left[-\frac{4}{s}-\frac{3}{s^2}\right]}$$

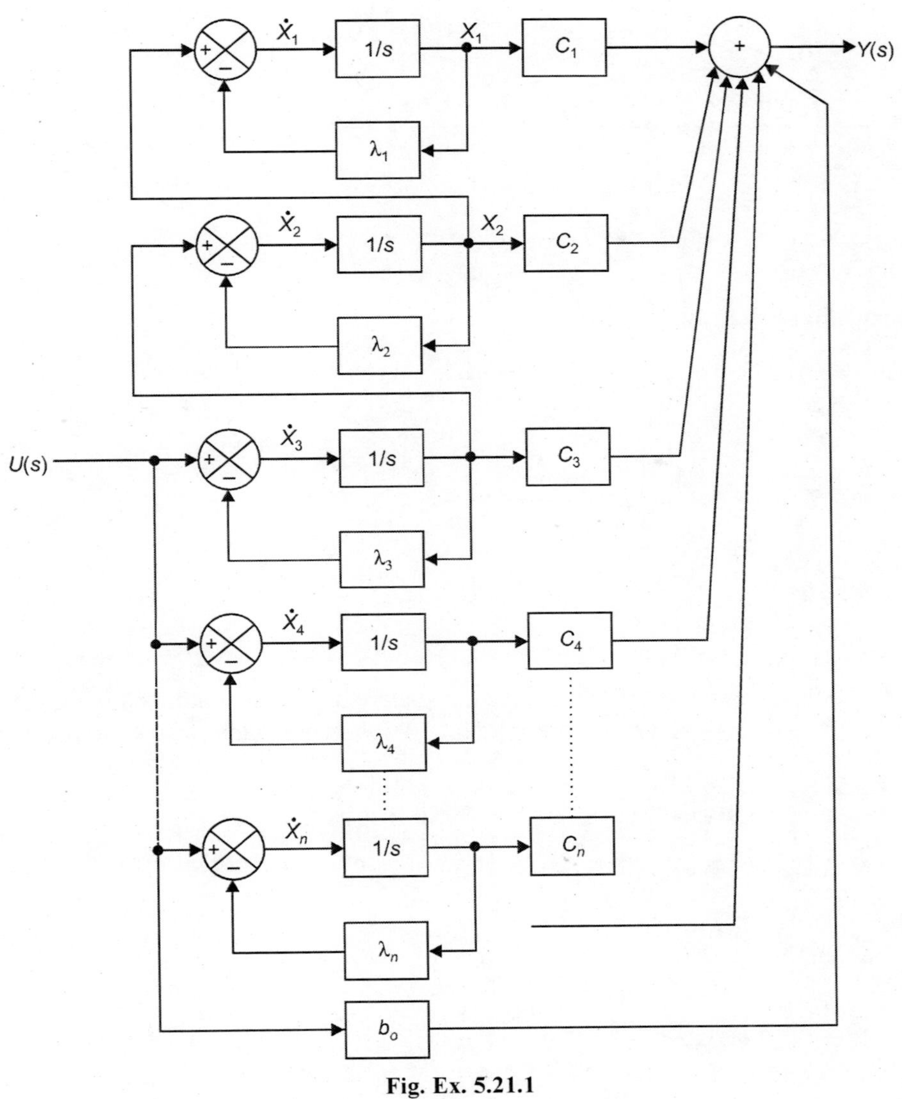

Fig. Ex. 5.21.1

A signal flow graph for the above transfer function can be constructed as shown Fig. (Ex. 5.21.1), with two forward paths and two individual loops. The forward paht gains are $\frac{10}{s^2}$ and $\frac{40}{s^3}$. The loop gains are $\frac{-4}{s}$ and $\frac{-3}{s^2}$.

Assign state variables at the output of integrator as shown in Fig. (5.11.1) an so the input of integrator is first derivative of state variable. The state equations are obtained by summing all the incoming signals to the integrator and equating to the corresponding first derivative of the state variable [refer Fig. Ex. 5.21.2 to 5.21.4.].

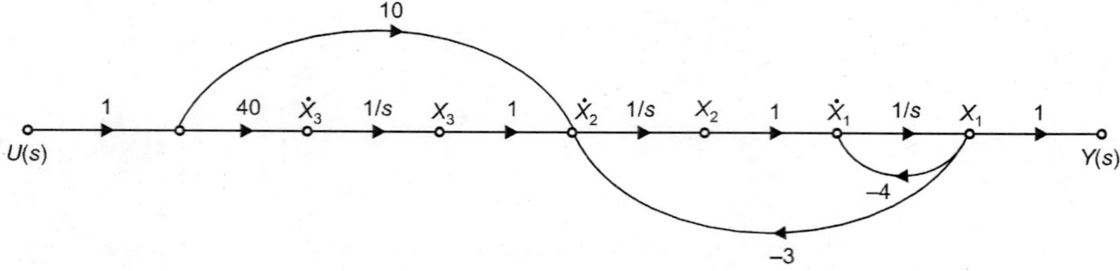

Fig. 5.21.1

The state equations are,

$$\dot{X}_1 = -4X_1 + X_2 \qquad \dot{X}_2 = -3X_1 + X_3 + 10u \qquad \dot{X}_3 = 40u$$

Fig. 5.21.2 Fig. 5.21.3 Fig. 5.21.4

The Output equation is, $y = X_1$

The state model is obtained by arranging the state equations and the output equation in the matrix form as shown below. The both diagram representation of this state model is shown in Fig. (Ex. 5.21.5)

$$\begin{bmatrix} \dot{X}_1 \\ \dot{X}_2 \\ \dot{X}_3 \end{bmatrix} = \begin{bmatrix} -4 & 1 & 0 \\ -3 & 0 & 1 \\ 0 & 0 & 0 \end{bmatrix} \begin{bmatrix} X_1 \\ X_2 \\ X_3 \end{bmatrix} + \begin{bmatrix} 0 \\ 10 \\ 40 \end{bmatrix} [u]$$

$$Y = \begin{bmatrix} 1 & 0 & 0 \end{bmatrix} \begin{bmatrix} X_1 \\ X_2 \\ X_3 \end{bmatrix}$$

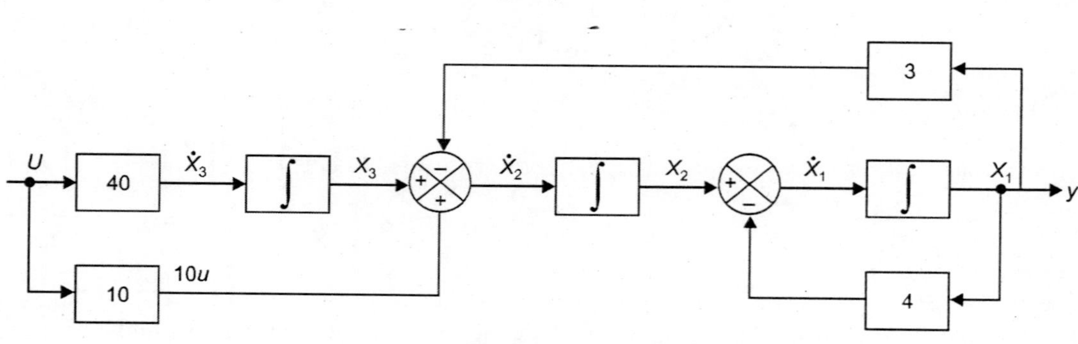

Fig. Ex. 5.21.5

Model 2: Given that,

$$\frac{Y(s)}{U(s)} = \frac{10(s+4)}{s(s+1)(s+3)}$$

Let, $$\frac{Y(s)}{U(s)} = \frac{X_1(s)}{U(s)} \cdot \frac{Y(s)}{X_1(s)} = \frac{10(s+4)}{s(s+1)(s+3)}$$

Let, $$\frac{X_1(s)}{u(s)} = \frac{1}{s(s+1)(s+3)}$$

and $$\frac{Y(s)}{X_1(s)} = 10(s+4)$$

$$\frac{X(s)}{u(s)} = \frac{1}{s(s+1)(s+3)} = \frac{1}{s(s^2+4s+3)} \quad (5.11.1)$$

$$\therefore \quad \frac{X_1(s)}{U(s)} = \frac{1}{s(s+1)(s+3)} = \frac{1}{s(s^2+4s+3)}; \frac{X_1(s)}{U(s)} = \frac{1}{s^3+4s^2+3s}$$

On cross multiplying the equation (5.11.1) we get,

$$X_1(s)\left[s^3+4s^2+3s\right] = U(s)$$

$$\therefore \quad s^3 X_1(s) + 4s^2 X_1(s) + 3s\, X_1(s) = U(s) \quad (5.11.2)$$

On taking inverse laplace transform of equation (5.11.2) we get,

$$\dddot{X}_1 + 4\ddot{X}_1 + 3\dot{X}_1 = u \quad (5.11.3)$$

Let the state variables be X_1, X_2 and X_3 ;

Where $X_2 = \dot{X}_1$ and $X_3 = \ddot{X}_1$

Put $\dot{X}_1 = X_2$; $\ddot{X}_1 = X_3$ and $\dddot{X}_1 = \dot{X}_3$ in equation (5.11.3)

$$\therefore \quad \dot{X}_3 + 4X_3 + 3X_2 = u$$

or $$\dot{X}_3 = 3X_2 - 4X_3 + u$$

The state equations are, $$\dot{X}_1 = X_2$$

$$\dot{X}_2 = X_3$$

$$\dot{X}_3 = -3X_2 - 4X_3 + u$$

Consider the second part of transfer function,

$$\frac{Y(s)}{X_1(s)} = 10(s+4) = 10s + 40 \quad (5.11.4)$$

On cross multiplying equation (5.11.4) we get,

$$Y(s) = 10s\, X_1(s) + 40\, X_1(s) \quad (5.11.5)$$

On taking inverse laplace transform of equation (5.11.5) we get,

$$y = 10\dot{X}_1 + 40X_1$$

Put $$\dot{X}_1 = X_2$$

$\therefore$ $$y = 10X_2 + 40\,X_1 = 40X_1 + 10X_2$$

Here, $y = 40\,X_1 + 10\,X_2$ is the output equation. The state model in the matrix form is shown below. The block diagram representation of this state model is shown in Fig. (Ex. 5.21.6)

$$\begin{bmatrix}\dot{X}_1\\ \dot{X}_2\\ \dot{X}_3\end{bmatrix} = \begin{bmatrix}0 & 1 & 0\\ 0 & 0 & 1\\ 0 & -3 & -4\end{bmatrix}\begin{bmatrix}X_1\\ X_2\\ X_3\end{bmatrix} + \begin{bmatrix}0\\ 0\\ 1\end{bmatrix}[u];$$

$$y = \begin{bmatrix}40 & 10 & 0\end{bmatrix}\begin{bmatrix}X_1\\ X_2\\ X_3\end{bmatrix}$$

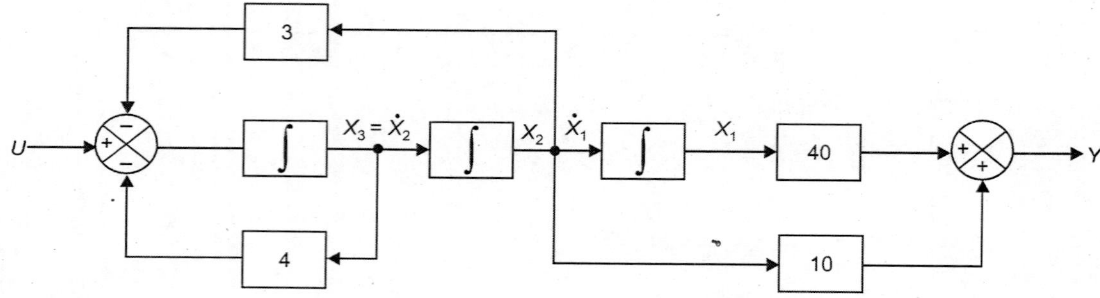

Fig. Ex. 5.21.6

Model 3: $$\frac{Y(s)}{U(s)} = \frac{10(s+4))}{s(s+1)(s+3)}$$

By applying partial fraction

$$\frac{Y(s)}{U(s)} = \frac{10(s+4))}{s(s+1)(s+3)} = \frac{A}{s} + \frac{B}{s+1} + \frac{C}{s+3}$$

$$A = \left.\frac{10(s+4)}{(s+1)(s+3)}\right|_{s=0} = \frac{10\times 4}{1\times 3} = \frac{40}{3}$$

$$B = \left.\frac{10(s+4)}{s(s+3)}\right|_{s=-1} = \frac{10(-1+4)}{-1(-1+3)} = \frac{10\times 3}{-1\times 2} = -15$$

$$C = \left.\frac{10(s+4)}{s(s+1)}\right|_{s=-3} = \frac{10(-3+4)}{-3(-3+1)} = \frac{10\times 1}{-3\times(-2)} = \frac{5}{3}$$

$$\frac{Y(s)}{U(s)} = \frac{\frac{40}{3}}{s} - \frac{15}{s+1} + \frac{\frac{5}{3}}{s+3} \qquad (5.11.6)$$

Equation (5.11.6) can be rearranged as shown below:

$$\frac{Y(s)}{U(s)} = \frac{40/3}{s} - \frac{15}{s\left(1+\frac{1}{s}\right)} + \frac{5/3}{s\left(1+\frac{3}{s}\right)}$$

$$\therefore \quad Y(s) = \left[\frac{1}{s} \times \frac{40}{3}\right] U(s) - \left[\frac{\frac{1}{s}}{1 - \frac{1}{s} \times 1} \times 15\right] U(s) + \left[\frac{\frac{1}{s}}{1 + \frac{1}{s} \times 3} \times \frac{5}{3}\right] U(s) \qquad (5.11.7)$$

The block diagram of the equation (5.11.7) is shown in Fig (Ex. 5.21.7)

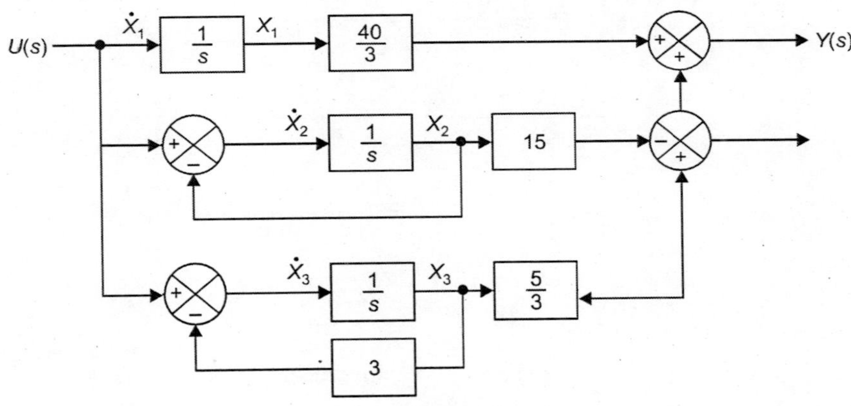

Fig. Ex. 5.21.7

Assign state variables at the output of the integrator as shown in Fig. (Ex. 5.21.7). At the input of the integrator, the first derivative of the state variables will be available. The state equations are obtained by adding in coming signals to the integrator and equating to the corresponding first derivative of the state variable.

$$\dot{X}_1 = u$$

$$\dot{X}_2 = -X_2 + u$$

$$\dot{X}_3 = -3X_3 + u$$

The output equation is $\quad y = \frac{40}{3}X_1 - 15X_2 + \frac{5}{3}X_3$

The state model is the matrix form is shown below.

The Fig (Ex. 5.21.7) is the block diagram representation of this state model.

$$\begin{bmatrix}\dot{X}_1\\ \dot{X}_2\\ \dot{X}_3\end{bmatrix} = \begin{bmatrix}0 & 0 & 0\\ 0 & -1 & 0\\ 0 & 0 & -3\end{bmatrix}\begin{bmatrix}X_1\\ X_2\\ X_3\end{bmatrix} + \begin{bmatrix}1\\ 1\\ 1\end{bmatrix}[u];$$

$$y = \begin{bmatrix}\frac{40}{3} & -15 & \frac{5}{3}\end{bmatrix}\begin{bmatrix}X_1\\ X_2\\ X_3\end{bmatrix}$$

EXAMPLE 5.22

Determine the Canonical state model of the system, whose transfer function is $T(s) = 2(s + 5)/[(s + 2)(s + 3)(s + 4)$.

Solution:

$$\frac{Y(s)}{U(s)} = \frac{2(s+5)}{(s+2)(s+3)(s+4)} = \frac{A}{s+2} + \frac{B}{s+3} + \frac{c}{s+4}$$

$$A = \left.\frac{2(s+5)}{(s+3)(s+4)}\right|_{s=-2} = \frac{2(-2+5)}{(-2+3)(-2+4)} = \frac{2\times 3}{1\times 2} = 3$$

$$B = \left.\frac{2(s+5)}{(s+2)(s+4)}\right|_{s=-3} = \frac{2(-3+5)}{(-3+2)(-3+4)} = \frac{2\times 2}{-1\times 1} = -4$$

$$C = \left.\frac{2(s+5)}{(s+2)(s+3)}\right|_{s=-4} = \frac{2(-4+5)}{(-4+2)(-4+3)} = \frac{2\times 1}{-2\times(-1)} = 1$$

$$\therefore \quad \frac{Y(s)}{U(s)} = \frac{3}{s+2} - \frac{4}{s+3} + \frac{1}{s+4} \tag{5.12.1}$$

The equation (5.12.1) can be rearranged as shown below:

$$\frac{Y(s)}{U(s)} = \frac{3}{s\left(1+\frac{2}{s}\right)} - \frac{4}{s\left(1+\frac{3}{s}\right)} + \frac{1}{s\left(1+\frac{4}{s}\right)}$$

$$\therefore \quad Y(s) = \left[\frac{\frac{1}{s}}{1+\frac{1}{s}\times 2}\times 3\right]U(s) - \left[\frac{\frac{1}{s}}{1+\frac{1}{s}\times 3}\times 4\right]U(s) + \left[\frac{\frac{1}{s}}{1+\frac{1}{s}\times 4}\times 4\right]U(s) \tag{5.12.2}$$

The equation (5.12.2) can be represented by the block diagram in Fig. (Ex. 5.22.1)

Assign state variables at the output of the integrators as shown in Fig (Ex. 5.22.1). At the input of the integrators we have first derivative of the state variables. The state equations are formed by adding all the incoming signals to the integrator and equating to the corresponding first derivative of state variable.

The state equations are,

$$\dot{X}_1 = -2X_1 + u$$

$$\dot{X}_2 = -3X_2 + u$$

$$\dot{X}_3 = -4X_3 + u$$

The output equation is, $y = 3X_1 - 4X_2 + X_3$

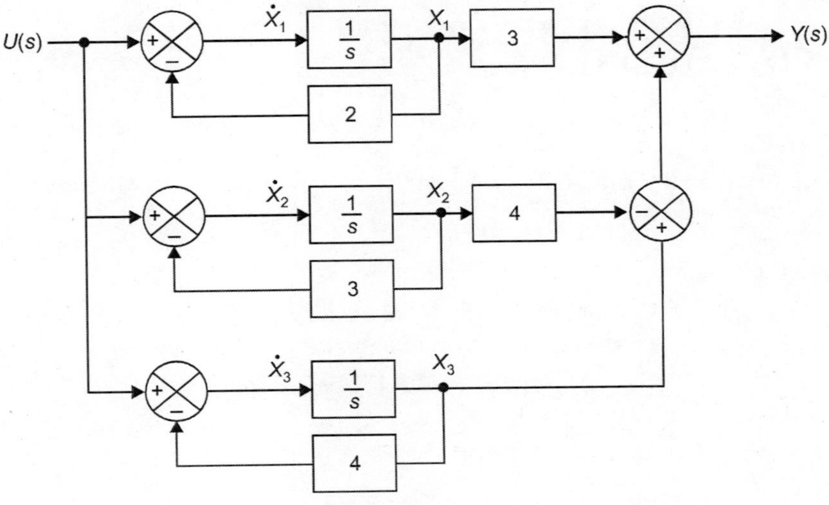

Fig. Ex. 5.22.1

The state model in matrix form is given below:

$$\begin{bmatrix} \dot{X}_1 \\ \dot{X}_2 \\ \dot{X}_3 \end{bmatrix} = \begin{bmatrix} -2 & 0 & 0 \\ 0 & -3 & 0 \\ 0 & 0 & -4 \end{bmatrix} \begin{bmatrix} X_1 \\ X_2 \\ X_3 \end{bmatrix} + \begin{bmatrix} 1 \\ 1 \\ 1 \end{bmatrix} [u];$$

$$Y = \begin{bmatrix} 3 & -4 & 1 \end{bmatrix} \begin{bmatrix} X_1 \\ X_2 \\ X_3 \end{bmatrix}$$

5.15 SOLUTION OF STATE EQUATIONS

Solution of Homogeneous State Equations (solution of state equations without input (or) excitation

Consider a first order differential equation, with initial condition, $X(o) = X_o$

$$\frac{dX}{dt} = ax; \quad X(o) = X_o \tag{5.43}$$

$$\frac{dX}{X} = a \cdot dt \text{ (From equation 5.4.3)} \tag{5.44}$$

integrating equation (5.44) we get,

$$\log X = at + C$$

$$\therefore \quad X = e^{(at+c)} = e^{at} . e^{c} \tag{5.45}$$

CHAPTER 5

We get the solution of first order differential equation as

$$X = e^{at} X_o \tag{5.46}$$

We know that,
$$e^X = \left[1 + X + \frac{1}{2!}X^2 + ..\frac{1}{n!}X^n + ..\right] \tag{5.47}$$

From equation (5.46) and (5.47) we get,

$$X = e^{at} X_O = \left(1 + at + \frac{1}{2!}a^2t^2 + \frac{1}{3!}a^3t^3 + .. + \frac{1}{i!}a^it^i + ..\right) X_o \tag{5.48}$$

Consider the state equations without input Vector, (i.e., homogeneous state equation)

$$\dot{X}(t) = A\,X(t)\,;\,X(o) = X_o \tag{5.49}$$

where $x(o)$ is the initial condition vector.

By analogy of the solution of first order differential equation [equation (5.48)], the solution of the matrix or vector equation can be assumed as shown in equation (5.50).

$$X(t) = A_o + A_1 t + A_2\, t^2 + A_3\, t^3 + ... + A_i\, t_i + ... \tag{5.50}$$

where A_o, A_1, A_2... A_t... are matrices and the elements of the matrices are constants. Differentiating equation (5.50) . We get,

$$\dot{X}(t) = A_1 + 2A_2 t + 3A_3 t^2 + ... + iA;\, t^{i-1} + ... \tag{5.51}$$

On multiplying equation (5.50) by A, we get,

$$A\,X(t) = A\left[A_o + A_1 t + A_2 t^2 + A_3 t_o^3 + ... + A_i t_i^i + ..\right] \tag{5.52}$$

From equation (5.49), we know that $\dot{X}(t) = A\,X(t)$.

Therefore we can equate the coefficient of equal powers of 't' in equation (5.51) and (5.52) as shown below:

On equating constant we get,

$$A_1 = A\,A_o$$

On equating coefficients of 't' we get,

$$2A_2 = AA_1$$

$$\therefore \quad A_2 = \frac{1}{2}AA_1$$

Put $A_1 = AA_o$

$$\therefore \quad A_2 = \frac{1}{2}A.A.A_o$$

$$A_2 = \frac{1}{2} \cdot A^2 A_o$$

On equating coefficients of t^2 we get,

$$3A_3 = AA_2$$

$$\therefore \quad A_3 = \frac{1}{3}AA_2$$

Put $A_2 = \frac{1}{2} \cdot A^2 A_o$

$$\therefore \quad A_3 = \frac{1}{3} A \text{Ч} \frac{1}{2} A^2 Ao$$

$$A_3 = \frac{1}{3!} A^3 A_o$$

Similarly, on equating coefficient of t^i we get,

$$A_i = \frac{1}{i!} A^i A_o$$

In the above analysis, the matrices A_1, A_2, A_3, etc., are expressed in terms of A and A_o. Hence replace the matrices $A_1, A_2, A_3 ... A_i$ in the assumed solution of $X(t)$ [i.e. equation (4.50)] by the equivalent terms obtained above..

$$\therefore \quad X(t) = A_o + AA_o t + \frac{1}{2!} A^2 A_o t^2 + \frac{1}{3!} A^3 A_o t^3 + .. + \frac{1}{i!} A^i A_o t^i + ..$$

$$= \left[1 + At + \frac{1}{2!} A^2 t^2 + \frac{1}{3!} A^3 t^3 + ... + \frac{1}{i!} A^i t^i + .. \right] A_o \quad (5.53)$$

where I is the unit Matrix.

It is given that, when $t = 0$, $X(t) = X(o) = X_o$

From equation (5.53) when $t = 0$, we get.

$$X(t)\big|_{t=0} = X(o) = A_o \quad (5.55)$$

From equation (5.54) and (5.55) we get,

$$A_o = X_o \quad (5.56)$$

Substituting (5.56) in (5.53) we get,

$$X(t) = \left[1 + At + \frac{1}{2!} A^2 t^2 + \frac{1}{3!} A^3 t^3 + ... + \frac{1}{i!} A^i t^i + .. \right] X_o \quad (5.57)$$

Each of the term inside the brackets is an n Ч m matrix. Because of the similarity of the entity inside the bracket with a scalar exponential of e^{at}, we call it a matrix exponential, which may be written as,

$$e^{at} = \left[1 + At + \frac{1}{2!} A^2 t^2 + \frac{1}{3!} A^3 t^3 + ... + \frac{1}{i!} A^i t^i + .. \right] \quad (5.58)$$

Hence the solution of the state equation is

$$\boxed{X(t) = e^{At} X_o} \quad (5.59)$$

The matrix e^{At} is called state transition matrix and denoted by $\phi(t)$. From the solution of the state equations it is observed that the initial state X_o at $t = 0$, is driven to state $X(t)$ at time t by state transition matrix.

Solution of Non Homogenous State Equations

(Solution of state equations with input or excitation)

The state equation of nth order system is given by,

$$\dot{X}(t) = A\,X(t) + B\,U(t); \quad X(o) = X_o \tag{5.60}$$

where X_o is initial condition vector.

The state equation of (5.60) can be rearranged as shown below:

$$\dot{X}(t) - AX(t) = B\,U(t) \tag{5.61}$$

Premultiply both sides of equation (5.61) by e^{-At};

$$e^{-At}\left[\dot{X}(t) - AX(t)\right] = e^{-At} B\,U(t) \tag{5.62}$$

$$e^{-At}\dot{X}(t) + e^{-At}(-A)\,X(t) = e^{-At} B\,U(t)$$

Consider the differential of $e^{-At} X(t)$

$$\frac{d}{dt}\left[e^{-At} X(t)\right] = e^{-At}\dot{X}(t) + e^{-At}(-A)\dot{X}(t) \tag{5.63}$$

Compare equation (5.62) and (5.63) we can write,

$$\frac{d}{dt}\left[e^{-At} X(t)\right] = e^{-At}\,B\,U(t)$$

$$d\left(e^{-At} X(t)\right) = e^{-At}\,B\,U(t)\,dt \tag{5.64}$$

Integrating equation (5.64) between limits 0 to t, we get,

$$e^{-At} X(t) = X_o + \int_0^t e^{-A\tau}\,BU(\tau)\,d\tau \tag{5.65}$$

where X_o = initial condition vector = integral constant

τ = Dummy variable substituted for t.

Premultiply both sides of equation (5.65) by e^{At},

$$e^{At} e^{-At} X(t) = e^{At} X_o + e^{At}\int_0^t e^{-A\tau} Bu(\tau)d\tau$$

$$X(t) = eA^t X_o + e^{At}\int_0^t e^{-A\tau}\,Bu(\tau)d\tau \tag{5.66}$$

The term e^{At} independent of the integral variable τ and *so* e^{At} can be brought inside the integral function.

$$\therefore \quad X(t) = e^{At} X_o + \int_0^t e^{At}\cdot e^{A\tau} BU(\tau)d\tau$$

$$X(t) = e^{At}X_o + \int_0^t e^{A(t-\tau)}BU(\tau)d\tau \quad (5.67)$$

The equation (5.67) is the solution of state equation, when the initial conditions are known at $t = 0$. If initial conditions are known at $t = t_o$ then the solution of state equation is given by equation (5.68)

$$X(t) = e^{A(t-t_o)}X(t_o) + \int_{t_o}^t e^{A(t-\tau)}B \cdot U(\tau)d\tau \quad (5.68)$$

The state transition matrix e^{At} is denoted by the symbol $\phi(t)$, i.e., $\phi(t) = e^{At}$

Hence, $e^{A(t-t_o)}$ can be expressed as,

$$e^{A(t-t_o)} = \phi(t - t_o) \quad (5.69)$$

and, $e^{A(t-\tau)}$ can be expressed as,

$$e^{A(t-\tau)} = \phi(t - \tau) \quad (5.70)$$

Equation (5.67) and (5.68) also can expressed as,

$X(t) = \phi(t)X(o) + \int_0^t \phi(t-\tau)B \cdot U(\tau) \cdot d\tau$. If the conditions are known $t = 0$ (5.71)

$X(t) = \phi(t - t_o)X(t_o) + \int_{t_o}^t \phi(t-\tau)B \cdot U(\tau)$. If the initial conditions are known $t = t_o$ (5.72)

Properties of State Transition Matrix

1. $\phi(o) = e^{AX_o} = I$ (Unit matrix)
2. $\phi(t) = e^{At} = (e^{-At})^{-1} = [\phi(-t)]^{-1}$ or $\phi^{-1}(t) = \phi(-t)$
3. $\phi(t_1 + t_2) = e^{A(t_1+t_2)} = (e^{At_1}) \cdot (e^{At_2}) = \phi(t_1)\,\phi(t_2) = \phi(t_2)\,.\,\phi(t_1)$

Computation of State Transition Matrix

The state transition matrix (e^{At}) can be computed by any one of the following methods.

Method 1: Computation of e^{At} using matrix exponential

Method 2: Computation of e^{At} using Laplace transform

Method 3: Computation of e^{At} by canonical transformation

Method 4: Computation of e^{At} using sylvester's interpretation formula (or) computation based on Cayley-Hamitton theorem.

Computation of state transition matrix using matrix exponential

In this method, the e^{At} is computed using the matrix exponential of equation (5.58), which is also given below:

$$e^{At} = I + At + \frac{1}{2!}A^2t^2 + \frac{1}{3!}A^3t^3 + \ldots + \frac{1}{1!}A^it^i + ..$$

where, e^{At} = state transition matrix of order n Ч n

A = system matrix of order n Ч n

I = unit matrix of order n Ч n

The disadvantage in this method is that each term of e^{At} will be an infinite series and the convergence of the infinite series are obtained by trial and error.

Computation of state transition matrix by laplace transform method

Consider the state equation without input vector,

$$\dot{X}(t) = A\,X(t) \tag{5.73}$$

Take laplace transform of equation (5.73) we get,

$$s\,X(s) - X(o) = A\text{Ч}(s)$$

$$s\,X(s) - AX(s) = X(o)$$

$$sI\,X(s) - AX(s) = X(o),$$

where I is a unit Matrix

$$(sI - A)\,X(s) = X(o)$$

Premultiply both sides by $(sI - A)^{-1}$

$$X(s) = (sI - A)^{-1}\,X(0)$$

take inverse laplace transform we get,

$$X(t) = L^{-1}[\,(sI - A)^{-1}]\,X(0)];$$

$$X(t) = L^{-1}[\,(sI - A)^{-1}]\,X(0) \tag{5.74}$$

Comparing equation (5.74) with the solution of state equation, $X(t) = e^{At}\,X(0)$ we get,

$$e^{At} = L^{-1}\,[\,(sI - A)^{-1}]\text{ (or) }L\,[e^{At}] = (sI - A)^{-1} \tag{5.75}$$

We know that, $e^{At} = \phi\,(t)$,

$$\therefore \quad L[e^{At}] = L\,[\phi(t)] = \phi\,(s) \tag{5.76}$$

where, $\phi\,(s) = (sI - A)^{-1}$ and it is called resolvant matrix.

From the system matrix, A the resolvant matrix, $\phi(s)$ can be computed. By taking inverse laplace transform of resolvent matrix, the state transition matrix is computed, from which the solution of state equation is obtained.

The solution of state equation is given by,

$$X(t) = e^{At}\,X(o)$$

$$\therefore \quad X(t) = L^{-1}\,[\,\phi\,(s)\,]\,X(o) \tag{5.77}$$

where, $\phi\,(s) = (sI - A)^{-1}$

Consider the state equation with forcing function [input (or) excitation],

$$\dot{X} = AX + BU \tag{5.78}$$

taking laplace of equation (5.78) we get,

$$s\,X(s) - X(o) = A\,X(s) + B\,U(s),$$
$$sI\,X(s) - A\,X(s) = X(o) + B\,U(s)$$
$$(sI - A)\,X(s) = X(o) + B\,U(s) \tag{5.79}$$

Premultiply the equation (5.79) by $(sI - A)^{-1}$

$$\therefore \quad X(s) = (sI - A)^{-1}\,X(o) + (sI - A)^{-1}\,B\,U(s)$$
$$= \phi\,(s)\,X(o) + \phi\,(s)\,B\,U(s) \tag{5.80}$$

take inverse laplace transform of equation (5.80) we get,

$$X(t) = \phi\,(t)\,X(o) + L^{-1}\,[\phi\,(s)\cdot B\,U(s)] \tag{5.81}$$

The equation (5.81) is the solution of state equation with forcing function.

EXAMPLE 5.23

Consider the matrix A. Compute e^{At} by two methods $A = \begin{bmatrix} 0 & 1 \\ -2 & -3 \end{bmatrix}$.

Solution: Method 1:

$$e^{At} = \left[I + At + \frac{1}{2!}A^2t^2 + \frac{1}{3!}A^3t^3 + .. \right].$$

$$A = \begin{bmatrix} 0 & 1 \\ -2 & -3 \end{bmatrix}$$

$$A^2 = A\cdot A = \begin{vmatrix} 0 & 1 \\ -2 & -3 \end{vmatrix}\begin{vmatrix} 0 & 1 \\ -2 & -3 \end{vmatrix} = \begin{vmatrix} -2 & -3 \\ 6 & 7 \end{vmatrix}$$

$$A^3 = A^2\cdot A = \begin{vmatrix} -2 & -3 \\ 6 & 7 \end{vmatrix}\begin{vmatrix} 0 & 1 \\ -2 & -3 \end{vmatrix} = \begin{vmatrix} 6 & 7 \\ -14 & -15 \end{vmatrix}$$

$$A^4 = A^3\cdot A = \begin{vmatrix} 6 & 7 \\ -14 & -15 \end{vmatrix}\begin{vmatrix} 0 & 1 \\ -2 & -3 \end{vmatrix} = \begin{vmatrix} -14 & -15 \\ 30 & 31 \end{vmatrix}$$

$$e^{At} = I + At + \frac{1}{2!}A^2t^2 + \frac{1}{3!}A^3t^3 + \frac{1}{4!}A^4t^4 + ..$$

$$= \begin{vmatrix} 1 & 0 \\ 0 & 1 \end{vmatrix} + \begin{vmatrix} 0 & 1 \\ -2 & -3 \end{vmatrix} t + \frac{1}{2}\begin{bmatrix} -2 & -3 \\ 6 & 7 \end{bmatrix} t^2 + \frac{1}{6}\begin{vmatrix} 6 & 7 \\ -14 & -15 \end{vmatrix} t^3 + \frac{1}{24}\begin{vmatrix} -14 & -15 \\ 30 & 31 \end{vmatrix} t^4 + ...$$

$$= \begin{vmatrix} 1 & 0 \\ 0 & 1 \end{vmatrix} + \begin{vmatrix} 0 & t \\ -2t & -3t \end{vmatrix} + \begin{vmatrix} t^2 & \frac{-3}{2}t^2 \\ 3t^2 & \frac{7}{2}t^2 \end{vmatrix} + \begin{vmatrix} t^3 & \frac{7}{6}t^3 \\ \frac{-7}{3}t^3 & \frac{-5}{2}t^3 \end{vmatrix} + \begin{vmatrix} \frac{-7}{12}t^4 & \frac{-5}{8}t^4 \\ \frac{5}{4}t^4 & \frac{31}{24}t^4 \end{vmatrix} + ...$$

$$= \begin{bmatrix} 1 - t^2 + t^3 - \frac{7}{12}t^4 + ... & t - \frac{3}{2}t^2 + \frac{7}{6}t^3 - \frac{5}{8}t^4 + ... \\ -2t + 3t^2 - \frac{7}{3}t^2 + \frac{5}{4}t^4 + .. & 1 - 3t + \frac{7}{2}t^2 - \frac{5}{2}t^3 + \frac{31}{24}t^4 + .. \end{bmatrix}$$

Each term in the matrix is an expansion of e^{at} · The conveyance of series is obtained by trial and error. Consider the expansion of e^{-t} and e^{-2t}.

$$e^{-t} = 1 - t + \frac{1}{2!}t^2 - \frac{1}{3!}t^3 + \frac{1}{4!}t^4 ..$$

$$= 1 - t + \frac{1}{2}t^2 - \frac{1}{6}t^3 + \frac{1}{24}t^4 ..$$

$$e^{-2t} = 1 - 2t + \frac{1}{2!}2^2t^2 - \frac{1}{3!}2^3t^3 + \frac{1}{4!}2^4t^4 ...$$

$$= 1 - 2t + 2t^2 - \frac{4}{3}t^3 + \frac{2}{3}t^4 ....$$

$$2e^{-t} - e^{-2t} = 2 - 2t + t^2 - \frac{1}{3}t^3 + \frac{1}{12} + t^4 .... - 1 + 2t - 2t^2 + \frac{4}{3}t^3 - \frac{2}{3}t^4 + ...$$

$$= 1 - t^2 + t^3 - \frac{7}{12}t^4 + ...$$

$$e^{-t} - e^{-2t} = 1 - t + \frac{1}{2}t^2 - \frac{1}{6}t^3 + \frac{1}{24}t^4 .. - 1 + 2t - 2t^2 + \frac{4}{3}t^3 - \frac{2}{3}t^4 + ...$$

$$= t - \frac{3}{2}t^2 + \frac{7}{6}t^3 - \frac{5}{8}t^4 + ...$$

$$-2e^{-t} + 2e^{-2t} = -2 + 2t - t^2 + \frac{1}{3}t^3 - \frac{1}{12}t^4 + ... + 2 - 4t + 4t^2 - \frac{8}{3}t^3 + \frac{4}{3}t^4.$$

$$= -2t + 3t^2 - \frac{7}{3}t^3 + \frac{5}{4}t^4$$

$$-e^{-t} + 2e^{-2t} = -1 + t - \frac{1}{2}t^2 + \frac{1}{6}t^3 - \frac{1}{24}t^4 + .. + 2 - 4t + 4t^2 - \frac{+8}{3}t^3 + \frac{4}{3}t^4 ..$$

$$= 1 - 3t + \frac{7}{2}t^2 - \frac{5}{2}t^3 + \frac{31}{24}t^4 + ..$$

$$\therefore \quad e^{At} = \begin{bmatrix} 2e^{-t} - e^{-2t} & e^{-t} - e^{-2t} \\ -2e^{-t} + 2e^{-2t} & -e^{-t} + 2e^{-2t} \end{bmatrix}$$

Method 2:

$$A = \begin{bmatrix} 0 & 1 \\ -2 & -3 \end{bmatrix}$$

$$e^{At}\, c = \phi(t) = L^{-1}[(sI - A)^{-1}]$$

$$sI - A = s\begin{bmatrix} 1 & 0 \\ 0 & 1 \end{bmatrix} - \begin{bmatrix} 0 & 1 \\ -2 & -3 \end{bmatrix} = \begin{bmatrix} s & o \\ o & s \end{bmatrix} - \begin{bmatrix} 0 & 1 \\ -2 & -3 \end{bmatrix} - \begin{bmatrix} s & -1 \\ 2 & s+3 \end{bmatrix}$$

Let $\Delta = |sI - A| = \text{determinant of } (sI - A)$

$\therefore$ $$\Delta = |sI - A| = \begin{vmatrix} s & -1 \\ 2 & s+3 \end{vmatrix} = s(s+3) + 2 = s^2 + 3s + 2 = (s+2)(s+1)$$

$$\phi(s) = [sI - A]^{-1} = \frac{[\text{Cofactor of } (sI - A)]^T}{\text{determinant of } (sI - A)} = \frac{[\text{Cofactor of } (sI - A)]^T}{\Delta}$$

$\therefore$ $$\phi(s) = \frac{1}{(s+1)(s+2)}\begin{bmatrix} s+3 & 1 \\ -2 & s \end{bmatrix}$$

$$\phi(s) = \begin{bmatrix} \dfrac{s+3}{(s+1)(s+2)} & \dfrac{1}{(s+1)(s+2)} \\ \dfrac{-2}{(s+1)(s+2)} & \dfrac{s}{(s+1)(s+2)} \end{bmatrix}$$

By partial fraction expansion, $\phi(s)$ can be written as,

$$\phi(s) = \begin{bmatrix} \dfrac{A_1}{s+1} + \dfrac{B_1}{s+2} & \dfrac{A_2}{s+1} + \dfrac{B_2}{s+2} \\ \dfrac{A_3}{s+1} + \dfrac{B_3}{s+2} & \dfrac{A_4}{s+1} + \dfrac{B_4}{s+2} \end{bmatrix}$$

$$\frac{s+3}{(s+1)(s+2)} = \frac{A_1}{s+1} + \frac{B_1}{s+2}$$

$$A_1 = \left.\frac{s+3}{s+2}\right|_{s=-1} = 2$$

$$B_1 = \left.\frac{s+3}{s+1}\right|_{s=-2} = -1$$

$$\frac{1}{(s+1)(s+2)} = \frac{A_2}{s+1} + \frac{B_2}{s+2}$$

$$A_2 = \left.\frac{1}{s+2}\right|_{s=-1} = 1$$

CHAPTER 5

$$B_2 = \left.\frac{1}{s+1}\right|_{s=-2} = -1$$

$$\frac{-2}{(s+1)(s+2} = \frac{A_3}{s+1} + \frac{B_3}{s+2}$$

$$A_3 = \left.\frac{-2}{s+2}\right|_{s=-1} = -2$$

$$B_3 = \left.\frac{-2}{s+1}\right|_{s=-2} = 2$$

$$\frac{s}{(s+1)(s+2)} = \frac{A_4}{s+1} + \frac{B_4}{s+2}$$

$$A_4 = \left.\frac{s}{s+2}\right|_{s=-1} = -1$$

$$B_4 = \left.\frac{s}{s+1}\right|_{s=-2} = 2$$

$$\therefore \quad \phi(s) = \begin{bmatrix} \frac{2}{s+1} - \frac{1}{s+2} & \frac{1}{s+1} - \frac{1}{s+2} \\ \frac{-2}{s+1} - \frac{2}{s+2} & \frac{-1}{s+1} + \frac{2}{s+2} \end{bmatrix}$$

Take inverse Laplace transform of $\phi(s)$ we get $\phi(t)$ where $\phi(t) = e^{At}$

$$\therefore \quad e^{At} = \phi(t) = \begin{bmatrix} 2e^{-t} - e^{-2t} & e^{-t} - e^{-2t} \\ -2e^{-t} + 2e^{-2t} & -e^{-t} + 2e^{-2t} \end{bmatrix}$$

Result: It is observed that the results of both the methods are same.

EXAMPLE 5.24

Given that $A_1 = \begin{vmatrix} \sigma & 0 \\ 0 & \sigma \end{vmatrix}$; $A_2 = \begin{vmatrix} 0 & \omega \\ -\omega & 0 \end{vmatrix}$; $A = \begin{vmatrix} \sigma & \omega \\ -\omega & \sigma \end{vmatrix}$ compute e^{At}.

Solution: Here

$$A = A_1 + A_2$$

$$e^{At} = e^{(A_1 + A_2)t} = e^{A_1 t} \cdot e^{A_2 t}$$

$$sI - A_I = s\begin{bmatrix} 1 & 0 \\ 0 & 1 \end{bmatrix} - \begin{bmatrix} \sigma & 0 \\ 0 & \sigma \end{bmatrix} = \begin{bmatrix} s & 0 \\ 0 & s \end{bmatrix} - \begin{bmatrix} \sigma & 0 \\ 0 & \sigma \end{bmatrix} - \begin{bmatrix} s-\sigma & 0 \\ 0 & s-\sigma \end{bmatrix}$$

$$\Delta_1 = |sI - A_1| = \begin{vmatrix} s-\sigma & 0 \\ 0 & s-\sigma \end{vmatrix} = (s-\sigma)^2$$

$$[sI - A_1]^{-1} = \frac{1}{\Delta_1}\begin{bmatrix} s-\sigma & 0 \\ 0 & s-\sigma \end{bmatrix} = \frac{1}{(s-\sigma)^2}\begin{bmatrix} s-\sigma & 0 \\ 0 & s-\sigma \end{bmatrix} = \begin{bmatrix} \frac{1}{s-\sigma} & 0 \\ 0 & \frac{1}{s-\sigma} \end{bmatrix}$$

$$e^{A_1 t} = L^{-1}\left[(sI - A_1)^{-1}\right] = \begin{bmatrix} e^{-\sigma t} & 0 \\ 0 & e^{-\sigma t} \end{bmatrix}$$

$$sI - A_2 = s\begin{vmatrix} 1 & 0 \\ 0 & 1 \end{vmatrix} - \begin{vmatrix} 0 & \omega \\ -\omega & 0 \end{vmatrix} = \begin{vmatrix} s & 0 \\ 0 & s \end{vmatrix} - \begin{vmatrix} 0 & \omega \\ -\omega & 0 \end{vmatrix} = \begin{vmatrix} s & -\omega \\ \omega & s \end{vmatrix}$$

$$\Delta_2 = |sI - A_2| = \begin{vmatrix} s & -\omega \\ \omega & s \end{vmatrix} = s^2 + \omega^2$$

$$[sI - A_2)^{-1} = \frac{1}{\Delta_2}\begin{bmatrix} s & \omega \\ -\omega & s \end{bmatrix} = \frac{1}{s^2+\omega^2}\begin{bmatrix} s & \omega \\ -\omega & s \end{bmatrix}$$

$$[sI - A_2]^{-1} = \begin{vmatrix} \frac{s}{s^2+\omega^2} & \frac{\omega}{s^2+\omega^2} \\ \frac{-\omega}{s^2+\omega^2} & \frac{s}{s^2+\omega^2} \end{vmatrix}$$

$$e^{A_2 t} = L^{-1}\left[(sI - A_2)^{-1}\right] = \begin{bmatrix} \cos\omega t & \sin\omega t \\ -\sin\omega t & \cos\omega t \end{bmatrix}$$

$$e^{At} = e^{A \cdot t} \cdot e^{A_2 t} = \begin{vmatrix} e^{-\sigma t} & 0 \\ 0 & e^{-\sigma t} \end{vmatrix}\begin{bmatrix} \cos\omega t & \sin\omega t \\ -\sin\omega t & \cos\omega t \end{bmatrix}$$

$$\therefore \quad e^{At} = \begin{bmatrix} e^{-\sigma t}\cos\omega t & e^{-\sigma t}\sin\omega t \\ -e^{-\sigma}t\sin\omega t & e^{-\sigma t}\cos\omega t \end{bmatrix}$$

EXAMPLE 5.25

For a system represented by state equation $\dot{X}(t) = A\,X(t)$. The response is $X(t) = \begin{bmatrix} e^{-2t} \\ -2e^{-2t} \end{bmatrix}$ when $X(o) = \begin{bmatrix} 1 \\ -2 \end{bmatrix}$ and $X(t) = \begin{vmatrix} e^{-t} \\ -e^{-t} \end{vmatrix}$ when $X(o) = \begin{bmatrix} 1 \\ -1 \end{bmatrix}$. Determine the system matrix A and the state transition matrix.

Solution: The solution of state equation is,

$$X(t) = e^{At}\,X(o) \tag{5.15.1}$$

Premultiply the equation (5.15.1) by e^{-At}

$$e^{-At}X(t) = e^{-At}\cdot e^{At}X(o)$$

$$\therefore \quad e^{-At}X(t) = X(o) \qquad (5.15.2)$$

One of the response is $X(t) = \begin{bmatrix} e^{-2t} \\ -2e^{-2t} \end{bmatrix}$ and $X(o) = \begin{bmatrix} 1 \\ -2 \end{bmatrix}$

On substituting the response in equation (5.15.2) we get,

$$e^{-At}\begin{bmatrix} e^{-2t} \\ -2e^{-2t} \end{bmatrix} = \begin{bmatrix} 1 \\ -2 \end{bmatrix} \qquad (5.15.3)$$

Let, $$e^{-At} = \begin{vmatrix} e_{11} & e_{12} \\ e_{21} & e_{22} \end{vmatrix} \qquad (5.15.4)$$

From equation (5.15.3) and (5.15.4) we can write

$$\begin{vmatrix} e_{11} & e_{12} \\ e_{21} & e_{22} \end{vmatrix}\begin{vmatrix} e^{-2t} \\ -2e^{-2t} \end{vmatrix} = \begin{vmatrix} 1 \\ -2 \end{vmatrix} \qquad (5.15.5)$$

On multiplying the equation (5.15.5) we get the following two equations,

$$e_{11}e^{-2t} - 2e_{12}e^{-2t} = 1 \qquad (5.15.6)$$

$$e_{21}e^{-2t} - 2e_{22}\,e^{-2t} = -2 \qquad (5.15.7)$$

The second solution of state equation is

$$X(t) = \begin{vmatrix} e^{-t} \\ -e^{-t} \end{vmatrix} \text{ and } X(o) = \begin{vmatrix} 1 \\ -1 \end{vmatrix}$$

substituting this solution in equation (5.15.2) we get,

$$e^{-At}\begin{bmatrix} e^{-t} \\ -e^{-t} \end{bmatrix} = \begin{vmatrix} 1 \\ -1 \end{vmatrix} \qquad (5.15.8)$$

From equation (5.15.4) and (5.15.8) we can write,

$$\begin{vmatrix} e_{11} & e_{12} \\ e_{21} & e_{22} \end{vmatrix}\begin{vmatrix} e^{-t} \\ -e^{-t} \end{vmatrix} = \begin{vmatrix} 1 \\ -1 \end{vmatrix} \qquad (5.15.9)$$

On multiplying the equation (5.15.9) we get the following two equations,

$$e_{11}e^{-t} - e_{12}e^{-t} = 1 \qquad (5.15.10)$$

$$e_{21}e^{-t} - e_{22}e^{-t} = -1 \qquad (5.15.11)$$

$$\begin{array}{ll} \text{equation (5.15.10)} \times e^{-t} & \Rightarrow e_{11}\,e^{-2t} - e_{12}e^{-2t} = e^{-t} \\ \text{equation (5.15.6)} \times 1 & \Rightarrow e_{11}\,e^{-2t} - 2e_{12}e^{-2t} = 1 \\ & \quad (-) \qquad\quad (+) \qquad (-) \\ \hline & \qquad e_{12}e^{-2t} = e^{-t} - 1 \end{array} \qquad (5.15.12)$$

From equation (5.15.12) we get,

$$e_{12} = \frac{e^{-t}-1}{e^{-2t}} = \frac{e^{-t}}{e^{-2t}} - \frac{1}{e^{-2t}} = e^{t} - e^{2t} \tag{5.15.13}$$

From equation (5.15.6),

$$e_{11} = \frac{1+2e_{12}e^{-2t}}{e^{-2t}}$$

Put $\quad e_{12} = e^{t} - e^{2t}$

$$\therefore \quad e_{11} = \frac{1+2\left(e^{t}-e^{2t}\right)e^{-2t}}{e^{-2t}} = \frac{1+2e^{-t}-2}{e^{-2t}}$$

$$= \frac{2e^{-t}-1}{e^{-2t}} = \frac{2e^{-t}}{e^{-2t}} - \frac{1}{e^{-2t}} = 2e^{t} - e^{2t}$$

equation (5.15.11) $Xe^{-t} \Rightarrow e_{21}\, e^{-2t} - e_{22}e^{-2t} = e^{-t}$

equation (5.15.7) $\;\; X1 \;\; \Rightarrow e_{21}\, e^{-2t} - 2e_{22}e^{-2t} = -2$

$\quad (+) \qquad (-) \qquad (+)$

$$e_{22}e^{-2t} = 2 - e^{-t} \tag{5.15.14}$$

From equation (5.15.14) we get,

$$e_{22} = \frac{-e^{-t}+2}{e^{-2t}} = \frac{-e^{-t}}{e^{-2t}} + \frac{2}{e^{-2t}} = -e^{t} + 2e^{2t}$$

From equation (5.15.11), $\quad e_{21} = \dfrac{-1+e_{22}e^{-t}}{e^{-t}}$

Put $\quad e_{22} = -e^{t} + 2e^{2t}$

$$\therefore \quad e_{21} = \frac{-1+\left(-e^{t}+2e^{2t}\right)e^{-t}}{e^{-t}}$$

$$e_{21} = \frac{-1-1+2e^{t}}{e^{-t}} = \frac{-2}{e^{-t}} + \frac{2e^{t}}{e^{-t}} = -2e^{t} + 2e^{2t}$$

$$\therefore \quad e^{-At} = \begin{vmatrix} e_{11} & e_{12} \\ e_{21} & e_{22} \end{vmatrix} = \begin{vmatrix} 2e^{t}-e^{2t} & e^{t}-e^{2t} \\ -2e^{t}+2e^{2t} & -e^{t}+2e^{2t} \end{vmatrix}$$

$$\therefore \quad e^{At} = \begin{vmatrix} 2e^{-t}-e^{-2t} & e^{-t}-e^{-2t} \\ -2e^{-t}+2e^{-2t} & -e^{-t}+2e^{-2t} \end{vmatrix}$$

e^{At} is the state transition matrix.

We know that, $L[e^{At}] = \phi(s)$

where $\phi(s) = (sI - A)^{-1};$

$\therefore$ $\phi(s)^{-1} = (sI - A)$ (or)

$$A = sI - \phi(s)^{-1}$$

$$\phi(s) = L\left[e^{At}\right]\begin{bmatrix} \frac{2}{s+1} - \frac{1}{s+2} & \frac{1}{s+1} - \frac{1}{s+2} \\ \frac{-2}{s+1} + \frac{2}{s+2} & \frac{-1}{s+1} + \frac{2}{s+2} \end{bmatrix}$$

$$= \begin{vmatrix} \frac{2(s+2)-(s+1)}{(s+1)(s+2)} & \frac{(s+2)-(s+1)}{(s+1)(s+2)} \\ \frac{-2(s+2)+2(s+1)}{(s+1)(s+2)} & \frac{-(s+2)+2(s+1)}{(s+1)(s+2} \end{vmatrix}$$

$$= \begin{vmatrix} \frac{s+3}{(s+1)(s+2)} & \frac{1}{(s+1)(s+2)} \\ \frac{-2}{(s+1)(s+2)} & \frac{s}{(s+1)(s+2)} \end{vmatrix}$$

Determinant of $\phi(s) = \dfrac{s(s+3)+2}{(s+1)^2(s+2)^2} = \dfrac{s^2+3s+2}{(s+1)^2(s+2)^2}$

$$= \frac{(s+1)(s+2)}{(s+1)^2(s+2)^2} = \frac{1}{(s+1)(s+2)}$$

$$\phi(s)^{-1} = \frac{1}{(s+1)(s+2)}\begin{bmatrix} \frac{s}{(s+1)(s+2)} & \frac{-1}{(s+1)(s+2)} \\ \frac{2}{(s+1)(s+2)} & \frac{s+3}{(s+1)(s+2)} \end{bmatrix} = \begin{bmatrix} s & -1 \\ 2 & s+3 \end{bmatrix}$$

$$A = sI\,\phi(s)^{-1}\begin{vmatrix} 1 & 0 \\ 0 & 1 \end{vmatrix} - \begin{vmatrix} s & -1 \\ 2 & s+3 \end{vmatrix} = \begin{vmatrix} s & 0 \\ 0 & s \end{vmatrix} - \begin{vmatrix} s & -1 \\ 2 & s+3 \end{vmatrix} = \begin{vmatrix} 0 & 1 \\ -1 & -3 \end{vmatrix}$$

Result: $A = \begin{vmatrix} 0 & 1 \\ -2 & -3 \end{vmatrix}; e^{At} = \begin{bmatrix} 2e^{-t} - e^{-2t} & e^{-t} - e^{-2t} \\ -2e^{-t} + 2e^{-2t} & -e^{-t} + 2e^{-2t} \end{bmatrix}$

EXAMPLE 5.26

A linear time-invariant system is characterized by homogeneous state equation. $\begin{vmatrix} \dot{X}_1 \\ \dot{X}_2 \end{vmatrix} = \begin{vmatrix} 1 & 0 \\ 1 & 1 \end{vmatrix}\begin{vmatrix} X_1 \\ X_2 \end{vmatrix}$. Compute the solution of the homogeneous equation, assuming the initial state vector $X_o = \begin{vmatrix} 1 \\ 0 \end{vmatrix}$.

Solution:

$$A = \begin{vmatrix} 1 & 0 \\ 1 & 1 \end{vmatrix};$$

$$sI - A = s\begin{vmatrix} 1 & 0 \\ 0 & 1 \end{vmatrix} - \begin{vmatrix} 1 & 0 \\ 1 & 1 \end{vmatrix} = \begin{vmatrix} s & 0 \\ 0 & s \end{vmatrix} - \begin{vmatrix} 1 & 0 \\ 1 & 1 \end{vmatrix} = \begin{vmatrix} s-1 & 0 \\ -1 & s-1 \end{vmatrix}$$

$$|sI - A| = \begin{vmatrix} s-1 & 0 \\ -1 & s-1 \end{vmatrix} = (s-1)^2 - 0 = (s-1)^2$$

$$(sI - A)^{-1} = \frac{1}{(s-1)^2}\begin{bmatrix} s-1 & 0 \\ 1 & s-1 \end{bmatrix}$$

$$= \begin{vmatrix} \frac{1}{s-1} & 0 \\ \frac{1}{(s-1)^2} & \frac{1}{s-1} \end{vmatrix}$$

$$e^{At} = \phi(t) = L^{-1}[\phi(s)] = L^{-1}\left[(sI - A)^{-1}\right] = \begin{vmatrix} e^t & 0 \\ te^t & e^t \end{vmatrix}$$

Result: The solution of the state equation is,

$$X(t) = e^{At} X_o = \begin{vmatrix} e^t & 0 \\ te^t & e^t \end{vmatrix}\begin{vmatrix} 1 \\ 0 \end{vmatrix} = \begin{vmatrix} e^t \\ te^t \end{vmatrix}$$

5.16 STATE SPACE REPRESENTATION OF DISCRETE TIME SYSTEMS

The state variable analysis techniques of continuous time systems can be extended to the discrete-time system. The discrete form of state space representation is analogous to the continuous form.

In the state variable formulation of a discrete time system, in general, a system consists of m-inputs P-outputs and n-state variables. The state space representation of discrete-time system may be drawn as shown in Fig. (5.6).

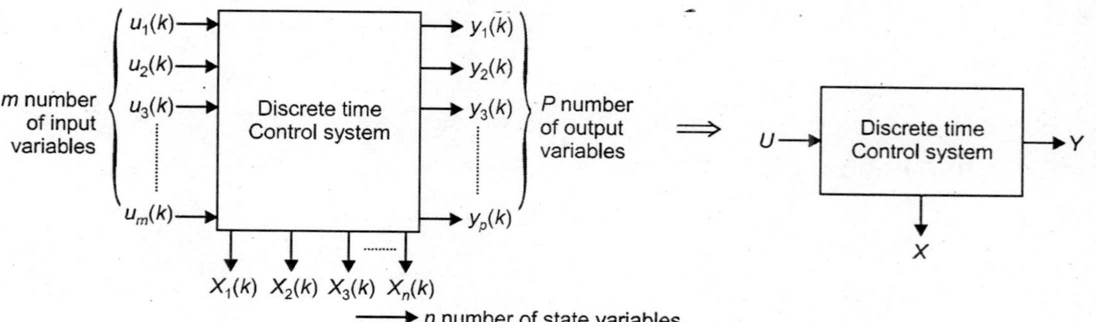

Fig. 5.6: State space representation of discrete time system.

Let, state variables = $X_1(k), X_2(k), X_3(k)... X_n(k)$
Input variables = $u_1(k), u_2(k), u_3(k)... u_m(k)$
Output variables = $y_1(k), y_2(k), y_3(k)... y_p(k)$

The different variables may be represented by the vectors (column matrix) as shown below.

Input Vector $$U(k) = \begin{vmatrix} u_1(k) \\ u_2(k) \\ \vdots \\ u_m(k) \end{vmatrix}$$

Output vector $$Y(k) = \begin{vmatrix} Y_1(k) \\ Y_2(k) \\ \vdots \\ Y_p(k) \end{vmatrix}$$

State variable vector $$X(k) = \begin{vmatrix} X_1(k) \\ X_2(k) \\ \vdots \\ X_n(k) \end{vmatrix}$$

The state equation of a discrete time system is a set of n-numbers of first order difference equations.

$$X_1(k+1) = f_1\,(X_1, X_2, ... X_n;\ u, u_2 ... u_m)$$
$$X_2(k+1) = f_2(X_1, d_2 .... X_n;\ u_1, u_2 .... u_m)$$
$$\vdots$$
$$X_n(k+1) = f_n(X_1, X_2 .... X_n;\ u_1, u_2 .... u_m)$$

For linear time invriant discrete time systems the above difference equations can be expressed as a linear combination of state variables and inpts.

$$X_1(k+1) = a_{11}\,X_1 + a_{12}X_2 + .. + a_{1n}\,X_n + b_{11}u_1 + b_{12}\,u_2 ... + b_{1m}\,u_m$$
$$X_2(k+1) = a_{21}\,X_1 + a_{22}\,X_2 + ... + a_{2n}X_n + b_{21}\,u_1 + b_{22}\,u_2 + ... + b_{2m}\,u_m$$
$$\vdots$$
$$X_n(k+1) = a_{n1}\,X_1 + a_{n2}\,X_2 + ... + a_{nn}X_n + b_{n1}\,u, + b_{n2}\,u_2 + .. + b_{nm}\,u_m$$

where the coefficients a_{ij} *and* b_{ij} are constants.

In the matrix form the above equations can be expressed as,

$$\begin{vmatrix} X_1(k+1) \\ X_2(k+1) \\ \vdots \\ X_n(k+1) \end{vmatrix} = \begin{vmatrix} a_{11} & a_{12} & \cdots & a_{1n} \\ a_{21} & a_{22} & \cdots & a_{2n} \\ \vdots & \vdots & & \vdots \\ a_{n1} & a_{n2} & \cdots & a_{nn} \end{vmatrix} \begin{vmatrix} X_1 \\ X_2 \\ \vdots \\ X_n \end{vmatrix} + \begin{vmatrix} b_{11} & b_{12} & \cdots & b_{1m} \\ b_{21} & b_{22} & \cdots & b_{2m} \\ \vdots & \vdots & & \vdots \\ b_{n1} & b_{n2} & \cdots & b_{nm} \end{vmatrix} \begin{vmatrix} u_1 \\ u_2 \\ \vdots \\ u_m \end{vmatrix} \quad (5.82)$$

The matrix equation (5.82) can be written in the vector notation as,

$$X(k+1) = A\,X(k) + B\,U(k) \qquad (5.83)$$

where,

$X(k)$ = state vector of order $(n \times 1)$

$U(k)$ = Input vector of order $(m \times 1)$

A = System matrix of order $(n \times n)$

B = Input matrix of order $(n \times m)$

The equation (5.83) is the state equation of (linear time invariant) discrete time system.

The output at any discrete time instant, K are functions of state variables and inputs. Hence the output variables of linear time invariant system can be expressed as a linear combination of state variables and inputs.

$$y_1 = c_{11}X_1 + c_{12}X_2 + .. + c_{1n}X_n + d_{11}u_1 + d_{12}u_2 + ... + d_{1m}u_m$$
$$y_2 = c_{21}X_1 + c_{22}X_2 + ... + c_{2n}X_n + d_{21}u_1 + d_{22}u_2 + ... + d_{2m}u_m$$
$$\vdots$$
$$y_p = c_{p1}X_1 + c_{P2}X_2 + ... + c_{pn}X_n + d_{p1}u_1 + dp_2u_2 + ... + d_{pm}u_m$$

where the coefficients c_{ij} and d_{ij} are constants.

In the matrix form the above equations can be expressed as,

$$\begin{vmatrix} y_1 \\ y_2 \\ y_3 \\ \vdots \\ y_p \end{vmatrix} = \begin{vmatrix} c_{11} & c_{12} & \cdots & c_{1n} \\ c_{21} & c_{22} & \cdots & c_{2n} \\ \vdots & \vdots & & \vdots \\ c_{p1} & c_{p2} & & c_{pm} \end{vmatrix} \begin{vmatrix} X_1 \\ X_2 \\ \vdots \\ X_n \end{vmatrix} + \begin{vmatrix} d_{11} & d_{12} & \cdots & d_{1m} \\ d_{21} & d_{22} & \cdots & d_{2m} \\ \vdots & \vdots & & \vdots \\ d_{p1} & d_{p2} & \cdots & d_{pn} \end{vmatrix} \begin{vmatrix} u_1 \\ u_2 \\ \vdots \\ u_m \end{vmatrix} \qquad (5.82)$$

The matrix equation (5.82) can be written in the vector notation as,

$$Y(k) = C\,X(k) + DU(k) \qquad (5.83)$$

Where,

$X(k)$ = state vector of order (n Ч 1)

$U(k)$ = Input vector of order(m Ч1)

A = System matrix of order (n Чm)

B = Input matrix of order (n Чm)

The equation (5.83) is the state equation of (linear time invariant) discrete time system.

The state equation and output equation together called as state model of the system. Hence the state model of discrete time system is given by the following equations.

$X(k+1) = A\,X(k) + B\,U(k)$ –state equation

$Y(k) = C\,X(k) + D\,U(K)$ –output equation

State Diagram of Discrete Time System

The state diagram of discrete time system can be either in block diagram form or signal flow graph form. The three fundamental elements, scalar, adder and unit delay elements are used to construct the state diagram. These basic elements are shown in Table (5.6).

Table 5.6: Basic elements of state diagram of discrete time system

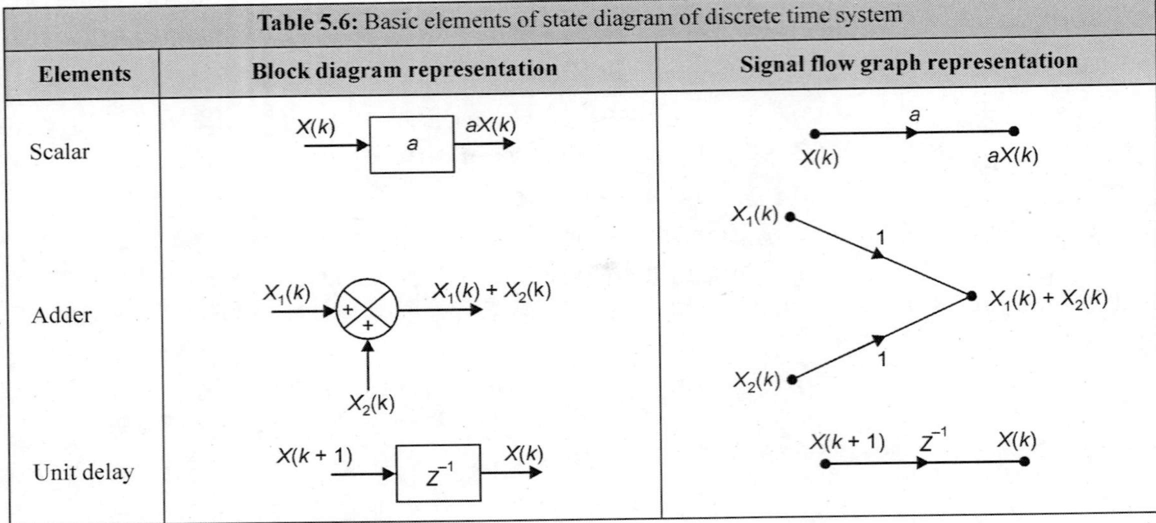

Elements	Block diagram representation	Signal flow graph representation
Scalar	$X(k)$ → a → $aX(k)$	$X(k)$ —a→ $aX(k)$
Adder	$X_1(k)$, $X_2(k)$ → (+, +) → $X_1(k) + X_2(k)$	$X_1(k)$ —1→, $X_2(k)$ —1→ $X_1(k) + X_2(k)$
Unit delay	$X(k + 1)$ → z^{-1} → $X(k)$	$X(k + 1)$ —z^{-1}→ $X(k)$

Scalar: The scalar is used to multiply a signal by constant

Adder: The adder is used to add two or more signals

Unit delay: The unit delay element will delay the signal passing through it by one sample time.

The block diagram representation of the state mud of discrete time system is shown in Fig. (5.7) and the signal flow graph representation is shown in Fig. (5.8)

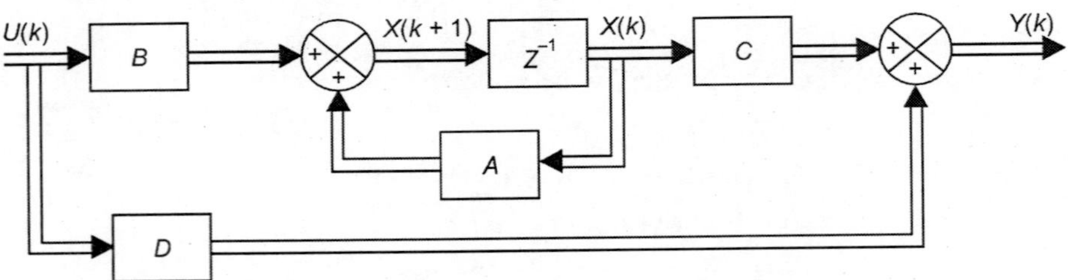

Fig. 5.7: Block diagram representation of discrete time system

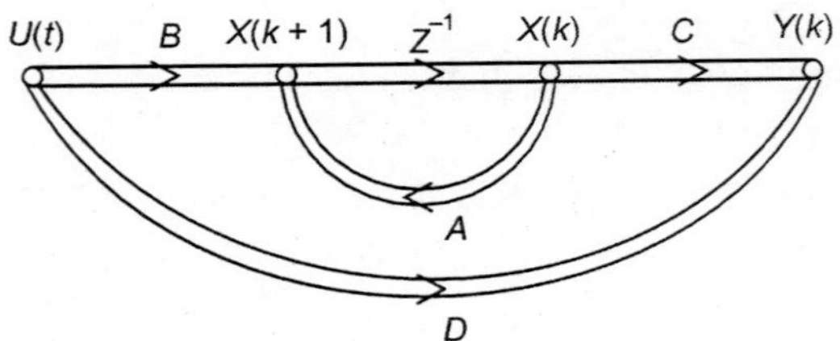

Fig. 5.8: Signal flowgraph representation of discrete time system.

Phase Variable Form of State Model

The discrete time system is governed by n^{th} order differential equation. The general form of nth order difference equation is,

$$y(k) = -\sum_{j=1}^{N} aj\ y(k-j) + \sum_{j=0}^{m} bj\ u(k-j) \tag{5.86}$$

Where aj and bj are constants for time invariant system.

On expanding the summation of equation (5.86) and rearranging we get,

$$y(k) + a_1\ y(k-1) + a_2 y(k-2) + \ldots + a_n\ Y(k-n) = b_o\ u(k) + b_1\ u(k-1) + b_2\ u(k-2) + \ldots + b_m\ u(k-m) \tag{5.87}$$

take Z-transform for equation (5.87) with zero initial conditions, we get,

$$Y(z) + a_1\ z^{-1}\ Y(z) + a_2\ z^{-2}\ Y(z) + .. + a_n\ z^{-n}\ Y(z) = b_o\ U(z) + b_1\ z^{-1}\ U(z) + b_2\ z^{-2}\ U(z) + .. + b_m\ z^{-m}\ U(z)$$

$$Y(z)\ [1 + a_1 z^{-1} + a_2 z^{-2} + .. + a_n z^{-n}] = [\ b_o + b_1\ z^{-1} + b_2\ z^{-2} + \ldots + b_m z^{-m}]\ U(z) \tag{5.88}$$

When $m = n$,

$$\frac{Y(z)}{U(z)} = \frac{b_o + b_1 z^{-1} + b_2 z^{-2} + .. + b_n z^{-n}}{1 + a_1 Z^{-1} + a_2 z^{-2} + \ldots + a_n z^{-n}} \tag{5.89}$$

The equation (5.88) is the transfer function of the discrete time system. The equation (5.88) can be expressed as shown below:

$$\frac{Y(z)}{U(z)} = \frac{b_o + b_1 z^{-1} + b_2 z^{-2} + .. + b_n z^{-n}}{1 - (-a_1 z^{-1} - a_2 z^{-2} \ldots - a_n z^{-n})} \tag{5.89}$$

On comparing equation (5.89) with Mason's gain formula, a signal flow graph can be constructed. Each numerator term of equation (5.89) represent a forward path gain and each denominator term of equation (5.89) represent an individual loop gain. The signal flow graph will not have any nontouching loops. The signal flow graph when $n = 3$ is shown in Fig. (5.9).

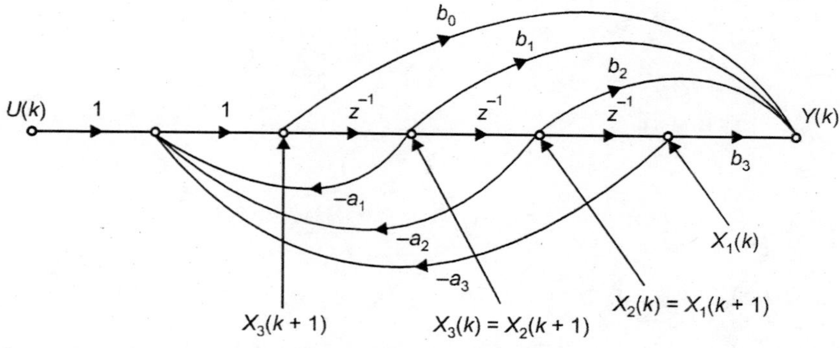

Fig. 5.9: Signal flow graph representation of discrete time system.

Let us assign state variables at the output of each unit delay element. hence the signal at the input of unit delay element will be a signal advanced by one sampling time. The state equations are obtained by summing up the signals at the input of each unit delay element.

In Fig. (5.9), at node X_2 and X_3 we get,

$$X_1\ (k+1) = X_2 \tag{5.90}$$

$$X_2\ (k+1) = X_3 \tag{5.91}$$

By summing up the incoming signals to node $X_3(k+1)$ we get,

$$X_3(k+1) = -a_3 X_1 - a_2 X_2 - a_1 X_3 + u \tag{5.92}$$

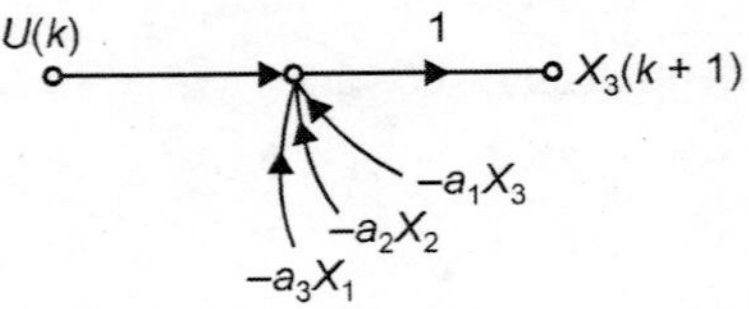

By summing up the incoming signals to node y we get, $b_o X_3(k+1)$

$$y = b_3 X_1 + b_2 X_2 + b_1 X_3 + b_0 X_3(k+1) \tag{5.93}$$

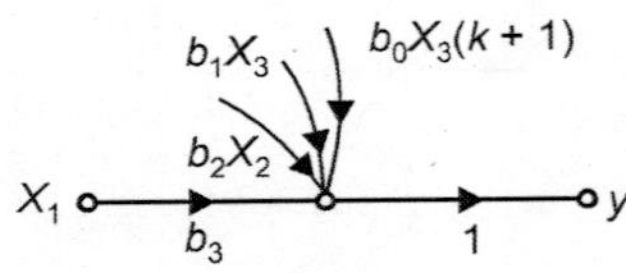

Fig. 5.93

Substitute for $XX_3(k+1)$ from equation (5.92) in equation (5.93)

$$y = b_3 X_1 + b_2 X_2 + b_1 X_3 + b_o(-a_3 X_1 - a_2 X_2 - a_1 X_3)$$

$\therefore$

$$y = (b_3 - b_o a_3) X_1 + (b_2 - b_o a_2) X_2 + (b_1 - b_o a_1) X_3 \tag{5.94}$$

The equations (5.90), (5.91) and (5.92) are state equations and equation (5.94) is the output equation. On arranging the state equations and output equations in the matrix form we get,

$$\begin{vmatrix} X_1(k+1) \\ X_2(k+1) \\ X_3(k+1) \end{vmatrix} = \begin{vmatrix} 0 & 1 & 0 \\ 0 & 0 & 1 \\ -a_3 & -a_2 & -a_1 \end{vmatrix} \begin{vmatrix} X_1 \\ X_2 \\ X_3 \end{vmatrix} + \begin{vmatrix} 0 \\ 0 \\ 1 \end{vmatrix} u \tag{5.95}$$

$$y = \left[(b_3 - b_o a_3)(b_2 - b_o a_2)(b_1 - b_o a_1)\right] \begin{bmatrix} X_1 \\ X_2 \\ X_3 \end{bmatrix} \tag{5.96}$$

The equations (5.95) and (5.96) can be extended to n^{th} order system as shown below:

$$\begin{vmatrix} X_1(k+1) \\ X_2(k+1) \\ X_3(k+1) \\ \vdots \\ X_n(k+1) \end{vmatrix} = \begin{vmatrix} 0 & 1 & 0 & \cdots & 0 \\ 0 & 0 & 1 & \cdots & 0 \\ 0 & 0 & 0 & \cdots & 0 \\ \vdots & \vdots & \vdots & & \\ -a_n & -a_{n-1} & -a_{n-2} & \cdots & -a_1 \end{vmatrix} \begin{vmatrix} X_1 \\ X_2 \\ X_3 \\ \vdots \\ X_n \end{vmatrix} + \begin{vmatrix} 0 \\ 0 \\ 0 \\ \vdots \\ 1 \end{vmatrix} u \tag{5.97}$$

$$y = \left[(b_n - b_o a_n)(b_{n-1} - b_o a_{n-1}) \ldots (b_1 - b_o a_1)\right] \begin{bmatrix} X_1 \\ X_2 \\ \vdots \\ X_n \end{bmatrix} \tag{5.98}$$

The equation (5.97) and (5.98) is the phase variable form of state model of discrete time n^{th} order system.

Canonical Form of State Model

The transfer function of equation (5.88) can be expressed as a summation using partial fraction as shown below:

$$\frac{Y(z)}{U(z)} = b_o + \frac{c_1}{z+\lambda_1} + \frac{c_2}{z+\lambda_2} + \ldots + \frac{c_n}{z+\lambda_n}$$

where c_1 , c_2 ... c_n are resides and λ_1, λ_2 .... λ_n are poles of the system.

$$\frac{Y(z)}{U(z)} = b_o + \frac{c_1}{z\left(1+\frac{\lambda_1}{z}\right)} + \frac{c_2}{z\left(1+\frac{\lambda_2}{z}\right)} + \ldots + \frac{c_n}{z\left(1+\frac{\lambda_3}{z}\right)}$$

$$\frac{Y(z)}{U(z)} = b_o + \frac{z^{-1}c_1}{1+z^{-1}\lambda_1} + \frac{z^{-1}c_2}{1+z^{-1}\lambda_2} + \ldots + \frac{z^{-1}c_n}{1+z^{-1}\lambda_n} \tag{5.99}$$

The equation (5.99) can be used to construct the block diagram of state model as shown in Fig. (5.10)

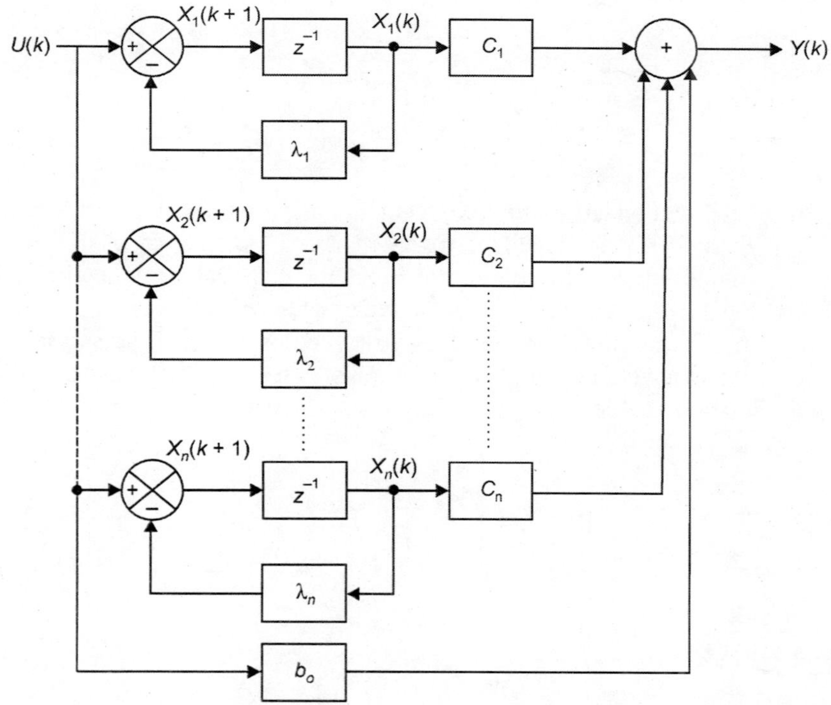

Fig. 5.10: Block diagram of canonical state model of discrete time system.

Let us assign the state variables at the output of unit delay elements. Hence the signal at the input of unit delay element will be a signal advanced by one sampling time. The state equations are obtained by summing of the signals at the input of each unit delay element.

The state equations are,

$$X_1(k+1) = -\lambda_1 X_1 + u$$
$$X_2(k+1) = -\lambda_2 X_2 + u$$
$$X_3(k+1) = -\lambda_{31} X_3 + u$$
$$\vdots \quad \vdots$$
$$Xx_n(k+1) = -\lambda_n X_n + u$$

The output equation is, $Y = c_1 X_1 + c_2 X_2 + c_3 X_3 + ... + c_n X_n + b_o u$

The state equations and output equations can be expressed in the matrix form as shown below. The equation (5.100) and (5.101) is the canonical form of state model of discrete time n^{th} order system.

$$\begin{vmatrix} X_1(k+1) \\ X_2(k+1) \\ X_3(k+1) \\ \vdots \\ X_n(k+1) \end{vmatrix} = \begin{vmatrix} -\lambda_1 & 0 & 0 & \cdots & 0 \\ 0 & -\lambda_2 & 0 & \cdots & 0 \\ 0 & 0 & -\lambda_3 & \cdots & 0 \\ \vdots & \vdots & \vdots & & \\ -a_n & -a_{n-1} & -a_{n-2} & \cdots & -\lambda_n \end{vmatrix} \begin{vmatrix} X_1 \\ X_2 \\ X_3 \\ \vdots \\ X_n \end{vmatrix} + \begin{vmatrix} 1 \\ 1 \\ 1 \\ \vdots \\ 1 \end{vmatrix} [u] \tag{5.100}$$

$$y = \begin{bmatrix} c_1 & c_2 & c_3 \ldots & c_n \end{bmatrix} \begin{bmatrix} X_1 \\ X_2 \\ X_3 \\ \vdots \\ X_n \end{bmatrix} + b_o u \tag{5.101}$$

when a pole of the transfer function has multiplicity then the state model will be in a special form called Jordan canonical form. In this form the system matrix A will have a Jordan block of size $q \times q$, correspond to a pole of value λ_i with multiplicity q. In the Jordan block the diagonal element will be the poles and the element just above the diagonal is one. Consider a system with poles $\lambda_1\ \lambda_2\ \lambda_3\ \lambda_4, \ldots\ \lambda_m$, where λ_1 has multiplicity of three. The input matrix and system matrix for this case will be as shown in equation (5.100a). The system matrix is also denoted as J. The transfer function of the system for this case is given by equation (5.99a) and the block diagram is shown in Fig. (5.10a)

$$B = \begin{vmatrix} 0 \\ 0 \\ 1 \\ 1 \\ \vdots \\ 1 \end{vmatrix}; A = J \begin{vmatrix} -\lambda_1 & 1 & 0 & 0 & \cdots & 0 \\ 0 & -\lambda_1 & 1 & 0 & \cdots & 0 \\ 0 & 0 & -\lambda_1 & 0 & \cdots & 0 \\ 0 & 0 & 0 & -\lambda_4 & \cdots & 0 \\ \vdots & \vdots & \vdots & \vdots & & \vdots \\ 0 & 0 & 0 & 0 & & -\lambda_n \end{vmatrix} \tag{5.100a}$$

$$\frac{Y(z)}{U(z)} = b_o + \frac{C_1}{(z+\lambda_1)^3} + \frac{C_2}{(z+\lambda_1)^2} + \frac{C_3}{z+\lambda_1} + \frac{C_4}{z+\lambda_4} + \ldots + \frac{C_n}{z+\lambda_n} \tag{5.99a}$$

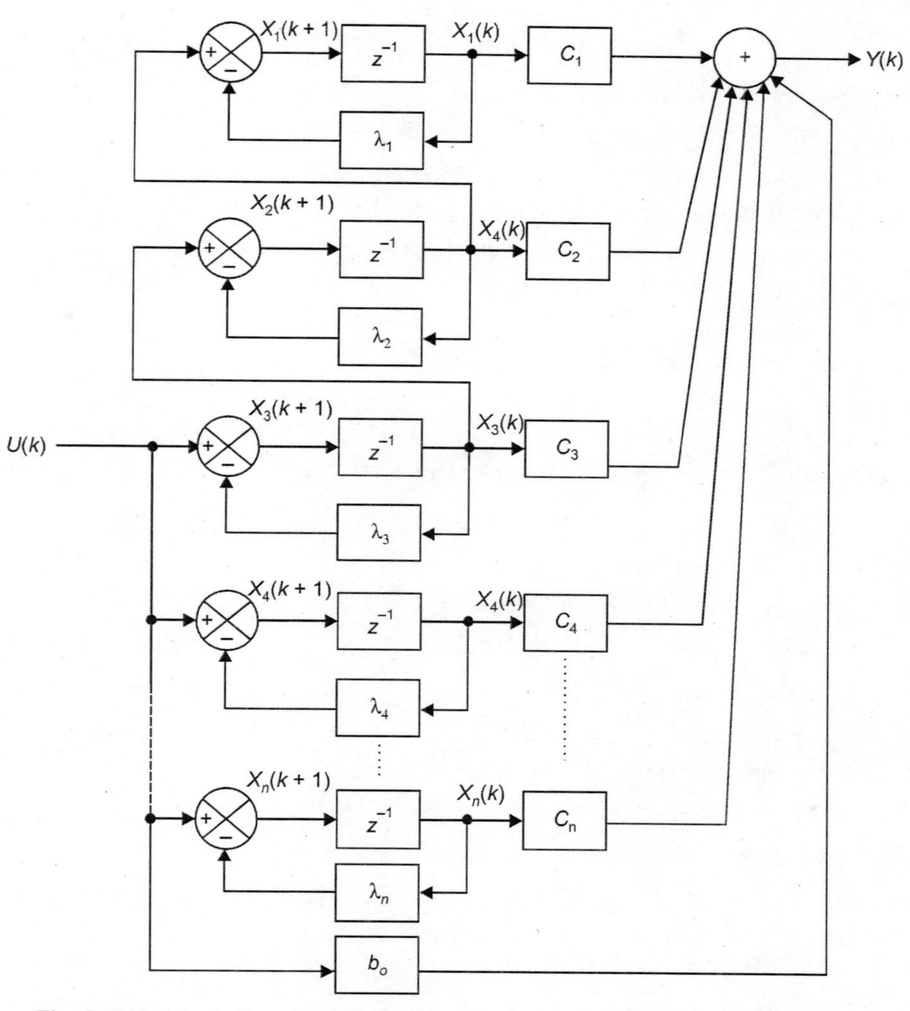

Fig. 5.10(a): Block diagram of Jordan Cannoical state model of discrete time system.

Solution of Discrete Time State Equation

The state equation of discrete time system is given by,

$$X(k+1) = A\,X(k) + B\,U(k) \quad (5.102)$$

when $k = 0$, the equation (5.102) can be written as

$$X(1) = A\,X(0) + B\,U(0) \quad (5.103)$$

when $k = 1$, the equation (5.102) can be written as

$$X(2) = A\,X(1) + B\,U(1) \quad (5.104)$$

On substituting for $X(1)$ from equation (5.103) in equation (5.104) we get,

$$X(2) = A[A\,X(0) + B\,U(0)\,] + B\,U(t)$$

$$\therefore \quad X(2) = A^2 X(0) + A B U(0) + B U(1) \tag{5.105}$$

On continuating this analysis till $X(k)$ we get,

$$X(k) = A^k X(0) + A^{(k-1)} B U(0) + A^{(k-2)} B U(1) + ... + AB U(k-2) + B U(k-1) \tag{5.106}$$

The matrix equation (5.106) is the solution of discrete time state equation. The matrix A^k is called the state transition matrix of discrete time system and it is also denoted by $\phi(k)$.

On substituting , $A^k = \phi(k)$ and $\phi(0) = I$ in equation (5.106) we get,

$$X(k) = \phi(k) X(0) + \phi(k-1) B U(0) + \phi(k-2) B U(1) + ... + \phi(1) B U(k-2) + \phi(0) B u(k-1)$$

$$\therefore \quad X(k) = \phi(k)X(0) + \sum_{j=0}^{k-1} \phi(k-1-j) BU(j) \tag{5.107}$$

Solution of Discrete Time State Equation using z-transform

Consider the state equation of the discrete time system

$$X(k+1) = A X(k) + B U(k) \tag{5.108}$$

Take z-transform of equation (5.108) we get,

$$z X(z) - X(0) = A X(z) + B U(z)$$

$$z X(z) - AX(z) = z X(0) + B U(z)$$

$$(zI - A)Xz = z X(0) + B U(z) \tag{5.109}$$

On premultiplying the equation (5.109) by $(zI - A)^{-1}$ we get,

$$X(z) = (zI - A)^{-1} z K(0) + (zI - A)^{-1} B U(z) \tag{5.110}$$

Take inverse z-transform of equation (5.110) we get $X(k)$

$$\therefore \quad X(k) = \mathcal{Z}^{-1} X\{x(z)\} = z^{-1}\left\{(zI - A)^{-1} zX(0) + (zI - A)^{-1} BU(z)\right\}$$

$$X(k) = \mathcal{Z}^{-1}\left\{(zI - A)^{-1} z\right\} X(0) + \mathcal{Z}^{-1}\left\{(zI - A)^{-1} BU(z)\right\} \tag{5.111}$$

Solution of discrete time state equation

Properties of State Transition Matrix of Discrete Time System

1. $\phi(0) = I$
2. $\phi^1(k) = \phi(-k)$
3. $\phi(k, k_o) = \phi(k - k_o) = A^{(k-k_o)}$; where $k > k_o$

Computation of State Transition Matrix

The state transition matrix A^k can be computed by any one of the following methods.

Method 1: Computation of A^k using z-transform

Method 2: Computation of A^k by canonical transformation

Method 3: Computation of A^k by cayley-Hamilton theorem.

The computation of A^k using z-transform have been dealt in this section.

On comparing equations (5.111) and (5.107) or [5.106] we can write,

State transition matrix, $$A^k = \mathcal{Z}^{-1}\left\{(zI - A)^{-1} z\right\} \qquad (5.112)$$

The equation (5.112) can be used to compute the state transition matrix, A^k.

EXAMPLE 5.27

A discrete time system has the transfer for function $\dfrac{Y(z)}{U(z)} = \dfrac{4z^3 - 12z^2 + 13z - 7}{(z-1)^2(z-2)}$.

Determine the state model of the system in

(a) Phase variable form and

(b) Jordan canonical form

Solution:

(a) Phase variable form of state model

Given that , $$\frac{Y(z)}{U(z)} = \frac{4z^3 - 12z^2 + 13z - 7}{(z-1)^2(z-2)}$$

$$= \frac{z^3\left(4 - 12z^{-1} + 13z^{-2} - 7z^{-3}\right)}{\left(z^2 - 2z + 1\right)(z-2)} = \frac{z^3(4 - 12z^{-1} + 13z^{-2} - 7z^{-3})}{z^3 - 2z^2 - 2z^2 + 4z + z - 2}$$

$$= \frac{z^3(4 - 12z^{-1} + 13z^{-2} - 7z^{-3})}{z^3(-4z^{-1} + 5z^{-2} - 2z^{-3})} = \frac{4 - 12z^{-1} + 13z^{-2} - 7z^{-3}}{1 - (4z^{-1} - 5z^{-2} + 2z^{-3})}$$

The equation (5.17.1) can be used to construct the signal flow graph of the discrete time system shown in Fig. (Ex. 5.27.1)

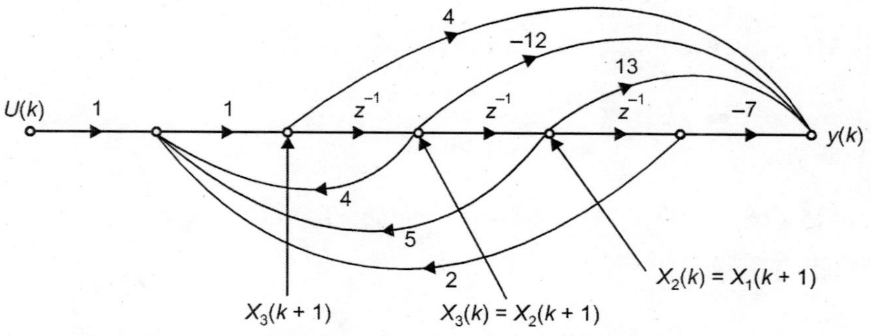

Fig. Ex. 5.27.1: Signal flow graph for transfer function of eq. (5.17.1).

From node X_2 and X_3 we get,

$$X_1(k+1) = X_2 \qquad (5.17.2)$$

CHAPTER 5

$$X_2(k+1) = X_3 \tag{5.17.3}$$

By summing up the incoming signals to node $X_3\,(k+1)$ we get,

$$X_3\,(k+1) = 2X_1 - 5X_2 + 4X_3 + u \tag{5.17.4}$$

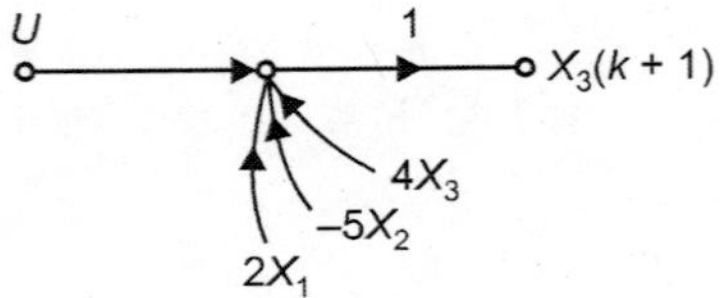

Fig. Ex. 5.27.2

The equations (5.17.2) to (5.17.4) are the state equations. The output equations is obtained by summing up the incoming signals to output node.

$$y = -7X_1 + 13\,X_2 - 12X_3 + 4X_3\,(\,k+1) \tag{5.17.5}$$

Fig. Ex. 5.27.3

On substituting for $X_3\,(k+1)$ from equation (5.17.4) in equation (5.17.5) we get,

$$y = -7X_1 + 13X_2 - 12X_3 + 4[2X_1 - 5X_2 + 4X_3 + u]$$

$$y = (-7+8)\,X_1 + (13-20)X_2 + (-12+16)\,X_3 + 4u$$

$$\therefore \quad y = X_1 - 7X_2 + 4X_3 + 4u \tag{5.17.6}$$

The state equations and output equation can be arranged in the matrix form as shown below:

$$\begin{vmatrix} X_1(k+1) \\ X_2(k+1) \\ X_3(k+1) \end{vmatrix} = \begin{vmatrix} 0 & 1 & 0 \\ 0 & 0 & 1 \\ 2 & -5 & 4 \end{vmatrix} \begin{vmatrix} X_1 \\ X_2 \\ X_3 \end{vmatrix} + \begin{vmatrix} 0 \\ 0 \\ 1 \end{vmatrix} u \tag{5.17.7}$$

$$Y = \begin{vmatrix} 1 & -7 & 4 \end{vmatrix} \begin{vmatrix} X_1 \\ X_2 \\ X_3 \end{vmatrix} + 4u \tag{5.17.8}$$

The equations (5.17.8) and (5.17.8) is the phase variable form of state model.

(b) Jordan Canonical form of state model

Given that,
$$\frac{Y(z)}{U(z)} = \frac{4z^3 - 12z^2 + 13z - 7}{(z-1)^2\,(z-2)}$$

$$= \frac{4z^3 - 12z^2 + 13z - 7}{z^3 - 4z^2 + 5z - 2}$$

$$\begin{array}{r|l} z^3 - 4z^2 + 5z - 2 & 4z^3 - 12z^2 + 13z - 7 \\ & \underline{4z^3 - 16z^2 + 20z - 8} \\ & \quad 4z^2 - 7z + 1 \end{array}$$

$$\frac{Y(z)}{U(z)} = 4 + \frac{4z^2 - 7z + 1}{z^3 - 4z^2 + 5z - 2} = 4 + \frac{4z^2 - 7z + 1}{(z-1)^2 (z-2)} \tag{5.17.9}$$

By partial fraction expansion the equation (5.17.9) can be written as,

$$\frac{Y(z)}{U(z)} = 4 + \frac{A_1}{(z-1)^2} + \frac{A_2}{z-1} + \frac{A_3}{z-2}$$

$$A_1 = (z-1)^2 \left.\frac{4z^2 - 7z + 1}{(z-1)^2 (z-2)}\right|_{z=1} = \left.\frac{4z^2 - 7z + 1}{z-2}\right|_{z=1}$$

$$= \frac{4 - 7 + 1}{1 - 2} = 2$$

$$A_2 = \left.\frac{d}{dz}\left[(z-1)^2 \frac{4z^2 - 7z + 1}{(z-1)^2 (z-2)}\right]\right|_{z=1} = \left.\frac{d}{dz}\left[\frac{4z^2 - 7z + 1}{z-2}\right]\right|_{z=1}$$

$$= \frac{(8z-7)(z-2) - (4z^2 - 7z + 1)}{(z-2)^2} = 1$$

$$A_3 = (z-2) \left.\frac{4z^2 - 7z + 1}{(z-1)^2 (z-2)}\right|_{z=2} = \left.\frac{4z^2 - 7z + 1}{(z-1)^2}\right|_{z=2}$$

$$= \frac{4 \times 2^2 - 7 \times 2 + 1}{(2-1)^2} = 3$$

$$\therefore \quad \frac{Y(z)}{U(z)} = 4 + \frac{2}{(z-1)^2} + \frac{1}{z-1} + \frac{3}{z-2} \tag{5.17.10}$$

The equation (5.17.10) can be used to construct the block diagram of the discrete time system shown in Fig. (Ex. 5.27.4)

Let us assign state variable at the output of each unit delay element. The state equations are formed by summing up the coming signals to the unit delay element.

The state equations are,

$$X_1(k+1) = X_1 + X_2$$

$$X_2(k+1) = X_2 + u$$

$$X_3(k+1) = 2X_3 + u$$

The output equation is,

$$y = 2X_1 + X_2 + 3X_3 + 4u$$

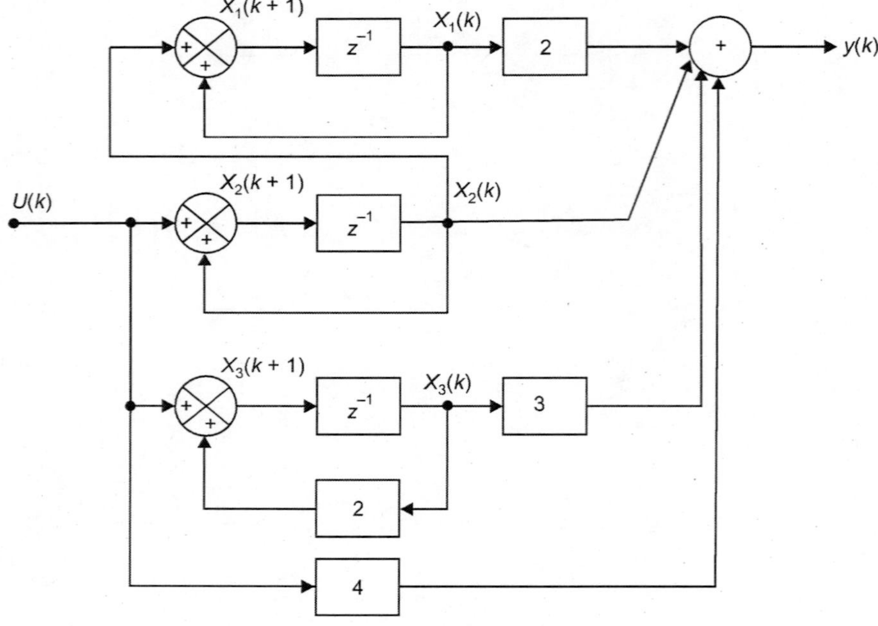

Fig. Ex. 5.27.4

Arrange the state equations and the output equation in the matrix form we get,

$$\begin{vmatrix} X_1(k+1) \\ X_2(k+2) \\ X_3(k+3) \end{vmatrix} = \begin{vmatrix} 1 & 1 & 0 \\ 0 & 1 & 0 \\ 0 & 0 & 2 \end{vmatrix} \begin{vmatrix} X_1 \\ X_2 \\ X_3 \end{vmatrix} + \begin{vmatrix} 0 \\ 1 \\ 1 \end{vmatrix} u \tag{5.17.11}$$

$$Y = \begin{vmatrix} 2 & 1 & 3 \end{vmatrix} \begin{vmatrix} X_1 \\ X_2 \\ X_3 \end{vmatrix} + 4u \tag{5.17.12}$$

The equation (5.17.11) and (5.17.12) constitute, Jordan canonical form of state model of the system.

EXAMPLE 5.28

A discrete time system is described by the difference equation,

$$y(k+2) + 5y(k+1) + 6(yk) = u(k)$$

$$Y(0) = Y(1) = 0 \; ; \; T = 1 \text{ sec.}$$

(a) Determine a state model in canonical form

(b) Find the state transition matrix

(c) For input $u(k) = 1$; $k \geq 1$, find the output $y(k)$

Solution:

(a) To determine the canonical form of state model

Given that, $y\,(k+2) + 5y(k+1) + 6(yk) = u(k)$ (5.18.1)

and $y(0) = y(1) = 0$; $T = 1$ sec.

Take z-transform of equation (5.18.1) with zero initial conditions we get,

$$z^2 y(z) + 5z\, y(z) + 6y(z) = U(z)$$

$$(z^2 + 5z + 6)y(z) = U(z)$$

$$\therefore \quad \frac{Y(z)}{U(z)} = \frac{1}{z^2 + 5z + 6} = \frac{1}{(z+3)(z+2)} \tag{5.18.2}$$

by partial fraction expansion the equation (5.18.2) can be expressed as,

$$\frac{Y(z)}{U(z)} = \frac{A_1}{z+3} + \frac{A_2}{z+2}$$

$$A_1 = (z+3)\frac{1}{(z+3)(z+2)}\bigg|_{z=-2} = \frac{1}{(z+2)}\bigg|_{z=-3} = \frac{1}{-3+2} = -1$$

$$\therefore \quad \frac{Y(z)}{U(z)} = \frac{-1}{z+3} + \frac{1}{z+2} \tag{5.18.3}$$

The equation (5.18.3) can be used to construct the block diagram shown in Fig. (Ex. 5.28.1)

From the block diagram, the state equations are,

$$X_1(k+1) = -3\, X_1 + u$$
$$X_2(k+1) = -2\, X_1 + u$$

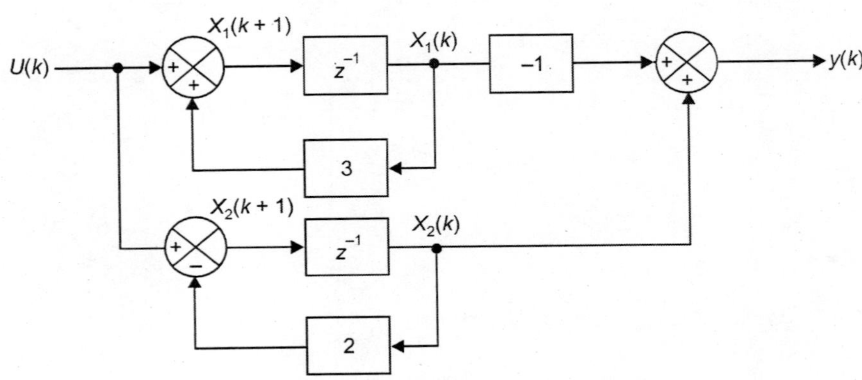

Fig. Ex. 5.28.1: Block diagram.

The output equation is

$$y = -X_1 + X_2$$

Arranging state equation and output equation in the matrix form, we get,

$$\begin{vmatrix} X_1(k+1) \\ X_3(k+1) \end{vmatrix} = \begin{vmatrix} -3 & 0 \\ 0 & -2 \end{vmatrix}\begin{vmatrix} X_1 \\ X_2 \end{vmatrix} + \begin{vmatrix} 1 \\ 1 \end{vmatrix} u \tag{5.18.4}$$

$$Y = \begin{bmatrix} -1 & 1 \end{bmatrix}\begin{vmatrix} X_1 \\ X_2 \end{vmatrix} \tag{5.18.5}$$

Equations (5.18.4) and (5.18.5) constitute the canonical form of state model.

(b) To find state transition matrix A^k:

The state transition matrix,

$$A^k = Z^{-1}\{[zI - A]^{-1}\, z\}$$

From the state equation we get,

$$A = \begin{vmatrix} -3 & 0 \\ 0 & -2 \end{vmatrix}$$

$$\therefore \quad zI - A = z\begin{vmatrix} 1 & 0 \\ 0 & 1 \end{vmatrix} - \begin{vmatrix} -3 & 0 \\ 0 & -2 \end{vmatrix} = \begin{vmatrix} z+3 & 0 \\ 0 & z+2 \end{vmatrix}$$

$$|zI - A| = \begin{vmatrix} z+3 & 0 \\ 0 & z+2 \end{vmatrix} = (z+2)(z+3)$$

$$|zI - A|^{-1} = \frac{1}{|zI - A|}\begin{vmatrix} z+2 & 0 \\ 0 & z+3 \end{vmatrix}^T$$

$$= \frac{1}{(z+2)(z+3)}\begin{vmatrix} z+2 & 0 \\ 0 & z+3 \end{vmatrix} = \begin{vmatrix} \dfrac{1}{z+3} & 0 \\ 0 & \dfrac{1}{z+2} \end{vmatrix} \tag{5.18.6}$$

$$A^K = Z^{-1}\{[zI - A]^{-1}\, z\} = Z^{-1}\begin{bmatrix} \dfrac{z}{z+3} & 0 \\ 0 & \dfrac{z}{z+2} \end{bmatrix}$$

we know that,

$$Z^{-1}\left\{\frac{z}{z-a}\right\} = a^k$$

$$\therefore \quad Z^{-1}\left\{\frac{z}{z+3}\right\} = (-3)^k \text{ and}$$

$$Z^{-1}\left\{\frac{z}{z+2}\right\} = (-2)^k$$

$\therefore$ state transition matrix,

$$A^k = \begin{bmatrix} (-3)^k & 0 \\ 0 & (-2)^k \end{bmatrix}$$

(c) To Find $y(k)$ when the input is unit step:

Consider the state equation,

$$X(k+1) = A\,X(k) + B\,U(k) \tag{5.18.8}$$

Or taking z-transform of equation (5.18.8) we get

$$z\,X(z) = z\,X(0) = A\,X(z) + B\,U(z)$$

Let us assume that, $X(0) = 0$

$$\therefore \quad z\,X(z) = A\,X(z)\,) = B\,U(z)$$

$$(z1 - A)X(z) = B\,U(z) \tag{5.18.9}$$

on premultiplying by $(z1 - A)^{-1}$ on both sides we get,

$$X(z) = (z1 - A)^{-1}\,B\,U(z) \tag{5.18.10}$$

Given that, $u(k) = 1$,

$$\therefore \quad U(z) = \mathcal{Z}\{u(k)\} = \frac{z}{z-1} \tag{5.18.11}$$

From equation (5.18.4) we get,

$$[z1 - A]^{1} = \begin{vmatrix} \dfrac{1}{z+3} & 0 \\ 0 & \dfrac{1}{z+2} \end{vmatrix}$$

From equation (5.18.4) we get,

$$B = \begin{bmatrix} 1 \\ 1 \end{bmatrix}$$

$$\therefore \quad X(z) = \begin{vmatrix} \dfrac{1}{z+3} & 0 \\ 0 & \dfrac{1}{z+2} \end{vmatrix} \begin{bmatrix} 1 \\ 1 \end{bmatrix} \begin{vmatrix} \dfrac{z}{z-1} \end{vmatrix}$$

$$= \begin{vmatrix} \dfrac{1}{z+3} \\ \dfrac{1}{z+2} \end{vmatrix} \begin{vmatrix} \dfrac{z}{z-1} \end{vmatrix} = \begin{vmatrix} \dfrac{z}{(z-1)(z+3)} \\ \dfrac{z}{(z-1)(z+2)} \end{vmatrix}$$

we know that $X(z) = \begin{vmatrix} X_1(z) \\ X_2(z) \end{vmatrix}$ and $X(k) = \begin{vmatrix} X_1(k) \\ X_2(k \end{vmatrix}$

Also $X(k) = \mathcal{Z}^{-1}\{X(z)\}$

$$\therefore \quad X(z) = \begin{vmatrix} X_1(z) \\ X_2(z) \end{vmatrix} = \begin{vmatrix} \dfrac{z}{(z-1)(z+3)} \\ \dfrac{z}{(z-1)(z+2)} \end{vmatrix}$$

and
$$X(k) = \begin{vmatrix} X_1(k) \\ X_2(k) \end{vmatrix} = \begin{vmatrix} \mathcal{Z}^{-1}\{X_1(z)\} \\ \mathcal{Z}^{-1}\{X_2(z)\} \end{vmatrix}$$

$$\therefore \quad X_1(k) = \mathcal{Z}^{-1}\left\{\frac{z}{(z-1)(z+3)}\right\} \text{ and } X_2(k) = \mathcal{Z}^{-1}\left\{\frac{z}{(z-1)(z+2)}\right\}$$

By partial fraction expansion,

$$\frac{z}{(z-1)(z+3)} = z\left[\frac{1}{(z-1)(z+3)}\right] = z\left[\frac{A_1}{z-1} + \frac{B_1}{z+3}\right]$$

$$A_1 = \frac{1}{(z-1)(z+3)}(z-1)\Big|_{z=1}$$

$$= \frac{1}{z+3}\Big|_{z=1} = \frac{1}{1+3} = \frac{1}{4}$$

$$B_1 = \frac{1}{(z-1)(z+3)}(z+3)\Big|_{z=-3} = \frac{1}{z-1}\Big|_{z=-3}$$

$$= \frac{1}{-3-1} = \frac{-1}{4}$$

$$\therefore \quad \frac{z}{(z-1)(z+3)} = z\left[\frac{A_1}{z-1} + \frac{B_1}{z+3}\right] = \left[\frac{1}{4}\frac{z}{z-1} - \frac{1}{4}\frac{z}{z+3}\right]$$

$$X_1(k) = \mathcal{Z}^{-1}\left\{\frac{z}{(z-1)(z+3)}\right\} = z^{-1}\left\{\frac{1}{4}\frac{z}{z-1} - \frac{1}{4}\frac{z}{2+3}\right\}$$

$$= \frac{1}{4}u(k) - \frac{1}{4}(-3)^k$$

By partial fraction expansion

$$\frac{z}{(z-1)(z+2)} = z\left[\frac{1}{(z-1)(z+2)}\right] = z\left[\frac{A_2}{z-1} + \frac{B_2}{z+2}\right]$$

$$A_2 = \frac{1}{(z-1)(z+2)}(z-1)\Big|_{z=1} = \frac{1}{z+2}\Big|_{z=1}$$

$$= \frac{1}{1+2} = \frac{1}{3}$$

$$B_2 = \frac{1}{(z-1)(z+2)}(z+2)\Big|_{z=-2} = \frac{1}{z-1}\Big|_{z=-2}$$

$$= \frac{1}{-2-1} = \frac{-1}{3}$$

$$\therefore \quad \frac{z}{(z-1)(z+2)} = z\left[\frac{A_2}{z-1} + \frac{B_2}{z+2}\right]$$

$$= \left[\frac{1}{3}\ \frac{z}{z-1} - \frac{1}{3}\ \frac{z}{z+2}\right]$$

$$X_2(k) = \mathcal{Z}^{-1}\left\{\frac{z}{(z-1)(z+2)}\right\}$$

$$= \mathcal{Z}^{-1}\left\{\frac{1}{3}\ \frac{z}{z-1} - \frac{1}{3}\ \frac{z}{z+2}\right\}$$

$$= \frac{1}{3}u(k) - \frac{1}{3}(-2)^k$$

From equation (5.18.5) we get,

Output, $$y(k) = \begin{bmatrix}-1 & 1\end{bmatrix}\begin{bmatrix}X_1(k)\\ X_2(k)\end{bmatrix}$$

$$\therefore \quad y(k) = -X_1(k) + X_2(k)$$

$$= \frac{1}{4}u(k) + \frac{1}{4}(-3)^k + \frac{1}{3}u(k) - \frac{1}{3}(-2)^k$$

$$= \frac{1}{4}(-3)^k - \frac{1}{3}(-2)^k + \frac{-3+4}{12}u(k)$$

$$= \frac{1}{4}(-3)^k - \frac{1}{3}(-2)^k + \frac{1}{12}u(k)$$

5.17 CONCEPT OF CONTROLLABILITY AND OBSERVABILITY CONTROLLABILITY

The controllability verifies the usefulness of a state variable. In the controllability test we can find, whether the state variable can be controlled to achieve the desired output. The choice of state variables is arbitrary while forming the state model. After determining the state model, the controllability of the state variable is verified. If the state variables is not controllable then we have to go for another choice of state variable.

Definition of Controllability

A system is said to be completely state controllable if it is possible to transfer the system state from any initial state $x(t_o)$ to any other desired state $x(t_d)$ in specified finite time by a control vector $U(t)$.

The controllability of a state model can be tested by Kalman's test (or) Gilbert's test.

Gilbert's method of testing controllability:

Case I

When the system matrix has distinct eigen values.

In this case the system matrix can be diagonalized and the state model can be converted to cannocial form.

Consider the state model of the system,

$$\dot{X} = AX + BU$$

$$Y = CX + DU$$

The state model can be converted to cannoical form by a transformation, $X = MZ$, when M is the model matrix and Z is the transformed state variable vector.

The transformed state model is given by

$$\dot{Z} = .\ Z + \tilde{B}U$$

$$Y = \tilde{C}Z + DU$$

where

$$. = M^{-1} AM$$

$$\tilde{B} = M^{-1}B$$

$$\tilde{C} = CM$$

In this case the necessary and sufficient condition for complete controllability is that, the matrix B must have no rows with all zeros. If any row of the matrix B is zero then the corresponding state variable is uncontrollable.

Case II

When the system matrix has repeated eigen values.

In this case, the system matrix cannot be diagegeneralized but can be transformed to Jordan canonical form.

Consider the state model of the system,

$$\dot{X} = AX + BU$$

$$Y = CX + DU$$

∴

The state model can be transformed to Jordan cannoical form by a transformation. $X = MZ$, where M is model matrix and Z is the transformed state variable vector.

The transformed state model is given by,

$$\dot{Z} = JZ + \tilde{B}U$$

$$Y = \tilde{C}Z + DU$$

where

$$J = M^{-1}AM$$

$$\tilde{B} = M^{-1}B$$

$$\tilde{C} = CM$$

In this case, the system is completely controllable if the elements of any row of $\tilde{B}$ that correspond to the last row of each Jordan block are not all zero and the row corresponding to other state variables must not have all zeros.

Kalman's method of testing controllability

Consider a system with state equation, $\dot{X} = AX + BU$. For this system, a composite matrix, Q_c can be formed,

$$Q_c = [B \quad AB \quad A^2B \quad ...A^{n-1} \quad B] \tag{5.10.1}$$

where n is the order of the system (n is also equal to number of state variables).

In this case the system is completely state controllable if the rank of the composite matrix, Q_c is n.

The rank of the matrix is n, if the determinant of $(n \times n)$ composite matrix Q_c is non-zero. i.e., $|Q_c| \neq 0$, then rank of $Q_c = n$ and the system is completely state controllable.

The advantage in Kalman's test is that the calculations are simpler. But the disadvantage in Kalman's test is that, we can't find the state variable which is uncontrollable. But in Gilbert's method we can find the uncontrollable state variable which is the state variable corresponding to the row of $\tilde{B}$ which has all zeros.

Condition for complete state controllability in the *s*-plane

A necessary and sufficient condition for complete state controllability is that no cancellation of poles and zeros occurs in the transfer function of the system. If cancellation occurs then the system cannot be controlled in the direction of the cancelled mode.

Observability

In observability test we can find whether the state variables is observable or measurable. The concept of observability is useful in solving the problem of reconstructing unmeasurable state variables form measurable ones in the minimum possible length of time. In state feedback control the estimation of unmeasurable state variables is essential in order to construct the control signals.

Definition of Observability

A system is said to be completely observable if every state $x(t)$ can be completely identified by measurements of the output $Y(t)$ over a finite time interval.

The observability of a system can be tested by either Gilbert's method (or) Kalman's method.

Gilbert's method of testing observability

Consider a state model of nth order system, $\dot{X} = AX + BU; Y = CX + DU$.

The state model can be transformed to a canonical Jordan canonical form by a transformation, $X = MZ$, where M is the modal matrix and Z is the transformed state variable vector.

The transformed state model is,

$$\dot{Z} = .\ Z + \tilde{B}U$$

$$Y = \tilde{C}Z + DU$$

or

$$\dot{Z} = JZ + \tilde{B}U$$

$$Y = \tilde{C}Z + DU$$

where

$$. = M^{-1}AM; \text{ if eigen values are distinct; } \tilde{B} = M^{-1}B$$

$$J = M^{-1}AM \text{ if eigne values have multiplicity; } \tilde{C} = CM$$

* The necessary and sufficient condition for complete observability is that none of the columns of the matrix $\tilde{C}$ be zero if any of the column's of $\tilde{C}$ has all zeros then the corresponding state variable is not observable.

Kalman's test for observability

Consider a system with state model,

$$\dot{X} = AX + BU;\ Y = CX + DU$$

For this system, a composite matrix, Q_0 can be formed such that,

$$Q_0 = [C^T \cdot A^TC^T \quad (A^T)^2C^T \quad (A^T)^3C^T \ldots (A^T)^{n-1}\,C^T] \tag{5.10.2}$$

where n is the order of the system (n is also equal to number of state variables).

* In this case, the system is completely observable if the rank of composite matrix, Q_0 is n.

The rank of the matrix is n, if the determinant of $n \times n$ composite matrix Q_0 is non-zero. The disadvantage in Kalman's test is that, the non observable state variables cannot be determined.

Condition for complete observability in the *s*-plane

The necessary and sufficient condition for complete observability is that no cancellation of poles and zeros occurs in the transfer function. If cancellation occurs, the cancelled mode cannot be observed in the output.

Relationships between controllability, observability and transfer functions

The concepts of controllability and observability is important in design of control system in state space. They give the existence of a complete solution to the control system design problem. The solution may not exist if the system is not controllable.

It is important to note that all physical systems are controllable and observable. However, the mathematical models of these systems may not posses the property of the controllability (or) observability. Then it is necessary to known the conditions under which a system is controllable and observable.

Duality property

The concepts of controllability and observability are dual concepts and it is proposed by Kalman's as principle of duality.

The principle of duality state s that a system is completely state controllable if and only if its dual system is completely. Observable or vice versa [i.e., if the system is observable then its dual is controllable]. Using the principle of duality, the observability of a given system can be checked by testing the state controllability of its dual or vice-versa.

Consider the systems S_1, described by the state model shown below.

$$\dot{X} = AX + BU$$

$$Y = CX$$

Let the dual of system S_1 be denoted as S_2 and the dual system S_2 is described the following state model.

$$\dot{Z} = A^T Z + C^T V$$

$$N = B^T Z$$

where, Z = state vector of dual system

V = Input vector of dual system;

N = Output vector of dual system

For the system S_1 the composite matrix, Q_{c1} for controllability is given by equation (5.10.3) and the composite matrix, Q_{o1} for observability is given by equation (5.10.4)

$$Q_{c1} = [B \quad AB \quad A^2B \quad \ldots A^{n-1}B] \tag{5.10.3}$$

$$Q_{o1} = [C^T \quad A^TC^T \quad (A^T)^2C^T \quad \ldots (A^T)^{n-1}C^T] \tag{5.12.4}$$

For the dual system S_2 the composite matrix, Q_{c2} for controllability is given by equation (5.1.0.5) and the composite matrix Q_{o2} for observability is given by equation (5.10.6).

$$Q_{c2} = [C^T \quad A^TC^T \quad (AT)^2C^T \quad \ldots (A^T)^{n-1}C^T] \tag{5.10.5}$$

$$Q_{o2} = [A \quad AB \quad A^2B \quad \ldots A^{n-1}B] \tag{5.10.6}$$

From equations (5.10.3) and (5.10.6) we get $Q_{c1} = Q_{o2}$, hence if the system S_1 is controllable then its dual system S_2 is observable.

From equations (5.10.4) and (5.10.5) we get $Q_{o1} = Q_{c2}$, hence if the system S_1 is observable then its dual system S_2 is controllable.

Effect of pole-zero cancellation in transfer function

The controllability and observability are closely related to the properties of the transfer function. Consider an nth order system with distinct eigen values. The transfer function of the system can be expressed as a ratio of two polynomials as shown in equation (5.10.7).

$$T(s) = \frac{Y(s)}{U(s)} = \frac{b_0 s^m + b_1 s^{m-1} + \ldots + b_{m-1}s_1 + b_m}{s^n + a_1 s^{n-1} + \ldots + a_{n-1}s + a_n}; m < n \tag{5.10.7}$$

$$= \frac{K(s+\beta_1)(s+\beta_2)\ldots(s+\beta_m)}{(s+\lambda_1)(s+\lambda_2)\ldots(s+\lambda_n)} \tag{5.10.8}$$

Apply partial fraction

$$\frac{Y(s)}{U(s)} = \frac{C_1}{s+\lambda_1} + \frac{C_2}{s+\lambda_2} + \ldots + \frac{C_1}{s+\lambda_1} + \ldots + \frac{C_n}{s+\lambda_n} \tag{5.10.9}$$

where, C_1, C_2, C_3, ... Cn are residues.

If the transfer function has identical pair of pole and zero at $\beta_i = \lambda_i$, then $C_i = 0$. The effect of this cancellation on controllability and observability properties depends on the choice of state variables.

In one method of state space modelling using cannoical of variables, the $C_i = 0$, will appear in input vector B and state x_i is uncontrollable. In other method of state space modelling using canonical variables, the $C_i = 0$, will appear in output vector C and the state x_i is shielded from observation.

The conclusion is that, if cancellation of pole-zero occurs in the transfer function of a system, then the system will be either not state controllable or unobservable, depending of how the state variables are defined. If the transfer function does not have pole-zero cancellation, the system can always be represented by completely controllable and observable state model.

EXAMPLE 5.29

Write the state equations for the system shown in Fig. (Ex. 5.29.1) in which x_1, x_2 and x_3 constitute the state vector. Determining whether the system is completely controllable and observable.

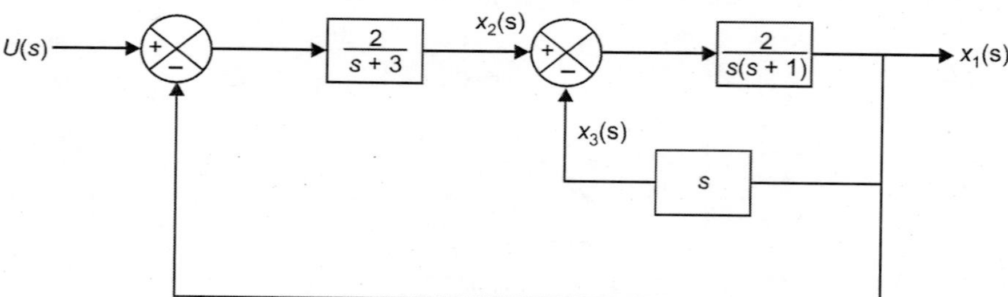

Fig. Ex. 5.29.1

Solution:

(i) State model

Consider the Fig. (5.10.1.1) for the block shown in Fig. (Ex. 5.29.2).

X2(s) − X3(s) → 2/s(s + 1) → X1(s)

Fig. Ex. 5.29.2

We can write state model equation for Fig. (Ex. 5.29.2)

$$X_1(s) = [X_2(s) - X_3(s)]\left[\frac{2}{s(s+1)}\right]$$

$$s(s+1)\,X_1(s) = 2\,X_2(s) - 2\,X_3(s)$$

$$s^2 X_1(s) + s X_1(s) = 2 X_2(s) - 2 X_3(s)$$

on taking inverse laplace transform,

$$\ddot{X}_1 + \dot{X}_1 = 2X_2 - 2X_3 \qquad (5.12.1.1)$$

with reference the Fig. (Ex. 5.29.3), we can write,

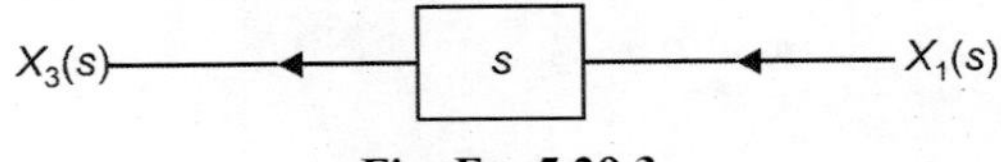

Fig. Ex. 5.29.3

$$X_3(s) = s X_1(s)$$

on taking inverse Laplace transform

$$X_3 = \dot{X}_1 \qquad (5.12.1.2)$$

with reference to Fig. (Ex. 5.29.4) we can write,

$U(s) - x_1(s)$ → $\frac{2}{s+3}$ → $x_2(s)$

Fig. Ex. 5.29.4

$$X_2(s) = [U(s) - X_1(s)]\left[\frac{2}{s+3}\right]$$

$$X_2(s)\,(s+3) = 2U(s) - 2\,X_1(s)$$

$$s\,X_2(s) + 3X_2(s) = 2U(s) - 2\,X_1(s)$$

on taking inverse Laplace transform

$$\dot{X}_2 + 3X_2 = 2u - 2X_1$$

$$\dot{X}_2 = -2X_1 - 3X_2 + 2u \qquad (5.10.1.3)$$

From equation (5.10.1.2), we get

$$\dot{X}_1 = X_3;$$

$$\therefore \qquad \ddot{X}_1 = \dot{X}_3$$

Put $\dot{X}_1 = X_3$ and $\ddot{X}_1 = \dot{X}_3$ in equation (5.10.1.1)

$$\therefore \qquad \dot{X}_3 + X_3 = 2X_2 - 2X_3$$

$$\dot{X}_3 = 2X_2 - 2X_3 - X_3$$

$$\dot{X}_3 = 2X_2 - 3X_2 \qquad (5.10.1.4)$$

The state equation are given by equation (5.10.1.2), (5.10.1.3) and (5.10.1.4)

$$\dot{X}_1 = X_3$$

$$\dot{X}_2 = -2X_1 - 3X_2 + 2u$$

$$\dot{X}_3 = 2X_2 - 3X_3$$

The output equation is $y = X_1$

The state model in the matrix from is

$$\begin{bmatrix} \dot{X}_1 \\ \dot{X}_2 \\ \dot{X}_3 \end{bmatrix} = \begin{bmatrix} 0 & 0 & 1 \\ -2 & -3 & 0 \\ 0 & 2 & -3 \end{bmatrix} \begin{bmatrix} X_1 \\ X_2 \\ X_3 \end{bmatrix} + \begin{bmatrix} 0 \\ 2 \\ 0 \end{bmatrix} u; \ y = [1 \ 0 \ 0] \begin{bmatrix} X_1 \\ X_2 \\ X_3 \end{bmatrix} \quad (5.10.1.5)$$

Kalman's Test for Controllability

$$A^2 = A \cdot A \begin{bmatrix} 0 & 0 & 1 \\ -2 & -3 & 0 \\ 0 & 2 & -3 \end{bmatrix} \begin{bmatrix} 0 & 0 & 1 \\ -2 & -3 & 0 \\ 0 & 2 & -3 \end{bmatrix} = \begin{bmatrix} 0 & 2 & -3 \\ 6 & 9 & -2 \\ -4 & -12 & 9 \end{bmatrix}$$

$$A \cdot B = \begin{bmatrix} 0 & 0 & 1 \\ -2 & -3 & 0 \\ 0 & 2 & -3 \end{bmatrix} \begin{bmatrix} 0 \\ 2 \\ 0 \end{bmatrix} = \begin{bmatrix} 0 \\ -6 \\ 4 \end{bmatrix}$$

$$A^2 \cdot B = \begin{bmatrix} 0 & 2 & -3 \\ 6 & 9 & -2 \\ -4 & -12 & 9 \end{bmatrix} \begin{bmatrix} 0 \\ 2 \\ 0 \end{bmatrix} = \begin{bmatrix} 4 \\ 18 \\ -24 \end{bmatrix}$$

The composite matrix for controllability,

$$Q_c = [B \quad AB \quad A^2B]$$

$$= \begin{bmatrix} 0 & 0 & 4 \\ 2 & -6 & 18 \\ 0 & 4 & -24 \end{bmatrix}$$

Determinant of $Q_c = \begin{vmatrix} 0 & 0 & 4 \\ 2 & -6 & 18 \\ 0 & 4 & -24 \end{vmatrix} = 4 \times 8 = 32$

Since $|Q_c| \neq 0$, the rank of $Q_c = 3$.

*Hence the system is completely state controllable.

Kalman's Test for Observability

$$A^T = \begin{bmatrix} 0 & 0 & 1 \\ -2 & -3 & 0 \\ 0 & 2 & -3 \end{bmatrix}^T = \begin{bmatrix} 0 & -2 & 0 \\ 0 & -3 & 2 \\ 1 & 0 & -3 \end{bmatrix}$$

$$C^T = [1\ 0\ 0]^T = \begin{bmatrix} 1 \\ 0 \\ 0 \end{bmatrix}$$

$$(A^T)^2 = \begin{bmatrix} 0 & -2 & 0 \\ 0 & -3 & 2 \\ 1 & 0 & -3 \end{bmatrix}\begin{bmatrix} 0 & -2 & 0 \\ 0 & -3 & 2 \\ 1 & 0 & -3 \end{bmatrix}$$

$$= \begin{bmatrix} 0 & 6 & -4 \\ 2 & 9 & -12 \\ -3 & -2 & 9 \end{bmatrix}$$

$$A^TC^T = \begin{bmatrix} 0 & -2 & 0 \\ 0 & -3 & 2 \\ 1 & 0 & -3 \end{bmatrix}\begin{bmatrix} 1 \\ 0 \\ 0 \end{bmatrix} = \begin{bmatrix} 0 \\ 0 \\ 1 \end{bmatrix}$$

$$(A^T)^2C^T = \begin{bmatrix} 0 & 6 & -4 \\ 2 & 9 & -12 \\ -3 & -2 & 9 \end{bmatrix}\begin{bmatrix} 1 \\ 0 \\ 0 \end{bmatrix} = \begin{bmatrix} 0 \\ 2 \\ -3 \end{bmatrix}$$

The composite matrix for observability

$$[C^T \quad A^TC^T \quad (A^T)^2C^T] = \begin{bmatrix} 1 & 0 & 0 \\ 0 & 0 & 2 \\ 0 & 1 & -3 \end{bmatrix}$$

$$\text{Determinant of } Q_0 \begin{vmatrix} 1 & 0 & 0 \\ 0 & 0 & 2 \\ 0 & 1 & -3 \end{vmatrix} = 1\,x - 2 = -2;$$

Since $Q_0 \neq 0$, the rank of $Q_0 = 3$.

* Hence the system is completely observable.

To find eigen values (Another method)

Here the system matrix, $A = \begin{bmatrix} 0 & 0 & 01 \\ -2 & -3 & 0 \\ 0 & 2 & -3 \end{bmatrix}$ from equation (5.10.1.5)

The characteristic equation is

$$|\lambda I - A| = 0$$

CHAPTER 5

$$(\lambda I - A) = \lambda \begin{bmatrix} 1 & 0 & 0 \\ 0 & 1 & 0 \\ 0 & 0 & 1 \end{bmatrix} - \begin{bmatrix} 0 & 0 & 1 \\ -2 & -3 & 0 \\ 0 & 2 & -3 \end{bmatrix} = \begin{bmatrix} \lambda & 0 & -1 \\ 2 & \lambda+3 & 0 \\ 0 & -2 & \lambda+3 \end{bmatrix}$$

$$|\lambda I - A| = \begin{vmatrix} \lambda & 0 & -1 \\ 2 & \lambda+3 & 0 \\ 0 & -2 & \lambda+3 \end{vmatrix} = \lambda(\lambda+3)^2 - 1(-4) = \lambda(\lambda^2 + 6\lambda + 9) + 4$$

$$= \lambda^3 + \lambda^2 + 9\lambda + 4 = (\lambda + 1)(\lambda^2 + 5\lambda + 4)$$

$$= (\lambda + 1)(\lambda + 1)(\lambda + 4) = (\lambda + 1)^2 (\lambda + 4)$$

The eigen values are $\lambda_1 = -1$, $\lambda_2 = -1$, and $\lambda_3 = -4$

To find eigen vectors

$$(\lambda_1 I - A) = \lambda_1 \begin{bmatrix} 1 & 0 & 0 \\ 0 & 1 & 0 \\ 0 & 0 & 1 \end{bmatrix} - \begin{bmatrix} 0 & 0 & 1 \\ -2 & -3 & 0 \\ 0 & 2 & -3 \end{bmatrix} = \begin{bmatrix} \lambda_1 & 0 & -1 \\ 2 & \lambda+3 & 0 \\ 0 & -2 & \lambda_1+3 \end{bmatrix}$$

Let C_{11}, C_{12} and C_{13} can be cofactors along 1st row of the matrix $(\lambda_1 I - A)$

$$C_{11} = (+1)\begin{vmatrix} \lambda_1+3 & 0 \\ -2 & \lambda_1+3 \end{vmatrix} = (\lambda_1+3)^2 = \lambda_1^2 + 6\lambda_1 + 9$$

$$C_{12} = (-1)\begin{vmatrix} 2 & 0 \\ 0 & \lambda_1+3 \end{vmatrix} = -[(\lambda_1+3)^2] = -2\lambda_1 - 6$$

$$C_{13} = (+1)\begin{vmatrix} 2 & \lambda_1+3 \\ 0 & -2 \end{vmatrix} = -4$$

$$m_1 = \begin{bmatrix} C_{11} \\ C_{12} \\ C_{13} \end{bmatrix} = \begin{bmatrix} \lambda_1^2 + 6\lambda_1 + 9 \\ -2\lambda_1 - 6 \\ -4 \end{bmatrix} = \begin{bmatrix} 1-6+9 \\ 2-6 \\ -4 \end{bmatrix} = \begin{bmatrix} 4 \\ -4 \\ -4 \end{bmatrix}$$

$$m_2 = \begin{bmatrix} \dfrac{d}{d\lambda_1} C_{11} \\ \dfrac{d}{d\lambda_1} C_{12} \\ \dfrac{d}{d\lambda_1} C_{13} \end{bmatrix} = \begin{bmatrix} 2\lambda_1 + 6 \\ -2 \\ 0 \end{bmatrix} = \begin{bmatrix} -2+6 \\ -2 \\ 0 \end{bmatrix} = \begin{bmatrix} 4 \\ -2 \\ 0 \end{bmatrix}$$

$$(\lambda_3 I - A) = (-4)\begin{bmatrix} 1 & 0 & 0 \\ 0 & 1 & 0 \\ 0 & 0 & 1 \end{bmatrix} - \begin{bmatrix} 0 & 0 & 1 \\ -2 & -3 & 0 \\ 0 & 2 & -3 \end{bmatrix} = \begin{bmatrix} -4 & 0 & -1 \\ 2 & -1 & 0 \\ 0 & -2 & -1 \end{bmatrix}$$

Let C_{11}, C_{12} and C_{13} be the cofactors along Ist row of the matrix $(\lambda_3 I - A)$.

$$C_{11} = (+1)\begin{vmatrix} -1 & 0 \\ -2 & -1 \end{vmatrix} = 1; \; C_{12} = (-1)\begin{vmatrix} 2 & 0 \\ 0 & -1 \end{vmatrix} = 2;$$

$$C_{13} = (+1)\begin{bmatrix} 2 & -1 \\ 0 & -2 \end{bmatrix} = -4$$

$$\therefore \qquad m_3 = \begin{bmatrix} C_{11} \\ C_{12} \\ C_{13} \end{bmatrix} = \begin{bmatrix} 1 \\ 2 \\ -4 \end{bmatrix}$$

To find canonical form of state model

The model matrix, M is given by

$$M = [m_1, m_2, m_3] = \begin{bmatrix} 4 & 4 & 1 \\ -4 & -2 & 2 \\ -4 & 0 & -4 \end{bmatrix}$$

$$M^{-1} = \frac{[\text{cofactor of } M]^{+}}{\text{Determinant of } M} = \frac{M^T co_f}{\Delta_M}$$

$$\Delta M = \begin{vmatrix} 4 & 4 & 1 \\ -4 & -2 & 2 \\ -4 & 0 & -4 \end{vmatrix} = 4(8) - 4(24) + 1(-8) = 32 - 96 - 8 = -72$$

$$M_{\text{cof}}^T = \begin{bmatrix} 8 & -24 & -8 \\ 16 & -12 & -16 \\ 10 & -12 & 8 \end{bmatrix}^T = \begin{bmatrix} 8 & 16 & 10 \\ -20 & -12 & -12 \\ -8 & -16 & 8 \end{bmatrix}$$

$$M^{-1} = \frac{1}{72}\begin{bmatrix} 8 & 16 & 10 \\ -24 & -12 & -12 \\ -8 & -16 & 8 \end{bmatrix} = \frac{1}{18}\begin{bmatrix} -2 & -4 & -2.5 \\ 6 & 3 & 3 \\ 2 & 4 & -2 \end{bmatrix}$$

$$J = M^{-1}AM = \frac{1}{18}\begin{bmatrix} -2 & -4 & -2.5 \\ 6 & 3 & 3 \\ 2 & 4 & -2 \end{bmatrix}\begin{bmatrix} 0 & 0 & 1 \\ -2- & 3 & 0 \\ 0 & 2 & -3 \end{bmatrix}\begin{bmatrix} 4 & 4 & 1 \\ -4 & -2 & 2 \\ -4 & 0 & -4 \end{bmatrix}$$

CHAPTER 5

$$= \frac{1}{18}\begin{bmatrix} 8 & 7 & 5.5 \\ -6 & -3 & -3 \\ -8 & -16 & 8 \end{bmatrix}\begin{bmatrix} 4 & 4 & 1 \\ -4 & -2 & 2 \\ -4 & 0 & -4 \end{bmatrix}$$

Jordan block

$$= \frac{1}{18}\begin{bmatrix} -18 & 18 & 0 \\ 0 & -18 & 0 \\ 0 & 0 & -72 \end{bmatrix} = \begin{bmatrix} -1 & 1 & 0 \\ 0 & -1 & 0 \\ 0 & 0 & -4 \end{bmatrix}$$

$$\tilde{B} = M^{-1}B = \frac{1}{18}\begin{bmatrix} -2 & -4 & -2.5 \\ 6 & 3 & 3 \\ 2 & 4 & -2 \end{bmatrix}\begin{bmatrix} 0 \\ 2 \\ 0 \end{bmatrix} = \begin{bmatrix} -4/9 \\ 3/9 \\ 4/9 \end{bmatrix}$$

$$\tilde{C} = CM = [1 \ 0 \ 0]\begin{bmatrix} 4 & 4 & 1 \\ -4 & -2 & 2 \\ -4 & 0 & -4 \end{bmatrix} = [4 \ 4 \ 1]$$

The Jordan canonical form of state model is shown below:

$\dot{Z} = JZ + \tilde{B}U$; $\quad Y = \tilde{C}Z + DU$ (Here DU is not defined)

$$\begin{bmatrix} \dot{Z}_1 \\ \dot{Z}_2 \\ \dot{Z}_3 \end{bmatrix} = \begin{bmatrix} -1 & 1 & 0 \\ 0 & -1 & 0 \\ 0 & 0 & -4 \end{bmatrix}\begin{bmatrix} Z_1 \\ Z_2 \\ Z_3 \end{bmatrix} + \begin{bmatrix} -4/9 \\ 3/9 \\ 4/9 \end{bmatrix}[u];$$

$$Y = [4 \ 4 \ 1]\begin{bmatrix} Z_1 \\ Z_2 \\ Z_3 \end{bmatrix}$$

Conclusion: It is observed that the elements of the now of $\tilde{B}$ are not all zeros. Hence the system is completely controllable. (or state controllable).

It is observed that the elements of the colours of $\tilde{C}$ are not all zeros. Hence the system is completely observable [i.e., all state variables are observable].

SOME WORKED EXAMPLES

EXAMPLE 1

Obtain the state model of the electrical network shown below Fig. (1) by choosing minimal number of state variables.

Solution: Let

$$i_1 = X_1$$
$$i_2 = X_2$$
$$V_3 = X_3$$

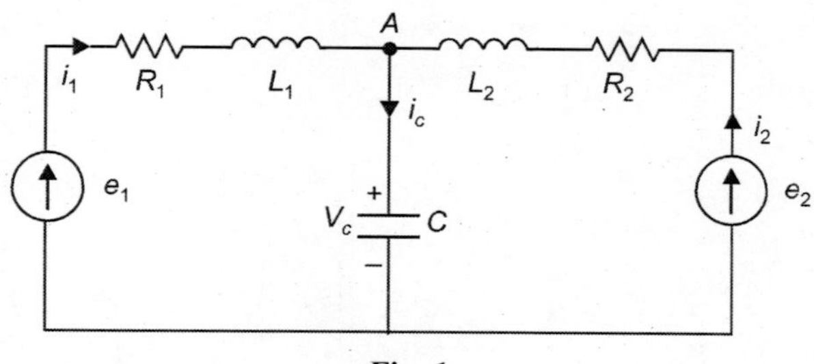

Fig. 1

$$L_1\dot{X}_1 = X_3 - X_1 R_1 - u$$

$$\dot{X}_1 = -\frac{R_1}{L_1}X_1 + \frac{1}{L_1}X_3 - \frac{1}{L_1}u \tag{1}$$

By applying KVL to loop 2,

$$V_l = L_2\frac{di_2}{dt} + R_2 i_2 + e(t)$$

$$X_3 = L_2\dot{X}_2 + R_2 X_2 + u$$

$$L_2\dot{X}_2 = X_3 - R_2 X_2 - u$$

$$\dot{X}_2 = -\frac{R_2}{L_2}X_2 + \frac{1}{L_2}X_3 - \frac{u}{L_2} \tag{2}$$

By applying KCl at node A,

$$i_1 + i_2 = i_c$$

$$X_1 + X_2 = C\frac{dV_c}{dt}$$

$$X_1 + X_2 = C\dot{X}_3$$

$$\dot{X}_3 = \frac{X_1}{C} + \frac{X_2}{C} \tag{3}$$

The equations (1), (2) and (3) are the state equations of the system. Hence

$$\dot{X}_1 = -\frac{R_1}{L_1}X_1 + \frac{1}{L_1}X_3 - \frac{1}{L_1}u$$

$$\dot{X}_2 = -\frac{R_2}{L_2}X_2 + \frac{1}{L_2}X_3 - \frac{u}{L_2}$$

$$\dot{X}_3 = \frac{X_1}{C} + \frac{X_2}{C}$$

on arranging the state equations is the matrix term we get,

$$\begin{bmatrix} \dot{X}_1 \\ \dot{X}_2 \\ \dot{X}_3 \end{bmatrix} = \begin{bmatrix} -\frac{R_1}{L_1} & 0 & \frac{1}{L_1} \\ 0 & -\frac{R_2}{L_2} & \frac{1}{L_2} \\ \frac{1}{C} & \frac{1}{C} & 0 \end{bmatrix} \begin{bmatrix} X_1 \\ X_2 \\ X_3 \end{bmatrix} + \begin{bmatrix} -\frac{1}{L_1} \\ -\frac{1}{L_2} \\ 0 \end{bmatrix} [u] \quad (4)$$

Equation (4) is the state equation

Let us choose the voltage across the resistances as output variables and the output variables are denoted by y_1 and y_2.

$$y_1 = i_1 R_1 \text{ and } y_2 = i_2 R_2 \quad (5)$$

on substituting the state variables in equation (3), we get,

$$y_1 = X_1 R_1; \quad y_2 = X_2 R_2 \quad (6)$$

on arranging the equations is the matrix form we get,

$$\begin{bmatrix} y_1 \\ y_2 \end{bmatrix} = \begin{bmatrix} R_1 & 0 \\ 0 & R_2 \end{bmatrix} \begin{bmatrix} x_1 \\ x_2 \end{bmatrix} \quad (7)$$

Equation (7) is the output equation. The equation (4) and (7) together constitute the state model of the system.

EXAMPLE 2

Construct the state model of mechanical system shown in Fig. (2).

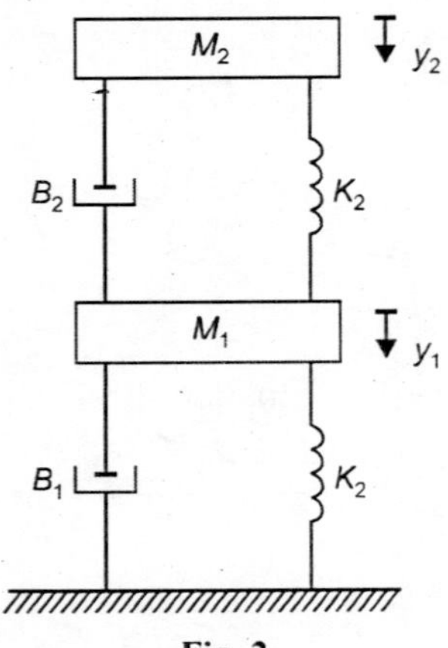

Fig. 2

Solution: Free body diagram of M_1

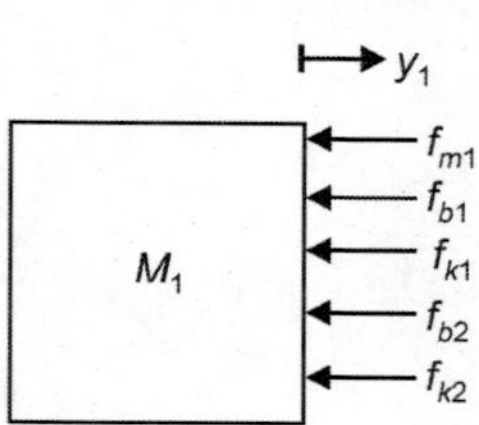

where,
$$f_{m1} = M\frac{d^2y_1}{dt^2}$$

$$f_{b1} = B_1\, dy_1/dt$$

$$f_{k1} = K_1y_1$$

$$f_{b2} = B_2\frac{d}{dt}(y_1 - y_2)$$

$$f_{K2} = K_2(y_1 - y_2)$$

By newton's second law, the torte balance equation at node M_1 is

$$0 = f_{m1} + f_{b1} + f_{k1} + f_{b2} + f_{k2}$$

$$0 = M_1\frac{d^2y_1}{dt^2} + B_1\frac{dy_1}{dt} + K_1y_1 + B_2\frac{d}{dt}(y_1 - y_2) + K_2(y_1 - y_2) \quad (1)$$

Free body diagram of M_2

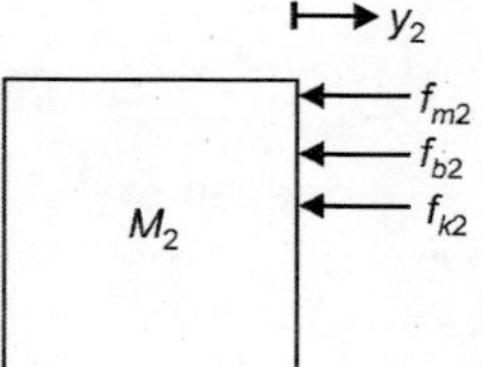

where
$$f_{m2} = M_2\frac{dy_2}{dt^2}$$

$$f_{b2} = B_2\frac{d}{dt}(y_2 - y_1)$$

$$f_{k2} = K_2(y_2 - y_1)$$

By newton's second law, the force balance equation at node M_2 is

$$0 = f_{m2} + f_{b2} + f_{k2}$$

$$0 = M_2\frac{d^2y_2}{dt^2} + B_2\frac{d(y_2 - y_1)}{dt} + K_2(y_2 - y_1) \quad (2)$$

Let us choose the state variable

$$y_1 = X_1, \quad y_2 = X_2, \quad \frac{dy_1}{dt} = X_3, \quad \frac{dy_2}{dt} = X_4$$

$$\frac{d^2y_1}{dt^2} = \dot{x}_3 \quad \text{and} \quad \frac{d^2y_2}{dt^2} = \dot{x}_4$$

on substituting state variables in equations (1) and (2)

$$0 = M_1\dot{X}_3 + B_1X_3 + K_1X_1 + B_2(X_3 - X_4) + K_2(X_1 - X_2)$$

$$0 = M_1\dot{X}_3 + X_3[B_1 + B_2] + X_1[K_1 + K_2] - B_2X_4 - K_2X_2$$

$$M_1\dot{X}_3 = -X_1[K_1 + K_2] + K_2X_2 - X_3[B_1 + B_2] + X_4B_2$$

$$\dot{X}_3 = \frac{-X_1[K_1 + K_2]}{M_1} + \frac{K_2}{M_1}X_2 - X_3\frac{[B_1 + B_2]}{M_1} + \frac{B_2}{M_1}X_4 \qquad (3)$$

similarly the equation (2) becomes,

$$0 = M_2\dot{X}_4 + B_2(X_4 - X_3) + K_2(X_2 - X_1)$$

$$M_2\dot{X}_4 = -B_2(X_4 - X_3) - K_2(X_2 - X_1)$$

$$\dot{X}_4 = +\frac{K_2}{M_2}X_1 - \frac{K_2}{M_2}X_2 + \frac{B_2}{M_2}X_3 - \frac{B_2}{M_2}X_4 \qquad (4)$$

The state variable $X_1 = y_1$

on differentiating $X_1 = y_1$ w.r.t. to t we get,

$$\frac{dX_1}{dt} = \frac{dy_1}{dt}$$

Let $\frac{dX_1}{dt} = \dot{X}_1$ and $\frac{dy_1}{dt} = X_3$, $\therefore\ \dot{X}_1 = X_3$ (5)

The equation (3), (4), (5) and (6) are called the state equations of mechanical system which is given in Fig. (2)

$$\dot{X}_1 = X_3$$

$$\dot{X}_2 = X_4$$

$$\dot{X}_3 = -\left(\frac{K_1 + K_2}{M_1}\right)X_1 + \frac{K_2}{M_1}X_2 - \frac{(B_1 + B_2)}{M_1}X_3 + \frac{B_2}{M_1}X_4$$

$$\dot{X}_4 = \frac{K_2}{M_2}X_1 - \frac{K_2}{M_2}X_2 + \frac{B_2}{M_2}X_3 - \frac{B_2}{M_2}X_4$$

on arranging the state equations.

$$\begin{bmatrix}\dot{X}_1\\ \dot{X}_2\\ \dot{X}_3\\ \dot{X}_4\end{bmatrix} = \begin{bmatrix} 0 & 0 & 1 & 0\\ 0 & 0 & 0 & 1\\ -\left[\frac{K_1 + K_2}{M_1}\right] & \frac{K_2}{M_1} & \frac{-(B_1 + B_2)}{M_1} & \frac{B_2}{M_1}\\ \frac{K_2}{M_2} & \frac{-K_2}{M_2} & \frac{B_2}{M_2} & -\frac{B_2}{M_2}\end{bmatrix}\begin{bmatrix}X_1\\ X_2\\ X_3\\ X_4\end{bmatrix} + \begin{bmatrix}0\\0\\0\\0\end{bmatrix}[u] \qquad (7)$$

Let the displacement y_1 and y_2 be the outputs of the system.

$\therefore$ $y_1 = X_1$ and $y_2 = X_2$

The output equation is matrix form is given by

$$\begin{bmatrix} y_1 \\ y_2 \end{bmatrix} = \begin{bmatrix} 1 & 0 & 0 & 0 \\ 0 & 1 & 0 & 0 \end{bmatrix} \begin{bmatrix} X_1 \\ X_2 \\ X_3 \\ X_4 \end{bmatrix} \tag{8}$$

The equations (7) we (8) together called state model of the system.

EXAMPLE 3

Construct a state model for the system. Characterized by the following differential equations. $\dddot{y} + 3\ddot{y} + 2\dot{y} = u$.

Solution: Let state variables X_1, X_2, X_3 are related to phase variables,

$$X_1 = y$$

$$X_2 = \frac{dy}{dt} = \dot{x}_1$$

$$X_3 = \frac{d^2y}{dt^2} = \dot{x}_2$$

and $$\frac{d^3y}{dt^3} = \dot{x}_3$$

The given equation can be rewritten as

$$\frac{d^3y}{dt^3} + 3\frac{d^2y}{dt^2} + 2\frac{dy}{dt} - u = 0 \tag{1}$$

on substituting state variables in equation (1)

$$\dot{X}_3 + 3X_3 + 2X_2 - u = 0$$

or $$\dot{x}_3 = -3x_3 - 2x_2 + u$$

The state equations are

$$\dot{X}_1 = X_2$$

$$\dot{X}_2 = X_3$$

$$\dot{X}_3 = -2X_2 - 3X_3 + u$$

on arranging the state equations in the matrix form we get,

$$\begin{bmatrix} \dot{X}_1 \\ \dot{X}_2 \\ \dot{X}_3 \end{bmatrix} = \begin{bmatrix} 0 & 1 & 0 \\ 0 & 0 & 1 \\ 0 & -2 & -3 \end{bmatrix} \begin{bmatrix} X_1 \\ X_2 \\ X_3 \end{bmatrix} + \begin{bmatrix} 0 \\ 0 \\ 1 \end{bmatrix} [u] \tag{2}$$

Here, y = output

But, $y = X_1$

$\therefore$ The output equation is $y = [1 \;\; 0 \;\; 0] \begin{bmatrix} X_1 \\ X_2 \\ Xs_3 \end{bmatrix}$

The equation (2) and (3) one called state model of the system.

EXAMPLE 4

In the state diagram of Fig. (3) Assign state variables and obtain state model of the system.

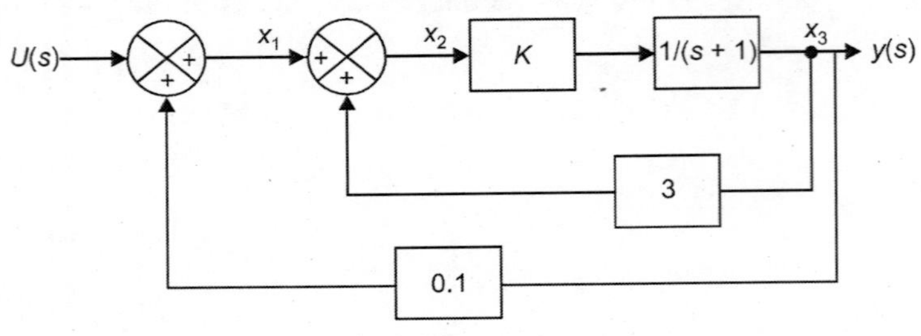

Fig. 3

Solution: The Fig. (3) can be redrawn as

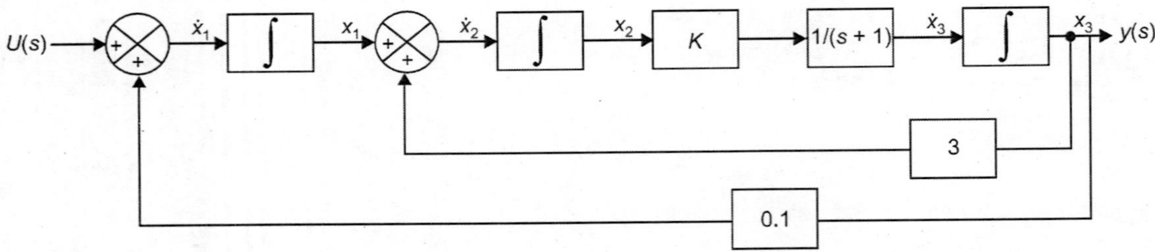

On adding the signal coming the 1st integrator we get,

$$X_1 + u + 0.1\, X_3$$

on adding the signals coming to the 2nd integrator we get,

$$\dot{X}_2 = X_1 + 3X_3$$

on adding the signals coming to the 3rd integrator we get,

$$\dot{X}_3 = \frac{K}{s+1} X_2$$

The state equations are

$$\dot{X}_1 = 0.1\, X_3 + u$$

$$\dot{X}_2 = X_1 + 3X_3$$

$$\dot{X}_3 = \left(\frac{K}{s+1}\right) X_2$$

The output equations are $y = X_3$

The state equations and output equations are arranged is the matrix form,

$$\begin{bmatrix} \dot{X}_1 \\ \dot{X}_2 \\ \dot{X}_3 \end{bmatrix} = \begin{bmatrix} 0 & 0 & 0.1 \\ 1 & 0 & s \\ 0 & \dfrac{K}{s+1} & 0 \end{bmatrix} \begin{bmatrix} X_1 \\ X_2 \\ X_3 \end{bmatrix} + \begin{bmatrix} 1 \\ 0 \\ 0 \end{bmatrix} [u] \quad (1)$$

$$[y] = [0 \quad 0 \quad 1] \begin{bmatrix} X_1 \\ X_2 \\ X_3 \end{bmatrix} \quad (2)$$

The equations (1) and (2) together called state model of the given system.

EXAMPLE 5

Obtain the state model of the system whose transfer function is given as, $\dfrac{Y(s)}{U(s)} = \dfrac{1}{s^2 + 3s + 2}$.

Solution:

$$\frac{Y(s)}{U(s)} = \frac{1}{s^2 + 3s + 2}$$

$$\frac{Y(s)}{U(s)} = \frac{1}{s^2\left[1 + \frac{3}{s} + \frac{1}{s^2}\right]} = \frac{1/s^2}{1 - \left[-\frac{3}{s} - \frac{2}{s^2}\right]}$$

The signal flow graph for the above transfer function is

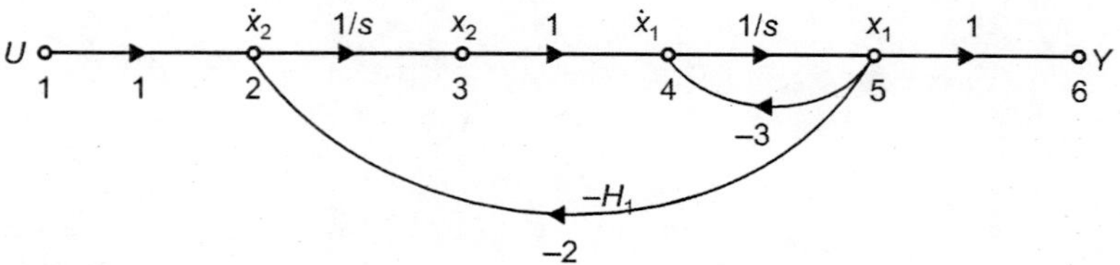

The state equations are, at node 2,

$$\dot{X}_2 = u + (-2\,X_1)$$

at node (4), $\dot{X}_1 = -3X_1 + X_2$

The output equation is $y = X_1$

The state model is the matrix form,

$$\begin{vmatrix} \dot{X}_1 \\ \dot{X}_2 \end{vmatrix} = \begin{bmatrix} -3 & 1 \\ -2 & 0 \end{bmatrix} \begin{bmatrix} X_1 \\ X_2 \end{bmatrix} + \begin{bmatrix} 0 \\ 1 \end{bmatrix} [u]$$

$$y = [1 \quad 0] \begin{bmatrix} x_1 \\ x_2 \end{bmatrix} \quad (2)$$

Equations (1) and (2) are called state model of the the given system.

EXAMPLE 6

Compute solution of the following equation. $\begin{bmatrix} \dot{X}_1 \\ \dot{X}_2 \end{bmatrix} = \begin{bmatrix} 1 & 1 \\ 0 & 1 \end{bmatrix} \begin{bmatrix} X_1 \\ X_2 \end{bmatrix}$ with $X_0 = \begin{vmatrix} 1 \\ 1 \end{vmatrix}$.

Solution: Here,

$$A = \begin{vmatrix} 1 & 1 \\ 0 & 1 \end{vmatrix};$$

$$sI = A = s\begin{vmatrix} 1 & 0 \\ 0 & 1 \end{vmatrix} - \begin{vmatrix} 1 & 1 \\ 0 & 1 \end{vmatrix} = \begin{vmatrix} s & 0 \\ 0 & s \end{vmatrix} - \begin{vmatrix} 1 & 1 \\ 0 & 1 \end{vmatrix}$$

$$= \begin{vmatrix} s-1 & -1 \\ 0 & s-1 \end{vmatrix}$$

$$|sI - A| = \begin{vmatrix} s-1 & -1 \\ 0 & s-1 \end{vmatrix} = (s-1)^2 - 0$$

$$|sI - A| = (s-1)^2$$

$$(sI - A)^{-1} = \frac{1}{(s-1)^2}\begin{bmatrix} s-1 & -1 \\ 0 & s-1 \end{bmatrix} = \begin{bmatrix} \frac{1}{s-1} & \frac{1}{(s-1)^2} \\ 0 & \frac{1}{(s-1)} \end{bmatrix}$$

$$e^{At} = \phi(t) = L^{-1}[\phi(s)] = L^{-1}[(sI - A)^{-1}] = \begin{bmatrix} e^t & te^t \\ 0 & e^t \end{bmatrix}$$

The solution of the state equation is

$$X_1(t) = e^{At}x_0 = \begin{bmatrix} e^t & te^t \\ 0 & e^t \end{bmatrix}\begin{bmatrix} 1 \\ 1 \end{bmatrix} = \begin{bmatrix} e^t te^t \\ e^t \end{bmatrix}$$

5.18 INTRODUCTION TO FUZZY LOGIC CONTROL

What is Fuzzy logic?

From the Fig. (5.11.1), we can understand the concept of fuzzy logic (or) what is fuzzy logic, Fuzzy logic can be defined as High-level processing of information through learning and experience. Knowledge representation and processing are the keys to any intelligent system. Knowledge is processed by the application of various laws of logic. We can see that the information gained from the Fig. (5.11.1) is that its same like a human brian. Fuzzy functions are similar to human brain it accept set of information and process through logics and law's to attain desired output. Thus fuzzy based products became highly competitive due to better performance, higher reliability, low power achievement.

To understand the concept of fuzzy logic let consider the following example. There is a set of 10 member in a team to perform some desired function.

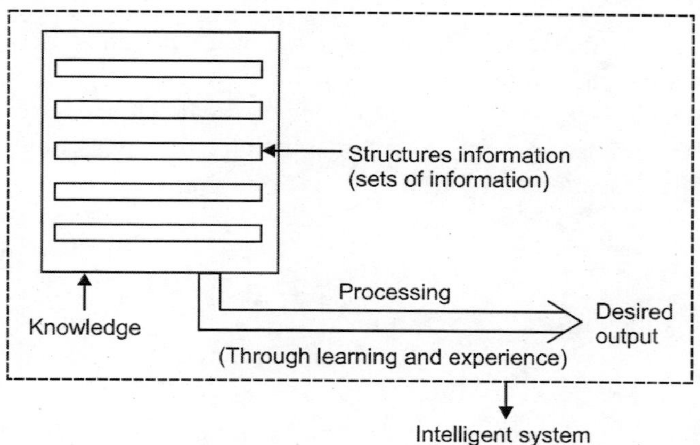

Fig. 5.11.1: Representation of fuzzy logic.

Team size = 10 nos.

$$X = \{x_1, x_2, x_3, ... x_{10}\}$$

$$x \in X$$

Let say, In that team the number of females are 4

Females = 4 nos.

$$A = \{x_1, x_2, x_3, x_4\}$$

A is sub set of *X* $\quad A\ \_\ X$

$$\mu_A : X \rightarrow \{0, 1\}$$

$$\mu_A : \begin{cases} 0; x = x_5, x_6, x_7, x_8, x_9, x_{10} \\ 1; x = x_1, x_2, x_3, x_4 \end{cases}$$

where,

μ_A is characteristic function of *A*. Let say x_1 and x_2 are below age 20, we may call them as "minors"

Minors = 2 nos.

$$B = \{x_1, x_2\}$$

B is subset of *A* $\quad B\ \_\ A$

$$\mu_B : x \rightarrow \{0, 1\}$$

$$\mu_B : \begin{cases} 1; x = x_1, x_2 \\ 0; x = \text{otherwise} \end{cases}$$

we have considered the "set of females *A*" and the "set of minors *B*" in *X*. If is also possible to consider a "set of young females *C*" If for our convenience we consider the "young" to be the same as "minor", then $C = B$, but in this case we have created a boundary, under which x_2 who is 19 is still young [$\mu_C(x_2) = 1$], but x_3 who just turned is no longer young [$\mu_C(x_3) = 0$]. In just one day, the value change from yes (1) to no. (0), and x_3 is now an old.

However, it is not possible that a young becomes an old over a period of 1 (or) two days, so we are introducing values such as 0.9 and 0.8 that are intermediate between 0 and 1, thus creating the concept of a "Fuzzy set" whereas a crisp set is defined by the characteristic function that can assume only the two values {0, 1}, a fuzzy set is defined by a "membership function" that can assume an infinite number of values, any real number in closed interval [0, 1].

$$\mu_C : X \rightarrow [0, 1]$$

such that

$$\mu_C : \begin{cases} 1 : x = x_1, x_2 \\ 0.9;\ x = x_3 \\ 0.2;\ x = x_4 \\ 0;\ \text{otherwise} \end{cases}$$

since [0, 1] incorporates {0, 1}, the concept of fuzzy set can be considered as an extended concept, which incorporates the concept of crisp set. For example, the crisp set of B of 'minors' can be regarded as a fuzzy set B with the membership function.

$$\mu_B(x) = \begin{cases} 1;\ x = x_1, x_2 \\ 0;\ \text{otherwise} \end{cases}$$

Fuzzy Logic Based Control System

A fuzzy logic based controller (FLC) is shown in Fig. (5.11.2). It consist of,

1. Fuzzification and fuzzifier,
2. Data base:
   (i) Input and output (ii) Fuzzy set and conditions
   (iii) Availability (iv) Completeness

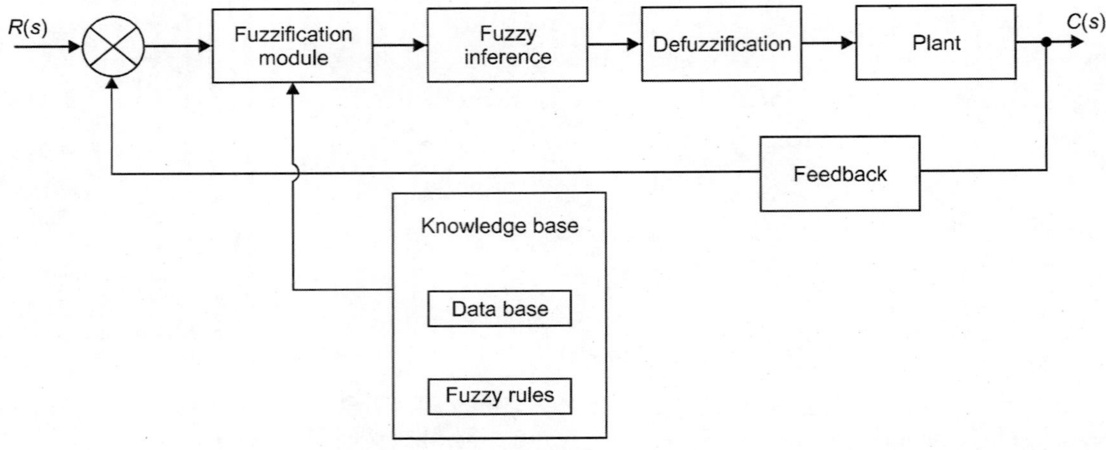

Fig. 5.11.2

3. Fuzzy rule:
   (i) State variables and control variables of fuzzy control rules.

(ii) Source and derivation fo fuzzy control rules.

(iii) Types of fuzzy control rules.

(iv) Consistency interactivity.

4. Fuzzy interference mechanism.
5. Defuzzication and defuzzifier.

* The section deals with the design of basic components of a FLC. The above listed components are necessary for the construction of any FLC. Fuzzification implies the process of transforming the crisp values of the inputs to a controller, to the fuzzy domain. The fuzzified values now fire the rule base and generate a so called fuzzified output. As a final step, the fuzzified output is defuzzified to yield the crisp controller output.

Let we will see the components of FLC one by one. Since the data manipulation in an FLC is based on fuzzy set theory, fuzzification is necessary in an earlier stage. Fuzzification module (FM) performs the following functions:

1. Measure the values of input variables.
2. Performs a scalar transformation which maps the physical values of the current process state variables into a normalized universe of discourse.
3. Performs the so called fuzzification which converts a point-wise (crisp), current value fo a process state variable into a fuzzy set, inorder to make it compatible with the fuzzy set representation of the process state variable in the rule-antecedent.

Knowledge base

The knowledge base of a FLC consists of a database and a rule base.

(i) Database: The basic function of a database is to provide necessary information for the proper functioning of the FM, the rule base, and the defuzzification module. This information includes;

* **Fuzzy sets:** representing the meaning of the values of the process and the control output variables. These concepts are defined based on experience and engineering judgement.

* Physical domains and their normalized counter parts together with the normalization or denormalization factors.

(ii) Rule base: The basic function of the rule base is to represent in a structured way with the control policy in the form of a set of rules.

Fuzzy Inference

There are two types of approaches employed in the design of the fuzzy inference of a FLC.

* Composition base inference.

* Individual rule-based inference.

Defuzzification module

The functions of defuzzification module (DM) are as follows:

- Performs the so called defuzzification which converts the set of modified control output values into a single point-wise values.

- Performs an output denormalization which maps the point-wise value of the control output onto its physical domain.

5.19 NEURAL NETWORKS

Artificial neural network is a technique which works as how brain performs. The human brain is made up of many millions of individual processing elements, called neurons, that are highly interconnected. A schematic diagram of a single biological neuron is shown in Fig. (5.13.1). Information from the outputs of the other neurons, in the form of electrical pulses, are received by the cell at connections called synapses. The synapses connect to the cell inputs, or dendrites, and the single output of the neuron appears at the atom. An electrical pulse is sent down the axon when the total input stimuli from all of the dendrites exceeds a certain threshold.

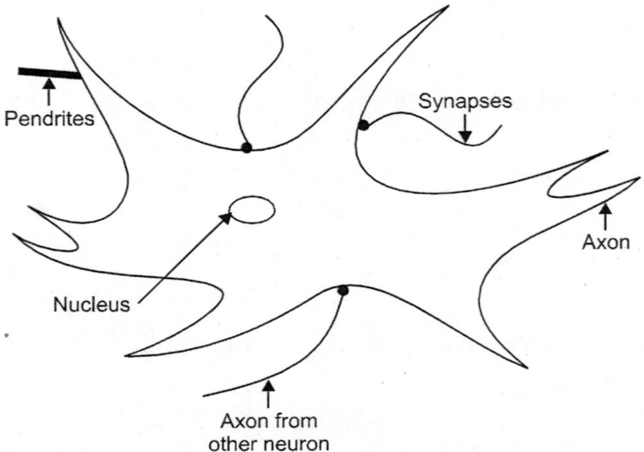

Fig. 5.13.1: A biological neuron.

Artificial neural networks are made up of simplified individual models of the biological neuron that are connected together to form a network. Information is stored in the network often in the form of different connection strengths, or weights, associated with the synapses in the artificial neuron models. Some of the various properties of neural networks are:

1. Inherent parallelism in the network architecture due to the repeated use of simple neuron processing elements. This leads to the possibility of very fast hardware implementations of neural networks.
2. Capability of learning information of example. The learning mechanism is often achieved by appropriate adjustment of the weights in the synapses of the artificial neuron models.
3. Ability to generalize to new inputs.
4. Fault tolerance. In general, network performance does not significantly degenerate if some of the network connections becomes faulty. Many different types of neural network are available and only the Multi Layer Network (MLN) which is most popular.

The most popular neural network architecture is the multi layer network. It consist of an input layer and an output layer as shown in Fig. (5.13.2). The output and the hidden layer are made up of a number of nodes. However the input layer is essentially a direct link to the inputs of the hidden layer and is included by convention. Sigmoidal activation functions for the nodes in the hidden and output layers are the most common choice, although variants on this are also possible. The outputs of each node in a layer are connected to the inputs of all the nodes in the subsequent layer. Data flows through the network in one direction only, from input to output; hence this type of network is called a feed-forward network.

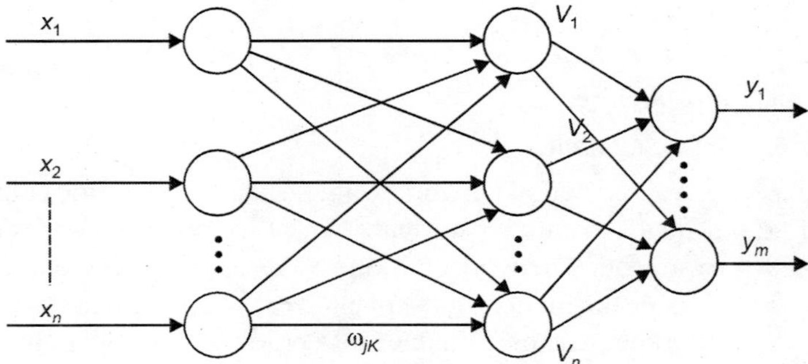

Fig. 5.13.2: A two layered network.

An important featane of MLN is that this network can accurately represent any continuous non-linear function relating the inputs end the outputs. Hence MLN exhibits potential for many application including modeling and control of real non-linear process.

Artificial Neuron Model

Neural networks can be considered as a large dimensional non-linear dynamic system, which is defined by a set of I[st] order non-linear differential equations. It acts as compensation elements in control systems. A neural network is a system with inputs and outputs and is composed of many simple and similar processing elements Fig. (5.13.3).

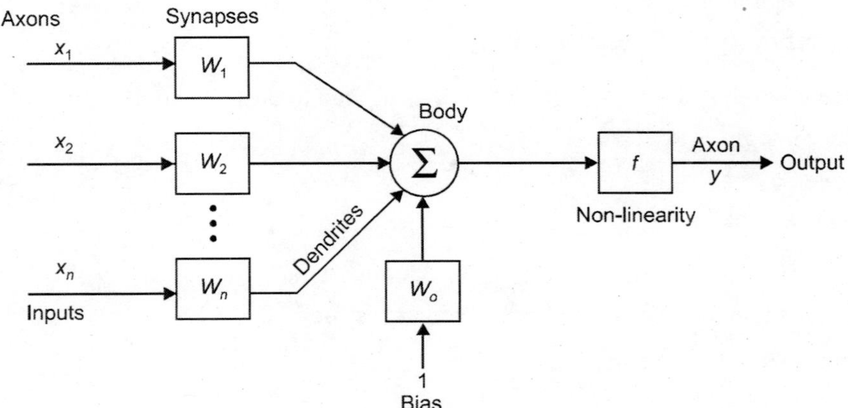

Fig. 5.13.3: Neuron model.

CHAPTER 5

Learning Control Architecture

The desired response of the plant is denoted by r and its actual output by y. The neural controller should act as the inverse of the plant, producing from the desired response r. A signal u that drives the output of the plant $y \equiv r$. The goal of the learning procedure, is to select the values of the weights of the network so that it produces $r \rightarrow u$ mappings at least over the range of r's in which the plant is to be operated. Four different learning methods are considered.

(i) Indirect learning architecture

(ii) General learning architecture

(iii) Specialized learning architecture

(iv) Forward inverse learning architecture

The conclusion can be made about neural network from the above discussions that, neural networks were used for controlling physical systems. Specifically four different methods for using error back propagation to train a feed-forward neural network controller to act as the inverse of the plant. The general learning method attempts to produce the inverse of the plant over the entire state space, but it can be very difficult to use it along to provide adequate performance in a practical control application. To avoid this kind of problem error is back-propagated through the plant, which allows to train the network exactly on the operational range of the plant. Forward inverse learning has very accurate learning architecture.

QUESTIONS AND ANSWERS

Q5.1 What are the time domain specifications which are needed to design a control system?

Ans. The time domain specifications needed to design a control system are

1. Rise time, t_r
2. Peak overshoot, M_P
3. Settling time, t_s
4. Damping ration, ζ_1
5. Natural frequency of oscillation, ω_n

Q5.2 What are the frequency domain specifications needed to design a control system?

Ans.

1. Phase margin
2. Gain margin
3. Resonant peak
4. Band width

Q5.3 What are the two methods of designing a control system:

Ans.

- Using root locus
- Using bode plot

Q5.4 What is compensation?

Ans. Compensation is the technique for designing system behaviour so as to meet the designed specifications, by introducing device called compensator.

Q5.5 What is series and parallel compensation?

Ans. (i) The series compensation is a design procedure in which a compensator is introduced in series with plant to alter the system behaviour and to provide satisfactory performance.

(ii) Some for parallel.

Q5.6 When lag/lead/lag-lead compensation are employed?

Ans. Lag compensator is employed for a stable system for improvement in steady state performance.

Lead compensatory is employed both for stable and unstable systems for improvement of transient state performance.

Lag-lead compensator is employed both for stable and unstable system for improvement is both steady state and transient state performance.

Q5.7 Discuss the effect of adding pole and zero to open loop transfer function f a system.

Ans. (i) Adding pole to open loop transfer function leads to reduce the steady state error. Thus steady state performance of the system is improved.

(ii) Adding pole will increase the order of the system which makes the system less stable then original system.

(iii) Adding zero to open loop transfer function leads to improve the transient response. It also reduces the rise time.

Q5.8 How root-loci are modified when a zero is added to open loop transfer function?

Ans. The addition of a zero to open loop transfer function will put the root locus to the left which make the system more stable and reduce the settling time.

Q5.9 How the root loci are modified when a pole is added to open loop transfer function of system?

Ans. The addition of a pole to the open-loop transfer function has the effect of pulling te root locus to the right, which reduce the relative stability of the system and increase the settling time.

Q5.10 How compensator is designed for control system using root locus?

Ans. Transient response specifications are translated into designed location for a pair of dominant closed loop poles. In order to satisfy the performance specifications the root loci should pass through these points.

Compensatory which is introduced will reshape the root loci and make them to pass through the points where the dominant closed loop poles are located.

Q5.11 Write the advantages of design using root locus?

Ans. Informations about both closed loop transient and frequency response can be obtained.

Q5.12 What are the informations that can be obtained from frequency response plots?

Ans.
- The low frequency region provides information regarding the steady-state performance.
- The high frequency region provides information regarding transient performance of the system.
- The mid frequency region provides information regarding relative stability of the system.

Q5.13 When maximum phase lag occurs in lag compensator? Give the expressions for maximum lag angle and the corresponding frequency?

Ans. The maximum phase lag = Geometric mean of two corner frequencies of the lag compensator.

Frequency corresponding to maximum phase angle

$$\omega_m = \sqrt{\omega c_1 \omega c_2} = \sqrt{\frac{1}{T}\frac{1}{\beta T}} = \frac{1}{T\sqrt{\beta}}$$

Q5.14 What is the relation between ϕ_m and β is lag compensator?

Ans. $\phi_m = \tan^{-1}\left(\frac{1-\beta}{2\sqrt{\beta}}\right)$ (or) $\beta = \frac{1-\sin\phi_m}{1+\sin\phi_m}$

Q5.15 What is lead compensator:

Ans. It is a device which is introduced in the system so as the meet the desired specifications.

Q5.16 What is the relation between ϕ_m and α in lead compensator?

Ans. $\phi_m = \tan^{-1}\left(\frac{1-\alpha}{2\sqrt{2}}\right)$ (or) $\alpha = \frac{1-\sin\phi_m}{1+\sin\phi_m}$.

Q5.17 When maximum phase lead occurs in lead compensator? Given the expressions for maximum lead angle and the corresponding frequency.

Ans. The maximum phase lead occurs at the geometric mean of two corner frequencies of the lead compensator.

Maximum phase lead angle,

$$\phi_m = \tan^{-1}\frac{(1-\alpha)}{2\sqrt{\alpha}}$$

Frequency corresponding to maximum phase lead angle

$$\omega_m = \frac{1}{T\sqrt{\alpha}}$$

Q5.18 What are the advantages of state space analysis?

Ans.
1. The state space analysis is applicable to any type of systems. They can be used for modeling and analysis of linear and non-linear systems, time invariant and time variant systems and multiple input/output systems.

2. The state space analysis can be performed with initial conditions.

3. The variables used to represent the system can be any variables in the system.

4. Using this analysis the internal states of the system at any instant at any time can be predicted

Q5.19 What are the drawbacks in transfer function model analysis?

Ans. 1. Transfer function is defined under zero initial conditions and it is applicable to only linear time invariant system.

2. Transfer function analysis is restricted to single input and output systems

3. Does not provides information regarding the internal state of the system

Q5.20 What is state and state variable?

Ans. The state is the condition of a system at any instant at any time, t. A set of variable which describes the state of the system at any instant at any time are called state variables.

Q5.21 What is a state vector?

Ans. The state vector is a (n Ч1) column matrix (or vector) whose elements are state variables of the system, (where in is the order of the system). It is denoted by $X(t)$.

Q5.22 Write the state model of nth order system?

Ans. The state model of a system consists of state equations and output equations. The state model of a nth model of a nth order system with m - inputs and p-outputs are

$$\dot{X}(t) = A X(t) + BU(t) \text{ - state equation}$$

$$Y(t) = C X(t) + DU(t) \text{ - Output equation}$$

Where,

$X(t)$ = State Vector of order ($n \times 1$)

A = System matrix of order ($n \times n$)

$Y(t)$ = Output Vector of order ($p \times 1$)

D = Transmission matrix of order ($p \times m$)

$U(t)$ = Input Vector of order ($m \times 1$)

B = Input matrix of order ($n \times$ m)

L = Output matrix of order ($p \times n$)

Q5.23 What is state space?

Ans. The set of all possible values which the state Vector $X(t)$ can have (or assume) at time t forms the state space of the system.

Q5.24 What is input and output space?

Ans. The set of all possible values which the input vector $U(t)$ can have (or assume) at time t forms the input space of the system.

The set of all possible values which the output vector $Y(t)$ can have (or assume) at time t forms the output space.

Q5.25 The state model of a linear time invariant system is given by

$\dot{X}(t) = A\,X(t) + B\,U(t)$

$Y(t) = C \times (t) + D\,U(t)$

Obtain the expression for transfer function of a system.

Ans. Given that

$$\dot{X}(t) = A\,X(t) + B\,U(t) \qquad (1)$$

$$Y(t) = C\,X(t) + D\,U(t) \qquad (2)$$

Take laplace transform of equation (1) with zero initial conditions we get,

$$s\,X(s) = A\,X(s) + B\,U(s)$$

$$s\,X(s) - A\,X(s) = B\,U(s)$$

$$(sI - A)\,X(s) = B\,U(s) \qquad (3)$$

Premultiplying equation (3) by $(sI - A)^{-1}$ we get

$$X(s) = (sI - A)^{-1}\,B\,U(s) \qquad (4)$$

Take laplace transform of equation (2) we get,

$$Y(s) = CX(s) + D\,U(s) \qquad (5)$$

Substitute for $X(s)$ from equation (4) in equation (5) we get,

$$Y(s) = c[(sI - A)^{-1}\,B\,U(s)\,] + D\,U(s)$$

$$= [C(sI - A)^{-1}\,B\,U(s) + D]\,U(s)$$

$$\therefore \quad \frac{Y(s)}{U(s)} = C\,(sI - A)^{-1}\,B\,U(s) + D \qquad (6)$$

Q5.26 What is state diagram?

Ans. The pictorial representation of the state model of the system is called state diagram. The state diagram of the system can be either in block diagram or in signal flow graph Form.

Q5.27 Draw the block diagram representation of state model?

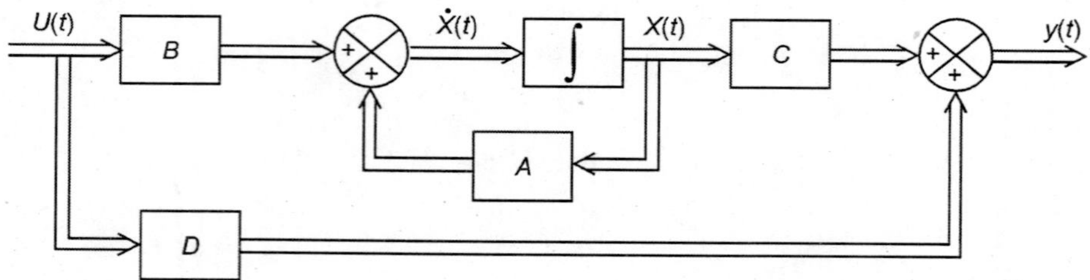

Block diagram of state model.

Q5.28 Draw the signal flow graph representation of state model?

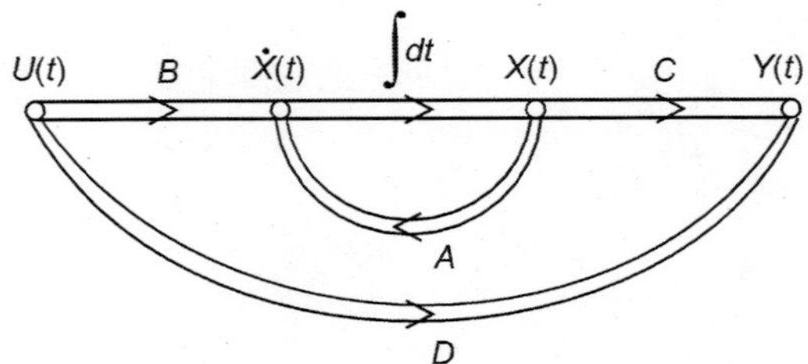

Signal flow graph of state model.

Q5.29 What are the basic elements used to construct the state diagram?

Ans. The basic elements used to construct the state diagram are scalar, Adder and integrator

Q5.30 Sketch the basis elements used to construct the block diagram of a state model?

Ans. The basic elements used to construct block diagram of a state model are shown in the following table.

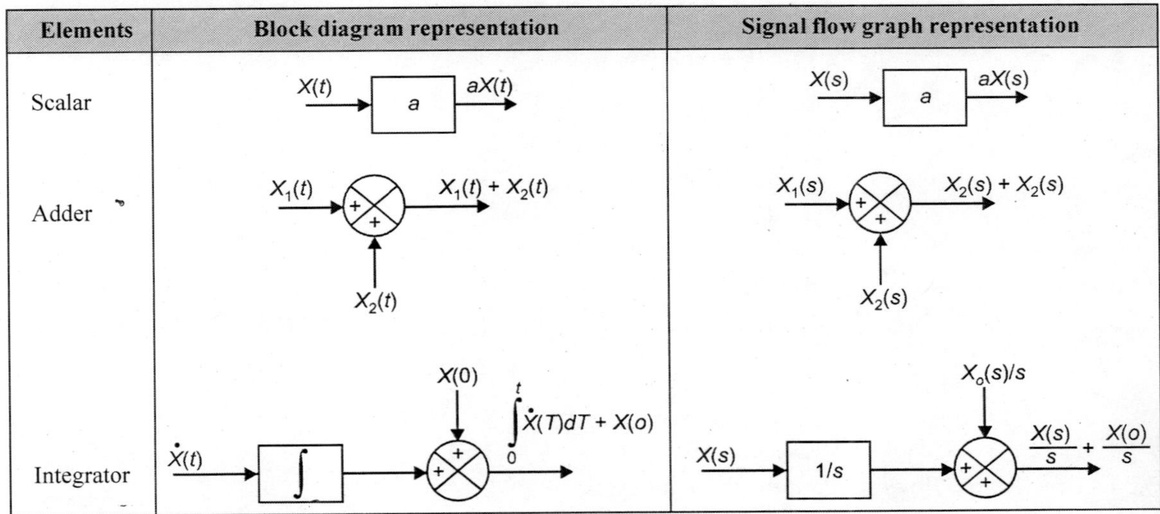

Elements	Block diagram representation	Signal flow graph representation
Scalar		
Adder		
Integrator		

CHAPTER 5

Q5.31 Sketch the basic elements used to construct the signal flow graph of a state model?

Ans. The basic elements used to construct the signal flow graph of a state model are shown in the following table.

Elements of signal flow graph		
Elements	**Time domain**	***s*-domain**
Scalar	X(t) —a→ aX(t)	X(s) —a→ aX(s)

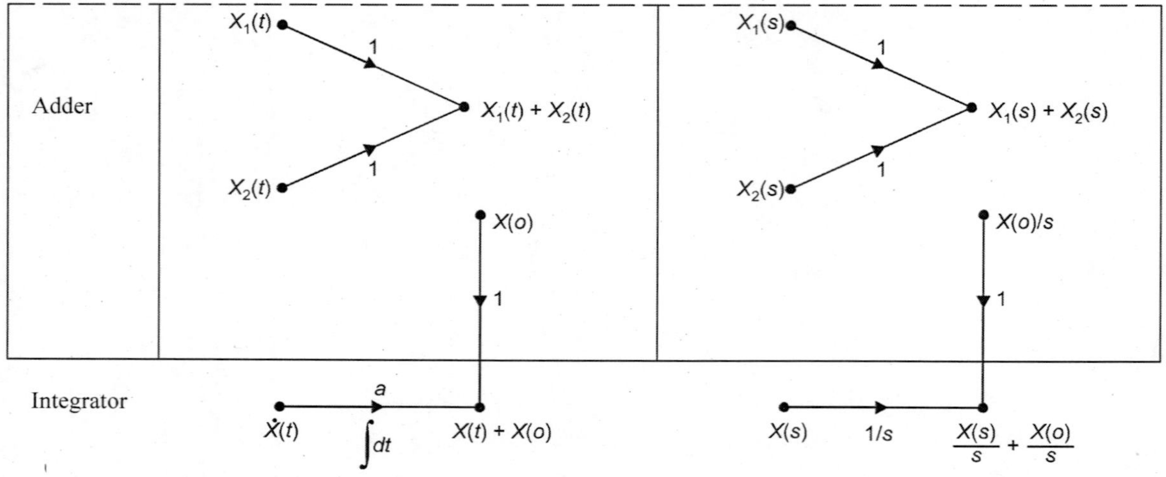

Q5.32 Draw the diagram of the system described by the state model

$$\begin{vmatrix} \dot{X}_1 \\ \dot{X}_2 \\ \dot{X}_3 \end{vmatrix} = \begin{vmatrix} 0 & 1 & 0 \\ 0 & 0 & 1 \\ 0 & a_2 & a_3 \end{vmatrix} \begin{vmatrix} X_1 \\ X_2 \\ X_3 \end{vmatrix} + \begin{vmatrix} 0 \\ 0 \\ 1 \end{vmatrix} u \text{ and } y = x_1$$

Ans.

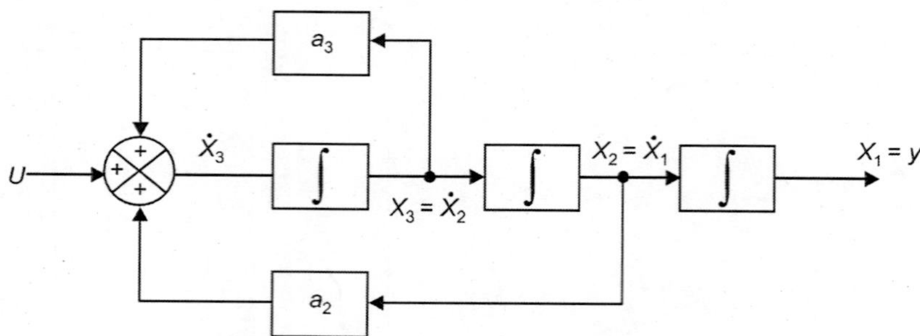

Q5.33 Draw signal flow graph of the system described by the state model.?

$$\begin{bmatrix} \dot{X}_1 \\ \dot{X}_2 \\ \dot{X}_3 \end{bmatrix} = \begin{vmatrix} a_1 & a_2 & 0 \\ 1 & 0 & 1 \\ 0 & 1 & 1 \end{vmatrix} \begin{vmatrix} X_1 \\ X_2 \\ X_3 \end{vmatrix} + \begin{vmatrix} 1 \\ 0 \\ 0 \end{vmatrix} u \text{ and } y = x_3$$

Ans.

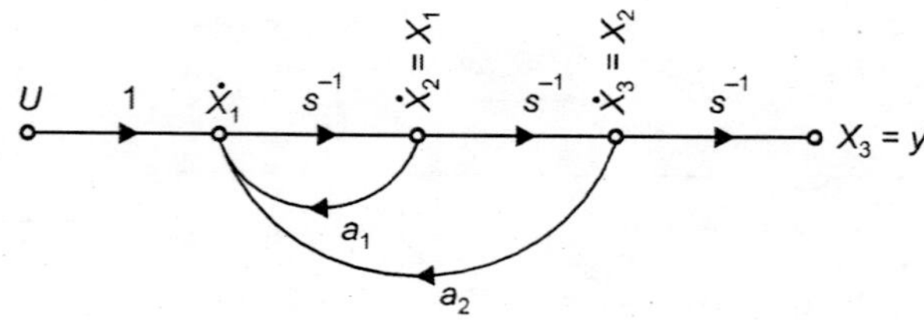

Q5.34 Determine the state model of the system represented by the block diagram.

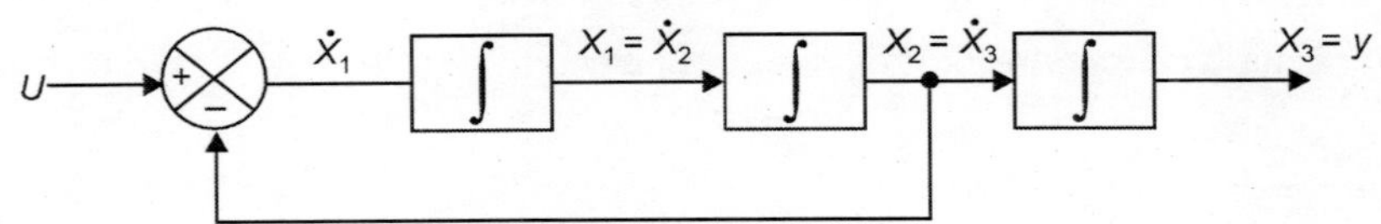

Ans. $$\begin{vmatrix}\dot{X}_1\\ \dot{X}_2\\ \dot{X}_3\end{vmatrix} = \begin{vmatrix}0 & 1 & 0\\ 1 & 0 & 0\\ 0 & 1 & 0\end{vmatrix}\begin{vmatrix}X_1\\ X_2\\ X_3\end{vmatrix} + \begin{vmatrix}1\\ 0\\ 0\end{vmatrix} u \text{ and } y = x_3$$

Q5.35 Determine the state model of the system represented by the signal flow graph.

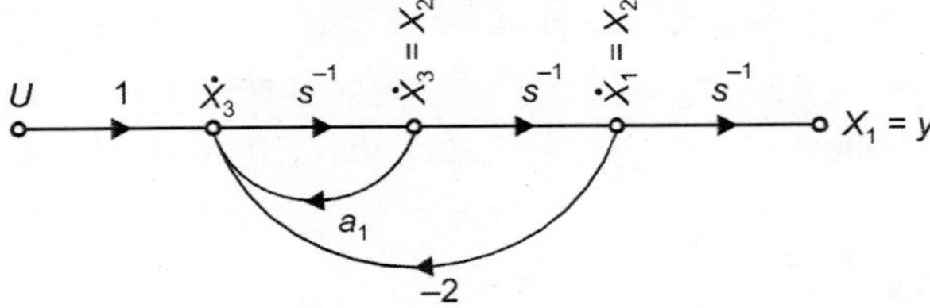

Ans. $$\begin{vmatrix}\dot{X}_1\\ \dot{X}_2\\ \dot{X}_3\end{vmatrix} = \begin{vmatrix}0 & 1 & 0\\ -2 & 0 & 1\\ 0 & 0 & 0\end{vmatrix}\begin{vmatrix}X_1\\ X_2\\ X_3\end{vmatrix} + \begin{vmatrix}0\\ 0\\ 1\end{vmatrix} u \text{ and } y = x_1$$

Q5.36 A system is characterized by the differential equation $\frac{d^2y}{dt^2}+10\frac{dy}{dt}+7y-u=0$. Determine its transfer function

Ans. Given that

$$\frac{d^2y}{dt^2}+10\frac{dy}{dt}+7y-u=0 \tag{1}$$

On taking laplace transform of equation (1) with zero initial conditions we get,

$$s^2Y(s) + 10s\,Y(s) + 7Y(s) - U(s) = 0$$

$$(s^2 + 10s + 7)Y(s) = U(s)$$

$$\therefore \quad \frac{Y(s)}{U(s)} = \frac{1}{s^2+10s+7} \tag{2}$$

The equation (2) is the transfer function of the system.

Q5.37 The transfer function of a system is given by $\frac{Y(s)}{U(s)} = \frac{10}{4s^2+2s+1}$, Determine the differential equation governing the system.

Ans. $$\frac{Y(s)}{U(s)} = \frac{10}{4s^2+2s+1}$$

$$(4s^2 + 2s + 1)\ Y(s) = 10\ U(s)$$

$$4s^2 Y(s) + 2(s)Y(s) + Y(s) = 10\ U(s) \quad (1)$$

Take inverse laplace transform of equation (1) we get,

$$4\frac{d^2y}{dt^2} + 2\frac{dy}{dt} + y = 10u$$

$$4\frac{d^2y}{dt^2} + 2\frac{dy}{dt} + y - 10u = 0 \quad (2)$$

The equation (2) is the differential equation governing the system

Q5.38 What are the advantages of state space model lingusing physical variable?

Ans. The advantages of choosing the physical variable are the following,

1. The state variable can be utilized for the purpose of feedback.
2. The implementation of design with state variable feedback becomes straight forward.
3. The solution of state equation gives time variation of variables which have direct relevance to the physical system.

Q5.39 What are phase variables?

Ans. The phase variables are defined as those particular state variables which are obtained from one of the system variable and its derivatives. Usually the variable used is the system output and the remaining state variables are then derivatives of the output.

Q5.40 What is bush form or companion form of state model?

Ans. In bush form (or) Companion form of state model, the system matrix, A has all 1's in the upper off-diagonal and its last row is comprised of the negative of the coefficients of the original differential equation and all other elements are zero. The companion form of state model is shown below:

$$\begin{vmatrix} \dot{X}_1 \\ \dot{X}_2 \\ \dot{X}_3 \\ \vdots \\ \dot{X}_3 \end{vmatrix} = \begin{vmatrix} 0 & 1 & 0 & 0 & \cdots\cdots & 0 \\ 0 & 0 & 1 & 0 & \cdots\cdots & 0 \\ 0 & 0 & 0 & 1 & \cdots\cdots & 0 \\ \vdots & & & & & \\ 0 & 0 & 0 & 0 & \cdots\cdots & 1 \\ -a_n & -a_{n-1} & -a_{n-2} & -a_{n-3} & \cdots & -a_1 \end{vmatrix} \begin{bmatrix} X_1 \\ X_2 \\ X_3 \\ \vdots \\ X_{n-1} \\ X_n \end{bmatrix}$$

$$Y = \begin{bmatrix} 1 & 0 & 0 & \cdots & 0 \end{bmatrix} \begin{vmatrix} X_1 \\ X_2 \\ X_3 \\ \vdots \\ X_n \end{vmatrix}$$

Q5.41 What are the advantages in choosing phase variables for state space modelling?

Ans.
1. Using phase variables the system state model can be written directly by inspection from the differential equation governing the system.
2. The phase variables provides a link between the transfer function design approach and time-domain design approach.

Q5.42 What is the disadvantages in choosing phase variable for state-space modelling?

Ans. The disadvantage in choosing phase variables is that the phase variables are not physical variables of the system and therefore are not available for measurement and control.

Q5.43 Write the canonical form of state model of nth order system.

Ans. In canonical form of state model, the system matrix, A will be a diagonal matrix. The canonical form of state model in the matrix form is given below.

$$\begin{vmatrix} \dot{X}_1 \\ \dot{X}_2 \\ \dot{X}_3 \\ \vdots \\ \dot{X}_n \end{vmatrix} = \begin{vmatrix} \lambda_1 & 0 & 0 & \cdots & 0 \\ 0 & \lambda_2 & 0 & \cdots & 0 \\ 0 & 0 & \lambda_3 & \cdots & 0 \\ \vdots & \vdots & \vdots & & \vdots \\ 0 & 0 & 0 & & \lambda_n \end{vmatrix} \begin{vmatrix} X_1 \\ X_2 \\ X_3 \\ \vdots \\ X_n \end{vmatrix} + \begin{vmatrix} 1 \\ 1 \\ 1 \\ \vdots \\ 1 \end{vmatrix} |u|$$

$$y = |c_1 \quad c_2 \quad c_3 \quad \cdots \quad c_n| \begin{vmatrix} X_1 \\ X_2 \\ X_3 \\ \vdots \\ X_n \end{vmatrix} + [b_o][u]$$

Q5.44 What is the advantage and the disadvantage in canonical form of state model.

Ans. The advantage of canonical form is that the state equations are independent of each other. The disadvantage is that the canonical variables are not available for measurement and control.

Q5.45 What is state transition matrix and how it is related to state of a system?

Ans. The matrix exponential e^{At} is called state transition matrix. In the expanded form,

$$e^{At} = I + At + \frac{1}{2!}A^2t^2 + \frac{1}{3!}A^3t^3 + \ldots + \frac{1}{1!}A^it^i + .$$

The state transition matrix is used to find the state of the system, at any time instant t, from the knowledge of the state at time, t_o. When the input is zero,

$$X(t) = e^{At} X(t_o)$$

when the input vector is

$$U(t), X(t) = e^{A(t-t_o)} \cdot X(t_0) + \int_{t_o}^{t} e^{A(t-\tau)} \cdot B \cdot U(\tau)d\tau$$

CHAPTER 5

where $X(t)$ = state vector at time, t

and $X(t_o)$ = state vector at time, t_o

Q5.46 Write the properties of state transition matrix?

Ans. The following are the properties of state transition matrix.

1. $\phi(o) = e^{AXo} = I$ (unit matrix)
2. $\phi(t) = e^{At} = (e^{-At})^{-1} = [\phi(-t)]^{-1}$
3. $\phi(t_1 + t_2) = e^{A(t_1 + t_2)} = e^{At_1} e^{At_2} = \phi(t_1)\, \phi(t_2) = \phi(t_2)\, \phi(t_1)$

Q5.47 Write the solution of homogeneous state equations?

Ans. The solution of homogeneous state equation is,

$$X(t) = e^{At} X_o$$

where, $X(t)$ = state vector at time, t

e^{At} = State transition matrix

X_o = Initial condition vector at $t = 0$

Q5.48 Write the solution of non-homogeneous state equations?

Ans. The solution of non-homogeneous state equation is,

$$X(t) = e^{A(t-t_o)} X(t_0) + \int_{t_o}^{t} e^{A(t-\tau)} \cdot B \cdot U(\tau) d\tau$$

where, $X(t)$ = state vector at time, t

$X(t_o)$ = Initial condition vector at $t = t_o$

B = Input matrix

$U(t)$ = Input vector

It initial conditions are known at $t = 0$, then put $t_o = 0$.

Q5.49 What is resolvant matrix?

Ans. The laplace transform of state transition matrix is called resolvant matrix.

Resolvant matrix, $\phi(s) = L[\phi(t)] = L[e^{-At}]$

Also $\phi(s) = [sI - A]^{-1}$

Q5.50 What are the different methods available for computing e^{AT}?

Ans. The following four methods are available for computing e^{At}

1. Computation of e^{At} using matrix exponential
2. Computation of e^{nt} using laplace transform
3. Computation of e^{At} by canonical transformation
4. Computation of e^{At} using cayley-Hamilton theorem

Q5.51 Write the state model of nth order discrete time system?

Ans. The state model of a system consists of state equation and output equation. The state model of a nth order discrete time system with m-inputs and p-outputs are,

$$X(k+1) = A\,X(k) + B\,U(k) \text{ - state equation}$$

$$Y(k) = C\,X(k) + D\,U(k) \text{ - Output equation}$$

Where, $X(k)$ = state vector of order $(n \times 1)$

$Y(k)$ = Output vector of order $(P \times 1)$

$U(k)$ = Input vector of order $(m \times 1)$

A = System matrix of order $(n \times n)$

B = Input matrix of order $(n \times m)$

C = Output matrix of order $(p \times n)$

D = Transmission matrix of order $(p \times m)$

Q5.52 What are the fundamental elements used to construct the state diagram of discrete time system?

Ans. The fundamental elements used to construct the state diagram of discrete time system are scalar, adder and unit delay element.

Q5.53 Sketch the basic elements used to construct the block diagram of discrete time system?

Ans. The basic elements used to construct the block diagram of discrete time system are shown below.

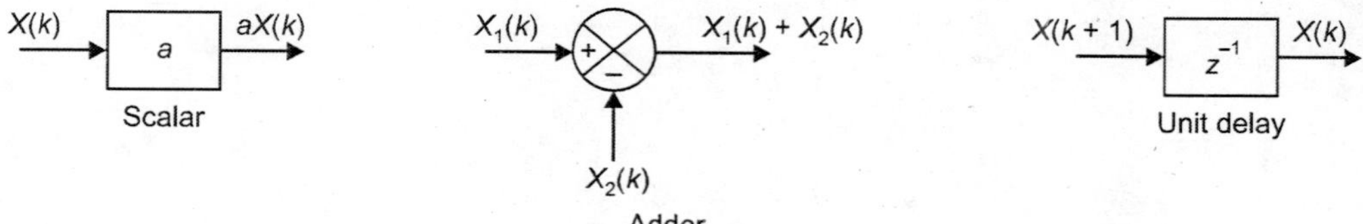

Q5.54 Draw the basic elements used to construct the signal flow graph of discrete time system

Ans. The basic elements used to construct the signal flow graph of discrete time system are shown below.

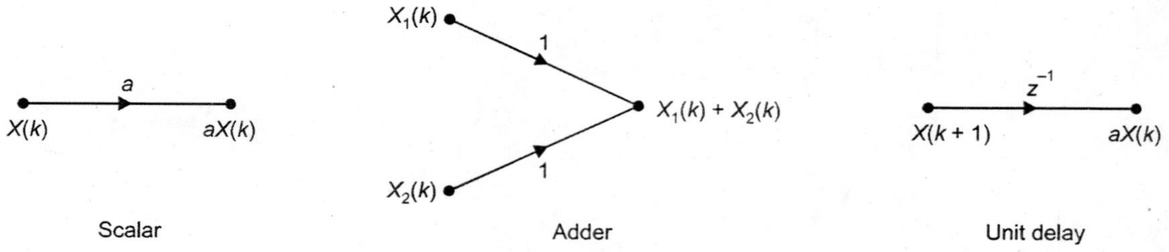

CHAPTER 5

Q5.55 Draw the block diagram representation of the state model of discrete time system?

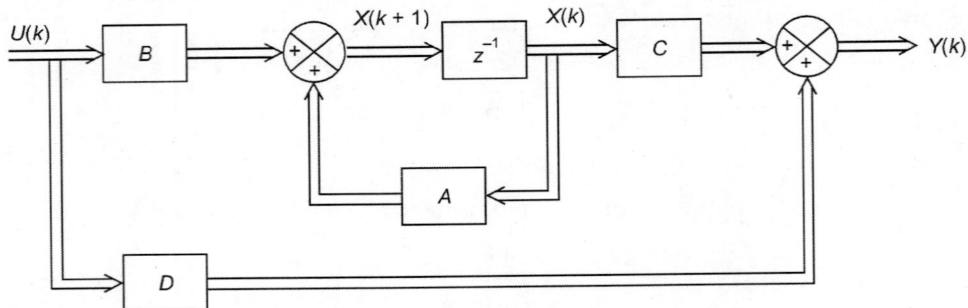

Q5.56 Draw the block diagram representation of the state model of discrete time system?

Ans.

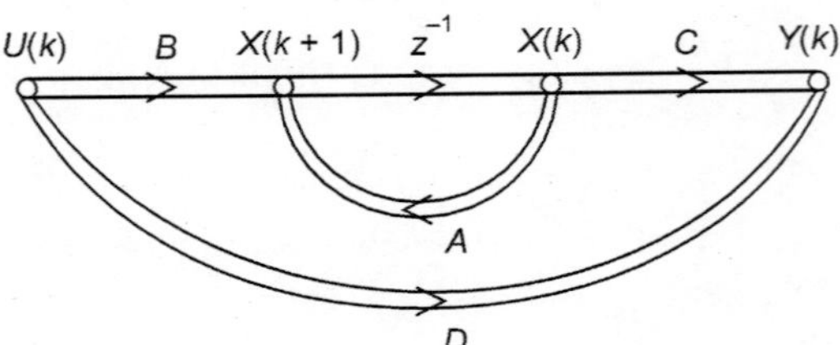

Q5.57 Draw the block diagram of the discrete time system described by the state model,

$$\begin{vmatrix} X_1(k+1) \\ X_2(k+1) \end{vmatrix} = \begin{vmatrix} 4 & 0 \\ 0 & 2 \end{vmatrix} \begin{vmatrix} X_1(k) \\ X_2(k) \end{vmatrix} + \begin{vmatrix} 1 \\ 1 \end{vmatrix} u(k)$$

$$Y(k) = \begin{vmatrix} 3 & 5 \end{vmatrix} \begin{vmatrix} x_1(k) \\ x_2(k) \end{vmatrix} + 10u(k)$$

Ans.

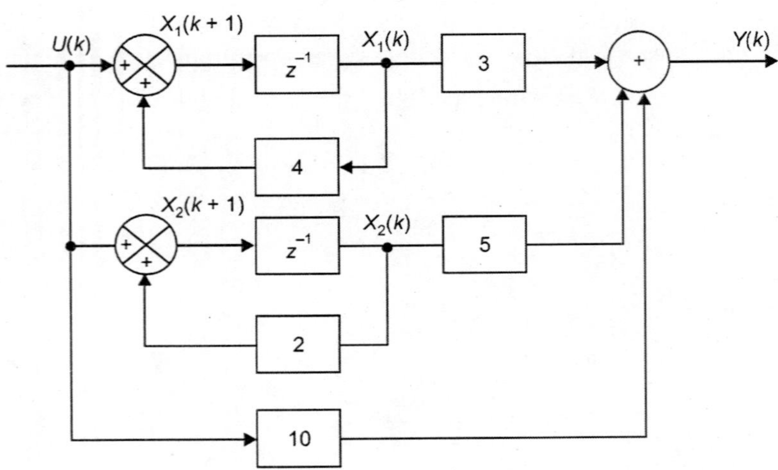

Q5.58 Draw the signal flow graph of the discrete time system described by the state model

$$\begin{vmatrix} X_1(k+1) \\ X_2(k+1) \end{vmatrix} = \begin{vmatrix} 0 & 1 \\ 2 & 3 \end{vmatrix} \begin{vmatrix} X_1(k) \\ X_2(k) \end{vmatrix} + \begin{vmatrix} 0 \\ 1 \end{vmatrix} u; Y(k) = X_1(k)$$

Ans.

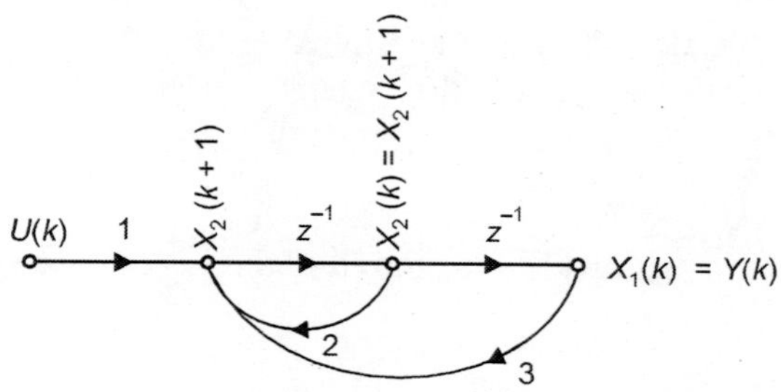

Q5.59 Determine the state model of the discrete time system represented by the block diagram of Fig. given below

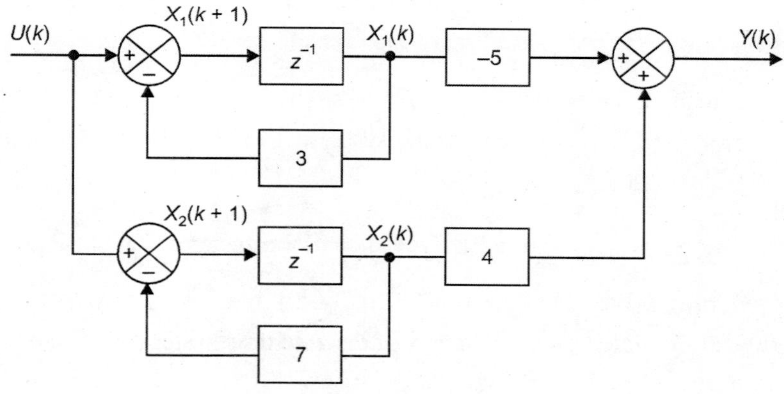

Ans.

$$\begin{vmatrix} X_1(k+1) \\ X_2(k+2) \end{vmatrix} = \begin{vmatrix} -3 & 0 \\ 0 & 7 \end{vmatrix} \begin{vmatrix} X_1(k) \\ X_2(k) \end{vmatrix} + \begin{vmatrix} 1 \\ 1 \end{vmatrix} |u|;$$

$$Y(k) = |-5 \quad 4| \begin{vmatrix} X_1(k) \\ X_2(k) \end{vmatrix}$$

Q5.60 Determine the state model of the discrete time system represented by the signal flow graph of Fig. given below.

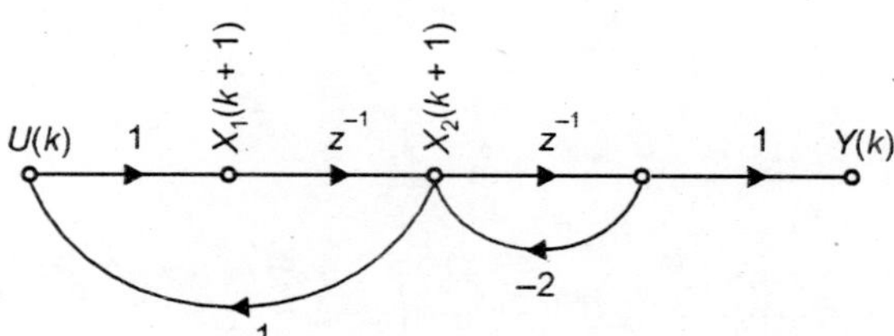

Ans.

$$\begin{vmatrix} X_1(k+1) \\ X_2(k+1) \end{vmatrix} = \begin{vmatrix} 0 & 1 \\ 1 & -2 \end{vmatrix} \begin{vmatrix} X_1(k) \\ X_2(k) \end{vmatrix} + \begin{vmatrix} 1 \\ 1 \end{vmatrix} u(k);$$

$$y(k) = X_2 k$$

Q5.61 A discrete time system is described by the difference equation, $y(k+2) + 3y(k+1) + 5y(k) = u(k)$. Determine the transfer function of the system.

Ans. Given that,

$$y(k+2) + 3y(k+1) + 5y(k) = u(k)$$

On taking z- transfer of equation (1) with zero initial conditions we get,

$$z^2 Y(z) + 3zY(z) + 6Y(z) = U(z)$$

$$(z^2 + 3z + 6)Y(z) = U(z)$$

$$\frac{Y(z)}{U(z)} = \frac{1}{z^2 + 3z + 6} \quad \text{Transfer function of the system}$$

The equation (2) is the transfer function of the system,

Q5.62 What is state transition matrix of discrete time system?

Ans. The matrix A^k [or $\phi(k)$] is called the state transition matrix of discrete time system.

It is given by, $A^k = \mathcal{Z}^{-1}\left\{(zI - A)^{-1} z\right\}$

The state transition matrix is used to find the state of the system, at any discrete time instant k.

Q5.63 Write the properties of the state transition matrix of discrete time system?

Ans. The properties of the state transition matrix of discrete time system are given below

1. $\phi(0) = 1$
2. $\phi^{-1}(k) = \phi(-k)$
3. $\phi(k, k_o) = \phi(k - k_0) = A^{(k - k_o)}$; where $k > k_o$

Q5.64 What are the different methods available for computing A^k?

Ans. The various methods available for computing A^k are,

1. Computation of A^k using z - transform
2. Computation of A^k by canonical transformation
3. Computation of A^k by caylay-Hamitton theorem.

Q5.65 Write the solution of discrete time state equation

Ans. The solution of discrete time state equation is

$$X(k) = \phi(k)X(0) + \sum_{j=0}^{k-1} \phi(k-1-j)\, Bu(j)$$

Q5.66 Write the expression to determine the solution of discrete time state equation using z-transform

Ans. The solution of discrete time state equation is given by

$$X(k) = \mathcal{Z}^{-1}\left\{(zI - A)^{-1} z\right\} X(o) + \mathcal{Z}^{-1}\left\{(zI - A)^{-1} BU(Z)\right\}$$

Q5.67 The state model of a discrete time system is given by,

$$X(k+1) = A\,X(k) + B\,U(k)$$
$$Y(k) = C\,X(k) + D\,U(k)$$

Determine its transfer function

Ans. The state equation is,

$$X(k+1) = A\,X(k) + B\,U(k) \quad (1)$$

On taking z-transform of equation (1) with zero initial conditions we get,

$$z\,X(z) = A\,X(z) + B\,U(k)$$
$$z\,X(z) = -AX(z) = B\,U(z)$$
$$(zI - A)X(z) = B\,u(z) \quad (2)$$

On Premultiplying equation (2) by $(zI - A)^{-1}$ we get,

$$X(z) = (zI - A)^{-1}\,B\,U(z) \quad (3)$$

The output equation is

$$Y(k) = C\,X(k) + D\,U(k) \quad (4)$$

On taking z-transform of equation (4) we get,

$$Y(z) = C\,X(z) + D\,U(z) \quad (5)$$

On substituting for $X(z)$ from equaation (3) in equation (5) we get,

$$Y(z) = C\,(zI - A)^{-1}\,B\,U(z) + D\,U(z)$$
$$Y(z) = [C(z1 - A)^{-1}\,B + D]\,U(z)$$
$$\frac{Y(z)}{U(z)} = C\,(z1 - A)^{-1}\,B + D \quad (6)$$

[transform function of discrete time system].

Q5.68 What is meant by diagonalization.

Ans. The process of converting the system matrix A into a diagonal matrix by a similarity transformation using the modal matrix M is called diagonalization.

Q5.69 What is canonical form of state model?

Ans. If the system matrix, A is the form of diagonal matrix then the state model is called canonical form.

The transformation used to diagonalize a system matrix is $X = MZ$.

where,
$X \Rightarrow$ original state vector
$Z \Rightarrow$ transformed state vector
$M \Rightarrow$ model matrix

Q5.70 What is Jordan Canonical form and Joradan Matrix?

Ans.
- When the eigen values have multiplicity the system matrix cannot be diagonalized. But the transformation, $X = MZ$ will transform the system matrix to a form called Jordan matrix, the transformed state model is this case is called Jordan canonical form.
- The Jordan matrix is the transformed system matrix using the transformation, $X = MZ$. The Jordan matrix, J will have a Jordan block of size $q \times q$ correspond to a eigen value of l_1 with multiplicity q. In the Jordan block the diagonal elements will be eigen values and elements just above the diagonal one.

Jordan block

$$J = \begin{bmatrix} \lambda_1 & 1 & 0 & \cdots & \cdots & 0 \\ 0 & \lambda_1 & 1 & \cdots & \cdots & 0 \\ 0 & 0 & \lambda_1 & \cdots & \cdots & 0 \\ 0 & 0 & 0 & \lambda_4 & \cdots & 0 \\ \vdots & \vdots & \vdots & \vdots & & \vdots \\ 0 & 0 & 0 & 0 & \cdots & \lambda_n \end{bmatrix}$$

Q5.71 Define controllability and what is the need for controllability test?

Ans.
- Possibility of transferring the system state from any initial state $x(t_0)$ to any other desired state $x(t)$, in specified finite time by a control vector $u(t)$.
- It is used to find the usefulness of a state variable.

Q5.72 Define observability and what is the need for observability?

Ans.
- A system is said to be completely observable if every state $x(t)$ can be completely identified by measurements of the output $y(t)$ ones a finite time interval.
- If is used to find whether the state variables are measurable (or) not.

Q5.73 State the duality between controllability and observability?

Ans. The principle of duality state that a system is completely state controllable if and only if its dual system is completely observable (or) viceversa.

Q5.74 What is the effect of pole-zero cancellation is transfer function.

Ans. If cancellation of pole-zero occurs in the transfer function of a system, then the system will be either not state controllable (or) unobservable depending on the state variables are defined.

Q5.75 What is pole placement by state feedback?

Ans. The pole placement by state feedback is a control system design technique, in which the state variables are used for feedback to achieve the desired closed loop poles.

Q5.76 How control system design is carried in state space?

Ans. In state space design of control system, any parameter (or) variable of a system are used for feedback to achieve the desired performance of the system.

The performance of the system is related to the location of closed loop poles. Hence in state space design the closed loop poles are placed at the desired locations by means of state feedback through an appropriate state feedback gain matrix, K.

EXERCISE PROBLEMS

Q.1 Design a phase lag network for a system having $G(s) = K/s(1 + .2s)2$ to have a phase margin of 30°.

Q.2 The unity feedback system is $G(s) = K/s\ (s + 3)$. It's velocity error constant of the system be equal to 15, while the damping ratio is 0.6 and velocity error is less than .25 rad per unit ramp input. Design a suitable lag compensator.

Q.3 Consider the unity feedback system with an open loop transfer function $G(s) = K/s(s + 2)2$. Design a suitable lead compensator so that phase margin is atleast 50° and velocity error constant is 20 s^{-1}.

Q.4 The open loop transfer function of certain unity feedback control system is given by $G(s) = \dfrac{K}{S(s+1)(.2s+1)}$ $K/0.1\ (s + 1)\ (.2s + 1)$. It is desired to have the phase margin to atleast 30°.

Design a suitable phase lead compensator.

Q.5 A unity feedback system with open loop transfer function $G(s) = K/s^2\ (s + 1.5)$ is to be read compensated to satisfy the following specifications.

(i) Damping ratio = .45

(ii) Undamped natural frequency, $\omega_n = 2.2$ rad/sec.

(iii) Velocity error constant, $K_v = 30$.

Q.6 The open loop transfer function of uncompensated system is $G(s) = K/s(s + 1)\ (s + 4)$. Design a lalg-lead compensator to meet the following specifications.

(i) $K_v \leq 5$

(ii) $z = -4$.

Q.7 Consider a unity feedback system with open loop transfer function, $G(s) = K/s\ (2s + 1)\ (.5s + 1)$. Design a suitable lag-lead compensator to meet the following specifications.

(i) $K_v = 30$

(ii) Phase margin ≥ 50.

Q.8 A linear time invariant system is described by the following state model. Obtain the canonical form of the state model given below.

(a) $$\begin{bmatrix} \dot{X}_1 \\ \dot{X}_2 \\ \dot{X}_3 \end{bmatrix} = \begin{bmatrix} 0 & 1 & 0 \\ -2 & -2 & -1 \\ 0 & 0 & -2 \end{bmatrix} \begin{bmatrix} X_1 \\ X_2 \\ X_3 \end{bmatrix} + \begin{bmatrix} 0 \\ 1 \\ 1 \end{bmatrix} [u] \text{ and } y = [3\ \ 1\ \ 0] \begin{bmatrix} X_1 \\ X_2 \\ X_3 \end{bmatrix}$$

CHAPTER 5

(b) $\begin{bmatrix} \dot{X}_1 \\ \dot{X}_2 \\ \dot{X}_3 \end{bmatrix} = \begin{bmatrix} 1 & 0 & +2 \\ -1 & 1 & 1 \\ 0 & 3 & -1 \end{bmatrix} \begin{bmatrix} X_1 \\ X_2 \\ X_3 \end{bmatrix} + \begin{bmatrix} 1 \\ 1 \\ 0 \end{bmatrix} [u]$ and $y = [1 \quad 0 \quad 1] \begin{bmatrix} X_1 \\ X_2 \\ X_3 \end{bmatrix}$

Q.9 Determine the state controllability for the systems represented by the following

(a) $\begin{bmatrix} \dot{X}_1 \\ \dot{X}_2 \end{bmatrix} = \begin{bmatrix} 2 & 1 \\ 0 & -1 \end{bmatrix} \begin{bmatrix} X_1 \\ X_2 \end{bmatrix} + \begin{bmatrix} 1 \\ 0 \end{bmatrix} [u]$

(b) $\begin{bmatrix} \dot{X}_1 \\ \dot{X}_2 \end{bmatrix} = \begin{bmatrix} 0 & 1 \\ -1 & 0 \end{bmatrix} \begin{bmatrix} X_1 \\ X_2 \end{bmatrix} + \begin{bmatrix} 0 \\ 1 \end{bmatrix} u$

Q.10 Examine the observability of the systems given below using canonical term.

(a) $\begin{bmatrix} \dot{X}_1 \\ \dot{X}_2 \\ \dot{X}_3 \end{bmatrix} = \begin{bmatrix} 2 & -2 & 3 \\ 1 & 1 & 1 \\ 1 & 3 & -1 \end{bmatrix} \begin{bmatrix} X_1 \\ X_2 \\ X_3 \end{bmatrix} + \begin{bmatrix} 1 \\ 1 \\ 1 \end{bmatrix} [u]$ and $y = [-3 \quad 5 \quad -2] \begin{bmatrix} x_1 \\ x_2 \\ x_3 \end{bmatrix}$

(b) $\begin{bmatrix} \dot{X}_1 \\ \dot{X}_2 \\ \dot{X}_3 \end{bmatrix} = \begin{bmatrix} 0 & 1 & 0 \\ 0 & 1 & 1 \\ 0 & -2 & -3 \end{bmatrix} \begin{bmatrix} X_1 \\ X_2 \\ X_3 \end{bmatrix} + \begin{bmatrix} 0 \\ 0 \\ 1 \end{bmatrix} [u]$ and $y = [3 \quad 4 \quad 1] \begin{bmatrix} x_1 \\ x_2 \\ x_3 \end{bmatrix}$

Q.11 Convert the following system matrix to cannoical form

(a) $A = \begin{bmatrix} -3 & 0 & 1 \\ 1 & 1 & 0 \\ 2 & -1 & 2 \end{bmatrix}$

(b) $A = \begin{bmatrix} 1 & 1 & -1 \\ 0 & 2 & 1 \\ 1 & 2 & 1 \end{bmatrix}$

Q.12 Obtain the state model of the electrical network shown below by choosing minimal member of state variables.

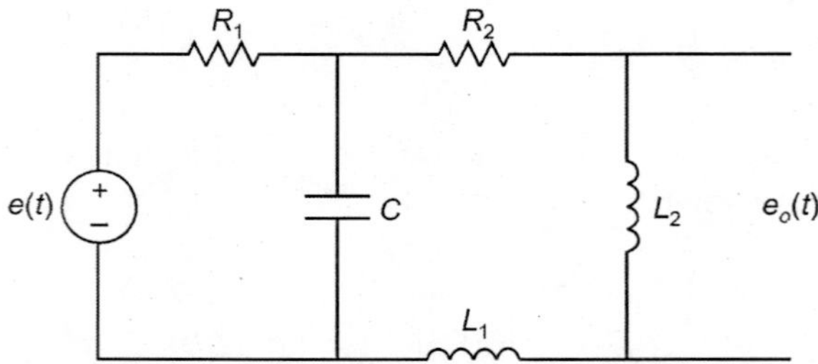

Q.13 Construct a state model for the system characterized by the following differential equation.

(a) $\dddot{y} + 6\ddot{y} + 11\dot{y} + 6y = u$

(b) $\ddot{y} + 2\dot{y} + y = \dot{u} + u$

Q.14 In the state diagram of the system shown below. Assign state variables and obtain state model of the system.

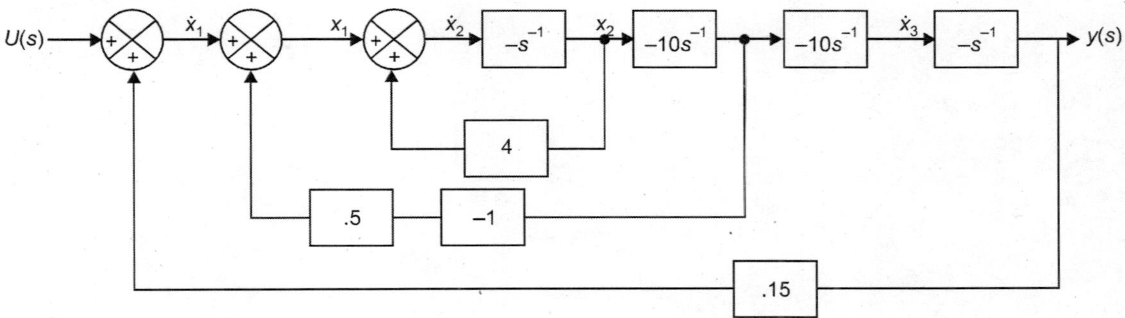

Q.15 The state diagram of a linear system is shown in Fig. below. Assign state variables and obtain the state model.

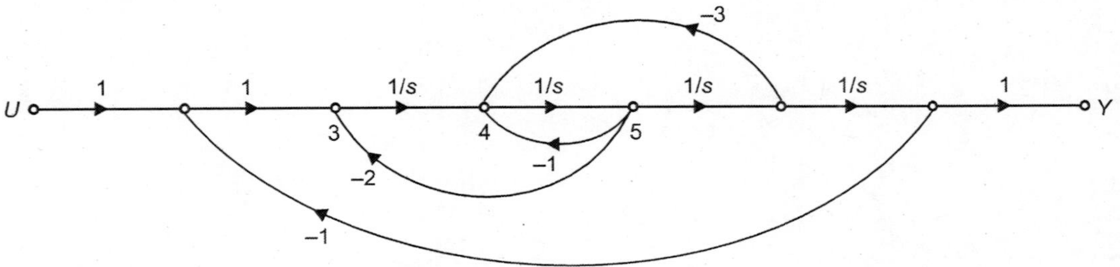

Q.16 Compute the solution of following equation

$$\begin{bmatrix} \dot{x}_1 \\ \dot{x}_2 \end{bmatrix} = \begin{bmatrix} 1 & 0 \\ 1 & 1 \end{bmatrix} \begin{bmatrix} x_1 \\ x_2 \end{bmatrix} + \begin{bmatrix} 1 \\ 1 \end{bmatrix} u(t) \text{ with } x_0 = \begin{bmatrix} 1 \\ 1 \end{bmatrix}.$$

CHAPTER 5